T0134750

Lecture Notes in Computer Science   13352

More information about this series at https://link.springer.com/bookseries/558

Derek Groen · Clélia de Mulatier ·
Maciej Paszynski · Valeria V. Krzhizhanovskaya ·
Jack J. Dongarra · Peter M. A. Sloot (Eds.)

Computational Science – ICCS 2022

22nd International Conference
London, UK, June 21–23, 2022
Proceedings, Part III

 Springer

Editors
Derek Groen ⓘ
Brunel University London
London, UK

Clélia de Mulatier ⓘ
University of Amsterdam
Amsterdam, The Netherlands

Maciej Paszynski ⓘ
AGH University of Science and Technology
Krakow, Poland

Valeria V. Krzhizhanovskaya ⓘ
University of Amsterdam
Amsterdam, The Netherlands

Jack J. Dongarra ⓘ
University of Tennessee at Knoxville
Knoxville, TN, USA

Peter M. A. Sloot ⓘ
University of Amsterdam
Amsterdam, The Netherlands

ISSN 0302-9743 ISSN 1611-3349 (electronic)
Lecture Notes in Computer Science
ISBN 978-3-031-08756-1 ISBN 978-3-031-08757-8 (eBook)
https://doi.org/10.1007/978-3-031-08757-8

Preface

Welcome to the 22nd annual International Conference on Computational Science (ICCS 2022 - https://www.iccs-meeting.org/iccs2022/), held during 21–23 June, 2022, at Brunel University London, UK. After more than two years of a pandemic that has changed so much of our world and daily lives, this edition marks our return to a – partially – in-person event. Those who were not yet able to join us in London had the option to participate online, as all conference sessions were streamed.

Although the challenges of such a hybrid format are manifold, we have tried our best to keep the ICCS community as dynamic, creative, and productive as always. We are proud to present the proceedings you are reading as a result of that.

Standing on the River Thames in southeast England, at the head of a 50-mile (80 km) estuary down to the North Sea, London is the capital and largest city of England and the UK. With a rich history spanning back to Roman times, modern London is one of the world's global cities, having a prominent role in areas ranging from arts and entertainment to commerce, finance, and education. London is the biggest urban economy in Europe and one of the major financial centres in the world. It also features Europe's largest concentration of higher education institutions.

ICCS 2022 was jointly organized by Brunel University London, the University of Amsterdam, NTU Singapore, and the University of Tennessee.

Brunel University London is a public research university located in the Uxbridge area of London. It was founded in 1966 and named after the Victorian engineer Isambard Kingdom Brunel, who managed to design and build a 214m long suspension bridge in Bristol back in 1831. Brunel is well-known for its excellent Engineering and Computer Science Departments, and its campus houses a dedicated conference centre (the Hamilton Centre) which was used to host ICCS. It is also one of the few universities to host a full-length athletics track, which has been used both for practice purposes by athletes such as Usain Bolt for the 2012 Olympics and for graduation ceremonies.

The International Conference on Computational Science is an annual conference that brings together researchers and scientists from mathematics and computer science as basic computing disciplines, as well as researchers from various application areas who are pioneering computational methods in sciences such as physics, chemistry, life sciences, engineering, arts, and humanitarian fields, to discuss problems and solutions in the area, identify new issues, and shape future directions for research.

Since its inception in 2001, ICCS has attracted increasing numbers of attendees and higher-quality papers, and this year – in spite of the ongoing pandemic—was not an exception, with over 300 registered participants. The proceedings series has become a primary intellectual resource for computational science researchers, defining and advancing the state of the art in this field.

The theme for 2022, "The Computational Planet," highlights the role of computational science in tackling the current challenges of the all-important quest for sustainable development. This conference aimed to be a unique event focusing on recent developments in scalable scientific algorithms, advanced software tools, computational grids, advanced numerical methods, and novel application areas. These innovative novel models, algorithms, and tools drive new science through efficient application in physical systems, computational and systems biology, environmental systems, finance, and other areas.

ICCS is well-known for its excellent lineup of keynote speakers. The keynotes for 2022 were as follows:

- Robert Axtell, George Mason University, USA
- Peter Coveney, University College London, UK
- Thomas Engels, Technische Universität Berlin, Germany
- Neil Ferguson, Imperial College London, UK
- Giulia Galli, University of Chicago, USA
- Rebecca Wade, Heidelberg Institute for Theoretical Studies, Germany

This year we had 474 submissions (169 submissions to the main track and 305 to the thematic tracks). In the main track, 55 full papers were accepted (32%), and in the thematic tracks, 120 full papers (39%). A higher acceptance rate in the thematic tracks is explained by the nature of these, where track organizers personally invite many experts in a particular field to participate in their sessions.

ICCS relies strongly on our thematic track organizers' vital contributions to attract high-quality papers in many subject areas. We would like to thank all committee members from the main and thematic tracks for their contribution to ensure a high standard for the accepted papers. We would also like to thank Springer, Elsevier, and Intellegibilis for their support. Finally, we appreciate all the local organizing committee members for their hard work to prepare for this conference.

We are proud to note that ICCS is an A-rank conference in the CORE classification.

We wish you good health in these troubled times and look forward to meeting you at the next conference, whether virtually or in-person.

June 2022

Derek Groen
Clélia de Mulatier
Maciej Paszynski
Valeria V. Krzhizhanovskaya
Jack J. Dongarra
Peter M. A. Sloot

Organization

General Chair

Valeria Krzhizhanovskaya   University of Amsterdam, The Netherlands

Main Track Chair

Clélia de Mulatier   University of Amsterdam, The Netherlands

Thematic Tracks Chair

Maciej Paszynski   AGH University of Science and Technology, Poland

Scientific Chairs

Peter M. A. Sloot   University of Amsterdam, The Netherlands | Complexity Institute NTU, Singapore

Jack Dongarra   University of Tennessee, USA

Local Organizing Committee

Chair

Derek Groen   Brunel University London, UK

Members

Simon Taylor	Brunel University London, UK
Anastasia Anagnostou	Brunel University London, UK
Diana Suleimenova	Brunel University London, UK
Xiaohui Liu	Brunel University London, UK
Zidong Wang	Brunel University London, UK
Steven Sam	Brunel University London, UK
Alireza Jahani	Brunel University London, UK
Yani Xue	Brunel University London, UK
Nadine Aburumman	Brunel University London, UK
Katie Mintram	Brunel University London, UK
Arindam Saha	Brunel University London, UK
Nura Abubakar	Brunel University London, UK

Thematic Tracks and Organizers

Advances in High-Performance Computational Earth Sciences: Applications and Frameworks – IHPCES

Takashi Shimokawabe	University of Tokyo, Japan
Kohei Fujita	University of Tokyo, Japan
Dominik Bartuschat	Friedrich-Alexander-Universität Erlangen-Nürnberg, Germany

Artificial Intelligence and High-Performance Computing for Advanced Simulations – AIHPC4AS

Maciej Paszynski	AGH University of Science and Technology, Poland

Biomedical and Bioinformatics Challenges for Computer Science – BBC

Mario Cannataro	Università Magna Graecia di Catanzaro, Italy
Giuseppe Agapito	Università Magna Graecia di Catanzaro, Italy
Mauro Castelli	Universidade Nova de Lisboa, Portugal
Riccardo Dondi	University of Bergamo, Italy
Rodrigo Weber dos Santos	Universidade Federal de Juiz de Fora, Brazil
Italo Zoppis	Università degli Studi di Milano-Bicocca, Italy

Computational Collective Intelligence – CCI

Marcin Maleszka	Wroclaw University of Science and Technology, Poland
Ngoc Thanh Nguyen	Wroclaw University of Science and Technology, Poland
Dosam Hwang	Yeungnam University, South Korea

Computational Health – CompHealth

Sergey Kovalchuk	ITMO University, Russia
Stefan Thurner	Medical University of Vienna, Austria
Georgiy Bobashev	RTI International, USA
Jude Hemanth	Karunya University, India
Anastasia Angelopoulou	University of Westminster, UK

Computational Optimization, Modelling, and Simulation – COMS

Xin-She Yang	Middlesex University London, UK
Leifur Leifsson	Purdue University, USA
Slawomir Koziel	Reykjavik University, Iceland

Computer Graphics, Image Processing, and Artificial Intelligence – CGIPAI

Andres Iglesias   Universidad de Cantabria, Spain

Machine Learning and Data Assimilation for Dynamical Systems – MLDADS

Rossella Arcucci   Imperial College London, UK

Multiscale Modelling and Simulation – MMS

Derek Groen   Brunel University London, UK
Diana Suleimenova   Brunel University London, UK
Bartosz Bosak   Poznan Supercomputing and Networking Center,
                                    Poland
Gabor Závodszky   University of Amsterdam, The Netherlands
Stefano Casarin   Houston Methodist Research Institute, USA
Ulf D. Schiller   Clemson University, USA
Wouter Edeling   Centrum Wiskunde & Informatica,
                                    The Netherlands

Quantum Computing – QCW

Katarzyna Rycerz   AGH University of Science and Technology,
                                    Poland
Marian Bubak   Sano Centre for Computational Medicine and
                                    AGH University of Science and Technology,
                                    Poland | University of Amsterdam,
                                    The Netherlands

Simulations of Flow and Transport: Modeling, Algorithms, and Computation – SOFTMAC

Shuyu Sun   King Abdullah University of Science and
                                    Technology, Saudi Arabia
Jingfa Li   Beijing Institute of Petrochemical Technology,
                                    China
James Liu   Colorado State University, USA

Smart Systems: Bringing Together Computer Vision, Sensor Networks, and Machine Learning – SmartSys

Pedro Cardoso   University of Algarve, Portugal
João Rodrigues   University of Algarve, Portugal
Jânio Monteiro   University of Algarve, Portugal
Roberto Lam   University of Algarve, Portugal

Software Engineering for Computational Science – SE4Science

Jeffrey Carver	University of Alabama, USA
Caroline Jay	University of Manchester, UK
Yochannah Yehudi	University of Manchester, UK
Neil Chue Hong	University of Edinburgh, UK

Solving Problems with Uncertainty – SPU

Vassil Alexandrov	Hartree Centre - STFC, UK
Aneta Karaivanova	Institute for Parallel Processing, Bulgarian Academy of Sciences, Bulgaria

Teaching Computational Science – WTCS

Angela Shiflet	Wofford College, USA
Nia Alexandrov	Hartree Centre - STFC, UK

Uncertainty Quantification for Computational Models – UNEQUIvOCAL

Wouter Edeling	Centrum Wiskunde & Informatica, The Netherlands
Anna Nikishova	SISSA, Italy

Reviewers

Tesfamariam Mulugeta Abuhay
Jaime Afonso Martins
Giuseppe Agapito
Shahbaz Ahmad
Elisabete Alberdi
Luis Alexandre
Nia Alexandrov
Vassil Alexandrov
Julen Alvarez-Aramberri
Domingos Alves
Sergey Alyaev
Anastasia Anagnostou
Anastasia Angelopoulou
Samuel Aning
Hideo Aochi
Rossella Arcucci
Costin Badica
Bartosz Balis
Daniel Balouek-Thomert
Krzysztof Banaś

Dariusz Barbucha
João Barroso
Valeria Bartsch
Dominik Bartuschat
Pouria Behnodfaur
Jörn Behrens
Adrian Bekasiewicz
Gebrail Bekdas
Mehmet Ali Belen
Stefano Beretta
Benjamin Berkels
Daniel Berrar
Georgiy Bobashev
Marcel Boersma
Tomasz Boiński
Carlos Bordons
Bartosz Bosak
Giuseppe Brandi
Lars Braubach
Marian Bubak

Jérémy Buisson
Aleksander Byrski
Cristiano Cabrita
Xing Cai
Barbara Calabrese
Nurullah Calik
Almudena Campuzano
Mario Cannataro
Pedro Cardoso
Alberto Carrassi
Alfonso Carriazo
Jeffrey Carver
Stefano Casarin
Manuel Castañón-Puga
Mauro Castelli
Nicholas Chancellor
Ehtzaz Chaudhry
Thierry Chaussalet
Sibo Cheng
Siew Ann Cheong
Andrei Chernykh
Lock-Yue Chew
Su-Fong Chien
Marta Chinnici
Amine Chohra
Neil Chue Hong
Svetlana Chuprina
Paola Cinnella
Noélia Correia
Adriano Cortes
Ana Cortes
Enrique Costa-Montenegro
David Coster
Carlos Cotta
Helene Coullon
Daan Crommelin
Attila Csikasz-Nagy
Javier Cuenca
António Cunha
Pawel Czarnul
Lisandro D. Dalcin
Bhaskar Dasgupta
Clélia de Mulatier
Charlotte Debus
Javier Delserlorente

Pasquale De-Luca
Quanling Deng
Vasily Desnitsky
Mittal Dhruv
Eric Dignum
Riccardo Dondi
Rafal Drezewski
Hans du Buf
Vitor Duarte
Richard Dwight
Wouter Edeling
Nasir Eisty
Kareem El-Safty
Nahid Emad
Gökhan Ertaylan
Roberto R. Expósito
Fangxin Fang
Antonino Fiannaca
Christos Filelis-Papadopoulos
Pawel Foszner
Piotr Frąckiewicz
Martin Frank
Alberto Freitas
Ruy Freitas Reis
Karl Frinkle
Kohei Fujita
Takeshi Fukaya
Wlodzimierz Funika
Takashi Furumura
Ernst Fusch
Leszek Gajecki
Ardelio Galletti
Marco Gallieri
Teresa Galvão
Akemi Galvez-Tomida
Maria Ganzha
Luis Garcia-Castillo
Bartłomiej Gardas
Delia Garijo
Frédéric Gava
Piotr Gawron
Bernhard Geiger
Alex Gerbessiotis
Philippe Giabbanelli
Konstantinos Giannoutakis

Adam Glos
Ivo Goncalves
Alexandrino Gonçalves
Jorge González-Domínguez
Yuriy Gorbachev
Pawel Gorecki
Markus Götz
Michael Gowanlock
George Gravvanis
Derek Groen
Lutz Gross
Lluis Guasch
Pedro Guerreiro
Tobias Guggemos
Xiaohu Guo
Manish Gupta
Piotr Gurgul
Zulfiqar Habib
Mohamed Hamada
Yue Hao
Habibollah Haron
Ali Hashemian
Carina Haupt
Claire Heaney
Alexander Heinecke
Jude Hemanth
Marcin Hernes
Bogumila Hnatkowska
Maximilian Höb
Jori Hoencamp
Rolf Hoffmann
Wladyslaw Homenda
Tzung-Pei Hong
Muhammad Hussain
Dosam Hwang
Mauro Iacono
David Iclanzan
Andres Iglesias
Mirjana Ivanovic
Takeshi Iwashita
Alireza Jahani
Peter Janků
Jiri Jaros
Agnieszka Jastrzebska
Caroline Jay

Piotr Jedrzejowicz
Gordan Jezic
Zhong Jin
David Johnson
Guido Juckeland
Piotr Kalita
Drona Kandhai
Epaminondas Kapetanios
Aneta Karaivanova
Artur Karczmarczyk
Takahiro Katagiri
Timo Kehrer
Christoph Kessler
Loo Chu Kiong
Harald Koestler
Ivana Kolingerova
Georgy Kopanitsa
Pavankumar Koratikere
Triston Kosloske
Sotiris Kotsiantis
Remous-Aris Koutsiamanis
Sergey Kovalchuk
Slawomir Koziel
Dariusz Krol
Marek Krótkiewicz
Valeria Krzhizhanovskaya
Marek Kubalcík
Sebastian Kuckuk
Eileen Kuehn
Michael Kuhn
Tomasz Kulpa
Julian Martin Kunkel
Krzysztof Kurowski
Marcin Kuta
Panagiotis Kyziropoulos
Roberto Lam
Anna-Lena Lamprecht
Kun-Chan Lan
Rubin Landau
Leon Lang
Johannes Langguth
Leifur Leifsson
Kenneth Leiter
Florin Leon
Vasiliy Leonenko

Jean-Hugues Lestang
Jake Lever
Andrew Lewis
Jingfa Li
Way Soong Lim
Denis Mayr Lima Martins
James Liu
Zhao Liu
Hong Liu
Che Liu
Yen-Chen Liu
Hui Liu
Marcelo Lobosco
Doina Logafatu
Marcin Los
Stephane Louise
Frederic Loulergue
Paul Lu
Stefan Luding
Laura Lyman
Lukasz Madej
Luca Magri
Peyman Mahouti
Marcin Maleszka
Bernadetta Maleszka
Alexander Malyshev
Livia Marcellino
Tomas Margalef
Tiziana Margaria
Svetozar Margenov
Osni Marques
Carmen Marquez
Paula Martins
Pawel Matuszyk
Valerie Maxville
Wagner Meira Jr.
Roderick Melnik
Pedro Mendes Guerreiro
Ivan Merelli
Lyudmila Mihaylova
Marianna Milano
Jaroslaw Miszczak
Janio Monteiro
Fernando Monteiro
Andrew Moore

Eugénia Moreira Bernardino
Anabela Moreira Bernardino
Peter Mueller
Ignacio Muga
Khan Muhammad
Daichi Mukunoki
Vivek Muniraj
Judit Munoz-Matute
Hiromichi Nagao
Jethro Nagawakar
Kengo Nakajima
Grzegorz J. Nalepa
Yves Nanfack
Pratik Nayak
Philipp Neumann
David Chek-Ling Ngo
Ngoc Thanh Nguyen
Nancy Nichols
Sinan Melih Nigdeli
Anna Nikishova
Hitoshi Nishizawa
Algirdas Noreika
Manuel Núñez
Frederike Oetker
Schenk Olaf
Javier Omella
Boon-Yaik Ooi
Eneko Osaba
Aziz Ouaarab
Raymond Padmos
Nikela Papadopoulou
Marcin Paprzycki
David Pardo
Diego Paredesconcha
Anna Paszynska
Maciej Paszynski
Ebo Peerbooms
Sara Perez-Carabaza
Dana Petcu
Serge Petiton
Frank Phillipson
Eugenio Piasini
Juan C. Pichel
Anna Pietrenko-Dabrowska
Laércio L. Pilla

Armando Pinho	Takashi Shimokawabe
Yuri Pirola	Alexander Shukhman
Mihail Popov	Marcin Sieniek
Cristina Portales	Nazareen Sikkandar-Basha
Roland Potthast	Robert Sinkovits
Małgorzata Przybyła-Kasperek	Mateusz Sitko
Ela Pustulka-Hunt	Haozhen Situ
Vladimir Puzyrev	Leszek Siwik
Rick Quax	Renata Słota
Cesar Quilodran-Casas	Oskar Slowik
Enrique S. Quintana-Orti	Grażyna Ślusarczyk
Issam Rais	Sucha Smanchat
Andrianirina Rakotoharisoa	Maciej Smołka
Raul Ramirez	Thiago Sobral
Celia Ramos	Isabel Sofia Brito
Vishwas Rao	Piotr Sowiński
Kurunathan Ratnavelu	Robert Speck
Lukasz Rauch	Christian Spieker
Robin Richardson	Michał Staniszewski
Miguel Ridao	Robert Staszewski
Heike Riel	Steve Stevenson
Sophie Robert	Tomasz Stopa
Joao Rodrigues	Achim Streit
Daniel Rodriguez	Barbara Strug
Albert Romkes	Patricia Suarez
Debraj Roy	Dante Suarez
Katarzyna Rycerz	Diana Suleimenova
Emmanuelle Saillard	Shuyu Sun
Ozlem Salehi	Martin Swain
Tarith Samson	Jerzy Świątek
Alberto Sanchez	Piotr Szczepaniak
Ayşin Sancı	Edward Szczerbicki
Gabriele Santin	Tadeusz Szuba
Vinicius Santos-Silva	Ryszard Tadeusiewicz
Allah Bux Sargano	Daisuke Takahashi
Robert Schaefer	Osamu Tatebe
Ulf D. Schiller	Carlos Tavares Calafate
Bertil Schmidt	Kasim Tersic
Martin Schreiber	Jannis Teunissen
Gabriela Schütz	Mau Luen Tham
Franciszek Seredynski	Stefan Thurner
Marzia Settino	Nestor Tiglao
Mostafa Shahriari	T. O. Ting
Zhendan Shang	Alfredo Tirado-Ramos
Angela Shiflet	Pawel Topa

Bogdan Trawinski
Jan Treur
Leonardo Trujillo
Paolo Trunfio
Hassan Ugail
Eirik Valseth
Casper van Elteren
Ben van Werkhoven
Vítor Vasconcelos
Alexandra Vatyan
Colin C. Venters
Milana Vuckovic
Shuangbu Wang
Jianwu Wang
Peng Wang
Katarzyna Wasielewska
Jaroslaw Watrobski
Rodrigo Weber dos Santos
Mei Wen
Lars Wienbrandt
Iza Wierzbowska
Maciej Woźniak
Dunhui Xiao

Huilin Xing
Yani Xue
Abuzer Yakaryilmaz
Xin-She Yang
Dongwei Ye
Yochannah Yehudi
Lihua You
Drago Žagar
Constantin-Bala Zamfirescu
Gabor Závodszky
Jian-Jun Zhang
Yao Zhang
Wenbin Zhang
Haoxi Zhang
Jinghui Zhong
Sotirios Ziavras
Zoltan Zimboras
Italo Zoppis
Chiara Zucco
Pavel Zun
Simon Portegies Zwart
Karol Życzkowski

Contents – Part III

Computational Optimization, Modelling and Simulation

Machine Learning and Data Assimilation for Dynamical Systems

Computational Health

Knowledge Discovery in Databases: Comorbidities in Tuberculosis Cases

Isabelle Carvalho[1]([✉])[iD], Mariane Barros Neiva[1][iD],
Newton Shydeo Brandão Miyoshi[2][iD], Nathalia Yukie Crepaldi[2][iD],
Filipe Andrade Bernardi[2,3][iD], Vinícius Costa Lima[2,3][iD],
Ketlin Fabri dos Santos[4][iD], Ana Clara de Andrade Mioto[3][iD],
Mariana Tavares Mozini[2][iD], Rafael Mello Galliez[5][iD],
Mauro Niskier Sanchez[6][iD], Afrânio Lineu Kritski[5][iD], and Domingos Alves[2][iD]

[1] Institute of Mathematical and Computer Sciences, University of Sao Paulo,
400 Trabalhador São Carlense Avenue, Sao Carlos, SP, Brazil
`isabelle.carvalho@alumni.usp.br, marianeneiva@usp.br`
[2] Ribeirao Preto Medical School, University of Sao Paulo,
3900 Bandeirantes Avenue, Ribeirao Preto, SP, Brazil
`{newton.sbm,nathaliayc,mtmozini}@usp.br, quiron@fmrp.usp.br`
[3] São Carlos School of Engineering - University of Sao Paulo,
400 Trabalhador São Carlense Avenue, São Carlos, SP, Brazil
`{filipepaulista12,viniciuslima,anaclara.mioto}@usp.br`
[4] Pontifical Catholic University of Minas Gerais,
500 Dom Jose Gaspar, 500, Belo Horizonte, MG, Brazil
`ketlin.fabri@gmail.com`
[5] Faculty of Medicine, Federal University of Rio Janeiro,
373 Carlos Chagas Filho Avenue, Rio de Janeiro, RJ, Brazil
`galliez77@ufrj.br, kristskia@gmail.com`
[6] School of Health Sciences, University of Brasilia,
Campos Univ. Darcy Ribeiro, Brasilia, DF, Brazil
`maurosanchez@unb.br`

Abstract. Unlike the primary condition under investigation, the term comorbidities define coexisting medical conditions that influence patient care during detection, therapy, and outcome. Tuberculosis continues to be one of the 10 leading causes of death globally. The aim of the study is to present the exploration of classic data mining techniques to find relationships between the outcome of TB cases (cure or death) and the comorbidities presented by the patient. The data are provided by TBWEB and represent TB cases in the territory of the state of São Paulo-Brazil, from 2006 to 2016. Techniques of feature selection and classification models were explored. As shown in the results, it was found high relevance for AIDS and alcoholism as comorbidities in the outcome of TB cases. Although the classifier performance did not present a significant statistical difference, there was a great reduction in the number of attributes and in the number of rules generated, showing, even more, the high relevance of the attributes: age group, AIDS, and other immunology in the classification of the outcome of TB cases. The explored techniques proved to be promising to support searching for unclear relationships in the TB context, providing, on average, a 73% accuracy in predicting the outcome of the cases according to characteristics that were analyzed.

D. Groen et al. (Eds.): ICCS 2022, LNCS 13352, pp. 3–13, 2022.
https://doi.org/10.1007/978-3-031-08757-8_1

Keywords: Comorbidity · Data mining · Knowledge discovery · Public health · Tuberculosis

1 Background

Tuberculosis (TB) is an infectious disease caused by a mycobacterium that mainly affects the lungs but can also appear in other organs of the body, such as bones, kidneys, and meninges (membranes that involve the brain). The infection has been treated for years as a public health problem and it has been the focus of many types of research. Besides the efforts, TB is still one of the main causes of death among infectious diseases, being the most important single infectious agent for mortality worldwide [27]. The problem is that TB compromises the patient's immune system, making them more susceptible to other diseases. Moreover, without adequate treatment, it might progress to more serious conditions as well as allowing for the development of drug resistance [8].

As mentioned before, TB is dangerous due to the ability to weaken the immune system, turning the body susceptible to other diseases. The term comorbidities define coexisting medical conditions, distinct from the primary condition under investigation, that influences patient care during detection, therapy, and outcome. The study of comorbidities associated with TB is extremely important to raise hypotheses about the relationship of other nosologies with the disease in order to help prevent and treat these patients [5].

1.1 The TB Scenario in Brazil and in the World

Tuberculosis continues to be one of the 10 leading causes of death globally nowadays. Since 1997, the World Health Organization (WHO) has been monitoring tuberculosis cases annually. It is estimated that TB caused 1.5 million deaths in 2020, including 208.000 deaths among HIV-positive people. Most of these cases occurred in emerging or underdeveloped countries. The combination of social and economic factors has contributed greatly to the reduction of these rates since effective treatment for TB already exists [27].

In Brazil, there are approximately 70,000 new TB cases per year, and it is one of the countries most affected by this issue [16]. Annually, the decrease in TB incidence does not exceed 2% while the ideal would be approximately 8%, to reach the goal of eliminating tuberculosis as a public health problem in Brazil [19]. In addition, treatment indicators do not reach 75% of cases depending on the region. Adherence to treatment is important because default contributes to disease transmission. According to the recommendation of the Brazilian Ministry of Health, cure rates below the 85% target and default rates above 5% demonstrate the need to increase the quality of treatment coverage [10,20]. All these conditions increase the vulnerability of Brazilian patients to be affected by comorbidities in TB and enhance the continuous necessity of research in this area.

1.2   The Comorbidities in TB Cases

The concern around the disease is even more relevant since TB shares socio-economic determinants with several other diseases. Thus, there are many studies that map the role of comorbidities in patients with tuberculosis [26]. Some of the tuberculosis-related comorbidities include HIV, diabetes, alcoholism, and drug addiction. The probability of developing TB disease is much higher among people infected with HIV and affected by risk factors such as diabetes, smoking, and alcohol consumption [21].

The relation between TB and HIV is one of the most studied topics in this context. Tuberculosis is one of the most common diseases among people with HIV, although the physical pathogenesis of TB acting as an immunosuppressant is not yet established. An HIV patient is 16 to 27 times more likely to develop TB than a person without HIV [12]. In the case of diabetes, the same relationship exists. Diabetes triples the risk of TB. This problem is still greater in emerging countries because of the increasing number of diabetes cases when compared to developed countries [11].

There is also evidence of alcohol use and the increased risk of developing tuberculosis. This risk increases in the case of alcoholic patients and/or people who drink more than 40g of alcohol per day [22]. Patients with TB also have conditions related to smoking. Population with a high smoking rate also presents a higher incidence of tuberculosis since it increases the risk of developing 2 to 3 folds [26]. In the case of drug abuse, users are considered risk groups for TB. This relationship has already been identified in several countries. One of the major problems associated with TB and drug abuse is treatment default [4].

1.3   Objectives

In the attempt to improve the knowledge about the patterns in TB, this article uses the power of data mining techniques to analyze the relationship among comorbidities and the final outcome (cure or death) in cases of tuberculosis.

2   Methods

In order to continue the study and to understand even more the relationship between TB and its related comorbidities, this work uses the power of machine learning (ML) techniques to statistically find the main aspects that can deal to cure or death in a case of TB. To summary the approach applied in the study, Fig. 1 presents the general steps of the research.

The data mining experiments were performed in three main steps (shown in Fig. 1 as 2a, 2b, and 2c). Each step of Fig. 1 is detailed as follows:

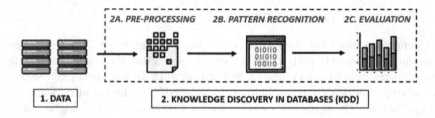

Fig. 1. Overview of methods.

1. **DATA:** The data used is provided by the Notification and Monitoring System of Tuberculosis of the Health Secretariat - São Paulo State Government (TBWeb) [2]. The database contains confirmed cases of tuberculosis in the territory of the state of São Paulo, from 2006 to 2016. In this step, a descriptive analysis of the data was explored.

2A. The key point of this step is to improve the representation and quality of raw data to provide effective analysis [25]. Two activities were explored in this step: class balancing and feature selection (FS).

   1. For class balancing, the method of sub-samples was applied to achieve equality in the number of instances for each class available in the set. This is a prime step in the process of knowledge discovery. The selected examples are the most relevant of the set, this is, the not selected examples are those with repeated or with little information and thus the least possible impairment in the analyzes.

   2. In the FS step, two different studies were evaluated:

      i. the use of a filter method that ranks each feature according to a degree of importance within the database computed by the filter algorithm. In the experiments, three different methodologies were applied: Correlation analysis (CFS), Information gain analysis (Gain Ratio), and Chi-square test.

      ii. the use of a wrapper method that involves the selection of attributes based on the classification algorithm to be used [9]. This analysis explored three algorithms: K-Nearest Neighbor (KNN), Bayesian Network (BN), and Decision Trees (DT).

2B. KDD - PATTERN RECOGNITION: This step aims the application of algorithms for the identification of relations in the data and the construction of mathematical models based on these relations [25]. Two analysis were explored in this step:

   1. For the FS filter method + rules extraction, three algorithms were applied: C4.5, PART, and RIPPER.

   2. For FS wrapper approaches + classification were applied to the same algorithms used for the construction of the relationships: KNN, BN, and DT.

2C. KDD - EVALUATION: The evaluation of the predicted patterns was performed through the accuracy of the exploited algorithms (KNN, BN, and DT). Accuracy presents the proportion of correct predictions.

This work is the first contact of the machine learning techniques with this dataset, which justifies the choice of classical algorithms and highly explored in the literature as objects of exploration. To initiate the discovery of knowledge in this context, we used a standard parameterization of the algorithms studied and the 10-cross-validation technique, to aid in the generalization of the results [25]. The tools Weka [9], Matlab [15], Python language [24] provided support in activities.

3   Results and Discussion

3.1   Data Characterization

The data set consists of 172,474 TB cases presented through 15 features (8 general features and 7 features about comorbidities). The general features are: id; race/color; age group; sex; are you pregnant?; naturalness; education and type of occupation. The comorbidities features are based on the presence or absence of AIDS, diabetes, alcoholism, mental disorders, drug addiction, other immunology, and tobacco use.

Figure 2 shows the distribution of TB cases with related comorbidities according to the output. We can observe that, proportionally, AIDS, and alcoholism are the two comorbidities that have greater cases of death compared to the other comorbidities.

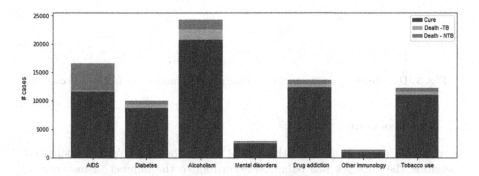

Fig. 2. Distribution of comorbidities in TB cases.

Figure 3 presents the presence or absence of comorbidities in TB cases, over the time studied. The first pattern that can be observed in Figs. 3(c), 3(e), and 3(g), starting around 2011, is the increase in the presence of alcohol, drugs, and tobacco, personal habits that influenced the incidence of TB cases. Two main classes are found in the TB database: 156,184 (91%) of the cases are cured and while 16,290 (9%) were deaths (with or without TB as the main cause of death).

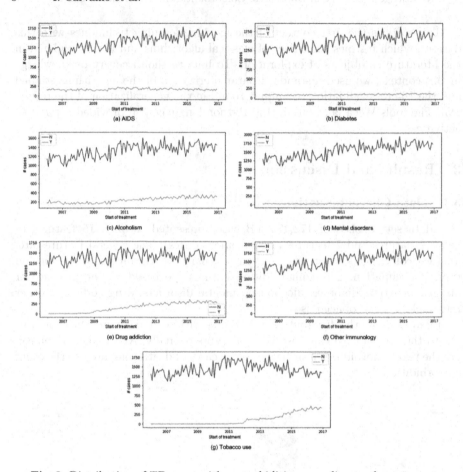

Fig. 3. Distribution of TB cases with comorbidities according to the outcome.

3.2   Pre-processing

As seen from the description of the classes above, the set is highly unbalanced. Therefore, the data set was balanced according to the number of instances of death set, which has lower samples. This resulted in 32,580 records, creating a 50/50 number of samples from each class in the dataset.

Furthermore, as the focus of the study is the relations between comorbidities and outcome of TB cases, four attributes were excluded from the analysis leaving us with 11 attributes composed the set for exploration: race/color, age group, sex, education, AIDS, diabetes, alcoholism, mental disorders, drug addiction, other immunology, and tobacco use. Feature selection results are to be described in the analysis sections below.

3.3   Analysis: FS Filter Method + Rules Extraction

Table 1 presents the most relevant characteristics selected in each FS method explored. The largest reduction was 73% in the number of attributes, from the original set to match set, i.e., the set with features selected by all methods.

Table 1. Description of selected characteristics - FS filter methods.

Dataset	#	Description
CFS	4	Age group, AIDS, mental disorders, other immunology
GainRatio	6	Age group, education, AIDS, alcoholism, mental disorders, other immunology
ChiSquare	6	Age group, sex, education, AIDS, alcoholism, other immunology
Match set	3	Age group, AIDS, other immunology

With the analysis of these ranks, a total of five subsets of features were analyzed in a first experiment (including the original set). Notice that age group, AIDS, and other immunology are presented as main features for all feature selection methods applied, showing the importance of these characteristics. For quantitative analysis, Table 2 presents the performance (accuracy) of the rule extraction models applied to the subsets. As presented in the table, one can see that there is no significant difference between the models, on average, 72% of correct predictions were maintained.

Table 2. Accuracy - Rule extraction models.

Dataset	Classifier		
	C4.5	PART	RIPPER
Original set	73.06%	72.96%	72.70%
CFS	71.76%	71.71%	70.99%
GainRatio	73.00%	73.21%	72.90%
ChiSquare	73.10%	73.16%	72.73%
Match set	71.83%	71.79%	71.35%

Table 3 shows the number of rules generated in each model and subset explored. As the objective is to evaluate the relationships among characteristics, the lower the set of characteristics, without affecting the performance of the classifier, the better the result. With this exploration, on average, there was an 85% reduction in the number of rules generated in relation to the original set.

Although the classifier performance did not present a significant statistical difference, there was a great reduction in the number of attributes and in the number of rules generated, showing, even more, the high relevance of the

Table 3. Number of generated rules.

Dataset	Classifier		
	C4.5	PART	RIPPER
Original set	135	266	23
CFS	22	19	8
GainRatio	57	49	10
ChiSquare	59	59	16
Match set	19	14	5

attributes: age group, AIDS, and other immunology in the classification of the outcome of TB cases.

The finding of the variable AIDS as relevant comorbidity for the outcome of cases TB is in agreement with discussions in the literature that directly correlate the two diseases [18]. TB is one of the most frequent opportunistic diseases in the HIV-infected patient and there is evidence that it is one of the main causes associated with death in this population [3].

3.4   Analysis: FS Wrapper Approaches + Classification Algorithms

For the second analysis, Table 4 presents the most relevant characteristics selected in each FS wrapper approach. In this case, the largest reduction was 64% in the number of attributes, from the original set to the match set.

Table 4. Description of selected characteristics - FS wrapper approaches.

Dataset	#	Description
1NN	6	Age group, education, AIDS, alcoholism, mental disorders, other immunology
BN	5	Age group, AIDS, alcoholism, mental disorders, other immunology
DT	8	Age group, sex, education, AIDS, diabetes, alcoholism, drug addiction, other immunology
Match set	4	Age group, AIDS, alcoholism, other immunology

Once more, there was also no significant difference between the models explored (Table 5), 72% of correct predictions were maintained (on average).

Table 5. Accuracy - Classification models.

Dataset	Classifier		
	1NN	BN	DT
Original set	71.92%	72.27%	73.06%
1NN	73.28%	72.23%	73.00%
BN	72.43%	72.38%	72.38%
DT	73.03%	72.34%	73.09%
Match set	72.50%	72.37%	72.43%

However, in this exploration, the alcoholism appears as comorbidity, not mentioned until then, was evaluated as an attribute of high relevance for the outcome of cases TB. The percentage of TB patients who have problems with alcoholism ranges from 10% to 50% [14, 22]. Furthermore, studies associate alcoholism with therapeutic failure, treatment abandonment, and death due to TB [13].

4 Conclusion

As a first approach of the KDD techniques to the Brazilian context of TB, the analyses performed showed to be promising when assessing the importance of comorbidities in TB cases. The findings are in agreement with recent studies in the health literature on TB, where AIDS and alcoholism comorbidities are being studied as important influencers for the treatment course of TB patients [4].

In future work, we intend to deepen the analyses, investigate the relationships found, investigate the differences among outcomes: death with main cause TB, death without main cause of TB, and treatment default, and also apply other methods of pattern recognition to the data set.

One aspect we intend to apply in this study is to use ontologies to contribute in every step of the KDD process: data selection, data preprocessing, data transformation, data mining, and interpretation and analysis of results. In the data selection step, ontologies can help to have a good understanding of the study domain and the data to be analyzed [1]. Data preprocessing can be done using restrictions and rules embedded in ontologies [7]. Feature engineering can be carried on using ontologies through semantic mapping enriching with additional knowledge [17, 23]. Obtained results can be analyzed using domain-specific ontologies. It is also possible to use ontologies to explain the results obtained from black-box ML algorithms [6].

Acknowledgements. We thank Freepik (www.freepik.com) to provide the icons used in the composition of Fig. 1. DA would like to thank the São Paulo Research Foundation for financial support (Process numbers: 2022/00020-0 | 2021/01961 | 2020/01975-9).

References

1. Abhishek, K., Singh, M.: An ontology based decision support for tuberculosis management and control in India. Int. J. Eng. Technol. **8**(6), 2860–2877 (2016)
2. Apunike, A.C., et al.: Analyses of public health databases via clinical pathway modelling: TBWEB. In: Krzhizhanovskaya, V.V., et al. (eds.) ICCS 2020. LNCS, vol. 12140, pp. 550–562. Springer, Cham (2020). https://doi.org/10.1007/978-3-030-50423-6_41
3. Bastos, S.H., Taminato, M., Fernandes, H., Figueiredo, T.M.R.M.d., Nichiata, L.Y.I., Hino, P.: Sociodemographic and health profile of TB/HIV co-infection in Brazil: a systematic review. Revista brasileira de enfermagem **72**(5), 1389–1396 (2019)
4. Deiss, R.G., Rodwell, T.C., Garfein, R.S.: Tuberculosis and illicit drug use: review and update. Clin. Infect. Dis. **48**(1), 72–82 (2009)
5. Farley, J.F., Harley, C.R., Devine, J.W.: A comparison of comorbidity measurements to predict healthcare expenditures. Am. J. Managed Care **12**(2), 110–118 (2006)
6. Faust, K., et al.: Intelligent feature engineering and ontological mapping of brain tumour histomorphologies by deep learning. Nat. Mach. Intell. **1**(7), 316–321 (2019)
7. Geisler, S., Quix, C., Weber, S., Jarke, M.: Ontology-based data quality management for data streams. J. Data Inf. Qual. (JDIQ) **7**(4), 1–34 (2016)
8. Glaziou, P., Floyd, K., Raviglione, M.C.: Global epidemiology of tuberculosis. In: Seminars in Respiratory and Critical Care Medicine, pp. 271–285. Thieme Medical Publishers (2018)
9. Johnston, A.H.: Practical machine learning: A beginner's guide to data mining with WEKA (2018)
10. Kritski, A., et al.: Tuberculosis: renewed challenge in Brazil. Revista da Sociedade Brasileira de Medicina Tropical **51**(1), 2–6 (2018)
11. Lönnroth, K., Roglic, G., Harries, A.D.: Improving tuberculosis prevention and care through addressing the global diabetes epidemic: from evidence to policy and practice. Lancet Diabetes Endocrinol. **2**(9), 730–739 (2014)
12. World Health Organization et al.: WHO policy on collaborative TB/HIV activities: guidelines for national programmes and other stakeholders. World Health Organization (2012)
13. Pelissari, D.M., et al.: Identifying socioeconomic, epidemiological and operational scenarios for tuberculosis control in Brazil: an ecological study. BMJ Open **8**(6), e018545 (2018)
14. Pereira, J.d.C., Silva, M.R., Costa, R.R.d., Guimarães, M.D.C., Leite, I.C.G.: Profile and follow-up of patients with tuberculosis in a priority city in Brazil. Revista de Saúde Pública **49**, 6 (2015)
15. Register, A.H.: A Guide to MATLAB Object-Oriented Programming. CRC Press (2007)
16. Reis-Santos, B., et al.: Tuberculosis in brazil and cash transfer programs: a longitudinal database study of the effect of cash transfer on cure rates. PLoS ONE **14**(2), e0212617 (2019)
17. Ristoski, P., Paulheim, H.: RDF2Vec: RDF graph embeddings for data mining. In: Groth, P., et al. (eds.) ISWC 2016. LNCS, vol. 9981, pp. 498–514. Springer, Cham (2016). https://doi.org/10.1007/978-3-319-46523-4_30
18. Samuels, J.P., Sood, A., Campbell, J.R., Khan, F.A., Johnston, J.C.: Comorbidities and treatment outcomes in multidrug resistant tuberculosis: a systematic review and meta-analysis. Sci. Rep. **8**(1), 1–13 (2018)

19. Secretariat of Health Surveillance, Department of Surveillance of Communicable Diseases, Ministry of Health: National plan to end tuberculosis as a public health problem. Virtual Health Library of the Brazilian Ministry of Health (2020)
20. Secretariat of Health Surveillance, Department of Surveillance of Communicable Diseases, Ministry of Health: Recommendations Manual For The Control Of Tuberculosis In Brazil. Virtual Health Library of the Brazilian Ministry of Health (2020)
21. Silva, D.R., et al.: Risk factors for tuberculosis: diabetes, smoking, alcohol use, and the use of other drugs. Jornal Brasileiro de Pneumologia **44**(2), 145–152 (2018)
22. Simou, E., Britton, J., Leonardi-Bee, J.: Alcohol consumption and risk of tuberculosis: a systematic review and meta-analysis. Int. J. Tuberc. Lung Dis. **22**(11), 1277–1285 (2018)
23. Unbehauen, J., Hellmann, S., Auer, S., Stadler, C.: Knowledge extraction from structured sources. In: Ceri, S., Brambilla, M. (eds.) Search Computing. LNCS, vol. 7538, pp. 34–52. Springer, Heidelberg (2012). https://doi.org/10.1007/978-3-642-34213-4_3
24. Van Rossum, G., Drake, F.L.: The python language reference manual. Network Theory Ltd. (2011)
25. Witten, I.H., Frank, E.: Data mining: practical machine learning tools and techniques with java implementations. ACM SIGMOD Rec. **31**(1), 76–77 (2002)
26. World Health Organization and others: TB comorbidities and risk factors (2019)
27. World Health Organization and others: Global tuberculosis report 2020 (2020)

Noninvasive Estimation of Mean Pulmonary Artery Pressure Using MRI, Computer Models, and Machine Learning

Michal K. Grzeszczyk[1](✉)(iD), Tadeusz Satława[1], Angela Lungu[2](iD),
Andrew Swift[3], Andrew Narracott[3,4](iD), Rod Hose[3], Tomasz Trzcinski[5,6,7](iD),
and Arkadiusz Sitek[1](iD)

[1] Sano Centre for Computational Medicine, Cracow, Poland
m.grzeszczyk@sanoscience.org
[2] Technical University of Cluj-Napoca, Cluj-Napoca, Romania
[3] The University of Sheffield, Sheffield, UK
[4] Insigneo Institute for in Silico Medicine, University of Sheffield, Sheffield, UK
[5] Warsaw University of Technology, Warsaw, Poland
[6] Tooploox, Wroclaw, Poland
[7] Jagiellonian University of Cracow, Cracow, Poland

Abstract. Pulmonary Hypertension (PH) is a severe disease characterized by an elevated pulmonary artery pressure. The gold standard for PH
diagnosis is measurement of mean Pulmonary Artery Pressure (mPAP)
during an invasive Right Heart Catheterization. In this paper, we investigate noninvasive approach to PH detection utilizing Magnetic Resonance
Imaging, Computer Models and Machine Learning. We show using the
ablation study, that physics-informed feature engineering based on models of blood circulation increases the performance of Gradient Boosting
Decision Trees-based algorithms for classification of PH and regression
of values of mPAP. We compare results of regression (with thresholding of estimated mPAP) and classification and demonstrate that metrics
achieved in both experiments are comparable. The predicted mPAP values are more informative to the physicians than the probability of PH
returned by classification models. They provide the intuitive explanation
of the outcome of the machine learning model (clinicians are accustomed
to the mPAP metric, contrary to the PH probability).

Keywords: Pulmonary hypertension · Regression · Gradient Boosting
Decision Trees · Mathematical modelling

1 Introduction

Pulmonary Hypertension is a severe disease difficult to diagnose with multiple
possible root causes [6]. For many years, PH was identified if a mean Pulmonary
Artery Pressure (mPAP) of a patient at rest was equal to or above 25 mmHg.
Recently, it has been suggested to lower the threshold to 20 mmHg [19]. The

D. Groen et al. (Eds.): ICCS 2022, LNCS 13352, pp. 14–27, 2022.
https://doi.org/10.1007/978-3-031-08757-8_2

precise measurement of mPAP is non-trivial and requires conducting an invasive Right Heart Catheterization (RHC) - the gold standard for diagnosing PH. This procedure carries risks, requires patient's preparation, trained staff, highly specialized equipment, it is expensive and time consuming. To lower the probability of complications it has to be performed at a specialized facility [5].

Non-invasive estimation of mPAP using medical imaging, mathematical modeling, and machine learning (ML) is an option to avoid issues related with RHC. Mathematical models, such as a Windkessel model, allow diagnosis of the vascular system parameters [23]. Different ML algorithms enable extracting knowledge about data samples and their performance usually increases with the addition of features from multiple domains.

In this paper, we present methods based on Gradient Boosting Decision Trees (GBDT) for non-invasive PH diagnosis. We use classic GBDT, DART (Dropouts meet Multiple Additive Regression Trees) [22] - a method utilizing dropouts of random trees during training - and GOSS (Gradient-based One-Side Sampling) [10] - a technique that uses different than GBDT process of training (retaining samples with large gradients and randomly dropping the ones with low gradients). We conduct analysis on data from 352-patient cohort and perform two tasks: classification of PH and regression of mPAP. As predictors, we use demographics features, measurements derived from Magnetic Resonance Imaging (MRI) and features obtained from 0D and 1D mathematical models [15].

Our main contribution is the demonstration of the ablation study, which shows, that physics-informed feature engineering based on mathematical models of blood circulation increases the performance of ML algorithms for classification and regression of PH and values of mPAP, respectively. Another significant contribution of this paper is comparison of utilities of classification and regression approaches for the detection of PH. While the regression achieves similar classification metrics (after thresholding of estimated mPAP), the values of predicted mPAP are more informative to the physicians than the probability of PH returned by classification models. As such, they provide the intuitive explanation of the outcome of the machine learning model (clinicians are accustomed to the mPAP metric, contrary to the PH probability).

2   Related Work

Multiple ML algorithms (utilizing features from various modalities like echocardiography, Computed Tomography (CT), or MRI) have been integrated for the purpose of the PH classification. In [14], five ML models were used and compared with each other. Boosted Classification Trees, Lasso Penalized Logistic Regression (LPLR), Random Forest (RF) for Regression, RF for Classification and Support Vector Machines (SVM) were adopted for mPAP prediction or PH classification basing on the echocardiographic measurements and basic patients characteristics (age, sex, BMI, body surface area). In [26], echocardiographic data was used to distinguish between pre- and post-capillary PH with one of the nine tested ML models (SVM, AdaBoost, LR, RF, Decision Trees (DT), K-Nearest Neighbours, GBDT, LogitBoost and Linear Discriminant Analysis (LDA)).

In [7], measurements derived from CT were used to train six ML classifiers to evaluate the probability of mPAP higher than 15 mmHg. Another approach was to record the heart sounds with a digital stethoscope to gather parameters for PH classification using LDA [2]. The analysis of the sounds revealed specific patterns in PH patients. In [1], it was noted that the sounds collected by phono-cardiogram can be applied for binary classification of PH with SVM. In [16], it was shown that MRI measurements combined with parameters from 0D and 1D computational models can be successfully used for PH and non-PH patients classification with DT. In our approach, we study the impact of mathematical models parameters on classification and regression. We also show the comparable performance of PH diagnosis with GBDT-based models in both tasks.

With the rise of Deep Learning (DL), multiple approaches of detecting PH directly from images, videos, or electrocardiography (ECG) signals were investigated. For example, chest X-Ray images can be utilized for binary classification of potential PH patients using Capsule Network with residual blocks [12]. In [27], three popular DL networks (ResNet50, Xception and Inception V3) were trained as predictors of PH. As shown in [13], an ensemble neural network can pose as a screening tool for PH from a 12-lead ECG signal.

ML can also be utilized for determining patients at risk of having Pulmonary Arterial Hypertension (PAH) from clinical records. In [11], it was shown that GBDT can help in screening for PAH based on their medical history. ML-based tools were also developed for the purpose of blood pressure estimation - in [25], Support Vector Machine Regression (SVR) models were applied for the prediction of the patient's blood pressure from the physiological data. Another example is an application of Multilayer Perceptron (MLP) for regression of systolic blood pressure using basic knowledge about patients (BMI, age, habits etc.) [24].

3   Methods

In this section, we describe our approaches to noninvasive PH diagnosis. We present the details of our dataset and introduce mathematical models which enabled the acquisition of physics-informed features. Finally, we train GBDT-based models on multiple feature sets to perform mPAP regression and PH classification experiments.

3.1   PH Dataset

Table 1 presents the available features of patients who were suspected with PH and underwent MRI and RHC within 48 h.

The medical procedures were performed at the Sheffield Pulmonary Vascular Disease Unit. The RHC procedure was conducted with a balloon-tipped 7.5-Fr thermodilution catheter. The PH was defined if measured mPAP $\geq$ 25 mmHg. Using these criteria out of the cohort of 352 patients 286 were diagnosed with PH. From 286 patients with PH, 142 had Pulmonary Arterial Hypertension, 86 had Chronic Thromboembolic PH, 35 PH cases were due to lung diseases

Table 1. PH dataset with patient related data, parameters derived from 0D and 1D models and measurements from MRI imaging. In the appendix (Sect. 7) we provide explanations for the feature names. P-value tests a null hypothesis that the coefficient of the univariate linear regression between a feature and mPAP is equal to zero.

Feature	No PH			PH			
	cnt	Mean	std	cnt	Mean	std	p-value
mPAP, mmHg	66	19.67	3.34	286	46.95	13.08	
Demographics							
Age, years	66	56.61	13.78	286	61.69	14.24	0.242
Gender, female/male	66	43/23		286	173/113		0.549
Who, no.	56	2.52	0.54	285	3.04	0.44	<0.001
bsa, m^2	65	1.88	0.25	286	1.82	0.22	0.24
0D and 1D models							
R_d, $kg/m^4 s$	66	6.08E+07	4.94E+07	286	1.46E+08	2.53E+08	<0.001
R_c, $kg/m^4 s$	66	7.94E+06	7.80E+06	286	9.17E+06	1.87E+07	0.072
C, $m^4 s^2/kg$	66	9.92E-09	6.71E-09	284	3.94E-04	6.65E-03	0.669
R_{tot}, $kg/m^4 s$	66	6.83E+07	5.38E+07	286	1.56E+08	2.62E+08	<0.001
W_b/W_{tot}	66	0.24	0.10	286	0.39	0.11	<0.001
MRI							
rac_fiesta, %	66	26.39	15.43	286	13.68	8.93	<0.001
syst_area_fiesta, cm^2	66	7.62	2.17	286	9.78	2.78	<0.001
diast_area_fiesta, cm^2	66	6.08	1.71	286	8.66	2.57	<0.001
rvedv, mL	66	118.93	36.00	286	159.58	58.27	<0.001
rvedv_index, mL/m^2	66	53.78	21.83	286	73.92	39.39	<0.001
rvesv, mL	66	55.41	20.68	286	102.48	49.92	<0.001
rvesv_index, mL/m^2	66	24.64	10.84	286	47.63	30.19	<0.001
rvef, %	66	53.32	9.86	286	38.05	13.59	<0.001
rvsv, mL	66	63.52	22.61	286	57.15	23.39	0.026
rvsv_index, mL/m^2	66	29.14	13.90	286	26.32	15.02	0.292
lvedv, mL	66	116.57	33.09	286	91.30	27.33	<0.001
lvedv_index, mL/m^2	66	53.16	21.90	286	41.25	19.20	<0.001
lvesv, mL	66	34.27	15.66	286	31.32	14.56	0.23
lvesv_index, mL/m^2	66	16.85	16.81	286	14.01	8.18	0.194
lvef, %	66	71.13	8.54	286	65.81	10.92	<0.001
lvsv, mL	66	82.30	23.30	286	59.97	19.93	<0.001
lvsv_index, mL/m^2	66	38.07	16.20	286	27.20	13.51	<0.001
rv_dia_mass, g	66	22.62	6.80	283	44.48	25.47	<0.001
lv_dia_mass, g	66	91.47	27.71	286	90.64	24.98	0.436
lv_syst_mass, g	66	111.74	32.17	286	99.83	26.39	<0.001
rv_mass_index, g/m^2	66	10.44	4.94	285	20.94	15.09	<0.001
lv_mass_index, g/m^2	59	40.90	17.87	243	39.84	18.99	0.442
sept_angle_syst, degrees	66	139.95	11.68	286	172.51	22.11	<0.001
sept_angle_diast, degrees	66	134.21	8.28	286	145.01	11.93	<0.001
4ch_la_area, mm^2	66	1921.95	387.56	286	1785.95	556.53	<0.001
4ch_la_length, mm^2	66	55.76	7.86	286	55.62	8.60	0.412
2ch_la_area, mm^2	66	1764.62	496.75	286	1901.67	545.35	0.855
2ch_la_length, mm^2	66	48.66	9.08	286	52.12	9.33	0.166
la_volume, mL	66	55.22	17.96	286	54.16	25.36	0.005
la_volume_index, mL/m^2	66	24.95	10.14	286	23.24	10.45	0.042
ao_qflowpos, L/min	65	6.09	1.50	285	5.29	1.50	<0.001
ao_qfp_ind, $L/min/m^2$	65	2.79	1.18	285	2.44	1.15	0.003
pa_qflowpos, L/min	66	5.50	1.84	284	5.00	1.97	0.006
pa_qflowneg, L/min	66	0.62	0.59	285	1.07	0.83	<0.001
pa_qfn_ind, $L/min/m^2$	66	9.70	7.19	284	17.49	9.85	<0.001
systolic_area_pc, mm^2	66	731.05	236.42	284	950.17	268.98	<0.001
diastolic_area_pc, mm^2	66	619.82	162.71	284	866.42	244.57	<0.001
rac_pc, %	66	17.02	13.70	284	10.01	8.14	<0.001

(e.g. Chronic Obstructive Pulmonary Disease), 15 cases were associated with left heart disease. The cause of PH in the rest of patients was either multifactorial or unknown. All of the available data samples are part of the ASPIRE Registry (Assessing the Severity of Pulmonary Hypertension In a Pulmonary Hypertension REferral Centre) [8].

MRI images were captured with 1.5-tesla whole-body scanner (GE HDx, GE Healthcare, Milwaukee) with an 8-channel cardiac coil. The images were acquired in the supine position during a breath hold. The balanced steady state free precession (bSSFP) sequences were spatially and temporally synchronized with the 2D phase contrast (PC) images of the Main Pulmonary Artery (MPA) using cardiac gating. Short-axis and four-chamber cardiac images were also collected. The features from MRI were obtained as in [21]. $A(t)$ area of the MPA was extracted from the semi-automatically segmented bSSFP images. The blood flow through MPA ($Q(t)$) was extracted from the segmented areas overlaid on PC images. Using those measurements 0D- and 1D-model features were derived.

To prepare the feature dataset for the training of ML models we fill the missing values using linear interpolation. We encode categorical features to numerical values and scale all the features to have means of 0 and variances of 1.

3.2   Features Derived from Models of Blood Circulation

The cardiovascular system (CVS) is a closed circuit with the main purpose of transporting oxygenated blood to organs and tissues [17]. It comprises especially from heart, blood and vessels. One of the main components of the CVS is the pulmonary circulation. The target of the pulmonary circulation is to transport the deoxygenated blood from the right ventricle through MPA and other arteries to lungs and deliver the oxygenated blood to the left ventricle [9]. Since CVS can be described by its haemodynamics and structure of heart and vessels, the computational models based on the simplified representation of CVS were introduced [18]. Those models range from 0D models simulating the global haemodynamics (e.g. resistance and compliance of the system) to 3D models representing the complex behaviour of vessels and the blood flow over time. In [15], two models (0D and 1D) based on MRI measurements for the diagnosis of PH were proposed.

0D Model. 0D models are often based on the hydraulic-electrical analogue - the blood flow and electrical circuits have many computational similarities [18]. For example, the friction in the vessel can be identified as resistance R, the blood pressure as voltage and the flow-rate as current. Thus, by applying electrical laws (e.g. Kirchhoff's law, Ohm's law), the simplified representation of the CVS can be achieved. 0D modelling of CVS started with the implementation of the two-element Windkessel model [23]. Different variants of this model appeared in the literature and it was applied to simulate pulmonary circulation [4,20]. The 3-element (R_cCR_d) Windkessel model comprises of the capacitor C characterizing the compliance of the pulmonary circulation and two resistors R_c and R_d representing the resistance proximal and distal to the capacitor respectively.

In [15], R_cCR_d model was applied to capture the characteristics of PH and non-PH patients. In this model, the sum of two resistors can be interpreted as the ratio between mean pressure and mean flow (pulmonary vascular resistance - PVR) and C indicates the compliance of the pulmonary arteries. To optimize the parameters of 0D model for the specific patient, two MRI imaging techniques of MPA were used: PC and bSSFP. The bSSFP images were segmented to find the area of MPA $(A(t))$ over time. Then, the segmented regions were overlayed over PC images to capture the blood flow through MPA $(Q(t))$. Having the $Q(t)$ and pressure $p(t)$ (derived from the measured MPA radius) the parameters of the Windkessel model which were best describing the relationship between $Q(t)$ and $p(t)$ over time could be derived.

1D Model. The simplified representation of the pulmonary vasculature is multiple elastic tubes with numerous branches. 1D models often analyse the propagation of the pressure and flow waves in such structures. The 1D equations of the waves travelling through elastic tubes are derived from Navier-Stokes equations. In [15], the analysis of the power of the pressure waves was performed. The pressure wave was broken down into forward and backward-travelling elements (since vessels are rugged and twisted, some waves are bouncing off the vessel walls and travel backward). It was assumed and confirmed that the power of the backward wave in relation to the total wave power was greatly higher in PH cases than in healthy ones. As diseased pulmonary vasculature contains more deposits and stenoses the ratio of the backward wave power to the total wave power (represented as W_b/W_{tot}) is higher than in the healthy one.

3.3   Machine Learning for PH Detection

mPAP Regression. The decision whether the patient is suffering from PH is more important to the doctors than the actual value of mPAP. However, the non-invasive prediction of the PH occurrence together with the predicted value of mPAP is more informative to the clinicians. Therefore, we decide to conduct two experiments: mPAP regression and PH classification.

To find the best ML algorithm for mPAP regression we train three models based on GBDT: classic GBDT, DART and GOSS. We use mPAP feature as the ground truth for our models. We find the best hyper parameters for the models using Bayes optimization with 8-fold cross-validation (CV). We optimize them for 200 iterations with minimizing Mean Squared Error (MSE) as the optimization target. Then, using the best found parameters we train the models with leave one out cross validation (LOOCV) and MSE as the objective function. We measure MSE, Root MSE (RMSE) and Mean Absolute Error (MAE) as regression metrics of the model.

We assume that mPAP ≥ 25 mmHg is a positive PH diagnosis. With this assumption, we compute the binary classification metrics after thresholding the predicted and measured mPAP with 25 value. We calculate accuracy, sensitivity, specificity, True Positives (TP), False Positives (FP), True Negatives (TN), False

Negatives (FN). To compare the impact of different feature sets on the results (demographics, MRI, mathematical models), we repeat the procedure of hyper parameter optimization, models training with LOOCV and metrics collection for different combinations of features. We compare results of all the approaches.

Additionally, we train four other than boosted tree ML models on all features and compare the metrics with GBDT-based methods using LOOCV-derived metrics. These additional methods are MLP, SVR, AdaBoost and RF.

PH Classification. We conduct the binary PH classification, similarly to mPAP regression. We binarize mPAP feature with 25 mmHg threshold and train three GBDT-based models on different variations of feature sets, previously optimizing the hyper parameters using Bayes optimization. The optimization is handled for 200 iterations with 8-fold stratified CV to ensure similar distribution of positive and negative samples over each fold. The optimization goal is the maximum area under the receiver operating characteristic (ROC) curve.

We train GBDT, DART and GOSS on best found parameters with LOOCV. We calculate binary classification metrics: area under ROC curve (AUC), sensitivity, specificity, accuracy, TP, FP, TN and FN. To compute the binary classification metrics we use multiple thresholding strategies: youden - maximization of specificity + sensitivity, f1 - maximization of f1 metric (harmonic mean of precision and recall), closest01 - the point which is closest to (0,1) point on the ROC curve, concordance - maximization of the product of sensitivity and specificity.

4 Results

In this section, we present results of our experiments. We analyze, through the ablation study, the impact of different feature sets on models performance and compare metrics achieved by regression and classification models. In our case, the ablation study means the removal of feature sets before the training to understand their contribution to the overall performance of ML models. We also show, that regression models can be utilized as a tool for PH classification.

4.1 mPAP Regression

Table 2 presents results of regression experiments. The lowest regression metrics are achieved by DART MAE = 5.94, RMSE = 7.85 and MSE = 61.66. GBDT has marginally better classification metrics with sensitivity = 0.96 (DART, 0.95), specificity = 0.74 (DART, 0.74) and accuracy = 0.92 (DART, 0.91). The difference between DART and GBDT results is not statistically significant (p-value = 0.93). Additionally, GBDT-based methods outperform other tested ML algorithms: RF, AdaBoost, SVR and MLP with RF achieving the lowest MAE (6.55) out of all compared methods (p-value = 0.003).

Table 3 shows results of the ablation study. For all models, MAE drops with different combinations of feature sets (demographics, mathematical models, MRI) as opposed to only one feature set. The lowest MAE when a single feature

Table 2. Results of mPAP value regression with LOOCV. Models trained on demographics, MRI-derived features and 0D and 1D models parameters. P-value is calculated based on MAE against DART model.

Method	MAE	RMSE	MSE	R^2	Sensitivity	Specificity	Accuracy	p-value
MLP	7.71	10.37	107.50	0.58	0.93	0.55	0.86	<0.001
SVR	7.29	9.39	88.14	0.65	0.95	0.55	0.88	<0.001
AdaBoost	6.92	8.92	79.59	0.69	0.97	0.41	0.87	<0.001
RandomForest	6.55	8.64	74.59	0.71	0.95	0.56	0.88	0.003
GOSS	6.44	8.38	70.22	0.72	0.96	0.67	0.90	<0.001
GBDT	5.95	7.91	62.55	0.75	0.96	0.74	0.92	0.93
DART	5.94	7.85	61.66	0.76	0.95	0.74	0.91	

Table 3. Ablation study over the combinations of available feature sets (demographics, MRI, 0D and 1D models). P-value is calculated against models trained on all features.

Demographics	✓			✓		✓	✓
0D and 1D models		✓		✓	✓		✓
MRI			✓		✓	✓	✓
Regression (MAE)							
GOSS	11.09	9.16	6.93	8.44	6.77	6.51	6.44
p-value	<0.001	<0.001	0.007	<0.001	0.012	0.645	
GBDT	10.85	9.14	6.69	8.33	6.49	6.34	5.95
p-value	<0.001	<0.001	<0.001	<0.001	<0.001	0.012	
DART	11.01	9.35	6.76	8.43	6.20	6.20	5.94
p-value	<0.001	<0.001	<0.001	<0.001	0.058	0.083	
Classification (AUC)							
GOSS	0.74	0.87	0.91	0.89	0.95	0.93	0.95
p-value	<0.001	<0.001	<0.001	<0.001	0.117	<0.001	
GBDT	0.77	0.85	0.93	0.88	0.94	0.93	0.94
p-value	<0.001	<0.001	0.593	<0.001	0.147	0.017	
DART	0.79	0.85	0.93	0.88	0.95	0.93	0.95
p-value	<0.001	<0.001	0.005	<0.001	0.028	<0.001	

set is used is achieved for MRI-derived measurements (GOSS, 6.93; GDBT 6.69; DART, 6.76). However, the combination of all available feature sets yields the best performance (GOSS, 6.44; GBDT 5.95; DART, 5.94). The physics-informed feature engineering performed by the addition of 0D and 1D models parameters improves metrics obtained in the regression.

The relations between predicted and measured mPAP values are shown in Fig. 1. The addition of mathematical models features decreases the number of FP and FN (calculated with 25 mmHg threshold) even though the models were

Table 4. Results of PH classification with LOOCV. Models trained on demographics, MRI-derived features and 0D and 1D models parameters. Metrics sens (sensitivity), spec (specificity), acc (accuracy) are given for multiple thresholding strategies: youden, concordance, 01 (closest01), f1 (maximizing f1 metric).

Method	AUC	Youden			Concordance			01			f1		
		sens	spec	acc	sens	spec	acc	sens	spec	acc	sens	spec	acc
GOSS	0.95	0.88	0.94	0.88	0.88	0.94	0.88	0.88	0.92	0.89	0.97	0.68	0.91
GBDT	0.94	0.84	0.95	0.86	0.84	0.95	0.86	0.84	0.95	0.86	0.94	0.76	0.91
DART	0.95	0.85	0.95	0.87	0.85	0.95	0.87	0.87	0.92	0.88	0.95	0.8	0.92

trained on the MSE which is a regression objective function. Only 17 predictions are FP and 11 are FN for GBDT (DART; 17 FP, 14 FN). For GOSS, GBDT and DART, only one FP sample was predicted as having higher mPAP than 40 mmHg. The measured value for this patient during RHC was 24 mmHg which by current indicators means PH positive patient. In case of mPAP above 45 mmHg, all samples are predicted positively meaning high confidence of this model above that value. All false negative samples have the predicted values above 20 mmHg.

4.2 PH Classification

In the PH classification experiments, the impact of 0D and 1D models parameters is also significant (Table 3). For the single feature set, the highest AUC is achieved for models trained on MRI-derived parameters (GOSS, 0.91; DART, 0.93; GDBT, 0.93). The addition of features from mathematical models improves the performance and acquires the same AUC as for the models trained on all parameters (GOSS, 0.95; GDBT 0.94; DART, 0.95). Table 4 shows the detailed results of PH classification models trained on all features. The highest AUC is achieved for GOSS and DART models. Those models have the highest specificity (GOSS, 0.94; GBDT, 0.95; DART, 0.95) when thresholding their predicted probabilities with youden or concordance strategies. However, the classification of PH patients is a task in which we would like to detect as many positive patients as possible (maximizing sensitivity) while retaining reasonably high specificity (the percentage of accurately stating that no PH is present). Such an approach is most closely achieved with maximizing f1 metric as the thresholding strategy. With this strategy, DART predictions yield best metrics with sensitivity of 0.95, specificity of 0.8 and accuracy of 0.92. The results are comparable with the best regression metrics (sensitivity $= 0.96$, specificity $= 0.74$ and accuracy $= 0.92$). The FN had mPAP close to 25 mmHg (with a maximum of 33 mmHg) and relatively small PVR, meaning, that no severe PH case was misclassified. Half of the FP had mPAP higher than 20 mmHg. The ROC curves for the three models are presented in Fig. 2.

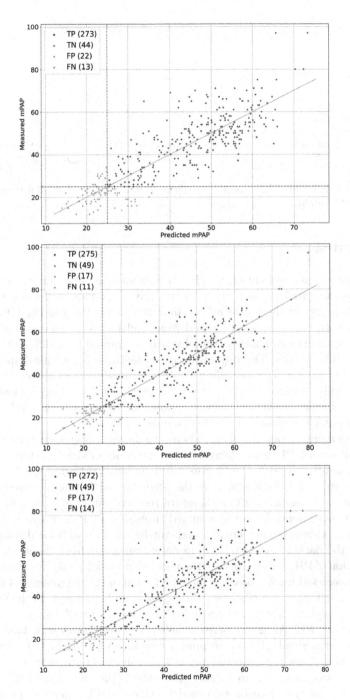

Fig. 1. Measured vs predicted values of mPAP with $\geq$ 25 mmHg thresholding for models trained on all parameters: GOSS (top, $R^2 = 0.72$, p-value < 0.001), GBDT (middle, $R^2 = 0.75$, p-value < 0.001), DART (bottom, $R^2 = 0.76$, p-value < 0.001).

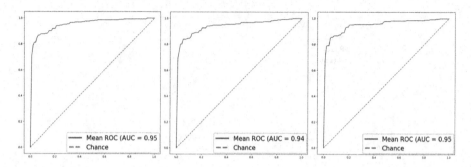

Fig. 2. ROC curves for GBDT-based classification models trained on all features: GOSS (left), GBDT (middle), DART (right).

5 Discussion

The noninvasive assessment of mPAP is a difficult task. In a clinical setting the pressure is measured through the invasive RHC. The models presented in this paper enable the prediction of mPAP in a noninvasive way using information about patients, measurements derived from multiple MRI images and mathematical models. The combination of all features acquired from different domains brings the best results. The physics-informed feature engineering improves the assessment of mPAP. The modelling of MPA haemodynamics enables the quantification of physiological markers that enhance the quality of predictions. While MRI is not a widely used test in PH diagnosis, we showed that it can be utilized for an accurate, noninvasive mPAP estimation.

What is more, as our knowledge about the disease progresses, the thresholds and definitions of PH may change. Our regression models are not restricted to the 25 mmHg threshold set before the training. Depending on the current and future state of PH classification, the predicted mPAP can be interpreted in different ways. Classification models return the probability of patient having PH - the probability depends on the assumed threshold for PH. In this setting, the regression models are more flexible and can be used as additional information regarding the patient's state to help in determining the final diagnosis, even, if the definition of PH changes. In case the regression model is used for classification only, the predicted mPAP poses as an explanation of the diagnoses. As shown in Fig. 1, the confidence in a positive PH diagnosis can be stronger as the predicted mPAP gets higher. Above the predicted value of 45 mmHg all patients were diagnosed with PH. All the positive samples that were misclassified as negative have the predicted mPAP over 20 mmHg which can be considered elevated. In other words we have not observed any critical failures of our models.

Nevertheless, clinicians are mostly interested in the final diagnosis of the ML models. We show that classification models achieve similar metrics as the regression models: sensitivity $= 0.95$, specificity $= 0.8$ and accuracy $= 0.92$ achieved by DART for classification, in comparison to sensitivity $= 0.96$, specificity $= 0.74$ and accuracy $= 0.92$ achieved by GBDT for regression. It is important to notice that the impact of the models performance by the features from mathematical models is more clearly represented in the classification task, because the described mathematical models were created for discrimination between PH and non-PH patients [15]. Parameters derived from those models pose as an accurate PH/non-PH differentiation mechanism and prediction of mPAP from those parameters may be a harder task. However, the addition of features derived from 0D and 1D models improves the regression metrics as well.

6 Conclusion

In this paper, we investigated the impact of physics-informed feature engineering on the performance of GBDT-based models for mPAP regression and PH classification. We showed that parameters from 0D and 1D mathematical models improve the metrics of tested models. Comparison of the results revealed that the PH diagnosis may be performed by regression models achieving similar metrics as the classification models. The provided, predicted mPAP value increases the confidence in the final diagnosis. Future works may include improvements in the feature engineering, utilizing deep learning to predict mPAP directly from MRI images or testing our methods on external datasets.

Acknowledgements. This publication is partly supported by the European Union's Horizon 2020 research and innovation programme under grant agreement Sano No. 857533 and the International Research Agendas programme of the Foundation for Polish Science, co-financed by the European Union under the European Regional Development Fund. This research was partly funded by Foundation for Polish Science (grant no POIR.04.04.00-00-14DE/18-00 carried out within the Team-Net program co-financed by the European Union under the European Regional Development Fund), National Science Centre, Poland (grant no 2020/39/B/ST6/01511). The authors have applied a CC BY license to any Author Accepted Manuscript (AAM) version arising from this submission, in accordance with the grants' open access conditions.

7 Appendix

Acronyms used in Table 1 and their explanations: mPAP: mean pulmonary arterial pressure measured during RHC procedure, who: WHO functional PAH score [3], bsa: body surface area, R_d: distal resistance calculated from 0D model, R_c: proximal resistance, C: total pulmonary compliance, R_{tot}: total resistance, W_b/W_{tot}: backward pressure wave to the total wave power, rac_fiesta: pulmonary arterial relative area change from bSSFP MRI, systolic_area_fiesta: syst area of MPA from bSSFP, diast_area_fiesta: diastolic area of MPA from bSSFP, rvedv: right ventricle end diastolic volume, rvedv_index: rv end diastolic volume index, rvesv: rv end systolic volume, rvesv_index: rv end systolic volume index, rvef: right ventricle ejection fraction, rvsv: rv stroke volume, rvsv_index: rvsv index, lvedv: left ventricle end diastolic volume, lvedv_index: lvedv index, lvesv: lv end systolic volume, lvesv_index: lvesv index, lvef: lv ejection fraction, lvsv: lv stroke volume, lvsv_index: lvsv index, rv_dia_mass: rv diastolic mass, lv_dia_mass: lv diastolic mass, lv_syst_mass: lv systolic mass, rv_mass_index: rv diastolic mass index, lv_mass_index: lv diastolic mass index, sept_angle_syst: systolic septal angle, sept_angle_diast: diastolic septal angle, 4ch_la_area: left atrium area 4 chamber, 4ch_la_length: la length 4 chamber, 2ch_la_area: left atrium area 2 chamber, 2ch_la_length: la length 2 chamber, la_volume: la volume, la_volume_index: la volume index, ao_qflowpos: aortic positive flow, ao_qfp_ind: aortic positive flow index, pa_qflowpos: PA positive flow, pa_qflowneg: PA negative flow, pa_qfn_ind: PA negative flow index, systolic_area_pc: systolic MPA area from PC, diastolic_area_pc: diastolic MPA area from PC, rac_pc: relative area change of MPA from PC.

References

1. Dennis, A., et al.: Noninvasive diagnosis of pulmonary hypertension using heart sound analysis. Comput. Biol. Med. **40**, 758–764 (2010). https://doi.org/10.1016/j.compbiomed.2010.07.003
2. Elgendi, M., et al.: The voice of the heart: vowel-like sound in pulmonary artery hypertension. Diseases **6** (2018). https://doi.org/10.3390/diseases6020026. www.mdpi.com/journal/diseases
3. Galie, N., et al.: Guidelines for the diagnosis and treatment of pulmonary hypertension: the task force for the diagnosis and treatment of pulmonary hypertension of the European society of cardiology (ESC) and the European respiratory society (ERS), endorsed by the international society of heart and lung transplantation (ISHLT). Eur. Heart J. **30**(20), 2493–2537 (2009)
4. Grant, B.J., Paradowski, L.J.: Characterization of pulmonary arterial input impedance with lumped parameter models. Am. J. Physiol.-Heart Circ. Physiol. **252**, H585–H593 (1987). https://doi.org/10.1152/ajpheart.1987.252.3.H585
5. Hoeper, M.M., Lee, S.H., Voswinckel, R., et al.: Complications of right heart catheterization procedures in patients with pulmonary hypertension in experienced centers. J. Am. Coll. Cardiol. **48**(12), 2546–2552 (2006)
6. Hoeper, M.M., et al.: Pulmonary hypertension. Dtsch Arztebl Int **114**, 73–84 (2017). https://doi.org/10.3238/arztebl.2017.0073
7. Huang, L., et al.: Prediction of pulmonary pressure after Glenn shunts by computed tomography-based machine learning models. Eur. Radiol. **30**, 1369–1377 (2020). https://doi.org/10.1007/s00330-019-06502-3
8. Hurdman, J., Condliffe, R., Elliot, C., Davies, C., Hill, C., et al.: Aspire registry: assessing the spectrum of pulmonary hypertension identified at a referral centre. Eur. Respir. J. **39**, 945–955 (2012). https://doi.org/10.1183/09031936.00078411
9. Jain, V., Bordes, S., Bhardwaj, A.: Physiology, Pulmonary Circulatory System. StatPearls Publishing (2021)
10. Ke, G., et al.: LightGBM: a highly efficient gradient boosting decision tree. Adv. Neural. Inf. Process. Syst. **30**, 3146–3154 (2017)
11. Kiely, D.G., et al.: Utilising artificial intelligence to determine patients at risk of a rare disease: idiopathic pulmonary arterial hypertension. Pulm. Circ. **9** (2019). https://doi.org/10.1177/2045894019890549
12. Kusunose, K., Hirata, Y., Tsuji, T., Kotoku, J., Sata, M.: Deep learning to predict elevated pulmonary artery pressure in patients with suspected pulmonary hypertension using standard chest X ray. Sci. Rep. **10** (2020). https://doi.org/10.1038/S41598-020-76359-W

13. Kwon, J.M., Kim, K.H., Inojosa, J.M., Jeon, K.H., Park, J., Oh, B.H.: Artificial intelligence for early prediction of pulmonary hypertension using electrocardiography. J. Heart Lung Transplant. **39**, 805–814 (2020). https://doi.org/10.1016/j.healun.2020.04.009
14. Leha, A., et al.: A machine learning approach for the prediction of pulmonary hypertension. PLoS ONE **14** (2019). https://doi.org/10.1371/journal.pone.0224453
15. Lungu, A., Wild, J.M., Capener, D., Kiely, D.G., Swift, A.J., Hose, D.R.: MRI model-based non-invasive differential diagnosis in pulmonary hypertension. J. Biomech. **47**, 2941–2947 (2014). https://doi.org/10.1016/j.jbiomech.2014.07.024
16. Lungu, A., Swift, A.J., Capener, D., Kiely, D., Hose, R., Wild, J.M.: Diagnosis of pulmonary hypertension from magnetic resonance imaging-based computational models and decision tree analysis. Pulm. Circ. **6**, 181–190 (2016). https://doi.org/10.1086/686020
17. Quarteroni, A., Manzoni, A., Vergara, C.: The cardiovascular system: mathematical modelling, numerical algorithms and clinical applications. Acta Numerica **26**, 365–590 (2017). https://doi.org/10.1017/S0962492917000046
18. Shi, Y., Lawford, P., Hose, R.: Review of zero-D and 1-D models of blood flow in the cardiovascular system. BioMedical Eng. OnLine **10**, 33 (2011). https://doi.org/10.1186/1475-925X-10-33
19. Simonneau, G., et al.: Haemodynamic definitions and updated clinical classification of pulmonary hypertension. Eur. Respir. J. **53** (2019). https://doi.org/10.1183/13993003.01913-2018
20. Slife, D.M., et al.: Pulmonary arterial compliance at rest and exercise in normal humans. Am. J. Physiol.-Heart Circ. Physiol. **258**, H1823–H1828 (1990). https://doi.org/10.1152/ajpheart.1990.258.6.H1823
21. Swift, A.J., Rajaram, S., Condliffe, R., et al.: Diagnostic accuracy of cardiovascular magnetic resonance imaging of right ventricular morphology and function in the assessment of suspected pulmonary hypertension results from the aspire registry. J. Cardiovasc. Magn. Reson. **14**(1), 1–10 (2012)
22. Vinayak, R.K., Gilad-Bachrach, R.: DART: dropouts meet multiple additive regression trees. In: Artificial Intelligence and Statistics, pp. 489–497. PMLR (2015)
23. Westerhof, N., Lankhaar, J.W., Westerhof, B.E.: The arterial Windkessel. Med. Biol. Eng. Comput. **47**, 131–141 (2008). https://doi.org/10.1007/s11517-008-0359-2
24. Wu, T.H., Pang, G.K.H., Kwong, E.W.Y.: Predicting systolic blood pressure using machine learning. In: 2014 7th International Conference on Information and Automation for Sustainability: "Sharpening the Future with Sustainable Technology", ICIAfS 2014, March 2014. https://doi.org/10.1109/ICIAFS.2014.7069529
25. Zhang, B., Ren, H., Huang, G., Cheng, Y., Hu, C.: Predicting blood pressure from physiological index data using the SVR algorithm. BMC Bioinform. **20** (2019). https://doi.org/10.1186/s12859-019-2667-y
26. Zhu, F., Xu, D., Liu, Y., Lou, K., He, Z., et al.: Machine learning for the diagnosis of pulmonary hypertension. Kardiologiya **60**, 96–101 (2020). https://doi.org/10.18087/cardio.2020.6.n953
27. Zou, X.L., et al.: A promising approach for screening pulmonary hypertension based on frontal chest radiographs using deep learning: a retrospective study. PloS One **15**(7) (2020). https://doi.org/10.1371/journal.pone.0236378

GAN-Based Data Augmentation for Prediction Improvement Using Gene Expression Data in Cancer

Francisco J. Moreno-Barea[(✉)] , José M. Jerez , and Leonardo Franco

Departamento de Lenguajes y Ciencias de la Computación, Escuela Técnica Superior
de Ingeniería Informática, Universidad de Málaga, Málaga, Spain
{fjmoreno,jja,lfranco}@lcc.uma.es

Abstract. Within the area of bioinformatics, Deep Learning (DL) models have shown exceptional results in applications in which histological images, scans and tomographies are used. However, when gene expression data is under analysis, the performance is often limited, further hampered by the complexity of these models that require several instances, in the order of thousands, to provide good results. Due to the difficulty and the costs involved in the collection of medical data, the application of Data Augmentation (DA) techniques to alleviate the lack of samples is a topic of great relevance. State-of-the-art models based on Conditional Generative Adversarial Networks (CGAN) and some introduced modifications are used in this work to investigate the effect of DA for prediction of the vital status of patients from RNA-Seq gene expression data. Experimental results on several real-world data sets demonstrate the effectiveness and efficiency of the proposed models. The application of DA methods significantly increase prediction accuracy, leading by 12% with respect to benchmark data sets and 3.15% with respect to data processed with feature selection. Results based on CGAN models outperform in most cases, alternative methods like the SMOTE or noise injection techniques.

Keywords: Data Augmentation · Gene expression · Bioinformatics · Deep Learning · CGAN

1 Introduction

Deep learning (DL) models have become the state-of-the-art prediction algorithms in several application tasks, translating into billions of dollars invested by industries towards its application. With the advancement of deep network architectures, the access to large databases and the use of powerful computing systems, DL models have made incredible progress in a large variety of problems. DL models have a more complex structure compared to traditional machine learning methods, as they include thousands of parameters and dozens of layers that must be adjusted during the training process, and because of this, its application requires the use of large data sets with thousands of instances in

© The Author(s), under exclusive license to Springer Nature Switzerland AG 2022
D. Groen et al. (Eds.): ICCS 2022, LNCS 13352, pp. 28–42, 2022.
https://doi.org/10.1007/978-3-031-08757-8_3

order achieve a better performance than traditional machine learning techniques (shallow ANNs, SVMs, RF, etc.) [7, 22].

In particular, in the area of bioinformatics large and readily available data sets are scarce. Medical records are sensitive data with associated privacy problems and a difficulty to obtain patient consent for massive dissemination. Further, gene expression data are significantly more difficult to obtain, they present a greater dispersion and are prone to suffer from the curse of dimensionality, as microarray data contains a greater number of features compared to the number of samples usually available. For these reasons, the application of data augmentation (DA) methods has become one of the relevant topics in the area, allowing for the addition of new synthetic generated samples. A revision of recent state-of-the-art works in the field related to DL models applied to genomic data sets showed that some advantages can be observed using these models [4, 26] but we have not found works applying DA as it is used in the present work with the aim of improving prediction capabilities.

Like it happens with DL models, DA best results are found in computer vision and image processing areas, where data possesses structure. Specifically, DA models have shown impressive results in generating synthetic realistic images, based on a framework called Generative Adversary Networks (GAN) [8, 19]. Essentially, a GAN model network generates new samples from a distribution learned from the original data set, and for this purpose, the GAN produces a confrontation between two competing neural networks that learn from each other. Apart from the achievements of GAN models obtained in image vision, they have proven to be useful also for the DA task with images [6, 9, 27]. Applying DA to non-image data sets is far more challenging. Experts in an specific domain can be asked to assess the quality of a generated image and to distinguish a synthetic from real samples. However, this type of human expert based evaluation is not feasible when applied to non image-sets, even less if we take into account gene expression data. Most common methods for applying DA to non-structured data are the SMOTE technique (synthetic minority oversampling technique) [3] designed to deal with imbalanced data sets, and the noise injection methods as a way to prevent overfitting and improve prediction accuracy [17, 18, 20, 29]. Nonetheless, in recent times GAN models have become one of the reference DA methods also with other types of structured data such as time series or signals [11, 23], with data sets without any type of spatial or temporal structure [5, 16, 21], and also in biomedical problems [1, 12, 14].

Taking into account all the aspects mentioned above, this work has several objectives. Current research attempts to add knowledge to the existing scientific literature related to the application of DA with GAN models in biomedical problems, and more specifically with gene expression data. On the other hand we analyse modifications to state-of-the-art DA methods in order to obtain an increase in the precision of the cancer prognosis prediction problem compared to the traditional SMOTE and noise injection methods, which will allow the efficient application of techniques of Deep learning-based DA to small and non-structured data sets across multiple domains. Finally, we want to verify the methods ability to replicate the gene expression data with the Fréchet Inception Distance (FID), and be able to provide support for the prediction results.

2   Methodology

We include in this section the Data Augmentation (DA) methods and models applied to a cancer prognosis problem with different gene expression data sets.

2.1   Noise Injection Method

To perform DA with image sets there are some methods whose execution and approach is simple, such as resampling, flipping, cropping, shifting, or noise injection. To perform DA with non-image sets, some of these methods can also be used, such as resampling, based on repeating random instances of the data, or noise injection, based simply on modifying instances with degrees of noise. Although the noise injection may have a simple approach, the application of a procedure based on this method can be modified to obtain effective results [17].

The noise injection method designed randomly selects training samples and modifies a maximum of 25% of the features. The noise is generated from a random normal distribution with a standard deviation of 0.2 and is added to the original value of the feature, being subsequently controlled so as not to exceed the range of $[0; 1]$. A standard deviation value of 0.2 is enough to create samples that does not stray too far from the real space of instances.

2.2   SMOTE Techniques

Apart from the addition of noise to perform DA with non image data sets, in the literature we can find some applications of SMOTE techniques (synthetic minority oversampling technique) [3] designed to generate synthetic data in data sets that present imbalanced classes. This oversampling technique uses a k-nearest neighbour algorithm, instead of random sampling with replacement. SMOTE performs a random interpolation of the instance of the selected minority class and its nearest neighbours, in order to balance the data set and operating in the feature space. The interpolation calculates the difference between the instance and each of the selected neighbours, multiplies the difference for each feature by a random normalisation and adds this value to the original feature of the sample. This process creates new instances of the minority class that are located within this space between the sample and its neighbours.

However, this technique has certain drawbacks due to random interpolation. One of the most notable disadvantages is the possible generation of samples that do not respect the geometry present in the data set. The generated samples can occupy positions in the feature space that belong to the majority class data. Other significant drawback is that SMOTE does not allow to control the amount of synthetic samples generated, only those necessary to balance the data set.

2.3 Conditional Generative Adversarial Networks

The standard GAN model [8] has a general structure composed by two neural networks, called the *generator* and the *discriminator*, that are trained simultaneously resulting in a confrontation process. In this way, the discriminator network (D) tries to distinguish whether a sample comes from the real distribution or is a synthetic sample, i.e., for the input sample x, the discriminator estimates the probability that it belongs to the real distribution or not. The generator network (G) gets as output a synthetic sample from a noisy random distribution z. The purpose of the generator is create new synthetic samples with features that approximate those present in the real samples, so that the discriminator network will not be able to distinguish these synthetic samples as samples not coming from the real distribution. Therefore, the generator process is opposite to that of the discriminator, giving rise to a competitive environment.

Specifically, the model considered was the Conditional GAN (CGAN) [15], a variant of the standard GAN model. In CGAN, the information concerning to a condition y, the sample label or other data information is taken into account in the network. In this way, the latent space z and the condition y are passed as input to the generator network. This condition can be created randomly when training the model and it can be controlled when generating synthetic samples. The condition is also added to the input of the discriminator network, being the same that has been used to create a synthetic sample by the generator or the label assigned to the real sample.

$$\min_G \max_D \; \mathbb{E}_{x \sim p_{data}(x)}[\log D(x|y)] + \mathbb{E}_{z \sim p_z(z)}[\log(1 - D(G(z|y)))] \qquad (1)$$

The objective cost function (Eq. 1) of the CGAN model presents the behaviours identified with the competitive process: one related to better recognise samples that belong to the real distribution and another related to better recognise samples created by the generator. In this way, the ability of the model to perceive whether the samples are real or fake is expressed in Eq. 2, and the error identified with the recognition of fake samples is expressed by Eq. 3.

$$\max_D \; \mathbb{E}_{x \sim p_{data}(x)}[\log D(x|y)] + \mathbb{E}_{z \sim p_z(z)}[\log(1 - D(G(z|y)))] \qquad (2)$$

$$\min_G \mathbb{E}_{z \sim p_z(z)}[\log(D(G(z|y)))] \qquad (3)$$

2.4 CGAN Modified Generative Process

Considering the DA process to deal with supervised benchmark problems, we implemented modifications to the standard CGAN generative process giving rise to the ModCGAN model. This modified model was developed in a previous work [16]. The most significant difference from the generative process performed with ModCGAN compared to CGAN is the use of an external classifier called "generative classifier". This generative classifier is used to label the synthetic samples

32   F. J. Moreno-Barea et al.

created by the generator and discard them if they do not present enough quality. The whole generative process is shown in Fig. 1. The generative classifier is
trained with the real samples from the training set, also adding noisy samples
from two different methods: a uniform random distribution and gaussian noise
injection. The use of these different noise sources teaches the classifier to distinguish real and fake samples, from pure noise (uniform random distribution) and
samples with similar aspects to the real distribution (gaussian noise injection).

In the ModCGAN generative process, the generator creates a synthetic sample from a noisy random distribution and a label, since its generative base is the
same as CGAN. However, instead of using the discriminant network, ModCGAN
uses the generative classifier to estimate the label for the synthetic sample. If
the 'noise' label is estimated, sample is considered fake and is discarded. On the
other hand, if the estimated label is different from 'noise', it means that classifier
has predicted a label from the real ones, so the sample is saved with the predicted
label. Applying this modified process, the synthetic sample may be assigned a
different label than the one used by the generator at the generative process.
Furthermore, it is possible that samples that the discriminator network could
consider fake are saved or, conversely, not save samples that the discriminator
could detect as real but that the generative classifier predicts as 'noise'.

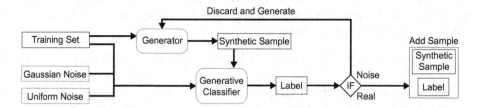

Fig. 1. Generative DA process for the creation of synthetic data in the ModCGAN
model. When the generative model and the generative classifier are trained, the synthetic sample is generated and the appropriate label is predicted. If the label represents
noise, the sample is discarded and another sample is generated, else the sample is save.

2.5   Modifications for Treating Unbalanced Data Set Distribution

A possible problem that arises from generating synthetic samples with GAN
models is the so-called collapse of the model. This problem arises when the
generator creates the same synthetic samples repeatedly. Also when it creates
synthetic samples from only one of the classes, usually from the majority class.
Two modifications, 'Multiclass' and 'Balanced Multiclass' [16], were applied to
avoid collapse of the model causing problems in the DA application. The modification was indicated in the models with the suffixes '_M' and '_BM' respectively.

The GAN-based models with these modifications use two independent models. Each independent model is trained with a set that presents all the samples
that belong to one of the classes with a random selection of samples from the

other class, trying to reach a ratio of 20%. The purpose is that each generator is able to focus on one class of the problem, always taking into account its differences with the samples of the other class. These generators are used in an ensemble methods process, each generating a specific proportion of samples that are joined to produce the final synthetic data set.

The specific number of samples generated by each generator depends on the desired number of samples, the proportion of each class in the original data set, and the implemented modification applied. The generative process with 'Multiclass' tries to keep the original classes proportion, while with 'Balanced Multiclass' it generates more samples for the minority class. It should be mentioned that the generative process followed in both modifications remains as the original GAN-based model, so if a sample is classified as fake or noise, the same generator model that created this sample generates a new one. These modifications are not intended to force the model to generate only synthetic samples that belong to one of the classes, but rather to better adapt the distribution of the samples to avoid collapse when the model trains with all the data.

2.6   Implementations of the Models

The CGAN and ModCGAN models, and the balance control modifications, present the same standard implementation with the exception of the specific implementation of the ModCGAN generative classifier. Generator network presents 4 hidden layers with Rectified Linear Unit (ReLU) [28] as activation function (de facto state-of-the-art activation functions in DL). Discriminator network also presents 4 hidden layers but with Leaky ReLU [13] activation function, since it provides more stability than ReLU in classification tasks. Both networks used batch normalisation as regularisation technique and Adam algorithm as optimisation algorithm with adaptive learning rate.

On the other hand, the classifier model used for the classification experiments and the generative classifier for synthetic process was a deep feedforward neural network. It presents 3 hidden layers with Leaky ReLU activation function and batch normalisation and dropout [24] at each hidden layer. The sigmoid activation function was used in the output neuron to classify patterns. Adam algorithm was also used as optimisation algorithm. The dropout rate applied was 0.1, 0.5 and 0.3 in the hidden layers of the generative classifier, and 0.3, 0.6 and 0.4 in the classifier used in test prediction experiments. The L2 norm was used in combination with dropout and batch normalisation to avoid overfitting.

2.7   Benchmark Data Sets

The benchmark data sets used for the experiments are freely available at The Cancer Genome Atlas (TCGA) website, provided by International Cancer Genome Consortium (ICGC). The data sets correspond to patients linked to 18 different cancer types: bladder carcinoma (blca), breast carcinoma (brca), colon adenocarcinoma (coad), glioblastoma multiforme (gbm), head and neck squamous cell carcinoma (hnsc), pan-kidney cohort (kipan), kidney renal cell carcinoma (kirc), brain lower grade glioma (lgg), liver hepatocellular carcinoma (lihc), lung adenocarcinoma (luad), lung squamous cell carcinoma (lusc), ovarian carcinoma (ov), prostate adenocarcinoma (prad), skin cutaneous melanoma (skcm), stomach adenocarcinoma (stad), stomach and esophageal carcinoma (stes), thyroid adenocarcinoma (thca) and uterine corpus endometrioid carcinoma (ucec).

The instances of these data sets represents patients affected of cancer, and for each patient it contains a row of 20531 variables than correspond to the expression level of a certain gene, so the data sets are RNA-Seq gene expression profiles after applying pre-processing procedures for batch correction and RSEM normalisation. A logarithmic (log2) transformation of the expression levels in the data was carried out, to approximate them to a normal distribution for its use with the predictive models. Additionally, a feature selection process was applied using the LASSO model [25] and the Gini importance from Random Forest method [2], reducing the number of genes. In order to perform a prediction analysis, vital status information for each patient has been collected, which is also freely available in TCGA. The vital status therefore supposes the label present in the data and the objective to be predicted.

$$\text{Balance} = \frac{H}{\log k} = \frac{-\sum_{i=1}^{k} \frac{c_i}{n} \log \frac{c_i}{n}}{\log k} \tag{4}$$

Table 1 shows some characteristics of the benchmark data sets, the columns show the name of the benchmark data set, the number of features after feature selection (Feat.) and instances (Inst.), the proportion of classes (Bal.), and the most significant gen according to the feature selection (Sig-Gen). Instead of showing the percentage of instances that belong to each class, we show a measure of balance (Eq. 4) based on the Shannon entropy (H). This measure is calculated given the number of instances n in the data set, the number of classes k, and the size of each class c_i. If the value of Balance is 1, the set is completely balanced, and if the value is 0, the set is completely unbalanced.

Table 1. Characteristics of the eighteen gene expression data sets studied.

Data	Feat.	Inst.	Bal.	Sig-Gen	Data	Feat.	Inst.	Bal.	Sig-Gen
blca	114	427	0.99	SPG7	luad	13	344	0.96	OR2T335
brca	74	1212	0.64	ZNF331	lusc	18	552	0.99	PYGB
coad	12	191	0.71	ALPK3	ov	33	307	0.97	PERP
gbm	19	171	0.73	ABCB8	prad	27	550	0.13	SNORA16A
hnsc	10	566	0.99	SLC25A43	skcm	21	473	1.00	INSR
kipan	102	1020	0.83	BANP	stad	3	450	0.96	LPPR2
kirc	88	606	0.92	DPAGT1	stes	47	646	0.97	PRTG
lgg	57	242	0.98	CCNI	thca	21	568	0.22	CXCL5
lihc	11	423	0.96	EIF5B	ucec	31	201	0.67	PEX11A

3   Experiments and Results

In order to keep complete independence between data generation, classifier model training, and prediction accuracy evaluation, we performed a division of the data set into training, validation and test sets. The synthetic data generation does not include any samples from the test set, which is kept separate for honest external performance testing. A 10-fold cross validation procedure is implemented in the prediction experiments and the training folds are augmented with synthetic samples. The result of the classification process is the average of the accuracy results obtained for the 10-folds. This process is further repeated with 10 different seeds to reduce possible random effects.

Table 2 shows the results obtained for the 18 data sets studied and described previously in Table 1. First column shows the test accuracy obtained with the original 'raw' data set, second column ('FS') shows the test accuracy when feature selection pre-processing is applied but not including any Data Augmentation process. Next group of columns show results obtained when DA is applied on the data sets after the feature selection process, showing the DA method used, the test accuracy obtained and the percentage of augmentation applied to the training data set. The model with the highest accuracy evaluated on the validation set is the indicated one. Last columns in the Table 2 show the relative difference (Eq. 5) obtained for each of the three DA methods applied: CGAN-based models, SMOTE and noise injection. The reference results for calculating the RD are the results obtained with the feature selection pre-processing. Last column, $\widehat{RD}$,

Table 2. Test accuracies obtained with the original data set (col. 2) and when a feature selection method is applied (col. 3). Cols. 4–6 show the test accuracy for the best case of the three implemented DA methods and the corresponding percentage of generated samples. Cols. 7–10 shows Test RD for the three used methods and last column ($\widehat{RD}$) the best RD obtained (see text for details).

Data	Original	FS	Data Augmentation			RD			$\widehat{RD}$
			Acc	Model	Perc	CGAN	SMOTE	NOISE	
gbm	0.6353	0.7794	0.8185	CGAN	200	**5.02**	−0.45	−1.21	5.02
coad	0.7015	0.7421	0.7529	CGAN	50	**1.45**	−5.82	−0.50	1.45
ucec	0.7250	0.7955	0.8405	NOISE	200	1.98	1.29	**5.66**	5.66
lgg	0.6802	0.8313	0.8413	CGAN_M	50	**1.20**	−1.03	0.70	1.20
ov	0.5192	0.7370	0.7921	ModCGAN_M	200	**7.47**	2.62	−6.56	7.47
luad	0.6071	0.6368	0.6694	CGAN_M	100	**5.13**	0.35	−1.48	5.13
lihc	0.6248	0.6690	0.6918	NOISE	200	1.19	1.64	**3.40**	3.40
blca	0.5566	0.7412	0.7636	CGAN_M	50	**3.03**	0.24	−0.76	3.03
stad	0.6093	0.6623	0.6708	SMOTE	None	0.79	**1.28**	−0.91	1.28
skcm	0.5213	0.6872	0.7377	ModCGAN_BM	200	**7.34**	−0.37	−2.94	7.34
prad	0.8918	0.9772	0.9810	SMOTE	None	0.31	**0.39**	0.04	0.39
lusc	0.5485	0.5624	0.5857	CGAN_BM	200	**4.15**	0.15	−4.98	4.15
hnsc	0.5919	0.7000	0.7388	ModCGAN_BM	200	**5.54**	1.50	−5.90	5.54
thca	0.9106	0.9640	0.9655	NOISE	200	0.05	−0.03	**0.16**	0.16
kirc	0.7239	0.7836	0.8041	ModCGAN_BM	100	**2.62**	0.01	2.04	2.62
stes	0.5747	0.5895	0.6033	CGAN_M	50	**2.35**	−0.43	0.03	2.35
kipan	0.7402	0.8265	0.8278	ModCGAN	200	**0.17**	−2.85	−1.31	0.17
brca	0.8072	0.8531	0.8567	ModCGAN	100	**0.42**	−0.90	0.09	0.42
Mean	0.6667	0.7521	0.7745			**2.79**	−0.13	−0.80	3.15

shows the maximum value for the relative difference over the results obtained previously (indicated with bold font).

$$RD = \frac{(\text{Acc\_Aug} - \text{Acc\_Ref})}{\text{Acc\_Ref}} \times 100 \qquad (5)$$

The results on Table 2 show an average improvement of 3.15% in $\widehat{RD}$ and 2% in test prediction accuracy using DA methods compared to the case when only feature selection is implemented. Using the FS pre-processing and DA methods a substantial improvement of approximately 11% is achieved over the original raw benchmark data sets, noting that the feature selected data set already permit to achieve almost a 9% increase in accuracy over the original data set, and thus obtaining a further increase with DA techniques is a relevant achievement.

The results indicate that for 13 out of 18 data sets best accuracy results are obtained through DA based on CGAN models, also obtaining in this case the best average results with a RD value of 2.79%. The noise injection method is the best one for 3 data sets and finally SMOTE leads in two cases; however both methods lead to negative RD values. In addition, 12 of 18 methods generate a percentage of samples greater than or equal to 100, so the models at least double the number of samples present in the training set. Regarding the efficacy of the different CGAN-based models (i.e. this analysis does not take into account noise injection or SMOTE methods), ModCGAN is the best option for 10 data sets while CGAN is the preferred method for the remaining 8.

Figure 2 shows the relationship between the test accuracy obtained for the data sets with FS and DA application and the obtained using the original

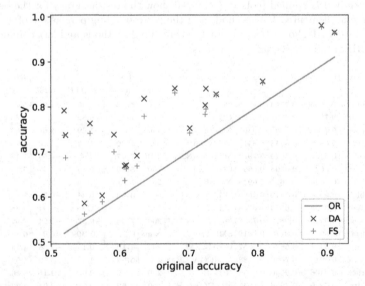

Fig. 2. Relationship between accuracy obtained with the processed data sets (DA and FS) and the original ones. Crosses represent the results obtained with augmented data sets and cross-hairs those obtained when only feature selection is applied. The continuous line represents the identity function.

raw benchmark data sets. The results show that the improvement in accuracy obtained from DA with respect to FS and the original results tends to be greater for those sets whose original accuracy is lower.

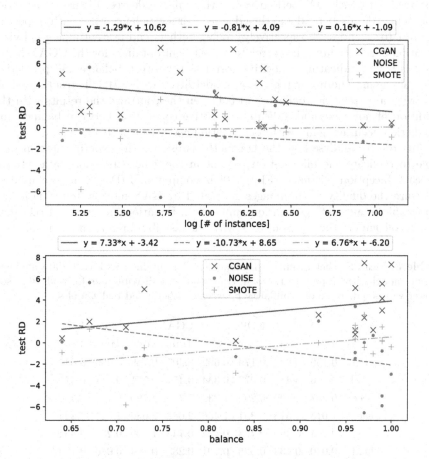

Fig. 3. Relative prediction accuracy difference (RD) *vs* the logarithm of the number of instances (top) and *vs* the balance (bottom). The lines are a linear regression adjusted to the data. Crosses and continuous line represent the results obtained with the CGAN model, dots and dashed line those obtained with the NOISE method, and cross-hairs and dashed-dotted line those obtained with SMOTE.

To analyse the influence of the data sets size on the precision obtained, we graph in Fig. 3 (top) the test relative difference RD obtained with each DA method versus the number of instances in a logarithmic scale on the x-axis. A linear regression model was fit to results for each type of DA method, obtaining for the CGAN-based models a correlation coefficient of -0.29, which indicates a moderate negative correlation between the number of instances and the prediction accuracy gain. Similar with the linear regression model fitted for the noise injection results, with a correlation coefficient of -0.14. The results obtained

with SMOTE show that the relationship between the number of instances and
the prediction accuracy gain remains stable, with a correlation coefficient of 0.04.

In a similar way, Fig. 3 (bottom) shows the test relative difference RD
obtained with each DA method versus the balance degree. Figure results does
not include data sets prad and thca due to their low balance value with respect to
the rest (0.13 and 0.22 respectively), being outliers and preventing a good visualisation of the results. The correlation coefficient obtained for the CGAN-based
models is 0.4, indicating a positive correlation between balance and prediction
accuracy gain. The linear regression model fitted with SMOTE results is similar, with a correlation coefficient of 0.44. On the contrary, the results with the
addition of noise method indicate a negative correlation between balance and
prediction accuracy gain, with a correlation coefficient of -0.43.

In order to analyse how the DA methods used in the experimentation were
able to replicate the information present in the gene expression data set, the
Fréchet Inception Distance (FID) [10] is computed. FID is a metric used to
measure the quality of the images generated by GAN models, but FID is also
applicable to any data generation application. Equation 6 shows the FID calculation to compare the distribution r with the distribution g from mean values of

Table 3. Mean Fréchet Inception Distance ($\widehat{FID}$) obtained with each data augmentation model with respect to the real Train and Test distributions for each data set.
Lower values indicate more similarity between synthetic and real samples.

Data	SMOTE		NOISE		CGAN		ModCGAN	
	Train	Test	Train	Test	Train	Test	Train	Test
gbm	0.150	0.893	0.179	0.602	0.407	0.769	0.206	0.644
coad	0.076	0.469	0.072	0.334	0.451	0.771	0.147	0.377
ucec	0.287	1.699	0.299	1.232	1.305	1.790	0.741	1.440
lgg	1.015	1.615	0.351	1.486	2.082	2.686	1.359	2.224
ov	0.264	0.490	0.103	0.434	0.745	1.099	0.256	0.554
luad	0.040	0.133	0.028	0.101	0.283	0.314	0.046	0.103
lihc	0.030	0.046	0.018	0.040	0.089	0.124	0.021	0.046
blca	1.441	2.037	0.541	2.060	3.170	3.820	2.057	2.471
stad	0.002	0.005	0.003	0.007	0.112	0.099	0.042	0.039
skcm	0.175	0.266	0.041	0.134	0.277	0.429	0.107	0.196
prad	0.269	0.000	0.047	0.094	1.020	1.795	1.142	2.687
lusc	0.086	0.135	0.029	0.093	0.280	0.326	0.051	0.105
hnsc	0.028	0.039	0.011	0.028	0.111	0.146	0.015	0.033
thca	0.223	1.083	0.222	0.623	0.801	1.265	0.411	0.804
kirc	0.378	0.949	0.341	1.127	3.413	3.657	1.359	1.708
stes	0.171	0.390	0.143	0.394	1.440	1.640	0.576	0.771
kipan	0.212	0.794	0.319	0.971	3.448	3.624	1.818	2.096
brca	0.128	0.700	0.206	0.682	2.064	2.336	1.052	1.291
Mean	0.276	0.652	0.164	0.580	1.194	1.483	0.634	0.977

the real (μ_r) and the generated (μ_g) vectors, the trace of the matrix (Tr) and the covariance matrix of the vectors (Σ_r, Σ_g).

$$\text{FID} = \|\mu_r - \mu_g\|^2 + \text{Tr}(\Sigma_r + \Sigma_g - 2(\Sigma_r\Sigma_g)^{1/2}) \qquad (6)$$

Mean FID values $(\widehat{FID})$ are obtained from the comparison of the synthetic class 0 distribution with the real class 0 distribution and the synthetic class 1 distribution with the real class 1 distribution. The $\widehat{FID}$ results for SMOTE only refer to the minority class comparison. Table 3 shows the $\widehat{FID}$ values obtained in the comparison with the real Train and Test distributions with the synthetic set generated with the different DA methods used, making a comparison between CGAN and ModCGAN.

The results reported in the Table 3 reveal that the addition of noise generates samples with great similarity to the original sample distribution. Samples generated by this method obtain the lowest $\widehat{FID}$ value, 0.164 on average measured on training samples. For the case of applying the SMOTE method, the analysis reveals a level of similarity slightly greater than the noise based one, with low $\widehat{FID}$ values (0.28 and 0.65 with respect to train and test). Regarding GAN-based methods, these add more variability to the augmented data sets, as samples generated with CGAN have less similarity for train and test data sets, reaching the highest average values of $\widehat{FID}$ (1.19 and 1.48 with respect to training and testing). On the other hand, the samples generated with ModCGAN model present lower values of $\widehat{FID}$, which indicates a greater similarity with the real samples. The $\widehat{FID}$ values are almost half of those obtained with CGAN (0.63 and 0.98).

4 Conclusion

In this work, we proposed the application of different state-of-the-art techniques for Data Augmentation (DA), with the aim of improving the prediction accuracy in patient prognosis analysis that can be obtained when data sets of RNA-Seq gene expression profiles are studied in different types of cancer. The results indicate that the application of DA methods can lead to an increase in prediction accuracy of approximately 3% (all the tested methods are evaluated, choosing the best one according to the validation error). This improvement has been achieved with respect to the data sets after applying a feature selection technique, as the improvement in prediction accuracy over the original raw data is approximately 11%. We observed also that conditional GAN models can greatly improve the generalisation results as a 2.79% increase was obtained, while alternative models like SMOTE and noise injection lead to negative results. Additionally, the quality of the generated samples was analysed to explain the performance achieved by each DA methods, and for this purpose the Fréchet Inception Distance (FID) was measured. From this analysis, we concluded that the noise addition method generates more similar samples, while CGAN-based models offers more variability. In the light of these results, we can draw the conclusion that greater variability in the augmented sets increases the potential of the prediction models to correctly classify test samples (never presented before to the classification model) that may not be similar to training ones.

In conclusion, DA techniques constitute a suitable approach to increase the prediction performance in patient prognosis analysis with data sets of RNA-Seq gene expression profiles. DA techniques based on CGAN models are capable of generating good quality synthetic data that lead on average to a 3% relative prediction increase. In relation to this, several future studies are planned, extending the application of DA methods to other gene expression data sets and bioinformatics tasks.

Acknowledgements. The authors acknowledge the support from MICINN (Spain) through grant TIN2017-88728-C2-1-R and PID2020-116898RB-I00, from Universidad de Málaga y Junta de Andalucía through grant UMA20-FEDERJA-045, and from Instituto de Investigación Biomédica de Málaga - IBIMA (all including FEDER funds). The results published here are based upon data generated by the TCGA Research Network: https://www.cancer.gov/tcga.

References

1. Barile, B., Marzullo, A., Stamile, C., Durand-Dubief, F., Sappey-Marinier, D.: Data augmentation using generative adversarial neural networks on brain structural connectivity in multiple sclerosis. Comput. Methods Programs Biomed. **206**, 106113 (2021). https://doi.org/10.1016/j.cmpb.2021.106113

2. Breiman, L.: Random forests. Mach. Learn. **45**(1), 5–32 (2001). https://doi.org/10.1023/a:1010933404324

3. Chawla, N.V., Bowyer, K.W., Hall, L.O., Kegelmeyer, W.P.: SMOTE: synthetic minority over-sampling technique. J. Artif. Intell. Res. **16**, 321–357 (2002). https://doi.org/10.1613/jair.953

4. Cheerla, A., Gevaert, O.: Deep learning with multimodal representation for pan-cancer prognosis prediction. Bioinformatics **35**(14), i446–i454 (2019). https://doi.org/10.1093/bioinformatics/btz342

5. Douzas, G., Bacao, F.: Effective data generation for imbalanced learning using conditional generative adversarial networks. Expert Syst. Appl. **91**, 464–471 (2018). https://doi.org/10.1016/j.eswa.2017.09.030

6. Frid-Adar, M., Diamant, I., Klang, E., Amitai, M., Goldberger, J., Greenspan, H.: GAN-based synthetic medical image augmentation for increased CNN performance in liver lesion classification. Neurocomputing **321**, 321–331 (2018). https://doi.org/10.1016/j.neucom.2018.09.013

7. Goodfellow, I., Bengio, Y., Courville, A.: Deep Learning. MIT Press, Cambridge (2016)

8. Goodfellow, I., et al.: Generative adversarial nets. In: Advances in Neural Information Processing Systems, pp. 2672–2680 (2014)

9. Han, C., et al.: GAN-based synthetic brain MR image generation. In: 2018 IEEE 15th International Symposium on Biomedical Imaging (ISBI 2018), pp. 734–738. IEEE (2018). https://doi.org/10.1109/isbi.2018.8363678

10. Heusel, M., Ramsauer, H., Unterthiner, T., Nessler, B., Hochreiter, S.: GANs trained by a two time-scale update rule converge to a local Nash equilibrium. In: Advances in Neural Information Processing Systems, vol. 30 (2017)
11. Hsu, W.N., Zhang, Y., Glass, J.: Unsupervised domain adaptation for robust speech recognition via variational autoencoder-based data augmentation. In: 2017 IEEE Automatic Speech Recognition and Understanding Workshop (ASRU), pp. 16–23. IEEE, December 2017. https://doi.org/10.1109/asru.2017.8268911
12. Liu, Y., Zhou, Y., Liu, X., Dong, F., Wang, C., Wang, Z.: Wasserstein GAN-based small-sample augmentation for new-generation artificial intelligence: a case study of cancer-staging data in biology. Engineering 5(1), 156–163 (2019). https://doi.org/10.1016/j.eng.2018.11.018
13. Maas, A.L., Hannun, A.Y., Ng, A.Y.: Rectifier nonlinearities improve neural network acoustic models. In: International Conference on Machine Learning, vol. 30, p. 3 (2013)
14. Marouf, M., et al.: Realistic in silico generation and augmentation of single-cell RNA-seq data using generative adversarial networks. Nat. Commun. 11(1), 1–12 (2020). https://doi.org/10.1038/s41467-019-14018-z
15. Mirza, M., Osindero, S.: Conditional generative adversarial nets (2014)
16. Moreno-Barea, F.J., Jerez, J.M., Franco, L.: Improving classification accuracy using data augmentation on small data sets. Expert Syst. Appl. 161, 113696 (2020). https://doi.org/10.1016/j.eswa.2020.113696
17. Moreno-Barea, F.J., Strazzera, F., Jerez, J.M., Urda, D., Franco, L.: Forward noise adjustment scheme for data augmentation. In: IEEE Symposium Series on Computational Intelligence (IEEE SSCI 2018) (2018). https://doi.org/10.1109/ssci.2018.8628917
18. Piotrowski, A.P., Napiorkowski, J.J.: A comparison of methods to avoid overfitting in neural networks training in the case of catchment runoff modelling. J. Hydrol. 476, 97–111 (2013). https://doi.org/10.1016/j.jhydrol.2012.10.019
19. Radford, A., Metz, L., Chintala, S.: Unsupervised representation learning with deep convolutional generative adversarial networks (2015)
20. Reed, R.D., Marks, R.J.: Neural Smithing: Supervised Learning in Feedforward Artificial Neural Networks. MIT Press, Cambridge (1998)
21. dos Santos Tanaka, F.H.K., Aranha, C.: Data augmentation using GANs. In: Proceedings of Machine Learning Research XXX 1, p. 16 (2019)
22. Schmidhuber, J.: Deep learning in neural networks: an overview. Neural Netw. 61, 85–117 (2015). https://doi.org/10.1016/j.neunet.2014.09.003
23. Shao, S., Wang, P., Yan, R.: Generative adversarial networks for data augmentation in machine fault diagnosis. Comput. Ind. 106, 85–93 (2019). https://doi.org/10.1016/j.compindJ.2019.01.001
24. Srivastava, N., Hinton, G., Krizhevsky, A., Sutskever, I., Salakhutdinov, R.: Dropout: a simple way to prevent neural networks from overfitting. J. Mach. Learn. Res. 15(1), 1929–1958 (2014)
25. Tibshirani, R.: Regression shrinkage and selection via the Lasso. J. R. Stat. Soc. Ser. B (Methodol.) 58(1), 267–288 (1996). https://doi.org/10.1111/j.2517-6161.1996.tb02080.x
26. Vale-Silva, L.A., Rohr, K.: Long-term cancer survival prediction using multimodal deep learning. Sci. Rep. 11(1), 1–12 (2021). https://doi.org/10.1038/s41598-021-92799-4

27. Waheed, A., Goyal, M., Gupta, D., Khanna, A., Al-Turjman, F., Pinheiro, P.R.: CovidGAN: data augmentation using auxiliary classifier GAN for improved COVID-19 detection. IEEE Access **8**, 91916–91923 (2020). https://doi.org/10.1109/access.2020.2994762
28. Xu, B., Wang, N., Chen, T., Li, M.: Empirical evaluation of rectified activations in convolutional network (2015)
29. Zur, R.M., Jiang, Y., Pesce, L., Drukker, K.: Noise injection for training artificial neural networks: a comparison with weight decay and early stopping. Med. Phys. **36**(10), 4810–4818 (2009). https://doi.org/10.1118/1.3213517

National Network for Rare Diseases in Brazil: The Computational Infrastructure and Preliminary Results

Diego Bettiol Yamada[1]([✉]), Filipe Andrade Bernardi[2], Márcio Eloi Colombo Filho[2], Mariane Barros Neiva[3], Vinícius Costa Lima[2], André Luiz Teixeira Vinci[1], Bibiana Mello de Oliveira[4,5], Têmis Maria Félix[5], and Domingos Alves[1]

[1] Ribeirao Preto Medical School, University of Sao Paulo, Ribeirao Preto, Brazil
diego.yamada@usp.br
[2] Bioengineering Postgraduate Program, University of Sao Paulo, Sao Carlos, Brazil
[3] Institute of Mathematics and Computer Sciences, University of São Paulo, São Carlos, Brazil
[4] Postgraduate Program in Genetics and Molecular Biology, Federal University of Rio Grande Do Sul, Porto Alegre, Brazil
[5] Medical Genetics Service, Porto Alegre Clinical Hospital, Porto Alegre, Brazil

Abstract. According to the World Health Organization, rare diseases currently represent a global public health priority. Although it has a low prevalence in the general population, this type of condition collectively affects up to 10% of the entire world population. Therefore, these pathologies are numerous and of a diverse nature, and some factors imply significant challenges for public health, such as the lack of structured and standardized knowledge about rare diseases in health units, the need for communication between multidisciplinary teams to understand phenomena and definition of accurate diagnoses, and the scarcity of experience on specific treatments. In addition, the often chronic and degenerative nature of these diseases generates a significant social and economic impact. This paper aims to present an initiative to develop a network of specialized reference centers for rare diseases in Brazil, covering all country regions. We propose collecting, mapping, analyzing data, and supporting effective communication between such centers to share clinical knowledge, evolution, and patient needs, through well-defined and standardized processes. We used validated structures to ensure data privacy and protection from participating health facilities to create this digital system. We also applied systems lifecycle methodologies, data modeling techniques, and quality management. Currently, the retrospective stage of the project is in its final phase, and some preliminary results can be verified. We developed an intuitive web portal for consulting the information collected, offering filters for personalized queries on rare diseases in Brazil to support evidence-based public decision-making.

Keywords: Rare disease · Health network · Digital health · Public health observatory

© The Author(s), under exclusive license to Springer Nature Switzerland AG 2022
D. Groen et al. (Eds.): ICCS 2022, LNCS 13352, pp. 43–49, 2022.
https://doi.org/10.1007/978-3-031-08757-8_4

1 Introduction

Rare Diseases (RDs) are conditions that, although individually presenting a low prevalence, together affect an expressive part of the world population. It is estimated that the set of all known RDs affects approximately 10% of all individuals on the planet [1]. Such pathologies are numerous and diverse, demanding the relationship of different fields of knowledge. These factors imply significant challenges for public health, such as the lack of structured learning in health units about each of these medical conditions, the difficulty and multidisciplinary nature involved in the accurate establishment of RD diagnoses, and the chronic and degenerative nature of these diseases, like malformation syndromes, morphological, and biological anomalies, which causes a significant impact and social burden [2].

According to the World Health Organization (WHO), to overcome the barriers, research and development of studies about RDs must involve the formation of multidisciplinary networks of collaboration between health professionals from different areas, reference centers, public health managers, and patient groups and associations [3, 4]. So, initiatives are developed to provide informational support to these networks, and the best known is Orphanet [5].

In Brazil, the Unified Health System ('SUS'; Portuguese: 'Sistema Único de Saúde') is responsible for offering comprehensive public care to all citizens and at all levels of complexity. Nevertheless, only in 2014 the Brazilian Policy for Comprehensive Care for People with Rare Diseases was established and instituted within the scope of the SUS [6].

However, governmental and socio-political financial support is not sufficient to guarantee the population's rights to health in practice. Problems with human, technological, and infrastructure resources are observed in health units in all Brazilian regions [7]. Therefore, the Brazilian Network of Rare Diseases ('RARAS'; Portuguese: 'Rede Nacional de Doenças Raras') project presents itself to reverse this scenario. This project aims to carry out a national representative survey about the epidemiology, clinical overview, diagnostic and therapeutic resources used, and costs for RD of genetic and non-genetic origin in Brazil and create a national RD surveillance network [1].

This paper aims to present the plan's preliminary results that involve all the computational and procedural infrastructure used to create the RARAS, following the WHO guidelines for developing digital health observatories [4].

2 Literature Review

RDs are a highly complex problem for health organizations all over the planet. The intrinsic multidisciplinarity of this area requires that health services articulate in an interoperable, transparent, and agile manner. In addition, communication between the various health units involved in the care of these patients must be precise and unambiguous to generate savings in resources and time, and consequently, greater patient satisfaction. Many collaborative RD health network initiatives exist, and others are emerging. Such networks are of great importance and value for the development of science and technology in this area [8].

The International Rare Diseases Research Consortium (IRDiRC) is a collaborative network that seeks new developments to generate knowledge and improvements in therapeutic, diagnostic methods, and consequently, quality of life for patients with RD [9]. The European Reference Networks (ERNs) were established to expand their scope of attention to rare and low-prevalence conditions in Europe [10]. At the same time, other initiatives already existed and were functioning at the national level.

Italy was one of the first European countries to create specific regulations for RDs. Its successful experience evidenced a reduction in the costs of health services, lower mobility of patients among the different health units, and better planning of public health policies [8]. In France, establishing the French National Plan for Rare Diseases contributed to structuring processes, integrating care models and epidemiological models for research [11].

Consequently, such projects promoted greater engagement in models of care for RDs. Similar contributions emerged in countries such as the United Kingdom, Germany, and the United States [12]. The Cooperation Rare Disease Action Plan was created based on the premises of the Asia-Pacific Economic Cooperation to improve the social and economic inclusion of people with RDs [13].

In Brazil, the Ministry of Health established the Brazilian Policy for Comprehensive Care for People with Rare Diseases. However, the non-standardization of processes, the heterogeneity of the health information systems, and the lack of terminologies make complex the task of guaranteeing the premises of the Brazilian plan [6, 7]. Therefore, we develop and offer the necessary computational and procedural infrastructure for what specialized centers in RDs from all regions of Brazil can collect, analyze, generate and share knowledge about RDs.

3 Methods

3.1 Study Design

This paper is part of a larger project, RARAS, which is an observational cohort study. The development of computing infrastructure is characterized as applied basic research because it uses validated scientific knowledge to develop methods and technologies to improve the understanding of events and phenomena, increasing our scientific knowledge base [14]. The RARAS project covers all Brazilian demographic regions, with 39 participating health centers. The inclusion criteria for the retrospective phase refers to patients seen at RDs services between 2018 and 2019.

For the prospective phase, the inclusion criterion refers to patients seen in these services in 2022, between April and September. Participating health centers include health units in 16 state capitals and other municipalities, totaling more than 47 million people living in these areas [15]. To centralize information from all participants, we used standardized electronic capture of health data techniques.

3.2 Electronic Data Captures Systems and Data Collection

The use of Electronic Data Capture (EDC) systems for data collection eliminates risks associated with paper-based instruments and enhances the collection of high-quality data necessary for conducting health research [16]. So, the EDCs systems REDCap [17] and KoBoToolbox [18] were selected in this study because they are free, stable, widely used, and well-documented software, with Application Programming Interfaces (API) available for integration with other systems.

We designed the electronic Case Report Forms (eCRFs) in REDCap to collect data from the health centers, and we structured processes to ensure the monitoring of data quality indicators through a methodology of auditory Early Hearing Detection and Intervention (EHDI) [19]. The classical database architecture Structured Query Language (SQL) was chosen to store the data, using the MYSQL database [20] once the REDCap supported it.

We define internationally validated standards for the interoperable structuring of the data collection stages. To identify a confirmed RD diagnosis, we use 3 selectable options: International Statistical Classification of Diseases and Related Health Problems-10 (ICD-10), Online Mendelian Inheritance in Man (OMIM), and ORPHA code. For the proper mapping of the signs and symptoms caused by pathologies, we use the Human Phenotype Ontology (HPO) [21].

Automatic and manual validations are performed to verify the consistency of the data entered in the data collection instruments. The automatic validation is done by a script that searches for outliers in critical fields on filled forms. For manual validation, we create a monitor process using the Business Process Model and Notation (BPMN), to standardize each step of the data collection process [22]. Six data managers are responsible for doing the manual validation.

3.3 Web Portal Development and Data Analysis

The processed information and analyzed data are made available in a web portal that is in development using open source technologies and languages, being PHP7 for the back-end side and HTML5, CSS3, and JavaScript for the front-end side, following the W3C guidelines [23]. The data is retrieved from an internal server, and Python scripts were developed to summarize the characteristics of the collected information.

We emphasize evaluating the distribution of human resources and laboratory procedures for each state and health center using statistical methodologies. Patients' data distributions such as age, initial symptoms, and diagnosis are also available. The preliminary web portal is available at the domain: https://raras.org.br/. However, as the study is still in progress, all tools have not yet been made available because we are following the project schedule.

4   Results and Discussion

The health center characterization form showed us that just over half of the participating centers (59%) have Electronic Health Records (EHRs) established in their work process. The rest of the centers use physical health records on paper. Less than half of the centers indicated that the data generated internally are shared with other organizations or systems (46.2%), showing the inadequacy of the SUS in guaranteeing adequate communication through the health network.

The retrospective data collection has 10,442 records entered from patients with RDs from 39 referral centers spread across all regions of Brazil, in a data collection time of 433 days. To verify the consistency of the data entered by the typists, we design a BPMN process. Figure 1 shows the clinical collection monitoring diagram.

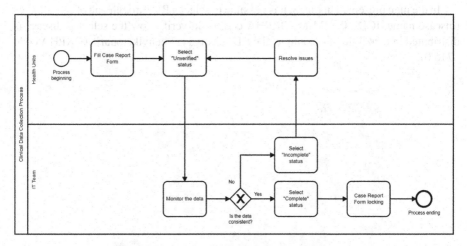

Fig. 1. Clinical data collection monitoring diagram

From the collected data, the distribution of the diagnostic status of all patients entered so far can be graphically visualized, as shown in Fig. 2. We can see that, although more than half of the patients entered have a confirmed diagnosis of RDs, 35.5% of the patients do not have an actual confirmed diagnosis.

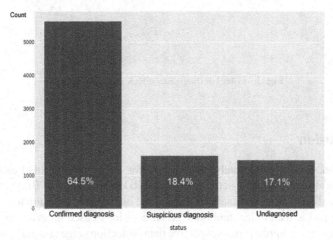

Fig. 2. Diagnostic status distribution

We also identify the moment of diagnosis of these patients, offering information of great use to direct public investments. This moment can be prenatal (1.01%), neonatal screening (8.79), postnatal (70.9), and 19.3% of the records did not have this information. Regarding the terminology used in these health units, we found that the ORPHA code was the most used (62.1%), followed by the ICD-10 (29.9%), and later, the OMIM (8%).

In the map presented in Fig. 3, it is possible to select a RD through one of its variables (disease name, ICD-10, OMIM, ORPHA code) and verify how the selected disease is distributed in Brazil. In this example, the RD selected is Phenylketonuria (ORPHA code –716.0).

Fig. 3. Filter for diagnoses check geographically

5 Conclusion

In conclusion, the computational infrastructure for implementing this health network promotes the ability to map the actual scenario of RD in all regions of Brazil. From these data, it is possible to carry out analyses to generate intelligence in health to support public planning and assist clinical and managerial decisions based on evidence. As future work, we intend to complete the prospective data collection stage and make all information available in a dynamic and user-friendly way on our web portal. Thus, we believe that improving health management, quality of care, and allocation of technological and human resources is possible.

Funding. This study was funded by the National Council for Scientific and Technological Development – CNPq and the Ministry of Health of Brazil – MoH.

References

1. Félix, T.M., et al.: Epidemiology of rare diseases in Brazil: protocol of the Brazilian rare diseases network (RARAS-BRDN). Orphanet J. Rare Dis. **17**(1), 1–13 (2022)
2. Nguengang Wakap, S., et al.: Estimating cumulative point prevalence of rare diseases: analysis of the Orphanet database. Eur. J. Hum. Genet. **28**(2), 165–173 (2020)
3. World Health Organization, Priority diseases and reasons for inclusion, https://www.who.int/medicines/areas/priority_medicines/Ch6_0Intro.pdf. Accessed 6 Feb 2022
4. World Health Organization - Guide for the establishment of health observatories. https://apps.who.int/iris/handle/10665/246123. Accessed 05 Apr 2022
5. Orphanet inventory of rare diseases. https://www.orpha.net/orphacom/cahiers/docs/GB/eproc_disease_inventory_PR_R1_Nom_04.pdf. Accessed 16 Feb 2022
6. Portaria nº 199, de 30 de Janeiro de 2014, Ministério da Saúde. https://bvsms.saude.gov.br/bvs/saudelegis/gm/2014/prt0199_30_01_2014.html. Accessed 16 Feb 2022
7. Iriart, J.A.B., Nucci, M.F., Muniz, T.P., Viana, G.B., Aureliano, W.D.A., Gibbon, S.: Da busca pelo diagnóstico às incertezas do tratamento: desafios do cuidado para as doenças genéticas raras no Brasil. Cien. Saude Colet. **24**, 3637–3650 (2019)
8. Simone Baldovino, M.D., Domenica Taruscio, M.D., Dario Roccatello, M.D.: Rare diseases in Europe: from a wide to a local perspective. Isr Med. Assoc. J. **18**, 359–363 (2016)
9. IRDiRC - Vision and Goals. https://irdirc.org/about-us/vision-goals/. Accessed 16 Feb 2022
10. Héon-Klin, V.: European reference networks for rare diseases: what is the conceptual framework? Orphanet J. Rare Dis. **12**(1), 1–9 (2017)
11. Choquet, R., Landais, P.: The French national registry for rare diseases: an integrated model from care to epidemiology and research. Orphanet J. Rare Dis. **9**(1), 1–2 (2014)
12. Alves, D., et al.: Mapping, infrastructure, and data analysis for the Brazilian network of rare diseases: protocol for the RARASnet observational cohort study. JMIR Res. Protoc. **10**(1), e24826 (2021)
13. APEC Rare Disease Network. Asia-Pacific Economic Cooperation. https://www.apec.org/rarediseases. Accessed 08 Jan 2021
14. Roll-Hansen, N.: Why the distinction between basic (theoretical) and applied (practical) research is important in the politics of science. London: London School of Economics and Political Science, Contingency and Dissent in Science Project (2009)
15. IBGE – Cidades. https://cidades.ibge.gov.br/. Accessed 16 Feb 2022
16. Pasalic, D., Reddy, J.P., Edwards, T., Pan, H.Y., Smith, B.D.: Implementing an electronic data capture system to improve clinical workflow in a large academic radiation oncology practice. JCO Clin. Canc. Inform. **2**, 1–12 (2018)
17. Wright, A.: REDCap: a tool for the electronic capture of research data. J. Electr. Res. Med. Libr. **13**(4), 197–201 (2016)
18. KoBoToolbox Website. https://www.kobotoolbox.org/. Accessed 16 Feb 2022
19. Chen, H., Hailey, D., Wang, N., Yu, P.: A review of data quality assessment methods for public health information systems. Int. J. Environ. Res. Public Health **11**(5), 5170–5207 (2014)
20. MySQL – Documentation. https://dev.mysql.com/doc/. Accessed 16 Feb 2022
21. Köhler, S., et al.: The human phenotype ontology project: linking molecular biology and disease through phenotype data. Nucleic Acids Res. **42**(D1), D966–D974 (2014)
22. Zarour, K., Benmerzoug, D., Guermouche, N., Drira, K.: A systematic literature review on BPMN extensions. Bus. Process Manag. J. (2019)
23. W3C Standards. https://www.w3.org/standards/. Accessed 04 Apr 2022

Classification of Uterine Fibroids in Ultrasound Images Using Deep Learning Model

K. T. Dilna[1,2], J. Anitha[1], A. Angelopoulou[3], E. Kapetanios[4], T. Chaussalet[3], and D. Jude Hemanth[1(✉)]

[1] Department of ECE, Karunya Institute of Technology and Sciences, Coimbatore, India
judehemanth@karunya.edu
[2] Department of ECE, College of Engineering and Technology, Payyanur, India
[3] School of Computer Science and Engineering, University of Westminster, London, UK
[4] School of Physics, Engineering and Computer Science, University of Hertfordshire, Herts, UK

Abstract. An abnormal growth develop in female uterus is uterus fibroids. Sometimes these fibroids may cause severe problems like miscarriage. If this fibroids are not detected it ultimately grows in size and numbers. Among different image modalities, ultrasound is more efficient to detect uterus fibroids. This paper proposes a model in deep learning for fibroid detection with many advantages. The proposed deep learning model overpowers the drawbacks of the existing methodologies of fibroid detection in all stages like noise removal, contrast enhancement, Classification. The preprocessed image is classified into two classes of data: fibroid and non-fibroid, which is done using the MBF-CDNN method. The method is validated using the parameters Sensitivity, specificity, accuracy, precision, F-measure. It is found that the sensitivity is 94.44%, specificity 95% and accuracy 94.736%.

Keywords: Monarch Butterfly (MB) · Optimization and Fuzzy bounding approach based convention · Deep Neural Network (MBF-CDNN)

1 Introduction

Uterus- the female reproductive system has hollow inside with thick muscular walls. Uterine fibroids (UF) are smooth muscle tumors that develop from the myometrium. The ultrasound (US) imaging technique is used, together with other imaging techniques, like X-ray, computerized tomography (CT) and Magnetic resonance imaging (MRI) for producing images of tissue for medical diagnosis. Image quality of scanned image is reduced mainly because of speckle noise. So, in US images, speckle noise filtering has an evident necessity in removing the speckle noise. There are many filters to reduce speckle noise, but while combining different technique, some important diagnostic information may misplace and it should be conserved.

N. Sriraam et al. [1] presented a backpropagation neural network (BPNN) for automated detection of ultrasonic uterine fibroid by using wavelet features. In order to distinguish the normal and fibroid imag- es, a feed-forward classifier was applied. But, in this method, the noisy data was not detected, which reduced the classification accuracy.

Yixuan Yuan et al. proposed a novel weighted locality-constrained linear coding (LLC) method for uterus image analysis [2]. Leonardo Rundo et al. elaborate a semi-automatic approach to detect fibroid which depends on region-growing segmentation technique [3]. Dynamic statistical shape model (SSM)-based segmentation was explained in [4] but it takes more running time. Alireza Fallahi et al. used FCM on MRI image [5] to segment uterine fibroid. It was two step process - segmentation using FCM and morphological operations and fuzzy alogorithm is used for refining the output. Properties of fibroids are not examined like infarct regions and calcified regions. T.Ratha Jeyalakshmi et al. provides mathematical morphology based methods for automated segmentation [6]. Fibroid in the inner wall of uterus is detected. Shivakumar K et al. has described GVF snake method for the Segmentation in uterus images [7]. Different ultrasound uterus image analysis methods are available in [8–10]. S. Prabakar et al. [11] defined morphological image cleaning (MMIC) algorithm to detect Uterine Fibroid. This algorithm had been developed, employed, and validated in LabVIEW vision assistant toolbox. Only a limited number of techniques have been developed using the deep learning model for fibroid detection. This paper proposes a deep learning model for fibroid detection with many advantages.

The rest of the paper is organized as follows: Sect. 2 provides a clear description of the proposed system. Section 3 provides discussions on the proposed method. Section 4 provides the conclusion and future direction of the proposed system.

2 Proposed Classification Method

Various methods developed for uterine fibroid detection, did not achieve a considerable level of accuracy for fibroid detection because they are not focusing on noise removal and efficient feature extraction. In this paper, an MBF-CDNN based classification algorithm is proposed for accurate detection of uterine fibroid by means of noise removal. The proposed methodology comprises of input data collection, preprocessing and classification. The preprocessed and contrast-enhanced image is given as the input to the MBF-CDNN classifier. The output of the classifier contains two classes of data as fibroid and non-fibroid. In this work, 256 * 256 Gy level ultrasound images are used with intensity ranges between 0 and 255.

The proposed method utilizes Monarch Butterfly (MB) Optimization and Fuzzy (F) bounding approach based Convolutional Deep Neural Network (MBF-CDNN).The CDNN is the hybridized form of CNN and DNN. The CDNN chooses a weight randomly that increases a training time and attained a maximum error in classification. To solve this problem, the Fuzzy bound method is used, which bounds the weights and selects the feasible one using the CQO-MBA method. In the CQO-MBA method, the slow convergence speed of the MBA method is resolved with chaos mechanism (C) and quasi-opposition (QO). Hence the proposed technique is named MBF-CDNN. The structure of MBF-CDNN is shown in Fig. 1,

The DNN network has number of hidden layers and it maps the input features from the fully connected layer with random weights and bias value. Based on the hidden layer and the weight vector which connects the hidden layer to the output layer the output vector is obtained. After obtaining the output response the error w_g is estimated for each

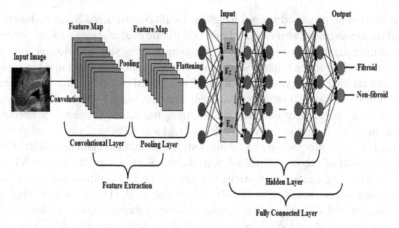

Fig. 1. Structure of MBF-CDNN classifier

input as,

$$w_g = \frac{1}{\hbar_n} \sum_{i=0}^{\hbar_n} X_m^o - g_j^m \tag{1}$$

where, $\hbar_n$ is the number of neurons in hidden layer, X_m^o is the output vector estimated at instance m, g_j^m is the ground truth vector at instance m (Table 1).

Table 1. Layer details of the above architecture MBF-CDNN

Layer	Kernels	Kernel Size	Output
Input	–	–	$32 \times 32 \times 1$
Convolution layer 1	6	5×5	$28 \times 28 \times 6$
Max pooling 1	–	2×2	$14 \times 14 \times 6$
Convolution layer 2	16	5×5	$10 \times 10 \times 16$
Max pooling 2	–	2×2	$5 \times 5 \times 16$
Convolution layer 3	120	5×5	120
Fully connected layer 1	–	–	84
Fully connected layer 2	–	–	10

When a new feature E_i^{m+1} added to the CDNN, the error w_{g+1} estimated, and the weights are to be updated without knowing the weights of the previous instance. If the error value attained is smaller than that estimated for the previous instance, the weight assigned to the network is that obtained using CDNN. Otherwise, a fuzzy bound approach is used. In the Fuzzy bound approach, it initialize the weight values and selects the appropriate one using CQO-MBA method. The Fuzzy bound approach follows a modification degree where the difference of the weights of previous instances are estimated and a hypothesis weight ψ^{hyp} is obtained as,

$$\psi^{hyp} = \psi_i^m \pm FB \tag{2}$$

where, FB is the fuzzy bound and ψ_i^m is the weight that need to be updated using MBA method. The objective of the MBA method is selecting the optimal weight value that gives the rate of minimum error value. By following the same behaviour of the butterfly in MBA method described in [12] the optimal weight value ψ_i^{opt} is updated as,

$$\psi_i^m = \psi_i^{opt} \tag{3}$$

Once the optimal weight value is selected it will be updated for the further classification.

3 Result and Discussion

In this section, the proposed deep learning method for uterine fibroid detection is evaluated by conducting the results of experiments on MATLAB. In this paper, results of MBF-DNN are compared with some traditional existing methods. The dataset consists of 259 images in which it has 119 fibroid and 133 non fibroid images. In this dataset, 80% of data is taken for training and 20% of data is taken for testing. By performing the statistical measures such as sensitivity, specificity, accuracy, precision, recall, F-measure, NPV, MCC, FPR, and FNR the performance of our proposed fibroid classification system is examined. The statistical metrics can be expressed in the terms of TP, FP, FN and TN values. The proposed MBF-CDNN classifier in fibroid detection is analysed by comparing with the existing methods such as, Monarch Butterfly Optimization Based Convolution Deep Neural Network (MB-CDNN), Convolution Deep Neural Network (CDNN), Convolutional Neural Network (CNN) (Table 2).

Table 2. Performance comparison of proposed method

Classification	TP	TN	FP	FN	Sensitivity	Specificity	Accuracy
MBF-CDNN	34	38	2	2	0.9444	0.95	0.9473
MB-CDNN	33	36	4	3	0.9166	0.9	0.9078
CDNN	33	35	5	3	0.9166	0.875	0.8947
CNN	31	33	7	5	0.8611	0.825	0.8421

Figure 2 analyses the performance with respect to the performance metrics precision, recall, and F-measure. The CDNN method have a medium level of values, such as 0.868421 precision, 0.916667 recall, and 0.891892 F-measure. The CNN method has the lowest level of precision and F-measure values and a medium level of recall. The

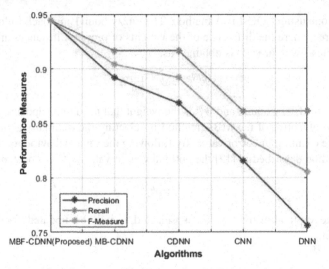

Fig. 2. Demonstrate the performance of the proposed and existing methods in terms of precision, recall, and F-measure

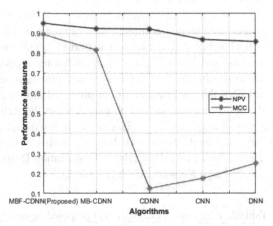

Fig. 3. Represents the performance based on NPV and MCC

value of precision, recall, and F-measure for the proposed method is 0.944444 higher than the existing methods.

The NPV and MCC metrics of the proposed and existing methods are analyzed in Fig. 3. The proposed method has an NPV of 0.95 where the other methods have 0.923077 for MB-CDNN, 0.921053 for CDNN and 0.857143 for CNN. The MCC of the proposed MBF-CDNN method is 0.8944. In the above figure, the MCC of the existing methods is lower than the proposed method.

Figure 4 evaluates the performance based on FPR and FRR. The FPR of the proposed method is 0.05 and MB-CDNN, CDNN and CNN methods have 0.1, 0.790569 and 0.612175 respectively. On the other hand, the lower FRR of the proposed method is

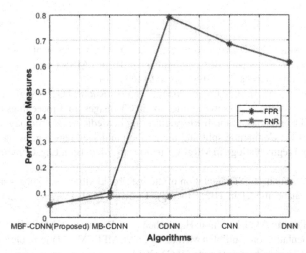

Fig. 4. Performance comparison in terms of FPR and FRR

0.055556 and the FRR of the existing MB-CDNN and CDNN is 0.083333. The FRR of CNN methods is 0.138889. From the overall analysis, it is proved that the proposed MBF-CDNN classifier is better than the other classifiers.

Table 3 shows the confusion matrix of the proposed and existing methods.

Table 3. Confusion matrix of the proposed and existing methods

Methods	MBF-CDNN		MB-CDNN		CDNN		CNN	
Categories	F	NF	F	NF	F	NF	F	NF
F	34	2	33	3	33	3	31	5
NF	2	38	4	36	5	35	7	33

4 Conclusion

The presence of fibroid can cause severe pain, infertility, and repeated miscarriages. So the detection of fibroid and treatment is the crucial factor in women health, US imaging is the most common modality for detecting fibroids. To detect fibroids, this paper proposed a deep learning model for the accurate detection of fibroids. In performance evaluation MBF-CDNN method weighted against several existing methods. The methods are compared based on some quality metrics. In the evaluation the proposed methods provide a better accuracy level than the existing methods. The proposed MBF-CDNN classifier has the accuracy level of 0. 947368%. These results proved that the proposed method is highly efficient for accurate detection of uterine fibroids.

References

1. Sriraam, N., Nithyashri, D., Vinodashri, L., Niranjan, P.M.: Detection of uterine fibroids using wavelet packet features with BPNN classifier. In: IEEE EMBS Conference on Biomedical Engineering and Sciences (IECBES), pp. 406–409 (2010)
2. Yuan, Y., Hoogi, A., Beaulieu, C.F., Meng, M.Q.H., Rubin, D.L.: Weighted locality–constrained linear coding for lesson classification in CT images. In: Proceedings of 37th Annual International Conference of the IEEE Engineering in Medicine and Biology Society (2015)
3. Rundo, L., et al.: Combining split-and-merge and multi-seed region growing algorithms for uterine fibroid segmentatiogn in MRgFUS treatments. Med. Biol. Eng. Comput. **54**(7), 1071–84 (2016)
4. Ni, B., He, F., Yuan, Z.: Segmentation of uterine fibroid ultrasound images using a dynamicstatistical shape model in HIFU therapy. Comput. Med. Imaging Graph. **46**, 302–314 (2015)
5. Fallahi, A., Pooyan, M., Khotanlou, H., Hashemi, H., Firouznia, K., Oghabian, M.A.: Uterine fibroid segmentation on Multiplan MRI using FCM, MPFCM and morphological operations. Biomed. Eng. Appl. Basis Commun. **26**(1) (2014)
6. Jeyalakshmi, R., Kadarkarai, R.: Segmentation and feature ex traction of fluid-filled uterine fibroid-a knowledge-based approach. Int. J. Sci. Technol. **4**, 405–416 (2010). ISSN 1905-7873
7. Harlapur, S.K., Hegadi, R.S.: Segmentation and analysis of fibroid from ultrasound images. Int. J. Comput. Appl. **0975**, 8887 (2015)
8. Yao, J., Chen, D., Lu, W., Premkumar, A.: Uterine fibroid segmentation and volume measurement on MRI. In: Proceedings of the SPIE, vol. 6143 (2006)
9. Alush, A., Greenspan, H., Goldberger, J.: Automated and interactive lesion detection and segmentation in uterine cervix images. IEEE Trans. Med. Imaging **29**(2) (2010)
10. Padghamod, M.J., Gawande, J.P.: Classification of ultrasonic uterine images. Adv. Res. Electr. Electron. Eng. **1**(3), 89–92 (2014)
11. Prabakar, S., Porkumaran, K., Guna Sundari, J.: Uterine fibroid segmentation and measurement based on morphological functions in graphical vision assistant tool. In: Sridhar, V., Sheshadri, H., Padma, M. (eds.) Emerging Research in Electronics, Computer Science and Technology. LNEE, vol. 248, pp. 357–366. Springer, New Delhi (2014). https://doi.org/10.1007/978-81-322-1157-0_36
12. Ghetas, M.: Learning-based monarch butterfly optimization algorithm for solving numerical optimization problems. Neural Comput. Appl. **34**, 3939–3957 (2022)

Explainable AI with Domain Adapted FastCAM for Endoscopy Images

Jan Stodt[1]([✉])[iD], Christoph Reich[1][iD], and Nathan Clarke[2][iD]

[1] Institute for Data Science Cloud Computing and IT -Security (IDACUS)
Furtwangen University for Applied Science,
78120 Furtwangen lm Schwarzwald, Germany
{Jan.Stodt,Christoph.Reich}@hs-furtwangen.de
[2] Centre for Security, Communications, and Networks Research (CSCAN)
Plymouth University, Portland Square, Plymouth PL4 8AA, UK
N.Clarke@plymouth.ac.uk

Abstract. Enormous potential of artificial intelligence (AI) exists in numerous products and services, especially in healthcare and medical technology. Explainability is a central prerequisite for certification procedures around the world and the fulfilment of transparency obligations. Explainability tools increase the comprehensibility of object recognition in images using Convolutional Neural Networks, but lack precision.

This paper adapts FastCAM for the domain of detection of medical instruments in endoscopy images. The results show that the Domain Adapted (DA)-FastCAM provides better results for the focus of the model than standard FastCAM weights.

Keywords: XAI · FastCAM · CNN · Healthcare · Endoscopy

1 Introduction

Explainable Artificial Intelligence (XAI) is essential for artificial intelligence products and services in healthcare and medical technology and is a central prerequisite for certification procedures around the world [2,5,7]. How black box models such as neural networks arrive at their results cannot be understood due to their complex processes. However, explainability tools can be used to increase comprehensibility at least for local explanation of individual decisions. In image processing with neural networks, one already speaks of an explanation of a decision when the areas in the input image that have led to the classification of an object are highlighted [2]. In this case, the inner workings of the model are not explained, only the data that is most significant for the decision-making process is highlighted, for example the focus of the model. The motivation of the work to optimize the weights of FastCAM for the endoscopy domain (DA-FastCAM), to achieve better results as the standard FastCAM. The paper works on endoscopy instrument recognition to perform plausibility tests to achieve trustable Convolutional Neural Networks (CNNs) based object detection. The aim is to support the certification of AI-based applications in medicine based on plausibility tests.

D. Groen et al. (Eds.): ICCS 2022, LNCS 13352, pp. 57–64, 2022.
https://doi.org/10.1007/978-3-031-08757-8_6

2   Related Work

In SHAP (SHapley Additive exPlanations) [6] each input feature is weighted regarding model output. All combinations of features are considered to determine the importance (positive and negative) of a single feature. Integrated Gradients [10] visualizes the importance of the input features of a CNN that contribute to the output of the model via heat maps. GradCAM (Gradient-weighted Class Activation Mapping) [9] allows generating visual explanations that highlight the most important regions of an image that predict the feature.

3   Domain Adapted (DA)-FastCAM

The goal is to optimize the weights of the XAI framework FastCAM [8] for the domain of endoscopy instrument recognition. FastCAM generates a focus area, which is an explanation for the results of object classification. The optimized FastCAM weights more accurately reflect the reality of the focus for the target domain compared to the original weights which are an average of the weights for the ImageNet, CSAIL Places and COWC datasets [8]. Areas that are important for the recognition are the focus of the model (Fig. 1 - Focus area). Focus areas that are not connected to main focus area are distraction area (Fig. 1 - Distraction area). Unimportant (out-of-focus) areas are masked (Fig. 1 - Masked area). Masked areas that occlude a part of the instrument are the occlusion areas (Fig. 1 - Occlusion area). Tests with the original FastCAM weights (Table 2) showed high occlusion, although the model recognized them correctly and a significant occurrence of distraction areas.

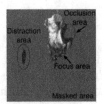

Fig. 1. Areas within an explanation frame

4   Optimization Approach for DA-FastCAM

The dataset used is CholecSeg8k [4] which provides segmentation for 8080 images of laparoscopic cholecystectomy. Each image annotated at the pixel level for thirteen classes common in laparoscopic cholecystectomy. The two tool classes grasper and L hook electrocautery (from here on referred to as hook) are the focus of this paper. The AI model utilized in this paper is the AlexNet based model by Ranem et al.[1] which has an average test accuracy of 67% for the Cholec80 [11] dataset. The AI model was used to classify frames of CholecSeg8k dataset.

4.1 Optimization Framework and Algorithms

The Optuna framework[2] was used to optimize the FastCAM mask weights of the five 2-D convolution layers of AlexNet. The metric used to evaluate the optimization is the Root-Mean-Square Difference (RMS). The RMS is calculated for a segmented frame and the explanation frame computed by FastCAM (for the input frame); see Fig. 2b, 2c and 2a. All RMS values are summed, and the average is calculated. The goal of the optimization is to minimize this average RMS value. The algorithms CMA-ES [3] and TPE [1] were used for optimization.

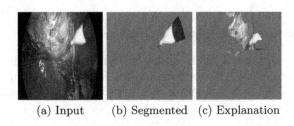

(a) Input   (b) Segmented   (c) Explanation

Fig. 2. Example frames of the RMS calculation

5 Evaluation of DA-FastCAM

A series of experiments were conducted for the optimization of the weights: a) selected frames of grasper and hook together, b) selected frames of grasper, c) selected frames of hook, d) grasper of frame 312 and e) hook of frame 28926. Table 1 provides an visual overview of the most important results. Numerical experiment results can be seen in Table 2, the best and worst RMS for the original and optimized weights can be seen in Table 3. 10 frames for the grasper and 10 frames for the hook were selected for the optimization. These 20 frames represent the typical views of the instruments and their position in the frame with different recognition rates by the model.

5.1 Optimization of Weights for Grasper and Hook Together

The weights are optimized for the selected frames of the grasper and hook.

---

[1] https://github.com/amrane99/CAI-Classification.
[2] https://optuna.org/.

Grasper & Hook CMA-ES. "312 - Grasper - Best": the focus includes the entire instrument (A) and does not cut off the upper half of the jaw (B). The distraction area changes its shape (C). "613 - Grasper - Worst": the focus includes the transition between the jaws and the outer tube (A) and a sharper edge to the masked area (B). However, a new occlusion area is created that covers the fenestrated opening in the jaws (C), considered a small disadvantage. "2850 - Hook - Best": the focus includes the transition between the white body and the outer tube (A). The distraction area changes its shape (B). "28911 - Hook - Worst", the focus is reduced so much that only a small section of to the white body of the instrument is visible (A). **Result:** Except for "28911 - Hook - Worst" a clearer focus was created.

Grasper & Hook TPE. The results are identical to Section Grasper & Hook CMA-ES; see Sect. 5.1 for discussion and visual presentation of the results.

5.2   Optimization of Weights for Grasper

The weights are optimized for the selected frames of the grasper. As seen in Table 3, the use of CMA-ES results in better RMS values compared to TPE. Therefore, for the following experiments the TPE algorithm is omitted.

Grasper CMA-ES. Despite the optimization for graspers, the results are visually identical to the results of Grasper & Hook CMA-ES, Grasper & Hook TPE and Grasper CMA-ES, therefore see Sect. 5.1 for discussion and visual presentation of the results.

5.3   Optimization of Weights for Hook

The weights for the selected frames are optimized for the Hook.

Hook CMA-ES. "293 - Grasper - Best" includes the transition between the jaws and the outer tube (A). The distraction area (B, C) changed their shape and the distraction area (C) is divided into two parts. "561 - Grasper - Worst": no occlusion area on the grasper mechanism (A) but grasper is more occluded (B). The distraction area changed its shape (D). An additional distraction area has been created (C). "28605 - Hook - Best" the focus includes more area of the outer tube (A). "2850 - Hook - Worst" the hook is completely occluded (A) and the focus area widens (B). **Result:** Except for "2850 - Hook - Worst", a clearer focus was created after optimization. Even if only hooks was optimized, the grasper focus was also optimized.

5.4 Optimization of Weights for Grasper of Frame 312

The weights are optimized for the frame 312. This frame was selected because it was the worst occluded explanation frame with the original weights for all selected graspers. The aim to see if it is possible to improve the masking of one grasper without degrading the results of the other graspers.

Grasper - 312 CMA-ES. "312 - Grasper - Optimized for", the focus includes both parts of the jaws (A) and the transition between the jaws and the outer tube (B). The two distraction areas (C, D) change their geometry and the distraction area (D) is divided into two parts. **Result:** Even if optimization was done on a specific grasper, for all graspers the focus after optimization is more precise on the respective instruments (also hook, albeit with limitations).

5.5 Optimization of Weights for Hook of Frame 28926

The weights are optimized for frame 28926. This frame was selected because it was the worst occluded explanation frame with the original weights for all selected hooks. The aim is to see if it is possible to improve the masking of one hook without degrading the results of the other hooks.

Hook - 28926 CMA-ES. "28926 - Hook - Optimized For" the focus contains the white body of the instrument (A). But also contains more of the black background (B). The distraction area is smaller but now divided into two parts (C). **Result:** Even if a specific hook is optimized, for all hooks (and grasper) except "2850 - Hook - Worst", the focus after optimization is more precise.

5.6 Optimization Result Overview

Table 2 shows the average RMS, the weights and at which epoch the optimum was achieved. Table 3 shows the best and worst RMS values per frame for the original weights and the optimized weights for grasper and hook. Interesting outcomes are: First, for grasper & hook the algorithm CMA-ES and TPE have the same RMS value and layer weights. Second, experiment grasper CMA-ES have the same RMS value as grasper & hook (CMA-ES, TPE). Third, the CMA-ES finds the optimal weights faster than TPE.

Table 1. Results of Grasper & Hook CMA-ES, Grasper & Hook TPE, Grasper CMA-ES, Grasper - 312 CMA-ES, Hook CMA-ES and Hook - 28926 CMA-ES

Original Weights	Optimized Weights	Original Weights	Optimized Weights
Frame: 312 - Tool: Grasper Type: Optimized For - RMS: 42.1729		Frame: 28926 - Tool: Hook Type: Optimized For - RMS: 46.0587	
Frame: 312 - Tool: Grasper Type: Best - RMS: 41.0978		Frame: 293 - Tool: Grasper Type: Best - RMS: 42.2980	
Frame: 613 - Tool: Grasper Type: Worst - RMS: 50.9357		Frame: 561 - Tool: Grasper Type: Worst - RMS: 60.3581	
Frame: 2850 - Tool: Hook Type: Best - RMS: 43.0347		Frame: 28605 - Tool: Hook Type: Best - RMS: 43.9202	
Frame: 28911 - Tool: Hook Type: Worst - RMS: 60.7620		Frame: 2850 - Tool: Hook Type: Worst - RMS: 60.7620	

Table 2. Optimization result overview

Index	Average RMS	Weight layer A	Weight layer B	Weight layer C	Weight layer D	Weight layer E	Epoch of best result
Original FastCAM weights	51.18	0.18	0.15	0.37	0.40	0.72	N/A
Grasper & Hook CMA-ES	48.62[a]	0.00	0.00	0.00	0.00	0.87	471
Grasper & Hook TPE	48.62	0.00	0.00	0.00	0.00	0.87	855
Grasper CMA-ES	48.62[b]	0.00	0.00	0.00	0.00	0.89	355
Hook CMA-ES	49.28	0.00	0.16	0.00	0.61	0.12	550
Grasper - 312 CMA-ES	49.81	0.01	0.95	0.01	0.06	0.34	396
Hook - 28926 CMA-ES	49.28	0.00	0.22	0.00	0.86	0.17	693

[a] Optimum reached earlier than TPE
[b] Same optimum RMS as for Grasper & Hook CMA-ES

Table 3. Overview of best and worst RMS for the original and optimized weights

Index	Original Weights RMS					Optimized Weights RMS				
	Average	Grasper		Hook		Average	Grasper		Hook	
		Best	Worst	Best	Worst		Best	Worst	Best	Worst
Grasper & Hook CMA-ES	51.14	44.20	59.42	47.19	59.09	48.62	41.09	53.56	43.03	60.76
Grasper & Hook TPE						48.62	41.09	53.56	43.03	60.76
Grasper CMA-ES						48.62	41.09	53.56	43.03	60.76
Hook CMA-ES						49.28	42.29	60.35	43.92	52.17
Grasper - 312 CMA-ES						49.81	42.15	58.54	41.31	60.76
Hook - 28926 CMA-ES						49.28	42.32	60.34	43.81	52.14

6 Conclusion

Experiments showed that the DA-FastCAM archives a general improvement of the original FastCAM weights via an automated process, validated by reduced RMS values after optimization compared to the RMS values for the original FastCAM weights. Through this optimization, the explanation frames had in average a smaller distraction area and a smaller occlusion area, and a more precise focus area. For a high accuracy of object recognition (e.g. graspers) the RMS values of the DA-FastCAM decreased significantly compared to a lower accuracy of object recognition (e.g. hooks). It is worth mentioning, that through the optimization of FastCAM, a bad CNN model gives bad XAI images. Optimization for all frames of the grasper and hook instruments was shown to be the best approach for optimization. It should be noted that even when optimizing only for a specific frame of grasper, optimization can be performed on average for all frames, whether grasper or hook, albeit with exceptions for some frames of hook. An expected result is that in general hooks are less recognized by the CNN model. For the choice of algorithm, the CMA-ES is recommended, as it not only finds the best weights, but also has the best performance compared to TPE. This is particularly important when numerous images (videos) have to be the basis of the optimization.

Acknowledgment. The authors would like to acknowledge the financial support from the German Federal Ministry of Research and Education (Bundesministerium für Bildung und Forschung) under grant CoHMed/PersonaMed A for this research. Thanks to the DigNest project the results will be disseminated at seminars and workshops.

References

1. Bergstra, J., Bardenet, R., Bengio, Y., Kégl, B.: Algorithms for hyper-parameter optimization. Adv. Neural Inf. Process. Syst. **24** (2011)
2. DIN: DIN SPEC 13288, Guideline for the development of deep learning image recognition systems in medicine. Tech. rep., https://dx.doi.org/10.31030/3235648
3. Hansen, N., Ostermeier, A.: Completely derandomized self-adaptation in evolution strategies. Evol. Comput. **9**(2), 159–195 (2001)
4. Hong, W.Y., Kao, C.L., Kuo, Y.H., Wang, J.R., Chang, W.L., Shih, C.S.: Cholec-Seg8k: a Semantic Segmentation Dataset for Laparoscopic Cholecystectomy Based on Cholec80. arXiv:2012.12453 [cs] (2020). http://arxiv.org/abs/2012.12453
5. ISO/IEC AWI TS 6254: Objectives and approaches for explainability of ml models and AI systems. Standard, ISO, Geneva, CH
6. Lundberg, S., Lee, S.I.: A unified approach to interpreting model predictions. Adv. Neural Inf. Process. Syst. **30** (2017)
7. Maack, S., Bertovic, M., Radtke, M.: Deutsche Normungsroadmap KI (2020)
8. Mundhenk, T.N., Chen, B.Y., Friedland, G.: Efficient saliency maps for explainable AI (2019)
9. Selvaraju, R.R., Cogswell, M., Das, A., Vedantam, R., Parikh, D., Batra, D.: Grad-cam: visual explanations from deep networks via gradient-based localization. In: Proceedings of the IEEE International Conference on Computer Vision, pp. 618–626 (2017)
10. Sundararajan, M., Taly, A., Yan, Q.: Axiomatic attribution for deep networks. In: International Conference on Machine Learning, pp. 3319–3328. PMLR (2017)
11. Twinanda, A.P., Shehata, S., Mutter, D., Marescaux, J., De Mathelin, M., Padoy, N.: Endonet: a deep architecture for recognition tasks on laparoscopic videos. IEEE Trans. Med. Imaging **36**(1), 86–97 (2016)

Sensitivity Analysis of a Model of Lower Limb Haemodynamics

Magdalena Otta[1,2,3]([✉])(iD), Ian Halliday[2,3], Janice Tsui[4,5], Chung Lim[4], Zbigniew R. Struzik[1,6,7](iD), and Andrew Narracott[2,3](iD)

[1] Sano Centre for Computational Medicine, Modelling and Simulation Research Team, Czarnowiejska 36, Building C5, 30-054 Kraków, Poland
`motta1@sheffield.ac.uk`
[2] Department of Infection, Immunity and Cardiovascular Disease, University of Sheffield, Sheffield, UK
[3] Insigneo Institute for in silico Medicine, University of Sheffield, Sheffield, UK
[4] University College London, London, UK
[5] Royal Free London NHS Foundation Trust, London, UK
[6] Faculty of Physics, University of Warsaw, Pasteura 5, 02-093 Warsaw, Poland
[7] Graduate School of Education, The University of Tokyo, 7-3-1 Hongo, Bunkyo-ku, Tokyo 113-0033, Japan
`https://sano.science/`

Abstract. Post-thrombotic syndrome (PTS) has variable clinical presentation with significant treatment costs and gaps in the evidence-base to support clinical decision making. The contribution of variations in venous anatomy to the risk of complications following treatment has yet to be characterized in detail. We report the development of a steady-state, 0D model of venous anatomy of the lower limb and assessments of local sensitivity (10% radius variation) and global sensitivity (50% radius variation) of the resulting flows to variability in venous anatomy. An analysis of orthogonal sensitivity was also performed. Local sensitivity analysis was repeated with four degrees of thrombosis in the left common iliac vein. The largest normalised sensitivities were observed in locations associated with the venous return. Both local and global approaches provided similar ranking of input parameters responsible for the variation of flow in a vessel where thrombosis is typically observed. When a thrombus was included in the model increase in absolute sensitivity was observed in the leg affected by the thrombosis. These results can be used to inform model reduction strategies and to target clinical data collection.

Keywords: Post-thrombotic syndrome · Venous model · Sensitivity analysis

1 Introduction

Deep vein thrombosis (DVT) of the lower limb is a health condition in which blood clots form in deep veins of the leg due to some pathological changes of the

D. Groen et al. (Eds.): ICCS 2022, LNCS 13352, pp. 65–77, 2022.
https://doi.org/10.1007/978-3-031-08757-8_7

blood vessels or the blood itself [1]. It is estimated to affect 1–2 per 1,000 people each year and between 20 and 50% of them will develop long term complications known as post-thrombotic syndrome, PTS [2,3]. The condition is not terminal, but it significantly impairs quality of life. The highly variable clinical presentation of PTS makes it difficult to treat and, due to extensive follow-up and repeat medical interventions, treatment pathways are associated with significant cost. The placement of a stent, a metallic scaffold used to restore the vessel diameter following disease, has increased in recent years [4], but there are significant gaps in the evidence-base to support clinical decision making around use of stents [5]. The contribution of venous anatomy to variation in blood flow in the region of the thrombosis and resulting risk of complications following stent placement has yet to be characterized in detail.

Reduced order modelling approaches using one dimensional (1D) and zero dimensional (0D) formulations to characterise pressure-flow distributions in blood vessels have been extensively reported in the literature, particularly in the context of research questions associated with the arterial circulation [6]. A 1D formulation considers continuous variation of variables along the circulation, whereas a 0D approach, also known as lumped-parameter, or compartmental modelling, represents elements of the circulation as lumped compartments. The status, challenges and prospects of the 0D method, again with emphasis on arterial applications, was recently provided by Hose et al. [13]. There have been relatively few studies which focus on the venous circulation, of particular note in the context of this study are reports by Müller and Toro [7] who describe a 1D model of both the arterial and venous circulation, with focus on the cerebral vasculature and Keijsers et al. [8] who employed a 1D formulation to study the interaction between the venous circulation in the lower limb and the activity of the calf muscle pump.

The assessment of model sensitivity and the quantification of the propagation of uncertainty from model inputs to model outputs has become acknowledged as an essential aspect of model development, particularly when model outputs are used to inform clinical decision making [9] or to identify biomarkers of disease states [10]. However, uncertainty quantification is often computationally expensive and is not always performed during the development of new modelling approaches. More complex models frequently lack sensitivity analysis.

This study reports the development of a model of venous haemodynamics with focus on the influence of variation in the venous anatomy on the distribution of flow within the veins of the lower limb. This represents the first step towards modelling venous flow of the lower limb to aid clinical decision making in treatment of PTS. The approach uses data reported by Müller and Toro [7] to characterise the lower limb circulation, combining this with a local and global sensitivity analysis.

2   Methods

This section summaries construction of the model under investigation followed by a description of the sensitivity analyses performed. The application of local and global sensitivity analyses to the venous network without thrombosis is described. This is followed by assessment of the change in local sensitivity of the network following the development of a thrombus. All analyses were performed using Python, using libraries including `numpy` and `pandas` to manipulate data and to solve the model, `matplotlib` to visualise results, and `SALib` to perform global sensitivity analysis.

2.1 Lower Limb Circulation Model

In this study a steady-state, 0D model was used to account for the complicated venous anatomy of the lower limb without considering the pulsatility of arterial or venous flow or vessel wall elasticity. The model topology and input parameters were taken from anatomical data reported by Müller and Toro (given in Table III, VI and VIII in [7]). The form of the model is shown in Fig. 1.

The mean radius and length of each vessel were used to compute the corresponding Poiseuille resistance, given by equation (1).

$$R = \frac{8\mu L}{\pi r^4} \tag{1}$$

where μ is the viscosity of blood, r is a radius representing the cross section of the considered vessel and L is its length.

To simplify the form of the model, where vessels were arranged in series in the vascular network, they were represented as a single resistive element. The arterial circulation was taken to start at the abdominal aorta (representing flow into the lower limb only) and three pathways were considered to contribute to venous return to the heart (the inferior vena cava, the azygos vein and the vertebral venous plexus). The full model consisted of 50 resistances (15 arterial, 8 capillary beds, 27 venous) informed by 42 input vessel radii and length parameters. The boundary conditions to the model consisted of the pressure gradient between the abdominal aorta and the venous return to the heart. The aortic pressure was set to 80 mmHg and the pressure at all outlets was assumed to be zero.

Vessel properties were assigned to the elements of the model, based on a prepared text file with one column listing these elements and another column listing all contributing vessel properties, to account for some of them being lumped into a single resistor. The resistance between the arterial and venous system was calculated by summing the resistance of arterioles, capillaries and venules and their distal resistance as provided by Müller and Toro.

To simulate thrombosis, resistance of specific blood vessels was recalculated and replaced in the dataset based on a percentage reduction in the mean radius. A system of equations describing the model was constructed using Kirchhoffs' laws. There are 30 unique pathways from inlet to outlet and the pressure drop along each must equal the specified boundary condition. At every junction, the sum of flows entering equals the sum of flows exiting that junction. This results in an overspecified system of 66 equations. A row-echelon reduction algorithm was implemented to reduce and solve the system. The solution of the model provides

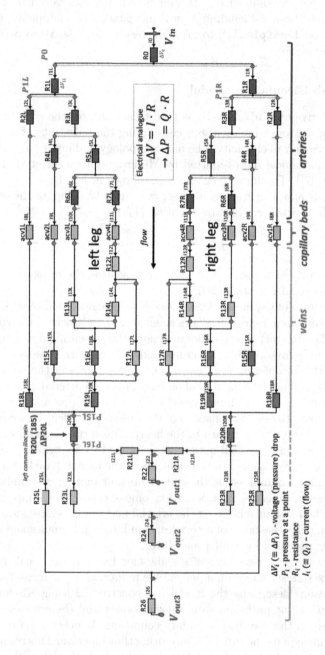

Fig. 1. Model formulation. Flow in the circulation is from right to left. Red elements represent large arteries, blue elements represent arterioles, capillaries and venules and green elements represent veins. All vessels are modelled as purely resistive elements. (Color figure online)

flow at all locations, which is then post-processed to provide the corresponding pressure everywhere in the model.

2.2 Sensitivity Analysis

Local Sensitivity: Local sensitivity indices are based upon the first-order partial derivative in (2), where y_i is the ith output of the model, p_j is the jth input parameter and Δp_j is a small perturbation in its value.

$$\frac{\partial y_i}{\partial p_j} = \lim_{\Delta p_j \to 0} \left(\frac{y_i(p_j + \Delta p_j) - y_i(p_j)}{\Delta p_j} \right) \tag{2}$$

Of course, all other input parameters p_k, $k \neq j$ are held constant when taking the above limit. Accordingly, parameter variation is designated OAT (*one at a time*), and takes place around a base state, representative of a given physiological state. For present purposes, the partial derivative in (2) is computed from the numerical quotient $\frac{\Delta y_i}{\Delta p_j}$, where Δy_i is the measured change in output y_i, caused by a change, Δp_j, in input parameter p_j, typically by a few percent. It is often convenient to contextualise this quotient, to account for the absolute value of y_i and p_j, as we describe shortly.

To assess the sensitivity of the model outputs to variability in venous anatomy a local sensitivity analysis was performed. Each of the 42 radii of the blood vessels was varied by $\pm 10\%$ from the reference value in turn, with all other radii kept constant. The model was solved 3 times for each vessel, with r_{base} - the original value, $r_{min} = 0.9 \cdot r_{base}$ and $r_{max} = 1.1 \cdot r_{base}$. This produced 3 sets of output flows, Q, for each of the 42 radii. It is important to note that because some vessels are lumped, two or three radii contribute to the same resistor.

Two matrices were constructed and compared to visualise the relationships between the change in radii (Δr) with the resulting change in flows (ΔQ). The first was a matrix of absolute changes, with elements described by Eq. (3).

$$a_{ij} = \frac{\Delta Q_j}{\Delta r_i} = \frac{[Q_{max} - Q_{min}]_j}{[r_{max} - r_{min}]_i} \tag{3}$$

Q_{max} corresponds to flows obtained by changing the radius of a vessel to r_{max}, and Q_{min} to r_{min}. Returning to the issue of contextualisation, our second matrix reports relative changes, with respect to the base values of the radii and flows, with elements described by Eq. (4). Changes ΔQ and Δr are calculated in the same way as in Eq. (3).

$$s_{ij} = \frac{\left[\frac{\Delta Q}{Q}\right]_j}{\left[\frac{\Delta r}{r}\right]_i} \tag{4}$$

The left common iliac vein, a deep vein of the ilio-femoral region, is a likely place for thrombus development. In the Müller and Toro dataset it corresponds to the vessel no. 185 (resistor $R20L$ in the current model) and is associated with flow $Q20L$. The relative influence of each input parameter on this flow value was evaluated using the sensitivity vector for this particular output value.

An analysis of orthogonal sensitivity was also performed. The vector in Eq. (5) expresses the relative sensitivity of all outputs (flows) to the ith input (radius).

$$\underline{s}_i = (s_{i1}, s_{i2}, s_{i3}, .., s_{iN}).$$ (5)

The normalised inner product in Eq. (6) therefore measures the orthogonality (effective independence) of the ith and j^{th} input parameters [11].

$$p_{ij} = \frac{\underline{s}_i \cdot \underline{s}_j}{|\underline{s}_i| \times |\underline{s}_i|}.$$ (6)

If the output scalar is close to ± 1, the input parameter pair has a similar effect on the system, if it is close to zero, their effects are independent.

Global Sensitivity - Sobol Analysis: Global sensitivity analysis differs from the local approach in that (i) all input parameters are varied (i.e. we no longer change input parameters OAT) and (ii) outputs (in contradistinction to their rates of change with respect to inputs) are considered directly. The type of analysis performed was the Sobol sensitivity analysis, which is based on variance decomposition. It allows one to measure the contributions of model inputs to variance of model outputs. First-order indices inform relative influence of every input, second order indices measure the contributions of the interactions between two input parameters, and so on [12].

The programming language used was Python with SALib library - following a procedure described in the SALib documentation. Interactions of first and second order were investigated. Ranges for parameter variation were set to $\pm 50\%$ from their nominal values. Data points were sampled from Saltelli sequence for six cases, $n = \{1, 3, 5, 7, 10\}$, with 2^n being an input to sobol.saltelli() function related to the number of samples, N_s. The specific relationship between n and N_s depends on the order of Sobol indices included in the analysis. For the case investigated, the smallest set of samples $N_s = 216$, and the biggest $N_s = 110,592$. Each sample is a 42-dimensional vector generating a 50-dimensional output flow vector. For each output, the vector containing the value of that output for all samples was provided to the sobol.analyze() function which returns a dictionary of Sobol indices and their confidence values. Convergence of first-order indices was checked to validate the sample size. The case of $n = 10$ was chosen for further analysis.

A matrix of first-order interactions between inputs and outputs as well as the corresponding confidence matrix were constructed to identify parameters of significant influence. Second-order interactions were first investigated by binning and assessing their index and confidence values for all parameter pairs. For each of the 50 output flows, a matrix of parameter interactions was constructed to identify interactions significant to the output variance.

Influence of Thrombus Formation: Four different degrees of thrombosis were introduced to the left common iliac vein (no. 185) by reducing its radius, in turn, by 30%, 40%, 50% and 60%. A local sensitivity matrix was constructed for each case, for ±10% changes to every radius, and compared to the original absolute sensitivity matrix for the no-thrombus case.

3   Results

This section summarises results obtained from the analysis of local sensitivity (varying one parameter at a time) and global sensitivity (varying all parameters simultaneously), as well as the influence of thrombus formation on absolute local sensitivity.

Local Sensitivity: A comparison of absolute and normalised (relative) sensitivity matrices is shown in Fig. 2. Input and output parameters corresponding to cells of highest and lowest sensitivity value are highlighted in yellow.

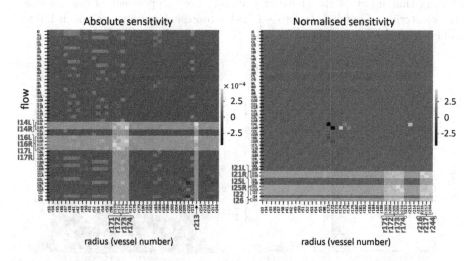

Fig. 2. Absolute vs normalised sensitivity. The most sensitive parameters and outputs are highlighted in both figures. Colour scales are adjusted such that distinctive features of both matrices are visible. (Color figure online)

The two matrices vary in that the most significant radii in the absolute sensitivity belong to veins in the middle of the model, whereas for relative sensitivity, vessels close to the venous return appear to be more significant. It is worth noting that cells of highest values in the absolute sensitivity matrix are still significant in the relative sensitivity matrix, but not as much as those in the bottom right corner.

Ranked normalised sensitivity for flow $Q20L$ is shown in Fig. 3. This is a flow in left common iliac vein - a potential site of thrombosis. Out of 42 radii, 13 display influence on the flow and the first 5 belong to the arterial network of the left leg.

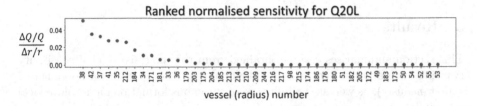

Fig. 3. Relative influence of input parameters on Q20L flow.

The distribution of inner product values and a heatmap of their absolute values are presented in Fig. 4. This distribution has a clear peak at zero, demonstrating that many of the sensitivity vectors are independent of one another. The clustering of entries close to -1 and 1 represent parameters which induce similar response of the system.

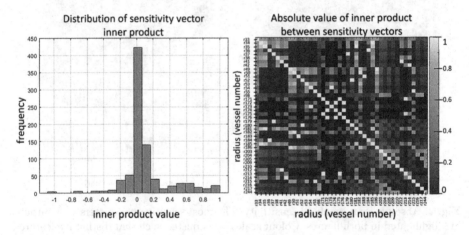

Fig. 4. Distribution of inner product values (left) and absolute value of the inner product (right). Absolute values range from 0 to 1.

Global Sensitivity - Sobol Analysis: Investigation of convergence in first-order indices and corresponding confidence values showed that, for small sample size (i.e., $n = \{1, 3\}$), many indices were negative and their confidence values were relatively large (≥ 1.0). For $n = 10$, all indices were ≥ -0.006. Corresponding confidence values were all ≤ 0.12 which indicates relatively good convergence.

The matrix of first-order indices for this sample size and corresponding confidence matrix are shown in Fig. 5. Both show similar patters, with visible symmetry between left and right leg.

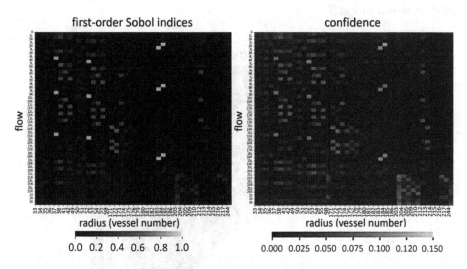

Fig. 5. Sobol indices of first order interactions between inputs and outputs (left) and corresponding confidence values (right). The colour scales are adjusted such that distinctive features of both matrices are visible. (Color figure online)

Ranked sensitivity for flow $Q20L$ extracted from the matrix of first-order Sobol indices is shown in Fig. 6. The first five input parameters are the same as those identified using the local sensitivity analysis, although they appear in a different order.

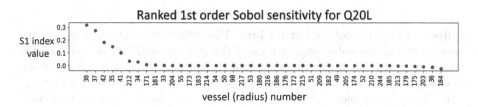

Fig. 6. Relative influence of input parameters (vessel radius) on Q20L flow based on 1st order Sobol indices.

Second-order indices generally contributed less to the output variance with values ranging from −0.06 to 0.20 with over 99% of these values being smaller than 0.10. Confidence values of S2 ranged between 0.0 and 0.12 with 97% lower than 0.05. An example of the matrix of second-order indices for output flow $Q20L$ is shown in Fig. 7.

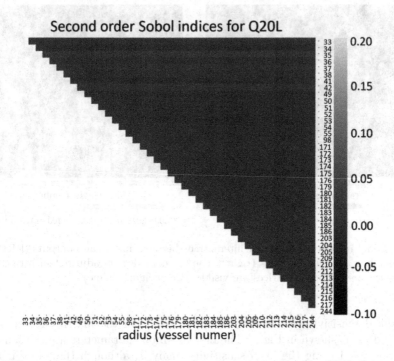

Fig. 7. Second order interactions between input parameters for flow Q20L. Interactions of a parameter with itself and duplicate interactions are excluded.

Influence of Thrombus Formation: The difference in absolute sensitivity between the no-thrombus sensitivity and the 30% and 60% thrombosis cases is shown in Fig. 8. The higher the degree of thrombosis, the bigger the difference in absolute sensitivity. This is observed for the flows of the left leg, marked by a red frame in the plot, and for the output flow Q_{24} in the *inferior vena cava*. These results demonstrate that sensitivity of the system is dominated by the effects of the thrombus when the reduction of the vessel lumen is $\geq 40\%$ while *anatomical* variation is 10%.

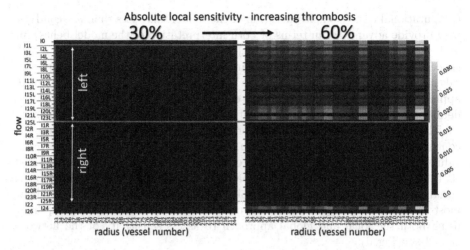

Fig. 8. Difference between absolute local sensitivity matrix for no-thrombosis case and two different degrees of thrombosis in the left common iliac vein. "Left" and "right" refer to legs. (Colour figure online)

4   Discussion

The focus of this study is on the assessment of the influence of anatomical variability on the distribution of flow within a model of the lower limb circulation with model parameters taken from Müller and Toro [7]. The sensitivity analyses reported here are an important step in examining the behaviour of the model, prior to further development including personalisation with clinical data, as discussed in detail by Huberts et al. [9].

The comparison of absolute and normalised sensitivity provided in Fig. 2 highlights the effect of variation of radii on the flow observed within the network. Both output measures provide useful information about the system and it is reasonable that the largest normalised change in flow is observed in locations associated with the venous return, as the flow in these vessels results from flow through all other vessels in the network.

The ranking of input parameters provided by the results shown in Fig. 3 is useful when the focus of the model operation is on predicting the flow in a specific location in the network (in this example the left common iliac vein). If such a model is personalised using clinical measurements of vascular anatomy then this ranking can be used to identify the most important radii for direct measurement and reduce the effort in assessing vasculature which has a small effect on the flow in the location of interest. Figure 3 suggests such an approach is likely to be feasible for the current network, as relatively few input parameters are associated with higher sensitivity values for this specific model output.

The visualisation of orthogonal sensitivity of the model input parameters shown in Fig. 4 has the potential to inform approaches to reduce the complexity of the model. Although this is not necessary in the current model, due to low computational cost (a single operation of the model takes less than a second), it may provide advantages in terms of both interpretation of the model behaviour and efficiency of model personalisation (by reducing the dimension of the search space). However, some care is required in interpretation of Fig. 4 as the approach taken here assigns the same significance to all model outputs. For specific clinical applications this may not be appropriate, as some model outputs will inform the detail of haemodynamics associated with patient outcomes more than others.

The global sensitivity analysis performed in this study was used to assess the contribution to variation of flow within the network over a wider range of radii values. Comparison of the magnitude of the first and second terms confirmed that interactions between input radii made a relatively small contribution to the variance in the output flows. It is notable that input radii which contributed the most to variation in a particular output flow were fairly consistently reported using both the local and global approach to assess sensitivity of the network (Fig. 3 and Fig. 6).

The results provided in Fig. 8 demonstrate the significance of the occlusion of a vessel due to the formation of a thrombus. Although thrombus formation is represented in the same manner as variation in anatomy, by varying the vessel radius, it is worth noting that this variation is of a larger range than that assumed for anatomical variation. The figure shows an increase in absolute sensitivity in the leg affected by the thrombosis. Changes in relative sensitivity were also observed, but only became significant for thrombosis greater than 60%.

5   Conclusion

This study demonstrates the value of local and global sensitivity analysis to inform development of a model of lower limb haemodynamics. The results obtained can be used to inform model reduction strategies and target clinical data collection, to maximise the accuracy of model estimates of flow in venous regions, prone to thrombus development. A suitably refined and parameterised model has clear potential to guide clinical decision-making by providing information beyond the level typically measured during patient follow-up.

Acknowledgements. This publication is supported by the European Union's Horizon 2020 research and innovation programme under grant agreement Sano No 857533 and carried out within the International Research Agendas programme of the Foundation for Polish Science, co-financed by the European Union under the European Regional Development Fund.

References

1. Stone, J.: Deep vein thrombosis: pathogenesis, diagnosis, and medical management. Cardiovasc. Diagn. Ther. **7**(3), 276–284 (2017)
2. Baldwin, M.J.: Post-thrombotic syndrome: a clinical review. J. Thromb. Haemost. **11**(5), 795–805 (2013)

3. Beckman, M.G.: Venous thromboembolism: a public health concern. Am. J. Prev. Med. **38**(4), 495–501 (2010)
4. Lim, C.S., et al.: A centralised complex venous service model in an NHS hospital. Br. J. Healthcare Manag. **26**(2), 2–15 (2022)
5. Black, S.A., et al.: Management of acute and chronic iliofemoral venous outflow obstruction: a multidisciplinary team consensus. Int. Angiol. **39**(1), 3–16 (2020)
6. Shi, Y., et al.: Review of zero-D and 1-D models of blood flow in the cardiovascular system. Biomed. Eng. Online **10**, 33 (2011)
7. Müller, L.O., Toro, E.F.: A global multiscale mathematical model for the human circulation with emphasis on the venous system. Int. J. Numer. Meth. Biomed. Eng. **30**, 681–725 (2014)
8. Keijsers, J.M., et al.: A 1D pulse wave propagation model of the hemodynamics of calf muscle pump function. Int. J. Numer. Method Biomed. Eng. **31**(7), e02716 (2015)
9. Huberts, W., et al.: What is needed to make cardiovascular models suitable for clinical decision support? A viewpoint paper. J. Comput. Sci. **24**, 68–84 (2018)
10. Benemerito, I., et al.: Determining clinically-viable biomarkers for ischaemic stroke through a mechanistic and machine learning approach. Ann. Biomed. Eng. (2022). https://doi.org/10.1007/s10439-022-02956-7
11. Li, R., et al.: Selection of model parameters for off-line parameter estimation. IEEE Trans. Control Syst. Technol. **12**(3), 402 (2004)
12. Saltelli, A., et al.: Variance based sensitivity analysis of model output. Design and estimator for the total sensitivity index. Comput. Phys. Commun. **181**(2), 259–270 (2010)
13. Hose, D.R., et al.: Cardiovascular models for personalised medicine: where now and where next? Med. Eng. Phys. **72**, 38–48 (2019)

Data Augmentation Techniques to Improve Metabolomic Analysis in Niemann-Pick Type C Disease

Francisco J. Moreno-Barea[1]([✉]) [iD], Leonardo Franco[1] [iD], David Elizondo[2] [iD], and Martin Grootveld[3] [iD]

[1] Escuela Técnica Superior de Ingeniería Informática Universidad de Málaga, Málaga, Spain
{fjmoreno,lfranco}@lcc.uma.es
[2] School of Computer Science and Informatics, Faculty of Technology, De Montfort University, The Gateway, Leicester, UK
elizondo@dmu.ac.uk
[3] Leicester School of Pharmacy, Faculty of Health and Life Sciences, De Montfort University, The Gateway, Leicester, UK
mgrootveld@dmu.ac.uk

Abstract. Niemann-Pick Class 1 (NPC1) disease is a rare and neurodegenerative disease, and often metabolomics datasets of NPC1 patients are limited in the number of samples and severely imbalanced. In order to improve the predictive capability and identify new biomarkers in an NPC1 disease urinary dataset, data augmentation (DA) techniques based on computational intelligence are employed to create additional synthetic samples. This paper presents DA techniques, based on the addition of noise, on oversampling techniques and using conditional generative adversarial networks, to evaluate their predictive capacities on a set of Nuclear Magnetic Resonance (NMR) profiles of urine samples. Prediction results obtained show increases in sensitivity (30%) and in F_1 score (20%). In addition, multivariate data analysis and variable importance in projection scores have been applied. These analyses show the ability of the DA methods to replicate the information of the metabolites and determined that selected metabolites (such as 3-aminoisobutyrate, 3-hidroxivaleric, quinolinate and trimethylamine) may be valuable biomarkers for the diagnosis of NPC1 disease.

Keywords: Metabolomics · Data augmentation · Niemann-Pick type C · Bioinformatics · Lysosomal storage disease

1 Introduction

Niemann-Pick type C disease (NPC, OMIM 257220) is a very rare neurodegenerative lysosomal storage disease caused by mutations in two genes NPC1 and NPC2 [22]. NPC affect approximately 1:100000 live births, although the NPC1

D. Groen et al. (Eds.): ICCS 2022, LNCS 13352, pp. 78–91, 2022.
https://doi.org/10.1007/978-3-031-08757-8_8

mutations account for 95% of cases observed. NPC involves the altered lysosomal storage of sphingosine, and leads to a loss of lysosomal calcium ions, a process accompanied by the accumulation of unesterified cholesterol and glycosphingolipids [19,24], along with decreased acidic store calcium levels [11]. Usually, NPC disease presents in childhood with clumsiness, ataxia, learning difficulties, vertical gaze paralysis, and dysphagia, together with cataplexy, epilepsy, and hepatosplenomegaly. Additionally, adult-onset illness may occur, and this may be associated with a neuropsychiatric presentation [22]. NPC disease also involves neuroinflammation, neuronal apoptosis, and oxidative stress [4].

For the diagnosis and prognostic monitoring of such diseases, metabolomics strategies are valuable because bioanalytical dataset systems can be analysed under pre-established conditions determined by the experimental design. The non-invasive nature of metabolomics and the close link of this type of data with the phenotype, make it an ideal tool for pharmaceutical and preventative health. Metabolomics is also applicable to the discovery of biomarkers, small molecules known as metabolites, as a support for decision making. Selected metabolites and their concentrations can be used to determine the status of different groups of samples based on their detection in control group samples, or in those collected from patients with a specified disease. Urine samples such as those analysed in this work contain informative metabolites that can be easily analysed for the purpose of discovering new biomarkers.

Notwithstanding, currently there is a clear lack of global, untargeted metabolomics studies focused on investigations of lysosomal storage diseases, with only a small number of studies being reported [18,20,21]. These studies justify the value offered by NMR-based metabolomics data analysis techniques, and the use of composites of both bioanalytical techniques and computational intelligence techniques is therefore further evolving and becoming more popular [13]. However, metabolomics datasets are often limited in the number of samples and heavily imbalanced. In this case, the lysosomal storage disorders are genetically-distinct and metabolically-related, rare inherited diseases. Because of this, the prior collection and the parental ethical consent required are often highly challenging hurdles to surmount. Additionally, obtaining a sufficient number of biofluid samples for NMR or other analyses adds to this complexity.

In this work, computational intelligence based Data Augmentation (DA) methods are used to generate more observations. DA has proven to be an effective technique to improve the performance of machine learning models, especially for applications related to problems involving datasets consisting of images [9], also in biomedical applications [6,14,23]. The application of DA techniques to datasets that are not images, signals or time series, is more complex. Experts find it easier to evaluate a generated image, being able to measure its quality and distinguish whether it is a 'synthetic' or a 'real' image. However, this type of evaluation conducted by human experts is not feasible when applications includes genomic or clinically-relevant metabolic data. DA techniques such as noise injection techniques [17,26] or the application of SMOTE techniques (synthetic minority oversampling technique) [2] are available to handle this type of dataset. A more recent technique known as Generative Adversarial Networks (GANs) has been proposed to be suitable for the analysis of these types of datasets [8]. GAN models have shown an impressive level of success in generating realistic images, and recently, it has been shown that they can also be applied as a DA method for datasets without any type of spatial or temporal structure [5,16], also in some biomedical applications [7,10,12]. To the best of the authors knowledge, there are no recent DA studies that show its application on metabolomics analysis.

Considering all the above aspects, the main objectives of this work are: (1) to apply different state-of-the-art DA methods to a small size metabolomics dataset aimed at obtaining an increase in the prediction performance of urine samples belonging to NPC1 disease patients, in order to demonstrate their usefulness in this research domain; (2) to analyse the ability of these DA methods to replicate the information of the metabolites using conventional forms of multivariate data analysis, such as partial least squares - discriminatory analysis (PLS-DA).

2 Materials

This study presents a UK-based clinical cohort consisting of 13 untreated NPC1 patients and 47 corresponding parental heterozygous carriers. The selection process for the NPC1 patient cohort was carefully conducted to select only patients not receiving any therapeutic agents. This process avoids any complications arising from the presence of urinary ^{1}H NMR resonances attributable to such drugs and their metabolites in the urinary metabolite profiles explored. The data for this study was collected with informed consent and previously approved by the appropriate Research Ethics Committee (06/MRE02/85). Urine samples were collected, thawed and centrifuged to remove any cells and debris. The sample mixtures were then transferred to NMR tubes for in-depth analysis.

Single-pulse ^{1}H NMR analysis of human urine samples were obtained using a Bruker Avance AV-600 spectrometer (Queen Mary University of London facility, London, UK) operating at a frequency of 600.13 MHz, as described in [21]. The intense H_2O/HOD signal ($\delta = 4.80$ ppm) was suppressed via gated decoupling during the delay between pulses. Chemical shift values were internally referenced to the methyl group resonances of acetate (s, $\delta = 1.920$ ppm), alanine (d, $\delta = 1.487$ ppm), creatinine (>NCH3 s, $\delta = 3.030$ ppm) and lactate (d, $\delta = 1.330$ ppm). Through a complete consideration of chemical shift values, coupling patterns and coupling constants, the identities of metabolite resonances present in spectra acquired were routinely assigned. These assignments were cross-checked with the *Human Metabolome Database (HMDB)* [25] and confirmed by one- (1D) and two-dimensional (2D) correlation (COSY) and total correlation (TOCSY) spectroscopic techniques.

The urinary dataset matrix consists of 60 spectra × 33 ^{1}H NMR-assigned metabolite predictor variables. This dataset was generated using macro procedures for line broadening, zero filling, Fourier-transformation and phase and baseline corrections, together with the subsequent application of a separate macro for the "intelligently-selected bucketing" (ISB) processing sub-routine. All procedures were performed using the ACD/Labs Spectrus Processor 2012 software package (ACD/Labs, Toronto, Ontario, Canada M5C 1T4). This ISB strategy ensured that all bucket edges featured did not coincide with ^{1}H NMR resonance maxima, and hence this approach avoided the splitting of signals across separate integral regions. Prior to data augmentation experiments, all sample ^{1}H NMR profiles were autoscaled column-(metabolite variable)-wise.

3   Data Augmentation Methods

Addition of Noise. The first of the methods used in this work for data augmentation is a simple and straightforward one, that can be easily applied, and has the ability to lead to competent results. Specifically, the method randomly selects samples and modifies a maximum of a 25% of the features present in the data. The process of generating a new feature value $\tilde{x}$ from the original value x is mathematically described in Eq. 1. The noise value obtained from a random normal distribution (denoted "RND") with a standard deviation/variance of 1.0, is added to the original value for the chosen feature. The resulting "noisy" value is controlled so that it does not exceed the real limit values established for its feature (MIN_Value and MAX_Value). A standard deviation of 1.0 at the random normal distribution is sufficient to generate a sample that is not too far from the actual sample.

$$\tilde{x} = \min(\text{MAX\_Value}, \max(\text{MIN\_Value}, x + \text{RND}(-1.0, 1.0))) \qquad (1)$$

A variation of the addition of noise method described above has been designed for balancing purposes. The method abbreviated as "Noise Bal", differs from the standard method in that it applies the random selection only to samples belonging to the minority class. Therefore, it only modifies and generates synthetic samples that belong to the minority class in an oversampling process. The rest of the method follows the same noise addition process described before.

SMOTE Technique. In clinical cohorts of rare diseases, it is easier to have more control samples available than samples from patients that present the disease. Therefore, medical datasets, as well as metabolomics datasets, are often imbalanced. The traditional oversampling method to reverse this situation by applying DA is SMOTE (Synthetic Minority Oversampling Technique) [2]. SMOTE uses a k-nearest neighbour algorithm on the minority class, rather than random sampling with replacement. In this way, the algorithm performs an interpolation between each sample x and its selected neighbours. The interpolation computes the difference between the sample x and each of the neighbours in the feature space, multiplies the difference of each feature by a random normalisation between 0 and 1, and adds this value to the feature of original sample x. This interpolation results in the synthetic samples generated by SMOTE being located within the space between the selected neighbours and the sample x. One disadvantage of the SMOTE algorithm application is the lack of control over the number of samples to generate. This technique is ineffective on well-balanced datasets, since oversampling aims to create a fully balance augmented dataset. Another disadvantage, derived from the interpolation process, is the creation of synthetic samples that do not follow the distribution of the original dataset.

Conditional GAN. The DA application of deep learning models known as Generative Adversarial Networks (GAN) [8] has shown an impressive success in the generation of realistic images. Specifically, the model considered in this work is the Conditional GAN (CGAN) [15], since a supervised task is performed. The CGAN model is a variant of the vanilla GAN model in which the information contained in the sample label y is taken into account. The generation of synthetic samples using GAN models occurs by learning the distribution of the original dataset. With this aim, GAN models have a structure divided into two neural networks trained simultaneously, the *generator* and the *discriminator*, yielding a confrontation between both so that they are able to learn from each other. In this manner, the objective of the discriminator network (D) is to estimate the probability of the sample arises from the real distribution or is a generated sample. However, the purpose of the generator network (G), which takes as input a noisy random distribution z and the condition y, is to produce a distribution $G(z)$ (synthetic sample) with features that approximate those present in the real samples. Therefore, the generator intends that the discriminator cannot distinguish these synthetic samples from the real ones.

$$\min_G \max_D \ \mathbb{E}_{x \sim p_{data}(x)}[\log D(x|y)] + \mathbb{E}_{z \sim p_z(z)}[\log(1 - D(G(z|y)))] \qquad (2)$$

The two behaviours described above within the competitive process can be distinguished in the CGAN objective cost function (Eq. 2). One part is related to achieving a better recognition of those samples that belong to the real distribution, while the other is related to achieving a better recognition of those synthetic samples created by the generator network. Thereby, the discriminator network is updated based on the error associated with the ability to perceive whether the samples are real or false, expressed in Eq. 3; and the generator network is updated from the error identified for false sample recognition, modelled by Eq. 4.

$$\max_D \ \mathbb{E}_{x \sim p_{data}(x)}[\log D(x|y)] + \mathbb{E}_{z \sim p_z(z)}[\log(1 - D(G(z|y)))] \qquad (3)$$

$$\min_G \ \mathbb{E}_{z \sim p_z(z)}[\log(D(G(z|y)))] \qquad (4)$$

4   Experiments and Results

The experimentation process followed is shown in Fig. 1. A stratified division of the original dataset into training and test sets is performed, allowing to maintain total independence between the synthetic data generation process and the evaluation of the experiments. Due to the reduced number of samples present in the benchmark dataset, a split of 60% of samples for training and 40% for testing is conducted. Depending on the DA method, the synthetic data generation process uses the training set to create the desired number of samples, following the procedures described in the previous section. Before the augmentation process, a principal component analysis (PCA) is performed on the training dataset. This process removes any high level of correlation (multicollinearity) between the variables of the metabolomics dataset. In this way, the score vectors are obtained and the training, test and synthetic datasets are transformed. Samples values are represented by their principal components instead of the original values from metabolomic variables.

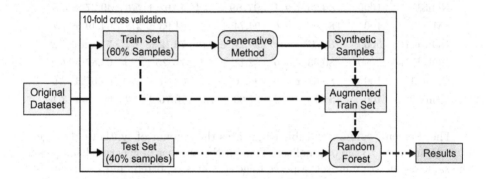

Fig. 1. Flow diagram of the whole experimentation process.

After this transformation process, the augmented training set is formed by adding synthetic generated data to the training one. A Random Forest system [1] is used for classifying the samples. Random Forest are essentially an ensemble of decision trees and establishes the outcome based on the individual tree predictions. The python scikit-learn package is used, with 1001 trees, bootstrap samples to build each tree, Gini impurity for tree splitting and 6 predictors selected at each node. The classifier model employs the augmented set to perform the training and the prediction is measured using the test set data. The entire process of data division, generation, training and evaluation is carried out through a cross-validation procedure, employed for obtaining a better estimate of the metrics, in order to avoid small dataset sampling biases.

4.1 Classification Performance

Test results obtained from the application of the different DA methods are shown in Table 1. To show whether the application of DA improves the classification performance, the results are compared with those obtained with the original non-augmented dataset, indicated in the table as 'None'. In order to be more exhaustive in the study, the results obtained with the combination of samples from two different DA models are included. 'Comb 1' refers to the results obtained with a combination between the CGAN model and the SMOTE method, and 'Comb 2' refers to the results obtained with a combination between the CGAN model and the Noise Bal strategy, when a 500% augmentation level is applied with Noise Bal.

Table 1. Test results acquired with a random forest system using each DA method and the percentage of augmentation applied.

Aug. model	Percent	Accuracy	Specificity	Sensitivity	F_1 score
None	None	85.42 ± 1.0	97.37 ± 0.6	40.10 ± 5.6	53.33 ± 2.4
CGAN	500	85.83 ± 1.1	90.53 ± 1.3	67.99 ± 4.5	63.47 ± 3.9
NOISE	1000	86.25 ± 1.1	**97.89 ± 0.4**	42.02 ± 5.0	49.27 ± 5.6
SMOTE	100	87.92 ± 1.0	94.74 ± 0.4	61.93 ± 4.8	64.49 ± 4.1
NOISE Bal	500	**89.17 ± 1.0**	94.21 ± 0.9	70.12 ± 3.7	**71.79 ± 2.8**
NOISE Bal	2000	80.83 ± 1.6	82.11 ± 2.5	**76.25 ± 2.9**	63.87 ± 1.9
Comb 1	100	85.63 ± 1.0	90.26 ± 1.2	68.01 ± 4.0	64.67 ± 3.0
Comb 2	100	84.79 ± 1.4	87.63 ± 1.4	73.76 ± 3.9	66.68 ± 3.1

The 'Percent' column of Table 1 indicates the amount of synthetic data generated compared to the original set. Thus, if the number of generated samples is the same as the training set, a percentage of 100 is reported; and if the number of training samples is multiplied by 10, a percentage of 1000 is reported. The remaining columns show the values (± 'between-validation performance' SE) obtained for each of the test metrics. The test metrics showed are the accuracy, specificity, sensitivity and F_1 score obtained. The F_1 score is the harmonic mean of the precision and sensitivity (Eq. 5), and allows a reliable measure of the prediction performance achieved in problems where sensitivity is more important.

$$F_1 \text{ score} = 2 \cdot \frac{\text{precision} \cdot \text{sensitivity}}{\text{precision} + \text{sensitivity}} = \frac{2TP}{2TP + FP + FN} \tag{5}$$

Results in Table 1 show that an improvement in test prediction accuracy, sensitivity and F_1 is achieved with almost all the DA methods compared to the values obtained with the non-augmented dataset ('None'). Using the Noise Bal method with 1000% DA, the highest accuracy (89.17%) and F_1 values (71.78%) are obtained. The Noise Bal approach with 2000% reaches the highest sensitivity values (78.8%), but with a lower accuracy (80.83%) and F_1 score (63.87%). These values show a substantial improvement compared to analysis of the dataset without augmentation.

Additional analyses were performed reviewing the impact of the number of samples generated with different DA methods on the test results obtained. Three test metrics (accuracy, specificity and sensitivity) obtained with three different DA methods (CGAN, Noise and Noise Bal) *versus* the number of instances on a logarithmic scale, are presented in Fig. 2. The results obtained with the CGAN

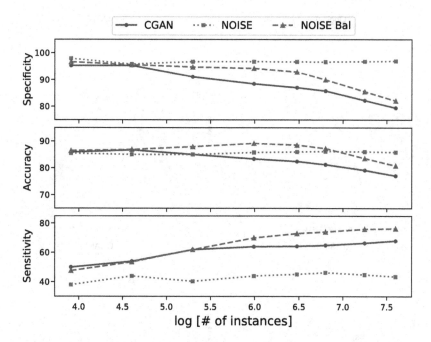

Fig. 2. Comparisons of the specificity, accuracy and sensitivity obtained with different DA methods *versus* the logarithm of the number of instances generated. Dots (solid) represent the results obtained with the CGAN model, squares (dotted) those with the NOISE method, and triangles (dashed) those with the NOISE Bal method.

model indicate a negative correlation between the number of instances created and the specificity and accuracy gain of the prediction. However, a positive correlation for the sensitivity gain was also observed. For the Noise Bal method, there is clearly a significant positive correlation between sensitivity and the number of instances created with this strategy. The specificity obtained decreases slightly until the abscissa axis reaches a value of 6.5, when it presents a significant negative correlation. The influence of specificity on the accuracy gain is noticeable, since they decreases at the same time when a large number of samples are generated. Finally, the values obtained for the metrics are approximately stable with respect to the number of instances generated with the standard Noise method.

4.2   Augmented Datasets Analysis

An important objective is to analyse how the DA methods were able to replicate the metabolomic information present in the dataset. The configuration of the augmented samples can be visualised in a two-dimensional space (component 2 *vs* component 1) through a partial least squares - discriminatory analysis (PLS-DA), using *MetaboAnalyst v4.0* software (University of Alberta and National Research Council, National Nanotechnology Institute (NINT), Edmonton, AB, Canada) [3]. This provided a means to check the information contained in the

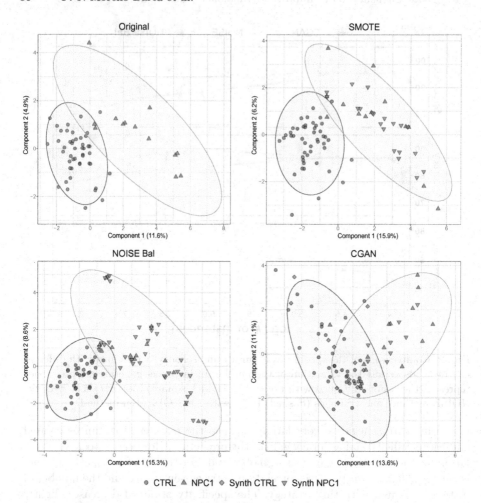

Fig. 3. PLS-DA component 2 *versus* component 1 scores plot for the original dataset, and the augmented datasets with SMOTE, the Noise Bal method and CGAN model. Colour codes: red circles, heterozygous carrier control urine; green diamond, synthetic control; blue triangles, NPC1 disease urine; purple inverse triangles, synthetic NPC1. (Color figure online)

augmented dataset and compare this with the information in the original samples, analyzing how the distribution and clustering is affected.

Figure 3 (top left) shows the PLS-DA results obtained by using the original NPC1 dataset. This reveals two significant groups for the samples that correspond to the possible classes of "disease state", with an area where both clusters converge. The cluster belonging to the control group appeared as a compact cluster, while the cluster conformed by the NPC1 disease samples was more dispersed. The results of PLS-DA after adding the samples generated with SMOTE are shown in Fig. 3 (top right). Here, it can be clearly seen how the creation of samples through SMOTE works. The synthetic samples are distributed throughout the 'real' NPC1 disease cluster, from the interpolation process. In this case, the small convergence zone between clusters avoids the SMOTE method disadvantages. Figure 3 (bottom left) shows the distribution of the augmented dataset created by using the Noise Bal method with respect to the original samples.The distribution of the samples is similar to the distribution observed with the original dataset, and with the augmented dataset produced by the SMOTE. Through the noise injection process, the synthetic samples belonging to the NPC1 disease class are found grouped around the original samples that they modify.

The PLS-DA scores plot when using the augmented dataset with the CGAN model is shown in Fig. 3 (bottom right). Contrary to previous DA methods, CGAN is not an oversampling technique, so the model creates samples belonging to both classes. The generated samples modify the dispersion and angle presented by the component analysis, causing the control group less compact. Although it is still possible to differentiate both groups, with a larger convergence zone. The generated synthetic samples fit satisfactorily the distribution of the 'real' samples for both groups, with some of them generated in the convergence zone.

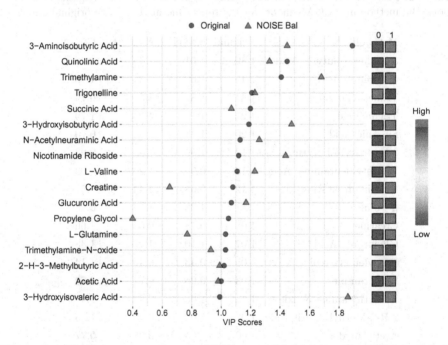

Fig. 4. Variable importance parameter (VIP) scores obtained from the PLS-DA applied to the original dataset and the application to the Noise Bal augmented dataset. Colour codes: blue circles, original VIP; red triangles, Noise Bal augmentation VIP. (Color figure online)

The variable importance in projection (VIP) scores with respect to component 1 were also obtained using PLS-DA. The VIP scores allow to analyse the differences between the heterozygous carrier group and NPC1 disease urine samples, measuring the importance of each metabolite in the differentiation process. Figure 4 shows the VIP scores for the top 17 metabolites obtained from the original dataset and a comparison with the VIP scores obtained from the augmented dataset with balanced addition of noise. The coloured boxes indicate the relative concentrations of the corresponding metabolite in each "disease state" group on the original dataset. Considering that the values >1.00 are significant, the most outstanding metabolite of the analysis in the original set was 3-aminoisobutyric acid with a VIP score equal to 1.89. With the Noise Bal augmentation, the most prominent metabolites were trimethylamine with a VIP score of 1.68 and 3-hydroxyisovaleric acid with a VIP score of 1.86. These values are higher than those obtained for the original dataset. Conversely, the 3-aminoisobutyric acid obtained a VIP score of 1.45 and propylene glycol obtained a VIP score of 0.4 with the Noise Bal dataset, values lower than the original ones. Notwithstanding, the analysis reveals a total of nine metabolites that obtain similar VIP values.

Table 2. PLS-DA VIP scores obtained using augmented datasets with the SMOTE, Noise Bal method and CGAN model for the top 17 metabolites with original dataset.

Metabolite	Original	SMOTE	Noise Bal	CGAN
3-Aminoisobutyric Acid	1.89	1.51	1.45	1.22
Quinolinic Acid	1.45	1.60	1.33	1.05
Trimethylamine	1.41	1.81	1.68	0.79
Trigonelline	1.21	1.12	1.23	1.79
Succinic Acid	1.20	1.13	1.07	0.70
3-Hydroxyisobutyric Acid	1.19	1.56	1.48	0.93
N-Acetylneuraminic Acid	1.13	0.93	1.26	0.93
Nicotinamide Riboside	1.12	0.68	1.44	1.51
L-Valine	1.11	1.38	1.23	0.66
Creatine	1.08	0.78	0.65	1.66
Glucuronic Acid	1.07	1.10	1.17	1.38
Propylene Glycol	1.05	0.43	0.40	1.14
L-Glutamine	1.03	0.96	0.77	0.70
Trimethylamine-N-oxide	1.03	1.05	0.93	1.49
2-H-3-Methylbutyric Acid	1.02	1.02	0.99	1.13
Acetic Acid	1.00	1.05	0.98	0.77
3-Hydroxyisovaleric Acid	0.99	1.69	1.86	0.58

The results obtained for each of the top 17 marker metabolites shown in Fig. 4 are summarised in Table 2. The analysis reveals certain metabolites showing analogous VIP scores for the SMOTE and Noise Bal approaches. Amongst these, the following metabolites should be highlighted: trimethylamine, 3-hydroxyisovaleric acid, 3-hydroxyisobutyric acid, 3-aminoisobutyric acid, and trigonelline. Both methods (SMOTE and Noise Bal) are oversampling ones, thus increasing the number of samples for the NPC1 disease class. This significantly influenced the analysis, which indicates a greater relevance of these metabolites to separate this group from the heterozygous carriers. Regarding the analysis using the augmented set with CGAN, the results show fewer similarities. The most differentiating metabolites with respect to the original dataset are trimethylamine with a VIP score of 0.79, compared to the original value equal to 1.41; and creatine with a VIP score of 1.66, and a value of 1.08 with the original dataset.

5 Conclusions

The different state-of-the-art techniques for Data Augmentation (DA) employed in this work clearly offer much potential regarding the analysis of metabolomics datasets, as these predominantly comprise small numbers of sample-donating participants, as it is the case of the NPC1 data examined here.

The results shown in Table 1 indicate a great improvement of test prediction, with an increase in predictive accuracy. This renders the balanced addition of noise (Noise Bal) the best DA method for this purpose. The augmented dataset reaches approximately a 4% improvement in accuracy compared to the analysis performed on the original dataset. Since the dataset is quite imbalanced, predictive accuracy is not the most representative metric, as it is more important that the largest number of patients with the disease be diagnosed as such. Therefore, most representative prediction metrics for this type of imbalanced problem are sensitivity and F_1 score. Table 1 shows that when performing data augmentation with the Noise Bal method and 500% DA, an approximate 30% improvement in sensitivity and a 20% improvement in F_1 score can be obtained.

In order to determine the ability of DA methods to replicate metabolic information, a PLS-DA was performed. The SMOTE and Noise Bal method show a good capacity to replicate the information of the metabolites from samples representing NPC1 disease. The results obtained from the analysis of the CGAN augmentation show the ability of this model to replicate information that fits the distribution of the 'real' samples. However, because CGAN can generate samples for both classes in the convergence zone of the clusters, the PLS-DA results differ from the original one. Finally, the VIP scores results obtained revealed a series of biomarkers which may be valuable for distinguishing between the urinary ^{1}H NMR profiles of NPC1 patients and their heterozygous healthy controls. These included the branched-chain amino acid valine, 3-aminoisobutyrate, 3-hidroxivaleric, quinolinate and trimethylamine. The selected metabolites and their relative importance rankings were found to be similar to those reported in a previously conducted study of the dataset analysed, and without any form of DA strategies [21].

In conclusion, DA techniques constitute a suitable approach to increase the prediction performance of Niemann-Pick Class C1 (NPC1) disease activity in patients when analysing ^{1}H NMR urinary metabolic datasets. DA techniques are capable of generating good quality synthetic data that lead to an increase in sensitivity of 30%, allowing the identification of urinary metabolomics biomarkers which will serve on the diagnosis and monitoring of the severity of patients with NPC1 disease. Future research directions will focus on testing different machine learning algorithms analysing their robustness in the prediction of rare diseases.

Acknowledgements. The authors acknowledge the support from MICINN (Spain) through grant TIN2017-88728-C2-1-R and PID2020-116898RB-I00, from Universidad de Málaga y Junta de Andalucía through grant UMA20-FEDERJA-045, and from IBIMA (all including FEDER funds).

References

1. Breiman, L.: Random forests. Mach. Learn. **45**(1), 5–32 (2001). https://doi.org/10.1023/a:1010933404324
2. Chawla, N.V., Bowyer, K.W., Hall, L.O., Kegelmeyer, W.P.: SMOTE: synthetic minority over-sampling technique. J. Artif. Intell. Res. **16**, 321–357 (2002). https://doi.org/10.1613/jair.953
3. Chong, J., et al.: MetaboAnalyst 4.0: towards more transparent and integrative metabolomics analysis. Nucl. Acids Res. **46**(W1), W486–W494 (2018). https://doi.org/10.1093/nar/gky310
4. Cougnoux, A., et al.: Necroptosis in Niemann-Pick disease, type C1: a potential therapeutic target. Cell Death Dis. **7**(3), e2147–e2147 (2016). https://doi.org/10.1038/cddis.2016.16
5. Douzas, G., Bacao, F.: Effective data generation for imbalanced learning using conditional generative adversarial networks. Expert Syst. Appl. **91**, 464–471 (2018). https://doi.org/10.1016/j.eswa.2017.09.030
6. Frid-Adar, M., Diamant, I., Klang, E., Amitai, M., Goldberger, J., Greenspan, H.: GAN-based synthetic medical image augmentation for increased CNN performance in liver lesion classification. Neurocomputing **321**, 321–331 (2018). https://doi.org/10.1016/j.neucom.2018.09.013
7. García-Ordás, M.T., Benavides, C., Benítez-Andrades, J.A., Alaiz-Moretón, H., García-Rodríguez, I.: Diabetes detection using deep learning techniques with oversampling and feature augmentation. Comput. Meth. Programs Biomed. **202**, 105968 (2021). https://doi.org/10.1016/j.cmpb.2021.105968
8. Goodfellow, I., et al.: Generative Adversarial Nets. In: Advances in Neural Information Processing Systems, vol. 3, pp. 2672–2680 (2014). https://doi.org/10.1145/3422622
9. He, K., Zhang, X., Ren, S., Sun, J.: Deep Residual Learning for Image Recognition. In: 2017 IEEE Conference on Computer Vision and Pattern Recognition (CVPR), pp. 770–778 (2016). https://doi.org/10.1109/cvpr.2016.90
10. Liu, Y., Zhou, Y., Liu, X., Dong, F., Wang, C., Wang, Z.: Wasserstein GAN-based small-sample augmentation for new-generation artificial intelligence: a case study of cancer-staging data in biology. Engineering **5**(1), 156–163 (2019). https://doi.org/10.1016/j.eng.2018.11.018

11. Lloyd-Evans, E., et al.: Niemann-Pick disease type C1 is a sphingosine storage disease that causes deregulation of lysosomal calcium. Nat. Med. **14**(11), 1247 (2008). https://doi.org/10.1038/nm.1876
12. Marouf, M., et al.: Realistic in silico generation and augmentation of single-cell RNA-seq data using generative adversarial networks. Nat. Commun. **11**(1), 1–12 (2020). https://doi.org/10.1038/s41467-019-14018-z
13. Marshall, D.D., Powers, R.: Beyond the paradigm: combining mass spectrometry and nuclear magnetic resonance for metabolomics. Prog. Nucl. Magn. Reson. Spectrosc. **100**, 1–16 (2017). https://doi.org/10.1016/j.pnmrs.2017.01.001
14. Marzullo, A., Moccia, S., Catellani, M., Calimeri, F., De Momi, E.: Towards realistic laparoscopic image generation using image-domain translation. Comput. Methods Programs Biomed. **200**, 105834 (2021). https://doi.org/10.1016/j.cmpb.2020.105834
15. Mirza, M., Osindero, S.: Conditional Generative Adversarial Nets. CoRR abs/1411.1784, November 2014. https://arxiv.org/abs/1411.1784
16. Moreno-Barea, F.J., Jerez, J.M., Franco, L.: Improving classification accuracy using data augmentation on small data sets. Expert Syst. Appl. **161**, 113696 (2020). https://doi.org/10.1016/j.eswa.2020.113696
17. Moreno-Barea, F.J., Strazzera, F., Jerez, J.M., Urda, D., Franco, L.: Forward Noise Adjustment Scheme for Data Augmentation. In: IEEE Symposium Series on Computational Intelligence (IEEE SSCI 2018), pp. 728–734 (2018). https://doi.org/10.1109/ssci.2018.8628917
18. Percival, B.C., Latour, Y.L., Tifft, C.J., Grootveld, M.: Rapid identification of new biomarkers for the classification of GM1 Type 2 Gangliosidosis using an unbiased 1H NMR-linked metabolomics strategy. Cells **10**(3), 572 (2021). https://doi.org/10.3390/cells10030572
19. Platt, F.M., d'Azzo, A., Davidson, B.L., Neufeld, E.F., Tifft, C.J.: Lysosomal storage diseases. Nat. Rev. Dis. Primers. **4**(1), 1–25 (2018). https://doi.org/10.1038/s41572-018-0025-4
20. Probert, F., et al.: NMR analysis reveals significant differences in the plasma metabolic profiles of Niemann Pick C1 patients, heterozygous carriers, and healthy controls. Sci. Rep. **7**(1), 1–12 (2017). https://doi.org/10.1038/s41598-017-06264-2
21. Ruiz-Rodado, V., et al.: 1H NMR-linked urinary metabolic profiling of Niemann-Pick Class C1 (NPC1) disease: identification of potential new biomarkers using correlated component regression (CCR) and genetic algorithm (GA) analysis strategies. Current Metabol. **2**(2), 88–121 (2014). https://doi.org/10.2174/2213235X02666141112215616
22. Vanier, M.T.: Niemann-Pick disease type C. Orphanet J. Rare Dis. **5**(1), 1–18 (2010). https://doi.org/10.1186/1750-1172-5-16
23. Waheed, A., Goyal, M., Gupta, D., Khanna, A., Al-Turjman, F., Pinheiro, P.R.: CovidGAN: data augmentation using auxiliary classifier GAN for improved COVID-19 detection. IEEE Access **8**, 91916–91923 (2020). https://doi.org/10.1109/access.2020.2994762
24. Winkler, M.B., et al.: Structural insight into eukaryotic sterol transport through Niemann-Pick type C proteins. Cell **179**(2), 485–497 (2019). https://doi.org/10.1016/j.cell.2019.08.038
25. Wishart, D.S., et al.: HMDB 4.0: the human metabolome database for 2018. Nucleic Acids Res. **46**(D1), D608–D617 (2018). https://doi.org/10.1093/nar/gkx1089
26. Zur, R.M., Jiang, Y., Pesce, L., Drukker, K.: Noise injection for training artificial neural networks: a comparison with weight decay and early stopping. Med. Phys. **36**(10), 4810–4818 (2009). https://doi.org/10.1118/1.3213517

Effect of Feature Discretization on Classification Performance of Explainable Scoring-Based Machine Learning Model

Arkadiusz Pajor[1,2]([×]) [iD], Jakub Żołnierek[3][iD], Bartlomiej Sniezynski[2][iD],
and Arkadiusz Sitek[1][iD]

[1] Sano Centre for Computational Medicine, Cracow, Poland
apajor@student.agh.edu.pl
[2] AGH University of Science and Technology, Cracow, Poland
[3] Maria Skłodowska-Curie Memorial Cancer Center, Warsaw, Poland

Abstract. We improve the utility of the Risk-calibrated Supersparse Linear Integer Model (RiskSLIM). It is a scoring system that is an interpretable machine learning classification model optimized for performance. Scoring systems are commonly used in healthcare and justice. We implement feature discretization (FD) in the hyperparameter optimization process to improve classification performance and refer to the new approach as FD-RiskSLIM. We test the approach using two medical applications. We compare the results of FD-RiskSLIM, RiskSLIM, and other machine learning (ML) models. We demonstrate that scoring models based on RiskSLIM, in addition to being interpretable, perform at least on par with the state-of-the-art ML models such as Gradient Boosting in terms of classification metrics. We show the superiority of FD-RiskSLIM over RiskSLIM.

1 Introduction

Machine Learning (ML) starts to play an important role in domains like healthcare where making high stake decisions is common. Examples of such domains include cancer prognosis [8], hypertension outcomes [5], heart diseases [16], and many others. Typically researchers focus on the predictive performance of the ML models. Gradient Boosting, Random Forest, or Artificial Neural Networks are considered state-of-the-art algorithms that show superior performance to simpler models like Decision Tree or Linear Regression. However, over the last few years regulations outlined in the General Protection Data Regulation (GDPR) and in particular "right to explanation" [13] emphasized the importance of ML-algorithm trustability, transparency, and fairness and have sparked a discussion around important needs for interpretability of ML models. In our research, we focus on algorithms that generate models with a high level of predictive performance and low complexity which are interpretable or explainable [25].

This work focuses on improvements of the Risk-calibrated Supersparse Linear Integer Model (RiskSLIM) introduced in [23] which is an interpretable ML model achieving accuracy comparable to black-box models mentioned before. We demonstrate that our improved interpretable RiskSLIM algorithm outperforms them for healthcare-related examples used here.

The main contributions of this paper: (1) we add feature discretization (FD) to RiskSLIM algorithm [23], (2) we compare FD-RiskSLIM with RiskSLIM and other classical ML techniques. We use two examples of medical applications, namely the prediction of heart-failure patient outcome and the prediction of the outcome of kidney cancer treatment. We demonstrate that the performance of our interpretable models (FD-RiskSLIM and RiskSLIM) is comparable or even superior to classical non-interpretable models such as Random Forest or Gradient Boosting. It is important to note that we optimize all algorithms used in this paper in the same way, using the algorithm for hyperparameter optimization (see Sect. 3.3).

2 Related Research

Our research was focused on medical decision support based on scoring systems. Scoring models are sparse linear models with integer coefficients that make them interpretable. Many popular scoring systems like Simplified Acute Physiology Score (SAPS) [15], Systemic Inflammatory Response Syndrome (SIRS) [4], Acute Physiologic Assessment and Chronic Health Evaluation (APACHE) [14], Stroke Risk Assessment in Atrial Fibrillation (CHADS$_2$) [12], Thrombolysis in Myocardial Infarction (TIMI) [3] to name a few. The models were built by domain experts based at least partially on their experience. In Fig. 1, a scoring model CHADS$_2$ is presented. The CHADS$_2$ index was created by including independent risk factors: prior cerebral ischemia, history of hypertension, diabetes mellitus, congestive heart failure, and age of 75 years. Other factors like high blood pressure or sex were not included based on domain expertise even though similar scoring systems developed by others take them into account. The points contributing to the overall score were assigned arbitrarily. However, they were validated with the use of an exponential survival model which measured how the rate of stroke was affected by 1-point increases in CHADS$_2$.

Other classical scoring systems such as SAPS [15] were built with the use of machine learning techniques. Features contributing to the overall SAPS score were selected with the use of cross-validation. The final model exposed twenty relevant features, but the complexity of the model was reduced by using logistic regression.

Our method is built on a novel approach to scoring systems introduced in [22]. The authors developed SLIM, the predecessor of RiskSLIM, which can build scoring systems directly from data with no necessity of using domain knowledge. With the use of SLIM, the authors managed to predict Obstructive Sleep Apnea using various information available for the patients using polysomnography results as the ground truth [24]. Data used for such prediction included standard medical information such as age, BMI, gender, diabetes, smoking, and past

Features	Points
Congestive heart failure	1
Hypertension: blood pressure consistently above 140/90 mmHg (or treated hypertension on medication)	1
Age >= 75 years	1
Diabetes mellitus	1
Prior Stroke or TIA or Thromboembolism	2

Score	0	1	2	3	4	5	6
Risk (%)	1.9	2.8	4.0	5.9	8.5	12.5	18.2

Fig. 1. Structure of CHADS$_2$ scoring system. Sum of points evaluates to the risk as presented.

problems with the heart reported by patients. Authors demonstrated that the scoring system built by them performed better than the commonly used STOP-BANG [7] scoring system, which in addition to features used by SLIM took into consideration symptoms reported by patients. The same scientific group used the successor of SLIM, RiskSLIM to predict seizures. The prediction was based on patterns in continuous electroencephalography (cEEG) [21]. Authors created a new scoring system 2HELPS2B and demonstrated that it performed as well as neural networks, but with important advantages of interpretability and transparency [20].

3   Methods

3.1   Risk-calibrated Supersparse Linear Integer Model (RiskSLIM)

RiskSLIM was introduced by Ustun and Rudin in 2019 [23]. It is a scoring system similar to the predictive models designed by humans over the last century (e.g. CHADS$_2$ [11]). However, contrary to the traditional models, RiskSLIM determines integer score points (which are coefficients of the linear model) relying solely on the data, using non-linear integer optimization, instead of obtaining score points from experts. Authors assumed that domain knowledge may be incorporated in a form of specific constraints to the input and output variables. It was shown that RiskSLIM allows the creation of scoring systems that give accurate and interpretable results for decisions related to medicine and criminal justice [21,24,26].

To train a RiskSLIM model, a mixed-integer, non-linear problem with hard computational complexity has to be solved. It follows that the computation for a highly-dimensional problem is time-consuming and often impractical in medical settings. The approach uses constraints to make score points to be small integers. The objective function is a log-loss (the same as used in the logistic regression) that is minimized during model training. As a result, we get risk-calibrated scoring systems, in which predicted risks agree with risks calculated directly from the data [19]. The formula which is used for the estimation of the risk of the event under consideration (stroke, cancer death) consists of an intercept value and a calculated integer score which is a sum of all score points provided by the model (score points are added to the score if the corresponding feature is present in the analyzed event):

$$\frac{1}{1 + \exp(-intercept - score)},\qquad(1)$$

where the intercept in the expression (1) is determined during the model training.

Table 1. Sample numerical feature discretized using 3 subspaces represented by columns with white background. Discretization is done with overlapping (top image) and with disjoint regions (bottom image). In FD-RiskSLIM a number of subspaces and threshold values are hyperparameters that are optimized (see Sect. 3.3). For presented experiments (see Sect. 4) we utilized disjoint regions.

feature	feature < 4.0	feature ≥ 4.0	feature ≥ 13.67
25	0	1	1
16	0	1	1
9	0	1	0
4	0	1	0
1	1	0	0
0	1	0	0
1	1	0	0
4	0	1	0
9	0	1	0
16	0	1	1
25	0	1	1

feature	0 ≤ feature < 4.0	4.0 ≤ feature < 13.67	13.67 ≤ feature < 25
25	0	0	1
16	0	0	1
9	0	1	0
4	0	1	0
1	1	0	0
0	1	0	0
1	1	0	0
4	0	1	0
9	0	1	0
16	0	0	1
25	0	0	1

3.2   RiskSLIM with Feature Discretization (FD-RiskSLIM)

The original RiskSLIM algorithm assigns at most a single coefficient (number of score points) to a given feature, so in a case when there is a potential non-linear relation between the feature and the output variable the model performance can be compromised. Such a single RiskSLIM coefficient is similar to a coefficient of a linear regression model where modeling non-linear relations is difficult. We designed the tool which allows us to divide the space of a numerical feature into subspaces (discretize it), so every subspace can have a different coefficient assigned.

The subspaces may be overlapping or disjoint from each other. For overlapping subspaces, one can define binary relation (lesser than, lesser or equal than, equal to, greater or equal than, greater than) between the actual feature value and a bin edge found by the selected discretization method. Table 1 presents a feature that is discretized in two ways, with overlapping and disjoint ranges as the output. For overlapping subspaces (upper part of the table) bin edges are values 4.0 and 13.67 and binary relations are *lesser than* and *greater or equal* for 4.0 and *greater or equal* for 13.67. For this kind of discretization, there can be many 1s in a row. For disjoint subspaces (lower part of the table) such binary relations are not applied. For this kind of discretization, there can be only a single 1 in a row.

To find appropriate discretization, an optimization algorithm is applied. The resulting bins depend on chosen discretization strategy (e.g. uniform, quantile, k-means, MDLP, etc.). In the optimization process hyperparameters like a number of bins for quantile discretization or maximum depth of a tree for MDLP discretizer are tuned. We also optimize RiskSLIM-specific hyperparameters such as a number and a magnitude of output coefficients and an intercept value. In our case, the goal of the optimization was to achieve maximal accuracy, but one can specify other metrics such as F1-score, Matthews correlation coefficient, etc. The full algorithm of FD-RiskSLIM is the following:

1. Define hyperparameters for RiskSLIM, as a grid of parameters,
2. Define hyperparameters for feature discretization, as a grid of parameters,
3. Pass these grids to the hyperparameter optimization framework,
4. Select the best model which is found by hyperparameter optimization,
5. Using a data test set compute the metrics of the model's performance.

3.3   Implementation

To perform a fair comparison of the performance of different machine learning algorithms, we optimized hyperparameters for all ML methods used in this paper using the Optuna hyperparameter optimization framework [2]. We wrapped the Optuna in a class to create a reusable tool to optimize any model, with any number and type of hyperparameters.

For feature discretization, we designed a solution that allows one to apply a discretizer of a choice. We provided the functionality that wraps the discretizer and exposes its ability through the constant interface. As a proof of concept, we used KBinsDiscretizer from scikit-learn library [17], which can use a few strategies of discretization (uniform, quantile, k-means) and the MDLP (Minimum Description Length Principle) discretizer [10].

For RiskSLIM we optimized hyperparameters that specify the number of coefficients (number of features used for risk calculation), their value (small non-zero integers), and the intercept (see Fig 1). For FD-RiskSLIM we also optimized hyperparameters of feature discretization, which involved finding an optimal number of bins and the position of their edges as described in Sect. 3.2.

We used RiskSLIM implementation by Ustun and Rudin [23] available on https://github.com/ustunb/risk-slim.

4   Experiments

We conducted experiments using two real-world medical datasets to demonstrate the application of FD-RiskSLIM. In the first experiment described in Sect. 4.1 we compare its performance to the performance of RiskSLIM. We also compared our results with results obtained by others on the same dataset. In the second experiment, we applied RiskSLIM, FD-RiskSLIM, and other ML methods that we implemented to the original dataset describing the survival of patients with kidney cancer (Sect. 4.2).

4.1   Prediction of Death of Patients with Heart Failure

The first set of experiments involves the dataset containing the medical records (13 features) of 299 heart failure patients collected at the Faisalabad Institute of Cardiology and the Allied Hospital in Faisalabad in Punjab, Pakistan in 2015 [1]. We test FD-RiskSLIM on the classification task of patient death during the observation period as in [6] where authors used the same heart-failure dataset. We followed the same methodology for model training and performance evaluation as in [6]. We split the dataset into 80% for the training set and 20% for the test set. As in [6], we used the following metrics: accuracy, F1 score, Matthews correlation coefficient (MCC), and the area under the ROC curve to measure the performance of RiskSLIM and FD-RiskSLIM.

We built the following four models. We created RiskSLIM and FD-RiskSLIM models by using all the features and RiskSLIM and FD-RiskSLIM models using only 2 out of 13 features. This choice was inspired by authors of [6] who predicted the survival only from *serum creatinine* and *ejection fraction* features ignoring the other 11 features. We scaled the features using a min-max (0–1) scaler as it ensures small final values of a score calculated with a ready scoring system. The models are presented in Fig. 2. Even if we passed all the features for training the output risk scoring systems A and B contain only part of them as the underlying algorithm apart from minimizing loss, it minimizes the number of outputted coefficients. The resulting risk formula differs between risk scoring systems as it also includes interception coefficient which can differ for different models.

Tables 2 and 3 show performance metrics. The results for models other than RiskSLIM and FD-RiskSLIM come from [6]. We followed the same methodology of performance evaluation. We assumed 50% of the risk threshold for positive evaluation to compute the accuracy. FD-RiskSLIM performed the best by far.

Features	Points
serum creatinine	6
age	4
sex	-1
ejection fraction	-5
Risk formula: $\dfrac{1}{1 + \exp(1 - score)}$	A

Features	Points
age between **63.65 (inc.)** and **95 (inc.)**	2
ejection fraction between **0 (inc.)** and **20**	2
serum creatinine between **1 (inc.)** and **9.4 (inc.)**	1
serum creatinine between **0 (inc.)** and **0.5**	-1
Risk formula: $\dfrac{1}{1 + \exp(2 - score)}$	B

Features	Points
serum creatinine	7
ejection fraction	-5
Risk formula: $\dfrac{1}{1 + \exp(-score)}$	C

Features	Points
serum creatinine between **7.7 (inc.)** and **9.4 (inc.)**	3
serum creatinine between **1 (inc.)** and **7.7**	-1
ejection fraction between **28 (inc.)** and **80 (inc.)**	-2
serum creatinine between **0 (inc.)** and **1**	-3
Risk formula: $\dfrac{1}{1 + \exp(3 - score)}$	D

Fig. 2. A and **B** are models trained on all features for RiskSLIM and FD-RiskSLIM, respectively. **C** and **D** are models trained on features including only serum creatinine and ejection fraction for RiskSLIM and FD-RiskSLIM, respectively. To obtain *score* we multiply the vector of *Features* by the vector of *Points*. A and C risk scoring systems contain continuous features only in the range 0–1. B and D risk scoring systems contain categorical features only with values 0 or 1.

Discussion. We found that RiskSLIM models perform superior to other machine learning models. This is a surprising finding and initially, we suspected that we may simply compute the metrics differently, on a different test set, etc., compared to [6]. However, in the second experiment (Sect. 4.2) we found similar superiority when we compared RiskSLIM with ML models that we implemented and optimized on identical test sets.

One of the greatest advantages of scoring systems is their clarity and interpretability. Only a small subset of features and simple formula are needed to quickly evaluate risk even using a piece of paper and a pen. This simplicity and interpretability cannot be achieved for models like Random Forest, Gradient Boosting, or Support Vector Machine. These scoring systems also directly expose feature importances. Essentially, deciding with the use of this system, one knows which feature contributes to the risk and by how much. The highest the absolute number of points assigned by a model to a given feature the biggest impact it has on the final risk.

When we consider interpretable ML models the Decision Tree model naturally comes to mind. We present the example in Fig. 3. It consists of four leaf nodes and it is simple and easy to interpret due to the shallow depth of the tree. However, the performance (averaged) of this model is substantially worse than RiskSLIM (accuracy: 0.735, F1: 0.532, ROC AUC: 0.675, MCC: 0.372). Another disadvantage of the Decision Tree model is that it forces conditions to be checked in a given order, while RiskSLIM allows to evaluate them in any order.

Table 2. Comparison of performances of the models trained on all the features. Results are sorted by MCC. All results except for RiskSLIM comes from [6] and represent means computed over the test set.

Method	MCC	F1 score	Accuracy	ROC AUC
FD-RiskSLIM	0.436	0.617	0.744	0.723
RiskSLIM	0.392	0.529	0.744	0.672
Random forests	0.384	0.547	0.740	0.800
Decision tree	0.376	0.554	0.737	0.681
Gradient boosting	0.367	0.527	0.738	0.754
Linear regression	0.332	0.475	0.730	0.643
One rule	0.319	0.465	0.729	0.637
Artificial neural network	0.262	0.483	0.680	0.559
Naive bayes	0.224	0.364	0.696	0.589
SVM radial	0.159	0.182	0.690	0.749
SVM linear	0.107	0.115	0.684	0.754
KNN	-0.025	0.148	0.624	0.493

Table 3. Comparison of performances of the models trained on serum creatinine and ejection fraction features only. Results are sorted by MCC. All results except for RiskSLIM and FD-RiskSLIM are reproduced from [6] and represent means computed over the test set.

Method	MCC	F1 score	Accuracy	ROC AUC
FD-RiskSLIM	0.435	0.616	0.746	0.719
Random forests	0.418	0.585	0.754	0.698
Gradient boosting	0.414	0.585	0.750	0.792
RiskSLIM	0.380	0.476	0.750	0.649
SVM radial	0.348	0.543	0.720	0.667

4.2 Prediction of Outcomes of Kidney Cancer Treatment

We also performed experiments on two datasets of medical records of patients with kidney cancer (metastatic renal-cell carcinoma) related to treatment using sunitinib and everolimus drugs. The datasets come from the National Institute of Oncology in Warsaw, Poland [9]. Most of the features in the datasets are binary or categorical (tumor grading, metastases, and other diseases), but there are also some numerical features such as age, weight, BMI, lymphocytes, leukocytes, and neutrophils. There are two outcome variables: (1) progression-free survival (PFS) time which is the length of time during and after the treatment of cancer, that a patient lives with cancer but it does not get worse, given in years, and overall survival (OS) time which is the length of time from the beginning of treatment to death, given in years. The data are right-censored but in this work, we ignored this as our main goal was to demonstrate the application of RiskSLIM algorithm.

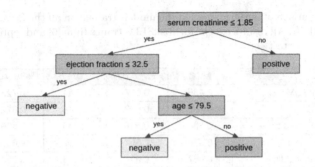

Fig. 3. Decision Tree model trained on all the features.

We selected two classification tasks to compare the performance of explainable methods (RiskSLIM and FD-RiskSLIM) with other state-of-the-art ML methods. We predict whether PFS and OS times are longer than their respective medians.

We transformed the original datasets by dropping irrelevant features and features with a substantial ratio of missing values (over 20%). We concatenated these two datasets to increase the total number of records. Only common features were kept in the final dataset. An additional categorical feature denoting the type of the treatment was added. The final dataset consisted of 149 events (rows) with 50 features.

For comparison, we also applied the following classification models: K-Nearest Neighbors, Decision Tree, Random Forest, Gradient Boosting, Support Vector Machine with the radial kernel, and Support Vector Machine with linear kernel. We divided the dataset into subsets for training and testing, in proportions 80:20, respectively. Splits were repeated 100 times and results are provided as averages.

Figure 4 shows scoring systems obtained for both classification tasks. Most of the features from the original dataset are categorical or binary therefore the discretization was applied only to four of them (BMI, lymphocytes, leukocytes, and neutrophils). All the numerical features were scaled with the use of a min-max (0–1) scaler. Scoring systems denoted as **A** and **B** allow calculating the risk of OS being longer than the median. Scoring systems denoted as **C** and **D** are classification models for PFS longer than the median.

Tables 4 and 5 present comparison of the performance of the different classification models. As for experiments presented in the Sect. 4.1 for RiskSLIM models we assumed that the risk greater or equal to 50% evaluates to the positive class. For models listed in both figures we performed hyperparameter optimization using Optuna implemented as described in Sect. 3.

Discussion. For both classification tasks, RiskSLIM performs as well as the state-of-the-art models like Gradient Boosting or Random Forest. FD-RiskSLIM performs better which is also consistent with findings of the experiment with the heart failure described in Sect. 4.1.

Features	Points
G1	2
NEUT > UT	-1
Heng1	-4
neutrophils	-5
MSKCC2	-5
Risk formula: $\dfrac{1}{1 + \exp(-6 - score)}$	**A**

Features	Points
lymphocytes between **2.13 (inc.)** and **5.21 (inc.)**	-1
NEUT > UT	-2
G3	-2
Heng1	-3
MSKCC2	-5
Risk formula: $\dfrac{1}{1 + \exp(-5 - score)}$	**B**

Features	Points
distant lymph nodes	1
number of other cancers	-1
AH	-1
T2	-1
NEUT > UT	-2
G3	-2
LDH > 1.5xUT	-7
Risk formula: $\dfrac{1}{1 + \exp(-9 - score)}$	**C**

Features	Points
leukocytes between **3.21 (inc.)** and **4.49 (inc.)**	1
AH	-1
HGB < LT	-1
LDH > 1.5xUT	-1
NEUT > UT	-1
G3	-1
Risk formula: $\dfrac{1}{1 + \exp(-3 - score)}$	**D**

Fig. 4. A and **B** are scoring models of risk for OS longer than median for RiskSLIM and FD-RiskSLIM. **C** and **D** are scoring models PFS longer than median for RiskSLIM and FD-RiskSLIM. To obtain the *score* we multiply the vector of *Features* by the vector of *Points*. Explanations for feature names are provided in the Appendix.

Most of the points present in risk scoring systems shown in Fig. 4 are negative. This indicates that they decrease the chance that someone would live without progression or overall longer than the median time. Interestingly, RiskSLIM/FD-RiskSLIM use G1 and G3 features (tumor grades 1 and 3) but do not use G2 and G4. We speculate that if G1 increases the chance and G3 decreases then G2 might have no contribution (e.g. coefficient equal to 0). Also, class imbalance may play a role as, for example, only less than 5% of the samples in the dataset have cancer grade 4 (G4).

As for previous experiments in the Sect. 4.1, apart from predictive accuracy, we are also interested in knowledge representation. RiskSLIM scoring systems have feature importance and their strength are provided explicitly. The evaluation of risk for new patients goes fast as there are just several coefficients that have to be summed and passed to the risk formula.

The clinical interpretation of the model is beyond the scope of this work. However, it is of great practical importance and we will pursue this direction of research in collaboration with our clinical colleagues in future work.

For comparison of explainability, we also built Decision Tree models for both classification tasks. They are presented in Fig. 5 and Fig. 6. These trees are not sparse and easy to work with due to a small number of rules. However, their performance is worse and they can not evaluate the risk, they allow for classification only.

102   A. Pajor et al.

Table 4. Comparison of performances of the models built for predicting whether overall survival time is longer than a median. Results are sorted by MCC.

Method	MCC	F1 score	Accuracy	ROC AUC
FD-RiskSLIM	0.225 (0.149)	0.744 (0.059)	0.648 (0.069)	0.594 (0.066)
RiskSLIM	0.224 (0.144)	0.726 (0.064)	0.638 (0.073)	0.601 (0.065)
Gradient boosting	0.211 (0.135)	0.715 (0.057)	0.632 (0.063)	0.598 (0.065)
Random forests	0.181 (0.148)	0.739 (0.058)	0.629 (0.074)	0.571 (0.063)
Decision tree	0.152 (0.152)	0.690 (0.076)	0.606 (0.074)	0.572 (0.072)
SVM linear	0.131 (0.128)	0.672 (0.063)	0.592 (0.061)	0.564 (0.063)
KNN	0.052 (0.152)	0.688 (0.062)	0.572 (0.073)	0.523 (0.065)
SVM radial	0.000 (0.000)	0.742 (0.058)	0.597 (0.074)	0.500 (0.000)

Table 5. Comparison of performances of the models built for predicting whether progression-free survival time is longer than a median. Results are sorted by MCC.

Method	MCC	F1 score	Accuracy	ROC AUC
Random forests	0.397 (0.127)	0.707 (0.075)	0.688 (0.069)	0.696 (0.064)
FD-RiskSLIM	0.389 (0.122)	0.745 (0.053)	0.689 (0.063)	0.679 (0.073)
Gradient boosting	0.379 (0.108)	0.703 (0.064)	0.684 (0.056)	0.689 (0.054)
RiskSLIM	0.340 (0.150)	0.705 (0.070)	0.668 (0.071)	0.665 (0.074)
Decision tree	0.313 (0.136)	0.654 (0.086)	0.649 (0.069)	0.655 (0.068)
SVM linear	0.229 (0.153)	0.635 (0.079)	0.613 (0.074)	0.614 (0.077)
KNN	0.162 (0.131)	0.616 (0.067)	0.581 (0.066)	0.580 (0.065)
SVM radial	-0.007 (0.030)	0.669 (0.110)	0.519 (0.072)	0.498 (0.007)

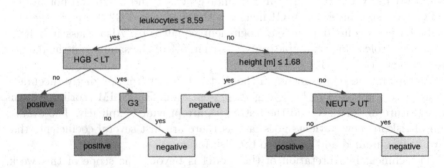

Fig. 5. Decision Tree model trained for predicting whether OS is longer than the median.

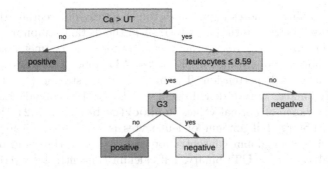

Fig. 6. Decision Tree model trained for predicting whether PFS is longer than the median.

5 Conclusions

RiskSLIM is capable of making accurate and interpretable predictions at the same time outperforming complex and non-interpretable black-box models in certain applications which we explored in this paper. Interpretability became one of the most important factors when it comes to making high stake decisions [18] e.g. in medical domains which underlines the importance of explainable models such as the ones considered here. Adding discretization makes this approach even better. The drawback of RiskSLIM is its high computational demand as model training is longer by orders of magnitude in comparison to models such as Decision Tree, Random Forest, or Gradient Boosting. Also feature discretization introduced by us here as a part of data preprocessing increases substantially model training time.

In future work, we plan to incorporate feature discretization into the model building stage to improve efficiency and reliability. Additionally, in the future work we will investigate results stability. We also plan to apply it to other domains.

Acknowledgements. This work is supported in part by the European Union's Horizon 2020 research and innovation programme under grant agreement Sano No. 857533 and the International Research Agendas programme of the Foundation for Polish Science, co-financed by the EU under the European Regional Development Fund.

7 Appendix

Explanations for the features listed in Fig. 4. **G1**: binary, from tumor grading, denotes low grade (tumor well differentiated), **NEUT > UT**: binary, 1 if concentration of neutrocytes exceeds upper threshold of a range of normal values, **Heng1**: binary, Heng scale, **neutrophils**: continuous, scaled with min-max (0–1) scaler, amount of neutrophils, **MSKCC2**: binary, MSKCC (Memorial Sloan Kettering Cancer Center) scale, **lymphocytes between 2.13 (inc.) and 5.21 (inc.)**: binary, 1 if amount of lymphocytes is inside this range, **G3**: binary, from tumor grading, denotes high grade (tumor poorly differentiated), **distant lymph nodes**: binary, 1 if tumor metastasis in the distant lymph nodes, **number of other cancers**: numeric, number of other cancers (metastasis) than: lungs, liver, bones and distant lymph nodes, **AH**: binary, denotes if someone suffers from arterial hypertension, **T2**: binary, cancer staging (T1–T4), **LDH > 1.5 x UT**: binary, lactate dehydrogenase activity, 1 if exceeds 1.5 x upper threshold of a range of normal values, **leukocytes between 3.21 (inc.) and 4.49 (inc.)**: binary, 1 if amount of leukocytes is inside this range, **HGB < LT**: binary, 1 if hemoglobin concentration exceeds lower threshold of a range of normal values, **Ca > UT**: binary, 1 if calcium concentration exceeds upper threshold of a range of normal values.

References

1. Ahmad, T., Munir, A., Bhatti, S.H., Aftab, M., Raza, M.A.: Survival analysis of heart failure patients: a case study. PLoS ONE **12**, e0181001 (2017)
2. Akiba, T., Sano, S., Yanase, T., Ohta, T., Koyama, M.: Optuna: a next-generation hyperparameter optimization framework. In: Proceedings of the 25rd ACM SIGKDD International Conference on Knowledge Discovery and Data Mining (2019)
3. Antman, E.M., et al.: The TIMI risk score for unstable angina/non-ST elevation MI: a method for prognostication and therapeutic decision making. JAMA **284**(7), 835–842 (2000)
4. Bone, R.C., et al.: Definitions for sepsis and organ failure and guidelines for the use of innovative therapies in sepsis. the ACCP/SCCM consensus conference committee. Am. Coll. Chest Phys./Soc. Crit. Care Med. Chest **101**(6), 1644–1655 (1992)
5. Chang, W., et al.: A machine-learning-based prediction method for hypertension outcomes based on medical data. Diagnostics **9**, 178 (2019)
6. Chicco, D., Jurman, G.: Machine learning can predict survival of patients with heart failure from serum creatinine and ejection fraction alone. BMC Med. Inform. Dec. Mak. **20**, 1-16 (2020)
7. Chung, F., Abdullah, H.R., Liao, P.: Stop-bang questionnaire: a practical approach to screen for obstructive sleep apnea. Chest **149**(3), 631–638 (2016). https://doi.org/10.1378/chest.15-0903
8. Cruz, J.A., Wishart, D.S.: Applications of machine learning in cancer prediction and prognosis. Cancer Inform. **2**, 59–77 (2007)
9. Dudek, A.Z., Żołnierek, J., Dham, A., Lindgren, B.R., Szczylik, C.: Sequential therapy with sorafenib and sunitinib in renal cell carcinoma. Cancer **115**, 61–67 (2009)
10. Fayyad, U.M., Irani, K.B.: Multi-interval discretization of continuous-valued attributes for classification learning. In: IJCAI (1993)
11. Gage, B.F., Waterman, A.D., Shannon, W., Boechler, M., Rich, M.W., Radford, M.J.: Validation of clinical classification schemes for predicting StrokeResults from the national registry of atrial fibrillation. JAMA **285**(22), 2864–2870 (2001). https://doi.org/10.1001/jama.285.22.2864
12. Gage, B.F., Waterman, A.D., Shannon, W.D., Boechler, M., Rich, M.W., Radford, M.J.: Validation of clinical classification schemes for predicting stroke: results from the national registry of atrial fibrillation. JAMA **285**, 2864–2870 (2001)

13. Goodman, B., Flaxman, S.: European union regulations on algorithmic decision-making and a "right to explanation" (2016). https://doi.org/10.1609/aimag.v38i3.2741, http://arxiv.org/abs/1606.08813
14. Knaus, W.A., et al.: The apache iii prognostic system. risk prediction of hospital mortality for critically ill hospitalized adults. Chest **100**(6), 1619–1636 (1991)
15. Metnitz, P.G.H., et al.: SAPs 3-from evaluation of the patient to evaluation of the intensive care unit. part 1: objectives, methods and cohort description. Intensive Care Med. **31**, 1336–1344 (2005)
16. Mohan, S., Thirumalai, C., Srivastava, G.: Effective heart disease prediction using hybrid machine learning techniques. IEEE Access **7**, 81542–81554 (2019)
17. Pedregosa, F., et al.: Scikit-learn: machine learning in Python. J. Mach. Learn. Res. **12**, 2825–2830 (2011)
18. Rudin, C.: Stop explaining black box machine learning models for high stakes decisions and use interpretable models instead. Nat. Mach. Intell. 1(5), 206–215 (2019)
19. Rudin, C., Ustun, B.: Optimized scoring systems: toward trust in machine learning for healthcare and criminal justice. Interfaces **48**, 449–466 (2018)
20. Struck, A.F., et al.: Comparison of machine learning models for seizure prediction in hospitalized patients, June 2019. https://onlinelibrary.wiley.com/doi/full/10.1002/acn3.50817
21. Struck, A.F., et al.: A practical risk score for EEG seizures in hospitalized patients (s11.002). Neurology **90** (2018). https://n.neurology.org/content/90/15_Supplement/S11.002
22. Ustun, B., Rudin, C.: Supersparse linear integer models for optimized medical scoring systems. Mach. Learn. **102**(3), 349–391 (2015). https://doi.org/10.1007/s10994-015-5528-6
23. Ustun, B., Rudin, C.: Learning optimized risk scores. J. Mach. Learn. Res. **20**(150), 1–75 (2019). http://jmlr.org/papers/v20/18-615.html
24. Ustun, B., Westover, M.B., Rudin, C., Bianchi, M.T.: Clinical prediction models for sleep apnea: The importance of medical history over symptoms. J. Clin. Sleep Med. JCSM Off. Publ. Am. Acad. Sleep Med. **12**(2), 161–8 (2016)
25. Vilone, G., Longo, L.: Explainable artificial intelligence: a systematic review. arXiv:2006.00093 (2020)
26. Wang, C.L., Han, B., Patel, B., Mohideen, F., Rudin, C.: In pursuit of interpretable, fair and accurate machine learning for criminal recidivism prediction. ArXiv:2005.04176 (2020)

Hybrid Modeling for Predicting Inpatient Treatment Outcome: COVID-19 Case

Simon D. Usoltsev and Ilia V. Derevitskii(✉)

ITMO University, Saint Petersburg, Russia
ivderevitckii@itmo.ru

Abstract. This study presents two methods to support the treatment process of inpatients with COVID-19. The first method is designed to predict treatment outcomes; this method is based on machine learning models and probabilistic graph models of patient clustering. The method demonstrates high quality in terms of predictive models, and the structure of the graph model is supported by knowledge from practical medicine and other studies. The method is used as a basis for finding the optimal intervention plan for severe patients. This plan is a set of interventions for patients that are optimal in terms of minimizing the probability of mortality. We tested the method for critically ill patients (item 4.5) and for 30% of all patients with lethal outcomes the methods found an intervention plan that leads to recovery as a treatment outcome as predicted. Both methods show high quality, and after validation by physicians, this method can be used as part of a decision support system for medical professionals working with COVID-19 patients.

Keywords: COVID-19 · Predict treatment outcome · Optimal interventions · Bayesian network

1 Introduction

Coronaviruses are a group of viruses that can cause acute respiratory disease in humans and can also be transmitted between species and cause various diseases. A new coronavirus, SARS-CoV-2 (Severe Acute Respiratory Syndrome Coronavirus), which was designated COVID-19 by the World Health Organization (WHO) on February 11, 2020. It causes an acute and deadly illness with a global mortality rate of 1.68%. In Russia, the mortality rate is 2.96% [1]. The level of morbidity and mortality in Russia from Covid-19 and the complications it causes is quite high, which creates a serious burden on the health care system [2], coronavirus can cause severe effects on the heart, brain, kidneys, blood vessels, and other vital organs. Therefore, it is especially important for patients who are hospitalized with this disease to prevent the development of serious complications [3]. The risk of complications can be reduced with the right therapy, as well as enough time that medical professionals can devote to heavy patients, in a heavy workload, to reduce the burden on the medical staff requires special methods to assess the severity of the patient admitted to the hospital, optimal in terms of accuracy and necessary data

D. Groen et al. (Eds.): ICCS 2022, LNCS 13352, pp. 106–112, 2022.
https://doi.org/10.1007/978-3-031-08757-8_10

for decision-making. This article presents a new hybrid approach for assessing the risk of hospital-acquired patient mortality, based on a highly interpretable Bayesian network model, with the possibility of editing the model based on expert data as well as predictive machine learning models. There are many approaches in the literature for assessing the risk of covid mortality. The methods are based on several different approaches: statistical, simulation, and dynamic modeling. Using linear prediction, some researchers have used regression modeling to calculate the risk of 15-day mortality and determine predictors of mortality in critically ill patients with covid [4–6]. The use of this approach has the following advantages: interpretability, ease of implementation, and availability of many software tools. Among the disadvantages of this approach, it is important to highlight the impossibility of detecting nonlinear relationships. Disadvantages of linear methods are eliminated in the simulation modeling approach. Blagojevic A et al. described the classification of clinical severity of patients based on blood biomarkers, which were selected as biomarkers that have the greatest impact on the classification of patients with Covid-19 [7]. Another approach is to model the course of the disease as a dynamic process. Several studies have used Markov chains Monte Carlo to estimate the mortality rate [8].

2   Model

The first step is data mining; our team structures, selects and aggregates information to build hypotheses and models, then we create special scripts to convert information to convert information from medical experts into matrices. The rows are cases of diseases, the columns are important indicators of the course of the disease. At this stage we use only expert modeling. This stage includes steps 1–5. The second step is to identify patterns from the data (steps 6–10), then the scheme. This stage involves processing noise, outliers, removing/substituting gaps in the data, categorical feature coding using one-hot-label-encoding/dummy encoding methods, scaling and protocolization. In the next step, we create a Bayesian network model and machine learning on the resulting data, then use the Bayesian network to select the therapy that reduces the probability of lethality. In the next step, we interpret the models using additive Shapley explanations (SHAP) as well as knowledge from medicine and other studies. We then validate the methods using the classical quality metric of the predictive problem and comparing the results of using the recommended therapy with real-world results (Fig. 1).

2.1   Data Mining

The study was based on a data set including medical records of 2445 patients treated for covid-19 in the hospital of the V.A. Almazov National Research Center, St. Petersburg, Russia (Table 1).

These medical indicators include information on medical history, test values, physical measurements, lifestyle, hospital department metadata, and medical professional opinions.

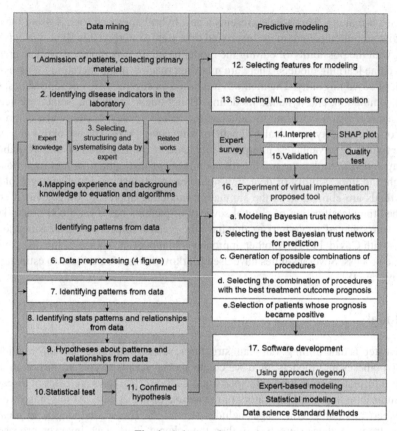

Fig. 1. Scheme of research.

3   Experimental Study

3.1   Mortality Prediction – Bayesian Networks

A Bayesian network is a graph-based probabilistic model, which is a set of variations and their Bayesian probability relations. The constructed Bayesian network was used to predict lethality. The network was constructed using the hill climb search approach with a k2 estimation metric; the approach implements a greedy local search that can be used to find the optimal network structure.

The model separates the metrics quite qualitatively according to groups of tests, which include attributes such as vein tests, blood test results, procedure correlation, and some patient characteristics. The red cluster localizes the traits that are related to the patient's procedures, the blue cluster localizes the area that is related to oxygen therapy. This area, through signs of weakness and fever above 37.5 is related to the purple area, in which headache and other general patient signs are localized. This area, through the sign of saturation, is related to the yellow sector, in which age, signs with the results of general blood tests and urine tests are localized (Fig. 2).

Table 1. Signs obtained during treatment in the data

Feature's group	Features
Measurements	Group includes height, weight, age, gender, SBP, DBP, heart rate, body mass index, hemoglobin, and other blood test data
Chronic diseases	Group includes a history of chronic diseases in the patient
Time information	Group includes time of onset, time of tests, date of complications, and other
Secondary attributes of the Covid-19	Group includes the presence of heaviness in the chest, lack of sense of smell and taste, fever, decreased consciousness, saturation, etc.
Signs of complications	The group includes the presence of pneumonia, anemia, hypothyroidism, bronchitis, and other
Drugs	The group includes 404 different drugs that were used to treat patients in the hospital. These drugs belong to the pharmacological group of glucocorticosteroids, insulins, anticoagulants, etc.
Procedures	The group includes procedures that were performed to maintain the patient's condition, such as blood transfusions, ventilation, oxygen therapy, and other

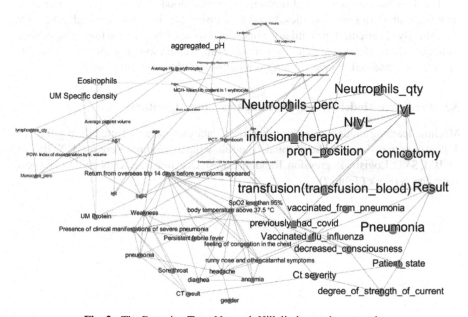

Fig. 2. The Bayesian Trust Network Hillclimb search approach

3.2 Prediction Experiment with Critically Ill Patients

For patients with a negative outcome, procedures were selected in which the prognosis of the model improved. For several patients and procedures, the model predicted a positive treatment outcome. The recommended procedures can be seen in Fig. 3.

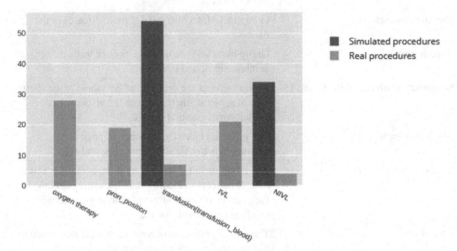

Fig. 3. Recommended procedures that improve prognosis for lethal patients

For these patients, the model recommends more blood transfusion and noninvasive lung ventilation. In several studies, blood transfusion has demonstrated high efficiency, for critically ill patients. On ventilator, after which some of them began to breathe on their own, also, the model recommends noninvasive lung ventilation, in some cases mortality of patients connected to invasive lung ventilation is rather high [9].

3.3 Predictive Modeling with Machine Learning Algorithms

Machine learning models were built based on data that were collected during the first 3 days of patients' stay in the hospital. The following algorithms were used: XGBoost, LGBM, SVC, Logistic Regression, Random Forest, Decision tree, KNN, SGD, Gaussian, Bayesian network (Table 2).

Table 2. The obtained cross-validation accuracy for predictive models

Model	Cross-validation-score, %
XGBoost	96.9
LGBM	96.84
KNN	96.8
SVC	96.6
Logistic Regression	96.5
Decision tree	95.91
SGD	91.97
Bayesian network	91.0
Gaussian	90.2
Random forest	87.2

The XGBoost model showed the highest accuracy score, consider the results of the model. We interpret the results of the model using SHAP values; with their help we can understand the importance of a single trait and its impact on the assessment of the severity of the patient's condition (Fig. 4).

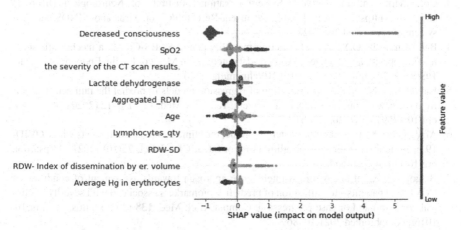

Fig. 4. Interpretation of obtained model values using SHAP values.

The graph shows the 10 indicators that most affect the assessment of the severity of the patient's condition. The most significant indicator for the model turned out to be the patient's decreased consciousness, which is often observed against the background of worsening lung condition. The second and the third significant indicator are the signs that are also related to the respiratory system: oxygen saturation level and CT scan results [10].

4 Conclusion and Future Work

This paper proposes a method for predicting treatment outcomes in critically ill patients and selecting optimal therapy. We used this method to create a practical tool for recommending intervention plans to reduce the probability of lethal outcome. We have tested the effectiveness of the method on critically ill patients with a negative treatment outcome, and according to the results of the model, for 30% of patients we can achieve a change in the results of treatment to recovery. All models show high quality, so they can be used as part of a decision support system for medical professionals working with patients.

Acknowledgments. This research is financially supported by The Russian Science Foundation, Agreement №17-71-30029 with co-financing of Bank Saint Petersburg.

References

1. Mortality Analyses - Johns Hopkins Coronavirus Resource Center.https://coronavirus.jhu.edu/data/mortality. Accessed 21 Jan 2022
2. Panda, P.K., Sharawat, I.K.: COVID-19 and/with dengue infection: a curse in an overburdened healthcare system. Trop. Doct. **51**(1), 106–108 (2021). https://doi.org/10.1177/0049475520975945
3. Coronavirus and the Nervous System | National Institute of Neurological Disorders and Stroke.https://www.ninds.nih.gov/Current-Research/Coronavirus-and-NINDS/nervous-system. Accessed 21 Jan 2022
4. Bello-Chavolla, O.Y., et al.: Predicting mortality due to SARS-CoV-2: a mechanistic score relating obesity and diabetes to COVID-19 outcomes in Mexico. J. Clin. Endocrinol. Metab. **105**(8) (2020). https://doi.org/10.1210/clinem/dgaa346
5. Karyakin, N.N., et al.: Modernization of regression models to predict the number of deaths from the new coronavirus infection. Sovrem. Tehnol. v Med. **12**(4), 6–12 (2020). https://doi.org/10.17691/stm2020.12.4.01
6. Araç, S., Özel, M.: A new parameter for predict the clinical outcome of patients with COVID-19 pneumonia: the direct/total bilirubin ratio. Int. J. Clin. Pract. **75**(10) (2021). https://doi.org/10.1111/ijcp.14557
7. Blagojević, A., et al.: Artificial intelligence approach towards assessment of condition of COVID-19 patients - Identification of predictive biomarkers associated with severity of clinical condition and disease progression. Comput. Biol. Med. **138** (2021). https://doi.org/10.1016/j.compbiomed.2021.104869
8. Zhicheng, D., et al.: Using Markov chain Monte Carlo methods to estimate the age-specific case fatality rate of COVID-19. Zhonghua Liu Xing Bing Xue Za Zhi **41**(11), 1777–1781 (2020). https://doi.org/10.3760/CMA.J.CN112338-20200609-00823
9. Duan, K., et al.: Effectiveness of convalescent plasma therapy in severe COVID-19 patients. Proc. Natl. Acad. Sci. U.S.A. **117**(17), 9490–9496 (2020). https://doi.org/10.1073/PNAS.2004168117/-/DCSUPPLEMENTAL
10. Chatterjee, N.A., et al.: Admission respiratory status predicts mortality in COVID-19. Influenza Other Respi. Viruses **15**(5), 569–572 (2021). https://doi.org/10.1111/IRV.12869

Neural Additive Models for Explainable Heart Attack Prediction

Ksenia Balabaeva[✉] and Sergey Kovalchuk

ITMO University, Saint Petersburg, Russia
kyubalabaeva@gmail.com

Abstract. Heart attack (HA) is a sudden health disorder when the flow of blood to the heart is blocked, causing damage to the heart. According to the World Health Organization (WHO), heart attack is one of the greatest causes of death and disability globally. Early recognition of the various warning signs of a HA can help reduce the severity. Different machine learning (ML) models have been developed to predict the heart attack. However, patients with arterial hypertension (AH) are especially prone to this disorder and have several features that distinguish them from other groups of patients. We apply these features to develop a special model for people suffering from AH. Moreover, we contribute to this field bringing more transparency to the modelling using interpretable machine learning. We also compare the patterns learned by methods with prior information used in heart attack scales and evaluate their efficiency.

Keywords: Heart attack prediction · XAI · Decision Tree · NAMs · Heart attack risk · Arterial hypertension · Interpretable machine learning

1 Introduction

Heart diseases are one of the main reasons of a death rate increase each year. Heart Attack (HA) is a sudden disruption of a cardiac system. In healthcare sector tons of data are generated in electronic health records (EHR) about patients. These records describe main patients' characteristics helping clinicians make decisions on diagnostics and treatment. According to one of the surveys conducted by WHO, the clinicians can accurately predict only 67% of heart diseases [1]. There are many factors possible influencing the outcome of heart disease and it's impossible to take everything in mind but still there is potential to increase the quality.

EHRs are valuable sources of data for machine learning models and medical decision support systems. Such systems already proved their accuracy, however, the pace of implementation and practical use remains low especially in medical institutions. One of the reasons of such refuse of AI technologies is a lack of trust and understanding among users.

We implement a machine learning based system that can detect and predict heart attack for patients using the medical records. The proposed solution is based on Neural Networks and generalized additive models (GAMs) [2, 3].

© The Author(s), under exclusive license to Springer Nature Switzerland AG 2022
D. Groen et al. (Eds.): ICCS 2022, LNCS 13352, pp. 113–121, 2022.
https://doi.org/10.1007/978-3-031-08757-8_11

The dataset used in our study was collected from Almazov Medical Research Center. It contains 17 features (such as age, gender, blood pressure, etc.) and 385 observations after preprocessing and data clearance. It was then split into 70% train sets and 30% test sets. Moreover, cross-validation was performed to compare methods and optimize hyperparameters. A series of experiments was conducted to examine the performance, accuracy, and interpretation stability of the proposed system. Experiments and were and modules were implemented in Python 3 programming language which predicts the risk of heart attack among patients with AH. The results show that the NAMs performance and accuracy is high enough for the task and it provides helpful interpretation.

2   Related Works

Different researchers have contributed to the development of digitalization and predictive analytics in medical domain. Prediction of heart disease based on machine learning algorithm is always curious case for researchers.

A concept of explainability and interpretation also plays an import role in healthcare [4]. Early works were devoted to IF-ELSE rules, where researchers modelled a set of diseases (lung cancer, asthma, and diabetes) based on electronic health records [5]. Researchers in [6] applied LIME to explain the prediction of heart failure by recurrent neural networks. They also provided explanations that allowed to identify the risk-factors such as kidney failure, anemia, and diabetes that increase the risk of heart failure.

In [7] authors used an algorithm named weighted association rule-based classifier (WAC), based on association rule mining to predict heart attack. Florence et al. [8] applied decision trees and artificial neural networks to the same task. In the work [9] authors used an algorithm based on graph association rules mining. One of the drawbacks of early works was poor accuracy and a lack of interpretation and transparency of the system. Therefore, recent advances of AI in medicine are achieved thanks to explainable artificial intelligence (XAI) using molecular data [10], deep meta-learning [11], and other methods [12]. In 2019 Madan M. et al. proposed a technique based on combination of Genetic Algorithm (GA) and Adaptive Neural Fuzzy Inference System (ANFIS) for explainable heart attack prediction [13]. However, the proposed approach wasn't compared with other ML algorithms to conclude about predictive efficiency.

In this work we're testing a novel transparent machine learning approach named Neural Additive Models (NAMs) for the task of heart attack prediction. NAMs were proposed by X et al. [3]. We also compare the predictive capacity of method with stability of explanations.

3   Methods

For many years neural nets outperformed other ML algorithms mostly while applied on unstructured data (images, text, etc.). However, in 2021 Agarwal et al. proposed a novel method bringing concept of neural networks (NN) to transparent class of generalized additive models (GAMs) [2, 3]. It allows to efficiently train neural nets on structured tabular data. GAMs have the form:

$$g\left(\mathbb{E}[y]\right) = \beta + f_1(x_1) + f_2(x_2) + \cdots + f_m(x_m) \tag{1}$$

where $x = (x_1, \ldots, x_m)$ is a vector of M features, y is a target variable, $g(.)$ is a link function (exponential, logistic, etc.) with f_i being univariate shape functions with $\mathbb{E}[f_i] = 0$.

Compared to GAMs, NAMs learn a linear combination of networks, where each separate network is trained on a single feature $x_i \in M$. These networks are trained jointly, using classic backpropagation mechanism.

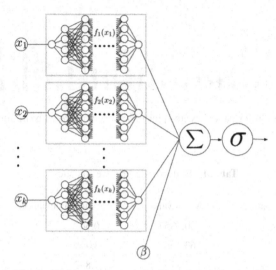

Fig. 1. NAMs architecture for binary classification [3]

Interpretation of NAMs is possible in the form of feature importance since each subnet is independent from other features and can be calculated separately. Moreover, each single subnet in the architecture can be represented as graph and "exactly describe how NAMs computes the prediction". In our study we compare NAMs to other self-explained ML methods, that can provide interpretation in the form of feature importances: Decision Tree, XGBoost and statistical model Logistic Regression. Moreover, we conduct additional experiments, to evaluate performance of widely-used clinical scale, named SCORE for evaluating heart disease risk development [14].

4 Experiments

The dataset was collected from Almazov Medical Research Center, Saint-Petersburg. It contains 385 observations representing patients with arterial hypertension. 281 have suffered from heart attack and 104 didn't. The dataset includes men and women from 37 to 87 years old (Fig. 1) (Fig. 2).

The feature set includes the following information: gender, age, height, weight, body mass index (BMI), body square area (BSA), smoking (0, 1), diabetes (0, 1). Systolic and diastolic blood pressure measurements: dad_min, dad_max, sad_min, sad_max. Laboratory test results: lpvp, lpnp, alanine transaminase (ALT), aspartate aminotransferase

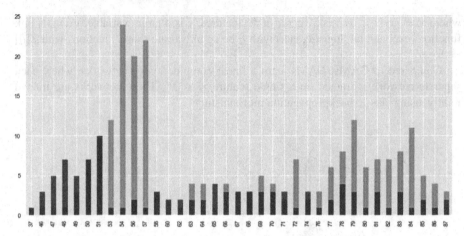

Fig. 2. Age distribution for patients with (pink) and without (green) heart attack (Color figure online)

Table 1. Features' descriptive statistics

Feature name	Mean value	Standard deviation
Gender	0.4857	0.50
Age	63.166	13.419
Height	167.71	7.872
Weight	80.65	14.32
BMI	29.17	8.815
BSA	1.943	0.2400
Smoking	0.127	0.333
Diabetes	0.945	0.227
DAD min	80.41	7.146
DAD max	89.48	7.589
SAD min	141.042	12.77
SAD max	151.22	15.109
LPVP	1.326	0.4105
LPNP	2.763	0.6609
ALT median	21.487	18.590
AST median	21.81	10.129
Urine	6.76	2.157
Stroke	0.729	0.44

(AST), urine. The descriptive statistics about features used for modelling is provided in Table 1.

The features for modelling were aggregated from several clinical episodes using descriptive statistics: mean, median, minimal, maximal and std values in cases where patients had previous episodes with necessary data collection. The outliers were detected with 3-sigma rule and clinical norms' limits. The missing values were filled with median.

The whole sample was randomly split into 70% training set and 30% test set. We selected 30% threshold for test to provide efficient data for model's testing and keep enough data for training.

Heart attack prediction can be solved as a binary classification task, with feature matrix X, target vector y, and model $f(X)$ that is trained on X to predict y. We selected a set of machine learning algorithms that can provide explanation in the form of feature importances: NAMs, Gradient boosting, Decision Tree, and Logistic Regression. In each approach, feature importances are calculated in different ways. Moreover, we compare ML algorithms to used scale SCORE for heart disease risk calculation that is widely used among clinicians. To adopt SCORE predictive capacity, we consider the highest possible risk (threshold >15%) as a positive class, and all other risks – as negative.

To compare predictive performance, we evaluate algorithms using F-score on 5-fold cross-validation and hold-out 30% test set. To test the interpretation stability, we train and test models on different random seeds (seed = 42; 2021; 2022) and evaluate the change in feature importance ranking using NDCG-metric. All experiments were conducted using Python 3.7.

5   Results and Discussion

Considering the predictive performance, best result both on test and validation was achieved by NAMs. Considering F1 on test set, XGB and logistic regression show similar but weaker results. That means that NAM outperforms other explainable ML algorithms in terms of accuracy and can be used for heart attack prediction in medical decision support systems (Table 2).

Table 2. Predictive performance of algorithms

Method	F1 cross-validation (STD)	Test
SCORE	0.3089 (±0.114)	0.419
XGB	0.7142 (±0.0183)	0.8349
Logistic Regression	0.7324 (±0.01)	0.8317
Decision Tree	0.6909 (±0.0528)	0.805
NAM	**0.8778 (±0.026)**	**0.8713**

Meanwhile, the SCORE scale showed the worst quality. Since this is a scale based on several rules, it can't observe dependencies in data. Moreover, since our dataset has bias (all patients have arterial hypertension), some of the rules in SCORE might not be valid. For instance, according to SCORE, the higher is the blood-pressure – the higher is the risk of a heart disease. But many patients take medical treatment to normalize systolic and diastolic blood pressure, which apparently doesn't guarantee the lower HA risk.

The consistency of interpretation was measured using NDCG -score (Table 3), calculating the difference in rank and weight of feature importance trained on the same models using different random seeds (42, 2021, 2022) and taking the mean value of such comparisons. The change in the seed also influenced train-test split, so the score includes data perturbations.

Table 3. Evaluation of interpretation stability

Method	Feature importance NDCG
NAM	0.7150
Logistic Regression	0.93937
XGB	0.9545
Decision Tree	0.7709

The highest stability was achieved by XGBoost and Logistic Regression, and the worst – by NAMs. We think that poor stability of NAM's interpretation might be caused by the lack of data, since usually neural nets require enormous data volumes for efficient and stable training.

Now let's discuss what exactly NAM's interpretation reveals. The most important risk-factors for prognosis were parameters related to blood pressure (SAD max, DAD max), height and age (Fig. 3). Gender and AST have the lowest importance and doesn't influence model's prediction.

Using NAM for modelling, offers visualization of a detailed feature contribution in the form of graphs (Fig. 4). Light-pink regions in graphs correspond to regions with low data density (few samples in a dataset), on the contrary, red regions correspond to high density in data. Blue line depicts shape functions, that tend to be smooth in regions with high density and serrated in regions of few data samples.

Thus, we see that low values of the blood pressure slightly increase the. It might seem counterintuitive, but extremely high values of maximum systolic and diastolic pressure may indicate the severe course of the disease and anti-AH treatment therapy. The high maximal value of arterial hypertension might be the evidence of a crisis when treatment fails to normalize the pressure. But at the same time, mean SAD and DAD might be much lower most of the time. Unfortunately, we can't test this hypothesis since blood pressure measurements were collected from anamnesis.

Mean Importance

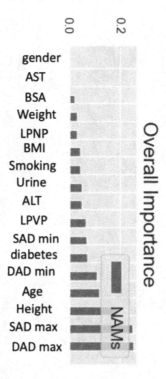

Fig. 3. NAMs feature importances

As for the patient's height, if it's lower that 178 cm, the height lowers the HA risk, however, if a person is higher than 180 cm the risk the positive feature contribution rises significantly. It is also clear that a model learned age patterns, the older a person – the more is a positive contribution of Age in a model.

Due to the transparency of NAM, we can find more insides concerning the connection between disease and risk factor, especially when it comes to specific patients' groups with different comorbidities, such as arterial hypertension or diabetes, where, as we saw, universal conservative methods of risk scoring (SCORE) may fail in terms of predictive performance.

120   K. Balabaeva and S. Kovalchuk

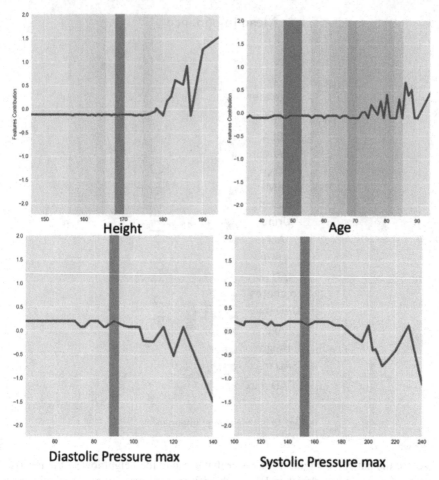

Fig. 4. NAMs intrinsic interpretation

6   Conclusion

To provide explainable and more transparent results for clinicians n a novel approach
NAM was tested to predict heart attack among patients with arterial hypertension. The
experiment confirmed that NAM predicts HA with a high accuracy and provide explanation in the form of graph and feature importances, showing redundant information on the
influence of each risk factor for the prediction. However, the stability of interpretation
might suffer from data perturbation and randomization parameter. This drawback can
be potentially solved with multiple training of NAM on different seeds and averaging
the feature importances.

Acknowledgement. This work was supported by the Ministry of Science and Higher Education
of Russian Federation, goszadanie no. 2019-1339.

References

1. Kirubha, V., Priya, S.M.: Survey on data mining algorithms in disease prediction. Int. J. Comput. Trends Technol. **38**(3), 124–128 (2016)
2. Hastie, T.J., Tibshirani, R.J.: Generalized Additive Models. Routledge, New York (2017)
3. Agarwal, R., et al.: Neural additive models: interpretable machine learning with neural nets. In: Advances in Neural Information Processing Systems 34 (2021)
4. Khedkar, S., Subramanian, V., Shinde, G., Gandhi, P.: Explainable AI in healthcare. In: Healthcare 2nd International Conference on Advances in Science & Technology (ICAST) (2019)
5. Lakkaraju, H., Bach, S.H., Leskovec, J.: Interpretable decision sets: a joint framework for description and pre-diction. In: Proceedings of the ACM SIGKDD International Conference on Knowledge Discovery and Data Mining, pp. 1675–1684, 13–17 (2016)
6. Khedkar, S., Subramanian, V., Shinde, G., Gandhi, P.: Explainable AI in healthcare. In: Healthcare (April 8, 2019). 2nd International Conference on Advances in Science & Technology (ICAST) (2019)
7. Soni, J., Ansari, U., Sharma, D., Soni, S.: Intelligent and effective heart disease prediction system using weighted associative classifiers. Int. J. Comput. Sci. Eng. **3**, 2385–2392 (2011)
8. Florence, S., Bhuvaneswari Amma, N.G., Annapoorani, G., Malathi, K.: Predicting the risk of heart attacks using neural network and decision tree. Int. J. Innov. Res. Comput. Commun. Eng. **2**, 7025–7030 (2014)
9. Jabbar, M.A., Deekshatulu, B.L., Chandra, P.: Graph based approach for heart disease pre-diction. In: Das, V. (eds.) Proceedings of ITC 2012. LNEE, vol. 150, pp. 465–474. Springer, New York (2012). https://doi.org/10.1007/978-1-4614-3363-7_54
10. Westerlund, A.M., Hawe, J.S., Heinig, M., Schunkert, H.: Risk prediction of cardiovascular events by exploration of molecular data with explainable artificial intelligence. Int. J. Mol. Sci. **22**(19), 10291 (2021)
11. Dağlarli, E.: Explainable artificial intelligence (xAI) approaches and deep meta-learning models. In: Advances and Applications in Deep Learning, vol. 79 (2020)
12. Duell, J., Fan, X., Burnett, B., Aarts, G., Zhou, S.M.: A comparison of explanations given by explainable artificial intelligence methods on analysing electronic health records. In: IEEE EMBS International Conference on Biomedical and Health Informatics (BHI), pp. 1–4. IEEE. (2021)
13. Aghamohammadi, M., Madan, M., Hong, J., Watson, I.: Predicting heart attack through explainable artificial intelligence. In: Rodrigues, J.M.F., et al. (eds.) ICCS 2019. LNCS, vol. 11537, pp. 633–645. Springer, Cham (2019). https://doi.org/10.1007/978-3-030-22741-8_45
14. Systematic COronary Risk Evaluation (SCORE). https://www.escardio.org/Education/Practice-Tools/CVD-prevention-toolbox/SCORE-Risk-Charts

Machine Learning Models for Predicting 30-Day Readmission of Elderly Patients Using Custom Target Encoding Approach

Nodira Nazyrova[(✉)], Thierry J. Chaussalet, and Salma Chahed

School of Computer Science and Engineering, University of Westminster,
London W1W 6UW, UK
W1169140@my.westminster.ac.uk

Abstract. The readmission rate is an important indicator of the hospital quality of care. With the upsetting increase in readmission rates worldwide, especially in geriatric patients, predicting unplanned readmissions becomes a very important task, that can help to improve the patient's well-being and reduce healthcare costs. With the aim of reducing hospital readmission, more attention is to be paid to home healthcare services, since home healthcare patients on average have more compromised health conditions. Machine Learning and Artificial intelligence algorithms were used to develop predictive models using MIMIC-IV repository. Developed predictive models account for various patient details, including demographical, administrative, disease-related and prescription-related data. Categorical features were encoded with a novel customized target encoding approach to improve the model performance avoiding data leakage and overfitting. This new risk-score based target encoding approach demonstrated similar performance to existing target encoding and Bayesian encoding approaches, with reduced data leakage, when assessed using Gini-importance. Developed models demonstrated good discriminative performance, AUC 0.75, TPR 0.69 TNR 0.67 for the best model. These encouraging results, as well as an effective feature engineering approach, can be used in further studies to develop more reliable 30-day readmission predictive models.

Keywords: 30-day readmission · Home care patient readmission · Machine learning model · Categorical feature encoding · Customized target encoding

1 Introduction

Readmission to the hospital within 30 days from discharge has been receiving growing attention due to its implications on cost and quality of care. In the UK, approximately one in six hospital admissions result in readmission, and elderly patients are more likely to be at risk of readmission [1]. Patients aged 65 and older account for 56% of readmission cases, which constitutes 60% of associated costs [2]. Moreover, there is a growing number of patients with multi-morbidities and in the future the comorbidities seem to take on greater importance due to the overall world's population ageing [3]. In the USA, 30-day readmissions were higher among elderly Medicare beneficiaries with chronic conditions (22.5%) than among those with acute conditions (19.3%) [4].

But not all readmission cases are unavoidable: the Medicare Payment Advisory Commission (MedPac) estimates 12% of readmission as potentially avoidable and prevention of 10% of these cases could save Medicare $1 billion [5]. Simple post-discharge calls and follow-up visits have proven to be effective measures to decrease early readmission cases in elderly patients [6]. With the aim of reducing ill health and preventing emergency admissions more attention should be paid to care home patients. Older people living in care homes are among the highest risk group for preventable ill health and the use of clinical services [7].

In the past decade, various efforts were invested in modelling the risk of 30-day readmission to hospitals. Several risk scoring systems are widely developed and adopted in hospitals to predict the risk of readmission or mortality. These scoring systems are based on baseline information obtained during the patient's hospital stay. But there is a lack of predictive models that consider the impact of prescribed medications along with more detailed clinical data.

In this paper we describe the development of the hospital readmission model which is based on the thorough health status of the patient created for a full cohort of elderly patients and a subset of home-care patients. In the feature engineering step, we developed a novel score-based target encoding approach. Our categorical feature encoding method increased the statistical performance of the model when compared to target agnostic encoding approaches. The models are built on the clinical data which is available before discharge, hence can be used to predict the risk of patient readmission and undertake preventive measures.

2 Study Design and Methodology

The study was a cross-sectional assessment of 63 557 geriatric patients which constitutes 140 518 hospital admissions between 2015 and 2019 from the Medical Information Mart for Intensive Care (MIMIC-IV) datasets to predict 30-day readmission.

We defined the criteria for readmission as an episode of unplanned hospitalization to an acute care hospital within 30 days of previous discharge. The unplanned hospitalizations are recorded under the emergency and urgent admission types, which can include both walk-in admissions and emergency department admissions. Whereas the planned hospitalizations are recorded as observation or elective admission types. Only unplanned hospitalizations were included in the analysis. Moreover, to exclude the episodes of observation stays in the emergency department, patients who spent less than one day in the emergency department are not considered readmitted.

Figure 1 shows the methodology adopted in this work. In the data preparation step, the clinical dataset for the study was extracted. The analysis was performed on the MIMIC-IV dataset - a large database with administrative, clinical and critical care data for patients admitted to the Beth Israel Deaconess Medical Centre [8]. It contains data about over 382 278 deidentified patients, which constitutes 523 740 hospital admissions. All patient identifiers are removed to comply with the Health Insurance Portability and Accountability (HIPAA) regulations [9].

In the data pre-processing step, the extracted dataset was split into training and test sets with 70% and 30% of records correspondingly. The missing value imputation and standardisation were applied to each set separately to avoid data leakage. The training set was modified with a state-of-the-art oversampling technique and both imbalanced and balanced datasets were used for the modelling using seven ML algorithms. In the feature engineering step, various categorical features were compared, and the novel score-based categorical feature encoder was proposed. Mean Decrease Impurity Filter was used for Feature Selection and feature importance levels were monitored with Gini Purity Scores. For the model evaluation, the AUC of the models and classes' accuracy rates were compared after the testing phase.

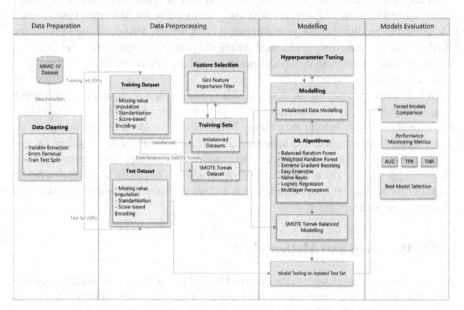

Fig. 1. The ML modelling methodology adopted in this work

2.1 Data Preparation

The full MIMIC-IV cohort had 523 740 admissions and 382 278 patients, which were further filtered out based on several criteria. For the analysis, only patients older than 65 years were included. Moreover, since the readmission rate is used to evaluate the quality of care in the hospitals, those patients who were discharged against advice and those who were discharged to hospices were excluded. The full sampling approach is demonstrated in Fig. 2. For the study, the final cohort of 140 518 admissions was used, of which 18 447 (13.12%) admissions were identified as readmission cases.

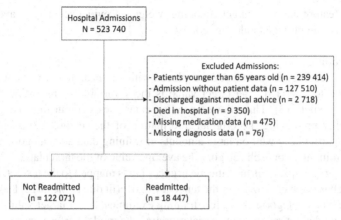

Fig. 2. Study flow diagram

2.2 Data Pre-processing

To prepare the datasets for the modelling and enhance performance several data transformations were performed.

Missing Value Imputation

Demographical variables often had a high number of missing values, therefore additional dimensions to indicate unknown features were added. A small number of patients (0.6%) did not have any associated diagnoses and medications records, therefore were excluded from the analysis. In 5% of admissions laboratory values for haemoglobin and sodium were missing. These variables were considered within normal limits, following the haemoglobin and sodium value imputation approach in similar studies [10, 11]. Most of the features in the newly formed dataset are dichotomous, hence only zero imputation was made for missing values.

Feature Scaling

All input variables were scaled to a common magnitude using a robust scaling approach [12]. A robust scaler removes the median and scales the data between the 25th quantile and 75th quantile range. Since the medical records often contain outliers, especially for geriatric patients with multiple comorbidities, the mean can be skewed by the extreme values, and typically these extreme values have a low probability of occurrence. Therefore, a robust scaling approach was adopted to avoid the negative impact of outliers on standardization.

Feature Engineering

Most machine learning algorithms required the input data to be a numeric matrix, hence it is required to encode categorical features, that do not have an intrinsic ordering. There are numerous ways to encode categorical features, however, not all of them preserve the original knowledge contained in categorical features. To select the most suitable categorical feature encoder, target encoding, weight of evidence encoder and custom-made score-based target encoder are compared.

Target Encoding
With target encoding, each category is replaced with the mean target value for samples having that category [13]. The target value is the y-variable or the value the model is trying to predict [13]. This allows encoding categories without increasing the data dimensionality preserving the original information of the features. This approach is performing particularly well on large amounts of training data and categorical features with low cardinality. For each category the average value of the target label is calculated on the training examples. Further, the mean encoding is mapped to the test set. However, this approach is often criticized for the tendency to overfit due to the target leakage [13]. In addition, when categories have few training examples, mean target values for these categories may be not representative, deteriorating the model performance.

Weight of Evidence Encoder
Weight of Evidence is a categorical feature encoder that measures the strength of a grouping technique that is used to separate one class from another in the following way: [14].

$$WoE = ln\left(\frac{\%of\ non-events}{\%of\ events}\right) \tag{1}$$

Similarly to Target Encoding, there is a potential for target leakage and overfitting of the model. To avoid this, random Gaussian noise may be injected to the variable during encoding.

Score-Based Target Encoding
Since the target encoding and weight of evidence encoding are often criticized for data leakage and overfitting, it was decided to adjust the encoder to avoid this behaviour. Similar to target encoding, categorical features are encoded using the target value in the training set. Features are replaced with the blend of the posterior probability of the target given a particular categorical value and the prior probability of the target over all the training data.

The readmission rate (target variable ratio) is used as a baseline rate, which is 13.12% for our dataset.

$$BaselineRate = \frac{Readmitted}{Readmitted + Not\ Readmitted} * 100 \tag{2}$$

When the feature readmission rate is within ½ standard deviation from the baseline readmission rate, this feature is encoded to 1, following the risk-scoring approach. Score boundaries were calculated using m standard deviations from the baseline readmission rate, where m is the incremental value.

If Encoded($\mathcal{X}^i$) == *Baseline Rate* ± 0.5 σ
  Then Score($\mathcal{X}^i$) = 1
If Encoded($\mathcal{X}^i$) == *Baseline Rate* + (m+0.5)σ
  Then Score ($\mathcal{X}^i$) feature = m+1
If Encoded($\mathcal{X}^i$) == *Baseline Rate* – (m+0.5)σ
  Then Score ($\mathcal{X}^i$) feature = 1-m

where: ($\mathcal{X}^i$) is given categorical feature, Encoded ($\mathcal{X}^i$) is the probability of the target (readmitted) given particular categorical value ($\mathcal{X}^i$) and the prior probability of the target over all the training data, σ - standard deviation of Encoded ($\mathcal{X}^i$) for the given categorical feature ($\mathcal{X}^i$), m – number of standard deviations;

To demonstrate how custom score-based target encoding is used, the example of encoding for 'Discharge Location' categorical value is provided in Table 1.

Table 1. Example of score-based encoding using 'Discharge Location' categorical feature

Discharge Location	Not readmitted	Readmitted	Readmission % for the feature	Encoded value
Psych facility	209	134	39.06	9
Chronic/long term acute care	3745	1039	21.71	4
Other facility	161	31	16.14	2
Acute hospital	631	109	14.72	1
Home health care	37620	6369	14.47	1
Skilled nursing facility	31409	5117	14.00	1
Assisted living	430	66	13.30	1
Rehab	5881	884	13.06	1
Home	41939	4687	10.05	0
Standard Deviation			**3.339**	

This approach can be strongly affected by the outliers in the standard deviation calculation. Therefore, the outliers should be omitted when calculating standard deviation. To avoid data leakage encoding should be based on the training set data and the obtained scores should be mapped to categorical features in the test set.

There was no significant difference in the statistical performance of the three encoders during the testing as can be seen in Table 2. However, since the problem of data leakage in target encoding and weight of evidence encoding was raised, we decided to observe the feature importance of these encoded categorical features (Fig. 3).

Both target encoding and weight of evidence encoding demonstrate a higher variance in feature importance than custom target encoding. Particularly, the care unit categorical

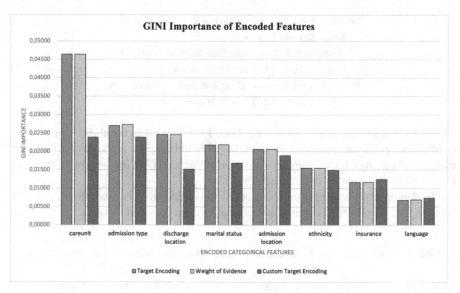

Fig. 3. Gini Importance of encoded features.

feature which contained a large number (31) of categories, has the highest feature importance of 0.46 for both target and weight of evidence encoders. This can be an indicator of data leakage for high cardinality features. Surprisingly, the majority of low cardinality features also have higher feature importance in target and weight of evidence encoders. The custom target encoding approach demonstrates a low variance of feature importance, at the same time models with this encoder demonstrate slightly higher AUC (0.06 increase) compared to the next best model built with the target encoder. The score-based target encoder makes the training algorithms put less emphasis on the categorical features, thus reducing target leakage. Table 2 demonstrates the performance of tree-based and linear models with the target, weight of evidence and score-based target encoding.

Table 2. Performance Metrics of Machine Learning Models with three different categorical feature encoding approaches: target, weight of evidence and score-based target encoding.

Models	Target encoding			Weight of evidence encoding			Score based target encoding		
	AUC	TPR	TNR	AUC	TPR	TNR	AUC	TPR	TNR
Balanced Random Forest	0.7414	0.68	0.67	0.7429	0.68	0.67	0.7409	0.68	0.67
XGBoost	0.7485	0.69	0.66	0.7493	0.69	0.67	0.7501	0.69	0.67
Logistic Regression	0.7213	0.64	0.69	0.7302	0.65	0.68	0.7212	0.64	0.69
Naïve Bayes	0.6913	0.62	0.66	0.6914	0.62	0.66	0.6913	0.62	0.66

Data Resampling

To tackle the class imbalance Hybrid SMOTE-Tomek Links resampling approach is used. The minority class is oversampled using SMOTE algorithm: creating artificial instances based on k-nearest neighbours. When the dataset contains ambiguous records, specifically, with two closest neighbour instances belonging to the opposite classes, such records are removed using the Tomek Links approach [15]. This approach helps to improve the class separation near the decision boundaries.

Feature Selection

Mean Decrease in Impurity was used to monitor the feature importance and select features for the model. Features with zero importance were excluded from the analysis. Figure 4. shows the top features with their Gini importance that were selected for the analysis.

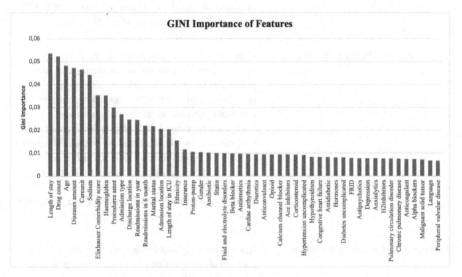

Fig. 4. Feature importance based on Gini Impurity

Since readmission modelling implies a higher level of uncertainty, more complete information about the patients should be used to improve the model discrimination. Several studies [16, 17] suggested that polypharmacy and medication noncompliance are one of the major reasons for readmission, followed by fall injuries when older patients are concerned. Therefore, it was decided to incorporate into the analysis these cases. Thirty-two groups of drugs that have the highest correlation to readmission were identified based on the existing studies [16, 17]. These drug groups included mainly drugs that increase the risk of falls, such as antihypertensives, antiarrhythmics, anticholinergics, antihistamines, sedatives, antipsychotics, and opioids. Moreover, some studies [18] outlined the increased risk of readmission for patients taking heart failure drugs, such as ace inhibitors, angiotensin receptor blockers, beta-blockers, and diuretics. The final model contains twenty-two groups of drugs chosen during the feature selection.

Even though most of the existing readmission studies use the Charlson Comorbidity index features, for this study, it was decided to use the Elixhauser Comorbidity index features, as it gives a more detailed view of patients' health conditions [19]. Elixhauser Comorbidity score incorporates 31 groups of diseases, including those in the Charlson comorbidity index. The full list of attributes used for modelling is provided in Table 3.

Table 3. Model variables

Attributes groups	Attributes
Demographic variables	Age, gender, ethnicity, marital status, language
Administrative variables	Length of stay, admission type, insurance, number of readmissions during the previous year, number of readmissions during the previous six month, number of ICD9/10-coded procedures, number of ICD9/10 encoded diseases, number of prescribed drugs, discharge care unit
Elixhauser comorbidity score diseases	Congestive heart failure, cardiac arrhythmias, valvular disease, pulmonary circulation disorders, peripheral vascular disorders, hypertension, paralysis, neurodegenerative disorders, chronic pulmonary disease, diabetes uncomplicated, diabetes complicated, hypothyroidism, renal failure, liver disease, peptic ulcer disease without bleeding, AIDS/HIV, lymphoma, metastatic cancer, solid tumour without metastasis, rheumatoid arthritis, coagulopathy, obesity, weight loss, fluid and electrolyte disorders, blood loss anaemia, deficiency anaemia, alcohol abuse, drug abuse, psychosis, depression
Prescribed medications (25 drug groups)	Anticoagulants, antibiotics, antipsychotics, anticonvulsants, anticholinergics, antiarrhythmics, antiemetics, anti-diabetic, antifungal, angiotensin, alpha-blockers, anxiolytics, ACE inhibitors, beta-blockers, calcium channel blockers, cardiac drugs, corticosteroids, diuretics, FRIDs (fall risk increasing drugs), h2 inhibitors, hormones, mineralocorticoids, PIMs (potentially inappropriate drugs), sedatives, h2 inhibitors

2.3 Modelling

Six machine learning and one deep learning algorithms were used to build predictive models: Logistic Regression, Naïve Bayes, Random Forest (balanced and weighted), Easy Ensemble, Extreme Gradient Boosting and Multilayer Perceptron. To account for the class imbalance, algorithms hyper-parameters to adjust class weights were applied. Models were validated on the training set using 5-fold validation. Models were tested on imbalanced dataset settings and Smote Tomek balanced dataset settings.

2.4 Model Evaluation

The selection of appropriate evaluation metrics is very important in clinical decision making and it often implies a trade-off between cost optimization and risk aversion. Acknowledging these risks, it was decided to use the following evaluation metrics: the True Positive Rate (Sensitivity), the True Negative Rate (Sensitivity) and the Area Under Receiver Operating Characteristic Curve (AUROC).

True Positive Rate (TPR) is used to measure the rate of correct prediction of the subjects from the readmission group. And **True Negative Rate (TNR)** is used to measure the rate of correct prediction of the subjects from the non-readmission group. ROC AUC metrics is a commonly used evaluation metric is a graph showing the performance of a classification model at all classification thresholds. This metric can show a trade-off between correctly predicted readmission cases and those who were falsely classified as readmitted, which in turn can impose additional costs on healthcare organizations.

To select the best models the highest AUC score was selected with the most balanced TPR and TNR. We assume equal classification importance for both classes.

3 Results

Imbalanced Classification Results
Table 4 shows the test performance of models built with the imbalanced dataset settings for all elderly patients and for those who were discharged to home health care settings. Multilayer Perceptron achieved the lowest AUC score of 0.61, and 0.26 TPR, 0.85 TNR due to the class imbalance. There are no hyper-parameters to account for the class imbalance. Ease Ensemble learner achieved 0.73 AUC, however the class accuracy difference achieved as much as 10%. Traditional ML algorithms showed less variance between TPR and TNR when used in a cost-sensitive approach with balanced class weights. The difference between the class accuracy was less than 5% for the remaining models. Naïve Bayes achieved the second lowest result with AUC 0.69, followed by the Logistic Regression. Tree-based ensemble learner demonstrated the best statistical performance, with the XGBoost achieving the highest 0.75 AUC with the most balanced class accuracy: 0.69 TPR and 0.67 TNR for the dataset with a full cohort of patients, and AUC 0.74, TPR 0.68 and TNR 0.65 for the home care patients.

Table 4. The test performance of ML/AI models trained with imbalanced datasets

ML models	Full cohort of patients			Home care patients		
	AUC	TPR	TNR	AUC	TPR	TNR
Balanced Random Forest	0.74	0.68	0.67	0.73	0.67	0.65
XGBoost	**0.75**	**0.69**	**0.67**	**0.74**	**0.68**	**0.65**
Weighted Random Forest	0.72	0.64	0.64	0.71	0.64	0.68
Easy ensemble	0.74	0.67	0.57	0.73	0.65	0.57
Logistic Regression	0.72	0.64	0.69	0.71	0.68	0.62
Naïve Bayes	0.69	0.62	0.66	0.68	0.59	0.66
Multilayer perceptron	0.61	0.26	0.85	0.61	0.25	0.85

SMOTE Re-sampled Datasets Modelling Results

SMOTE Tomek resampling technique applied to the imbalanced dataset did not result in improved model performance. Overall, there is a slight AUC score decrease for most of the models with the resampled dataset settings. The class variance was reduced as expected. Multilayer Perceptron has less class accuracy imbalance, and still demonstrates the poorest AUC and TPR. Other predictors demonstrate a balanced classification between the two classes. XGBoost again demonstrates the highest AUC and TPR for both dataset settings: full cohort of patients and home care patients (see Table 5).

Table 5. The test performance of ML/AI models trained with resampled datasets

ML models	Full cohort of patients			Home care patients		
	AUC	TPR	TNR	AUC	TPR	TNR
Balanced Random Forest	0.73	0.60	0.67	0.72	0.65	0.65
XGBoost	**0.74**	**0.71**	**0.64**	**0.73**	**0.69**	**0.62**
Weighted Random Forest	0.72	0.64	0.64	0.71	0.64	0.68
Easy ensemble	0.71	0.66	0.57	0.70	0.65	0.59
Logistic Regression	0.71	0.63	0.68	0.70	0.66	0.62
Naïve Bayes	0.69	0.63	0.66	0.69	0.67	0.64
Multilayer perceptron	0.65	0.30	0.79	0.64	0.30	0.78

Baseline Characteristics Analysis

Analysis of baseline characteristics of the baseline population demonstrated that over-all home care patients have more aggravated health conditions, with the average age higher than the full cohorts', they have more comorbidities and have a longer length of stay in the hospital. They are prescribed more medications and have higher number of registered diseases. Moreover, this cohort of patients more often live without partner (widowed or divorced). There are more women admitted to the hospital, however, men more often have readmission cases. Detailed baseline characteristics are provided in the Table 6.

Table 6. Baseline characteristics of the study population by 30-day readmission status.

Characteristics	Full cohort		Home care patients	
Readmission (%)	Not readmitted 122,071 (87%)	Readmitted 18,447 (13%)	Not readmitted 69,042 (86%)	Readmitted 11,490 (14%)
Age, (mean, SD)	77.06 (8.23)	76.89 (8.18)	78.70 (8.42)	78.14 (8.36)
Length of stay (mean, SD)	5.13 (6.28)	6.43 (8.3)	5.80 (6.02)	6.46 (7.73)
Emergency department visits in last 12 month (mean, SD)	0.29 (0.82)	0.91 (1.79)	0.34 (0.90)	0.98 (1.88)
Emergency department visits in last 6 month (mean, SD)	0.23 (0.66)	0.73 (1.35)	0.27 (0.73)	0.77 (1.41)
Gender:				
Male	58 752 (48%)	9 398 (51%)	30 868 (45%)	5 388 (48%)
Female	63 319 (52%)	9 049 (49%)	38 429 (55%)	5 865 (52%)
Ethnicity:				
White	92 743 (75%)	13 547 (76%)	51 930 (75%)	8 644 (75%)
Asian	3 451 (3%)	527 (3%)	1 792 (2%)	323 (2%)
Hispanic/Latino	3 811 (3%)	606 (3%)	2 089 (3%)	421 (3%)
Black/African American	14 065 (11%)	2 293 (13%)	8 245 (12%)	1 620 (14%)
American Indian/Alaska Native	214 (0.2%)	37 (0.2%)	117 (0.16%)	26 (0.23%)
Unknown	8931 (7%)	768 (4%)	4869(7%)	456 (4%)
Insurance:				

(continued)

Table 6. (*continued*)

Characteristics	Full cohort		Home care patients	
Readmission (%)	Not readmitted 122,071 (87%)	Readmitted 18,447 (13%)	Not readmitted 69,042 (86%)	Readmitted 11,490 (14%)
Medicaid	2,241 (2%)	352 (2%)	938 (1.5%)	186 (2%)
Medicare	83 291 (67.5%)	12 374 (70%)	47 942 (69.5%)	8 115 (71%)
Other	37 683 (30%)	5 052 (28%)	20 162(29%)	3 119 (27%)
Marital Status:				
Married	60 022 (49%)	8 678 (49%)	29 832 (43%)	5 074 (44%)
Single	23 021 (18%)	3374 (19%	13 390 (19%)	2 268 (20%)
Widowed	27 991 (22%)	4 368 (24%)	18 712 (27%)	3 258 (28%)
Divorced	9 139 (7%)	1 234 (7%)	5,323 (8%)	824 (7%)
Unknown	3 042 (2%)	124 (0.7%)	1,785 (2%)	66 (0.5%)
Language:				
English speaking	107 584 (87.4%)	15 342 (86%)	59 814 (87%)	9 772 (85%)
Non-English Speaking	15 631 (12.6%)	2 436 (14%)	9 228 (13%)	1 718 (15%)
Admission type:				
Emergency	65 628 (54%)	13 147 (73%)	36 324 (53%)	8 265 (72%)
Urgent	11 651 (9%)	1 944 (11%)	7 058 (10%)	1 166 (10%)
Elective	44 792 (37%)	3 356 (17%)	25 650 (37%)	2 059 (18%)
Charlson Comorbidity Index (mean, SD)	3.19 (2.99)	4.10 (3.13)	3.45 (3.08)	4.36 (3.16)
Diseases Amount (mean, SD)	14.44 (6.97)	15.96 (7.20)	15.67 (6.91)	16.72 (7.00)
Prescribed drugs amount (mean, SD)	23.11 (11.22)	26.24 (12.41)	25.11 (10.87)	26.61 (10.90)

4   Discussion

We developed machine learning and artificial intelligence models to predict the 30-day hospital readmission. The developed model was validated on the full cohort of elderly patients and a subset of patients, who use home care services.

We adopted a novel customized target encoding approach in the feature engineering step. Categorical variables were encoded using a score-based target encoder. This approach demonstrated similar performance with the target encoder and the weight of evidence encoder. However, a comparison of Gini feature importance for all three types of encoders upholds the well-known problem of data leakage in target encoders, including weight of evidence encoder. Whereas score-based target encoding demonstrated less emphasis on encoded categorical features, preserving the good model performance. Developed models were modified to account for the high-class imbalance. While traditional algorithms benefited from the hybrid oversampling approach, tree-based models performed better in ensemble learner mode with the class weight adjustment.

Among all developed and analysed models, ensemble learners and specifically XGB dominated the highest ranking in TPR, TNR and AUC scores for both imbalanced and re-sampled data states for the proposed dataset with AUC 0.75, TPR 0.69, TNR 0.67.

During our analysis we found that polypharmacy is one of the most important predictors of readmission. A higher number of prescribed medications had a positive correlation with readmission cases. Moreover, some drug groups were identified to have an impact on classification results, such as diuretics, corticosteroids, anticoagulants and fall risk-increasing drugs.

In concordance with previous studies, the number of comorbidities directly correlates with the frequency of 30-day readmission after hospital discharge [20, 21]. Particularly such conditions as fluid and electrolyte disorders, cardiac arrhythmia, hypertension, congestive heart failure, pulmonary circulation disorders, depression, and malignant cancer. Interestingly, some of these comorbidities are not covered in the widely used Charlson Comorbidity index, such as fluid and electrolyte disorder, cardiac arrhythmia, hypertension and depression. Whereas most of the existing studies utilize Charlson Comorbidity Index, Elixhauser Comorbidity Index could provide a better overview of the patients' health conditions [19]. Future work will take into consideration drug-drug interactions as an important readmission predictor. Polypharmacy and drug interactions should be thoroughly examined when analysing geriatric patient readmission. Moreover, future work should consider the impact of post-discharge services on the 30-day readmission when the data is available.

References

1. NHS Digital. Hospital Admitted Patient Care Activity 2019-20 (2020). https://digital.nhs. uk/data-and-information/publications/statistical/hospital-admitted-patient-care-activity/201 9-20. Accessed 12 Feb 2022
2. Glans, M., Kragh Ekstam, A., Jakobsson, U., et al.: Risk factors for hospital readmission in older adults within 30 days of discharge – a comparative retrospective study. BMC Geriatr. **20**, 467 (2020). https://doi.org/10.1186/s12877-020-01867-3
3. Fabbri, E., Zoli, M., Gonzalez-Freire, M., Salive, M.E., Studenski, S.A., Ferrucci, L.: Aging and multimorbidity: new tasks, priorities, and frontiers for integrated gerontological and clinical research. J. Am. Med. Dir. Assoc. **16**(8), 640–647 (2015). https://doi.org/10.1016/j. jamda.2015.03.013
4. Raval, A.D., Zhou, S., Wei, W., Bhattacharjee, S., Miao, R., Sambamoorthi, U.: 30-day readmission among elderly medicare beneficiaries with type 2 diabetes. Popul. Health Manag. **18**(4), 256–264 (2015). https://doi.org/10.1089/pop.2014.0116
5. McIlvennan, C.K., Eapen, Z.J., Allen, L.A.: Hospital readmissions reduction program. Circulation **131**(20), 1796–1803 (2015). https://doi.org/10.1161/CIRCULATIONAHA.114. 010270
6. Vernon, D., Brown, J.E., Griffiths, E., Nevill, A.M., Pinkney, M.: Reducing readmission rates through a discharge follow-up service. Future Healthcare J. **6**(2), 114–117 (2019). https://doi. org/10.7861/futurehosp.6-2-114

7. Dixon, J.: Reducing emergency admissions from care homes: a measure of success for the NHS long-term plan (2019). https://www.hsj.co.uk/emergency-care/reducing-emergency-admissions-from-care-homes-a-measure-of-success-for-the-nhs-long-term-plan/7025675
8. Johnson, A., Bulgarelli, L., Pollard, T., Horng, S., Celi, L.A., Mark, R.: MIMIC-IV (version 0.4). PhysioNet (2020). https://doi.org/10.13026/a3wn-hq05
9. Institute of Medicine (US) Committee on Health Research and the Privacy of Health Information, The HIPAA Privacy Rule, Nass, S.J., Levit, L.A., Gostin, L.O. (eds.) Beyond the HIPAA Privacy Rule: Enhancing Privacy, Improving Health Through Research. National Academies Press (US), Washington DC (2009). 4, HIPAA, the Privacy Rule, and Its Application to Health Research. https://www.ncbi.nlm.nih.gov/books/NBK9573/
10. Donzé, J., Aujesky, D., Williams, D., Schnipper, J.L.: Potentially avoidable 30-day hospital readmissions in medical patients: derivation and validation of a prediction model. JAMA Intern. Med. **173**(8), 632–638 (2013). https://doi.org/10.1001/jamainternmed.2013.3023
11. Robinson, R., Hudali, T.: The HOSPITAL score and LACE index as predictors of 30 day readmission in a retrospective study at a university-affiliated community hospital. PeerJ **5**, e3137 (2017). https://doi.org/10.7717/peerj.3137
12. Iglewicz, B.: Robust scale estimators and confidence intervals for location. In: Hoaglin, D.C., Mosteller, M., Tukey, J.W. (eds.) Understanding Robust and Exploratory Data Analysis. Wiley, NY (1983)
13. Pargent, F., Pfisterer, F., Thomas, J., Bischl, B.: Regularized target encoding outperforms traditional methods in supervised machine learning with high cardinality features. arXiv preprint arXiv:2104.00629 (2021)
14. Lund, B.: Weight of Evidence Coding and Binning of Predictors in Logistic Regression (2016)
15. Batista, G., Prati, R., Monard, M.: A study of the behavior of several methods for balancing machine learning training data. SIGKDD Explor. **6**(1), 20–29 (2004). https://doi.org/10.1145/1007730.1007735
16. Glans, M., Kragh Ekstam, A., Jakobsson, U., Bondesson, Å., Midlöv, P.: Medication-related hospital readmissions within 30 days of discharge—a retrospective study of risk factors in older adults. PLoS ONE **16**(6), e0253024 (2021). https://doi.org/10.1371/journal.pone.0253024
17. Uitvlugt, E., et al.: Medication-related hospital readmissions within 30 days of discharge: prevalence, preventability, type of medication errors and risk factors. Front. Pharmacol. **12**, 567424 (2021). https://doi.org/10.3389/fphar.2021.567424
18. Parajuli, P., et al.: Heart failure drug class effects on 30-day readmission rates in patients with heart failure with preserved ejection fraction: a retrospective single center study. Medicines (Basel) **7**(5), 30 (2020). https://doi.org/10.3390/medicines7050030. PMID:32443705; PMCID:PMC7281589
19. Buhr, R.G., Jackson, N.J., Kominski, G.F., Dubinett, S.M., Ong, M.K., Mangione, C.M.: Comorbidity and thirty-day hospital readmission odds in chronic obstructive pulmonary disease: a comparison of the Charlson and Elixhauser comorbidity indices. BMC Health Serv. Res. **19**(1), 701 (2019). https://doi.org/10.1186/s12913-019-4549-4. PMID:31615508; PMCID:PMC6794890
20. Picker, D., Heard, K., Bailey, T.C., et al.: The number of discharge medications predicts thirty-day hospital readmission: a cohort study. BMC Health Serv. Res. **15**, 282 (2015). https://doi.org/10.1186/s12913-015-0950-9
21. Pereira, F., Verloo, H., Zhivko, T., et al.: Risk of 30-day hospital readmission associated with medical conditions and drug regimens of polymedicated, older inpatients discharged home: a registry-based cohort study. BMJ Open **11**, e052755 (2021). https://doi.org/10.1136/bmjopen-2021-052755

Patient- and Ventilator-Specific Modeling to Drive the Use and Development of 3D Printed Devices for Rapid Ventilator Splitting During the COVID-19 Pandemic

Muath Bishawi[1,2], Michael Kaplan[2], Simbarashe Chidyagwai[2(✉)],
Jhaymie Cappiello[3], Anne Cherry[4], David MacLeod[4], Ken Gall[5,6],
Nathan Evans[6], Michael Kim[6], Rajib Shaha[6], John Whittle[4],
Melanie Hollidge[4], George Truskey[2], and Amanda Randles[2]

[1] Department of Surgery, Duke University Medical Center, Durham, NC, USA
[2] Department of Biomedical Engineering, Pratt School of Engineering,
Duke University, Durham, NC, USA
sgc22@duke.edu
[3] Department of Respiratory Therapy, Duke University Medical Center,
Durham, NC, USA
[4] Department of Anesthesiology, Duke University Medical Center, Durham, NC, USA
[5] Department of Mechanical Engineering, Pratt School of Engineering, Duke
University, Durham, NC, USA
[6] restor3d Inc, Durham, NC, USA

Abstract. In the early days of the COVID-19 pandemic, there was a pressing need for an expansion of the ventilator capacity in response to the COVID19 pandemic. Reserved for dire situations, ventilator splitting is complex, and has previously been limited to patients with similar pulmonary compliances and tidal volume requirements. To address this need, we developed a system to enable rapid and efficacious splitting between two or more patients with varying lung compliances and tidal volume requirements. We present here a computational framework to both drive device design and inform patient-specific device tuning. By creating a patient- and ventilator-specific airflow model, we were able to identify pressure-controlled splitting as preferable to volume-controlled as well create a simulation-guided framework to identify the optimal airflow resistor for a given patient pairing. In this work, we present the computational model, validation of the model against benchtop test lungs and standard-of-care ventilators, and the methods that enabled simulation of over 200 million patient scenarios using 800,000 compute hours in a 72 h period.

Keywords: Airflow · Cloud computing · Ventilator modeling

D. Groen et al. (Eds.): ICCS 2022, LNCS 13352, pp. 137–149, 2022.
https://doi.org/10.1007/978-3-031-08757-8_13

1 Introduction

The COVID19 pandemic has shed light on the need for emergency ventilator systems which can be rapidly deployed when the demand for ventilators surpasses their supply [1], such as during regional emergencies[2], global pandemics [3], and in low-resource ICUs [4]. These various scenarios require a ventilator sharing strategy that maximizes the number of patients able to receive potentially life-saving treatment from a limited number of ventilators. Ventilator splitting has been introduced as a strategy to support multiple patients on the same ventilator and has been implemented at a number of institutions during dire situations [5–7]. Recent advances, such as the addition of resistors[8], clamps [9], and valves [10], has allowed ventilator splitting to be useful for carefully matched patients [7]. However, ventilator splitting cannot be safely and rapidly implemented for patients with significantly differing pulmonary compliances [11] or minute ventilation requirements [10], as this could lead to volutrauma, barotrauma, and/or hypoventilation of one or both of the patients. Concerns related to the safety of ventilator splitting has prevented it from being recommended as a general solution for ventilator shortages in the most extreme of circumstances [12].

To address this problem, we developed a rapidly deployable, simple, and low-cost ventilator splitter and resistor system (VSRS) with 3D-printed interchangeable airflow resistors and a clinical support mobile app informed by over 200 million individual simulations to allow for patients with differing pulmonary mechanics to share the same ventilator. A standard ventilator is able to be retrofitted to a split ventilator configuration with the addition of only two essential 3D-printed components, the splitter, which connects one standard ventilator tubing inlet to two tubing outlets, and the resistor. The resistors allow independent differential control over the tidal volumes and pressures delivered to each patient and a computational model was created to quantify how ventilator settings, endotracheal tube diameters, and patient pulmonary compliances affect delivered tidal volumes and pressures with the VSRS. When used together, the 3D printed components and the numerical modelling allow clinicians to quickly, but safely ventilate multiple patients, even when the patients have differing ventilatory requirements and pulmonary compliances. While we present the novel VSRS here, we specifically focus on the role of the computational model. In this work, we present the development of a computational model to simulate air flow in the VSRS and guide usage decisions, validation of the model against benchtop experiments with test lungs, use of the model to determine that safest splitting is pressure-driven as opposed to volume-driven, and finally use of a massively parallel cloud-based framework to pre-compute a wide expance of clinically relevant scenarios.

2   Methods

2.1   Design of the Ventilator Splitter and Resistor System

The VSRS we developed consists of two primary components: the splitter and the resistor, shown in Fig. 1 (b). The splitter component is a Y-shaped adaptor that splits a single airflow into two separate channels (used for the inspiratory limb splitting). When used in reverse, the splitter can combine airflow from two channels into a single channel (used for the expiratory limb). The splitter has a continuously graded diameter such that the interior diameter of the splitter's single-channel end fits over the exterior diameter of standard ventilator tubing, while the exterior diameters of the splitter's dual-channel ends fit within the interior diameter of standard ventilator tubing. The splitter features a 60° junction between the two dual-channel ends. The splitter fits standard ventilator tubing and at least two splitters would be required within a shared ventilator circuit.

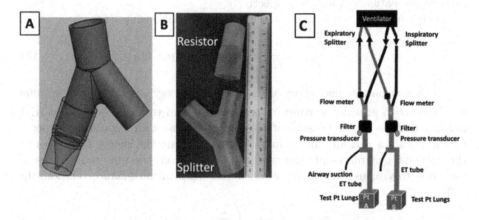

Fig. 1. Splitter and resistor design and test circuit configuration: (1A) Overall design, (1B) 3D printed components and (1C) Benchtop circuit setup to capture two patients with different lung compliances

The components of the VSRS are manufactured from a commercially available photopolymer via highly accurate stereolithography (SLA), using a commercially available printer (FormLabs 2, Formlabs Inc, Somerville, MA). This method creates highly reproducible air flow tubes, with a printing dimensional accuracy of approximately 0.5% of the diameter. All prints were done at restor3d (Durham, NC) following their internal quality system for resin-based 3D printing.

2.2   Reduced Order Ventilator Computational Model

<u>Numerical Model.</u> A pressure-controlled and volume-controlled ventilator were considered for the computational model. A range of Reynolds numbers were

simulated as a function of the ventilator settings, with higher Reynolds numbers approaching 4000. Air flow from the ventilator source to the patient was modeled using pipe flow dynamics in a gas network. The gas flow through the pipes is governed by the laws of mass, momentum and energy from which the pressure, velocity, density and temperature of the gas volume were solved. The mass conservation relates the mass flow rates to the pressure and temperature of the gas volume by the following relationship[15]:

$$\frac{\partial M}{\partial p} \cdot \frac{dp_I}{dt} + \frac{\partial M}{\partial T} \cdot \frac{dT_I}{dt} = \dot{m}_A + \dot{m}_B \tag{1}$$

were $\frac{\partial M}{\partial p}$ is the partial derivative of the mass of the gas flow with respect to pressure at constant temperature and volume. $\frac{\partial M}{\partial T}$ is the partial derivative of the mass of the gas volume with respect to temperature at constant pressure and volume, p_I is the pressure of the gas volume, T_I is the temperature and t is time. M is the mas of gas entering the device. $\dot{m}_A$ and $\dot{m}_B$ are the mass flow rates to the patients A and B, respectively.

Energy conservation is given by the following relationship [16]:

$$\frac{\partial U}{\partial p} \cdot \frac{dp_I}{dt} + \frac{\partial U}{\partial T} \cdot \frac{dT_I}{dt} = \Phi_A + \Phi_B + Q_H \tag{2}$$

$\frac{\partial U}{\partial p}$ is the partial derivative of the internal energy of the control volume with respect to pressure at constant temperature and volume. $\frac{\partial U}{\partial T}$ is the partial derivative of the internal energy of the control volume with respect to temperature at constant pressure and volume. Φ_A and Φ_B are the energy flow rates to the patients. Q_H represents the energy flow rate from the pipe wall. Pressure loses due to viscous friction are given by the momentum balance relationship [17]:

$$p_A - p_I = \left(\frac{\dot{m}_A}{S}\right)^2 \cdot \left(\frac{1}{\rho_I} - \frac{1}{\rho_A}\right) + \Delta p_{AI}$$

$$p_B - p_I = \left(\frac{\dot{m}_B}{S}\right)^2 \cdot \left(\frac{1}{\rho_I} - \frac{1}{\rho_B}\right) + \Delta p_{BI} \tag{3}$$

p_A and p_B are the pressures at the inlet and the outlet of the pipe respectively. ρ_A and ρ_B represent densities at the inlet and outlet of the pipe. S is the cross sectional area and Δp_{AI} and Δp_{BI} are the pressure loses due to friction.

The two patients are connected to each other by way of a junction, with a resistor connected distal to one of the branches to enable controlling the differential flow. The lungs are modeled as a Hookean spring (modelling the inverse of the compliance of the lungs) and a viscous dashpot (modelling the resistance of the upper respiratory tract) in parallel. This lung model presented in this study is consistent with other studies that have used a resistor - capacitor model to represent the lungs[13]. The simulations were performed using MathWork's Simscape (Simulink v4.8) Foundational Blocks. The ventilator is simulated by a pulse wave form generator with a period corresponding to the respiratory rate and inspiratory time, with a maximum value corresponding to the PIP and minimum value corresponding to the PEEP. In order for the system to reach steady state, at least 5 breath cycles were simulated for each set of parameters. The input parameters used for the numerical model are shown in Table 1 and 2.

Parallelization. It was important to not only develop a method that could accurately calculate resulting tidal volume for any two given patients and set ventilator to inform the decision of resistor to choose, but, moreover, overall time-to-solution was critical. It was important to pre-calculate all data so that it could be provided to doctors **in an offline capacity** and easily accessible manner without having to wait on a simulation. The goal was to provide clinicians with the ability to enter minimal information about the patients and the ventilator and receive immediate feedback regarding how to setup the VSRS. To that end, all of the combinations needed to be simulated *a priori*. To accomplish this, we parallelized the model and deployed it in a Microsoft Azure instance. The use of a cloud-based architecture was a strong fit for our needs based on the ability to configure the infrastructural aggregate to suit the problem at hand. We used 24,000 cores of the Azure individual node type HB60 in a datacenter in Western Europe. While the individual pairings were fundamentally framed as an embarrassingly parallel problem, a tighter coupling was imposed to reduce overall time-to-solution. As the code was built with MATLAB which relies on runtime compilation, the time to compile the code was actually significant compared to the simulation. In order to run the required 270 million simulations to span our search space (Table 2), it was important to reduce the time spent in compilation. We set up a systematic job hierarchy to maximize use of a built-in functionality for sweeping across parameters. We were thus able to pair down 270 million simulations to 146 thousand jobs, with only one compilation per job. All post-processing was completed on-node after the simulation completed to minimize data communication and storage. Only steady-state tidal volume, maximum delivered pressure, and minimum delivered pressure were stored. Lessons learned regarding fast deployment of such models are described in [14].

Table 1. Range of parameters simulated with the computational model

	Minimum value	Maximum value	Step size
Pulmonary Compliance (ml/cmH_2O)	10	100	1
Endotracheal Tube Diameter (mm)	6	8.5	0.5
Peak Inspiratory Pressure (cmH_2O)	20	50	1
Positive End - Expiratory Pressure (cmH_2O)	5	20	1
Inspiratory to Expiratory Ratio	1:3	1:1	Fractional
Respiratory Rate (breaths/minute)	10	30	1
Resistor Radii (mm)	2.5	5.5	0.5

2.3   Benchtop Testing of VSRS Using Test Lungs

Benchtop testing was used to confirm that the VSRS could be used to ventilate
two test lungs (Linear Test Lung, Ingmar Medical, Pittsburgh, PA) properly.
The measurements from this circuit were used to validate the computational
model. The benchtop circuit was used to assess applicability to both standard
ICU ventilators and anesthesia machines, both of which could be required during times of ventilator shortages. As shown in Fig. 1(c), the circuit incorporated a number of one-way valves at the inspiratory and expiratory limbs to
limit mixing and viral/bacterial filters were placed at locations of possible cross
contamination. The circuit consisted of two test lungs in which two different
compliances were tested during our experiments, as presented in Table 1. Using
ventilator settings with a respiratory rate of 20 bpm (breaths per minute), a
PEEP (positive end-expiratory pressure) of 5 cmH2O, and PIP (peak inspiratory pressure) of 20 cmH2O, the delivered tidal volumes to the low compliance
artificial lung was 352–359 ml and 566–567 ml to the medium compliance lung
(Table 2). The precise compliance of the test lungs in 'low' configuration was
determined to be 18 ml/cmH2O, and 34–36 ml/cmH2O for the 'medium' configuration (Table 2). Ventilator parameters recorded were: peak inspiratory pressure
(cmH2O), mean pressure (cmH2O), tidal volume (mL), and dynamic compliance
(mL/cmH2O). Individual test lung circuit parameters recorded (each circuit, A
and B) were: distal circuit maximum and trough pressure (mmHg), using a dry
disposable pressure transducer (Transpac IV, ICU Medical, San Clemente, CA)
connected directly to the gas sample port of each distal circuit filter. Pressures
were recorded using a GE Carescape Monitor. Each distal circuit tidal volume
was measured using in-line volume monitors (Ohmeda 6800 Volume Monitor,
Bird Products, Palm Springs, CA). A similar method of validating the benchtop
setup for use on two test lungs with different compliances was demonstrated by,
[10].

Table 2. Individual test lung mechanics using the single circuit set up

	Patient A low compliance	Patient B low compliance	Patient A medium compliance	Patient B medium compliance
Compliance	18	18	34	36
Pressure (Peak/Low)	26/13	26/13	25/13	25/13
Tidal volume	352	359	566	567

3   Results

3.1   The Computational Model Predicts Benchtop Measurements

The computational model was validated against the test lung data for multiple different pulmonary compliance values, respiratory rates, and resistor sizes.

There is excellent agreement between the computational model and the test lung data, which is evidenced by a Pearson correlation coefficient of 0.9697 and a p-value of less than 0.0001. Figure 2 illustrates the comparison between simulated and benchtop data for one set of lung compliances and respiratory rates, while varying the resistor sizes. Both the benchtop data and the computational model agree that a 3.5 mm resistor on the higher compliance limb will result in equivalent delivered tidal volumes to both lungs for this specific configuration.

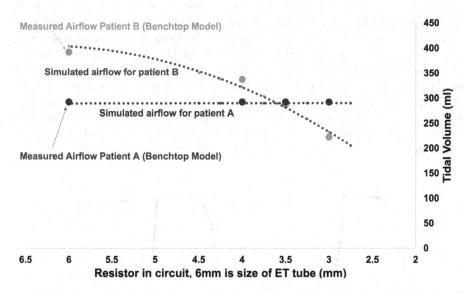

Fig. 2. Comparison of the numerical results to the benchtop experimental data Comparison of the simulation results (dotted lines) with the benchtop results (large circles) for patient A (medium compliance) and patient B (low compliance). Both are in agreement over a range of resistor sizes and demonstrate similar delivered tidal volumes to both patients using a 3.5 mm resistor.

More generally, the computational model supports the findings from the benchtop model that adding a resistor to the split ventilator circuit significantly alters delivered tidal volumes and pressures. Figure 3 illustrates the model outputs of predicted pressure, flow rate, and volume waveforms for a patient with a lower pulmonary compliance (Patient A) sharing a ventilator with a patient with higher pulmonary compliance and a resistor (Patient B). The characteristic waveform of pressure-controlled ventilation is observed for patient A, where the ventilator reaches the ventilator-specified peak inspiratory pressure and then a plateau occurs. However, for patient B, the resistor slows the buildup of pressure and the ventilator-specified peak inspiratory pressure is never reached. As a result, the flow rate waveform for patients A and B markedly differ, and as demonstrated, the computational model is able to accurately capture the effect of different resistor sizes on the resulting tidal volume for each patient.

3.2 Pressure-controlled Ventilation Protects Patients from Changes in the Opposing Patients Circuit

The computational model was also used to improve the design of the VSRS. In particular, there was an outstanding question regarding whether the ventilator splitting should be pressure-controlled or volume-controlled. The validated computational model allowed this question to be resolved through comparison of simulations in each case. Figure 4(a) illustrates how in pressure-controlled ventilation both the tidal volumes and pressures to patient A are not affected by changes to patient B's circuit, such as the addition of a resistor, which was a trend also observed in the benchtop data. In both pressure-controlled and volume-controlled ventilation, the pressure delivered to the bifurcation at the circuit is always equal to each branch, and therefore the delivered tidal volumes to each patient are independently a function of how their pulmonary characteristics respond to this pressure. However, in volume-controlled ventilation, we demonstrate that a coupling of the delivered tidal volumes between the two patients occurs. When the resistance of one limb of the circuit increases, such as through a kink in tubing or secretions, the ventilator senses the resulting decrease in com-

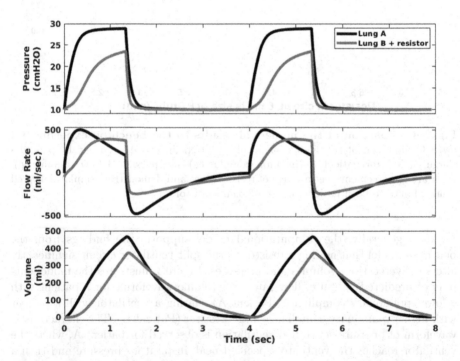

Fig. 3. Model output waveforms for pressure, flow rate, and tidal volume Example model output for simulated patient A (compliance of 30 ml/cmH2O) and simulated patient B (compliance of 75 ml/cmH2O) using a 4 mm resistor. While Patient A experiences the peak inspiratory pressure set by the ventilator, the presence of the resistor results in decreased peak pressures to Patient B.

bined delivered tidal volume and subsequently increases the delivered pressures in an attempt to achieve the desired tidal volume, which results in increased pressures and volumes to both patients. This differs from pressure-controlled ventilation, where patient A is protected from experiencing dangerously elevated pressures and tidal volumes in response to changes in resistance in patient B's circuit (Fig. 4(a)). Therefore, we show that pressure-controlled ventilator splitting offers a markedly improved safety profile due to a reduced risk of barotrauma or volutrauma to one patient from changes in the opposite patient's circuit.

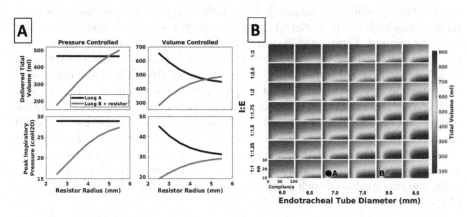

Fig. 4. (A) Comparing pressure-controlled and volume-controlled ventilation in the split ventilator configuration Computational model output of tidal volume (top row) and peak inspiratory pressure (bottom row) for different resistor sizes under pressure-controlled ventilation (left column) and volume-controlled ventilation (right column). (B) Multiple dimensions of nonlinearity for predicted tidal volumes The predicted volumes for a 8.5 mm endotracheal tube compared to a 6 mm endotracheal tube as a function of pulmonary compliance and respiratory rate with other ventilator settings held constant (PIP of 28, PEEP of 8, I:E of 1).

While this finding was significant in driving VSRS usage setup in a manner to better protect the patient, it also had implications for our computational model. Because the patients could be viewed as decoupled under a pressure-controlled setup, we could assume no interaction between patients on the same ventilator and subsequently decouple the simulations as well. Functionally, this finding led to a drastic reduction in the number of parameter pairings that needed to be simulated and allowed us to restrict our search space.

3.3 Tidal Volumes from Pressure-Controlled Ventilation Are Highly Sensitive to Small Changes in Ventilator Settings, Endotracheal Tube Sizes, and Patient Characteristics

Predicted tidal volumes from pressure-controlled ventilation are sensitive to small changes in multiple parameters (Table 1) in a non-linear fashion (Fig. 4(b)).

At lower compliance values, different endotracheal tube sizes result in only minimal changes in tidal volumes and tidal volumes increase roughly linearly with compliance.

However, as Fig. 4(b) illustrates, for larger compliance values there is a marked change in the increase in tidal volume as a function of pulmonary compliance and significant differences in tidal volumes due to different endotracheal tube sizes can occur. This is an important characteristic to consider as pulmonary compliance is expected to change due to disease progression or recovery. For example, safe tidal volumes for a given set of patient characteristics and ventilator settings while the patient has low pulmonary compliance would become dangerously high as the patient improves, and therefore a different resistor size needs to be used. Figure 4(b) also demonstrates the non-linear effect of respiratory rate (RR) on tidal volumes, where the sensitivity of tidal volumes to changes in RR is much greater for lower RRs than for higher RRs. These and other non-linearities necessitate a high-resolution parameter exploration in order to safely and precisely quantify the effect of multiple patient parameters and ventilator settings on delivered tidal volumes. These experiments emphasize how small changes in patient-specific parameters can lead to large changes in delivered tidal volumes. Given the large parameter space required to explore how these patient-specific parameters affect tidal volumes, a computational model was needed to make the exploration of such a large parameter space feasible in a short period of time.

3.4   The Computational Model Solves the Large Parameter Space of Potential Patient Pairings

To properly account for the range of different factors influencing patient-delivered tidal volumes in a split-ventilator configuration, we performed the largest-to-date computational effort to simulate the necessary number of different ventilator settings and patient-specific parameters that clinicians may encounter. Figure 5 displays the results of exploring the seven-dimensional parameter space (Table 1) which were found to significantly affect predicted tidal volumes and pressures. 270 million different simulations were required to explore the parameter space at a sufficient resolution such that the step size for a given parameter resulted in less than a 5% change in tidal volume.

Figure 5 depicts the scale of the parameter sweep that was performed. The left panel of Fig. 5 shows how increasing the driving pressure (PIP-PEEP), as well as increasing the I:E and therefore the inspiratory time, acts to increase tidal volumes. The right panel of Fig. 5 is an expansion of the predicted tidal volumes of the upper left panel by selecting a PIP of 28, PEEP of 8, and I:E of 1 and illustrating tidal volumes as a function of endotracheal tube sizes, RR, and compliance. The interplay of multiple parameters is observed as the effect of changing RR on tidal volumes is itself dependent on the patient's pulmonary compliance and endotracheal tube size. To complete these simulations, 24,000 cores were used over a 72 h period using more then one million core hours in Microsoft Azure.

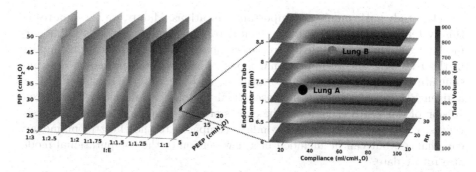

Fig. 5. Illustration of the results from computational model's parameter sweep Left: The average tidal volume without a resistor for all simulations as a function of PIP, PEEP, and I:E. Right: Average tidal volume without a resistor for all simulations originating from the black square on the upper left figure (PIP of 28, PEEP of 8, I:E of 1) as a function of RR, compliances, and endotracheal tube diameters.

4   Discussion and Conclusion

Shared ventilatory support still poses significant risk of harm and should not be undertaken unless there are no other viable options. However, the dire circumstances during the COVID-19 pandemic has brought renewed interest [15] and innovation [10,11] in ventilator sharing, which has applications in future respiratory outbreaks, the battlefield, and in low resource ICU settings, as well as during the current global pandemic. The proposed VSRS relies exclusively on simple 3D printed components that can be readily created in locations with a 3D printer and shipped to nearby hospitals, as well as a free mobile app that removes the guess work from deciding how to pair patients and what resistor size to use. In this work, we demonstrate the role that a computational fluid dynamics model can play in guiding system design and VSRS use.

The computational model was able to demonstrate the large difference in delivered airflow to two lungs of different compliances sharing the same ventilator without use of the VSRS. This confirms the feared clinical scenario that would lead to one patient potentially experiencing volutrauma and/or the other having inadequate ventilation. Using 3D printed airflow resistors in the circuit for the patient with the higher compliance lungs allows for control over the delivered tidal volume to the higher compliance lung. We showed that a pressure-controlled splitting using the VSRS can reduce the effects of one patient's ventilation on the other in a shared ventilator setting. Moreover, we confirmed that the airflow can be predicted using the computational model we have developed, which allows clinicians to select the resistor that will result in the desired tidal volumes for each patient, even for patients with very different ventilation needs and pulmonary compliances. By leveraging large-scale cloud resources, we were able to pre-compute results for 270 million potential clinical scenarios to be fed into a mobile app that has already been created to assist clinicians with this process, [14].

This study has a number of important limitations. Mainly, we have not tested this system on real patients during ventilator splitting. A patient's pulmonary mechanics might have subtle differences compared to the simplifications implicit in the test lung and computational model. For example, the model does not simulate the effect of alveolar recruitment, and therefore increasing PEEP at a given PIP never results in increased tidal volumes. Additionally, the current version of the VSRS excludes of the effects of differing patient pulmonary resistances, which we found to be secondary compared to the effect of pulmonary compliances, but can be included in a new version of the computational model at a future date.

The VSRS is designed to initially have the patient on a single ventilator in order to determine their patient-specific pulmonary compliance and, in the case of volume-controlled ventilation, resistance. Once the patient is stabilized, they then can be moved to a split ventilator configuration using the VSRS for longer term ventilatory support with their now known patient-specific pulmonary characteristics entered into the app in order to choose the proper resistor size. It will be important to redetermine the optimal resistor and ventilator settings as the patients' conditions can change rapidly with time, which can be accomplished done by temporarily occluding air flow to one of the patients for a few breaths to determine the pulmonary compliance for the other patient. The developed computational model can then be used to rapidly recalculate optimal resistor and ventilator settings.

However, the goal behind ventilator splitting is to address periods of short supply, especially given reports of prolonged ventilator support times for COVID-19 patients. Reserved for dire situations, ventilator splitting is complex, and introduces many safety concerns related to the lack of control over individual patient's respiratory support, some of which are alleviated by the VSRS. By precomputing the hundreds of millions of different possible combinations of ventilator settings and patient-specific characteristics and by taking advantage of simple 3D printable geometries, the VSRS can be rapidly deployed at minimal cost wherever the need for ventilators surpasses their supply.

Acknowledgements. The authors would like to thank the COVID-19 HPC Consortium for both the compute hours and the broader Microsoft team for all of the support. They would like to thank T. Milledge, C. Kneifel, V. Orlikowski, J. Dorff, and M. Newton for critical computing support.

References

1. Optimizing ventilator use during the COVID-19 pandemic. https://www.hhs.gov/sites/default/files/optimizing-ventilator-use-during-covid19-pandemic.pdf. Accessed 20 Apr 2020
2. Wilgis, J.: Strategies for providing mechanical ventilation in a mass casualty incident: distribution versus stockpiling. Respir. Care **53**(1), 96–103 (2008)
3. Truog, R.D., Mitchell, C., Daley, G.Q.: The toughest triage-allocating ventilators in a pandemic. N. Engl. J. Med. **382**(21), 1973–1975 (2020)

4. Guérin, C., Lévy, P.: Easier access to mechanical ventilation worldwide: an urgent need for low income countries, especially in face of the growing COVID-19 crisis (2020)
5. Hadar, S.S., et al.: Adaptive split ventilator system enables parallel ventilation, individual monitoring and ventilation pressures control for each lung simulators. medRxiv (2020)
6. Tronstad, C., et al.: Splitting one ventilator for multiple patients-a technical assessment (2020)
7. Beitler, J.R., et al.: Ventilator sharing during an acute shortage caused by the COVID-19 pandemic. Am. J. Respir. Crit. Care Med. **202**(4), 600–604 (2020)
8. Lai, B.K., Erian, J.L., Pew, S.H., Eckmann, M.S.: Emergency open-source three-dimensional printable ventilator circuit splitter and flow regulator during the COVID-19 pandemic. Anesthesiology **133**(1), 246–248 (2020)
9. Clarke, A., Stephens, A., Liao, S., Byrne, T., Gregory, S.: Coping with COVID-19: ventilator splitting with differential driving pressures using standard hospital equipment. Anaesthesia **75**(7), 872–880 (2020)
10. Srinivasan, S.S., et al.: A rapidly deployable individualized system for augmenting ventilator capacity. Sci. Transl. Med. **12**(549), eabb9401 (2020)
11. Clarke, A.: 3D printed circuit splitter and flow restriction devices for multiple patient lung ventilation using one anaesthesia workstation or ventilator. Anaesthesia **75**(6), 819–820 (2020)
12. Joint statement on multiple patients per ventilator. https://www.asahq.org/about-asa/newsroom/news-releases/2020/03/joint-statement-on-multiple-patients-per-ventilator. Accessed 25 Apr 2020
13. Schmidt, M., Foitzik, B., Hochmuth, O., Schmalisch, G.: Computer simulation of the measured respiratory impedance in newborn infants and the effect of the measurement equipment. Med. Eng. Phys. **20**(3), 220–228 (1998)
14. Kaplan, M., et al.: Cloud computing for COVID-19: lessons learned from massively parallel models of ventilator splitting. Comput. Sci. Eng. **22**(6), 37–47 (2020)
15. Cherry, A.D., Cappiello, J., Bishawi, M., Hollidge, M.G., MacLeod, D.B.: Shared ventilation: toward safer ventilator splitting in resource emergencies. Anesthesiology **133**(3), 681–683 (2020)

AI Classifications Applied to Neuropsychological Trials in Normal Individuals that Predict Progression to Cognitive Decline

Andrzej W. Przybyszewski[1,2](✉) ⓘ and the BIOCARD Study Team*

[1] Polish-Japanese Academy of Information Technology, 02-008 Warsaw, Poland
przy@pjwstk.edu.pl

[2] Department Neurology, University of Massachusetts Medical School, Worcester, MA 01655, USA
Andrzej.Przybyszewski@umassmed.edu

Abstract. The processes of neurodegeneration related to Alzheimer's disease (AD) begin several decades before the first symptoms. We have used granular computing rules (rough set theory) to classify cognitive data from BIOCARD study that have been started over 20 years ago with 354 normal subjects. Patients were evaluated every year by team of neuropsychologists and neurologists and classified as normal, with MCI (mild cognitive impairments), or with dementia. As the decision attribute we have used CDRSUM (Clinical Dementia Rating Sum of Boxes) as more quantitative measure than above classification. Based on 150 stable subjects with different stages of AD we have found rules (granules) that classify cognitive attributes with disease stages (CDRSUM). By applying these rules to normal (CDRSUM = 0) 21 subjects we have predicted that one subject might get mild dementia (CDRSUM > 4.5), one very mild dementia (CDRSUM > 2.25), and five other might get questionable impairment (CDRSUM > 0.75). AI methods can find, invisible for neuropsychologists, patterns in cognitive attributes of normal subjects that might indicate their pre-dementia stage.

Keywords: Granular computing · Rough set · Rules · Cognition

1 Introduction

As our population is aging, it causes that the prevalence of AD related dementia is fast increasing [1]. About 5.7 million Americans have actually AD, and the prevalence worldwide is estimated to be as high as 24 million. By 2050, AD number could potentially rise to 14 million in the US [1], and dementia (60–70% AD) to 139 millions worldwide (World Health Organization, 2021). Because AD biomarkers were identified in recent years, AD related changes might be found in the preclinical AD phase that opens possibilities of the new preventive methods developments.

Cognitive changes are dominant symptoms in the Alzheimer's disease (AD). In the most cases of AD neurodegeneration starts from the hippocampus and frontal cortex, and it related to memory and orientation problems. With the disease progression, other brain regions become also affected. There is no cure for AD, as during the diagnosis of the first clinical symptoms many parts of the brain are already dead.

As each patient has dissimilar neurodegeneration developments, their compensation (brain plasticity) and in the consequence symptoms might be various; finding partial optimal treatment is an art for an experienced neurologist.

The neurodegeneration developments that start several decades before first symptoms, and they were registered as changes in the Cerebral Spinal Fluid (CSF) t-tau. Whereas the cognitive tests had changepoints in about a decade before symptoms onset [2, 3]. As cognitive changes can be easy and in the noninvasive way measured online, in this project, we have predicted disease onset with sets of psychophysical attributes found as the most meaningful in patients from the BIOCARD study publications [4, 5]. In addition, we have combined them with results of the apolipoprotein E ApoE genotype [4]. Albert et al. [5] have successfully predicted conversion from normal to MCI (Mild Cognitive Impairment) due to AD, 5 years after baseline, for 224 subjects by using the following parameters: CSF β-amyloid and p-tau, MRI hippocampal and entorhinal cortex volumes, cognitive tests scores, and APOE genotype. However, their predictions were for the whole populations and with many different parameters [5], and ours are for each individual subject based on APOE genotype and only cognitive attributes.

This study is the continuation of the rough set theory application to follow predominantly the cognitive changes in the neurodegenerative diseases (ND) such as Parkinson's [6] and now in Alzheimer's diseases.

2 Methods

We have analyzed cognitive and APOE data of 150 subjects consist of: 40 normal subjects, 70 MCI (Mild Cognitive Impairment), and 40 subjects with dementias (AD). These data were basis of our general basic model (G Model) connecting cognitive attributes with different disease stages related to CDRSUM (Clinical Dementia Rating Sum of Boxes). We have also used 40 AD subjects from this group as another model for advanced patients - AD Model. We have tested using above two Models, on 21 of classified by clinicians as normal subjects (N Group), with the purpose to estimate their stages (CDRSUM) on similarities to our models.

In all subjects with recorded their age, had the following neuropsychological tests performed every year: Logical Memory Immediate (LOGMEM1A), Logical Memory Delayed (LOGMEM2A), Trail Making, Part A (TrailA - connecting time in sec of random placed numbers), Trail Making Part B (TrailB - connecting time in sec of random placed numbers and letters), Digit Symbol Substitution Test (DSST), Verbal Fluency Letter F (FCORR), Rey Figure Recall (REYRECAL), Paired Associate Immediate (PAIRED1), Paired Associate Delayed (PAIRED2), Boston Naming Test (BOSTON), and CVLT (California Verbal Learning Test). In addition, we have subjects' age (years), APOE genotype; individuals who are *ApoE-4* carriers vs. non-carriers (digitized as 1 vs. 0), and CDRSUM (sum of boxes) as precise and quantitative general index of the Clinical Dementia Rating [7]. There are the following CDRSUM values related to different stages of normal, pre-, and clinical confirmed AD patients: for prodromal patients are: (0.0) – normal; (0.5–4.0) – questionable cognitive impairment; (0.5–2.5) – questionable impairment; (3.0–4.0) – very mild dementia; (4.5–9.0) – mild dementia [7].

*Data used in preparation of this article were derived from BIOCARD study, supported by grant U19 - AG033655 from the National Institute on Aging. The BIOCARD study team did not participate in the analysis or writing of this report, however, they contributed to the design and implementation of the study. A listing of BIOCARD investigators can be found on the BIOCARD website (on the 'BIOCARD Data Access Procedures' page, 'Acknowledgement Agreement' document).

2.1 Rough Set Theory

Our data mining analysis follows rough set theory (RST) discovered by Prof. Zdzislaw Pawlak [8], whose solutions of the vague concept of boundaries were approximated by sharp sets of the upper and lower approximations [8]. It was demonstrated previously that RST gave the best results in the PD symptoms classifications in comparison to other methodologies [9]. Details of RST were described in the previous ICCS conference [10]. We have used Rough Set Exploration System RSES 2.2 as a toolset for analyzing data with rough set methods [11].

3 Results

3.1 Statistics

We have performed statistical analysis for all 15 attributes, and we found that 7 attributes had stat. sig. difference of means: FCORR, REYRECAL, PAIRED1, PAIRED2, BOSTON, CVLT, CDRSUM. It was found for different groups of subjects: normal (N), mixture of normal MCI, and AD (G Model), and AD (AD Model).

3.2 Rules from General Model (G Model)

We have placed G Model data in the following information table (Table 1):

Table 1. Part of the decision table for Model1 subjects

P#	age	Lgm1A	Lgm2A	TrailA	TrailB	DSST	Fcorr	Reyrcl	APOE	...	CDRSUM
67643	74	9	8	40	208	35	14	18	1	...	0.5
70407	88	8	5	66	150	21	21	10	0	...	4.5
102541	71	15	25	25	202	52	17	23.5	0	...	1
119156	92	7	34	34	386	40	20	10.5	0	...	3.5
139134	81	6	51	51	60	49	13	6	1	...	2.5
142376	76	18	54	54	50	19	14	12	0	...	0

The complete Table 1 has 150 rows, and 15 columns, there are shown the following condition attributes: P# - number given to each patients, age –age of subject, Lgm1A -Logical Memory Immediate, Lgm2A - Logical Memory Delayed, TrailA -Trail Making Part A, TrailB -Trail Making Part B, DSST - Digit Symbol Substitution Test, Fcorr -Verbal Fluency Letter F, TrailA and TrailB are growing from N to AD, DSST is decreasing from N to AD in a similar way as Fcorr (FCORR). Reyrcl - Rey Figure Recall, APOE

- *ApoE* genotype, ... CDRSUM -sum of boxes- index of the Clinical Dementia Rating. We have used RSES 2.2 for G Model group discretization with the global cuts (RSES 2.2) [13]. There were the following 3 ranges of the decision attribute CDRSUM: "(-Inf, 0.75)", "(0.75, 1.25)", "(1.25, Inf)". We had obtained 2581 rules using the exhaustive algorithm for G Model subjects. There are two rules below:

$$(FCORR="(-Inf,10.5)")\&(REYRECAL="(-Inf,15.75)")\&(APOE=1) => (CDRSUM = "(1.25,Inf)"[7]) \; 7$$

$$\text{(1)}$$

$$(LOGMEM1A="(16.0,20.5)")\&(BOSTON="(-Inf,26.5)")\&(age="(73.5,86.5)") => (CDRSUM="(0.75,1.25)"[5]) \; 5$$

$$\text{(2)}$$

We read above equations (Eq. 1) as following: it fulfils 7 cases that if FCORR is below 10.5 and REYRECAL is below 15.75 and APOE is 1 then CDRSUM is above 1.25 that means questionable impairment. Equation 2 is for CDRSUM between 0.77 and 1.25 and is based on not very good Boston naming (BOSTON) results.

By rules obtained from the G Model we have predicted the CDRSUM of each subject the N Group. There were 21 normal (with CDRSUM = 0) subjects.

Table 2. Confusion matrix for CDRSUM of N Group by rules obtained from G Model by local cuts [11].

Predicted				
Actual	"(-Inf, 0.75)	"(1.25, Inf)"	"(0.75, 1.25)"	ACC
"(-Inf, 0.75)"	17.0	2.0	2.0	0.81
"(1.25, Inf)"	0.0	0.0	0.0	0.0
"(0.75, 1.25)"	0.0	0.0	0.0	0.0
TPR	1.0	0.0	0.0	

TPR: True positive rates for decision classes; ACC: Accuracy for decision classes: the global coverage was 1.0 and the global accuracy was 0.81, the coverage for decision classes was 1.0, 0.0, 0.0.

We were interested in those normal subjects who had predicted values of the CDR-SUM > 0. It Table 2 states that 17 are normal, there were two subjects with predicted values of CDRSUM = (0.75, 1.25), and two others with CDRSUM > 1.25. All four might have cognitive impairments.

In Table 3 we have also used RSES 2.2 for G Model group discretization by the *global cuts* [11]. There were the following 3 ranges of the decision attribute CDRSUM: "(-Inf, 0.75)", "(0.75, 2.25)", "(2.25, Inf)". We have obtained 324 rules with the genetic algorithm for G Model subjects.

Table 3. Confusion matrix for CDRSUM of N Group by rules obtained from G Model by the *global cuts* [11]. Predicted.

Actual	"(-Inf, 0.75)"	"(2.25, Inf)"	"(0.75, 2.25)"	ACC
"(-Inf, 0.75)"	15.0	2.0	4.0	0.71
"(2.25, Inf)"	0.0	0.0	0.0	0.0
"(0.75, 2.25)"	0.0	0.0	0.0	0.0
TPR	1.0	0.0	0.0	

TPR: True positive rates for decision classes; ACC: Accuracy for decision classes: the global coverage was 1.0 and the global accuracy was 0.714, the coverage for decision classes was 1.0, 0.0, 0.0.

We were interested in those normal subjects who had predicted values of the CDR-SUM > 0. From Table 3 there were four subjects with CDRSUM = (0.75, 2.25) that with values between (0.5–2.5) might have a questionable impairment [9], and two subjects with CDRSUM = (2.25, Inf)): *401297* and *164087* that means that they might have a very mild dementia or mild dementia [7] as below in Eqs. 3 and 4.

(Pat=401297)& (LOGMEM1A= "(-Inf,15.5)"))&(LOGMEM2A = "(-Inf,16.5)")&
(TRAILA= "(-Inf,23.5)")& (TRAILB= "(-Inf,74.5)")&(FCORR="(-Inf,16.5)")&
(REYRECAL= "(-Inf,15.75)")&(PAIRD2= "(6.5,Inf)")&(age= "(-Inf,76.5)") &
(APOE=1) =>(CDRSUM= "(2.25,Inf)"

$$(3)$$

The first patient *(Pat = 401297)* as states in Eq. 3 has the low *FCORR* (below 16.5) and *REYRECAL* (below 15.75) values, as well as bad APOE genotype that mainly caused his CDRSUM above 2.25. That might suggest very mild dementia.

By using rules from the AD Model group, we have also found that *(Pat = 164087)* has even larger CDRSUM that is related to the execution function timing (long *TrailB*), and the low *FCORR* and *PAIRD2* values (Eq. 4).

(Pat=164087)&(LOGMEM1A="(14.5,15.5)")&LOGMEM2A="(7.0,Inf)")&(TrailB
"(74.5,153.0)")&(FCOR="(-Inf,12.5)")&(REYRECAL="(-Inf,21.5)")&(PAIRD2=
"(-Inf,6.5)")&((BOSTON="(25.5,Inf)=>(CDRSUM="("(4.5,7.0)"

$$(4)$$

As CDRSUM of *(Pat = 164087)* was predicted above as to be higher than 4.5 it means that this patient's cognitive results suggested that he might have mild dementia.

4 Discussion

Alzheimer's disease has long prodromal phase, with neurodegeneration beginning decades before symptoms onset (first clinical manifestation). This creates a challenge to the development of therapeutics since it is much more difficult to reverse the disease process and recover normal neuronal function without the ability to detect changes earlier.

Brain plasticity may partially explain why individuals can have no or minimal symptoms despite several decades of extensive neurodegeneration. During this long period, individual compensatory processes may develop differently between subjects. In this study, we aim to detect the beginning of compensatory changes reflective of underlying neurodegeneration in those developing dementia. We have developed novel tool to more easily and accuracy monitor ongoing progression by looking into patterns of cognitive attributes' values and comparing them with our Models (general and AD).

We have applied rough set theory and its rules as the granular computing to estimate a possible disease progression in normal subjects from the BIOCARD study. We used the intelligent granular computing with the rough set rules to investigates tests results set as granules for individual patients. To estimate their properties, we need to have a Model that has the meaning and tells us what the importance of the pattern (granule) is. In fact, our granules are complex (c-granules) as they are changing their properties with time of the neurodegeneration development till become like granules of the patients with dementia or PD [12]. In this work, we have limited our test to the static granules (in one time moment) and we have tried to estimate what is the meaning of a particular, individual granule. We have used two models: G Model (general model) have granules related to normal subjects, MCI and AD patients. On its basis we have obtained a large set of rules that have represented subjects' different stages of the disease from the normal to dementia. We have tested several of such models mostly changing normal subjects and getting different rules, which we have applied to other normal subjects and estimated what 'normal' means. Also, rules can be created with different granularity and algorithms that might give different classifications.

Therefore, we were looking for classifications that are universal e.g., they give similar results with different sets of rules. G Model has given us rules that are subtle and determine the beginning of possible symptoms. In the next step, we have used a model based on the more advanced patients in the progression of the disease– AD Model that gave rules based on AD patients. We got higher values of the CRDSUM that gave us only classifications of the possibly subjects with the mild dementia. Looking into different rules, some of them is easy to interpret, but other patients' granules look relatively normal. As our rules are applied to different subjects there are not certain, and we have confirmed our classifications by using different set of rules with different granularity and algorithms that may give different consistent or inconsistent classifications. Therefore, they are only indications for the clinician to test certain patients more carefully as they might already have some unnoticed dementia related symptoms.

5 Conclusions

Our main assumption was to have a universal *dementia related Model* that represents expertise of the clinical doctors: neurologists and neuropsychologists. We have used the supervised learning to get granules that connect patterns of 13 cognitive tests with the clinical symptoms measured as the CDRSUM (quantitative measure related to different dementia stages [7]). In our population there are 42 patients with dementia (two of them did not have cognitive tests), therefore, in our Model we have used 40 AD, 40 normal subjects, and we found 70 MCI that have consistent symptom.

We have concentrated on the predictions of the conversion from normal to prodromal AD of the individual subjects in contrast to the population of patients as in many of the studies e.g., [2–5]. We have applied rules form our Model to the cognitive test results of each patient with the purpose to find similarities indicating dementia. We have obtained some consistent results, but the core of our model (AD patients) is relatively small (40 patients) that does not give power (number of rules) to cover many individual cases, and therefore gives us, in part inconsistent classifications. However, classifying individual subjects for the prodromal stage of AD seems encouraging.

References

1. Alzheimer's Association 2018 Alzheimer's disease facts and figures. Alzheimer's & Dementia **14,** 367–429 (2018)
2. Sperling, R.A., Aisen, P.S., Beckett, L.A., Bennett, D.A., et al.: Toward defining the pre-clinical stages of Alzheimer's disease: recommendations from the National Institute on Aging-Alzheimer's Association workgroups on diagnostic guidelines for Alzheimer's disease. Alzheimers Dement. **7,** 280–292 (2011)
3. Younes, L., Albert, M., Moghekar, A., et al.: Identifying changepoints in biomarkers during the preclinical phase of Alzheimer's disease. Front. Aging Neurosci. **11,** 74 (2019)
4. Albert, M., Soldan, A., Gottesman, R., et al.: The BIOCARD research team, cognitive changes preceding clinical symptom onset of mild cognitive impairment and relationship to *ApoE* genotype. Curr. Alzheimer Res. **11**(8), 773–784 (2014)
5. Albert, M., Zhu,Y., Moghekar, et al.: Predicting progression from normal cognition to mild cognitive impairment for individuals at 5years Brain. **141**(3), 877–887 (2018)
6. Przybyszewski, A.W., Nowacki, J.P., Drabik, A., Szlufik, S., Koziorowski, D.M.: Concept of Parkinson leading to understanding mechanisms of the disease. In: Nguyen, N.T., Iliadis, L., Maglogiannis, I., Trawiński, B. (eds.) ICCCI 2021. LNCS (LNAI), vol. 12876, pp. 456–466. Springer, Cham (2021). https://doi.org/10.1007/978-3-030-88081-1_34
7. O'Bryant, S.E., Waring, S.C., Cullum, C.M., et al.: Staging dementia using clinical dementia rating scale sum of boxes scores: a texas Alzheimer's research consortium study. Arch Neurol. **65**(8), 1091–1095 (2008)
8. Pawlak, Z.: Rough Sets: Theoretical Aspects of Reasoning About Data. Kluwer, Dordrecht (1991)
9. Przybyszewski, A.W., Kon, M., Szlufik, S., Szymanski, A., Koziorowski, D.M.: Multimodal learning and intelligent prediction of symptom development in individual parkinson's patients. Sensors **16**(9), 1498 (2016). https://doi.org/10.3390/s16091498
10. Przybyszewski, A.W.: Theory of mind helps to predict neurodegenerative processes in Parkinson's disease. In: Paszynski, M., Kranzlmüller, D., Krzhizhanovskaya, V.V., Dongarra, J.J., Sloot, P.M.A. (eds.) ICCS 2021. LNCS, vol. 12744, pp. 542–555. Springer, Cham (2021). https://doi.org/10.1007/978-3-030-77967-2_45
11. Bazan, J.G., Szczuka, M.: RSES and RSESlib - a collection of tools for rough set computations. In: Ziarko, W., Yao, Y. (eds.) RSCTC 2000. LNCS (LNAI), vol. 2005, pp. 106–113. Springer, Heidelberg (2001). https://doi.org/10.1007/3-540-45554-X_12
12. Przybyszewski, A.W.: Parkinson's disease development prediction by c-granule computing. In: Nguyen, N.T., Chbeir, R., Exposito, E., Aniorté, P., Trawiński, B. (eds.) ICCCI 2019. LNCS (LNAI), vol. 11683, pp. 296–306. Springer, Cham (2019). https://doi.org/10.1007/978-3-030-28377-3_24

Super-Resolution Convolutional Network for Image Quality Enhancement in Remote Photoplethysmography Based Heart Rate Estimation

K. Smera Premkumar[1], A. Angelopoulou[2], E. Kapetanios[3], T. Chaussalet[2], and D. Jude Hemanth[1(✉)]

[1] Department of ECE, Karunya Institute of Technology and Sciences, Coimbatore, India
judehemanth@karunya.edu
[2] School of Computer Science and Engineering, University of Westminster, London, UK
[3] School of Physics, Engineering and Computer Science, University of Hertfordshire, Hatfield, UK

Abstract. Heart rate (HR) is one of the important vital parameters of the human body and understanding this vital sign provides key insights into human wellness. Imaging photoplethysmography (iPPG) allows HR detection from video recordings and its unbeatable compliance over the state of art methods has made much attention among researchers. Since it is a camera-based technique, measurement accuracy depends on the quality of input images. In this paper, we present a pipeline for efficient measurement of HR that includes a learning-based super-resolution preprocessing step. This preprocessing image enhancement step has shown promising results on low-resolution input images and works better on iPPG algorithms. The experimental results verified the reliability of this method.

Keywords: Remote Photoplethysmography · Image enhancement · Heart Rate (HR) Detection · Convolutional neural network

1 Introduction

Heart Rate Measurement (HRM) is a crucial physiological regulator of a person's total cardiac output. The far-flung heart rate tracking devices have led to substantial interest in Heart Rate (HR) as a prospective clinical diagnosis tool. The gold standard to analyze cardiac measurements is an Electrocardiogram (ECG) which measures the electrical activity of the heart through attached sensors (called electrodes) to the skin. Another favoured optical-based technique is Photoplethysmography (PPG) which detects the changes in blood volume pulse (BVP) via contact sensors attached to anatomical locations such as wrists, fingers, and toes. Commercial wearable devices such as fitness trackers, and smart watches make use of this principle where the sensor emits light to the skin and measures the reflected intensity due to optical absorption of blood.

These state-of-art methods of measuring cardiac activities either need physical contact in clinical settings or a sensor attached to the body and it is not suitable for long time monitoring as they can cause discomfort, especially in neonates, and elderly care. In recent years, Imaging Photoplethysmography (iPPG) or Remote Photoplethysmography (rPPG) has been a prominent topic among researchers that measures HR from face images by tracking volumetric variations of blood circulation, while it is invisible to the human eye. Even though it is a progressive method of PPG technology, the iPPG method does not require any kind of physical contact. This relies on the signals obtained from the video streams that can predict not only heart rate but also other vital information of the body like heart rate variability, blood pressure etc. and thereby infers mental stress, variations in cardiovascular functions, quality of sleep, drowsiness.

Recently, Covid 19 Pandemic made a huge transition in healthcare from in-person care to telehealth. This practical application of the PPG principle can measure the vital parameters well-nigh possible from any mobile camera/webcam. It provides a static, contactless health monitoring of patients and no longer needs any clinical settings. Although appreciable progress has been made in rPPG methods, still a few challenges remain open such as motion, skin tone, compression etc. Since rPPG is a camera-based technique Compression artifact is an inevitable challenge, especially in real-time scenarios such as telehealth. Remote measurement of HR relies on very fine details of the input video and a small artifact can make a huge impact on the accuracy of results especially when it comes to processing low-resolution images/videos. This paper aims at a fast super-resolution network (FSRCNN) for the low-resolution video input which enhances the input frames that ensure vital sign measurement more efficiently. We present a pipeline for iPPG measurement with a preprocessing super-resolution step which provides an improved signal to noise ratio (SNR) and combat the effects of temporal compression of video to an extent.

2 Related Works

Photoplethysmography (PPG) is a non-invasive, optical technique that is used to detect volumetric changes of blood in the microvascular bed of tissue [1]. The possibility of non-contact physiological computation using a thermal camera has been introduced in [2] and [3] demonstrated that plethysmography signals could be measured from the human face from simple consumer-level camera recordings with ambient light conditions. Since then, a substantial number of researches have been conducted in remote Photoplethysmography.An ICA based algorithm has been explained in [4] as an optimal combination of raw signals in which the raw signals is separated into independent non-Gaussian channels. In this method, authors arbitrate that the second component produced after the ICA is considered as a periodic one and used for further processing.

With the emergence of the deep learning end to end method, extensive opportunities are opening up for performing tasks more efficiently in a better way. Chen et.al [5] was introduced the first end to end Learning Model 'DeepPhys' which is based on a Convolutional Attention Network (CAN) and enables spatiotemporal visualization of signals. This paper proposed a skin reflection model that is exceptionally robust in different illumination conditions. Subsequently, many learning-based methods have been proposed in the literature for HR estimation which includes illustrated ETA-rPPGNet illustrated by [6], meta phys model demonstrated by [7] and the neural architecture, AutoHR proposed [8].

Although appreciable progress has been made in rPPG methods in the last few years, still a few challenges remain open such as motion robustness, illumination, and skin tone compression artifacts etc. [9]. It should be noted that the compression artifacts are also an inevitable challenge in rPPG environment, but only a few approaches are there in literature to succeed in dealing with compression artifacts. Most video datasets are captured as raw images to avoid lossy compression and it demands immense storage. But it is a major challenge in real-time scenarios such as telemedicine, as it requires large memory requirements. Since rPPG measurement leverages very fine details of input video, compression makes a huge impact on the accuracy and robustness of physiological signal measurement. A significant research gap can be seen in literature to incorporate compression of low-resolution input videos into consideration. McDuff et al. [10] explained the impacts of video compression and Song et al. [11] have shown the resolution of the input videos affects the quality of output measurement.

3 Methods

We present a preprocessing step that can improve the accuracy of the output physiological signal even if the input image is with low resolution. The motivation behind this work is the great success of learning based super-resolution networks on face images/videos. The process of enhancing a low resolution (LR) image is called super-resolution (SR) while an interpolation method takes the weighted average of the neighboring pixels. A neural network can hallucinate details based on some prior information it collects from a large set of images and the details are then added to the image to create an SR image. An enhanced, sharper image can lead to reliable skin segmentation and thereby missing colour signal information can be extracted more efficiently.

3.1 Super Resolution Preprocessing

We use a Fast SRCNN proposed by C.Dong et al. [12] as the preprocessing enhancement network. We compared the Fast SRCNN with interpolation methods and normal SR convolutional network to check the computational complexity and speed. The main advantage of Fast SRCNN is that it directly accepts a low-resolution input and no longer needs a bicubic interpolation step. If we are resizing or moving an image, the new locations we have got will not necessarily match up with the previous ones. Bilinear interprets the missing values by using linear interpolation between the values whereas a bicubic interpolation takes 16 pixels (4 × 4) into the account. We have tried a super-resolution network proposed by c.Dong [13] for the preprocessing step. This method uses a bicubic interpolation preprocessing step to up sample the LR image and then a convolution operation is performed to improve image quality.

We found that this method has high computational complexity and low speed. To overcome this, we have tried FSRCNN, the advanced version of this method. and it works better in our work. Fast SRCNN capable of producing a super scaled image directly from the LR image. As shown in Fig. 2, the FSRCNN has four steps before a deconvolution process which include feature extraction, shrinking, non-linear mapping and expanding process. The feature extraction process replaces bicubic interpolation with 5×5 convolutions and a feature map reduction has happened in shrinking. Then 3×3 multiple layers are applied followed by a 1×1 convolution expansion. It is used to reduce the number of parameters and it could help to speed up the network. PReLU is used as an activation function. The network was pretrained and this network shows faster performance on standard image datasets for image SR Then using a 9×9 filter, a high-resolution image is reconstructed. This network shows higher image quality as well as a shorter run time when compared to interpolation and SRCNN. The network architecture is illustrated in Fig. 1.

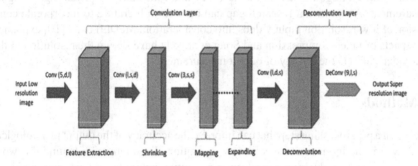

Fig. 1. Architecture of fast super-resolution convolutional neural network

We can represent an FSRCNN network as FSRCNN (d, s, m). The computational complexity can be calculated as

$$O\left\{\left(25d + sd + 9ms^2 + ds + 81d\right)S_{LR}\right\} = O\{(9ms2 + 2sd + 106d)S_{LR}\}$$

We exclude the parameters of PReLU, which introduce a negligible computational cost. To compare the cost function, we use the mean square error (MSE). The optimization objective is represented as

$$\min \theta \frac{1}{n} \sum_{i=1}^{n} \left\| F\left(Y_s^i; \theta\right) - Xi_2^2 \right\|,$$

where Y_s^i and Xi is the ith LR and HR sub-image pair in the training data and $F\left(Y_s^i; \theta\right)$ is the network output for Y_s^i with parameters θ. Using a standard backpropagation, parameters are optimized using stochastic gradient descent.

To compare the efficiency of FSRCNN, the LR images are first up sampled using bicubic interpolation, SRCNN and then FSRCNN. From our experiments, it is evident that FSRCNN uses small filter sizes and a deeper network, and it has better PSNR (image quality) and less computational complexity compared to the other two methods. Because of its efficiency in training and testing over upscaling factors, it could be beneficial in real-time scenarios on generic CPU.

4   Results and Discussions

The results of PSNR (dB) and test time have been verified on two publicly available datasets, MAHNOB-HCI and VicarVision. We use a pre-trained SRCNN network and used two public benchmark datasets for testing. The inference time is tested with the C++ implementation on an intel core i5 processor. The qualitative results based on image quality (PSNR) and test time are listed in Table 1.

Table 1. Test results

Test Data Set	Scaling Factor	Bicubic Interpolation PSNR/Time	SRCNN PSNR/Time(sec on CPU)	SRCNN PSNR /Time (sec on CPU)
MAHNOB-HCI	2	30.14/ -	32.35/1.3	33.25/.098
	3	26.45/ -	29.26/1.3	29.43/ .074
VICARVISION	2	29.81/ -	31.53/1.8	31.82/0.088
	3	26.41/ -	28.47/1.7	30.43/0.097

We use a pre-trained model for an upscaling factor in advance. During testing convolution operations are performed once and then, the image is up sampled to different sizes and the corresponding deconvolution layer is processed.

Fig. 2. Visualization of enhanced images

From the results, it is evident that the noticeable improvement in image quality achieves high running speed over interpolation and SRCNN methods. An example of visualization of image enhancement using interpolation and SRCNN can see in Fig. 2. It has fast training and testing speed across different upscaling factors and this model can be adapted for real-time video SR.

4.1 HR Estimation

To test the reliability of this SR network in iPPG environment, we used two previously published iPPG methods to calculate the HR. To achieve this, the resulting pixels from the SRCNN network were spatially averaged for each frame to obtain the test signals for HR measurement. We used ICA method explained in [4] and the POS method for

PPG signal recovery. Spatially averaged colour signals were normalized and multiplied by the projection matrix P. A Butterworth filter is then applied to the model output. To extract the heart rate from the corresponding pulse signal, we used FFT, and HR was chosen from the peak with the greatest power. The HR was calculated using the average interbeat signal interval in seconds for a 30 s window.

Table 2. The effectiveness of the FSRCNN method on iPPG algorithms in terms of PSNR and RMSE

Method	ICA		POS	
	RMSE	SNR	RMSE	SNR
Bicubic	5.79	0.067	4.12	-0.043
SRCNN	5.21	0.012	5.12	-0.028
FSRCNN	4.74	0.027	4.31	-0.015

We perform HR estimation 30 s windows for each video. To check the quality, We used performance metric root mean square error (RMSE),

$$RMSE = \sqrt{\frac{\sum_{i=1}^{N}\left|HR_i - \widehat{HR_i}\right|^2}{N}}$$

where N is the total number of observation windows. HR_i is the ith measurement and $\widehat{HR_i}$ is the corresponding prediction. We evaluate the heart rate estimate using performance metrics RMSE and the blood volume pulse SNR.

The SRCNN network was trained using a dataset that is pre-trained using a different dataset and the evaluation is conducted using two publicly available datasets – VicarVision and MAHNOB-HCI. We compared the performance of bicubic interpolation, SRCNN and FSRCNN methods. We used 30 s of each video to measure the Heart Rate. We have chosen 0.6 s time interval. Table 2 shows the results using RMSE and SNR for each method. The FSRCNN method outperforms the bicubic interpolation method. The FSRCNN has reduced error compared to the bicubic interpolation method. The interpolation and FSRCNN method have reduced RMSE when compared to the low-resolution frames. Despite its reduced error, the FSRCNN network has less computational complexity and high speed compared to other interpolation methods.

5 Conclusion

In this paper, we present a pipeline with a Fast Super-resolution CNN based image enhancement method as preprocessing step for HR measurement from face videos. From the results, the proposed method has more accuracy than other interpolation methods.

We have observed that this network can provide better results for low resolution and compressed input frames. This method can help to combine the advantage of super-resolution with the iPPG measurements. In our future work, we would like to investigate the possibilities of advancements of remote methods using neural models to alleviate the challenges in rPPG measurement.

References

1. Allen, J.: Photoplethysmography and its application in clinical physiological measurement. Physiol. Meas. **28**(3) (2007). https://doi.org/10.1088/0967-3334/28/3/R01
2. Pavlidis, I., Dowdall, J., Sun, N., Puri, C., Fei, J., Garbey, M.: Interacting with human physiology. Comput. Vis. Image Underst. (2007). https://doi.org/10.1016/j.cviu.2006.11.018
3. Verkruysse, W., Svaasand, L.O., Nelson, J.S.: Remote plethysmographic imaging using ambient light. Opt. Express **16**(26), 21 434–21 445 (2008)
4. Poh, M.Z., McDuff, D.J., Picard, R.W.: Non-contact, automated cardiac pulse measurements using video imaging and blind source separation. Opt. Express **18**(10), 10762–10774 (2010). https://doi.org/10.1364/OE.18.010762
5. Chen, W., McDuff, D.: DeepPhys: video-based physiological measurement using convolutional attention networks. In: Ferrari, V., Hebert, M., Sminchisescu, C., Weiss, Y. (eds.) ECCV 2018. LNCS, vol. 11206, pp. 356–373. Springer, Cham (2018). https://doi.org/10.1007/978-3-030-01216-8_22
6. Hu, M., Qian, F., Guo, D., Wang, X., He, L., Ren, F.: ETA-rPPGNet: effective time-domain attention network for remote heart rate measurement. IEEE Trans. Instrum. Meas. **70** (2021). https://doi.org/10.1109/TIM.2021.3058983
7. Liu, X., Jiang, Z., Fromm, J., Xu, X., Patel, S., McDuff, D.: MetaPhys: few-shot adaptation for non-contact physiological measurement. In: ACM CHIL 2021–Proceedings of the 2021 ACM Conference on Health, Inference, and Learning, vol. 1, Issue 1. Association for Computing Machinery (2021). https://doi.org/10.1145/3450439.3451870
8. Yu, Z., Li, X., Niu, X., Shi, J., Zhao, G.: AutoHR: a strong end-to-end baseline for remote heart rate measurement with neural searching. IEEE Signal Process. Lett. **27**, 1245–1249 (2020). https://doi.org/10.1109/LSP.2020.3007086
9. Hoffman, W.F.C., Lakens, D.: Addressing reproducibility issues in remote Photoplethysmography (rPPG) research: an investigation of current challenges and release of a public algorithm benchmarking dataset, 25 June 2021. https://doi.org/10.17605/OSF.IO/XJF7U
10. McDuff, D.J., Blackford, E.B., Estepp, J.R.: The impact of video compression on remote cardiac pulse measurement using imaging Photoplethysmography. In: 2017 12th IEEE International Conference on Automatic Face & Gesture Recognition (FG 2017), pp. 63–70 (2017). https://doi.org/10.1109/FG.2017.17
11. Song, R., Zhang, S., Cheng, J., Li, C., Chen, X.: New insights on super-high resolution for video-based heart rate estimation with a semi-blind source separation method. Comput. Biol. Med. **116**(Complete) (2020). https://doi.org/10.1016/j.compbiomed.2019.103535
12. Dong, C., Loy, C.C., He, K., Tang, X.: Image super-resolution using deep convolutional networks. IEEE Trans. Pattern Anal. Mach. Intell. **38**(2), 295–307 (2016). https://doi.org/10.1109/TPAMI.2015.2439281
13. Dong, C., Loy, C.C., Tang, X.: Accelerating the super-resolution convolutional neural network. In: Leibe, B., Matas, J., Sebe, N., Welling, M. (eds.) ECCV 2016. LNCS, vol. 9906, pp. 391–407. Springer, Cham (2016). https://doi.org/10.1007/978-3-319-46475-6_25

A Hybrid Modeling Framework for City-Scale Dynamics of Multi-strain Influenza Epidemics

Vasiliy Leonenko(✉) (iD)

ITMO University, 49 Kronverksky Pr., St. Petersburg 197101, Russia
vnleonenko@yandex.ru

Abstract. In the current paper we present a hybrid modeling framework which allows to simulate co-circulation of influenza strains in urban settings. It comprises a detailed agent–based model coupled with SEIR-type compartmental model. While the former makes it possible to simulate the initial phase of an outbreak when the heterogeneity of the contact network is crucial, the latter approximates the disease dynamics after the occurrence of mass infection thus dramatically increasing the framework performance. The numerical experiments with the model are presented and their results are discussed.

Keywords: Python · Influenza · Co-circulation · Agent-based models · Compartmental models

1 Introduction

Outbreaks of influenza, one of the oldest and the most widely spread human infectious diseases, result in 3 to 5 million cases of severe illness annually worldwide, and the mortality rate is from 250 to 640 thousand individuals per year [15]. In addition to induced mortality, influenza causes an increase of heart attacks and strokes [6], as well as other disease complications. To enhance the capabilities of influenza surveillance and, as a consequence, to find means of restraining influenza epidemics and reducing the mortality attributed to influenza complications, the healthcare organs widely use statistical and mechanistic models. Among the factors of influenza dynamics, that are considered influential and thus should be included into the models, are contact patterns in the population [1,20,22], [30], and the immunity levels to various influenza strains [2,12,17,27]. The latter is connected with the former, as the heterogeneity of networks of disease transmission might cause uneven distribution of the infected and consequently the immune people, leading to intricate prevalence dynamics and the inability of simple models to predict it. As an example, in a deterministic SEIR

This research was supported by The Russian Science Foundation, Agreement #20-71-00142.

model it is assumed that the population immunity level directly defines the outbreak incidence dynamics and ultimately the outbreak size. At the same time, is it known that in real life the infection prevalence dynamics is very dependent on the stochastic effects inherent to the initial stages of the epidemic onset and the contact network clustering [5,9].

The modeling technique which makes it possible to account for the influence of contact network heterogeneity on the disease transmission is multi-agent modeling. There is a number of known publications on the topic, including such articles as [21], where multi-agent modeling of vaccination scenarios in heterogeneous populations based on social network incidence data. Another examples include a multi-component stochastic model that reproduces the dynamics of influenza in certain regions of England and Wales for 14 years [2] and the works of research teams that use a multi-agent approach to predict the dynamics of influenza based on synthetic populations—these are the teams of the University Pittsburgh ([19,20,28]), RTI International [7,8], and Wake Forest University School of Medicine [11]. The same concept was recently applied to COVID-19 modeling, with the examples such as COVASIM [16,18]. The author of this article employed an agent-based model to replicate the 2010–2011 outbreak in St Petersburg by means of the synthetic population of this city [22,24] and analyzed the co-circulation of several influenza strains in the same population depending on the initial immunity levels [26].

One of the main drawbacks of the multi-agent approach which seriously limits its application is related to excessive demand of computational power to handle the experiments with the model. Even in the simplest case, when the aim is to calculate the disease trajectory for one outbreak at the city scale with a predefined set of parameter values, several simulation runs are required to address stochastic uncertainty. As a result, the experiment may last from hours to days, depending on the model employed and the computational resources available. Obviously, in these circumstances the tasks which require many repetitive launches with different parameter values, such as model calibration to data or uncertainty/sensitivity analysis, may not be performed in reasonable time. There exist different methods to overcome this obstacle, namely, those related to preliminary data modification (for example, using a representative sample of individuals rather than the whole population in the simulations), to algorithm optimization and parallel computing (particularly, GPGPU-compatible framework implementations), and to simplification of some of the processes within the dynamics of the regarded system (for instance, by training neural networks on the output of multi-agent models to replicate disease incidence and prevalence trajectories without actual simulation). Among the last group of approaches hybrid modeling of disease dynamics can be named [4,13,14]. The mentioned method is based on the idea that in some sets of conditions the difference in the outputs of the detailed multi-agent models and much simpler compartmental ones may be negligible [29] which makes it possible to locally replace the former approach by the latter without dramatic loss of disease dynamic reproduction accuracy. With the mentioned benefits come the drawbacks. Particularly, the necessity of using two different modeling techniques for describing a single infection process instance raises a question of compatibility of those techniques. How smooth the transitions between the two models are and what is the influence of switching condition on the regarded disease dynamics? We try to address these questions in the current study, using a created hybrid modeling framework which allows to replicate artificial outbreaks caused by co-circulation of influenza strains in a synthetic population. To the author's knowledge, it is the first attempt of hybrid simulation of virus co-circulation. The presented study is a part of the ongoing research, the ultimate aim of which consists in quantifying the interplay between the immunity formation dynamics and the circulation of influenza strains in Russian cities. The study results are also applicable to modeling the circulation of arbitrary acute respiratory infections, particularly, COVID-19.

2   Methods

2.1   The Multi-agent Model

Overall Description. The original model used as a base for the framework is an agent-based model of co-circulation of different influenza strains in a synthetic population which is described in detail in [26]. It has discrete time with the modeling step equal to one day. The epidemic process is initiated by assigning randomly an infectious status to some individuals in the synthetic population at the beginning of the simulation. The model output includes generation of spatial distributions of the incidence cases via independent simulation runs, calculation of cumulative incidence and prevalence in the area under study and assessing the levels of herd immunity in the population after the outbreak. It is possible to collect additional data, such as places of infection (school, workplace, home and its immediate vicinity) for each incidence case, which allows to assess the contribution of contacts in each type of place to the spread of infection.

Population. The population-related parameters used for the model are organized in a form of a synthetic population of St Petersburg for the year 2010. The population includes 40213 households with the cumulative number of dwellers being 4,865,118 individuals. The residential buildings are regarded as a bunch of separate dwellings, and the individual can contact only with the people they share a dwelling with (e.g., with family members). Following the statistics of the governmental service "Open data of Saint Petersburg" [10], the average number of people per dwelling as 2.57, which is used as a mean for the generation of number of dwellers in each household (Poisson distribution is employed). We assumed that all the young people aged 7 to 17 attend schools, and the adults of working age (18 to 55 for males and 18 to 60 for females) can work. The workplaces are split into small compartments within which the daily contacts occur. The average workplace size was chosen equal to the average daily number of workplace contacts. We generate the workplace compartment sizes using the Poisson distribution with the corresponding mean. We consider that workplaces for the adults and schools for the school–age children are selected randomly from the available positions within a certain radius from the household (based on general knowledge, we took 15 km and 5 km correspondingly, which seems adequate for St Petersburg). If there are no vacancies in schools/vacant workplaces within this radius, the closest vacancy is assigned disregarding the distance. The remained individuals without schools are assigned to closest schools in disregard of the school capacity, while the remained individuals without workplaces are considered jobless.

Contacts. We assume that there exist the following patterns of a daily activity depending on the individual:

- stay in the household with a fixed id during the whole day (pre–school children, retired, unemployed)
- go to the school with a fixed id (students)
- go to the workplace with the particular id (working adults)

Hence, each day one individual has 1 to 2 places of potential contacts which are not changed over time. The contact numbers were derived from the data used by the author in the compartmental influenza model for St. Petersburg [23]. These data were calculated from the contact matrices for Russian cities [1]. We assumed that, in average, the dwellers of St. Petersburg have 1.57 contacts within the household and 8.5 outside it (at school or at work, which are mutually exclusive). Taking into account the differences in activity patterns of people (some of them do not work or study), the average calculated number of daily contacts in a model is around 6.51, which is close to the average number introduced in [23] (6.528). The role of public transport in spreading the infection is not considered. In the current version of the framework, the weekends are not regarded separately, i.e. the behavior of the individuals is the same during all the days.

Disease Onset and Recovery. The rate of effective contacts in a particular activity location (that is, the contacts between a susceptible and an infected individual which result in new infection cases) depends on the average number of contacts per person per day and the infection transmission coefficient, which are parameters of the model. We take a simplifying assumption that the infection transmission coefficients are not dependent on the strain. If various strains are instantaneously transmitted to an individual at the place of contact, one of them is selected at random as the one causing the infection. Each agent in the population potentially interacts with other agents if they attend the same school (for schoolchildren), workplace (for working age adults), or lives in the same household.

The infectivity of each individual depends on their day of infection. The fraction of infectious individuals in the group of individuals infected τ days before the current moment t is defined by a piecewise constant function g_τ which reflects the change of individual infectiousness over time from the moment of acquiring influenza. It is assumed that there exists some moment $\bar{t}$: $\forall t \geq \bar{t}\ g_\tau = 0$, which corresponds to the moment of recovery. The values of $g(\tau)$ were set according to [3], with τ measured in days: $g(0) = g(1) = 0$, $g(2) = 0.9$, $g(3) = 0.9$, $g(4) = 0.55$, $g(5) = 0.3$, $g(6) = 0.15$, $g(7) = 0.05$, $g(8) = g(9) = \cdots = 0$. We assume that the fraction of infectious individuals over time is not dependent on the strain. Individuals recovered from the disease are considered immune to the particular influenza strain, that caused it, until the end of the simulation. Cross-immunity is not considered, i.e. the mentioned recovered individuals do not acquire immunity to other influenza strains (Table 1).

Table 1. Multi-agent model parameters

Parameter name	Description	Value
α_m	A fraction of the individuals which are non-immune to the virus strain m	{0.78, 0.74, 0.6} [26]
λ	Infection transmission coefficient	0.3 [26]
c_{sch}	Average daily number of contacts in schools	8.5
c_{wp}	Average daily number of contacts in workplaces	8.5
c_{hh}	Average daily number of contacts in households	1.57
$I_0^{(m)}$	Initial number of individuals infected by a given strain m	5

2.2 The Compartmental Model

As a simplified substitute for the multi-agent model, a multi-strain compartmental model is used based on a deterministic system of difference equations, with the time step equal to one day. The thorough model description can be found in [25]. Analogous to a multi-agent model, we consider the co-circulation of three influenza strains, A(H1N1)pdm09, A(H3N2) and B, thus, we assume $n_s = 3$, where n_s is the total number of regarded strains. Different strains of influenza B type are not distinguished and the dominant B type strain is regarded during each epidemic season. Let $x_t^{(h)}$ be the fraction of susceptible individuals in the population with exposure history $h \in \overline{1, n_s + 1}$, $y_t^{(m)}$ be the number of individuals newly infected at the moment t by the virus strain m and $\overline{y}_t^{(m)}$ – the cumulative number of infectious persons by the time t transmitting the virus strain m, $m \in \overline{1, n_s}$. A possibility of co-infection by multiple strains in the course of one season is not regarded, hence, the individuals recovered from the influenza caused by any of the circulating strains are considered immune.

However, this assumed cross-immunity between virus strains is not transferred to the next epidemic season.

The susceptible individuals are divided into subgroups based on their exposure history h, $h \in \overline{1, n_s + 1}$. A group of susceptible individuals with exposure history state $h \in \overline{1, n_s}$ is composed of those individuals who were subjected to infection by the strain m in the previous epidemic season, whereas a group with exposure history state $h = n_s + 1$ is regarded as naive to the infection caused by any strain. The variable $\mu \in [0; 1)$ reflects the fraction of population which do not participate in infection transmission. In the default case, $\mu = 0$. Due to immunity waning, the individuals with the history of exposure to a fixed influenza strain in the preceding season might lose immunity to the same strain in the following epidemic season. We assume that the fraction a of those individuals, $a \in (0; 1)$, becomes susceptible, whereas $1 - a$ individuals retain their immunity during the modeled epidemic season. As a result, a function $f(h, m)$ is introduced into the model which defines the proportion of the individuals with exposure history state h, who are susceptible to virus strain m:

$$f(h, m) = \begin{cases} a, & m = h, \\ 1, & m \neq h. \end{cases} \tag{1}$$

The modeling equation system is formulated in the following way:

$$x_{t+1}^{(h)} = \max \left\{ 0, \left(1 - \sum_{m=1}^{n_s} \frac{\beta^{(m)}}{\rho} \overline{y}_t^{(m)} f(h, m) \right) x_t^{(h)} \right\}, h \in \overline{1, n_s + 1}, \tag{2}$$

$$y_{t+1}^{(m)} = \frac{\beta^{(m)}}{\rho} \overline{y}_t^{(m)} \sum_{h=1}^{n_s+1} f(h, m) x_t^{(h)}, m \in \overline{1, n_s},$$

$$\overline{y}_t^{(m)} = \sum_{\tau=0}^{T} y_{t-\tau}^{(m)} g_\tau^{(m)}, m \in \overline{1, n_s},$$

$$x_0^{(h)} = \alpha^{(h)} ((1 - \mu)\rho - \sum_{m=1}^{n_s} y_0^{(m)}) \geq 0, h \in \overline{1, n_s + 1},$$

$$y_0^{(m)} = \varphi_0^{(m)} \geq 0, m \in \overline{1, n_s}. \tag{3}$$

The piecewise constant function g_τ gives a fraction of infectious individuals in the group of individuals infected τ days before the current moment t and is defined in the same way as in the multi-agent model. An intricate model of contacts in a synthetic population is replaced by a mass action law with the intensity of effective contacts $\beta^{(m)}$, defined separately for each influenza strain m:

$$\beta^{(m)} = \lambda^{(m)} \delta,$$

where $\lambda^{(m)}$ is virulence of the strain m, δ is the average number of contacts in the population [23].

While in the original model from [25] it was assumed that all the initially infected individuals are in their first infective day (for the employed $g(\tau)$ it is the day 2 after the infection), the compartmental submodel from this study is able to handle the disease prehistory, i.e. the initially infected people are distinguished by the day of their infection. The number of infected is transferred from the multi-agent model at the moment of switching (Table 2).

Table 2. Compartmental model parameters

Variable	Description	Values
ρ	Population size, people	$4,865,118$
$\alpha^{(h)}$	A fraction of population exposed to the strain m in the preceding epidemic season, $h \in \overline{1,m}$	$\{0.78, 0.74, 0.6\}$
$\lambda^{(m)}$	Virulence of the virus strain m	0.3
a	The fraction of people who lost immunity after being exposed to the virus strain in the preceding epidemic season	0.3 [25]
δ	Average daily number of contacts in the population for a fixed individual	6.528 [23]

2.3   Switching Algorithm

One of the important aspects of a hybrid modeling is to properly decide how to define the switching conditions, when a detailed multi-agent model should be replaced by a compartmental model during a simulation run. As it can be seen from Fig. 1, the calculation time for a single algorithm step (a modeling day) is growing very fast with the step number due to the increase in the number of infected people in the population. Thus, it is beneficial to approximate the infection process by a simpler model starting from the moment, when the number of infected people becomes large.

In [4] two switching condition types were proposed:

- Switch to a compartmental model when a certain number of infected individuals in the population is reached;
- Switch to a compartmental model when the effective reproduction number of the infection is stabilized (the difference between the corresponding values for the subsequent simulation days becomes lower than a certain threshold).

While both switching conditions are quite effective and interpretable in the case of a disease dynamics caused by a single virus, they cannot be easily adopted for our case due to the existence of multiple viruses in the population. The typical situation in virus co-circulation modeling corresponds to high prevalence caused by one strain and low prevalence caused by other strains. As a result, if we make a switch based on the cumulative numbers of the infected people or the cumulative reproduction number, the disease dynamics of the virus with

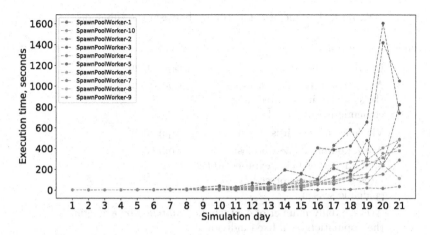

Fig. 1. Calculation time for a single time step depending on the simulation day.

low prevalence might be altered compared to the original model. On the other hand, if the switch is to be performed when the threshold is reached by the rarest strain's infection number, it will lead to no switching at all or to the late switching, because minor epidemics with the prevalence never exceeding a threshold are typical for almost every simulation run.

In this study, we perform numerical experiments with a hybrid model based on a time-related switching condition, i.e. the moment of the switch is tied to the simulation day. Table 3 contains execution time corresponding to simulation runs from Sect. 3.1. The table demonstrates that earlier switching gives an immense economy of execution time. In fact, the share of computational time for the compartmental submodel in the overall simulation process might be considered negligible, because it is much faster than its multi-agent counterpart, thus, the hybrid simulation with the switching moment t^* is very close in computation time to performing t^* time steps of the original multi-agent model.

Table 3. Hybrid modeling algorithm performance

Moment of switching, t^*	5	10	15	20
Execution time, seconds	267.6	1185.5	4748.5	131784.8

A crucial aspect of the switching is to ensure that the parameters of both submodels align, otherwise instead of a single simulated prevalence trajectory with varied level of detail we might obtain two independent epidemic processes. In Table 4, the parameter matching is described.

As it is clear from the table, the following two parameters are the main source of potential bias between the outputs of the two submodels:

Table 4. Parameter compatibility between the submodels

Variable	Description	Compatibility between submodels
ρ	Population size, people	Equivalent
$\alpha^{(h)}$	A fraction of population exposed to the strain m in the preceding epidemic season, $h \in \overline{1, m^*}$	Equivalent
$\lambda^{(m)}$	Virulence of the virus strain m	Equivalent
a	A fraction of people who lost immunity after being exposed to the virus strain in the preceding epidemic season	Equivalent
δ	Average daily number of contacts in the population for a fixed individual	Matched by averaging
μ	A fraction of the individuals with the protection from infection by any influenza strains	Population submodel only

- Average daily number of contacts δ. Obviously, the contact process in the compartmental model lacks much detail compared to the explicit modeling of the contacts in a synthetic population, thus even if the average number of contacts is correctly calculated from the corresponding multi-agent submodel data, the discrepancy in the actual number of contacts is inevitable.
- A fraction μ of the individuals with the protection from infection by any influenza strains. This parameter was somewhat artificially added to a model to make it possible to calibrate the compartmental model to real data. In [25] it was shown that with $\mu = 0$, when all the individuals without expection are prone to the infection, the compartmental model gives implausible prevalence curves—they are either too high (unrealistic epidemic intensity) or too wide (unrealistic epidemic duration) compared to the real observed epidemic outbreaks. We assumed thus that μ is dependent on the fraction of cases which is missed in the statistics due to under-reporting (hence, lower prevalence peaks in data) and also on the topology of the contact network (unlike it is assumed in SEIR-type compartmental models, not all of the individuals participate in the epidemic transmission).

The analysis of the dependence of hybrid model output on t^* and μ was performed in the numerical experiments described in the following section.

3   Simulation

The simulation framework implementing the described hybrid model was developed using Python 3.8 programming language. The simulation runs were executed in parallel using `multiprocessing` library. The hardware used was Intel

Xeon cluster with 24 virtual (12 physical) cores. A single experiment included 10–20 repetitive simulation runs. The result of one experiment is thus 30–60 trajectories in total (10–20 for each of three co-circulating influenza strains).

3.1 Switching Moment Influence

The first set of experiments with the hybrid model was conducted to compare prevalence trajectories obtained with different values of t^*. The value $\mu = 0.9$ was set based on the previous experience of compartmental model calibration [25]. On the Fig. 2 below, from left to right in a fixed row one can see output prevalence curves for three circulating virus strains. From top to bottom, it is shown how the output corresponding to a fixed virus strain changes depending on the input parameters. The moment of switching is shown on each graph by a gray dotted line.

Figure 2 shows that the resulting output is generally similar throughout the experiments, however, there exists a visible increase in the maximum total number of infected individuals with the increase of t^* (the images top to bottom for a fixed strain). Also, the very moment of the switch is clearly visible on the graphs, which indicates that the submodel matching by simple equaling of the parameter values does not allow for smooth transition. Thus the framework might require modifications to assure correct coupling of the submodels. That observation is coherent with the findings demonstrated by other research groups [29].

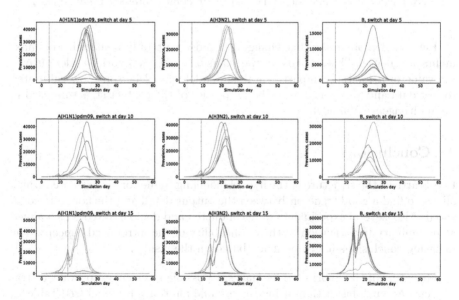

Fig. 2. Prevalence dynamics for different switching moments t^*, the fraction of protected individuals $\mu = 0.9$ set according to [25]. A row of three figures corresponds to an output of a single experiment while a column shows the comparative prevalence of infection caused by one virus strain depending on t^*.

3.2   Protected Individuals Fraction Influence

The second set of experiments with the hybrid model was conducted to find out how the value of μ might influence the resulting output.

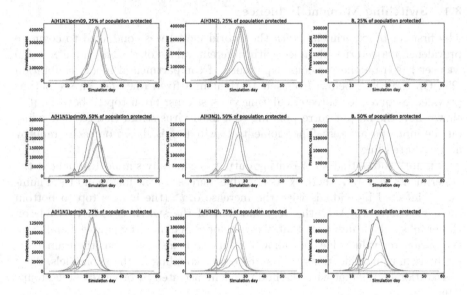

Fig. 3. Incidence dynamics depending on the fraction of protected individuals μ.

Figure 3 demonstrates that changing μ indeed alters dramatically the maximum prevalence. Thus, to ensure the plausibility of the hybrid model output, the value of μ for the compartmental submodel should be somehow calibrated to real data and/or aligned with the properties of the contact structure used in the multi-agent submodel.

4   Conclusions

In this article, the structure of the hybrid modeling framework is presented which allows to find a good trade-off between the output detail and the computational speed. Also, several experiments were conducted and demonstrated which aimed at preliminary investigation of the applicability of the introduced concept. The following conclusions might be made based on the results:

- The usage of the hybrid model makes it possible to calculate prevalence trajectories much faster than using the original multi-agent model (see Table 3), and to add more detail to the disease transmission description (e.g., by considering supermarkets and public transport as potential places of contacts). However, since the choice of the switching moment t^* alters the output, and, obviously, the switch made too early renders a multi-agent component of the hybrid network useless, the problem of careful selection of this value should be properly addressed in the forthcoming research. One of the things which is to be done in that direction is to compare the presented outputs of the hybrid model with the output of the original multi-agent model and to assess how fast the bias between the trajectories decreases with the increase of t^*. This investigation will make it possible to find a dependence between the output accuracy and the simulation execution time under a fixed set of input parameters.

- The demonstrated discontinuity of prevalence trajectories at the moment of switching calls for refining the switching algorithm. Particularly, it is necessary to establish whether this problem arises due to incompatibility of modeling structures or poor matching of parameter values. According to [29], while there is an equivalence between the averaged and network model, the parameters still need to be adjusted. Since the networks are not homogeneous, one needs to calculate the revised "effective" value of disease transmission intensity. This approach, along with the switching based on stabilized transmission detection, will be implemented in the near future.
- The presence of the parameter μ in the compartmental submodel, which does not have a clear interpretation, complicates the alignment of the two submodels. A possible solution of the issue is to find a way how to derive the value of μ from real data or from the parameter values of the multi-agent submodel. Since this result might be obtained only after clarifying the parameter meaning, a full-fledged separate research is required.
- In the current article, separate trajectories were shown on the graphs, since the aim was to demonstrate how a given trajectory behaves before and after switching the submodels. However, for the practical aims of using the hybrid modeling framework, particularly, for the calibration to real data and for the uncertainty analysis, the confidence intervals should be assessed and compared depending on switching moment t^*. That will allow to understand how the usage of the hybrid model alters the uncertainty of the prevalence estimation compared to the original multi-agent framework.

The author believes that the mentioned steps will make the described hybrid modeling framework a valueable tool for short-term infection prediction of influenza and COVID-19 prevalence, along with the analysis of their possible co-circulation, which is stated to pose potential danger for the population well-being.

References

1. Ajelli, M., Litvinova, M.: Estimating contact patterns relevant to the spread of infectious diseases in Russia. J. Theor. Biol. **419**, 1–7 (2017)
2. Baguelin, M., Flasche, S., Camacho, A., Demiris, N., Miller, E., Edmunds, W.J.: Assessing optimal target populations for influenza vaccination programmes: an evidence synthesis and modelling study. PLoS Med. **10**(10), e1001527 (2013)

3. Baroyan, O., Basilevsky, U., Ermakov, V., Frank, K., Rvachev, L., Shashkov, V.: Computer modelling of influenza epidemics for large-scale systems of cities and territories. In: Proceedings of the WHO Symposium on Quantitative Epidemiology, Moscow (1970)
4. Bobashev, G.V., Goedecke, D.M., Yu, F., Epstein, J.M.: A hybrid epidemic model: combining the advantages of agent-based and equation-based approaches. In: 2007 Winter Simulation Conference, pp. 1532–1537. IEEE (2007)
5. Brett, T., et al.: Detecting critical slowing down in high-dimensional epidemiological systems. PLOS Comput. Biol. **16**(3), 1–19 (2020)
6. CDC: People with heart disease and those who have had a stroke are at high risk of developing complications from influenza (the flu). http://www.cdc.gov/flu/heartdisease/
7. Cooley, P., et al.: The role of subway travel in an influenza epidemic: a New York City simulation. J. Urban Health **88**(5), 982 (2011)
8. Cooley, P.C., Bartsch, S.M., Brown, S.T., Wheaton, W.D., Wagener, D.K., Lee, B.Y.: Weekends as social distancing and their effect on the spread of influenza. Comput. Math. Organ. Theory **22**(1), 71–87 (2015). https://doi.org/10.1007/s10588-015-9198-5
9. Drake, J.M., et al.: The statistics of epidemic transitions. PLOS Comput. Biol. **15**(5), 1–14 (2019)
10. Government of Saint Petersburg: Otkritie dannie Sankt-Peterburga [Open data of Saint-Petersburg], (In Russian). https://data.gov.spb.ru/opendata/7840013199-passports_houses/versions/9/. Accessed 19 Apr 2020
11. Guo, D., Li, K.C., Peters, T.R., Snively, B.M., Poehling, K.A., Zhou, X.: Multi-scale modeling for the transmission of influenza and the evaluation of interventions toward it. Sci. Rep. **5**(1), 1–9 (2015)
12. Hill, E.M., Petrou, S., De Lusignan, S., Yonova, I., Keeling, M.J.: seasonal influenza: modelling approaches to capture immunity propagation. PLoS Comput. Biol. **15**(10), e1007096 (2019)
13. Hunter, E., Kelleher, J.D.: Adapting an agent-based model of infectious disease spread in an Irish county to COVID-19. Systems **9**(2), 41 (2021)
14. Hunter, E., Mac Namee, B., Kelleher, J.: A hybrid agent-based and equation based model for the spread of infectious diseases. J. Artif. Soc. Soc. Simul. **23**(4) (2020)
15. Iuliano, A.D., et al.: Estimates of global seasonal influenza-associated respiratory mortality: a modelling study. Lancet **391**(10127), 1285–1300 (2018)
16. Kerr, C.C., et al.: COVASIM: an agent-based model of COVID-19 dynamics and interventions. PLoS Comput. Biol. **17**(7), e1009149 (2021)
17. Konshina, O., Sominina, A., Smorodintseva, E., Stolyarov, K., Nikonorov, I.: Population immunity to influenza virus A(H1N1)pdm09, A(H3N2) and B in the adult population of the Russian Federation long-term research results. Russ. J. Infect. Immun. **7**(1), 27–33 (2017). in Russian, https://doi.org/10.15789/2220-7619-2017-1-27-33
18. Krivorotko, O., Sosnovskaia, M., Vashchenko, I., Kerr, C., Lesnic, D.: Agent-based modeling of COVID-19 outbreaks for New York state and UK: Parameter identification algorithm. Infect. Dis. Model. **7**(1), 30–44 (2022)
19. Kumar, S., Grefenstette, J.J., Galloway, D., Albert, S.M., Burke, D.S.: Policies to reduce influenza in the workplace: impact assessments using an agent-based model. Am. J. Public Health **103**(8), 1406–1411 (2013)
20. Kumar, S., Piper, K., Galloway, D.D., Hadler, J.L., Grefenstette, J.J.: Is population structure sufficient to generate area-level inequalities in influenza rates? An examination using agent-based models. BMC Public Health **15**(1), 947 (2015)

21. Lee, B.Y., et al.: A computer simulation of vaccine prioritization, allocation, and rationing during the 2009 h1n1 influenza pandemic. Vaccine **28**(31), 4875–4879 (2010)
22. Leonenko, V., Arzamastsev, S., Bobashev, G.: Contact patterns and influenza outbreaks in Russian cities: a proof-of-concept study via agent-based modeling. J. Comput. Sci. **44**, 101156 (2020)
23. Leonenko, V., Bobashev, G.: Analyzing influenza outbreaks in Russia using an age-structured dynamic transmission model. Epidemics **29**, 100358 (2019)
24. Leonenko, V., Lobachev, A., Bobashev, G.: Spatial modeling of influenza outbreaks in Saint Petersburg using synthetic populations. In: Rodrigues, J.M.F., et al. (eds.) ICCS 2019. LNCS, vol. 11536, pp. 492–505. Springer, Cham (2019). https://doi.org/10.1007/978-3-030-22734-0_36
25. Leonenko, V.N.: Herd immunity levels and multi-strain influenza epidemics in Russia: a modelling study. Russ. J. Numer. Anal. Math. Model. **36**(5), 279–291 (2021)
26. Leonenko, V.N.: Modeling co-circulation of influenza strains in heterogeneous urban populations: the role of herd immunity and uncertainty factors. In: Paszynski, M., Kranzlmüller, D., Krzhizhanovskaya, V.V., Dongarra, J.J., Sloot, P.M.A. (eds.) ICCS 2021. LNCS, vol. 12744, pp. 663–669. Springer, Cham (2021). https://doi.org/10.1007/978-3-030-77967-2_55
27. Leonenko, V.N., Danilenko, D.M.: Modeling the dynamics of population immunity to influenza in Russian cities. In: ITM Web of Conferences, vol. 31, p. 03001 (2020)
28. Lukens, S., et al.: A large-scale immuno-epidemiological simulation of influenza a epidemics. BMC Public Health **14**(1), 1–15 (2014)
29. Rahmandad, H., Sterman, J.: Heterogeneity and network structure in the dynamics of diffusion: comparing agent-based and differential equation models. Manag. Sci. **54**(5), 998–1014 (2008)
30. Vlad, A.I., Sannikova, T.E., Romanyukha, A.A.: Transmission of acute respiratory infections in a city: Agent-based approach. Math. Biol. Bioinform. **15**(2), 338–356 (2020)

Computational Optimization, Modelling and Simulation

Numerical Stability of Tangents and Adjoints of Implicit Functions

Uwe Naumann[✉][iD]

Department of Computer Science, RWTH Aachen University,
52056 Aachen, Germany
naumann@stce.rwth-aachen.de,
http://www.stce.rwth-aachen.de/

Abstract. We investigate errors in tangents and adjoints of implicit functions resulting from errors in the primal solution due to approximations computed by a numerical solver.

Adjoints of systems of linear equations turn out to be unconditionally numerically stable. Tangents of systems of linear equations can become instable as well as both tangents and adjoints of systems of nonlinear equations, which extends to optima of convex unconstrained objectives. Sufficient conditions for numerical stability are derived.

Keywords: Algorithmic differentiation · Implicit function

1 Introduction

We consider twice differentiable implicit functions

$$F : \mathbb{R}^m \to \mathbb{R}^n : \mathbf{p} \mapsto \mathbf{x} = F(\mathbf{p}) \tag{1}$$

defined by the roots of residuals

$$R : \mathbb{R}^n \times \mathbb{R}^m \to \mathbb{R}^n : (\mathbf{x}, \mathbf{p}) \mapsto R(\mathbf{x}, \mathbf{p}) . \tag{2}$$

R is referred to as the primal residual as opposed to tangent and adjoint residuals to be considered later. Primal roots of the residual satisfying

$$R(\mathbf{x}, \mathbf{p}) = 0 \tag{3}$$

are assumed to be approximated by numerical solvers

$$S : \mathbb{R}^m \to \mathbb{R}^n : \mathbf{p} \mapsto \mathbf{x} + \Delta\mathbf{x} = S(\mathbf{p})$$

with an absolute error $\Delta\mathbf{x}$ yielding a relative error $\delta\mathbf{x}$ of norm

$$\|\delta\mathbf{x}\| = \frac{\|\Delta\mathbf{x}\|}{\|\mathbf{x}\|} = \frac{\|S(\mathbf{p}) - F(\mathbf{p})\|}{\|F(\mathbf{p})\|} .$$

© The Author(s), under exclusive license to Springer Nature Switzerland AG 2022
D. Groen et al. (Eds.): ICCS 2022, LNCS 13352, pp. 181–187, 2022.
https://doi.org/10.1007/978-3-031-08757-8_17

We investigate (relative) errors in corresponding tangents

$$\dot{\mathbf{x}} = \dot{F}(\mathbf{x}, \dot{\mathbf{p}}) \equiv \frac{dF}{d\mathbf{p}} \cdot \dot{\mathbf{p}} \tag{4}$$

and adjoints

$$\bar{\mathbf{p}} = \bar{F}(\mathbf{x}, \bar{\mathbf{x}}) \equiv \frac{dF}{d\mathbf{p}}^T \cdot \bar{\mathbf{x}} \tag{5}$$

due to $\Delta\mathbf{x}$. *Algorithmic* tangents and adjoints result from the application of algorithmic differentiation (AD) [3,4] to the solver S. *Symbolic* tangents and adjoints can be derived at the solution of Eq. (3) in terms of tangents and adjoints of the residual [2,5]. AD of the solver can thus be avoided which typically results in a considerably lower computational complexity.

2   Prerequisites

We perform standard first-order error analysis. For a given absolute error $\Delta\mathbf{p}$ in the input of a function F the absolute error in the result is estimated as

$$\Delta\mathbf{x} \approx \frac{dF}{d\mathbf{p}} \cdot \Delta\mathbf{p} . \tag{6}$$

Equation (1) is differentiated with respect to $\mathbf{p}$ in the direction of the absolute error $\Delta\mathbf{p}$. From the Taylor series expansion of

$$\mathbf{x} + \Delta\mathbf{x} = \mathbf{x} + \frac{dF}{d\mathbf{p}} \cdot \Delta\mathbf{p} + O(\|\Delta\mathbf{p}\|^2)$$

it follows that negligence of the remainder within a neighborhood of $\mathbf{x}$ containing $\Delta\mathbf{x}$ is reasonable for $\|\Delta\mathbf{p}\| \to 0$ and assuming convergence of the Taylor series to the correct function value. For linear F we get $\Delta\mathbf{x} = \frac{dF}{d\mathbf{p}} \cdot \Delta\mathbf{p}$ due to the vanishing remainder.

Tangents and adjoints of Eq. (1) can be expressed as matrix equations over derivatives of the residual. The fundamental operations involved are scalar multiplications and additions, outer vector products, matrix-vector products and solutions of systems of linear equations.

It is well-known that scalar multiplication $y = x_1 \cdot x_2$ is numerically stable with relative error $|\delta y| = |\delta x_1| + |\delta x_2|$. A similar result holds for scalar division. It generalizes naturally to element-wise multiplication and division of vectors, matrices, and higher-order tensors as well as to the outer product of two vectors.

Scalar addition $y = x_1 + x_2$ on the other hand is known to be numerically unstable $|\delta y|\frac{|\Delta x_1 + \Delta x_2|}{|x_1 + x_2|} \to \infty$ for $\Delta x_1 \neq -\Delta x_2$ and $x_1 \to -x_2$. A similar result holds for scalar subtraction.

Numerical instability of scalar addition prevents unconditional numerical stability of inner vector products as well as matrix-vector/matrix products and solutions of systems of linear equations. Sufficient conditions for numerical stability need to be formulated.

The relative error of a matrix-vector product $\mathbf{x} = A \cdot \mathbf{b}$ for $A \in \mathbb{R}^{n \times n}$ and $\mathbf{x}, \mathbf{b} \in \mathbb{R}^n$ is easily shown to be equal to

$$\|\delta\mathbf{x}\| \approx \kappa(A) \cdot (\|\delta A\| + \|\delta\mathbf{b}\|) . \tag{7}$$

Depending on the magnitude of the condition number $\kappa(A) \equiv \|A^{-1}\| \cdot \|A\|$ of A the relative error of the matrix-vector product can suffer from a potentially dramatic amplification of the relative errors in the arguments.

We take a closer look at the derivation of a similar result for systems of linear equations $A \cdot \mathbf{x} = \mathbf{b}$ for $A \in \mathbb{R}^{n \times n}$ and $\mathbf{x}, \mathbf{b} \in \mathbb{R}^n$. Differentiation in the direction of non-vanishing absolute errors $\Delta A \in \mathbb{R}^{n \times n}$ and $\Delta\mathbf{x}, \Delta\mathbf{b} \in \mathbb{R}^n$ yields

$$\Delta\mathbf{x} = A^{-1} \cdot (\Delta\mathbf{b} - \Delta A \cdot \mathbf{x})$$

and hence the first-order error estimate

$$
\begin{aligned}
\frac{\|\Delta\mathbf{x}\|}{\|\mathbf{x}\|} &= \frac{\|A^{-1} \cdot (\Delta\mathbf{b} - \Delta A \cdot \mathbf{x})\|}{\|\mathbf{x}\|} \\
&\leq \frac{\|A^{-1} \cdot \Delta\mathbf{b}\|}{\|\mathbf{x}\|} + \frac{\|A^{-1} \cdot \Delta A \cdot \mathbf{x}\|}{\|\mathbf{x}\|} \\
&\leq \frac{\|A^{-1}\| \cdot \|\Delta\mathbf{b}\|}{\|\mathbf{x}\|} + \frac{\|A^{-1} \cdot \Delta A \cdot \mathbf{x}\|}{\|\mathbf{x}\|} = \frac{\|A\| \cdot \|A^{-1}\| \cdot \|\Delta\mathbf{b}\|}{\|A\| \cdot \|\mathbf{x}\|} + \frac{\|A^{-1} \cdot \Delta A \cdot \mathbf{x}\|}{\|\mathbf{x}\|} \\
&\leq \kappa(A) \cdot \frac{\|\Delta\mathbf{b}\|}{\|A \cdot \mathbf{x}\|} + \frac{\|A^{-1} \cdot \Delta A \cdot \mathbf{x}\|}{\|\mathbf{x}\|} = \kappa(A) \cdot \frac{\|\Delta\mathbf{b}\|}{\|\mathbf{b}\|} + \frac{\|A^{-1} \cdot \Delta A \cdot \mathbf{x}\|}{\|\mathbf{x}\|} \\
&\leq \kappa(A) \cdot \frac{\|\Delta\mathbf{b}\|}{\|\mathbf{b}\|} + \frac{\|A^{-1}\| \cdot \|\Delta A\| \cdot \|\mathbf{x}\|}{\|\mathbf{x}\|} = \kappa(A) \cdot \frac{\|\Delta\mathbf{b}\|}{\|\mathbf{b}\|} + \|A^{-1}\| \cdot \|\Delta A\| \\
&= \kappa(A) \cdot \frac{\|\Delta\mathbf{b}\|}{\|\mathbf{b}\|} + \frac{\|A\| \cdot \|A^{-1}\| \cdot \|\Delta A\|}{\|A\|} = \kappa(A) \cdot \frac{\|\Delta\mathbf{b}\|}{\|\mathbf{b}\|} + \kappa(A) \cdot \frac{\|\Delta A\|}{\|A\|} \\
&= \kappa(A) \cdot \left(\frac{\|\Delta\mathbf{b}\|}{\|\mathbf{b}\|} + \frac{\|\Delta A\|}{\|A\|} \right) .
\end{aligned}
$$

As for matrix-vector products we get

$$\|\delta\mathbf{x}\| \approx \kappa(A) \cdot (\|\delta A\| + \|\delta\mathbf{b}\|) . \tag{8}$$

Again, a low condition number of A is sufficient for numerical stability.

3   Errors in Tangents and Adjoints of Implicit Functions

Differentiation of Eq. (3) with respect to $\mathbf{p}$ yields

$$\frac{\partial R}{\partial \mathbf{x}} \cdot \frac{d\mathbf{x}}{d\mathbf{p}} + \frac{\partial R}{\partial \mathbf{p}} = R_{\mathbf{x}} \cdot \frac{d\mathbf{x}}{d\mathbf{p}} + R_{\mathbf{p}} = 0 , \tag{9}$$

where ∂ denotes partial differentiation. Multiplication with $\dot{\mathbf{p}}$ from the right yields the tangent residual

$$R_{\mathbf{x}} \cdot \frac{d\mathbf{x}}{d\mathbf{p}} \cdot \dot{\mathbf{p}} + R_{\mathbf{p}} \cdot \dot{\mathbf{p}} = R_{\mathbf{x}} \cdot \dot{\mathbf{x}} + R_{\mathbf{p}} \cdot \dot{\mathbf{p}} = 0 . \tag{10}$$

The tangent $\dot{\mathbf{x}}$ can be computed as the solution of the system of linear equations

$$R_{\mathbf{x}} \cdot \dot{\mathbf{x}} = -R_{\mathbf{p}} \cdot \dot{\mathbf{p}} .$$

The right-hand side is obtained by a single evaluation of the tangent residual. Tangents in the directions of the Cartesian basis of $\mathbb{R}^n$ yields $R_{\mathbf{x}}$. Potential sparsity can and should be exploited [1]. An error $\Delta\mathbf{x}$ in the primal solution yields a corresponding error in the tangent for $R_{\mathbf{x}} = R_{\mathbf{x}}(\mathbf{x})$ and/or $R_{\mathbf{p}} = R_{\mathbf{p}}(\mathbf{x})$.

From Eq. (9) it follows that for regular $R_{\mathbf{x}}$

$$\frac{d\mathbf{x}}{d\mathbf{p}} = -R_{\mathbf{x}}^{-1} \cdot R_{\mathbf{p}} .$$

Transposition of the latter followed by multiplication with $\bar{\mathbf{x}}$ from the right yields

$$\bar{\mathbf{p}} = \frac{d\mathbf{x}}{d\mathbf{p}}^T \cdot \bar{\mathbf{x}} = -R_{\mathbf{p}}^T \cdot R_{\mathbf{x}}^{-T} \cdot \bar{\mathbf{x}} . \tag{11}$$

The adjoint $\bar{\mathbf{p}}$ can be computed as the solution of the system of linear equations

$$R_{\mathbf{x}}^T \cdot \mathbf{z} = -\bar{\mathbf{x}}$$

followed by the evaluation of the adjoint residual yielding

$$\bar{\mathbf{p}} = R_{\mathbf{p}}^T \cdot \mathbf{z}$$

. Again, an error $\Delta\mathbf{x}$ in the primal solution yields a corresponding error in the adjoint.

3.1 Systems of Linear Equations

The tangent of the solution of the primal system of linear equations

$$A \cdot \mathbf{x} = \mathbf{b} \tag{12}$$

is defined as $\dot{\mathbf{x}} = \dot{\mathbf{x}}_A + \dot{\mathbf{x}}_{\mathbf{b}}$, where

$$A \cdot \dot{\mathbf{x}}_{\mathbf{b}} = \dot{\mathbf{b}} \tag{13}$$

and

$$A \cdot \dot{\mathbf{x}}_A = -\dot{A} \cdot \mathbf{x} \tag{14}$$

[2]. An error $\Delta\mathbf{x}$ in the primal solution which, for example, might result from the use of an indirect solver yields an erroneous tangent

$$\dot{\mathbf{x}} + \Delta\dot{\mathbf{x}} = (\dot{\mathbf{x}}_A + \Delta\dot{\mathbf{x}}_A) + (\dot{\mathbf{x}}_{\mathbf{b}} + \Delta\dot{\mathbf{x}}_{\mathbf{b}}) .$$

Application of Eq. (8) to Eq. (13) yields

$$\|\delta\dot{\mathbf{x}}_{\mathbf{b}}\| \approx \kappa(A) \cdot (\|\delta A\| + \|\delta\dot{\mathbf{b}}\|) .$$

Independence of $\dot{\mathbf{x}}_\mathbf{b}$ from $\mathbf{x}$ (and hence from $\varDelta\mathbf{x}$) implies $\delta\dot{\mathbf{x}}_\mathbf{b} = 0$ for error-free A and $\dot{\mathbf{b}}$, that is $\varDelta\dot{\mathbf{x}} = \varDelta\dot{\mathbf{x}}_A$, respectively $\delta\dot{\mathbf{x}} = \delta\dot{\mathbf{x}}_A$. Let $\mathbf{c} = -\dot{A}\cdot\mathbf{x}$. With Eq. (7) it follows that

$$\|\delta\mathbf{c}\| \approx \kappa(\dot{A})\cdot\|\delta\mathbf{x}\|$$

as $\delta\dot{A} = 0$ Moreover, application of Eq. (8) to $A\cdot\dot{\mathbf{x}}_A = \mathbf{c}$ yields

$$\|\delta\dot{\mathbf{x}}_A\| \approx \kappa(A)\cdot\|\delta\mathbf{c}\| \ .$$

Consequently,

$$\|\delta\dot{\mathbf{x}}_A\| \approx \kappa(A)\cdot\kappa(\dot{A})\cdot\|\delta\mathbf{x}\| \ . \tag{15}$$

Low condition numbers of both A and $\dot{A}$ ensure numerical stability of tangent systems of linear equations.

The adjoint of the primal linear system in Eq. (12) is defined as

$$A^T\cdot\bar{\mathbf{b}} = \bar{\mathbf{x}} \tag{16}$$

and

$$\bar{A} = -\bar{\mathbf{b}}\cdot\mathbf{x}^T \tag{17}$$

[2]. Application of Eq. (8) to Eq. (16) yields

$$\delta\bar{\mathbf{b}} \approx \kappa(A)\cdot(\delta A + \delta\bar{\mathbf{x}}) \ .$$

Independence of $\bar{\mathbf{b}}$ from $\mathbf{x}$ (and hence from $\varDelta\mathbf{x}$) implies $\delta\bar{\mathbf{b}} = 0$ for error-free A and $\bar{\mathbf{x}}$. The outer product $\bar{A} = -\bar{\mathbf{b}}\cdot\mathbf{x}^T$ is numerically stable as scalar multiplication is. Consequently, adjoint systems of linear equations are numerically stable.

3.2   Systems of Nonlinear Equations

Differentiation of Eq. (10) in the direction of absolute errors $\varDelta R_\mathbf{x} \in \mathbb{R}^{n\times n}$, $\varDelta\dot{\mathbf{x}} \in \mathbb{R}^n$, $\varDelta R_\mathbf{p} \in \mathbb{R}^{n\times m}$ and $\varDelta\dot{\mathbf{p}} \in \mathbb{R}^m$ yields

$$\varDelta R_\mathbf{x}\cdot\dot{\mathbf{x}} + R_\mathbf{x}\cdot\varDelta\dot{\mathbf{x}} + \varDelta R_\mathbf{p}\cdot\dot{\mathbf{p}}\ [\underbrace{+R_\mathbf{p}\cdot\varDelta\dot{\mathbf{p}}}_{=0}] = 0$$

as $\varDelta\dot{\mathbf{p}} = 0$ and hence

$$\varDelta\dot{\mathbf{x}} = R_\mathbf{x}^{-1}\cdot(\varDelta R_\mathbf{x}\cdot\dot{\mathbf{x}} + \varDelta R_\mathbf{p}\cdot\dot{\mathbf{p}}) \ .$$

First-order estimates for

$$\varDelta R_\mathbf{x}\cdot\dot{\mathbf{x}} = [\varDelta R_\mathbf{x}\cdot\dot{\mathbf{x}}]_i \approx [R_{\mathbf{x},\mathbf{x}}]_{i,j,k}\cdot[\dot{\mathbf{x}}]_j\cdot[\varDelta\mathbf{x}]_k \equiv \varDelta\dot{R}_\mathbf{x}\cdot\varDelta\mathbf{x}$$

and

$$\varDelta R_\mathbf{p}\cdot\dot{\mathbf{p}} = [\varDelta R_\mathbf{p}\cdot\dot{\mathbf{p}}]_i \approx [R_{\mathbf{p},\mathbf{x}}]_{i,j,k}\cdot[\dot{\mathbf{p}}]_j\cdot[\varDelta\mathbf{x}]_k \equiv \varDelta\dot{R}_\mathbf{p}\cdot\varDelta\mathbf{x}$$

in index notation (summation over the shared index) yield

$$\Delta \dot{\mathbf{x}} \approx R_{\mathbf{x}}^{-1} \cdot (\Delta \dot{R}_{\mathbf{x}} + \Delta \dot{R}_{\mathbf{p}}) \cdot \Delta \mathbf{x}$$

and hence, with Eq. (7),

$$\|\delta \dot{\mathbf{x}}\| \approx \kappa(R_{\mathbf{x}}) \cdot \kappa(\Delta \dot{R}_{\mathbf{x}} + \Delta \dot{R}_{\mathbf{p}}) \cdot \|\delta \mathbf{x}\| . \tag{18}$$

Low condition numbers of the respective first and second derivatives of the residual ensure numerical stability of tangent systems of nonlinear equations. Both $\Delta \dot{R}_{\mathbf{x}}$ and $\Delta \dot{R}_{\mathbf{p}}$ can be computed by algorithmic differentiation (AD) [3,4].

Application of Eq. (8) to the system of linear equations

$$R_{\mathbf{x}}^T \cdot \mathbf{z} = -\bar{\mathbf{x}}$$

for $\Delta \bar{\mathbf{x}} = 0$ yields

$$\Delta \mathbf{z} = R_{\mathbf{x}}^{-T} \cdot \Delta R_{\mathbf{x}}^T \cdot \mathbf{z}$$

and hence

$$\|\delta \mathbf{z}\| \approx \kappa(R_{\mathbf{x}}) \cdot \kappa(\Delta \bar{R}_{\mathbf{x}}) \cdot \|\delta \mathbf{x}\| ,$$

where

$$[\Delta R_{\mathbf{x}}^T \cdot \mathbf{z}]_j \approx [R_{\mathbf{x},\mathbf{x}}]_{i,j,k} \cdot [\mathbf{z}]_i \cdot [\Delta \mathbf{x}]_k \equiv \Delta \bar{R}_{\mathbf{x}} \cdot \Delta \mathbf{x} .$$

Differentiation of $\bar{\mathbf{p}} = R_{\mathbf{p}}^T \cdot \mathbf{z}$ in the direction of the non-vanishing absolute errors $\Delta R_{\mathbf{p}}^T \in \mathbb{R}^{m \times n}$ and $\Delta \mathbf{z} \in \mathbb{R}^n$ yields

$$\Delta \bar{\mathbf{p}} = \Delta R_{\mathbf{p}}^T \cdot \mathbf{z} + R_{\mathbf{p}}^T \cdot \Delta \mathbf{z} = \Delta R_{\mathbf{p}}^T \cdot \mathbf{z} + R_{\mathbf{p}}^T \cdot R_{\mathbf{x}}^{-T} \cdot \Delta R_{\mathbf{x}}^T \cdot \mathbf{z}$$

and hence

$$\|\delta \bar{\mathbf{p}}\| \approx \left(\kappa(\Delta \bar{R}_{\mathbf{p}}) + \kappa(R_{\mathbf{p}}) \cdot \kappa(R_{\mathbf{x}}) \cdot \kappa(\Delta \bar{R}_{\mathbf{x}}) \right) \cdot \|\delta \mathbf{x}\| , \tag{19}$$

where

$$[\Delta R_{\mathbf{p}}^T \cdot \mathbf{z}]_j \approx [R_{\mathbf{p},\mathbf{x}}]_{i,j,k} \cdot [\mathbf{z}]_i \cdot [\Delta \mathbf{x}]_k \equiv \Delta \bar{R}_{\mathbf{p}} \cdot \Delta \mathbf{x} .$$

Low condition numbers of the respective first and second derivatives of the residual ensure numerical stability of adjoint systems of nonlinear equations. Both $\Delta \bar{R}_{\mathbf{x}}$ and $\Delta \bar{R}_{\mathbf{p}}$ can be computed by AD.

3.3   Convex Unconstrained Objectives

The first-order optimality condition for a parameterized convex unconstrained objective

$$f : \mathbb{R}^n \times \mathbb{R}^m \to \mathbb{R} : (\mathbf{x}, \mathbf{p}) \mapsto y = f(\mathbf{x}, \mathbf{p})$$

yields the residual $f_{\mathbf{x}}(\mathbf{x}, \mathbf{p}) = 0$. Consequently, assuming f to be three times differentiable,

$$\|\delta \dot{\mathbf{x}}\| \approx \kappa(f_{\mathbf{x},\mathbf{x}}) \cdot \kappa(\Delta \dot{f}_{\mathbf{x},\mathbf{x}} + \Delta \dot{f}_{\mathbf{x},\mathbf{p}}) \cdot \|\delta \mathbf{x}\| , \tag{20}$$

where

$$\Delta f_{\mathbf{x},\mathbf{x}} \cdot \dot{\mathbf{x}} = [\Delta f_{\mathbf{x},\mathbf{x}} \cdot \dot{\mathbf{x}}]_i \approx [f_{\mathbf{x},\mathbf{x},\mathbf{x}}]_{i,j,k} \cdot [\dot{\mathbf{x}}]_j \cdot [\Delta \mathbf{x}]_k \equiv \Delta \dot{f}_{\mathbf{x},\mathbf{x}} \cdot \Delta \mathbf{x}$$

and

$$\Delta f_{\mathbf{x},\mathbf{p}} \cdot \dot{\mathbf{p}} = [\Delta f_{\mathbf{x},\mathbf{p}} \cdot \dot{\mathbf{p}}]_i \approx [f_{\mathbf{x},\mathbf{p},\mathbf{x}}]_{i,j,k} \cdot [\dot{\mathbf{p}}]_j \cdot [\Delta \mathbf{x}]_k \equiv \Delta \dot{f}_{\mathbf{x},\mathbf{p}} \cdot \Delta \mathbf{x} \, .$$

Similarly,

$$\|\delta \bar{\mathbf{p}}\| \approx \left(\kappa(\Delta \bar{f}_{\mathbf{x},\mathbf{p}}) + \kappa(f_{\mathbf{x},\mathbf{p}}) \cdot \kappa(f_{\mathbf{x},\mathbf{x}}) \cdot \kappa(\Delta \bar{f}_{\mathbf{x},\mathbf{x}}) \right) \cdot \|\delta \mathbf{x}\| \, , \tag{21}$$

where

$$[\Delta f_{\mathbf{x},\mathbf{x}}^T \cdot \mathbf{z}]_j = [\Delta f_{\mathbf{x},\mathbf{x}} \cdot \mathbf{z}]_j \approx [f_{\mathbf{x},\mathbf{x},\mathbf{x}}]_{i,j,k} \cdot [\mathbf{z}]_i \cdot [\Delta \mathbf{x}]_k \equiv \Delta \bar{f}_{\mathbf{x},\mathbf{x}} \cdot \Delta \mathbf{x}$$

and

$$[\Delta f_{\mathbf{x},\mathbf{p}}^T \cdot \mathbf{z}]_j \approx [f_{\mathbf{x},\mathbf{p},\mathbf{x}}]_{i,j,k} \cdot [\mathbf{z}]_i \cdot [\Delta \mathbf{x}]_k \equiv \Delta \bar{f}_{\mathbf{x},\mathbf{p}} \cdot \Delta \mathbf{x} \, .$$

Low condition numbers of the respective second and third derivatives of the objective ensure numerical stability of tangent and adjoint optima of convex unconstrained objectives. Both $\Delta \dot{f}_{\mathbf{x},\mathbf{x}}$ and $\Delta \dot{f}_{\mathbf{x},\mathbf{p}}$ as well as $\Delta \bar{f}_{\mathbf{x},\mathbf{x}}$ and $\Delta \bar{f}_{\mathbf{x},\mathbf{p}}$ can be computed by AD.

4   Conclusion

Adjoint systems of linear equations are numerically stable with respect to errors in the primal solution. However, numerical stability of tangents and adjoints of implicit functions cannot be guaranteed in general. Sufficient conditions in terms of derivatives of the residual are given by Eqs. (15), (18), (19), (20) and (21). AD can be used to compute these derivatives. Corresponding symbolic tangents and adjoints should be augmented with optional estimation of conditions of the relevant derivatives.

References

1. Gebremedhin, A., Manne, F., Pothen, A.: What color is your Jacobian? Graph coloring for computing derivatives. SIAM Rev. **47**(4), 629–705 (2005)
2. Giles, M.: Collected matrix derivative results for forward and reverse mode algorithmic differentiation. In: Bischof, C., Bücker, M., Hovland, P., Naumann, U., Utke, J. (eds.) Advances in Automatic Differentiation. Lecture Notes in Computational Science and Engineering, vol. 64. Springer, Heidelberg (2008). https://doi.org/10.1007/978-3-540-68942-3_4
3. Griewank, A., Walther, A.: Evaluating Derivatives: Principles and Techniques of Algorithmic Differentiation. Number 105 in Other Titles in Applied Mathematics, 2nd edn. SIAM, Philadelphia (2008)
4. Naumann, U.: The Art of Differentiating Computer Programs. An Introduction to Algorithmic Differentiation. Number SE24 in Software, Environments, and Tools. SIAM (2012)
5. Naumann, U., Lotz, J., Leppkes, K., Towara, M.: Algorithmic differentiation of numerical methods: tangent and adjoint solvers for parameterized systems of nonlinear equations. ACM Trans. Math. Softw. **41**, 26 (2015)

Analysis of Parameters Distribution of EEG Signals for Five Epileptic Seizure Phases Modeled by Duffing Van Der Pol Oscillator

Beata Szuflitowska$^{(\boxtimes)}$ and Przemyslaw Orlowski$^{(\boxtimes)}$

West Pomeranian University of Technology in Szczecin, Sikorskiego 37, 70-313 Szczecin, Poland
bszuflitowska@zut.edu.pl

Abstract. Complex temporal epilepsy belongs to the most common type of brain disorder. Nevertheless, the wave patterns of this type of seizure, especially associated with behavioral changes, are difficult to interpret clinically. A helpful tool seems to be the statistical and time-frequency analysis of modeled epilepsy signals. The main goal of the study is the application of the Van der Pol model oscillator to study brain activity and intra-individual variability during complex temporal seizures registered in one patient. The achievement of the article is the confirmation that the statistical analysis of optimal values of three pairs of parameters of the duffing Van der Pol oscillator model enables the differentiation of the individual phases of the seizure in short-period seizure waves. In addition, the article attempts to compare the real signals recorded during the attack and modeled using frequency and time-frequency analysis. Similarities of power spectra and entropy samples of real and generated signals in low-frequency values are noted, and differences in higher values are explained about the clinical interpretation of the records.

Keywords: Van der Pol oscillator · EEG · Parameter estimation · Biological process model

1 Introduction

The electroencephalogram (EEG) is a representative signal informing about the state of the brain [1]. The shape of the wave may contain useful information about brain pathology. EEG records during some epileptic seizures, Alzheimer's, and Parkinson's disease are much more ordered oscillatory than in healthy records [1–4]. The bigger problem is distinguishing subtle changes in brain wave patterns as the seizure spreads. Some changes are very subtle highly subjective, the symptoms may appear at random in the time scale. Therefore, the EEG signal parameters, extracted, analyzed, and modeled using computers, are highly useful in diagnostics. The analysis of EEG relies mainly on time-frequency analysis [5–9], wavelet analysis [8, 10, 11]. In the case of newborns due to strong non-stationary properties of EEG signals, Nabeel and Sadiq [12] proposed

© The Author(s), under exclusive license to Springer Nature Switzerland AG 2022
D. Groen et al. (Eds.): ICCS 2022, LNCS 13352, pp. 188–201, 2022.
https://doi.org/10.1007/978-3-031-08757-8_18

Adaptive Directional Time-Frequency Distribution (ADTFD) can lead to better classification of ictal EEG signals. The ADTFD gives a highly concentrated time-frequency representation of spikes and sinusoids. In turn, in many hardware implementations of automatic epilepsy detection, wavelet transforms, principal component analysis, Hilbert-Huang transform, and support vector machines [13, 14]. Several types of entropy, i.e. sample, multiscale, and permutation entropy are used in the analysis of seizure spread. The experimental results obtained for 2 s EEG sequences show that the mean value of permutation entropy gradually decreases from the seizure-free (pre- and post-ictal) to the seizure phase [13]. Equally important parameters that measure the complexity of signal are Hjorth's parameters: activity, mobility, and complexity, which are useful for the quantitative description of the EEG. EEG modeling is also based on its non-stationary nature and includes i.e. random and backpropagation (BP) neural networks and coupled oscillators [15–21]. The system described in [16] determines the areas with the highest activity of spikes/poly-spikes of the signal received for one channel. In addition to the automatic selection, these discharges are also possible to verify and possibly change manually.

In the article, we have presented a modified variant of the deterministic duffing Van der Pol oscillator model, proposed in previous works by Ghorbanian et al., to model healthy and Alzheimer's disease signals [2, 3]. In our earlier works, we have used this model to analyze epileptic ictal signals for the first time [20, 21]. The current study is a continuation of this research.

An important goal of the article is to determine the relationship between the parameters, model sizes, and patterns of seizure waves. The study is an attempt to learn and understand the mechanisms accompanying the following five epileptic conditions: the onset of seizures, sequences with leg movement, records during automatic movements, related to the state of confusion, and final state of the seizure, registered in one patient. Another purpose of the paper is to evaluate the possibilities of differentiating epileptic states as above based on the parameters of the duffing Van der Pol oscillator model determined for the real EEG signal. We are driven by the motivation to use the obtained results in an expert system to support the process of medical diagnostics.

Similar to the articles [2, 3], the possibility of extracting frequency bands for which the dominant values of the power spectra of real and generated are examined. Application of the deterministic duffing Van der Pol oscillator model to differentiate the same five phases of seizures was described in [21], however, without a statistical analysis of the optimal parameter values. In turn, the statistical analysis of the optimal values of the model parameters for the differentiation of only three groups of signals: pre-, ictal, and post-ictal, was carried out by us in [20]. The values of sample entropy of real EEG signals registered during five ictal phases were presented in [21], without comparison with the values of sample entropy obtained for the modeled signals. To our knowledge, so far in the literature, spectrograms have not been used to analyze signals from such precisely separated seizure phases: the onset, the confusion, and automatic movements.

The main achievement of the article is the confirmation that the statistical analysis of optimal values of three pairs of parameters of the duffing Van der Pol oscillator model: linear stiffness coefficient ς, nonlinear stiffness coefficient ρ, Van der Pol damping coefficient ε enable the differentiation of the individual phases of the seizure in short period seizure waves, e.g. fast and slow waves. In addition, it has been shown that the power spectra of the real and generated signal are dominated by certain components in the frequency bands δ, θ, and α in each considered ictal stage.

The article analyzes the distribution of the duffing Van der Pol oscillator parameters for five seizure states: the onset of seizures, sequences with leg movement, records during automatic movements, related to the state of confusion, and final state of the seizure, derived from 15 registered EEG sequences. The determined distributions are presented in the form of box plots, independently for each of the 6 parameters of the duffing Van der Pol model: linear stiffness coefficients ς_1 and ς_2, nonlinear stiffness coefficients ρ_1 and ρ_2, Van der Pol damping coefficients ε_1 and ε_2. We made a comparative time-frequency analysis of epileptic real signals and the corresponding signals generated by the duffing Van der Pol oscillator model. The analysis of average values of power spectra is performed for three real and three generated signals in given frequency ranges in each considered phase of the seizure and the relative error between the average power values of the real and generated signal spectrum in three determined frequency bands: σ, θ, and α are calculated. The sample entropy values of real and generated signals obtained for the onset, the tangled stage, and automatic movements are compared. The achieved results are verified based on spectrograms made for real signals.

2 Materials and Methods

2.1 EEG Signals

EEG signals presented in this paper were recorded from a right-handed 55 aged female who takes Phenytoin, at Temple University Hospital and is seizure-free for 7 months [22]. A more detailed description of the patient and the test conditions can be found in [21]. We considered sequences 10 s (number of samples $N = 2500$) registered by electrode T3: 3 sequences corresponding to the onset of the seizure, 3 sequences of the phase with leg movements, sequences with registered automatic movements, 3 sequences of the entanglement, and 3 sequences associated with the end of the seizure.

2.2 Duffing Van Der Pol Oscillator

We based on the oscillator model described in the literature for the first time by Ghorbanian et al. The deterministic coupled system of duffing Van der Pol oscillators can be written as [2, 3]:

$$\dot{x}_1 = x_3$$
$$\dot{x}_2 = x_4$$
$$\dot{x}_3 = -(\varsigma_1 + \varsigma_2)x_1 + \varsigma_2 x_2 - \varsigma_1(x_1)^3 - \rho_2(x_1 - x_2)^3 + \varepsilon_1 x_3(1 - x_1)$$
$$\dot{x}_4 = \varsigma_2 x_1 - \varsigma_2(x_1 - x_2)^3 + \varepsilon_2 x_4\left(1 - (x_2)^2\right) \quad (1)$$

where ς is the linear stiffness coefficient, ρ is the nonlinear stiffness coefficient, associated with the strength of the duffing nonlinearity resulting in multiple resonant frequencies, ε is the Van der Pol damping coefficient related to the strength of Van der

Pol nonlinearity. Parameters ς_1, ρ_1, ε_1 and ς_2, ρ_2, ε_2 belong to the first and second oscillator, respectively.

The velocity of the second oscillator is selected as the model output. The initial conditions are equal to: $x_1(0) = 0$, $x_2(0) = 1$, $x_3(0) = 0$, $x_4(0) = 0$. Equation (1) is solved by Runge-Kutta iterative method.

2.3 Optimization Scheme

Figure 1 presents a block diagram of the optimization scheme proposed in the paper. The optimization process is repeated for each signal of mentioned earlier set of 15 recorded EEG signals. To determine the number of the sequence we introduced the parameter d, where $d = 1, 2,...,.15$.

Before calculating the discrete Fourier transform of each sample of real and generates EEG sequence has been multiplied by the appropriate Blackman's window coefficient [2]. The discrete Fourier transform (DFT) can be rewritten as:

$$X^d(k) = \sum_{n=0}^{N-1} x_w^d(n)\omega_N(n, k) \tag{2}$$

where:

$\omega_N = \exp(-j2\pi nk/N)$ is the Nth root of unity.

According to Fig. 1, the amplitude of DFT of the signal is normalized in the range of [0, 1]:

$$\hat{X}_d(k) = \frac{|X_d(k)|}{\max|X_d(k)|} \tag{3}$$

Next, the power of normalized DFT amplitude sequences in five major frequency bands are calculated according to the formula:

$$\widehat{P}_d^v = \frac{1}{|S_v|} \sum_{k\in Sv} \left(\hat{X}_d(k)\right)^2 \tag{4}$$

where: $v = 1,...,5$ is the number of the frequency band, Sv - set of discrete frequencies, corresponding to five major frequency bands [2]. The indexes v and Sv are presented in Table 1.

Table 1. The major EEG frequency bands.

Number of frequency band v	EEG band	Hz	S_v of discrete frequencies (k)
1	δ	1–4	$\{1,\ldots,15\}$
2	θ	4–8	$\{16,\ldots,31\}$
3	α	8–13	$\{32,\ldots,51\}$
4	β	13–30	$\{52,\ldots,119\}$
5	γ	30–60	$\{120,\ldots,239\}$

The power of normalized DFT amplitude of the generated signal is calculated in the same manner in the same way as shown in formula (4). Having the power averaged power for each frequency band of the real and generated signal with the determined values, we can determine the objective function L (5):

$$L(\Omega, d) = \sum_{v=1}^{5} \left(\hat{P}_d^v - \breve{P}^v P^v(\varsigma_1, \varsigma_2, \rho_1, \rho_2, \varepsilon_1, \varepsilon_2) \right)^2$$

$$\Omega = [\varsigma_1, \varsigma_2, \rho_1, \rho_2, \varepsilon_1, \varepsilon_2] \tag{5}$$

where: L is the cost function, Ω is the vector of design model variables, $0 \leq \varsigma_{1,2} \leq 200$, $0 \leq \rho_{1,2} \leq 100, 0 \leq \varepsilon_{1,2}$ are the decision variables of the optimization.

The optimization goal is error minimization:

$$\min_{\Omega} \ L(\Omega, d)$$

The optimization has been carried out using a genetic algorithm (GA) in a Matlab environment with bult-in function ga. The optimization scheme is repeated for each pair of real sequences and the corresponding generated sequence.

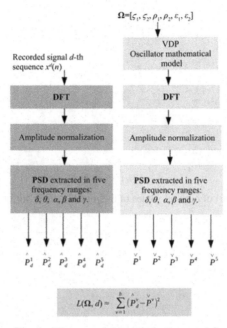

Fig. 1. Diagram of the optimization model.

3   Results

The analysis based on the optimal values of parameters, average values of the power spectrum of real and generated signals, the maximal difference between the power spectrum of generated and real signal, and the sample entropy of them is performed taking into account types of sequences. The estimated values of parameters are presented in the form of box plots, where the plots are made separately for each parameter of the model. Figure 2 shows the results obtained for the linear stiffness coefficient produces by the first oscillator. On the x-axis graph, separate states of seizures are marked as sequence no: 1 indicates the onset of seizures, 2 - sequences with leg movement, 3 - automatic movements, 4 is related to the state of confusion, and 5 determines the final state of the seizure. Low and comparable values of the parameter are obtained for the phases in which there are no changes in behavior, i.e. the onset of the seizure (marked as 1 in the graph and the end of the seizure marked as 5). High values of the parameter and large dispersion of the results, which reflect the height of the box, are achieved for the moments when there are large changes in the patient's behavior: leg movement and confusion.

Figure 3 has been created for the linear stiffness parameter generated by the second oscillator. Similar to the first diagram, we can distinguish two analogous phases with significantly greater dispersion of parameter values. Additionally, comparable median values optimal parameter values (marked with a red line inside the box) are obtained for the initial phase and with the accompanying automatic movements occurring in the patient. The low median value of the parameter also makes it possible to distinguish the

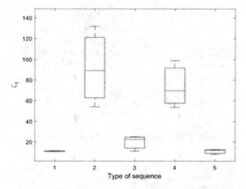

Fig. 2. The box chart of optimal values of the model's linear stiffness parameter ς_1.

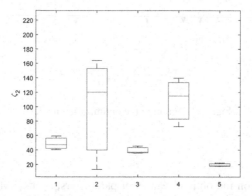

Fig. 3. The box chart of optimal values of the model's linear stiffness parameter ς_2.

final phase of the seizure. Low and comparable values of the parameter are obtained for the phases in which there are no changes in behavior, i.e. the onset of the seizure (marked as 1 in the graph and the end of the seizure marked as 5). High values of the parameter and large dispersion of the results, which reflect the height of the box, are achieved for the moments when there are large changes in the patient's behavior: leg movement and confusion.

Figures 4–5 refer to the nonlinear stiffness coefficients, ρ_1 and ρ_2, respectively. In the initial phase of the seizure, high values of the non-linear stiffness parameter generated by the first oscillator are achieved. A negative correlation is also obtained for the state in which the patient moves her leg, i.e. low values of nonlinear parameters.

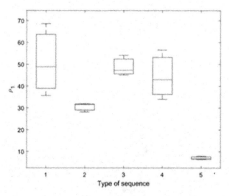

Fig. 4. The box chart of optimal values of the model's non-linear stiffness parameter ρ_1.

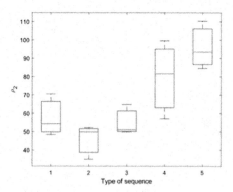

Fig. 5. The box chart of optimal values of the model's non-linear stiffness parameter ρ_2.

The final stage of the seizure is characterized only by a high parameter value of ρ_2. Figures 6–7 present results achieved for the Van der Pol damping coefficients. For states marked as 1 and 3, the median of ε_1 values hovers around 5 and is different from the other two phases (4 and 5). Based on Fig. 7, median, first and third quartile values of ε_2 are comparable in all phases.

In a further step power spectra of real signals (marked in blue) and generated by the oscillator model (red dotted line) are compared. Figure 8 shows the power spectra of real and generated EEG signals recorded during the entanglement state. Differences in the values of the power spectrum of the real and generated signal are obtained for α and β frequencies. From about 30 Hz, a rightward shift in the spectrum of the generated signal is observed.

Table 2 shows the average values of power spectra of three real and three generated signals in given frequency ranges in each considered phase of the seizure, where R denotes recorded data and G refers to generated signal. According to Fig. 8, the results obtained in Table 2 show that the power spectra of the real and generated signal are dominated by components in the frequency bands: δ, θ, and α (high average values of the power spectrum in these bands for each considered phase from Table 2).

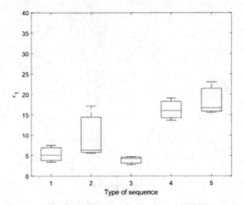

Fig. 6. The box chart of optimal values of the model's Van der Pol damping parameter ε_1.

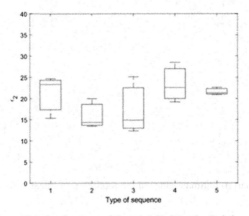

Fig. 7. The box chart of optimal values of the model's Van der Pol damping parameter ε_2.

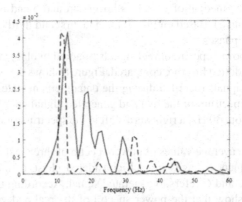

Fig. 8. Comparison of the power spectrum of the real and the generated signal corresponding to the entangled state.

Table 3 presents the average relative error between the average power values of the real and generated signal spectrum in the three dominant frequency bands of spectrograms: δ, θ, and α obtained in each determined phase of seizures. It can be seen clearly, that the stage of the seizure of low values of the determined ratio is the final state for each described EEG band. Interestingly, high values of average percentage power error for δ (equal to 50%) and low (5.6%) for θ are achieved for the seizure phases related to the patient's movements (marked in Table as 2 and 3). The α band is characterized by high power error values of the real and generated signal spectrum in the phases (1–4) close to 30% without the end of the seizure.

Table 2. The average power spectrum values of the real and generated signal in all discussed phases of seizure.

Phase		EEG band				
		δ	θ	α	β	γ
1 Onset	R	0.1860	0.2300	0.1200	0.0180	0.0010
	G	0.1970	0.2100	0.0900	0.0060	0.0006
2 Leg movement	R	0.0015	0.0038	0.0016	0.0004	0.0004
	G	0.0017	0.0039	0.0014	0.0005	0.0003
3 Automatic movements	R	0.0006	0.0024	0.0008	0.0003	0.0006
	G	0.0008	0.0028	0.0010	0.0004	0.0001
4 Confusion	R	0.0006	0.0046	0.0007	0.0005	0.0003
	G	0.0005	0.0050	0.0004	0.0002	0.0002
5 Final seizure state	R	0.1867	0.2150	0.1301	0.0067	0.0005
	G	0.1837	0.2167	0.1300	0.0080	0.0001

Table 3. The average percentage power error values of the real and generated signal spectrum in three determined frequency bands: δ, θ, and α.

Phase	EEG band		
	δ	θ	α
1 Onset	7.9%	16.7%	30.0%
2 Leg movement	50.0%	5.6%	30.0%
3 Automatic movements	50.0%	5.6%	30.0%
4 Confusion	24.5%	22.0%	36.9%
5 Final seizure state	7.0%	10.9%	14.0%

To examine the nature of the EEG signal more closely, spectrograms are made for a rectangular window of 512 sequence samples. The y axis of the spectrogram shows frequency and the x axis-time.

Figure 9 is made for sequences when the first panic pattern was registered. The spectrogram is dominated by low frequencies corresponding to waves δ, θ, and α. In the ranges of time 0–2 and 19–20 s of the seizure, spectrum building high frequencies (γ) is observed. In the time interval between 5 and 13 s the spectrum stabilizes to δ, θ, α, and β frequencies. From 14 to 18 s the components associated with complex slow waves dominate in the spectrum (rhythms δ, θ) and fast amplitude peak waves of low amplitude (α rhythm), which may be due to the patient's hyperventilation.

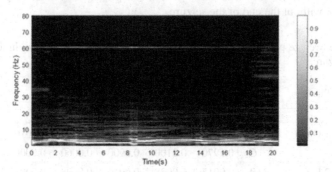

Fig. 9. A spectrogram of real EEG signal presented the onset of the seizure (marked in Tables 2–3 as phase 1).

Figures 10 and 11 present spectrograms obtained for the real signals during the confusion and automatic movements. The spectrogram presented in Fig. 10 shows a stable spectrum of all considered frequency bands. In the case of automatic movements, three phases can be distinguished in the spectrogram: in the 0–5 s time interval, δ, θ, and α frequencies dominate, between 6 and 10 s, also higher β and γ bands are present, from 10 to 20 s in the spectroscope there are frequencies up to 30 Hz. Next, a comparison of the calculated entropy value for real and generated signals for three phases in Figs. 9–11 is analysed. The results are summarized in Table 4.

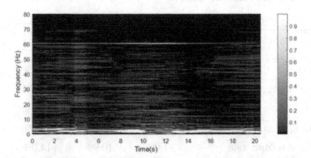

Fig. 10. A spectrogram of real sequences for entangled stage (marked in Tables 2–3 as phase 4).

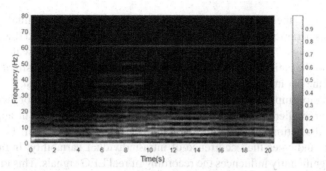

Fig. 11. A spectrogram of real sequences achieved during automatic movements (presented in Tables 2–3 as phase 3).

Table 4. The sample entropy of real and corresponding generated signals presented in Figs. 9–11.

Phase	Sample entropy	
	R	G
1	0.24	0.07
3	0.08	0.05
4	0.12	0.06

4   Discussion and Conclusion

Based on Table 2, it can see clearly that the dominant frequency bands in the power spectrum of signals presenting the complex temporal seizure are δ, θ, and α in all considered phases of the seizure. The plot obtained with the classical FFT analysis of the real signal is comparable with the plot obtained for the generated signal (see Fig. 8). In β and γ frequency ranges, the differences between the real signals and the corresponding generated signals occur in the average value of the spectra, i.e. average power spectrum in the onset obtained for real signal in β range is equal to 0.018 vs. 0.0006 for generated signal, and for the entanglement 0.0005 vs. 0.0002, respectively (Table 2). The results from Table 3 show that in states where the patient's anxious behavior affects the recording of brain waves (i.e. movements, the entanglement), high values of the coefficient are obtained, calculated as the average relative power error of generated and real signal in two determined frequency bands: σ and α, and low values-for θ band. The ratio values from Table 3 allow to distinguish sequences related to movements (marked as 2 and 3), in which similar average percentage power error values are obtained. Based on the low value of the error coefficients obtained for the δ band, the initial and final stages of the seizure (marked as 1 and 5) can be also eminent from those in which the propagation of the seizure waves took place over time. The proposed deterministic model encounters difficulties with chaotic, difficult to interpret waveforms related to the patient's behavior. An experienced clinician also had a problem with interpreting these states, who considered the record of the signal related to the movement of the leg difficult to assess. Based on the box plots, high dispersions can be seen in the optimal values of the linear stiffness parameters for the phases in which the patient is aroused (the second and fourth boxes in Figs. 2–3). On the other hand, for the states occurring without changing the behavior of the patient (the first and fifth box in Figs. 4–5), there is a slight variation in the values of the parameters. Considering the nonlinear parameters of stiffness, we note an inverse relationship, greater differentiation in the parameter values for the initial state (first box), and smaller for the leg movement (third box) on Figs. 4–5. The appearance of dispersion in the optimal values of the parameters of signals generated in the same phase of the seizure indicates the occurrence of intra-individual variability. Intra-individual variability significantly influences the recording of real EEG signals. This indicates that the proposed model, in this respect, maintains the physiological dependencies occurring in real signals.

Despite the presence of dispersion in the optimal values of the model parameters, their median, maximum or minimum values, as discussed in Sect. 3, allow determining the seizure phases and wave patterns associated with these states.

Low optimal values of linear and nonlinear stiffness parameters are noted in the case of recording low waves, the terminal phase of the seizure. Small optimal values of linear parameters and higher nonlinear parameters accompany the occurrence of wave waves with a complex of slow waves (the onset and low optimal values of linear parameters and higher nonlinear parameters accompany the occurrence of wave waves with a complex of slow waves). High values of linear and nonlinear parameters generate fast waves of high amplitude (hand movement, entanglement).

Besides it allows us to determine the parameters of the oscillator model, with the help of which we could differentiate the individual phases of the seizure in a short period, and thus differentiate seizure waves, e.g. fast and slow waves. In depth study of generated signals, including time-frequency analysis, is to add a stochastic component to the model.

References

1. Acharya, U.R., Molinari, F., Vinitha, S., Chattopadhyay, S.: Automated diagnosis of epileptic EEG using entropies. Biomed. Signal Process. Control **4**(7), 401–408 (2012)
2. Ghorbanian, P.: Non-Stationary Time Series Analysis and Stochastic Modeling of EEG and its Application to Alzheimer's Disease. [Doctoral dissertation, Villanova University] (2014)
3. Ghorbanian, P., Ramakrishnan, S., Ashrafiuon, H.: Stochastic non-linear oscillator models of EEG: the Alzheimer's disease case. Front. Comput. Neurosci. **9**(48) (2015). https://doi.org/10.3389/fncom.2015.00048
4. Botcharova, M.: Modelling and analysis of amplitude, phase and synchrony in human brain activity patterns. [Doctoral dissertation, University College London] (2014)
5. Yuan, Y., Xun, G., Jia, K., Zhang, A.: A multi-view deep learning method for epileptic seizure detection using short-time fourier transform. In: ACM-BCB Proceedings of the 8th ACM International Conference on Bioinformatics, Computational Biology, and Health Informatics, August 2017, pp. 213–222 (2017). https://doi.org/10.1145/3107411.3107419
6. Szuflitowska, B., Orlowski, P.: Comparison of the EEG signal classifiers LDA, NBC and GNBC based on time-frequency features. Pomiary Automatyka Robotyka **2**(21), 39–45 (2017)

7. Li, M., Chen, W.: FFT-based deep feature learning method for EEG classification. Biomed. Signal Process. Control **66**, 102492 (2021)
8. Chen, G., Xie, W., Bui, T.D., Krzyżak, A.: Automatic epileptic seizure detection in EEG using nonsubsampled wavelet–fourier features. J. Med. Biol. Eng. **37**, 123–131(2017)
9. Khan, N.A., Ali, S.: Classification of EEG signals using adaptive time-frequency distributions. Metrol. Meas. Syst. **23**(2), 251–260 (2016)
10. Kocadaglia, O., Langarib, R.: Classification of EEG signals for epileptic seizures using hybrid artificial neural networks based wavelet transforms and fuzzy relations. Expert Syst. Appl. **88**, 419–434 (2017)
11. Alturki, F.A., AlSharabi, K., Abdurraqeeb, A.M., Aljalal, M.: EEG signal analysis for diagnosing neurological disorders using discrete wavelet transform and intelligent techniques. Sensors **21**(20), 6932 (2021). https://doi.org/10.3390/s21206932
12. Zhang, Q., Hu, Y., Potter, T., Li, R., Quach, M., Zhang, Y.: Establishing functional brain networks using a nonlinear partial directed coherence method to predict epileptic seizures. J. Neurosci. Methods **329**, 108447 (2020)
13. Shriram, R., Baskar, V.V., Martin, B., Sundhararajan, M., Daimiwal, N.: Energy distribution and coherence-based changes in normal and epileptic electroencephalogram. In: Satapathy, S.C., Bhateja, V., Das, S. (eds.) Smart Intelligent Computing and Applications. SIST, vol. 104, pp. 625–635. Springer, Singapore (2019). https://doi.org/10.1007/978-981-13-1921-1_61
14. Albera, I., et al.: ICA-based EEG denoising: a comparative analysis of fifteen methods. Bull. Pol. Acad. Sci.: Tech. Sci. **60**(3) (2012). https://doi.org/10.2478/v10175-012-0052-3
15. Rafiammal, S.S., et al.: A low power and high performance hardware design for automatic epilepsy seizure detection. Int. J. Electron. Telecommun. **65**(4), 707–712 (2019)
16. Gaidar, V., Sudakov, O.: Design of wearable EEG device for seizures early detection. Int. J. Electron. Telecommun. **67**(2), 187–192 (2021)
17. Liu, L.: Recognition and analysis of motor imagery EEG signal based on improved BP neural network. IEEE Access **7**, 47794–47803 (2019)
18. Gandhi, T., et al.: Epilepsy diagnosis using combined duffing oscillator and PNN model. J. Bioinform. Intell. Control **1**(1), 64–70 (2012)
19. Tabi, C.B.: Dynamical analysis of the FitzHugh-Nagumo oscillatons through a modified Van der Pol equation with fractional-order derivative term. Int. J. Non-Linear Mech. **105**, 173–178 (2018)
20. Szuflitowska, B., Orlowski, P.: Statistical and physiologically analysis of using a Duffing-van der Pol oscillator to modeled ictal signals. In: Proceedings of the 16th International Conference on Control, Automation, Robotics and Vision (ICARCV), pp. 1137–1142 (2020). ieee.org/document/9305339
21. Szuflitowska, B., Orlowski, P.: Analysis of complex partial seizure using non-linear duffing van der pol oscillator model. In: Paszynski, M., Kranzlmüller, D., Krzhizhanovskaya, V.V., Dongarra, J.J., Sloot, P.M.A. (eds.) ICCS 2021. LNCS, vol. 12745, pp. 433–440. Springer, Cham (2021). https://doi.org/10.1007/978-3-030-77970-2_33
22. Obeid, I., Picone, J., Harabagiu, S.: Automatic discovery and processing of EEG cohorts from clinical records. In: Big Data to Knowledge All Hands Grantee Meeting, p. 1. Bethesda, Maryland, USA: National Institutes of Health (2016). https://pubmed.ncbi.nlm.nih.gov/24509598/

Multi-criterial Design of Antennas with Tolerance Analysis Using Response-Feature Predictors

Anna Pietrenko-Dabrowska[1] [ID], Slawomir Koziel[1,2(✉)] [ID], and Leifur Leifsson[3] [ID]

[1] Faculty of Electronics Telecommunications and Informatics, Gdansk University of Technology, Narutowicza 11/12, 80-233 Gdansk, Poland
anna.dabrowska@pg.edu.pl
[2] Engineering Optimization and Modeling Center, Department of Engineering, Reykjavík University, Menntavegur 1, 102 Reykjavík, Iceland
koziel@ru.is
[3] School of Aeronautics and Astronautics, Purdue University, West Lafayette, IN 47907, USA
leifur@purdue.edu

Abstract. Imperfect manufacturing is one of the factors affecting the performance of antenna systems. It is particularly important when design specifications are strict and leave a minimum leeway for a degradation caused by geometry or material parameter deviations from their nominal values. At the same time, conventional antenna design procedures routinely neglect to take the fabrication tolerances into account, which is mainly a result of a challenging nature of uncertainty quantification. Nevertheless, the ability to assess the effects of parameter deviations and to mitigate thereof is instrumental in achieving truly robust antenna designs. Furthermore, identifying the antenna-specific relationships between nominal requirements and tolerance immunity is essential to determine the necessary levels of fabrication accuracy, which affects the both the reliability and the manufacturing costs. This paper proposes a technique for multi-criterial optimization of antenna structures oriented towards rendering a family of designs representing trade-offs between the nominal performance and the robustness. The fundamental components of our procedure are feature-based regression models constructed at the level of selected characteristics points of the antenna outputs. The trade-off designs are generated sequentially, using local search carried out for gradually relaxed nominal requirements. Numerical experiments conducted for two microstrip antennas demonstrate that the proposed algorithm is capable of yielding the performance/robustness Pareto set at the cost of only a few dozens of EM analysis of the antenna at hand per design, while ensuring reliability, as validated by means of EM-based Monte Carlo simulation.

Keywords: Antenna design · Multi-criterial optimization · Simulation-based design · Manufacturing tolerances · Statistical analysis · Response features

1 Introduction

Manufacturing processes such as chemical etching or mechanical milling are of finite resolution and accuracy, whereas our knowledge of material parameters (e.g., substrate

D. Groen et al. (Eds.): ICCS 2022, LNCS 13352, pp. 202–216, 2022.
https://doi.org/10.1007/978-3-031-08757-8_19

permittivity) and operating conditions (e.g., input power level, temperature) is always limited. At the same time, the aforementioned uncertainties, especially deviations of geometry parameters from their nominal values, may be detrimental to the electrical and field characteristics of antennas [1]. In order to meet the stringent performance demands imposed on contemporary radiating structures, the design process should account for the effects of uncertainties to ensure that the system operates properly even under the most pessimistic scenarios.

In practical terms, the improvement of antenna performance requires utilization of numerical optimization methods [2–4]. At the same time, for the sake of reliability, parameter tuning is normally carried out using full-wave electromagnetic (EM) simulation models, which incurs considerable computational expenses. These are especially high in the case of global search procedures [5–7]. If the design process is to account for both the nominal performance and the effects of fabrication tolerances, in particular, if performance-robustness trade-offs are of interest, multi-objective optimization (MO) becomes imperative. MO is a CPU-heavy endeavor. The most popular acceleration methods involve surrogate modeling techniques [8, 9], both data-driven (kriging [10], support-vector regression [11]) or physics-based (e.g., space mapping [12], sequential domain patching [13]), often combined with machine-learning methodologies [14].

Quantification of the effects of fabrication tolerances requires appropriate statistical performance metrics. In the case of antennas, it is usually the yield [15], which is a likelihood of satisfying given performance requirements under the assumed probability distributions that govern deviations of the antenna parameters. Consequently, robust design techniques are mainly concerned with yield improvement [16, 17]. The alternative is to seek for the maximum allowed levels of input tolerances, for which the system outputs remain acceptable (maximum input tolerance hypervolume, MITH [18]). Unfortunately, estimation of the yield is a computationally expensive process. For example, EM-driven Monte Carlo (MC) simulation typically requires hundreds of EM analyses. Most of state-of-the-art statistical analysis methods rely on surrogate modeling methods [19–21], with a notable example of polynomial chaos expansion (PCE) [22]. Yet, handling higher-dimensional problems is still an issue due to considerable initial cost of surrogate model construction. A possible way of alleviating these difficulties is performance-driven modeling [23].

The literature offers few methods for multi-objective antenna design with tolerance analysis. For example, in [24], kriging surrogates are employed along with the worst-case analysis at the Pareto-optimal designs found by means of the particle swarm optimization algorithm. Machine-learning approach involving Gaussian Process Regression surrogates has been reported in [25], whereas [26] is the only methods that explicitly handles input tolerance hypervolume as one of the design objectives. In all cases, low-dimensional parameter spaces are considered.

This paper introduces a novel surrogate-based algorithm for low-cost tolerance-aware multi-objective design of antenna structures. In our methodology, maximization of the input tolerance levels for which the design specifications are still met is treated as one of the explicit objectives, the other being nominal performance of the antenna at hand. The optimization process is expedited through the employment of feature-based regression models, rendered at the level of suitably chosen characteristic points of

antenna responses. The Pareto-optimal designs are identified sequentially for the selected
values of relevant performance figures, using local gradient-based tuning. The presented
technique is demonstrated using two microstrip antennas, and shown to be both reliable
and computationally efficient with the CPU cost of generating trade-off designs as low
as a few dozens of EM simulations per point.

2 Multi-criterial Antenna Optimization with Tolerance Analysis

This section introduces the proposed multi-objective optimization strategy with tolerance
analysis. Formulation of the design task is followed by an exposition of the statistical
analysis approach, a description of the procedure for identifying the trade-off designs,
as well as a summary of the entire MO framework.

2.1 Problem Formulation

We denote by $R(x)$ the antenna responses corresponding to the parameter vector $x = [x_1 \ldots x_n]^T$, and obtained through full-wave EM analysis. We will also use additional
symbols to denote specific frequency characteristics such as reflection $S_{11}(x,f)$, axial
ratio $AR(x,f)$, or gain $G(x,f)$, where f stands for the frequency. The function $F_p(x)$ will
be used to denote the nominal performance for the antenna, i.e., assuming no fabrication
tolerances.

Consider a multi-band antenna with the target operating frequencies f_{0k}, $k = 1, \ldots,$
N, and target bandwidths B_k. The design specifications are defined for a performance
parameter $P(x,f)$, which should not exceed the value of $P_{\max}$ over the bandwidths of
interest. In other words, the specifications are satisfied if

$$\max\left\{f \in \bigcup_{k=1}^{N} \left[f_{0k} - B_k, f_{0k} + B_k\right] : |P(x,f)|\right\} \le P_{\max} \tag{1}$$

For example, if $P(x,f) = |S_{11}(x,f)|$ (antenna input characteristics), the acceptable
level is typically set to $P_{\max} = S_{11.\max} = -10$ dB.

The best nominal design x^p is obtained by improving the performance parameter P
as much as possible over the target bandwidths, i.e., we have

$$x^p = \arg\min_{x}\{\max\{f \in \bigcup_{k=1}^{N} \left[f_{0k} - B_k, f_{0k} + B_k\right] : P(x,f)\}\} \tag{2}$$

According to (2), the target nominal antenna performance $F_p(x)$ is simply $P_{\max}$.

Let $F_r(x)$ be a function representing the antenna design robustness. In this work,
we assume that parameter deviations follow independent Gaussian distributions of zero
mean and a variance σ (the same for all parameters); generalization of arbitrary distributions is straightforward. We define $F_r(x) = \sigma(x)$, where the dependence on the design x
emphasizes the fact that the maximum allowed variance is a function of antenna parameters. The meaning of F_r is that—at design x—it is the maximum value of the variance
σ, for which the performance specifications are still satisfied for any design perturbed
with respect to x, with the perturbations not larger than 3σ.

The tolerance-aware multi-objective optimization task can be then formulated as

$$\mathbf{x}^* = \arg\min_{\mathbf{x}}\left[F_p(\mathbf{x}) \quad - F_r(\mathbf{x})\right] \tag{3}$$

Thus, the objective is to improve both the nominal performance $F_p(x)$ and the robustness $F_r(x)$. Note that both objectives are conflicting as imposing more demanding target nominal performance leads to a reduced robustness, because there is a smaller margin for parameter deviations left. We also have two extreme designs: (i) the best nominal design x^p, and (ii) the minimum acceptable performance design x^r. The latter corresponds to the highest target value of F_p that can be accepted for a given application (e.g., -10 dB in the case of reflection response). Also, x^r is found by maximizing F_r given the aforementioned highest value of F_p. Here is a brief characteristic of the two designs:

- x^p: as this design corresponds to the best nominal performance (e.g., the lowest in-band reflection of the antenna), it features the minimum robustness. In particular, the level of parameter deviations ensuring the fulfilment of performance specifications is zero because any deviation from the nominal values results in worsening of F_p;
- x^r: at this design, we have the largest performance margin w.r.t. the best nominal design x^p. Thus, x^r exhibits the largest robustness as the feasible region for the highest considered value of F_p is the largest.

The designs that are globally non-dominated in the Pareto sense [27] w.r.t. to F_p and F_r form the Pareto front X_P [27]. These are the best possible trade-offs between the nominal performance and the robustness. Our goal is to identify a discrete subset of X_P, distributed uniformly along the front. The concepts considered in this sections have been illustrated in Fig. 1.

2.2 Yield Estimation by Means of Response Features

The robustness $F_r(x)$ is defined to be the maximum value of the variance σ of Gaussian probability distributions characterizing the geometry parameter deviations, for which the fabrication yield retains 100%. The yield is defined as [28]

$$Y(\mathbf{x}) = \int_{X_f} p(\mathbf{y}, \mathbf{x})d\mathbf{y} \tag{4}$$

In (4), $p(y, x)$ is a probability density function describing statistical variations of the design y w.r.t. the vector x. The feasible space X_f contains the designs that meet the performance specifications (cf. (1)). As X_f is unknown explicitly, the yield is normally approximated through Monte Carlo (MC) simulation as

$$Y(\mathbf{x}) = N_r^{-1} \sum_{k=1}^{N_r} H(\mathbf{x} + d\mathbf{x}^{(k)}) \tag{5}$$

where $dx^{(k)}$, $k = 1, ..., N_r$, are generated using the function p. The function H equals 1 if the design specifications are satisfied, and zero otherwise. Evaluation of (5) is computationally expensive, therefore, surrogate modeling techniques are often used for the sake of accelerating the process [19–21]. The robustness metric $F_r(x) = \sigma(x)$ is computed as

$$F_r(\mathbf{x}) = -\arg \max_{\sigma} \{Y(\mathbf{x}, \sigma) = 1\} \qquad (6)$$

where the explicit dependence of Y on σ is to emphasize that the variance determines the input tolerance levels, which affect the yield.

Here, efficient evaluation of (6) is ensured by employing feature-based regression models described below. The response feature technology has been introduced in [29] to accelerate antenna parameter tuning. The key idea is to reformulate the design task in terms of suitable chosen characteristic (or feature) points, e.g., frequency and level coordinates of antenna resonances, the coordinates of which are in weakly-nonlinear relationship with the antenna geometry parameters. This reformulation leads to a faster convergence of the optimization algorithms [30], as well as a reduced number of training data points when constructing surrogate models [31].

An illustration of the feature points for a reflection response of a triple band antenna can be found in Fig. 2. In the considered example, the points are defined to verify satisfaction of the performance specs imposed on the impedance matching of the device.

In relation to the performance requirements of (1), the feature vector $\mathbf{P}(x)$ at design x is defined as

$$\mathbf{P}(\mathbf{x}) = [p_1(\mathbf{x})\, p_2(\mathbf{x}) \dots p_{2N}(\mathbf{x})]^T = [f_1(\mathbf{x})\, f_2(\mathbf{x})\ \dots\ f_{2N}(\mathbf{x})]^T \qquad (7)$$

Here, the frequencies f_{2k-1} and f_{2k} are such that $P(x, f_{2k-1}) = P(x, f_{2k}) = P_{\max}$ for the kth operating band, $k = 1, \dots, N$. The condition (1) can be reformulated using $\mathbf{P}$ as.

$$p_{2k-1}(\mathbf{x}) \leq f_{0.k} - B_k, \quad p_{2k}(\mathbf{x}) \geq f_{0.k} + B_k, k = 1, \dots, N \qquad (8)$$

As the dependence between the feature points and antenna geometry parameters is weakly nonlinear, the feature vector $\mathbf{P}(x)$ at the design x located in a small vicinity of the current design $\mathbf{x}^{(i)}$ (produced in the course of the optimization process) can be predicted using a simple regression model. Here, we use a linear model $L_P^{(i)}(x)$

$$L_P^{(i)}(\mathbf{x}) = [p_{L.1}(\mathbf{x}) \dots p_{L.2N}(\mathbf{x})]^T = \begin{bmatrix} l_{0.1} + \mathbf{L}_1^T(\mathbf{x} - \mathbf{x}^{(i)}) \\ \vdots \\ l_{0.2N} + \mathbf{L}_{2N}^T(\mathbf{x} - \mathbf{x}^{(i)}) \end{bmatrix} \qquad (9)$$

The model is identified using $n + 1$ training pairs $\{\mathbf{x}_B^{(j)}, \mathbf{P}(\mathbf{x}_B^{(j)})\}, j = 1, \dots, n + 1$, arranged as follows: $\mathbf{x}_B^{(1)} = \mathbf{x}^{(i)}$, and $\mathbf{x}_B^{(j)} = \mathbf{x}^{(i)} + [0 \dots 0\, d\, 0 \dots 0]^T$ (d on the (j– 1)th position). The distance parameter $d = 3\sigma$, where σ is the variance of the Gaussian distribution governing the antenna parameter deviations. The coefficients of $L_P^{(i)}$ are found as.

$$\begin{bmatrix} l_{0.j} \\ \mathbf{L}_j \end{bmatrix} = \begin{bmatrix} 1 & (\mathbf{x}_B^{(1)} - \mathbf{x}^{(i)})^T \\ \vdots & \vdots \\ 1 & (\mathbf{x}_B^{(n+1)} - \mathbf{x}^{(i)})^T \end{bmatrix}^{-1} \begin{bmatrix} p_j(\mathbf{x}_B^{(1)}) \\ \vdots \\ p_j(\mathbf{x}_B^{(n+1)}) \end{bmatrix}, j = 1, \dots, 2N \qquad (10)$$

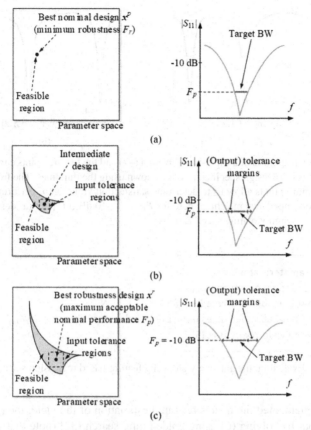

Fig. 1. Tolerance-aware multi-objective antenna design. A feasible region, shaded grey, contains designs satisfying performance requirements for a given F_p; the region becomes larger when relaxing F_p. The right-hand-side plots show exemplary reflection responses vs. target impedance bandwidth for different levels of target nominal performance threshold F_p: (a) at the best nominal design x^p, the feasible region is a single point x^p, hence, the input tolerance level is zero; (b) for an intermediate design, the feasible region is larger and the most robust design is centred therein to maximize the input tolerance ranges ensuring performance requirements satisfaction; (c) for the most robust design, corresponding to the maximum acceptable level of F_p (e.g., –10 dB for $|S_{11}|$), the input tolerance levels are the largest upon concluding the optimization process. The family of designs obtained for different values of F_p form a Pareto set (performance vs. robustness trade-offs).

The robustness metric $F_r(x)$ of (6) is evaluated by numerically integrating (4) with the use of the regression model (9). As the condition (1) is equivalent to (8), the yield $Y(x,\sigma)$ can be estimated using random observables $x_r^{(j)}$ allocated using the probability distribution characterized by the variance σ. The yield evaluation procedure has been summarized in Fig. 3. All steps in the above algorithm of Fig. 3 are vectorized to speed up the process.

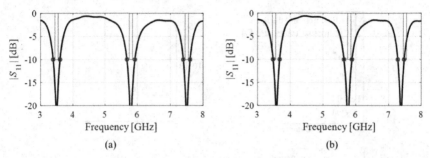

Fig. 2. Reflection responses of a triple-band antenna (—) and the feature points corresponding to −10 dB |S_{11}| levels (o). Design specifications are shown using the thin lines. The frequency coordinates of the feature points allow us to determine satisfaction/violation of performance requirements imposed on impedance matching, here for $P_{max} = -10$ dB (cf. (1)): (a) design satisfying specifications, (b) design violating specifications.

---

1. Input parameter: variance σ;
2. Generate random observables $\{x_r^{(j)}\}_{j=1,\,...,\,N_r}$;
3. Evaluate regression surrogate $L_P^{(i)}(x_r^{(j)})$ for $j = 1, ..., N_r$;
4. Evaluate (8) for all observables using predicted feature points $p_{Lk}(x_r^{(j)})$, $j = 1, ..., N_r$;
5. Estimate the yield $Y(x,\sigma)$ as in (5).

Fig. 3. Evaluating the antenna yield using feature-based regression surrogate.

Having implemented the means for rapid estimation of the yield, the evaluation of F_r is carried out by solving (6) using golden ratio search [32] (note that given a joint variance σ, the task (6) is a one-dimensional problem). Should the probability distribution be determined by multiple parameters (e.g., a covariance matrix), the problem (6) can be solved by means of other methods, e.g., gradient-based algorithms.

2.3 Generating Trade-off Designs

Our goal is to generate N_P trade-off designs, which form a discrete set of Pareto-optimal vectors w.r.t. the objectives F_p and F_r. Note that the Pareto front is spanned between the best nominal design x^p, and the most robust design x^r (cf. Sect. 2.1). The first trade-off design is therefore assigned as $x^{(1)} = x^p$, and it is obtained using (2). Consequently, the nominal objective function value $F_p(x^{(1)})$, denoted as $P_{max.1}$, is

$$P_{max.1} = \max\left\{ f \in \bigcup_{k=1}^{N} \left[f_{0k} - B_k, f_{0k} + B_k \right] : P(x^p, f) \right\} \quad (11)$$

In the next step, we set $P_{max.NP} = P_{max}$ (the maximum acceptable target level), which will determine the opposite end of the Pareto set. The $N_P - 1$ trade-off designs $x^{(j)}$, $j = 2, ..., N_P$, that remain to be generated will be obtained for the target levels $P_{max.j}$, $j =$

$1, \ldots, N_P$, uniformly distributed between $P_{\max.1}$ and $P_{\max.NP}$, i.e., we set

$$F_p(\mathbf{x}^{(j)}) = P_{\max.j} = P_{\max.1} + \left[P_{\max.Np} - P_{\max.1}\right]\frac{j-1}{N_P-1} \tag{12}$$

This arrangement leads to equally-spaced trade-off designs w.r.t. the nominal performance objective F_p.

The designs $\mathbf{x}^{(j)}, j = 2, \ldots, N_P$, are found as

$$\mathbf{x}^{(j)} = \arg \min_{\mathbf{x}} F_r(\mathbf{x}) \tag{13}$$

with $P_{\max}$ in (8) set to $P_{\max.j}$. Solving (13) means that the antenna robustness, understood as in (6), is maximized for the (nominal) target value set to $P_{\max.j}$.

The problem (13) is solved using the trust-region (TR) framework [33], which produces a series of approximation $\mathbf{x}^{(j.i+1)}$ to $\mathbf{x}^{(j)}$ as

$$\mathbf{x}^{(j.i+1)} = \arg \min_{||\mathbf{x}-\mathbf{x}^{(j.i)}||\leq d^{(i)}} F_r(\mathbf{x}) \tag{14}$$

starting from $\mathbf{x}^{(j.0)} = \mathbf{x}^{(j-1)}$. The search region size $d^{(i)}$ is adjusted based on standard TR rules [33]. The acceptance of a design $\mathbf{x}^{(j.i+1)}$ depends on the gain ratio

$$r = \frac{F_r^{\#}(\mathbf{x}^{(j.i+1)}) - F_r(\mathbf{x}^{(j.i)})}{F_r(\mathbf{x}^{(j.i+1)}) - F_r(\mathbf{x}^{(j.i)})} \tag{15}$$

which compares the actual improvement of the antenna robustness with the prediction of the feature-based regression surrogate. Note that $F_r^{\#}$ in the numerator of (15) is calculated as in Sect. 2.2, but with $L_P^{(j.i)}$ replaced by the linear model $L_P^{\#(j.i)}$. The latter is constructed as in (9), (10) but with the coefficient vector $[l_{0.1} \ldots l_{0.2N}]^T$ replaced by $P(\mathbf{x}^{(j.i+1)})$. Using $F_r^{\#}$ enables low-cost evaluation of the gain ratio (only one EM simulation is involved). Although this is an approximation, it is justified by a typically small distance between $\mathbf{x}^{(j.i)}$ and $\mathbf{x}^{(j.i+1)}$, comparable to σ.

The vector $\mathbf{x}^{(j.i+1)}$ is accepted if $r > 0$. Otherwise, the search radius $d^{(i)}$ is reduced, and the iteration is repeated. The termination condition is $||\mathbf{x}^{(i+1)} - \mathbf{x}^{(i)}|| < \varepsilon$ OR $d(i) < \varepsilon$, where ε is the required resolution of the search process (e.g., $\varepsilon = 10^{-3}$). The concept of iterative generation of the trade-off designs has been shown in Fig. 4.

2.4 Optimization Algorithm

The flow diagram of the proposed tolerance-aware MO procedure has been shown in Fig. 5. As mentioned earlier, the best nominal design $\mathbf{x}^p$ is first identified, followed by the establishment of the target performance levels $P_{\max.j}$. The latter are determined using the assumed maximum acceptable performance level $P_{\max}$, $F_p(\mathbf{x}^p)$, and the number of designs N_P. The performance-robustness trade-off designs are then obtained by sequentially solving (13) as described in Sect. 2.3.

3 Verification Case Studies

This section provides numerical verification of the multi-objective optimization procedure of Sect. 2. It is based on two microstrip antennas, a dual-band dipole, and a quasi-Yagi structure. In both cases, the goal is to find the trade-offs between the nominal performance defined through maximum in-band reflection, and the robustness, defined as the maximum level of parameter deviations that still ensure 100% fabrication yield.

3.1 Case I: Dual-Band Dipole Antenna

Consider a dual-band dipole with truncated substrate [34] shown in Fig. 6(a), implemented on RO4003 substrate ($\varepsilon_r = 3.38$, $h = 0.81$ mm). The parameter vector is $x = [L_{rr}\ d\ W_s\ W_d\ S\ L_d\ L_{gr}\ W_{gr}]^T$ (dimensions in mm, except the relative ones ending with the subscript r). Other parameters are: $W_r = 5$ mm, $L_s = 5$ mm, $L_0 = 25$ mm, $W_0 = 1.9$ mm, $L_r = L_{rr}((W_s - W_0)/2 - W_d - d)$, $L_g = L_{gr}(L_0 - W_g/2 + W_0/2)$, $W_g = W_{gr}W_s$, and $g = W_d$. The EM model is implemented in CST Microwave Studio and evaluated using the time-domain solver.

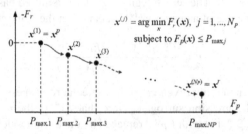

Fig. 4. Conceptual illustration of sequential generation of performance-robustness trade-off designs.

The target operating frequencies and bandwidths are $f_{01} = 3.5$ GHz, $f_{02} = 4.2$ GHz, and $B_1 = B_2 = 80$ MHz, respectively. The best nominal performance design is $x^p = [0.91\ 1.45\ 48.01\ 3.66\ 1.80\ 4.97\ 1.00\ 0.38]^T$. At this design, we have $F_p(x^p) = -15.1$ dB, and $F_r(x^p) = 0$ (cf. Sect. 2.1). Five more trade-off designs have been obtained, corresponding to $P_{max.2} = -14$ dB, $P_{max.3} = -13$ dB, through $P_{max.6} = -10$ dB (the highest acceptable in-band reflection level), as shown in Table 1 and Fig. 6(b). Figure 7 visualizes EM-driven Monte Carlo (MC) simulation for the selected designs. MC confirms that the fabrication yield is indeed close to 100% for all pairs $\{F_p(x^{(j)}), F_r(x^{(j)})\}$. The actual yield is between 98 and 100% (design dependent), yet, it should be noted that MC itself is characterized by a relatively large yield estimation variance due to using only 500 samples. The proposed MO approach is computationally efficient. The average cost of generating one trade-off design is only about 62 EM simulations of the antenna structure, which is possible by utilization of the feature-based surrogates (cf. Sect. 2.2).

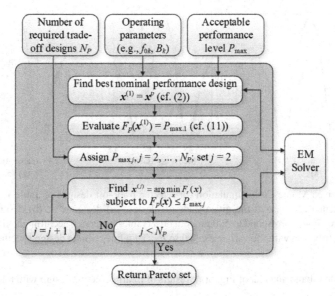

Fig. 5. Flow diagram of the proposed tolerance-aware multi-objective optimization algorithm using the feature-based regression surrogates and trust-region parameter adjustment process.

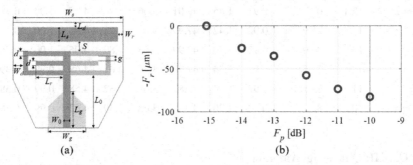

Fig. 6. Dual-band dipole antenna with truncated substrate [34]: (a) antenna geometry, the light-gray shade marks the ground plane, (b) performance-robustness trade-off designs obtained using the proposed procedure. The vertical line marks the maximum acceptable in-band reflection level.

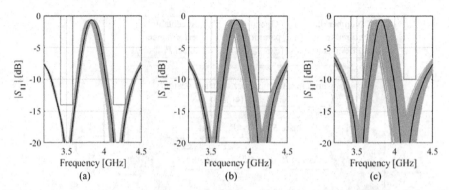

Fig. 7. EM-driven Monte Carlo simulation for selected trade-off designs of Table 1. Black line shows the antenna response at the given trade-off design: (a) design $x^{(2)}$, (b) design $x^{(4)}$, (c) design $x^{(6)}$, grey lines correspond to 500 random observables generated according to the assumed probability distribution with the variance equal to F_r. Thin lines denote design specifications.

Table 1. Dual-band antenna of Fig. 6(a): results of multi-objective design with tolerances

Design	Objectives		Geometry parameters [absolute in mm, relative unitless]							
	F_p [dB]	F_r [μm]	L_{rr}	d	W_s	W_d	S	L_d	L_{gr}	W_{gr}
$x^{(1)} = x^r$	− 15.1	0	0.91	1.45	48.01	3.66	1.80	4.97	1.00	0.38
$x^{(2)}$	− 14	25.9	0.92	1.42	47.99	3.62	1.78	4.93	1.00	0.38
$x^{(3)}$	− 13	35.0	0.92	1.40	47.99	3.65	1.80	4.95	1.00	0.38
$x^{(4)}$	− 12	57.5	0.92	1.34	47.98	3.72	2.01	4.68	0.99	0.39
$x^{(5)}$	− 11	73.5	0.92	1.24	47.87	3.65	2.32	4.58	0.99	0.38
$x^{(6)} = x^r$	− 10	82.7	0.92	1.20	47.85	3.68	2.39	4.62	0.99	0.38

3.2 Case II: Quasi-yagi Antenna

The center frequency and bandwidth are $f_{01} = 2.5$ GHz and $B_1 = 50$ MHz, respectively. Furthermore, the realized gain at 2.5 GHz is to be at least 7.9 (i.e., 8 dB with the tolerance of 0.1 dB). The best nominal performance design $x^p = [20.21\ 12.33\ 16.47\ 26.09\ 52.06$ $1.83\ 1.02\ 4.39\ 4.26\ 0.37\ 0.44\ 0.98\ 0.71\ 0.72]^T$ corresponds to $F_p(x^p) = -17.0$ dB. Seven additional trade-off designs have been found, corresponding to $P_{max.2} = -16$ dB, $P_{max.3}$ $= -15$ dB, through $P_{max.7} = -10$ dB.

The results have been gathered in Table 2 and Fig. 8(b). Figure 9 visualizes EM-based Monte Carlo simulation for the selected trade-off designs. The average value of the estimated yield is 97%, which is sufficiently close to 100 given the challenging nature of the problem (fourteen parameters, and limited predictive power of the surrogate model for larger values of input tolerances). Again, the proposed algorithm exhibits excellent computational efficiency with the average cost of identifying the trade-off designs of only 82 EM simulations per point.

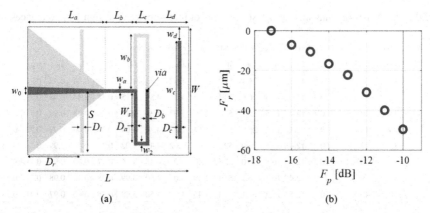

(a)   (b)

Fig. 8. Quasi-Yagi antenna with integrated balun [35]: (a) antenna geometry, light-gray shade indicates ground-plane metallization, (b) performance-robustness trade-off designs obtained using the proposed procedure for multi-objective optimization with tolerances. The vertical line marks the maximum acceptable in-band reflection level.

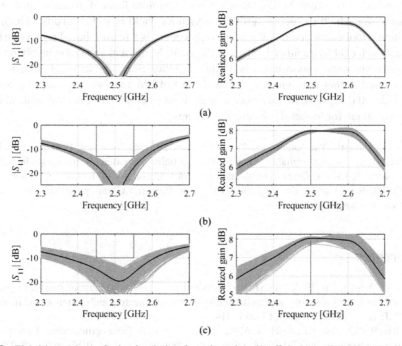

(a)

(b)

(c)

Fig. 9. EM-driven Monte Carlo simulation for selected trade-off designs of Table 2. Black line shows the antenna response at the given trade-off design: (a) design $x^{(2)}$, (b) design $x^{(5)}$, (c) design $x^{(8)}$, grey lines correspond to 500 random observables generated according to the assumed probability distribution with the variance equal to F_r. Thin lines denote design specifications.

Table 2. Quasi-Yagi antenna of Fig. 8(a): results of multi-objective design with tolerances

Design	Objectives		Geometry parameters [absolute in mm, relative unitless]													
	F_p [dB]	F_r [μm]	L_a	L_b	L_c	L_d	W	w_a	D_a	D_b	D_c	D_{lr}	D_{rr}	S_r	w_{br}	D_a
1	− 17	0	20.2	12.3	16.5	26.1	52.1	1.83	1.02	4.39	4.26	0.37	0.44	0.98	0.71	0.72
2	− 16	7.1	20.2	12.3	16.5	26.0	52.1	1.73	1.02	4.34	4.15	0.38	0.44	0.98	0.71	0.72
3	− 15	10.5	20.3	12.4	16.5	26.0	52.1	1.73	1.03	4.34	4.11	0.38	0.43	0.98	0.71	0.72
4	− 14	16.6	20.2	12.4	16.5	26.0	52.1	1.64	1.02	4.36	4.04	0.38	0.40	0.99	0.71	0.72
5	− 13	22.2	20.2	12.4	16.5	26.0	52.1	1.69	1.02	4.41	4.02	0.38	0.40	0.99	0.71	0.73
6	− 12	30.9	20.3	12.4	16.5	26.0	52.1	1.66	1.01	4.48	4.06	0.37	0.40	0.98	0.71	0.73
7	− 11	39.9	20.5	12.4	16.4	26.1	52.1	1.49	1.01	4.50	3.99	0.38	0.40	0.98	0.71	0.73
8	− 10	49.5	20.6	12.5	16.3	26.1	52.1	1.45	1.01	4.50	3.98	0.35	0.40	0.97	0.71	0.73

4 Conclusions

This paper proposed a novel technique for multi-objective optimization of antenna structures with tolerance analysis. Our approach allows for low-cost identification of the designs representing the best possible trade-offs between the nominal performance and the robustness. The latter is understood as the maximum level of geometry parameter deviations for which the perfect (100%) fabrication yield is still attainable. The main algorithmic tool employed in the presented procedure is a feature-based regression surrogate, which enables rapid estimation of the yield. Numerical validation involving two microstrip antennas demonstrates both the reliability and computational efficiency of the proposed framework, with only a few dozens of EM analyses required to generate each trade-off design. The future work will focus on extending the range of applicability of the technique for higher-dimensional problems.

Acknowledgement. The authors would like to thank Dassault Systemes, France, for making CST Microwave Studio available. This work is partially supported by the Icelandic Centre for Research (RANNIS) Grant 206606 and by Gdańsk University of Technology Grant DEC-41/2020/IDUB/I.3.3 under the Argentum Triggering Research Grants program - 'Excellence Initiative - Research University'.

References

1. Li, Y., Ding, Y., Zio, E.: Random fuzzy extension of the universal generating function approach for the reliability assessment of multi-state systems under aleatory and epistemic uncertainties. IEEE Trans. Reliabil. **63**(1), 13–25 (2014)
2. Elias, B.B.Q., Soh, P.J., Al-Hadi, A.A., Akkaraekthalin, P.: Gain optimization of low-profile textile antennas using CMA and active mode subtraction method. IEEE Access **9**, 23691–23704 (2021)
3. Zhang, Z., Chen, H., Jiang, F., Yu, Y., Cheng, Q.S.: A benchmark test suite for antenna S-parameter optimization. IEEE Trans. Ant. Propag. **69**, 6635–6650 (2021)
4. Koziel, S., Pietrenko-Dabrowska, A.: Robust parameter tuning of antenna structures by means of design specification adaptation. IEEE Trans. Ant. Propag. **69**, 8790–8798 (2021)

5. Xu, Q., Zeng, S., Zhao, F., Jiao, R., Li, C.: On formulating and designing antenna arrays by evolutionary algorithms. IEEE Trans. Ant. Propag. **69**(2), 1118–1129 (2021)
6. Al-Azza, A.A., Al-Jodah, A.A., Harackiewicz, F.J.: Spider monkey optimization: a novel technique for antenna optimization. IEEE Ant. Wirel. Propag. Lett. **15**, 1016–1019 (2016)
7. Darvish, A., Ebrahimzadeh, A.: Improved fruit-fly optimization algorithm and its applications in antenna arrays synthesis. IEEE Trans. Ant. Propag. **66**(4), 1756–1766 (2018)
8. Koziel, S., Sigurdsson, A.T.: Multi-fidelity EM simulations and constrained surrogate modeling for low-cost multi-objective design optimization of antennas. IET Microw. Ant. Prop. **12**(13), 2025–2029 (2018)
9. Xiao, S., et al.: Multi-objective Pareto optimization of electromagnetic devices exploiting kriging with Lipschitzian optimized expected improvement. IEEE Trans. Magn. **54**(3), Art no. 7001704 (2018)
10. De Villiers, D.I.L., Couckuyt, I., Dhaene, T.: Multi-objective optimization of reflector antennas using kriging and probability of improvement. In: International Symposium on Antennas and Propagation, San Diego, USA, pp. 985–986 (2017)
11. Lv, Z., Wang, L., Han, Z., Zhao, J., Wang, W.: Surrogate-assisted particle swarm optimization algorithm with Pareto active learning for expensive multi-objective optimization. IEEE J. Automatica Sinica **6**(3), 838–849 (2019)
12. Koziel, S., Ogurtsov, S.: Multi-objective design of antennas using variable-fidelity simulations and surrogate models. IEEE Trans. Ant. Prop. **61**(12), 5931–5939 (2013)
13. Liu, Y., Cheng, Q.S., Koziel, S.: A generalized SDP multi-objective optimization method for EM-based microwave device design. Sensors **19**(14), 3065 (2019)
14. Toktas, A., Ustun, D., Tekbas, M.: Multi-objective design of multi-layer radar absorber using surrogate-based optimization. IEEE Trans. Microw. Theory Techn. **67**(8), 3318–3329 (2019)
15. Du, J., Roblin, C.: Stochastic surrogate models of deformable antennas based on vector spherical harmonics and polynomial chaos expansions: application to textile antennas. IEEE Trans. Ant. Propag. **66**(7), 3610–3622 (2018)
16. Rayas-Sanchez, J.E., Gutierrez-Ayala, V.: EM-based statistical analysis and yield estimation using linear-input and neural-output space mapping. In: IEEE MTT-S International Microwave Symposium Digest (IMS), pp. 1597–1600 (2006)
17. Zhang, J., Zhang, C., Feng, F., Zhang, W., Ma, J., Zhang, Q.J.: Polynomial chaos-based approach to yield-driven EM optimization. IEEE Trans. Microw. Theory Tech. **66**(7), 3186–3199 (2018)
18. Wu, Q., Chen, W., Yu, C., Wang, H., Hong, W.: Multilayer machine learning-assisted optimization-based robust design and its applications to antennas and arrays. IEEE Trans. Ant. Prop. **69**, 6052–6057 (2021)
19. Abdel-Malek, H.L., Hassan, A.S.O., Soliman, E.A., Dakroury, S.A.: The ellipsoidal technique for design centering of microwave circuits exploiting space-mapping interpolating surrogates. IEEE Trans. Microw. Theory Techn. **54**(10), 3731–3738 (2006)
20. Ma, B., Lei, G., Liu, C., Zhu,. J., Guo, Y.: Robust tolerance design optimization of a PM claw pole motor with soft magnetic composite cores. IEEE Trans. Magn., **54**(3), Art no. 8102404 (2018)
21. Ren, Z., He, S., Zhang, D., Zhang, Y., Koh, C.S.: A possibility-based robust optimal design algorithm in preliminary design state of electromagnetic devices. IEEE Trans. Magn., **52**(3), Art no. 7001504 (2016)
22. Spina, D., Ferranti, F., Antonini, G., Dhaene, T., Knockaert, L.: Efficient variability analysis of electromagnetic systems via polynomial chaos and model order reduction. IEEE Trans. Comp. Packaging Manuf. Techn. **4**(6), 1038–1051 (2014)
23. Pietrenko-Dabrowska, A., Koziel, S., Al-Hasan, M.: Expedited yield optimization of narrow- and multi-band antennas using performance-driven surrogates. IEEE Access **8**, 143104–143113 (2020)

24. Xia, B., Ren, Z., Koh, C.S.: Utilizing Kriging surrogate models for multi-objective robust optimization of electromagnetic devices. IEEE Trans. Magn., **50**(2), Art no. 7017104 (2014)
25. Liu, B., Aliakbarian, H., Ma, Z., Vandenbosch, G.A., Gielen, G., Excell, P.: An efficient method for antenna design optimization based on evolutionary computation and machine learning techniques. IEEE Trans. Ant. Propag. **62**, 7–18 (2014)
26. Easum, J.A., Nagar, J., Werner, P.L., Werner, D.H.: Efficient multi-objective antenna optimization with tolerance analysis through the use of surrogate models. IEEE Trans. Ant. Prop. **66**(12), 6706–6715 (2018)
27. Deb, K.: Multi-Objective Optimization Using Evolutionary Algorithms. Wiley, New York (2001)
28. Hassan, A.S.O., Abdel-Malek, H.L., Mohamed, A.S.A., Abuelfadl, T.M., Elqenawy, A.E.: Statistical design centering of RF cavity linear accelerator via non-derivative trust region optimization. In: IEEE International Conference on Numerical EM Multiphysics Modeling Optics (NEMO), pp. 1–3 (2015)
29. Koziel, S.: Fast simulation-driven antenna design using response-feature surrogates. Int. J. RF & Microwave CAE **25**(5), 394–402 (2015)
30. Pietrenko-Dabrowska, A., Koziel, S.: Fast design closure of compact microwave components by means of feature-based metamodels. Electronics, **10**(1), Art no. 10 (2021)
31. Pietrenko-Dabrowska, A., Koziel, S.: Simulation-driven antenna modeling by means of response features and confined domains of reduced dimensionality. IEEE Access **8**, 228942–228954 (2020)
32. Press, W.H., Teukolsky, S.A., Vetterling, W.T., Flannery, B.P.: Golden section search in one dimension. In: Numerical Recipes: The Art of Scientific Computing, 3rd edn. Cambridge University Press, New York (2007)
33. Conn, A.R., Gould, N.I.M., Toint, P.L.: Trust Region Methods. MPS-SIAM Series on Optimization (2000)
34. Qudrat-E-Maula, M., Shafai, L.: A dual band microstrip dipole antenna. In: International Symposium Antennas, Technology and Applied Electronics (ANTEM), Victoria, BC, Canada, 13–16 July 2014, pp. 1–2 (2014)
35. Farran, M., et al.: Compact quasi-Yagi antenna with folded dipole fed by tapered integrated balun. Electron. Lett. **52**(10), 789–790 (2016)

Expedited Optimization of Passive Microwave Devices Using Gradient Search and Principal Directions

Slawomir Koziel[1,2]([⊠]) [ID], Anna Pietrenko-Dabrowska[2] [ID], and Leifur Leifsson[3] [ID]

[1] Department of Engineering, Engineering Optimization and Modeling Center, Reykjavík University, Menntavegur 1, 102 Reykjavík, Iceland
`koziel@ru.is`
[2] Faculty of Electronics Telecommunications and Informatics, Gdansk University of Technology, Narutowicza 11/12, 80-233 Gdansk, Poland
`anna.dabrowska@pg.edu.pl`
[3] School of Aeronautics and Astronautics, Purdue University, West Lafayette, IN 47907, USA
`leifur@purdue.edu`

Abstract. Over the recent years, utilization of numerical optimization techniques has become ubiquitous in the design of high-frequency systems, including microwave passive components. The primary reason is that the circuits become increasingly complex to meet ever growing performance demands concerning their electrical performance, additional functionalities, as well as miniaturization. Nonetheless, as reliable evaluation of microwave device characteristics requires full-wave electromagnetic (EM) analysis, optimization procedures tend to be computationally expensive, to the extent of being prohibitive when using conventional algorithms. Accelerating EM-driven optimization is therefore a matter of practical necessity. This paper proposes a novel approach to reduced-cost gradient-based parameter tuning of passive microwave circuits with numerical derivatives. Our technique is based on restricting the finite-differentiation (FD)-based sensitivity updates to a small set of principal directions, identified as having the most significant effect on the circuit responses over the frequency bands of interest. The principal directions are found in the form of an orthonormal basis, using an auxiliary optimization process repeated before each iteration of the optimization algorithm. Extensive verification experiments conducted using a compact branch-line coupler and a dual-band power divider demonstrate up to fifty percent speedup obtained over the reference algorithm (using full FD sensitivity updates), as well as a considerable improvement over several accelerated algorithms. The computational savings are obtained with negligible degradation of the design quality.

Keywords: Microwave design · Simulation-driven optimization · Principal directions · Gradient-based search · Sparse sensitivity updates

D. Groen et al. (Eds.): ICCS 2022, LNCS 13352, pp. 217–233, 2022.
https://doi.org/10.1007/978-3-031-08757-8_20

1 Introduction

Design of modern microwave components is a challenging endeavor. On the one hand, performance demands imposed on high-frequency systems become increasingly stringent, especially those associated with the emerging application areas (internet of things, IoT [1], wearable and implantable devices [2], 5G wireless communications). On the other hand, there is a growing trend to make the circuits more versatile (reconfigurability [3], multi-band operation [4], unconventional phase characteristics [5]). Another issue is miniaturization, which has become imperative in many cases due to the limitations on the physical space assigned for the passive circuitry [6, 7]. All these factors contribute to the increasing complexity of microwave components, which are described by many parameters, whereas their accurate evaluation requires costly full-wave electromagnetic (EM) analysis. As simpler models (e.g., equivalent networks) are no longer adequate, EM-driven optimization has become a necessity for parameter tuning. Yet, it is a computationally expensive procedure [8], which may require dozens, hundreds (gradient-based optimization over high-dimensional spaces), or even thousands of circuit simulations (global optimization [9], uncertainty quantification [10]).

The literature offers a large number of techniques for accelerating simulation-driven optimization of high-frequency components. In the context of local search, these include the employment of adjoint sensitivities [11], mesh deformation for fast sensitivity evaluation [12], feature-based methods [13], as well as cognition-driven design [14]. Surrogate-based optimization (SBO) is another and rapidly growing class of methods, which may involve both physics-based [15], and data-driven (approximation) models [16]. The latter (kriging [17], radial basis functions [18], support vector regression, neural networks [19]) are popular in global [20] and multi-criterial optimization [21], as well as statistical design [22]. Physics-based methods (space mapping [23], response correction techniques [15, 24]) are typically used in local search [25]. Unfortunately, SBO is affected by a number of issues, e.g., related to availability and setup of low-fidelity representations (physics-based models), or curse of dimensionality (data-driven models). Therefore, successful application examples of surrogate-assisted frameworks, are often limited to components described by a few parameters [26, 27].

Among the various optimization tasks, it is local parameter tuning that is by far the most often undertaken procedure in the case of high-frequency components, including microwave devices. The reason is the availability of reasonably good initial designs that are found through theoretical analysis or EM-based parametric studies. Local optimization is typically realized using gradient-based methods. Their computational efficiency mainly depends on the cost of estimating the gradients of the system characteristics. If adjoint sensitivities [11] are not accessible, the gradients are estimated through finite differentiation (FD). Acceleration is possible by restricting FD to selected system parameters, which may be decided upon based on investigating design relocation [28], detecting sensitivity patterns [29], or selective employment of updating formulas [30]. The aforementioned methods typically lead to at least forty percent speedup (in some cases, up to sixty) over full-FD updating schemes, with usually minor quality degradation. Yet, the efficacy depends on appropriate setup of the control parameters, which may be problem-dependent [28–30]. Also, sparse sensitivity updates are still limited to the coordinate system axes. Performing the updates along arbitrary directions seems to be potentially more advantageous.

In this paper, we propose a novel technique for accelerated gradient-based design optimization of passive microwave devices. Our methodology restricts the finite-differentiation (FD)-based sensitivity updates to a small set of so-called dominant directions that are associated with the most significant variability of the system responses over the frequency bands of interest. The dominant directions form an orthonormal basis updated before each iteration of the optimization algorithm, and obtained by solving an auxiliary optimization sub-problem. In practice, only a few directions are used for sensitivity updating, which results in considerable computational savings. For the two microwave circuits used as verification case studies, the cost reduction is as high as fifty percent over the reference algorithm involving full-FD updates, without compromising the design quality. At the same time, the proposed procedure is faster than several accelerated versions previously reported in the literature.

2 Microwave Design Optimization Using EM Models

Here, we recall the formulation of microwave design optimization task, as well as discuss the standard trust-region gradient-based algorithm, which is the foundation for the accelerated procedure introduced in Sect. 3, as well as one of the benchmark algorithms considered in Sect. 4.

2.1 Optimization Task Formulation

Optimization of microwave circuits often requires handling of several character-istics (reflection, transmission, phase, etc.), performance figures (operating fre-quency/bandwidth, power split ratio), as well as constraints. For the sake of simplic-ity, the design task is most often formulated to minimize a scalar objective function, which aggregates the goals and constraints in a problem-dependent manner. Here, the parameter tuning problem is defined as

$$x^* = \arg \min_{x} U(x) \tag{1}$$

where U is the merit (objective) function quantifying the design quality, x is a vector of adjustable parameters, and x^* is the optimum design to be found. Some examples of design tasks, and the corresponding objective functions can be found in Table 1. Therein, the following notation is used for the circuit S-parameters: $S_{kl}(x,f)$, where f is the frequency, whereas k and l denote respective circuit ports.

2.2 Gradient-Based Search with Numerical Derivatives

In this work, we are concerned with local, gradient-based optimization of microwave components. As mentioned earlier, the major contributor to the computational cost is the evaluation of the circuit response gradients, which, in the absence of faster methods (e.g., adjoint sensitivities [17]), is realized through finite differentiation.

Our reference algorithm is the trust-region (TR) procedure [31]. Therein, the optimum design x^* is approximated using a sequence of designs $x^{(i)}$, $i = 0, 1, ...,$ found by solving

$$x^{(i+1)} = \arg \min_{x; \, -d^{(i)} \leq x - x^{(i)} \leq d^{(i)}} U_L^{(i)}(x) \tag{2}$$

Table 1. Examples of microwave design optimization tasks

Design problem	Objective function	Comments
Design of an impedance matching transformer for minimum in-band reflection	$U(x) = \max\{f \in F : |S_{11}(x,f)|\}$	$|S_{11}(x,f)|$ is the circuit reflection; F is the frequency range of interest
Design of a dual-band coupler for the operating frequencies f_1 and f_2. The objectives are to ensure equal power split, and to minimize the circuit matching and isolation at f_1 and f_2	$U(x) = \max\{|S_{11}(x,f_1)|, |S_{11}(x,f_2)|,$ $|S_{41}(x,f_1)|, |S_{41}(x,f_1)|\}$ $+\beta[(|S_{21}(x,f_1)| - |S_{31}(x,f_1)|)^2$ $+(|S_{21}(x,f_2)| - |S_{31}(x,f_2)|)^2]$	Minimization of matching and isolation is the primary objective. The second term is a penalty function enforcing equal power split
Design of a compact coupler for size reduction. The objective is to minimize the circuit footprint area $A(x)$, while maintaining a sufficient -20 dB bandwidth for $|S_{11}|$ and $|S_{41}|$, and equal power split at the operating frequency f_0. The bandwidth is defined as $[f_0 - B, f_0 + B]$	$U(x) = A(x)+$ $+\beta_1[\max\{c(x) + 20, 0\}/20]^2$ $+\beta_2[|S_{21}(x,f_0)| - |S_{31}(x,f_0)|]^2$ where $c(x) = \max\{f \in [f_0 - B, f_0 + B] :$ $\max\{|S_{11}(x,f)|, |S_{41}(x,f)|\}$	Function c quantifies a possible violation of the -20 dB level within the bandwidth of interest. The third term is similar to the second term in the second example

(continued)

Table 1. *(continued)*

Design problem	Objective function	Comments
Design of a triple-band power divider to operate at the frequencies f_1, f_2, and f_3. The objectives include ensuring equal power split, minimization of the input matching $|S_{11}|$, output matching $|S_{22}|$ and $|S_{33}|$, as well as minimization of isolation $|S_{23}|$ (all at f_1, f_2, and f_3)	$U(\boldsymbol{x}) = \max\{ \max_{k,l\in\{1,2,3\}} |S_{kk}(\boldsymbol{x}, f_l)|,$ $\max_{l\in\{1,2,3\}} |S_{23}(\boldsymbol{x}, f_l)|\}$ $+\beta \sum_{l=1}^{3} (|S_{21}(\boldsymbol{x}, f_l)| - |S_{31}(\boldsymbol{x}, f_l)|)^2$	The second term is a penalty function enforcing equal power split at all operating frequencies

In (2), the function $U_L^{(i)}$ is defined just as the original objective function U; however, the circuit characteristics $S_{kl}(\boldsymbol{x}, f)$ are approximated using the linear expansion models

$$S_{L.kl}^{(i)}(\boldsymbol{x}, f) = S_{kl}(\boldsymbol{x}^{(i)}, f) + \nabla_{kl}(\boldsymbol{x}^{(i)}, f) \cdot (\boldsymbol{x} - \boldsymbol{x}^{(i)}) \qquad (3)$$

The gradients of S_{kl} at the design $\boldsymbol{x}$ and frequency f are defined as

$$\nabla_{kl}(\boldsymbol{x}, f) = \left[\frac{\partial S_{kl}(\boldsymbol{x}, f)}{\partial x_1} \quad \cdots \quad \frac{\partial S_{kl}(\boldsymbol{x}, f)}{\partial x_n} \right] \qquad (4)$$

It should be noted that the solution to sub-problem (2) is found in the interval $[\boldsymbol{x}^{(i)} - \boldsymbol{d}^{(i)}, \boldsymbol{x}^{(i)} + \boldsymbol{d}^{(i)}]$, the size of which is adjusted using the TR rules [31]. The computational cost of (2) is at least $n + 1$ EM simulations due to evaluation of (4) using finite differentiation. If the iteration fails, i.e., if $U(\boldsymbol{x}^{(i+1)}) \geq U(\boldsymbol{x}^{(i)})$, it is repeated with reduced $\boldsymbol{d}^{(i)}$. As mentioned earlier, the above procedure will be a benchmark algorithm in Sect. 4. It is also a basis for the procedure proposed in this work, as elaborated on in Sect. 3.

3 Accelerated Optimization by Means of Principal Directions

This section describes the proposed approach to accelerated parameter tuning of passive microwave devices. It is based on the concept of principal directions introduced in Sect. 3.1, and sensitivity updates restricted to a small set thereof, as elaborated on in Sect. 3.2. The complete optimization algorithm is summarized in Sect. 3.3.

3.1 Principal Directions and Their Identification

The proposed approach is based on restricting the finite-differentiation (FD) updates of the system sensitivity matrix to a few principal directions, which are the most influential in terms of response variability. This translates into a reduced operating cost of the optimization process.

In order to find the principal directions, we need to first decide upon the response variability metric F_v. It is defined by considering the frequency range of interest F, determined according to the design problem at hand. F can be a discrete set of target operating frequencies, or a continuous frequency interval, if the circuit characteristics are of interest over a specified bandwidth. The variability of the S-parameter S_{kl} is defined as

$$F_{v.kl}(S_{kl}(\boldsymbol{x}_1), S_{kl}(\boldsymbol{x}_2)) = \sqrt{\int_F \left[|S_{kl}(\boldsymbol{x}_1,f)| - |S_{kl}(\boldsymbol{x}_2,f)|\right]^2 df} \qquad (5)$$

which, in the case of a discrete set of frequencies $f_j, j = 1, \ldots, p$, becomes

$$F_{v.kl}(S_{kl}(\boldsymbol{x}_1), S_{kl}(\boldsymbol{x}_2)) = \sqrt{\sum_{j=1}^{p} \left[|S_{kl}(\boldsymbol{x}_1,f_j)| - |S_{kl}(\boldsymbol{x}_2,f_j)|\right]^2} \qquad (6)$$

This can be generalized for multiple circuit responses S_{kl}, $\{k,l\} \in \{\{k_1,l_1\}, \ldots, \{k_r,l_r\}\}$ (e.g., $k = 1, 2, 3, 4$, and $l = 1$, for a coupler structure), as the average of $F_{v.kl}$ for all characteristics involved

$$F_v(\boldsymbol{x}_1, \boldsymbol{x}_2) = \frac{1}{r} \sum_{j=1}^{r} F_{v.k_jl_j}\left(S_{k_jl_j}(\boldsymbol{x}_1), S_{k_jl_j}(\boldsymbol{x}_2)\right) \qquad (7)$$

Our goal is to identify an orthonormal basis of vectors $\{v^{(j)}\}_{j=1,\ldots,n}$, ordered in a descending manner with respect to F_v. A small subset thereof will govern the sensitivity updates, cf. Sect. 3.2. Let $\boldsymbol{x}^{(i)}$ be the design found in the ith iteration of the optimization process, and let $S_{L.kl}^{(i)}(\boldsymbol{x},f)$ be the linear model of S_{kl} at $\boldsymbol{x}^{(i)}$ (cf. (3)). We define

$$v^{(1)} = \arg \max_{v;\ ||v||=1} F_{L.v}\left(\boldsymbol{x}^{(i)} + v, \boldsymbol{x}^{(i)}\right) \qquad (8)$$

where

$$F_{L.v}\left(\boldsymbol{x}^{(i)} + v, \boldsymbol{x}^{(i)}\right) = \frac{1}{r} \sum_{j=1}^{r} F_{v.k_jl_j}\left(S_{L.k_jl_j}^{(i)}(\boldsymbol{x}^{(i)} + v), S_{k_jl_j}(\boldsymbol{x}^{(i)})\right) \qquad (9)$$

According to (8), $v^{(1)}$ maximizes the variability metric F_v in the vicinity of $\boldsymbol{x}^{(i)}$. The S-parameters at $\boldsymbol{x}^{(i)} + v$ are estimated using the respective linear models in order to make the solution to (8) computationally feasible.

In order to find the remaining $n-1$ directions, $v^{(2)}, v^{(3)}, \ldots$, a process similar to (8) is executed with additional constraints imposed to ensure orthogonality of the vectors. Having $v^{(k)}, k = 1, \ldots, j, v^{(j+1)}$ is identified as

$$v^{(j+1)} = \arg \max_{\bar{v}} F_{L.v}\left(\boldsymbol{x}^{(i)} + \bar{v}, \boldsymbol{x}^{(i)}\right) \qquad (10)$$

where $\bar{v}$ has the form of

$$\bar{v} = \frac{P^{(j)}(v)}{||P^{(j)}(v)||} \tag{11}$$

in which

$$P^{(j)}(v) = v - \sum_{k=1}^{j} v^{(k)} \left[(v^{(k)})^T v \right] \tag{12}$$

It can be noted that (11) and (12) ensure that $v^{(j+1)}$ has a unity length and it is orthogonal to $v^{(k)}$, $k = 1, ..., j$.

Figure 1 provides a graphical illustration of the aforementioned concepts for a compact branch-line coupler. In the considered case, the majority of response changes occur along the first three directions, therefore, restricting the sensitivity updates to this subset seems reasonable.

3.2 Restricted Sensitivity Updates

The procedure for generating the principal directions ensures that $F_{L.v}(x^{(i)} + v^{(1)}, x^{(i)}) > F_{L.v}(x^{(i)} + v^{(2)}, x^{(i)}) > ... > F_{L.v}(x^{(i)} + v^{(n)}, x^{(i)})$, with only a few directions responsible for the majority of response changes in the vicinity of $x^{(i)}$ (cf. Fig. 1). Consider the variability factors C_j determining the (relative) contribution of the first j directions to the overall response variation.

$$C_j = \frac{\sqrt{\sum_{k=1}^{i} \left[F_{L.v}(x^{(i)} + v^{(k)}, x^{(i)}) \right]^2}}{\sum_{k=1}^{n} \left[F_{L.v}(x^{(i)} + v^{(k)}, x^{(i)}) \right]^2} \tag{13}$$

The number j_{update} of directions utilized for the sensitivity updates can be computed based on the user-defined threshold C_{th}, typically set to 0.9 or higher. We have

$$j_{update} = \arg \min \{ j \in \{1, 2, ..., n\} : C_j \geq C_{th} \} \tag{14}$$

Going back to Fig. 1, we would have $j_{update} = 2$ given $C_{th} = 0.95$. This means that the first two directions $v^{(j)}$ contribute at least 95% of response variability as defined by (13). However, C_j are calculated using the linear models (3), which is an approximation. Therefore, in practice, it is recommended to introduce a lower bound on j_{update} at the level of about one third of the parameter space dimensionality, so that (14) is modified to $j_{update} = \max\{\arg\min\{j \in \{1, 2, ..., n\}: C_j \geq C_{th}\}, \lceil n/3 \rceil\}$, where $\lceil \cdot \rceil$ stands for the ceiling function.

Having j_{update}, the S-parameter sensitivity is updated using EM simulations results at the design perturbed along the selected principal directions $v^{(j)}, j = 1, ..., j_{update}$. The perturbation data is incorporated using the Broyden formula [32], as illustrated in Fig. 2.

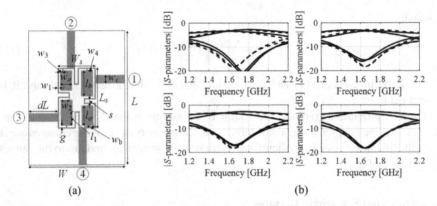

Fig. 1. Compact branch-line coupler and its S-parameter variability along the principal directions obtained using the procedure of Sect. 3.1: (a) circuit geometry, (b) EM-simulated S-parameters at the selected design $x^{(i)}$ (- - -) and at $x^{(i)} + hv^{(j)}$ (——) for $j = 1, 2, 3,$ and 4 (from top-left to bottom-right). Normalized response variability values are 1.00, 0.76, 0.19, and 0.07.

3.3 Optimization Algorithm

Figure 3 shows the flow diagram of the proposed optimization procedure with sparse sensitivity updates. The core optimization algorithm is the trust-region routine outlined in Sect. 2.2. The only control parameters are the threshold C_{th} (cf. Sect. 3.2), and the termination threshold ε that determines the resolution of the search process (set to $\varepsilon = 10^{-3}$ in the verification experiments of Sect. 4).

The sensitivities of the scattering parameters $S_{kl}(x^{(i)})$, $k, l = 1, ..., p$, are evaluated using FD in the first iteration of the algorithm, so that the principal directions can be identified with a sufficient accuracy. In further iterations, the gradients are updated using the EM-simulated circuit characteristics at the new design $x^{(i+1)}$, and along the selected principal directions. The TR size vector $d^{(i+1)}$ is updated based on the gain ratio r. The rules are as follows: if $r > 0.75$, then $d^{(i+1)} = 2d^{(i)}$; if $r < 0.25$, then $d^{(i+1)} = d^{(i)}/3$ if $r < 0.5$.

4 Demonstration Case Studies and Benchmarking

This section summarizes numerical verification of the optimization procedure introduced in Sect. 3. It is based on two microstrip components, and includes evaluation of the reliability and computational efficiency of the optimization process, as well as comparisons with several benchmark methods characterized in Sect. 4.2.

4.1 Verification Circuits

Verification experiments are based on the circuits shown in Fig. 4. These are a dual-band power divider operating at 2.4 GHz and 3.8 GHz (Circuit I) [33], and a compact branch-line coupler (BLC) operating at 1.5 GHz (Circuit II) [34]. The relevant circuit details have been gathered in Table 2. The EM simulation models of both circuits are implemented in CST Microwave Studio and evaluated using the time-domain solver.

---

1. Set $j = 1$;
2. Obtain temporary design $\boldsymbol{x}_{tmp} = \boldsymbol{x}^{(i)} + h\boldsymbol{v}^{(j)}$, where $h > 0$ is the step size;
3. Obtain EM-simulated circuit characteristics $S_{kl}(\boldsymbol{x}_{tmp})$, $\{k,l\} \in \{\{k_1,l_1\}, \ldots, \{k_p,l_p\}\}$;
4. Update the sensitivity vectors ∇_{kl} for $\{k,l\} \in \{\{k_1,l_1\}, \ldots, \{k_p,l_p\}\}$ (here, $\boldsymbol{h}^{(j)} = h\boldsymbol{v}^{(j)}$) as

$$\nabla_{kl}(\boldsymbol{x}^{(i)}) \Leftarrow \nabla_{kl}(\boldsymbol{x}^{(i)}) + \frac{\left(\left[S_{kl}(\boldsymbol{x}_{tmp}) - S_{kl}(\boldsymbol{x}^{(i)})\right] - \nabla_{kl}(\boldsymbol{x}^{(i)}) \cdot \boldsymbol{h}^{(j)}\right) \cdot \boldsymbol{h}^{(j)T}}{\boldsymbol{h}^{(j)T}\boldsymbol{h}^{(j)}} ;$$

5. If $j < j_{update}$
      Go to 2;
   else
      END
   end

Fig. 2. Sparse sensitivity updates using principal directions. The step size h is set to a fraction of mm (between 0.02 to 0.1), a typical value for FD carried out on EM-simulated responses.

226   S. Koziel et al.

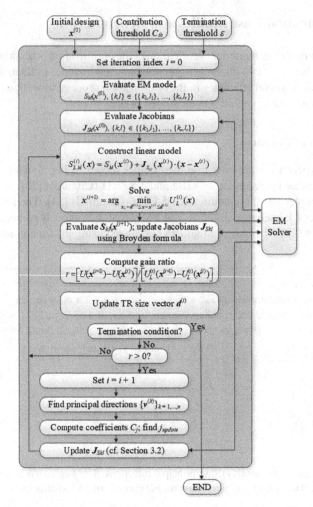

Fig. 3. Flow diagram of the proposed accelerated optimization algorithm with sparse sensitivity updates using principal directions.

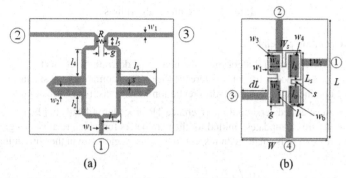

Fig. 4. Verification circuits: (a) dual-band power divider (Circuit I) [33], (b) miniaturized branch-line coupler (Circuit II) [34].

Table 2. Verification circuits

	Case study	
	Circuit I	Circuit II
Substrate	AD250 ($\varepsilon_r = 2.5$, $h = 0.81$ mm, tanδ = 0.0018)	RO4003 ($\varepsilon_r = 3.5$, $h = 0.76$ mm, tanδ = 0.0027)
Design parameters	$x = [l_1\ l_2\ l_3\ l_4\ l_5\ s\ w_2]^T$	$x = [g\ l_{1r}\ l_a\ l_b\ w_1\ w_{2r}\ w_{3r}\ w_{4r}\ w_a\ w_b]^T$
Other parameters	$w_1 = 2.2$, $g = 1$ mm	$L = 2dL + L_s$, $L_s = 4w_1 + 4g + s + l_a + l_b$, $W = 2dL + W_s$, $W_s = 4w_1 + 4g + s + 2w_a$, $l_1 = l_b l_{1r}$, $w_2 = w_a w_{2r}$, $w_3 = w_{3r} w_a$, and $w_4 = w_{4r} w_a$
Design specifications	• Operating frequencies: $f_1 =$ 2.4 GHz and $f_2 = 3.8$ GHz • Goal 1: minimize input matching $|S_{11}|$, and output matching $|S_{22}|$, $|S_{33}|$, at both f_1 and f_2 • Goal 2: minimize isolation $|S_{23}|$ at both f_1 and f_2 • Goal 3: ensure equal power split (note: this is implied by the structure symmetry)	• Operating frequency $f_0 =$ 1.5 GHz • Goal 1: minimize matching $|S_{11}|$ and isolation $|S_{41}|$ at f_0 • Goal 2: ensure equal power split, i.e., $|S_{21}| = |S_{31}|$ at f_0

4.2 Experimental Setup

The algorithm of Sect. 3 has been compared to four benchmark routines briefly character-ized in Table 3. Each procedure has been executed ten times starting from a random initial design. The results statistics discussed in Sect. 4.3 account for the average performance of the search process.

Table 3. Benchmark algorithms

Algorithm	Operating principles
I	• Conventional TR algorithm with numerical derivatives, cf. Sect. 2.2 • This method is used as a reference to compute computational savings that can be obtained w.r.t. full finite-differentiation-based sensitivity estimation
II	• Accelerated version of the reference TR procedure (details in [29]) • Sensitivity updates omitted for the parameters for which relative design relocation is small (with respect to the trust-region size in the current iteration), i.e., $$\varphi_k^i = \left| x_k^{(i+1)} - x_k^{(i)} \right| \Big/ d_k^{(i)}, k = 1, \ldots, n,$$ where $x_k^{(i)}$ and $x_k^{(i+1)}$ are the kth entries of the parameter vectors $x^{(i)}$ and $x^{(i+1)}$ from the last two iterations, respectively; $d_k^{(i)}$ refers to the kth element of the TR size vector $d^{(i)}$ • FD is omitted for the parameters, for which φ_k^i factors are below a user-defined threshold • For each gradient vector, FD update is enforced at least once every few iterations • Control parameter N_{iter}: the maximum admissible number of iterations without FD (typically from 3 to 5; in this work we set $N_{iter} = 3$); increasing N_{iter} likely leads to cost savings but may be detrimental for the design quality
III	• Accelerated version of the reference TR procedure (details in [30]) • Sensitivity updates realized with a Broyden formula (BF) for parameters that are sufficiently aligned with the direction of the recent design relocation. The alignment is quantified as $$\gamma_k^{(i)} = \left| h^{(i)T} e^{(k)} \right| \Big/ \left\| h^{(i)} \right\|, k = 1, \ldots, n$$ where $e^{(k)} = [0 \ldots 0\ 1\ 0 \ldots 0]^T$ with one on the kth position, is the kth standard basis vector, and $h^{(i+1)} = x^{(i+1)} - x^{(i)}$ • The kth gradient vector is updated using BF if $\gamma_k^{(i)}$ is larger than a user-defined threshold γ_{min} • Control parameter $0 \le \gamma_{min} \le 1$ allows for adjusting the trade-offs between the CPU cost and design quality. Higher γ_{min} makes the condition for using BF more rigorous, and a larger number of gradient vectors is updated using finite differentiation leading to improved quality but lower savings • In this work, $\gamma_{min} = 0.9$, as recommended in [30]

(continued)

Expedited Optimization of Passive Microwave Devices   229

Table 3. (continued)

Algorithm	Operating principles
IV	• Accelerated version of the reference TR algorithm (details in [35]) • Sensitivity updates based on detected stable sensitivity patterns across the algorithm iterations. Let $\nabla_S = [\nabla_1 \ldots \nabla_n]^T$ be the gradient vector of relevant system responses, where ∇_k stands for sensitivity w.r.t the k-th parameter, $k = 1, \ldots, n$. Also, let $\nabla_k^{(i)}(f)$ and $\nabla_k^{(i-1)}(f)$ refer to the k-th component of the gradient ∇_S in the ith and $(i-1)$th iteration. The gradient change factors are defined as $$d_k^{(i+1)} = \operatorname*{mean}_{f \in F}\left(2 \cdot \frac{\left|\nabla_k^{(i)}(f)\right| - \left|\nabla_k^{(i-1)}(f)\right|}{\left|\nabla_k^{(i)}(f)\right| + \left|\nabla_k^{(i-1)}(f)\right|}\right)$$ i.e., represent relative gradient changes averaged over the frequency range of interest • Let $d_{\min}^{(i)} = \min\{k = 1, \ldots, n: d_k^{(i)}\}$ and $d_{\max}^{(i)} = \max\{k = 1, \ldots, n: d_k^{(i)}\}$. Using these, the numbers $N_k^{(i)}$ of future iterations without FD-based updates for the k-th parameter are set to $$N_k^{(i)} = \left[\left[N_{\max} + a^{(i)}(d_k^{(i)} - d_{\min}^{(i)})\right]\right]$$ where $a^{(i)} = (N_{\max} - N_{\min})/(d_{\min}^{(i)} - d_{\max}^{(i)})$ ([[.]] is the nearest integer function). $N_{\min}/N_{\max}$ are the minimum/maximum number of iterations without FD (control parameters) • Remark: sensitivity update using FD is executed at least once per $N_{\max}$ iterations, and not more often than every $N_{\min}$ iterations. Here, we use $N_{\max} = 5$ and $N_{\min} = 1$ (cf. [35])

4.3 Results and Discussion

Tables 4 and 5 provide the results obtained for Circuits I and II, respectively. The circuit frequency characteristics at the initial and optimized designs obtained for the selected runs of our algorithm can be found in Figs. 5 and 6. As mentioned earlier, the results are in the form of statistics based on ten independent runs of the proposed and benchmark algorithms.

The performance of the presented algorithm in comparison to the benchmark can be characterized as follows:

- The algorithm of Sect. 3 performs consistently for both circuits in terms of all considered factors (computational complexity, reliability, and solution repeatability). The quality of optimized designs is comparable to Algorithm I (the reference);
- The achieved CPU savings are as high as forty percent for Circuit I and over fifty percent for Circuit II. These figures are higher for Algorithms II through IV (which are all accelerated procedures). The quality of the optimized designs is better than for the benchmark procedures.

Another important advantage of the proposed technique is the simplicity of its setup. Apart from the termination threshold, there is only one control parameter C_{th}, the meaning of which is intuitive as explained in Sect. 3.2.

Table 4. Optimization results for Circuit I

Algorithm	Performance figure				
	CPU Cost[1]	Cost savings[2]	$\max|S_{11}|$[3]	$\Delta \max|S_{11}|$[4]	Std $\max|S_{11}|$[5]
Algorithm I	99.8	–	−31.6 dB	–	4.5 dB
Algorithm II	77.5	22.4%	−28.6 dB	3.0 dB	6.0 dB
Algorithm III	81.9	17.9%	−28.1 dB	3.5 dB	5.8 dB
Algorithm IV	67.9	32.1%	−29.5 dB	2.1 dB	5.7 dB
This work ($C_{th} = 0.8$)[6]	59.8	40.1%	−30.3 dB	1.3 dB	4.3 dB
This work ($C_{th} = 0.9$)[6]	60.8	39.1%	−30.6 dB	1.0 dB	4.2 dB

[1] Number of equivalent EM evaluation of the circuit (averaged over ten algorithm runs)
[2] Relative savings in percent w.r.t. Algorithm I.
[3] Objective function value, averaged over ten algorithm runs.
[4] Degradation of the objective function value w.r.t. the TR algorithm in dB, averaged over ten algorithm runs.
[5] Standard deviation of the objective function value in dB across the set of ten algorithm runs.
[6] Index j_{update} selected as in (14).

Table 5. Optimization results for Circuit II

Algorithm	Performance figure				
	CPU Cost[1]	Cost savings[2]	$\max|S_{11}|$[3]	$\Delta \max|S_{11}|$[4]	Std $\max|S_{11}|$[5]
Algorithm I	84.6	–	−20.9 dB	–	1.5 dB
Algorithm II	67.4	20.0%	−20.7 dB	0.2 dB	2.1 dB
Algorithm III	43.6	48.4%	−18.8 dB	2.1 dB	3.8 dB
Algorithm IV	64.2	24.1%	−20.7 dB	0.2 dB	1.8 dB
This work ($C_{th} = 0.8$)[6]	38.3	54.7%	−21.9 dB	−1.0 dB	2.9 dB
This work ($C_{th} = 0.9$)[6]	51.5	39.1%	−21.5 dB	−0.6 dB	3.4 dB

[1-5] As in Table 4.

Also, as shown in Tables 4 and 5, the algorithm performance is only weakly dependent on the value C_{th}. On the other hand, the operation of the benchmark techniques depends on a larger number of control parameters, the setup of which is more intricate (cf. [29, 30, 35]).

The improvements achieved by the presented method over the benchmark is mainly related to the fact that the sensitivity updates are not restricted to the coordinate system axes, which was the case for the earlier algorithms.

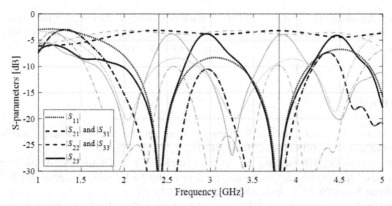

Fig. 5. S-parameters of Circuit I for the selected run of the proposed algorithm. Initial and optimized designs are shown using grey and black lines, respectively. The vertical lines mark the target operating frequencies.

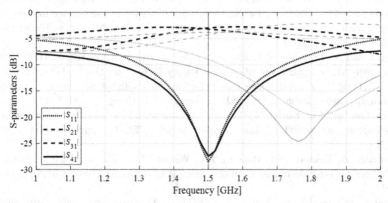

Fig. 6. S-parameters of Circuit II for the selected run of the proposed algorithm. Initial and optimized designs are shown using grey and black lines, respectively. The vertical line marks the target operating frequency.

5 Conclusions

This paper proposed a novel technique for accelerated local parameter tuning of microwave components. Our methodology capitalizes on restricting the finite-differentiation-based sensitivity updates to a selected set of principal directions established as having the major influence on the system response variability. Extensive numerical experiments indicate superiority of the proposed technique over several benchmark methods, including previously reported expedited optimization frameworks. The future work will be focused on achieving further acceleration through the incorporation of variable-resolution simulation models.

Acknowledgement. The authors would like to thank Dassault Systemes, France, for making CST Microwave Studio available. This work is partially supported by the Icelandic Centre for Research (RANNIS) Grant 206606 and by National Science Centre of Poland Grant 2020/37/B/ST7/01448.

References

1. Gumber, K., Dejous, C., Hemour, S.: Harmonic reflection amplifier for widespread backscatter internet-of-things. IEEE Trans. Microwave Theory Techn. **69**(1), 774–785 (2021)
2. Gugliandolo, G., Naishadham, K., Neri, G., Fernicola, V.C., Donato, N.: A novel sensor-integrated aperture coupled microwave patch resonator for humidity detection. IEEE Trans. Instrum. Meas. **70**, 1–11, Article No. 9506611 (2021)
3. Pradhan, N.C., Subramanian, K.S., Barik, R.K., Cheng, Q.S.: Design of compact substrate integrated waveguide based triple- and quad-band power dividers. IEEE Microwave Wirel. Comp. Lett. **31**(4), 365–368 (2021)
4. Li, Q., Yang, T.: Compact UWB half-mode SIW bandpass filter with fully reconfigurable single and dual notched bands. IEEE Trans. Microwave Theory Technol. **69**(1), 65–74 (2021)
5. Liu, H., Fang, S., Wang, Z., Fu, S.: Design of arbitrary-phase-difference transdirectional coupler and its application to a flexible Butler matrix. IEEE Trans. Microwave Theory Technol. **67**(10), 4175–4185 (2019)
6. Sieganschin, A., Tegowski, B., Jaschke, T., Jacob, A.F.: Compact diplexers with folded circular SIW cavity filters. IEEE Trans. Microwave Theory Technol. **69**(1), 111–118 (2021)
7. Zhu, Y., Dong, Y.: A novel compact wide-stopband filter with hybrid structure by combining SIW and microstrip technologies. IEEE Microwave Wirel. Comp. Lett. **31**(7), 841–844 (2021)
8. Rayas-Sanchez, J.E., Koziel, S., Bandler, J.W.: Advanced RF and microwave design optimization: a journey and a vision of future trends. IEEE J. Microwaves **1**(1), 481–493 (2021)
9. Bayraktar, Z., Komurcu, M., Bossard, J.A., Werner, D.H.: The wind driven optimization technique and its application in electromagnetics. IEEE Trans. Ant. Propag. **61**(5), 2745–2757 (2013)
10. Pietrenko-Dabrowska, A., Koziel, S., Al-Hasan, M.: Expedited yield optimization of narrow- and multi-band antennas using performance-driven surrogates. IEEE Access **9**, 143104–143113 (2020)
11. Koziel, S., Mosler, F., Reitzinger, S., Thoma, P.: Robust microwave design optimization using adjoint sensitivity and trust regions. Int. J. RF Microwave CAE **22**(1), 10–19 (2012)
12. Feng, F., Zhang, J., Zhang, W., Zhao, Z., Jin, J., Zhang, Q.: Coarse- and fine-mesh space mapping for EM optimization incorporating mesh deformation. IEEE Microwave Wirel. Comp. Lett. **29**(8), 510–512 (2019)
13. Koziel, S.: Fast simulation-driven antenna design using response-feature surrogates. Int. J. RF Microwave CAE **25**(5), 394–402 (2015)
14. Jin, J., et al.: Advanced cognition-driven EM optimization incorporating transfer function-based feature surrogate for microwave filters. IEEE Trans. Microwave Theory Technol. **69**(1), 15–28 (2021)
15. Koziel, S., Leifsson, L.: Simulation-Driven Design by Knowledge-Based Response Correction Techniques. Springer, New York (2016). https://doi.org/10.1007/978-3-319-301 15-0
16. Van Nechel, E., Ferranti, F., Rolain, Y., Lataire, J.: Model-driven design of microwave filters based on scalable circuit models. IEEE Trans. Microwave Theory Technol. **66**(10), 4390–4396 (2018)

17. Li, Y., Xiao, S., Rotaru, M., Sykulski, J.K.: A dual kriging approach with improved points selection algorithm for memory efficient surrogate optimization in electromagnetics. IEEE Trans. Magn. **52**(3), 1–4, Article No. 7000504 (2016)
18. Barmuta, P., Ferranti, F., Gibiino, G.P., Lewandowski, A., Schreurs, D.M.M.P.: Compact behavioral models of nonlinear active devices using response surface methodology. IEEE Trans. Microwave Theory Techol. **63**(1), 56–64 (2015)
19. Brihuega, A., Anttila, L., Valkama, M.: Neural-network-based digital predistortion for active antenna arrays under load modulation. IEEE Microwave Wirel. Comp. Lett. **30**(8), 843–846 (2020)
20. Liu, B., Yang, H., Lancaster, M.J.: Global optimization of microwave filters based on a surrogate model-assisted evolutionary algorithm. IEEE Trans. Microwave Theory Technol. **65**(6), 1976–1985 (2017)
21. Toktas, A., Ustun, D., Tekbas, M.: Multi-objective design of multi-layer radar absorber using surrogate-based optimization. IEEE Trans. Microwave Theory Technol. **67**(8), 3318–3329 (2019)
22. Zhang, J., Zhang, C., Feng, F., Zhang, W., Ma, J., Zhang, Q.J.: Polynomial chaos-based approach to yield-driven EM optimization. IEEE Trans. Microwave Theory Techol. **66**(7), 3186–3199 (20180
23. Zhang, J., Feng, F., Jin, J., Zhang, Q.-J.: Efficient yield estimation of microwave structures using mesh deformation-incorporated space mapping surrogates. IEEE Microwave Wirel. Comp. Lett. **30**(10), 937–940 (2020)
24. Koziel, S., Unnsteinsson, S.D.: Expedited design closure of antennas by means of trust-region-based adaptive response scaling. IEEE Ant. Wirel. Prop. Lett. **17**(6), 1099–1103 (2018)
25. Li, S., Fan, X., Laforge, P.D., Cheng, Q.S.: Surrogate model-based space mapping in postfabrication bandpass filters' tuning. IEEE Trans. Microwave Theory Technol. **68**(6), 2172–2182 (2020)
26. Lim, D.K., Woo, D.K., Yeo, H.K., Jung, S.Y., Ro, J.S., Jung, H.K.: A novel surrogate-assisted multi-objective optimization algorithm for an electromagnetic machine design. IEEE Trans. Magn. **51**(3), 8200804 (2015)
27. An, S., Yang, S., Mohammed, O.A.: A Kriging-assisted light beam search method for multi-objective electromagnetic inverse problems. IEEE Trans. Magn. **54**(3), 7001104 (2018)
28. Koziel, S.: Improved trust-region gradient-search algorithm for accelerated optimization of wideband antenna input characteristics. Int. J. RF Microwave CAE **29**(4), e21567 (2019)
29. Pietrenko-Dabrowska, A., Koziel, S.: Computationally-efficient design optimization of antennas by accelerated gradient search with sensitivity and design change monitoring. IET Microwaves Ant. Prop. **14**(2), 165–170 (2020)
30. Pietrenko-Dabrowska, A., Koziel, S.: Numerically efficient algorithm for compact microwave device optimization with flexible sensitivity updating scheme. Int. J. RF Microwave CAE **29**(7) (2019)
31. Conn, A.R., Gould, N.I.M., Toint, P.L.: Trust region methods, MPS-SIAM Series on Optimization (2000)
32. Broyden, C.G.: A class of methods for solving nonlinear simultaneous equations. Math. Comp. **19**, 577–593 (1965)
33. Lin, Z., Chu, Q.-X.: A novel approach to the design of dual-band power divider with variable power dividing ratio based on coupled-lines. Prog. Electromagn. Res. **103**, 271–284 (2010)
34. Tseng, C., Chang, C.: A rigorous design methodology for compact planar branch-line and rat-race couplers with asymmetrical T-structures. IEEE Trans. Microwave Theory Technol. **60**(7), 2085–2092 (2012)
35. Pietrenko-Dabrowska, A., Koziel, S.: Expedited antenna optimization with numerical derivatives and gradient change tracking. Eng. Comp. **37**(4), 1179–1193 (2019)

Approach to Imputation Multivariate Missing Data of Urban Buildings by Chained Equations Based on Geospatial Information

Alexander A. Khrulkov(✉), Margarita E. Mishina, and Sergey A. Mityagin

Saint-Petersburg National Research University of Information Technologies,
Mechanics and Optics (ITMO University), 197101 Saint-Petersburg, Russia
oneonwar@gmail.com, marg.mished@gmail.com, mityagin.spb@gmail.com

Abstract. Accurate information about real estate in the city, and about residential buildings in particular, is the most important resource for managing the development of the urban environment. Information about residential buildings, for example, the number of residents, is used in the inventory and digitization of the urban economy and subsequently becomes the basis of digital platforms for managing urban processes. Inventory of urban property can be carried out independently by different departments within the framework of official functions, which leads to the problem of conflicting information and missing values in urban data, in building data in particular. These problems are especially pronounced when information from different sources is combined to create centralized repositories and digital twins of the city. This leads to the need to develop approaches to filling missing values and correcting distorted information about residential buildings. As part of this work, the authors propose an approach to data imputation of residential buildings, including additional information about the environment. The analysis of the effectiveness of the approach is based on data collected for St. Petersburg (Russia).

Keywords: Urban data · Residential buildings · Multiple imputation · Machine learning · Geospatial information

1 Introduction

One of the main information resources of the city is information about residential buildings. The importance of this resource is determined by the fact that residential real estate accounts for a significant proportion of all real estate in the city and largely forms the urban environment. Buildings, as atomic city objects, have spatial and attribute characteristics. Spatial characteristics determine the location of a building in space in a given coordinate system, and attributive characteristics describe such features as number of storeys, population, area, etc.

Information about buildings posted in open Internet resources (for example, OpenStreetMap) forms the basis of many parametric methods for assessing the territory used in urban studies [1–3]. In addition, information about residential buildings and the citizens living in them is used simultaneously in many business processes of the city: from managing public transport services to the population to providing citizens with social infrastructure facilities. This leads to the formation of parallel processes of inventory and monitoring of the parameters of residential buildings, and therefore with which errors inevitably arise when combining these data. The most common manifestation is the incompleteness and inaccuracy of the information used [4]. Urban building data often contains omissions and erroneous values for one or more of the considered features, this makes it difficult to process and analyze these data. This leads to problems at the level of intersectoral interaction and management of complex projects for the development of territories, as well as creating or integrating city management systems based on digital twins and models of the urban environment.

Such a situation requires the development of approaches to imputation and correction of information about residential buildings and other objects of the urban environment. Within the framework of this article, we consider a general approach to filling missing data in the residential building features.

2   Techniques for Handling the Missing Data

2.1   Missing Data Types

The existing approaches to imputation missing data are based on the fundamental theory of Donald Rubin [5], which classifies such data into three categories depending on the reason for their absence: missing completely at random (MCAR), missing at random (MAR), missing not at random (MNAR).

The reason for the absence of data of the MCAR category is a random event that does not depend on any factors. For example, some information may be lost due to a failure in the storage system or as a result of incorrect actions when working with data. In practice, missing data of the MCAR category lead to a decrease in the analyzed sample without introducing a systematic error.

On the contrary, data of the MAR category are characterized by the dependence of the missing value on the known value of another feature of the object. This case is the most common, in particular, it includes the processing of missing data about urban buildings - missing values are often observed in relation to certain types of buildings, for example, new residential complexes or individual housing construction projects. As a rule, more advanced techniques are used to obtain unbiased statistical estimates when imputing MAR data than simply excluding them from the sample.

In the data feature of the MNAR category there is the systematic association of missing values with the reason for their absence. Thus, in surveys conducted in the course of urban studies, the presence or absence of a respondent's answer often depends on his position on the issue under consideration [6]. The processing of data of the MNAR category is the most problematic case, leading to an analysis of the reasons for the occurrence of missing values and to a change in the data collection strategy.

Determining the category of missing data (MCAR, MAR, or MNAR) allows us to choose the most appropriate imputation method that provides correct aggregated results and reliable statistical inferences.

2.2 Overview of Techniques for Handling the Missing Data

Standard way for missing data processing is to exclude them from the sample. In case of this approach the following methods are used: listwise deletion and pairwise deletion, which could be used for MCAR data. In addition, applying these methods for strongly damaged samples leads to significant data losses.

In most cases filling the omissions are more rational. Common ways to fill the missing data consist in imputation with the: median, mean, random value from the sample or constant value such as zero. More complicated approach is to use an EM-algorithm based on iterative computation of maximum likelihood estimates. For a wide class of problems, the maximum likelihood method is consistent and asymptotically efficient, however, to use it, it is necessary to have an idea of the laws of distribution of values in the observed features of an object. This shortcoming is devoid of widely used methods of regression and classification, which involve fitting a model based on known data.

Multiple imputation is the most universal approach to processing missing data, providing unbiased estimates and reliable confidence intervals, including for data of the MAR and MNAR category [7]. The concept of multiple imputation of missing data is based on the idea of combining several results obtained by a single imputation, reflecting the uncertainty of the imputed values. An increase in the number of iterations at which a single data imputation is carried out contributes to the achievement of convergence and stability of the final values. However, given the limited computing resources, in practice, in most cases, 3 to 5 iterations are sufficient.

2.3 Machine Learning Methods Used in Data Imputation

Since the data contains both quantitative and categorical features of objects, impute missing values involves solving both regression and classification problems. At the moment, there are many software-implemented mathematical methods, the success of which is determined by the available data and the dependencies that exist in them. The most common methods are: cluster analysis, K-nearest neighbors, Decision Trees and Random Forests [8].

An important step in the use of the described mathematical methods of data prediction is the hyperparameter tuning that are set before starting training, which affect the accuracy of the results and the cost of computing resources. Hyperparameter optimization methods are conditionally divided into two groups.

The first group includes methods based on enumeration of various combinations of hyperparameters. A reliable way to identify the optimal combination is to enumerate all possible hyperparameter values. A significant disadvantage of this method is its exponential complexity, which limits its use on large grids of parameters. Random and Successive Halving [9, 10] are considered effective modifications of the classical Grid Search algorithm, based on the reduction of the considered combinations due to various heuristics.

Another group includes sequential optimization methods based on a surrogate model that take into the results obtained after running in previous iterations (also called Bayesian optimization methods). The most common surrogate models are Gaussian processes and Parzen trees. The efficiency of Bayesian optimization based on Gaussian processes depends on the space of hyperparameters and decreases greatly when the dimension of the space increases, equating to the efficiency of Random Grid Search [11]. Bayesian optimization based on the Parzen window is considered to be less expensive in terms of the required computational resources, but this approach does not take into account the possible interaction of the hyperparameters under consideration [12].

The choice of an appropriate hyperparameter optimization method largely depends on the problem being solved and the available computing resources. For simple models built on a small amount of data, Grid Search guarantees accurate and unambiguous results in a reasonable time. On the contrary, in the case of scalable models, it is advisable to use Bayesian optimization or Successive Halving.

3 Approach to Imputation Multivariate Missing Urban Buildings Data Based on Geospatial Information

3.1 Description of the Developed Approach

Considering that city buildings data has multivariate missing values, when more than one attribute has missing data, applying single imputation methods cannot provide a predictable result. In this regard to impute missing data of residential buildings multiple imputation by chained equations is used [13]. The imputation of missing values at each iteration is carried out by regression and classification methods.

Cities consist of territorial units characterized by the prevailing evolutionary-morphological type of development, and having characteristic values for a number of features (number of storeys of tasks, density characteristics, etc.) [14]. The developed approach assumes that the use of additional information about the features of neighboring objects, as well as, if available, other aggregated information about the spatial environment (for example, about services located nearby or calculated spatial indices [15, 16]) provides more accurate results for data imputation process.

According to the given block diagram (Fig. 1):

1. At the first stage, the input data is preprocessed: categorical features are binarized and the known values are reduced to a numerical type. Positions of missing values are written to a new variable, and initial values are initialized in their place, which are subsequently imputed by predicted values. The initial values are the distance-weighted average values of other objects (1).

$$x_{init}^t = \frac{\sum_{i=1}^{k-1} w_i \cdot x_i^t}{\sum_{i=1}^{k-1} w_i}, \; w_i = \frac{1}{d} \tag{1}$$

where x_{init}^t—the calculated initial value of the object on the feature t; x_i^t—the initial value of another object on the feature t; w_i—coefficient of remoteness of another object; k—the number of objects in the dataset; d—the Euclidean distance between objects (2).

$$d = \sqrt{(x_1 - x_2)^2 + (y_1 - y_2)^2} \tag{2}$$

where x, y—coordinates of two points.

2. The next step is to expand the original feature space. For each object from the input dataset, a search is made in the metric space for K-nearest neighbors in a given radius r (3), based on the Euclidean distance measured between the centers of objects (2). The result of the search is two matrices containing the indices of the nearest neighbors and the distance to them for each sample object. The obtained indices are used to calculate the average values of neighboring objects, which supplement the input data array.

$$kNN(q) = \{R \subseteq X, |R| = k \cap \forall n \in R, o \in X - R : d(q, n) \leq d(q, o), d(q, n) \leq r\} \tag{3}$$

where X—the set of all objects in the dataset; k—the number of required neighbors; q—the object for which the search for neighbors is carried out; d—a function of distance; r—the specified neighbor search radius.

   If additional data is used (for example, on the calculated spatial indices of territorial units), they are aggregated and attached to the initial dataset by a given attribute (for example, by a building or block identifier).

3. Imputation of missing values at each iteration begins with the selection of dependent and independent variables in the prepared data. The dependent variable is the first in order feature that has missing values in the initial dataset for any objects of the sample (the positions of missing values before the initialization of the initial values were written to the variable at step 1). Then, a training sample is selected in the data, consisting of objects that do not have missing values in the original dataset according to the attribute assigned to the dependent variable (4).

$$X^{raw}_{m \times n} = X^{obs_t}_{p \times n} \cup X^{mis_t}_{k \times n} \ u \ \varnothing = X^{obs_t}_{p \times n} \cap X^{mis_t}_{k \times n} \tag{4}$$

where $X^{raw}_{m \times n}$—the prepared dataset; $X^{obs_t}_{p \times n}$—data that does not contain missing values in the initial dataset on the feature t; $X^{mis_t}_{k \times n}$—data containing missing values in the initial dataset on the feature t.

4. On the selected sample, the regression and classification models are trained. The selection of a combination of hyperparameters that provide the best predictive ability of the method is performed by the Grid Search with Successive Halving [10]. The assessment of the quality of a model trained on a certain combination of hyperparameters is carried out using the k-fold cross-validation. The loss function for regression models is the Mean Square Error and for classification models is the Log Loss function. Prediction of missing values on the rest of the data is made with the best combination of hyperparameters. The calculated values impute the missing values.

5. Having imputed the missing values of the dependent variable, steps 3 and 4 are repeated cyclically for each variable that has omissions in the values.

6. Steps 3–7 constitute one complete iteration of data imputation, the total number of which is specified by the user, based on available computing resources. According to [17], upon imputation of at least 2 iterations, the coefficients of the models become stable, which makes it possible to avoid the dependence of the result obtained on the order in which the target variables are assigned.

7. The result of steps 1–6 is a complete dataset containing the imputed values. According to the concept of multiple data imputation, steps 1–6 are repeated several times (from 3 to 5 are considered sufficient [18]), after which the values in independently imputed datasets are averaged. The final dataset and the average quality score obtained by the k-fold cross-validation in step 4 are the output of the method.

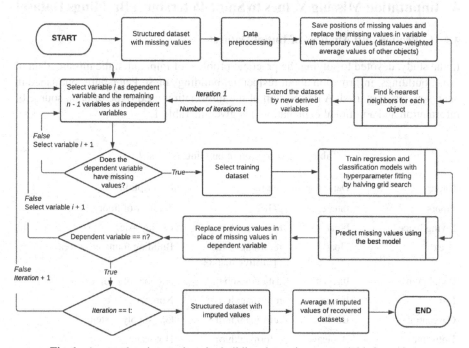

Fig. 1. Approach to imputation city building data using geospatial information

3.2 Findings of the Developed Approach

The issue of incomplete data imputation has a fundamental theoretical basis on which the developed approach to imputation missing data of urban buildings is based. The developed approach implements the concept of multiple data imputation, which takes into account the uncertainty of the imputed values by predicting several values for each missing value using the selected model. Since the missing values are found in data that have different internal dependencies, the presented approach preliminarily selects the best predictive model sequentially for each considered feature.

The advantage of the developed approach in the framework of solving the applied problem of building data imputation is the presence of the stage of expanding the initial dataset with derived features containing information extracted from spatial dependencies that are initially inaccessible to the model, but known to the researcher. At the current stage, additional derived features are information about neighboring objects.

In the future, it is planned to identify new spatial dependencies that exist in the urban environment and use them to obtain the most accurate imputed values.

A potential obstacle to the application of the presented approach is its resource consumption. Due to the iterative process, it takes a significant amount of time to impute a large number of missing values.

4   Imputation Missing Values to Saint-Petersburg Buildings Dataset

4.1   Collection and Analysis of Initial Dataset

In the study, devoted to solving the practical problem of imputation the missing data of urban buildings, information about residential buildings in St. Petersburg, published in open sources, was used. A description of the information used (type of data, source of information and additional explanations) is given in Table 1.

Table 1. Description of building data used.

Name	Data type	Source	Description
Floors	Integer	2GIS	Number of floors
LivingSpace	Float	reformagkh.ru,	Living space
Area	Float	reformagkh.ru, openstreetmap.org	Building foundation area
Population	Integer	data.gov.spb.ru	Number of people
Lift	Integer	reformagkh.ru	Number of lifts
Gascentral	Boolean	reformagkh.ru	Gas supply system
Hotwater	Boolean	reformagkh.ru	Hot water
Electricity	Boolean	reformagkh.ru	Electricity
Coordinates (x, y)	Float	openstreetmap.org	X, Y in 3857 ESPG projection

When collecting initial data, emphasis was placed on the residential development of the city, since at the time of the study there was the most complete database of this type of buildings in open sources of information, of which the reformagkh.ru resource was considered the most reliable. The collected data contained information on 12,867 residential buildings. The collected raw data contained no missing values. To conduct the study, the modeling of missing values was carried out by random removal of feature values. The degree of damage to the initial dataset was determined by the percentage of "sown" omissions from the total number of feature values and ranged from 10 to 90% in 10% increments. In the figure (Fig. 2), the buildings of the Dekabristov Island municipality are marked in red, the data on which are deliberately distorted in this way. An example of such distorted data is given in the Table 2.

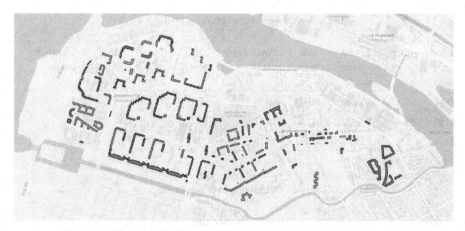

Fig. 2. Buildings with distorted data (red polygons) (Color figure online)

Table 2. Distorted data samples of three buildings.

ID	Floors	Living space	Area	Population	Lift	Gas central	Hot water	Electricity
1511	16	NULL	9378	509	8	NULL	True	True
1356	12	28873	20341	NULL	15	NULL	True	True
2314	NULL	1197	1988	0	0	True	False	True

4.2 Investigating the Effectiveness of an Approach to City Building Data Imputation

The approach was implemented in the python programming language, using open libraries - numpy, geopandas and scikit-learn. At the first stage of imputation, the input data was preprocessed: all values of the features were reduced to a numeric type. For features Gascentral, Hotwater, Electricity, the binarization procedure was carried out (True value - 1, False value - 0). Initial values were initialized in place of missing values. To expand the input dataset, information about neighbors was used - the average values of the features of the three nearest neighboring buildings located within a radius of 500 m. An example of such derived data is presented in Table 3.

The selection of the best model for predicting missing values was carried out using the Grid Search with Successive Halving. The use of the algorithm of Successive Halving for reduction of the hyperparameter combinations made it possible to get away from the exponential complexity of the basic grid search and obtain the optimal combination in less time. At the stage of determining the best predictive model, the three most common methods of regression and classification in the context of data imputation (according to Sect. 2.2) were investigated: K-nearest neighbors, Random Forests, and Gradient Boosting Decision Trees. The optimized hyperparameters are presented in Table 4. Z-normalization of the data was preliminarily carried out.

Table 3. Average characteristics of neighbors.

ID	Floors neigh	Living space neigh	Area neigh	Population neigh	Lift neigh	Gas central neigh	Hot water neigh	Electricity neigh
1511	16	12646	4591	506	7	0	1	1
1356	20	28302	14740	1101	13	1	1	1
2314	5	1920	2121	16	0	1	0	1

Table 4. Optimizing hyperparameters.

Method	Hyperparameter grid
K-nearest neighbors	The number of neighbors is 3, 5, 7, 9
Random forests	The number of trees is 50, 100, 150, 200 The maximum depth of trees is 2, 3, 4, 5, 6, 7, 8 The minimum number of objects to split a node is 2, 5, 10 The subset of features for node splitting is all, sqrt
Gradient Boosting Decision Trees	The number of trees is 50, 100, 150, 200 The maximum depth of trees is 2, 3, 4, 5, 6, 7, 8 The minimum number of objects to split a node is 2, 5, 10 The subset of features for node splitting is all, sqrt The learning rate is 0.01, 0.1, 1

The evaluation of the generalizing abilities of models with various combinations of hyperparameters is carried out by the k-fold cross-validation method on 5 groups of samples. The following were used as loss functions: the Mean Square Error (for regression problems) and the Log Loss function (for classification problems).

The quality metrics of the imputed data were: the coefficient of determination (for regression problems) and the F-measure (for classification problems). The quality metrics calculated for a dataset with a degree of damage of 10% are shown in Table 5. According to the results, the best generalizing ability is possessed by the Gradient Boosting Decision Trees. With an increase in the percentage of deleted values, similar results were observed (Fig. 3).

In order to identify the impact on the accuracy of data imputation of additional derived features, quality metrics were also calculated for data imputed using the developed approach, but without expanding the initial feature space with derived information - information about neighboring objects. The results obtained for a data set with a degree of damage of 10% are presented in Table 5. According to the results of calculations, for any degree of damage to the input set, the accuracy of the restored values turned out to be higher with a preliminary expansion of the feature space. With an increase in the percentage of missing values, the influence of derived information increased (Fig. 4).

Based on the results of experiments conducted on 9 datasets with varying degrees of damage, it was found that predicting missing values using the best model based on

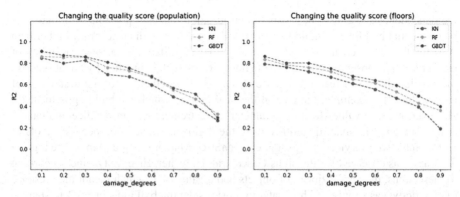

Fig. 3. Comparison of method quality metrics for the features Floors and Population

Table 5. Compare accuracy results for different predictive methods.

Variable	With neighbor information			Without neighbor information
	Gradient boosting	Random forest	KNeighbors	Gradient boosting
Floors	0.85	0.84	0.79	0.82
Living space	0.95	0.94	0.89	0.93
Area	0.54	0.55	0.47	0.54
Population	0.91	0.84	0.84	0.77
Lift	0.89	0.89	0.82	0.88
Gascentral	0.96	0.96	0.95	0.96
Hotwater	0.93	0.93	0.92	0.80
Electricity	0.99	0.99	0.99	0.99

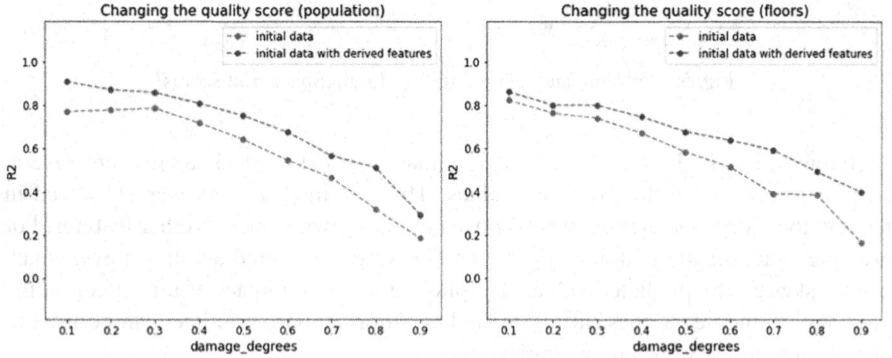

Fig. 4. Comparison of data imputation methods by characteristics Floors and population

the dependencies available in geospatial information makes it possible to impute data on residential buildings with up to 30% of missing values with an accuracy higher than 0.80 for the specified metrics. In the case under consideration, a lower accuracy of predictions was observed only for the feature Area, which is justified by the absence in the existing dataset of features that are closely related to the area of the foundation of the building (for example, the area of the land plot on which the building is located). However, missing values for this parameter are extremely rare in practice and can be quickly restored in small quantities using the digitization and geoprocessing tools of cartographic web services (for example, OpenStreetMap or Google Maps). The highest accuracy was observed for the signs Gascentral, Hotwater, Electricity, since according to these signs, the vast majority of objects had a value of 1, which meant the presence of electricity, gas supply and hot water in most residential buildings in St. Petersburg.

For the analysis of outliers in the imputed values, absolute and relative shifts from the true values were calculated (5)–(6). As an example, Fig. 5 plots the absolute and relative offsets of LivingSpace predictions for a 10% damaged dataset.

$$ab_i = y_{true} - y_{pred} \tag{5}$$

$$pb_i = 100 \cdot \left| \frac{(y_{true} - y_{pred})}{y_{true}} \right| \tag{6}$$

where y_{true}—the true value of the feature; y_{pred}—the predicted value of the feature.

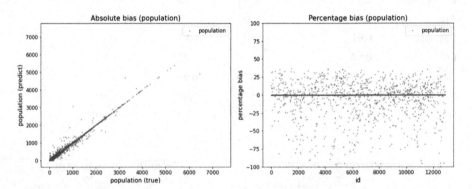

Fig. 5. Absolute and relative shift of LivingSpace trait scores

Even with a high coefficient of determination equal to 0.95, there were several significant outliers in the predicted values. The identified outliers were observed in relation to objects containing obvious errors in the known values of other features. For example, some of the buildings in the dataset were considered single-storeys, which greatly skewed the predicted values. Despite the negative impact of this effect on the accuracy of imputed values, in future studies, information on outliers can be used to identify inaccurate values in the initial data.

4.3 Comparative Analysis of Approaches to Data Imputation

An analysis of existing approaches to data imputation showed that the implemented solutions in the presence of missing values in the dataset, with the exception of their removal, can be used:

1. Filling in missing values with a point estimate (for example, an average).
2. The experimental method of the scikit-learn library is IterativeImputer.

Singular mean filling is a common solution to the problem of missing data, which, however, tends to skew any statistical estimate of a feature other than the mean.

The IterativeImputer method implements a multiple data imputation strategy using predictive models. As input, IterativeImputer accepts one specified regression or classification method, a method for initializing initial values in positions of unknown features (mean, median, mode, or constant), the number of imputation and a number of other parameters that determine the output of data. This method is a general tool for solving the problem of missing data of various directions and does not allow considering the specifics of urban data, including identifying new dependencies by expanding the feature space. Another disadvantage is the lack of an implemented algorithm for choosing the best predictive model, which, as a result, leads to the impossibility of simultaneously imputing quantitative and qualitative values, as well as automated selection of the best combination of hyperparameters.

The comparison of the alternative solutions with the developed approach was carried out for the input dataset with different degrees of damage by calculating the coefficient of determination and the F-measure. In the IterativeImputer method, gradient boosting decision trees was used as a predictive model with default parameters in the scikit-learn library. A demonstration of the results obtained is shown in Fig. 6 (similar results were observed on other grounds).

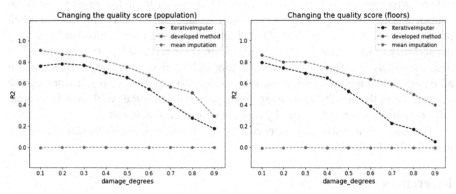

Fig. 6. Comparison of data imputation methods on the example of the features Floors and Population

The results of the calculations showed that filling the missing values of quantitative features with the average is a nominal solution that does not improve the situation -

the accuracy of the imputed values remains close to zero. Using the implemented IterativeImputer method of the scikit-learn library made it possible to consider existing data dependencies and obtain an acceptable accuracy of the imputed values at the level of 0.80 for the specified metrics (with a data corruption degree of no more than 10% and considering that the optimal regression or classification method is known). The developed data imputation approach showed its effectiveness in comparison with alternative solutions in all cases considered (with any degree of data damage). The advantage of the developed approach was achieved by extracting additional information from the data, which is then used to impute missing values.

5 Conclusion

Complete and accurate city building data is a critical resource for city management. However, such data often contain omissions and erroneous values for one or more features, which makes it difficult to process and further analyze them. This problem on a large scale leads to a decrease in the effectiveness of intersectoral interaction, management of complex projects for the development of territories and the use of integrating systems for urban management.

The developed approach for urban building data imputation is based on multiple data imputation, as well as regression and classification methods. The novelty of the developed approach lies in the presence of stages of expanding the initial feature space and determining the best model for predicting missing values. The extension of the feature space is carried out through calculations and transformations performed on the available geospatial information. The determination of the best predictive method is carried out according to the Grid Search algorithm with sequential halving of the specified combinations of hyperparameters of regression and classification methods.

It should be noted that at this stage, the presented method does not allow making confident judgments about values for individual objects, but it makes it possible to obtain correct aggregated values for territorial units and formulate reliable statistical conclusions when making decisions in the field of management and optimization of urban processes. In further studies, it is planned to conduct experiments aimed at improving the accuracy of the imputed values by identifying new relationships that exist in urban data, as well as to explore the developed approach on data containing information on all types of buildings (residential and non-residential). A separate area of work will be the analysis of the possibility of using this approach to solve the problem of inaccurate information about urban facilities by recalculating all available values.

This research is financially supported by The Russian Science Foundation, Agreement №. 20-11-20264.

References

1. Khrulkov, A., Mityagin, S.A., Repkin, A.I.: Multi-factor approach to investment attractiveness assessment of urban spaces. Proc. Comput. Sci. **178**, 94–102 (2020)
2. van Nes, A., Pont, M.B., Mashhoodi, B.: Combination of Space syntax with spacematrix and the mixed-use index: the Rotterdam South test case. In: 8th International Space Syntax Symposium, Santiago de Chile, 3–6 January (2012)

3. Boiko, D., Parygin, D., Savina, O., Golubev, A., Zelenskiy, I., Mityagin, S.: Approaches to analysis of factors affecting the residential real estate bid prices in case of open data use. In: Chugunov, A., Khodachek, I., Misnikov, Y., Trutnev, D. (eds.) EGOSE 2019. CCIS, vol. 1135, pp. 360–375. Springer, Cham (2020). https://doi.org/10.1007/978-3-030-39296-3_27
4. Wang, R.Y., Strong, D.M.: Beyond accuracy: what data quality means to data consumers. J. Manag. Inf. Syst. **12**, 5–33 (1996)
5. Rubin, D.B.: Inference and missing data. Biometrika **63**, 581–592 (1976)
6. Curley, C., Krause, R.M., Feiock, R., Hawkins, C.: Dealing with missing data: a comparative exploration of approaches using the integrated city sustainability database. Urban Affairs Rev. **55**, 591–615 (2019)
7. van Buuren, S.: Flexible Imputation of Missing Data. CRC Press (2018)
8. Lin, W.-C., Tsai, C.-F.: Missing value imputation: a review and analysis of the literature (2006–2017). Artif. Intell. Rev. **53**(2), 1487–1509 (2019). https://doi.org/10.1007/s10462-019-09709-4
9. Bergstra, J., Bengio, Y.: Random search for hyper-parameter optimization. J. Mach. Learn. Res. **13**, 281–305 (2012)
10. Jamieson, K., Talwalkar, A.: Non-stochastic best arm identification and hyperparameter optimization. In: Artificial Intelligence and Statistics, pp 240–248 (2016)
11. Li, L., Jamieson, K., DeSalvo, G., et al.: Hyperband: a novel bandit-based approach to hyperparameter optimization. J. Mach. Learn. Res. **18**, 6765–6816 (2017)
12. Bergstra, J., Bardenet, R., Bengio, Y., Kégl, B.: Algorithms for hyper-parameter optimization. Adv. Neural Inf. Process. Syst. **24** (2011)
13. van Buuren, S., Groothuis-Oudshoorn, K.: mice: Multivariate imputation by chained equations in R. J. Stat. Softw. **45**, 1–67 (2011)
14. Conzen, M.R.G.: Alnwick, Northumberland: a study in town-plan analysis. Transactions and Papers (Institute of British Geographers) iii–122 (1960)
15. Pont, M.Y.B., Haupt, P.A.: Spacematrix. Space, Density and Urban Form. NAi Publishers (2010)
16. van den Hoek, J.W.: Towards a Mixed-use Index (MXI) as a tool for urban planning and analysis. Urbanism: PhD Research 2008–2012 65 (2009)
17. Templ, M., Kowarik, A., Filzmoser, P.: Iterative stepwise regression imputation using standard and robust methods. Comput. Stat. Data Anal. **55**, 2793–2806 (2011)
18. Schafer, J.L., Graham, J.W.: Missing data our view of the state of the art. Psychol. Methods **7**, 147 (2002)

Global Design Optimization of Microwave Circuits Using Response Feature Inverse Surrogates

Anna Pietrenko-Dabrowska[1] ⓘ, Slawomir Koziel[1,2(✉)] ⓘ, and Leifur Leifsson[3] ⓘ

[1] Faculty of Electronics Telecommunications and Informatics, Gdansk University of Technology, Narutowicza 11/12, 80-233 Gdansk, Poland
anna.dabrowska@pg.edu.pl
[2] Department of Engineering, Engineering Optimization and Modeling Center, Reykjavík University, Menntavegur 1, 102 Reykjavík, Iceland
koziel@ru.is
[3] School of Aeronautics and Astronautics, Purdue University, West Lafayette, IN 47907, USA
leifur@purdue.edu

Abstract. Modern microwave design has become heavily reliant on full-wave electromagnetic (EM) simulation tools, which are necessary for accurate evaluation of microwave components. Consequently, it is also indispensable for their development, especially the adjustment of geometry parameters, oriented towards performance improvement. However, EM-driven optimization procedures incur considerable computational expenses, which may become impractical even in the case of local tuning, and prohibitive whenever global search is vital (e.g., multi-model tasks, simulation-based miniaturization, circuit re-design within extended ranges of operating frequencies). This work presents a novel approach to a computationally-efficient globalized parameter tuning of microwave components. Our framework employs the response feature technology, along with the inverse surrogate models. The latter permit low-cost exploration of the parameter space, and identification of the most advantageous regions that contain designs featuring performance parameters sufficiently close to the assumed target. The initial parameter vectors rendered in such a way undergo then local, gradient-based tuning. The incorporation of response features allows for constructing the inverse model using small training data sets due to simple (weakly-nonlinear) relationships between the operating parameters and dimensions of the circuit under design. Global optimization of the two microstrip components (a coupler and a power divider) is carried out for the sake of verification. The results demonstrate global search capability, excellent success rate, as well as remarkable efficiency with the average optimization cost of about a hundred of EM simulations of the circuit necessary to conclude the search process.

Keywords: Microwave design · Global optimization · Surrogate modeling · Surrogate-assisted design · Inverse models · Response features

D. Groen et al. (Eds.): ICCS 2022, LNCS 13352, pp. 248–262, 2022.
https://doi.org/10.1007/978-3-031-08757-8_22

1 Introduction

Over the years, microwave devices become increasingly complex [1, 2], which is mainly caused by stringent performance requirements imposed thereon [3–5]. Another reason is miniaturization [6], which requires the employment of transmission line (TL) meandering [7] or compact microwave resonant cells (CMRCs) [8], leading to densely arranged layouts [9]. Reliable evaluation of such circuits is only possible using full-wave EM analysis. Furthermore, achieving top electrical performance requires careful tuning of all system parameters. Unfortunately, EM-driven optimization is computationally expensive, even in the case of local search. Meanwhile, in a number of situations (multimodal problems [10], multi-objective design [11], unavailability of reasonable initial designs [12]), global optimization is required, the cost of which is considerably higher.

Nowadays, global search has been dominated by nature-inspired methods [13]-[15], which come in many variations [16–21]. The operating principles of these methods are rooted in relaying information between the sets (populations) of competing solutions, arranged by partially stochastic operators [13–17]. Nature-inspired algorithms are easy to apply and handle but their computational complexity is high, with a one-time optimization run involving as many as hundreds and thousands of objective function evaluations. Consequently, practical application of nature-inspired method in high-frequency design requires utilization of acceleration techniques, the most popular of which is surrogate modeling [22, 23]. Some of specific modeling techniques being in use in this context include kriging [24], Gaussian process regression [25], and neural networks [26].

Despite its advantages, surrogate-assisted nature-inspired global search has some limitations, the most serious of which is caused by the curse of dimensionality. As far as microwave design is concerned, yet another difficulty poses high nonlinearity of frequency characteristics. Both factors make a construction of reliable surrogates difficult. In practice, only relatively simple components parameterized using a few variables can be efficiently handled [27, 28]. A recent performance-driven modeling [29, 30] can somewhat mitigate this issue by confining the modeling process to a region containing designs of superior quality w.r.t. to the performance figures of choice [29]. This concept can be generalized to variable-resolution setup [31]. Another expedient strategy is the response feature technology [32], where the design process is reformulated in terms of appositely defined characteristics points [33]. This technology capitalizes on close-to-linear relationships between the feature coordinates and designable parameters of the system at hand [34].

In this paper, we propose a novel technique for globalized parameter tuning of passive microwave devices. Our methodology is based on inverse regression surrogates, established at the tier of response features of the component under design. The surrogates allow for low-cost and rapid exploration of the parameter space, and to identify the most encouraging regions thereof. The design found with the aid of the regression model is further enhanced using the conventional gradient-based procedure. The proposed framework has been validated using two microstrip devices, and its global search capability has been demonstrated along with the low execution cost of a few dozens of EM analyses of the structure under design. Extensive benchmarking also corroborates superiority over nature-inspired optimization, as well as multiple-start gradient search.

2   Global Design by Response-Feature Inverse Surrogates

This section outlines the proposed global optimization procedure, which employs response-feature surrogates for fast detection of the part of the design space where high-quality designs likely reside, and to render satisfactory starting points for local fine-tuning. The section begins by formulating the optimization task (Sect. 2.1). Section 2.2 recalls response features methodology. Global search and local tuning stages are delineated in Sects. 2.3 and 2.4, respectively. Section 2.5 briefly summarizes the proposed optimization technique.

2.1   Simulation-Based Optimization. Problem Formulation

For quantifying of the performance of the microwave component under design, we adopt a merit function U. The optimization task is formulated as

$$x^* = \arg\min_{x} U(x, F_t) \tag{1}$$

where $F_t = [F_{t.1} \ldots F_{t.K}]^T$ refers to a target vector whose entries are device operating parameters (such as operating frequencies or power split), whereas x denotes a design parameters vector (typically, circuit dimensions). The function U is evaluated based on EM-simulated characteristics of a given structure, typically, scattering parameters $S_{kl}(x, f)$, with k and l being the circuit ports, and f denoting the frequency.

2.2   Response Features for Design Assessment

Global optimization may be indispensable in the absence of a quality initial design or when given optimization task is multimodal. Yet, due to highly nonlinear relationships between the component characteristics and both design variables and frequency, as well as dimensionality issues, global search is burdensome. Moreover, finding a globally optimal design when using simulation-driven procedures is usually uneconomic. This is especially the case for nature-inspired algorithms, the cost of which may reach up to several thousands of EM-analyses of the considered component. Surrogate-based optimization of microwave passives is also encumbered by the aforementioned issues. This makes construction of reliable surrogates for structures described by an extended number of design variables and, at the same time, within sufficiently broad parameter ranges virtually unrealizable using conventional metamodeling techniques.

One of the available approaches for circumventing the aforementioned difficulties is response features technology [32]. The said technique operates at the tier of the characteristic points of the circuit response (in contrast to traditional procedures which handle the response in its entirety). The rationale behind adopting this strategy is the fact that the relationship between the features and design variables is nonlinear to a considerably smaller degree than that of the entire characteristic, as shown in Fig. 1 for an example microwave component (miniaturized compact rat-race coupler). The feature points are defined with regard to the particular formulation of the optimization task at hand, as well as the actual shape of the circuit response. Various definitions

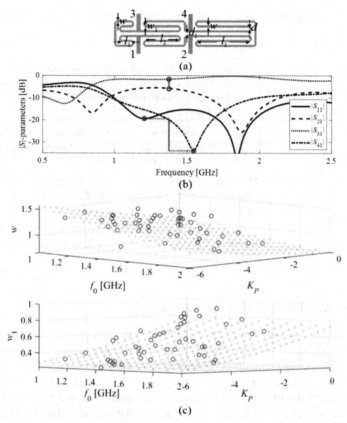

Fig. 1. Visual illustration of the weakly-nonlinear relationship between response features and geometry parameters of a circuit: a) an example structure (compact rat-race coupler), b) characteristic points corresponding to the minima of $|S_{11}|$ and $|S_{41}|$, and power split ratio K_P (marked with circles); the vertical line indicates the approximate circuit operating frequency f_0 (assessed as the minima average); c) relationship between the f_0 and K_P and the selected design variables; the designs are marked with the circles, whereas the regression model is shown with gray points.

include: frequency and level coordinates of the resonances [34], or points that delineate a bandwidth [35].

In Fig. 1(b), the operating frequency of the circuit (denoted as f_0) is roughly estimated as the midpoint between the minima of the matching and isolation characteristic, and the power split is assessed as $K_P = |S_{21}(f_0)| - |S_{31}(f_0)|$. Clearly, for certain designs, some of (or even all) the feature points may not be discernible. Figure 1(c) depicts the relationship between f_0 and K_P and two design variables of the coupler of Fig. 1(a). The circuit designs are shown against the inverse regression model $\alpha_0 + \alpha_1 \exp(\alpha_2 f_0 + \alpha_3 K_P)$ representing the trends between the circuit operating and geometry parameters. This form of a regression model is sufficient to describe weakly nonlinear dependence of the device dimensions on its operating parameters, which characterizes many real-world components and devices.

Table 1. Global optimization using inverse feature-based surrogates: notation

Notation	Description
$F(x) = [F_1(x) \dots F_K(x)]^T$	Operating parameter vector at the design x (estimated based on response features extracted from EM-simulated response); if some features are indiscernible and the parameters cannot be assessed: $F(x) = [0 \dots 0]^T$
$L(x) = [l_1(x) \dots l_K(x)]^T$	Auxiliary vector whose entries are design quality coefficients (corresponding to the entries of $F(x)$); if some elements of $L(x)$ cannot be derived: $L(x) = [0 \dots 0]^T$
$D(F,F_t)$	Distance function that measures the misalignment between the target vector F_t and the current one F; here: $D(F,F_t) = \|F - F_t\|$ (i.e., we use L_2-norm-based distance)
D_{accept}	Termination parameter (user-defined) of the global search stage: if $D(F, F_t) \leq D_{accept}$ the design x is acceptably close to the target F_t

2.3 Global Search Stage Using Inverse Models

In the presented approach, parameters of the inverse regression model are estimated based on randomly acquired observables. As the observable quality with respect to the target performance requirements is not known beforehand, we need a measure to assess it and to decide whether a particular observable is to be accepted or not. Here, this quality is quantified with the use of the response features derived from the frequency characteristics, which are subsequently compared with the assumed targets. The accepted observables are then utilized to build an inverse regression surrogate, which is exploited to delimit the promising region of the design space, and also to render infill points for the metamodel refinement. In each iteration, a single infill point is reinserted to the observable set replacing that of the poorest quality.

The adopted notation is provided in Table 1. Some additional clarification is required on the vector $L(x)$: its entries are the design quality coefficients, definition of which is problem-specific. If the design process aims at minimizing the levels of the reflection and isolation characteristics at the operating frequency, the respective coefficient l_k may be defined as the mean of $|S_{11}|$ and $|S_{41}|$ at the said minima. If, however, power split ratio is of interest, the correspondent l_k may be a discrepancy between the actual power split and the target one. In this work, the coefficients l_k are normalized, i.e., $0 \leq l_k \leq 1$: we have $l_k = 0$ for the best design, and $l_k = 1$ for the design of the lowest quality.

The essential steps of the discussed global optimization procedure include: rendering observables intertwined by the construction and refinement of the inverse surrogate, followed by rendering a candidate design, and concluded by its tuning in the case it is close enough to the target. The random observables, i.e., parameter vectors $x^{(j)} = [x_1^{(j)} \dots x_n^{(j)}]^T, j = 1, \dots, N$, are uniformly spaced within the box-constrained design space X, delimited by the lower and upper variable boundaries. The observable generation stops when N designs of $\|F(x^{(j)})\| > 0, j = 1, \dots, N$, have been gathered. The triplets $\{F(x^{(j)}), L(x^{(j)}), x^{(j)}\}_{j=1,\dots,N}$, constitute a training data for an inverse regressive metamodel $r_I(F)$, which assesses the trend between circuit operating parameters and dimensions. Once the inverse surrogate is constructed, it is utilized for prediction of the candidate design $x_{tmp} = r_I(F_t)$ for a given operating parameter vector F_t. If the two following conditions are satisfied: $\|F(x_{tmp})\| > 0$ and $D(F(x_{tmp}),F_t) < \max\{j = 1, ..., N: D(F(x^{(j)}),F_t)\}$, the design x_{tmp} substitutes that of the largest distance $D(F(x^{(j)}),F_t)\}$. Ultimately, the inverse surrogate is reset with a new training set.

The inverse model (re)construction and design prediction are reprised until a design sufficiently close to the assumed target is detected, i.e., the design for which $D(F(x_{tmp}),F_t) < D_{\max}$, where $D_{\max}$ denotes an acceptance threshold (set by the user). This design is subsequently refined in the local tuning stage. In the case such a design has not been reached, the procedure is executed until the assumed computational budget has ran out, then, the best design is returned. Figure 2 explains graphically the aforementioned concepts: starting from observable generation, through inverse model construction and refinement, until local tuning of the globally optimal design. Note that Figs. 2 (b)–(d) show the procedure for a single variable x, which, in fact, is carried out for all the parameters concurrently.

The analytical form of the inverse surrogate $r_I(F)$ does not need to be intricate, as the relation between operating parameters and circuit dimensions is weakly nonlinear. Still, it is essential to ensure a sufficient flexibility, since this dependence may in some cases be inversely proportional. Therefore, we adopt a regression model of the form:

$$r_I(F) = r_I\left(\begin{bmatrix} f_1 \\ \cdots \\ f_K \end{bmatrix}\right) = \begin{bmatrix} r_{I.1}(F) \\ \cdots \\ r_{I.n}(F) \end{bmatrix} = \begin{bmatrix} p_{1.0} + p_{1.1}e^{\sum_{k=1}^{K} p_{1.k+1}f_k} \\ \cdots \\ p_{n.0} + p_{n.1}e^{\sum_{k=1}^{K} p_{n.k+1}f_k} \end{bmatrix} \quad (2)$$

In order to identify the surrogate $r_I(F)$, one needs to solve

$$[p_{j.0}\ p_{j.1}\ \cdots\ p_{j.K+1}] = \arg \min_{[b_0\ b_1\ \cdots\ b_{K+1}]} \sum_{k=1}^{N} w_k \left[r_{I.j}\left(F(x^{(k)})\right) - x_j^{(k)} \right]^2, j = 1, \ldots, n \quad (3)$$

using the triplets $\{F(x^{(j)}), L(x^{(j)}), x^{(j)}\}_{j=1,...,N}$, as training data samples. The weights $w_k = [1 - \max\{l_1(x^{(j)}),..., l_k(x^{(j)})\}]^2$, $k = 1, ..., N$, are to ensure that the high-quality observables influence r_I in a more significant manner.

2.4 Local Tuning

The design $x^{(0)}$ rendered by the global optimization procedure of Sect. 2.3 satisfies the condition $D(F(x^{(0)}),F_t) \leq D_{accept}$, where the value of the threshold D_{accept} is set so as to ensure that the operating parameters at $x^{(0)}$ are in a sufficient proximity to F_t. This is to enable reaching the target through a local search procedure. Here, we exploit the trust-region (TR) optimization algorithm with numerical derivatives [36], which renders a series designs $x^{(i)}$, $i = 0, 1, ...$ that approximate the optimal solution x^*, by solving

$$x^{(i+1)} = \arg \min_{x;\ -d^{(i)} \leq x - x^{(i)} \leq d^{(i)}} U_L(x, F_t) \quad (4)$$

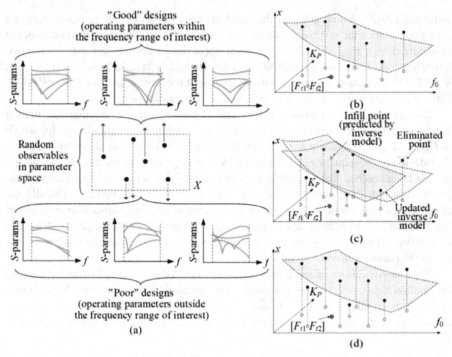

Fig. 2. Major stages of the presented global optimization framework: (a) selection of high-quality observables, (b) two-dimensional objective space: observables (•) and their projections onto the f_0-K_P plane, the initial inverse surrogate (gray surface); the target vector of operating parameters (blue circle); (c) first iteration: the infill point rendered by r_I (gray circle) substitutes the poorest observable and the model is reset; (d) final iteration: the observables are focused in the proximity of the target operating parameters and r_I predicts the design close enough to the target. (Color figure online)

within the trust region of the perimeter $d^{(i)}$. The TR region size is determined according to the routine setup of the TR algorithm [36]. The merit function U_L (4) is of the same form as U, yet, it is evaluated with the use of the first-order Taylor model $G^{(i)}(x,f)$ of the circuit responses at the vector $x^{(i)}$: $G^{(i)}(x,f) = S_{kl}(x^{(i)},f) + \nabla S_{kl}(x^{(i)},f) \cdot (x - x^{(i)})$. The linear model is defined for the scattering parameter S_{kl}, with the gradients estimated through finite differentiation (FD). This adds n EM simulations for each iteration to the overall optimization cost. The local search terminates if $\|x^{(i+1)} - x^{(i)}\| < \varepsilon$ (convergence in argument), or the TR size has shrunk below ε, i.e., $\|d^{(i)}\| < \varepsilon$. In this work, we employ $\varepsilon = 10^{-3}$. For computational efficiency, when approaching convergence (i.e., for $\|x^{(i+1)} - x^{(i)}\| < 10\varepsilon$) the rank-one Broyden formula [36] takes the place of FD for gradient evaluation.

2.5 Algorithm Summary

This section summarizes the proposed global search procedure with feature-based inverse surrogates, the main stages of which include the global exploration stage described in Sect. 2.3, in which the starting point $x^{(0)}$ is found, and the local refinement procedure (Sect. 2.4) that yields the design x^* using the trust-region algorithm as a search engine. The control parameters of our technique include: the number N of the observables required for inverse surrogate construction and the threshold D_{accept}. Typically, N of the same order as parameter space dimensionality is a reasonable choice; yet, N should also account for the number of design objectives so as to adequately represent the curvature of the set comprising high-quality designs. The purpose of the second parameter D_{accept} is to ensure attainability of the assumed targets through a local search.

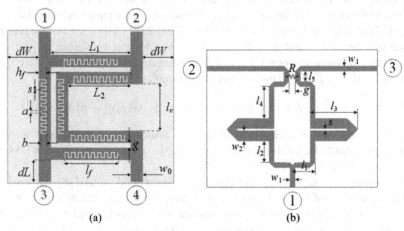

Fig. 3. Verification case strictures: (a) rat-race coupler with defected microstrip structure (RRC) [37], and (b) dual-band power divider (PD), with the lumped resistor marked as R [38].

If the operating parameters are circuit operating frequency (or frequencies), which is often the case, D_{accept} should be equal to around 50% of the target bandwidth. The parameters $N_{max.k}$, $k = 1, 2, 3$, (maximal number of EM analyses for initial sampling, global and local search phase, respectively) serve to define the computational budget. Their values should give enough room for the algorithm to converge with a sufficient accuracy, typically it suffices if $N_{max.1}$ and $N_{max.2}$ equal around $10N$, and $N_{max.3}$ is around five times higher.

3   Verification Experiments

This section reports the results of a numerical verification of the proposed global optimization procedure. The performance of our methodology is demonstrated using a miniaturized coupler and a dual-band power divider. The obtained results are benchmarked against: (i) local optimization starting from multiple random initial deigns (to justify the necessity for a global search), and (ii) particle swarm optimizer (commonly choice for performing global search). The control parameters of our framework are: $N = 10$, $D_{accept} = 0.2$, $N_{max.1} = N_{max.2} = 100$, $N_{max.3} = 500$, $\varepsilon = 10^{-3}$. For the benchmark PSO procedure we have: population size 10, $\chi = 0.73$, and $c_1 = c_2 = 2.05$. We also adopt standard setup [36] of the TR algorithm.

3.1 Verification Circuits

The verification structures utilized to validate the proposed approach are shown in Fig. 3. The first circuit is a miniaturized rat-race coupler (RRC), in which a defected meander spurline inside a folded transmission line is employed [37]. RRC is fabricated on a substrate of the height $h = 0.15$ mm. The vector of design variables is $x = [L_1 \, b_r \, g \, h_{fr} \, s \, l_{fr}]^T$ (dimensions in mm apart from unitless relative parameters indicated with the subscript r). We also have: $L_2 = L_1 - g - w_0$, $a = (l_f - 17s)/16$, $b = (h_f - s)b_r$, $l_f = L_2 \, l_{fr}$, $l_v = L_1 - 2g - 2w_0$, and $h_f = s + (w_0 - s)h_{fr}$; $dW = dL = 10$ mm. The width w_0 of input line is evaluated taking into account substrate permittivity ε_r to obtain 50 Ω input impedance. The lower and upper boundaries of the parameters are: $l = [20.0 \, 0.1 \, 1.0 \, 0.2 \, 0.2 \, 0.2]^T$ and $u = [15.0 \, 30.0 \, 50.0 \, 2.0 \, 2.0 \, 2.0]^T$. The design objectives are: (i) minimization of $|S_{11}|$ and $|S_{41}|$ characteristics (matching and isolation, respectively) at the target operating frequency f_0, and (ii) obtaining equal power split $d_S(x, f_0) = |S_{21}(x, f_0)| - |S_{31}(x, f_0)|$. The merit function is formulated as

$$U(x, F_t) = U(x, [f_0 \, K_P]^T) = \max\{|S_{11}(x, f_0)|, |S_{41}(x, f_0)|\} + \beta[d_S(x, f_0)]^2 \quad (5)$$

The second component of (5) is a penalty function ensuring equal power split ratio, and β denotes the penalty factor. Two design cases have been considered: (i) $f_0 = 1.5$ GHz, $\varepsilon_r = 2.5$, and (ii) (i) $f_0 = 1.2$ GHz, $\varepsilon_r = 4.4$.

The second structure is a dual-band equal-split power divider (PD) [38], implemented on AD250 substrate of $\varepsilon_r = 2.5$, and $h = 0.81$ mm. The circuit variables is is $x = [l_1 \, l_2 \, l_3 \, l_4 \, l_5 \, s \, w_2]^T$ (all in mm); fixed dimensions: $w_1 = 2.2$ mm and $g = 1$ mm. The lower and upper parameter boundaries are: $l = [10.0 \, 1.0 \, 10.0 \, 0.5 \, 1.0 \, 0.1 \, 1.5]^T$ and $u = [40.0 \, 20.0 \, 40.0 \, 15.0 \, 6.0 \, 1.5 \, 8.0]^T$, respectively. The goal is to minimize $|S_{11}|$, $|S_{22}|$, $|S_{33}|$, and $|S_{23}|$ (input matching, output matching, and isolation, respectively) at the operating frequencies $f_{0.1}$ and $f_{0.2}$. The objective function is formulated as

$$U(x, F_t) = U(x, [f_{0.1} \, f_{0.2}]^T) = \max\{|S_{11}(x, f_{0.1})|, |S_{22}(x, f_{0.1})|, |S_{33}(x, f_{0.1})|,$$
$$|S_{23}(x, f_{0.1})|, |S_{11}(x, f_{0.2})|, |S_{22}(x, f_{0.2})|, |S_{33}x, f_{0.2})|, |S_{23}(x, f_{0.2})|\} \quad (6)$$

Table 2. Optimization results for circuits of Fig. 3

Circuit	Target operating parameters	Optimization method	Proposed algorithm	PSO		TR-based local search
				50 iterations	100 iterations	
RRC	$f_0 =$ 1.5 GHz, $\varepsilon_r = 2.5$	Average objective function value [dB]	−18.6	−17.6	−19.2	−1.8

(continued)

Table 2. (*continued*)

Circuit	Target operating parameters	Optimization method	Proposed algorithm	PSO 50 iterations	PSO 100 iterations	TR-based local search
		Computational cost[a]	85.5	500	1,000	77.0
		Success rate[b]	10/10	10/10	10/10	5/10
	$f_0 =$ 1.2 GHz, $\varepsilon_r = 4.4$	Average objective function value [dB]	−21.5	−19.4	−22.5	7.6
		Computational cost[a]	90.2	500	1,000	83.8
		Success rate[b]	10/10	9/10	10/10	5/10
PD	$f_1 =$ 3.0 GHz, $f_2 =$ 4.8 GHz	Average objective function value [dB]	−33.9	−19.6	−18.8	−12.3
		Computational cost[a]	99.1	500	1,000	95.1
		Success rate[b]	10/10	8/10	9/10	2/10
	$f_1 =$ 2.0 GHz, $f_2 =$ 3.3 GHz	Average objective function value [dB]	−23.6	−18.8	−9.7	−20.6
		Computational cost[a]	99.2	500	1,000	93.8
		Success rate[b]	10/10	8/10	9/10	7/10

[a]The cost expressed in terms of the number of EM analyses of the circuit under design.
[b]Number of algorithm runs with operating parameters meeting the condition $D(F(x^*), f_t) \leq D_{accept}$.

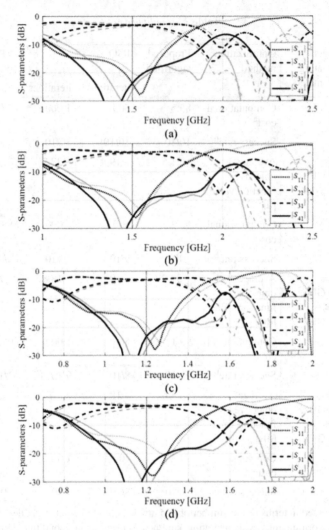

Fig. 4. Global search with inverse surrogates, S-parameters of RRC: (i) $f_0 = 1.5$ GHz, $\varepsilon_r = 2.5$ (a), (b) designs 1 and 2, respectively, (ii) $f_0 = 1.2$ GHz, $\varepsilon_r = 4.4$: (c), (d) designs 1 and 2, respectively. The initial design rendered within the global search stage is marked gray, the refined design is marked black. Vertical line marks the target operating frequency.

Here, the equal power split condition is enforced by the structure symmetry, thus, it does not need to be dealt with in the optimization process. The following design cases have been considered: (i) $f_{0.1} = 3.0$ GHz, $f_{0.2} = 4.8$ GHz, and (ii) $f_{0.1} = 2.0$ GHz, $f_{0.2} = 3.3$ GHz. The simulation models of both devices are realized in CST Microwave Studio, and evaluated using its transient solver.

3.2 Results

Table 2 provides the numerical results for both RRC and PD. The circuit responses at the designs optimized by the discussed global optimization framework are shown in Figs. 4 and 5. The main observations concerning the presented optimization strategy may be summarized as follows. Our approach exhibits a global search capability as high-grade designs have been yielded in all the ten runs of the algorithm. Moreover,

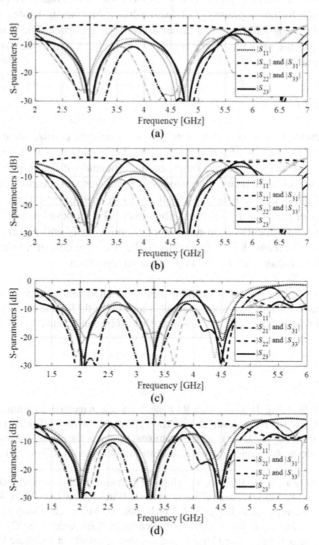

Fig. 5. Global search with inverse surrogates, S-parameters of PD: (i) $f_{0.1} = 3.0$ GHz, $f_{0.2} = 4.8$ GHz (a), (b) designs 1 and 2, respectively, (ii) $f_{0.1} = 2.0$ GHz, $f_{0.2} = 3.3$ GHz: (c), (d) designs 1 and 2, respectively. The initial design rendered within the global search stage is marked gray, the refined design is marked black. Vertical lines mark the target operating frequencies.

Figs. 4 and 5 show that the global search stage renders high-quality designs (with the operating frequency allocated close to the target). As for the benchmark TR algorithm from random initial designs, it fails in half of the cases. Although PSO optimizer performs much better, yet, its CPU cost is considerably higher that of the proposed procedure.

As far as the design quality is concerned, for our technique, it is significantly higher than for both benchmark procedures. Moreover, our approach is virtually immune to the initial design quality, which proves its robustness. The proposed methodology is only somewhat more expensive than gradient-based local search (of around ten percent). This efficiency is due to the employment of feature-based inverse surrogates, particularly, constructing the metamodel over low-dimensional objective space. The aforementioned advantages make the presented technique a low-cost alternative to mainstream global optimization methods.

4   Conclusions

The paper presented a low-cost procedure for global design optimization of microwave passives. The key concept of our framework is the employment of response feature inverse regression surrogates. The computational efficiency of the proposed technique capitalizes on weakly nonlinear relationship between the response features and geometry parameters of the system at hand, which allows for rendering the inverse model using only a handful of pre-selected random observables. The proposed procedure has been validated using a miniaturized rat-race coupler, and a dual-band power divider. The CPU cost of our framework is only slightly higher than the cost of a local search and significantly lower than that of a PSO optimization algorithm.

Acknowledgement. The authors would like to thank Dassault Systemes, France, for making CST Microwave Studio available. This work is partially supported by the Icelandic Centre for Research (RANNIS) Grant 206606 and by National Science Centre of Poland Grant 2020/37/B/ST7/01448.

References

1. Ma, P., Wei, B., Hong, J., Xu, Z., Cao, B., Jiang, L.: A design method of multimode multiband bandpass filters. IEEE Trans. Microwave Theory Technol. **66**(6), 2791–2799 (2018)
2. Yang, Q., Jiao, Y., Zhang, Z.: Compact multiband bandpass filter using low-pass filter combined with open stub-loaded shorted stub. IEEE Trans. Microwave Theory Technol. **66**(4), 1926–1938 (2018)
3. Hagag, M.F., Zhang, R., Peroulis, D.: High-performance tunable narrowband SIW cavity-based quadrature hybrid coupler. IEEE Microwave Wirel. Comp. Lett. **29**(1), 41–43 (2019)
4. Gómez-García, R., Rosario-De Jesus, J., Psychogiou, D.: Multi-band bandpass and bandstop RF filtering couplers with dynamically-controlled bands. IEEE Access **6**, 32321–32327 (2018)
5. Zhang, R., Peroulis, D.: Mixed lumped and distributed circuits in wideband bandpass filter application for spurious-response suppression. IEEE Microwave Wirel. Comp. Lett. **28**(11), 978–980 (2018)
6. Sheikhi, A., Alipour, A., Mir, A.: Design and fabrication of an ultra-wide stopband compact bandpass filter. IEEE Trans. Circuits Syst. II **67**(2), 265–269 (2020)

7. Firmansyah, T., Alaydrus, M., Wahyu, Y., Rahardjo, E.T., Wibisono, G.: A highly independent multiband bandpass filter using a multi-coupled line stub-SIR with folding structure. IEEE Access **8**, 83009–83026 (2020)

8. Chen, S., Guo, M., Xu, K., Zhao, P., Dong, L., Wang, G.: A frequency synthesizer based microwave permittivity sensor using CMRC structure. IEEE Access **6**, 8556–8563 (2018)

9. Chi, J.G., Kim, Y.J.: A compact wideband millimeter-wave quadrature hybrid coupler using artificial transmission lines on a glass substrate. IEEE Microwave Wirel. Comp. Lett. **30**(11), 1037–1040 (2020)

10. Koziel, S., Abdullah, M.: Machine-learning-powered EM-based framework for efficient and reliable design of low scattering metasurfaces. IEEE Trans. Microwave Theory Technol. **69**(4), 2028–2041 (2021)

11. Rayas-Sanchez, J.E., Koziel, S., Bandler, J.W.: Advanced RF and microwave design optimization: a journey and a vision of future trends. IEEE J. Microwaves **1**(1), 481–493 (2021)

12. Jin, H., Zhou, Y., Huang, Y.M., Ding, S., Wu, K.: Miniaturized broadband coupler made of slow-wave half-mode substrate integrated waveguide. IEEE Microwave Wirel. Comp. Lett. **27**(2), 132–134 (2017)

13. Li, X., Luk, K.M.: The grey wolf optimizer and its applications in electromagnetics. IEEE Trans. Ant. Prop. **68**(3), 2186–2197 (2020)

14. Luo, X., Yang, B., Qian, H.J.: Adaptive synthesis for resonator-coupled filters based on particle swarm optimization. IEEE Trans. Microwave Theory Technol. **67**(2), 712–725 (2019)

15. Majumder, A., Chatterjee, S., Chatterjee, S., Sinha Chaudhari, S., Poddar, D.R.: Optimization of small-signal model of GaN HEMT by using evolutionary algorithms. IEEE Microwave Wirel. Comp. Lett. **27**(4), 362–364 (2017)

16. Ding, D., Zhang, Q., Xia, J., Zhou, A., Yang, L.: Wiggly parallel-coupled line design by using multiobjective evolutionay algorithm. IEEE Microwave Wirel. Comp. Lett. **28**(8), 648–650 (2018)

17. Zhu, D.Z., Werner, P.L., Werner, D.H.: Design and optimization of 3-D frequency-selective surfaces based on a multiobjective lazy ant colony optimization algorithm. IEEE Trans. Ant. Propag. **65**(12), 7137–7149 (2017)

18. Greda, L.A., Winterstein, A., Lemes, D.L., Heckler, M.V.T.: Beamsteering and beamshaping using a linear antenna array based on particle swarm optimization. IEEE Access **7**, 141562–141573 (2019)

19. Cui, C., Jiao, Y., Zhang, L.: Synthesis of some low sidelobe linear arrays using hybrid differential evolution algorithm integrated with convex programming. IEEE Ant. Wirel. Propag. Lett. **16**, 2444–2448 (2017)

20. Baumgartner, P., et al.: Multi-objective optimization of Yagi-Uda antenna applying enhanced firefly algorithm with adaptive cost function. IEEE Trans. Magn. **54**(3), 8000504 (2018)

21. Yang, S.H., Kiang, J.F.: Optimization of sparse linear arrays using harmony search algorithms. IEEE Trans. Ant. Prop. **63**(11), 4732–4738 (2015)

22. Zhang, Z., Cheng, Q.S., Chen, H., Jiang, F.: An efficient hybrid sampling method for neural network-based microwave component modeling and optimization. IEEE Microwave Wirel. Comp. Lett. **30**(7), 625–662 (2020)

23. Van Nechel, E., Ferranti, F., Rolain, Y., Lataire., J.: Model-driven design of microwave filters based on scalable circuit models. IEEE Trans. Microwave Theory Technol. **66**(10), 4390–4396 (2018)

24. Li, Y., Xiao, S., Rotaru, M., Sykulski, J.K.: A dual kriging approach with improved points selection algorithm for memory efficient surrogate optimization in electromagnetics. IEEE Trans. Magn. **52**(3), 1–4, Article No. 7000504 (2016)

25. Jacobs, J.P.: Characterization by Gaussian processes of finite substrate size effects on gain patterns of microstrip antennas. IET Microwaves Ant. Prop. **10**(11), 1189–1195 (2016)

26. Ogut, M., Bosch-Lluis, X., Reising, S.C.: A deep learning approach for microwave and millimeter-wave radiometer calibration. IEEE Trans. Geosci. Remote Sens. **57**(8), 5344–5355 (2019)
27. Lim, D.K., Yi, K.P., Jung, S.Y., Jung, H.K., Ro, J.S.: Optimal design of an interior permanent magnet synchronous motor by using a new surrogate-assisted multi-objective optimization. IEEE Trans. Magn. **51**(11), 1–4, Article No. 8207504 (2015)
28. Taran, N., Ionel, D.M., Dorrell, D.G.: Two-level surrogate-assisted differential evolution multi-objective optimization of electric machines using 3-D FEA. IEEE Trans. Magn. **54**(11), 1–5, Article No. 8107605 (2018)
29. Koziel, S., Pietrenko-Dabrowska, A.: Performance-Driven Surrogate Modeling of High-Frequency Structures. Springer, New York (2020). https://doi.org/10.1007/978-3-030-389 26-0
30. Koziel, S.: Low-cost data-driven surrogate modeling of antenna structures by constrained sampling. IEEE Antennas Wirel. Prop. Lett. **16**, 461–464 (2017)
31. Pietrenko-Dabrowska, A., Koziel, S.: Antenna modeling using variable-fidelity EM simulations and constrained co-kriging. IEEE Access **8**(1), 91048–91056 (2020)
32. Koziel, S.: Fast simulation-driven antenna design using response-feature surrogates. Int. J. RF Microwave CAE **25**(5), 394–402 (2015)
33. Koziel, S., Pietrenko-Dabrowska, A.: Expedited feature-based quasi-global optimization of multi-band antennas with Jacobian variability tracking. IEEE Access **8**, 83907–83915 (2020)
34. Koziel, S., Bandler, J.W.: Reliable microwave modeling by means of variable-fidelity response features. IEEE Trans. Microwave Theory Technol. **63**(12), 4247–4254 (2015)
35. Pietrenko-Dabrowska, A., Koziel, S.: Fast design closure of compact microwave components by means of feature-based metamodels. Electronics **10**, 10 (2021)
36. Conn, A.R., Gould, N.I.M., Toint, P.L.: Trust Region Methods. MPS-SIAM Series on Optimization, SIAM, Philadelphia (2000)
37. Phani Kumar, K.V., Karthikeyan, S.S.: A novel design of ratrace coupler using defected microstrip structure and folding technique. In: IEEE Applied Electromagnetics Conf. (AEMC), Bhubaneswar, India, pp. 1–2 (2013)
38. Lin, Z., Chu, Q.-X.: A novel approach to the design of dual-band power divider with variable power dividing ratio based on coupled-lines. Prog. Electromagn. Res. **103**, 271–284 (2010)
39. Kennedy, J., Eberhart, R.C.: Swarm Intelligence. Morgan Kaufmann, San Francisco (2001)

Classification of Soil Bacteria Based on Machine Learning and Image Processing

Aleksandra Konopka[1], Karol Struniawski[1]([✉]), Ryszard Kozera[1,2],
Paweł Trzciński[3], Lidia Sas-Paszt[3], Anna Lisek[3], Krzysztof Górnik[3],
Edyta Derkowska[3], Sławomir Głuszek[3], Beata Sumorok[3],
and Magdalena Frąc[4]

[1] Institute of Information Technology, Warsaw University of Life Sciences - SGGW,
ul. Nowoursynowska 159, 02-776 Warsaw, Poland
{aleksandra_konopka,karol_struniawski,ryszard_kozera}@sggw.edu.pl
[2] School of Physics, Mathematics and Computing, The University of Western
Australia, 35 Stirling Highway, Perth, Crawley, WA 6009, Australia
ryszard.kozera@uwa.edu.au
[3] Department of Microbiology and Rhizosphere, The National Institute
of Horticultural Research, ul. Pomologiczna 18, 96-100 Skierniewice, Poland
{pawel.trzcinski,lidia.sas,anna.lisek,krzysztof.gornik,edyta.derkowska,
slawomir.gluszek,beata.sumorok}@inhort.pl
[4] Institute of Agrophysics, Polish Academy of Sciences, ul. Doświadczalna 4,
20-290 Lublin, Poland
m.frac@ipan.lublin.pl

Abstract. Soil bacteria play a fundamental role in plant growth. This paper focuses on developing and testing some techniques designed to identify automatically such microorganisms. More specifically, the recognition performed here deals with the specific five genera of soil bacteria. Their microscopic images are classified with machine learning methods using shape and image texture descriptors. Feature determination based on shape relies on interpolation and curvature estimation whereas feature recognition based on image texture resorts to the spatial relationships between chrominance and luminance of pixels using co-occurrence matrices. From the variety of modelling methods applied here the best reported result amounts to 97% of accuracy. This outcome is obtained upon incorporating the set of features from both groups and subsequently merging classification and feature selection methods: Extreme Learning Machine - Radial Basis Function with Sparse Multinomial Logistic Regression with Bayesian Regularization and also k-Nearest Neighbors classifier with Fast Correlation Based Filter. The optimal parameters involved in merged classifiers are obtained upon computational testing and simulation.

Keywords: Soil bacteria · Machine learning · Image analysis · Shape and image texture extraction · Spline interpolation · Modelling and simulation · Computational optimization

D. Groen et al. (Eds.): ICCS 2022, LNCS 13352, pp. 263–277, 2022.
https://doi.org/10.1007/978-3-031-08757-8_23

1   Introduction

Soil bacteria despite of their small size may have a large impact on plant growth. Some of them are beneficial to agricultural sector, while the others are either harmless or pathogenic causing a vast diversity of plant diseases. Consequently, bacteria recognition becomes an important task for scientists equally as a research and agricultural problem. Bacteria identification is usually carried out using specific markers changing their color as a reaction to specific chemical compounds. The morphology of the bacteria colony is also usually analyzed by examining its shape, edges, color, colony distribution, consistency and surface structure [7]. This approach is usually laborious and depends on the subjective perceptiveness of the scientist. A natural step accelerating and facilitating the latter is to automate the process of microscopic image analysis. This paper[1,2] resorts to machine learning and image processing methods applied to soil bacteria recognition. In general, comparing images of bacteria belonging to certain species is difficult since they adopt similar morphologies [22]. Due to this reason, it is decided to distinguish here the input bacteria on the genera level. The microscopic images of soil bacteria examined in this paper (which are part of our data-set available in full resolution under the URL link: https://bit.ly/3qdDuHo) include pictures of *Enterobacter*, *Rhizobium*, *Pantoea*, *Bradyrhizobium* and *Pseudomonas* (see Fig. 1). The pictures of investigated bacteria are obtained from Symbio-Bank - the collection of microorganisms of The National Institute of Horticultural Research in Skierniewice. Some of *Enterobacter* are considered plant pathogens, whereas the others are conducive for plant growth [13]. The bacteria of the genus *Rhizobium* have a positive effect on increasing the yield of grains and the protein content in pea grains [23]. *Rhizobium* and *Bradyrhizobium* are nitrogen-fixing soil bacteria that live in symbiosis with legumes [3]. On the other hand, *Pantoea* causing plant infections [24] is also used in the production of antibiotics [2]. Some *Pseudomonas* are plant pathogens, while the others are used to stimulate plant growth and to remediate contaminated soil [30]. This paper discusses the identification of bacteria genera based on their morphological features. The calculated traits refer to bacteria shape and image texture. In order to automatize the entire recognition process a variety of feature selection and class recognition methods adapting the concept of machine learning are applied. On the basis of the supplied training data-set a classification model is built permitting to automatically categorize soil microorganisms.

2   Work-flow Scheme

The work-flow scheme adopted in this work consists of the following four consecutive steps: *segmentation of the Region of Interest, feature generation, feature selection* and *class recognition*.

---
[1] This research is financed by The National Centre for Research and Development of the BIOSTRATEG Project (Eco-Fruits) BIOSTRATEG3/344433/16/NCBR/2018.
[2] This work is a part of Polish National Centre of Research and Development research project POIR.01.02.00-00-0160/20.

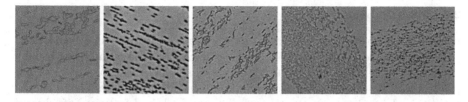

Fig. 1. Microscopic images of (from left): *Enterobacter, Rhizobium, Pantoea, Bradyrhizobium* and *Pseudomonas*. For more pictures see URL link: https://bit.ly/3qdDuHo.

2.1 Segmentation of the Region of Interest

The aim of this step is to extract bacteria and background image regions. Binary mask filter is applied yielding white pixels representing bacteria zones and black pixels corresponding to the background. To achieve the latter the image is first converted into gray-scale and then Otsu automatic image thresholding [18] with open and close morphological operations [28] is applied (see Fig. 2).

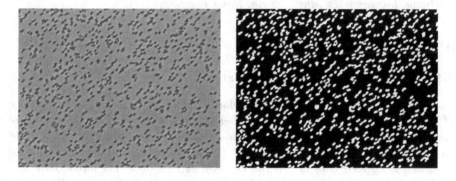

Fig. 2. Microscopic image of *Rhizobium* and its binary mask.

2.2 Feature Generation

Features that are considered in this paper refer to bacteria shape and texture of the input image. The determination of bacteria shape relies on estimating its boundary with the aid of cubic spline interpolation [5]. The latter permits to estimate the curvature of bacteria's boundary and to extract some correlation between selected distances and angles concerning the shape of bacteria in question. On the other hand, image texture features contain information about spatial relations between chrominance and luminance of the image pixels. To exploit such information, a statistical approach based on computation of the co-occurrence matrices is used [27]. The latter permits to estimate an image texture as a quantitative measure of luminance over the entire input image.

2.3 Feature Selection

It is common that processing large set of generated features may yield some of them highly correlated with one another. Such potential redundancy usually impacts on the classification accuracy. In contrast, the other group of extracted features can be poorly correlated to the dependent variable affiliated to the respective class. Consequently, the reduced set of selected features (sifted from the full set of initially determined features) should consist of those which are strongly correlated to the image class and weakly associated with the remaining features. In order to accomplish the latter the following methods for improving feature selection are used here: FCBF (*Fast Correlation Based Filter*), SBMLR (*Sparse Multinomial Logistic Regression with Bayesian Regularization*) and CFS (*Correlation-based Feature Selection*) - see e.g. [21].

2.4 Class Recognition

This paper resorts to the machine learning classifiers such as RF (*Random Forest*), SVM (*Support Vector Machine*), kNN (*k-Nearest Neighbors*), MLP (*Multilayer Perceptron*) [17], ELM (*Extreme Learning Machine*) [26] and ELM-RBF (*Extreme Learning Machine - Radial Basis Function*) [10].

3 Features Based on Shape

3.1 Planar Cubic Spline Interpolation

Consider now the ordered set of $m+1$ planar points $\mathcal{Q}_m = \{q_k\}_{k=0}^m$ i.e. sequence of points $q_k = (x_k, y_k)$ contained in 2D-Euclidean space $\mathbb{E}^2$. In the context of this work $\mathcal{Q}_m$ represents selected points of bacteria's boundary $\partial\Gamma$. In a quest to extract some shape information of $\partial\Gamma$ (or its estimate) an interpolation based approach is applied here [5]. In the classical setting of fitting input data, $\mathcal{Q}_m$ is also supplemented with the associated parameters, called interpolation knots $\mathcal{T}_m = \{t_k\}_{k=0}^m$ subject to $t_k < t_{k+1}$, $t_0 = 0$, $t_m = T$ and $t_k \in [0, T]$. Here the unknown function $\gamma : [0, T] \rightarrow \mathbb{E}^2$ meeting the constraints $\gamma(t_k) = q_k$ is assumed to satisfy $graph(\gamma) = \partial\Gamma$. For a given pair $(\mathcal{Q}_m, \mathcal{T}_m)$ there is a variety of interpolation schemes $\gamma_I : [0, T] \rightarrow \mathbb{E}^2$ fulfilling $\gamma_I(t_k) = q_k$ - see e.g. [5] or [9]. Since the selected interpolant γ_I to fit $(\mathcal{Q}_m, \mathcal{T}_m)$ should be both twice-differentiable (for curvature calculation) and should not render too excessive variations of $graph(\gamma_I)$ (for arbitrary m) a cubic spline $\gamma_I = \gamma^{cs}$ is a natural choice [5]. The interpolant γ^{cs} is defined as a track-sum of cubics $\{\gamma_k^{cs}\}_{k=0}^{m-1}$ with each cubic $\gamma_k^{cs} : [t_k, t_{k+1}] \rightarrow \mathbb{E}^2$ depending on four 2D-parameters (as $a_k, b_k, c_k, d_k \in \mathbb{R}^2$)

$$\gamma_k^{cs}(t) = a_k + b_k(t - t_k) + c_k(t - t_k)^2 + d_k(t - t_k)^3. \tag{1}$$

Here $4 \times m$ coefficients $\{(a_k, b_k, c_k, d_k)\}_{k=0}^{m-1}$ are calculable from $4 \times m$ constraints:

1. $2 \times m$ interpolation conditions for $k \in \{0, 1, \ldots, m-1\}$:

$$\gamma_k^{cs}(t_k) = q_k \quad \text{and} \quad \gamma_k^{cs}(t_{k+1}) = q_{k+1}. \tag{2}$$

2. $m - 1$ internal points' first-order smoothness for $k \in \{0, 1, \ldots, m-2\}$:

$$\dot{\gamma}_k^{cs}(t_{k+1}) = \dot{\gamma}_{k+1}^{cs}(t_{k+1}). \tag{3}$$

3. $m - 1$ internal points' second-order smoothness for $k \in \{0, 1, \ldots, m-2\}$:

$$\ddot{\gamma}_k^{cs}(t_{k+1}) = \ddot{\gamma}_{k+1}^{cs}(t_{k+1}). \tag{4}$$

4. 2 boundary conditions complementing (2), (3), (4) to yield $4 \times m$ equations.

Usually, the last two equations are obtainable from extra conditions such as e.g. $\dot{\gamma}(0) = v_0$ and $\dot{\gamma}(T) = v_m$. Indeed, the latter yields two missing equations:

$$\dot{\gamma}_0^{cs}(t_0 = 0) = v_0 \quad \text{and} \quad \dot{\gamma}_{m-1}^{cs}(t_m = T) = v_m. \tag{5}$$

Although, in our setting both velocities v_0 and v_m are not *a priori* given, they can be still estimated from $(\mathcal{Q}_m, \mathcal{T}_m)$ following the concept of modified Hermite interpolation [14]. Indeed, a unique Lagrange cubic $\gamma_0^{L(3)} : [0, t_3] \to \mathbb{E}^2$ interpolating the first four points $\{q_k\}_{k=0}^3$ at $\{t_k\}_{k=0}^3$ (see [5]) yields some estimate of $v_0 \approx \hat{v}_0 = \dot{\gamma}_3^{L(3)}(0)$. Similarly, a Lagrange cubic $\gamma_{m-3}^{L(3)} : [t_{m-3}, t_m] \to \mathbb{E}^2$ interpolating the last four points $\{q_k\}_{k=m-3}^m$ at $\{t_k\}_{k=m-3}^m$ renders some approximation of terminal velocity $v_m \approx \hat{v}_m = \dot{\gamma}_{m-3}^{L(3)}(t_m)$. Consequently, taking into account the latter, condition (5) modifies into:

$$\dot{\gamma}_0^{cs}(t_0 = 0) = \hat{v}_0 \quad \text{and} \quad \dot{\gamma}_{m-1}^{cs}(t_m = T) = \hat{v}_m. \tag{6}$$

The scheme for selection $\mathcal{Q}_m$ from $\partial\Gamma$ is described in Subsect. 3.3. Note that in our setting to approximate $\partial\Gamma$ with the closed curve as $q_0 \neq q_m$ we extend $\mathcal{Q}_m$ to $\hat{\mathcal{Q}}_{m+1} = \{\hat{q}_k\}_{k=0}^{m+1}$ so that $\hat{q}_k = q_k$ (for $k = 0, 1, \ldots, m$) and $\hat{q}_{m+1} = q_0$.

Upon selecting the interpolation points $\mathcal{Q}_m$ (and thus $\hat{\mathcal{Q}}_{m+1}$) from the bacteria's boundary $\partial\Gamma$ the next step is to estimate the accompanying knots $\mathcal{T}_{m+1} \approx \hat{\mathcal{T}}_{m+1} = \{\hat{t}_k\}_{k=0}^{m+1}$ (as $\mathcal{T}_{m+1}$ is not available out of input images) from the distribution of $\hat{\mathcal{Q}}_{m+1}$. This permits to construct the interpolant $\hat{\gamma}^{cs} : [0, \hat{T}] \to \mathbb{E}^2$ as a track-sum $\hat{\gamma}^{cs} = \{\hat{\gamma}_k^{cs}\}_{k=0}^{m+1}$, with $\hat{\gamma}_k^{cs} : [\hat{t}_k, \hat{t}_{k+1}] \to \mathbb{E}^2$ satisfying (1), (2), (3), (4) and (6) along $\hat{\mathcal{Q}}_{m+1}$ with somehow estimated knots $\hat{\mathcal{T}}_{m+1}$. Addressing the latter, we resort here to the so-called *exponential parameterization* commonly used in computer graphics [19] and defined in accordance with:

$$\hat{t}_0 = 0, \qquad \hat{t}_{k+1} = \hat{t}_k + \|q_{k+1} - q_k\|^\lambda, \quad k = 0, 1, \ldots, m, \tag{7}$$

for some parameter $\lambda \in [0, 1]$, where $\|\cdot\|$ is a standard Euclidean norm. This paper selects $\lambda = 0.5$ in (7) yielding the so-called *centripetal parameterization* with $\hat{T} = \sum_{k=0}^m \|q_{k+1} - q_k\|^{1/2}$ (see [19]). Note that in order to preserve $t_k < t_{k+1}$ it is also assumed that $q_k \neq q_{k+1}$. More information on exponential parameterization (7) and other knots selection schemes can be found e.g. in [14–16,19].

3.2   Curvature Calculation

Having found a cubic spline $\hat{\gamma}^{cs}$ approximating bacteria's boundary $\partial\Gamma$ one may extract some shape information of $\partial\Gamma$ by analyzing the geometry of $graph(\hat{\gamma}^{cs})$ forming the planar curve assumed also to estimate $\partial\Gamma$ (for m sufficiently big). In this work a curvature of $\hat{\gamma}^{cs}$ is computed to form a geometrical marker of $\partial\Gamma$ used later as one of the differentiating ingredients in classification process. Recall, that the curvature $\kappa(t)$ of a planar curve $\gamma : [a, b] \rightarrow \mathbb{E}^2$ at a given point $t \in [a, b]$ measures the amount by which such curve deviates from a tangent line at point $\gamma(t)$ - see [29]. The respective formula for the curvature $\kappa(t)$ of regular curve γ (i.e. for γ for which $\dot{\gamma}(t) \neq \mathbf{0}$ over $t \in [a, b]$) reads as:

$$\kappa(t) = \frac{\|\boldsymbol{T}'(t)\|}{\|\boldsymbol{r}'(t)\|}, \tag{8}$$

where $\boldsymbol{r}(t) = \dot{\gamma}(t)$ is a tangent vector to γ at t with its normalized vector $\boldsymbol{T}(t) = \boldsymbol{r}(t)/\|\boldsymbol{r}(t)\|$. In particular, for arc-length parameterization expressed as $s = \phi(t) = \int_a^t \|\boldsymbol{r}'(u)\| \, du$, for which reparameterized curve $\bar{\gamma}(s) = (\gamma \circ \phi^{-1})(s)$ satisfies $\|\dot{\bar{\gamma}}(s)\| = 1$ (yielding $\|\boldsymbol{T}(s)\| = 1$ - see [8]), the equation (8) reformulates into (with the respective derivative calculated for s-variable) $\kappa(s) = \|\boldsymbol{T}'(s)\|$.

3.3   Features Calculation

To estimate bacteria's shape (assumed here to be "more or less" convex), Region of Interest (ROI) mask is applied. In doing so, the following Matlab functions are exploited: *rgb2gray, imbinarize, imfill, bwareaopen* and *multithresh*. Upon localizing a single bacteria with ROI mask, the Laplacian filter is used to extract $\partial\Gamma$ of the analyzed object [4]. Next all computed boundary points $\mathcal{Q}_{\hat{m}} = \{q_j\}_{j=0}^{\hat{m}}$ are sorted out clock-wisely. To achieve the latter, we calculate and compare the angle between a given point $q_j \in \mathcal{Q}_{\hat{m}}$ and mean location of $\mathcal{Q}_{\hat{m}}$ i.e. the point $(\bar{x} = (1/(\hat{m}+1)) \sum_{k=0}^{\hat{m}} x_k, \bar{y} = (1/(\hat{m}+1)) \sum_{k=0}^{\hat{m}} y_k)$. It is assumed here that (x_k, y_k) represent Cartesian coordinates of the centers of bacterial boundary pixels (the center of the coordinate system is set in the upper left corner of an image). As a result the boundary of each single bacteria $\partial\Gamma$ is represented by a large set of points $\mathcal{Q}_{\hat{m}}$ which in turn is reduced to terser set $\mathcal{Q}_m$ with $m + 1 = 10$ or $m + 1 = 20$ points (and thus to $\hat{\mathcal{Q}}_{m+1}$ - see Subsect. 3.1) to be fitted with $\hat{\gamma}^{cs}$ and $\hat{T}$ governed by (7). Such reduction is carried out upon selecting from $\mathcal{Q}_{\hat{m}}$ "more or less" equally spaced points with respect to their index distribution taken in clockwise order (e.g. for $\hat{m} = 54$ a possible reduction leads to $\{q_0, q_8, q_{17}, q_{26}, q_{35}, q_{44}, q_{54}\}$).

The feature extraction process aimed to determine some bacteria's shape information relies on curvature calculation from the estimated bacteria's boundary $\partial\Gamma$ - see [1]. In doing so, formula (8) is applied to cubic spline $\hat{\gamma}_k^{cs}$ (see Subsect. 3.1). More precisely, with the aid of (8) for each $\hat{\gamma}_k^{cs}$ we compute over $[\hat{t}_k, \hat{t}_{k+1}]$ the maximal and minimal values of the curvature function $\kappa(\hat{t})$ (i.e. κ^{max} and κ^{min}) yielding the corresponding knots $\hat{t}_k^{max}, \hat{t}_k^{min} \in [\hat{t}_k, \hat{t}_{k+1}]$ obtained from $\kappa^{max} = \kappa(\hat{t}_k^{max})$ and $\kappa^{min} = \kappa(\hat{t}_k^{min})$, respectively. Note that if the pair of knots $(\hat{t}_k^{max}, \hat{t}_k^{min})$ is not uniquely determined one can choose e.g. the smallest two knots $t_k^{max,min} \in [\hat{t}_k, \hat{t}_{k+1}]$, respectively. This in turn, permits to determine two points $q_k^{max} = \hat{\gamma}_k^{cs}(\hat{t}_k^{max})$ and $q_k^{min} = \hat{\gamma}_k^{cs}(\hat{t}_k^{min})$ having maximal and minimal curvature $\hat{\gamma}_k^{cs}$ over the segment $[\hat{t}_k, \hat{t}_{k+1}]$. According to the order of all knots $\hat{t}_k^{max}$, $\hat{t}_k^{min}$ we re-index points q_k^{min} and q_k^{max} placing them into one sequence formula $\{q_k^{ext}\}_{k=0}^{2m+1}$ (where either $ext = max$ or $ext = min$). In the next step we determine the center of mass $q_c = (1/2(m+1))\sum_{k=0}^{2m+1} q_k^{ext}$ needed as a reference point to compute the $2(m+1)$ distance values $a_k = \|q_k^{ext} - q_c\|$. In sequel, a family of triangles $\Delta_k(q_k^{ext}, q_c, q_{k+1}^{ext})$ with common apex at q_c (with the respective lengths of Δ_k sides: a_k, $b_k = \|q_{k+1}^{ext} - q_k^{ext}\|$ and a_{k+1}) form a polygonal approximation of $\partial\Gamma$. Given the lengths of all sides of triangle Δ_k, its respective angles $\alpha_k, \beta_k, \gamma_k = \pi - (\alpha_k + \beta_k)$ are easily computable from the cosine theorem - here $\alpha_k = \angle(q_k^{ext}, q_c, q_{k+1}^{ext})$, $\beta_k = \angle(q_k^{ext}, q_{k+1}^{ext}, q_c)$ and $\gamma_k = \angle(q_{k+1}^{ext}, q_k^{ext}, q_c)$. Having determined the above distances and angles a given bacteria can be represented by the following four vectors (forming de facto its *polygonal shape descriptors*): $\boldsymbol{a} = (a_0, \ldots, a_{2m+1})$, $\boldsymbol{b} = (b_0, \ldots, b_{2m+1})$, $\boldsymbol{\alpha} = (\alpha_0, \ldots, \alpha_{2m+1})$ and $\boldsymbol{\beta} = (\beta_0, \ldots, \beta_{2m+1})$. At this point, we assume that we are given a reference bacteria (a kind of "geodetic benchmark" not necessarily belonging to any investigated herein soil microorganisms' classes) to which different five classes of examined bacteria are compared accordingly. Experiments carried out so far based on 5 generic representatives - one for each bacteria class - did not improve the results over selecting one reference bacteria. To juxtapose vectors representing an examined bacteria with the reference bacteria vectors $\boldsymbol{a}^{ref}$, $\boldsymbol{b}^{ref}$, $\boldsymbol{\alpha}^{ref}$, $\boldsymbol{\beta}^{ref}$ we calculate the cross correlation coefficient [6] between the respective pairs (i.e. $xcorr(\boldsymbol{a}, \boldsymbol{a}^{ref})$). For four cross correlation vectors one chooses their respective greatest values $a^{max}, b^{max}, \alpha^{max}$ and β^{max}. In each picture, we select from 5 to 50 bacteria whose surface area is the closest to the median surface area of all the bacteria in the input picture. The latter permits to select bacteria characterized by the average size and stage of growth. The less bacteria we select the less likely we qualify a group of overlapping bacteria as a single object. We considered 6 features (listed below) based on shape calculated for a fixed amount of points on one bacteria and the number of bacteria analyzed in a single image. We estimated the edge of bacteria using $m+1 = 10$ or $m+1 = 20$ points on one bacteria and compared $l = 5, 10, 20, 25, 30, 40$ and 50 bacteria on one image. The following 6 features based on shape information are considered:

1. *Mean bacteria arc-length* - which is a sum of all arc-lengths representing the perimeters of all selected bacteria divided by l.
2. *Mean curvature of l bacteria* - is a sum $(1/l)\sum_{k=1}^{l}\sum_{j=0}^{m}\int_{\hat{t}_j}^{\hat{t}_{j+1}} \kappa_j^k(\hat{t})d\hat{t}$, where κ_j^k represents the curvature of k-th bacteria along j-th segment (see (8)).
3. *Mean maximal first distance correlation* - $(1/l)\sum_{k=1}^{l} a_k^{max}$.
4. *Mean maximal second distance correlation* - $(1/l)\sum_{k=1}^{l} b_k^{max}$.
5. *Mean maximal first angle correlation* - $(1/l)\sum_{k=1}^{l} \alpha_k^{max}$.
6. *Mean maximal second angle correlation* - $(1/l)\sum_{k=1}^{l} \beta_k^{max}$.

4   Features Based on Texture

The second group of examined features relies on image texture analysis. In [12] Haralick introduced statistical measures resorting to the second order image histogram called GLCM (*Grey-Level Co-Occurrence Matrix*). In this paper we also used GLRLM (*Gray-Level Run-Length Matrix*) measures - see [11].

4.1   GLCM Features

GLCM is calculated for the following directional angles α: $0°$, $45°$, $90°$, $135°$ and distance d on quantized image Ω to n levels that are represented in gray-scale. The co-occurrence matrix M of size $n \times n$ is initialized with all its coefficients set to zero. Assume the image Ω is represented by the pixel table $\bar{M}$ having m_1 rows and n_1 columns. Note that here pixel $(1,1)$ represents the top-left pixel in Ω, whereas pixel (m_1, n_1) corresponds to the bottom-right image pixel. In addition, let matrix $W^{k,l}$ have m_w rows and n_w columns. $W^{k,l}$ is used iteratively to extract the following pixels of $\bar{M}[i,j]$: $m_w(k-1) < i \leq km_w$ and $n_w(l-1) < j \leq ln_w$. The latter can be geometrically viewed as positioning top-left corner of $W^{k,l}$ at (k,l) pixel of Ω. The coverage of Ω with $W^{k,l}$ abides the following pattern. First Ω is horizontally covered by windows $W^{1,1}, W^{1,n_w+1}, W^{1,2n_w+1}, \ldots, W^{1,n_1-n_w+1}$, respectively. Next after vertical shift to $W^{m_w+1,1}$ we move horizontally up to W^{m_w+1,n_1-n_w+1}. This procedure of disjoint coverage of Ω is continued up until reaching $W^{m_1-m_w+1,n_1-n_w+1}$ window. Note that if either $m_1(mod\ m_w) \neq 0$ or $n_1(mod\ n_w) \neq 0$ (since in practice $n_w \ll n_1$ and $m_w \ll m_1$) one can supply extra missing pixels for the most right or bottom part of Ω by extrapolation techniques. Additionally for each of $W^{k,l}$ we iterate over the pixels in that window incrementing values in M based on the correlations between pixels for direction α and distance d that is explained below (for more details see [12]).

Assume we use a window of size $m_w = 5$ and $n_w = 5$. For every $W^{k,l}$ we go through each pixel in that window. Let $n = 5$, $d = 2$, $\alpha = 45°$ and $w_{11}^{k,l}, \ldots, w_{55}^{k,l}$ be certain pixel values in $W^{k,l}$ (see Fig. 3). Here we are in the iterative step that analyzes pixel $w_{42}^{k,l}$ (that is marked as dark gray), its value is equal to 0. Then we know that we increment co-occurrences in GLCM matrix in first row that is responsible for relationships between values of pixels 0 and $n_i = 0, \ldots, 4$. Next step is to check values of pixels located in direction $\alpha = 45°$ and maximum distance of $d = 2$. There are two pixels meeting these requirements: $w_{33}^{k,l}$ and $w_{24}^{k,l}$. Since $w_{33}^{k,l} = 0$ we have to increment value in first row and in first column, and since $w_{24}^{k,l} = 3$ we need to increment value in first row and in fourth column in GLCM. Finally after we moved our window through the entire image registering co-occurrences of the pixels we divide each of the values in $W^{k,l}$ by n^2 that gives us probabilities of co-occurrences between gray levels of pixels for direction α and maximum distance d. Based on GLCM computation, the following 8 statistical measures are calculated [32]: *Contrast, Correlation, Energy, Homogeneity, Autocorrelation, Cluster Prominence, Inverse Difference and Dissimilarity*. Note that as GLCM is calculated for 4 different directions, we obtain measures such as Contrast $0°$, Contrast $45°$, Contrast $90°$ and Contrast $135°$. The final value of Contrast is taken as mean of these four values. The 8 statistical measures from above determine 8 texture features based on GLCM.

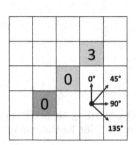

Fig. 3. Increment of the values in GLCM (right) based on the window $W^{k,l}$ (left) of size 5×5, pixel $w_{42}^{k,l}$ for $n = 5$, $d = 2$, $\alpha = 45°$ and a presentation of all possible directions α from the pixel $w_{44}^{k,l}$ marked with a red dot. (Color figure online)

4.2   GLRLM Features

The calculations of GLRLM based features are very similar to determining GLCM. The computations involved are also carried out for same directional angles α: $0°$, $45°$, $90°$, $135°$ and for the maximum distance d on quantized image to n levels that is represented in gray-scale. We initialize GLRLM of size $n \times d$ with zeros. Using window $m_w \times n_w$ we move through the image in the same manner as in GLCM and increment values in matrix according to the run-length of co-occurrence pixels for direction α and length that is equal to $d_i = 1, \ldots, d$. In the next step each of the values in GLRLM is divided by dn. The latter introduces matrix with probabilities of the respective co-occurrences of the gray level $n_i = 0, \ldots, n-1$ and run-length d_i.

As an example assume an input image is quantized to 5 gray levels, maximum run-length $d = 3$, window's size is 5×5, $\alpha = 90°$ and that we analyze pixels with values equal to zero. Let $W^{k,l}$ have sample entries as shown in left window of Fig. 4. The algorithm represents searching sequences of length d_i of pixels that have values equal to zero in $W^{k,l}$ e.g. two sequences of length 3 (marked as green - see Fig. 4) are found filling GLRLM for $n = 0$ with value $d_3 = 2$. For more details on GLRLM see also [11].

Again based on probabilities stored in GLRLM the following statistical measures are calculated [11]: *Short Run Emphasis, Long Run Emphasis, Grey Level Non-uniformity, Run Length Non-uniformity, Run Percentage, Low Grey Level Run Emphasis, High Grey Level Run Emphasis Short Run Low Gray Level Emphasis, Short Run High Gray Level Emphasis, Long Run Low Gray Level Emphasis* and *Long Run High Gray Level Emphasis*. As previously, these 11 measures are computed along 4 different directions and analogously to the case of GLCM the respective mean values of each measures are determined. Ultimately,

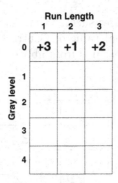

Fig. 4. Incrementing values in GLRLM (right) according to pixels in window $W^{k,l}$ (left) for $\alpha = 90°$ and the gray level equal to zero. (Color figure online)

this approach determines additional group of 11 features based on texture information. The latter together with the previously introduced GLCM based group yields 19 texture based features considered in this paper. They complement previously discussed 6 shape based features introduced to accomplish bacteria classification.

5   Selected Class Recognition Methods

5.1   KNN

K-Nearest Neighbors is a classification method [17] that permits to assign a new object to one of the constructed classes. We are supplied here with a data-set attributed with the existing features and its membership to the respective class. Based on the latter, a new object needs to be classified with respect to the same set of attached features. In doing so, the calculation of Euclidean distances between a new object and every single object from the data-set is performed. We sort these objects by distance in ascending order. Then we choose k objects (k value is set arbitrarily) whose distances are the smallest and conduct majority voting (see [17]) to decide to which class the new object should be attached.

5.2   ELM-RBF

Extreme Learning Machine - Radial Basis Function is feed-forward neural network with two hidden layers [10]. First one is responsible for the input vectors conversion to the distances based on Gaussian radial function to the closest centroid. Their amount is selected arbitrarily and they are computed using k-means method, where k defines number of centroids to be calculated. This procedure is very similar to RBN (*Radial Basis Network*) and brings much more robustness to the prediction [20]. Upon converting the input vectors into distances in the first hidden layer, we treat them as input vector in ELM network that contains one hidden layer with experimentally chosen number of neurons and their activation function. The bias and weights between hidden layers are assigned randomly. Weights between hidden and output layer are calculated using Moore-Penrose matrix pseudo-inverse operation [25].

6   Experiments and Results

We report now on generated experimental results based either on combined set of shape and texture features or solely relying on shape or texture information.

First, the experiments incorporating a full set of shape and texture attributes juxtapose different feature selection and classification methods to reach satisfactory classification accuracy. Additionally, the tests f determining the optimal number of interpolation points $\hat{Q}_{m+1}$ along $\partial\Gamma$ together with gauging the amount of selected bacteria are carried out. The best classification results obtained yield $m + 1 = 10$ and 50 bacteria. We present now the list of implemented methods (see also [18,31]) with the experimentally picked up optimal parameters guaranteeing the highest possible classification accuracy: SVM (using *radial basis function*), RF (using 200 trees), kNN (using $k = 1$), MLP (using back-propagation learning method, topology of the net $22 - 20 - 22$ and *tanh* as an activation function on all hidden layers), ELM (with 2800 neurons in hidden layer units with *tanh* activation function) and ELM-RBF (900 neurons with linear activation function in hidden layer units and 40 centroids).

Table 1. Mean accuracy percentage of 50 tests using 10% cross validation and feature selection with classification method performed on shape and texture features.

Feature selection method	SVM	RF	kNN	ELM	MLP	ELM-RBF
None	93.84	92.76	95.30	92.50	92.03	94.95
FCBF	95.69	93.62	97.07	78.80	89.73	94.05
SBMLR	96.61	93.92	96.00	91.61	92.65	97.03
CFS	93.61	91.23	95.00	85.34	91.03	94.74

As shown in Table 1, the best result in bacteria classification amounts to 97% in accuracy recognition which is obtained upon either applying ELM-RBF with SBMLR or using kNN with FCBF. In this case ELM-RBF shows superiority over ELM method increasing accuracy by over 5% and due to a smaller number of neurons in hidden layer has a vastly shorter training and testing time. Here SBLMR selects 5 shape and 15 texture features whereas FCBF relies on using 3 shape and 4 texture traits. In order to justify merging features from both classes of examined attributes (i.e. shape and texture), we subsequently tested the bacteria classification accuracy when either only the set of shape or the set of texture features is admitted, respectively.

Table 2. Mean accuracy percentage of 50 tests using 10% cross validation and feature selection with classification method performed on features based on shape.

Feature selection method	SVM	RF	kNN	ELM	MLP	ELM-RBF
None	75.69	72.61	77.00	49.84	73.84	76.92
FCBF	78.07	74.07	74.76	29.38	70.84	76.00
SBMLR	76.06	76.40	77.56	48.00	73.47	77.96
CFS	78.92	73.30	76.69	42.92	74.38	77.69

Table 3. Mean accuracy percentage of 50 tests using 10% cross validation and feature selection with classification method performed on features based on texture.

Feature selection method	SVM	RF	kNN	ELM	MLP	ELM-RBF
None	80.07	76.30	81.61	64.84	67.46	68.23
FCBF	75.76	79.76	76.61	41.61	68.30	74.76
SBMLR	78.93	77.06	82.27	63.46	67.81	69.64
CFS	79.00	78.84	78.53	55.46	70.53	74.38

The best experimental result relying exclusively on shape features amounts to 78.92% in classification accuracy (see Table 2). It is derived with the aid of SVM coupled with CFS. In contrast, the best accuracy using solely texture based features equals 82.27% (see Table 3) and is achieved upon combining kNN with SBMLR. Having juxtaposed results from Table 1, 2 and 3 it is transparent that merging shape and texture features improves classification accuracy by 15%. As shown in Table 1 the mean accuracy from 50 tests using 10% cross validation amounts to 97% matching state of the art results.

7   Conclusions

Experiments based on 6 shape features render (for our data) the best accuracy reaching 78.92% correct classification for SVM combined with CFS. On the other hand class recognition based on 19 image texture traits yields up to 82.27% for kNN and SBMLR. In contrast, gathering together both shape and texture information (totalling 25 conjugated features) leads up to 97% correct classification upon coupling either kNN with FCBF or ELM-RBF with SBMLR. The iterative optimization of the classification model parameters including selection of the number of knots and the amount of bacteria analyzed in one picture, improves accuracy and reduces time execution of the implemented congregated classifier. These results seem to be unexpectedly satisfactory for our proposed *aggregated bacteria classifier* in the absence of incorporating color information. Still, within the setting of this work, there is a natural scope for further improvements. In particular, any method selecting characteristic benchmark bacteria for a given genus permitting to compare bacteria's curvature with the reference bacteria would be desirable. In this work originally, such five exemplary bacteria were selected arbitrarily but the results obtained did not improve significantly the case of fixing one reference bacteria for five considered genera. Another related issue refers to the task of selecting all significant points (and knots) on the bacteria's boundary (see e.g. [5,14–16,19]). This work assumes "more or less" equally spaced points Q_m. The impact of convexity or non-convexity of the bacteria should also be analyzed with respect to ordering $Q_{\hat{m}}$. Furthermore, the comparison of standard classification methods with deep learning methods and extending admissible set of features incorporating color and dispersion information forms potential research topics within the field of soil bacteria classification. Lastly, the robustness of all examined methods may also be tested against the varying number of bacteria genera (or their respective representatives) and possibly in regard to other dynamic factors such as time aspect impacting on shape, size or a color of the examined bacteria and/or its bacterial colony distribution.

References

1. Amirani, M.C., Gol, Z.S., Shirazi, A.A.B.: Efficient feature extraction for shape-based image retrieval. J. Appl. Sci. **8**, 2378–2386 (2008). https://doi.org/10.3923/jas.2008.2378.2386
2. Anderson, L.M., Stockwell, V.O., Loper, J.E.: An extracellular protease of Pseudomonas fluorescens inactivates antibiotics of Pantoea agglomerans. Phytopathology **94**(11), 1228–1234 (2004). https://doi.org/10.1094/PHYTO.2004.94.11.1228
3. Beeckmans, S., Xie, J.: Glyoxylate cycle. In: Reference Module in Biomedical Sciences. Elsevier (2015). https://doi.org/10.1016/B978-0-12-801238-3.02440-5
4. Bhairannawar, S.S.: Chapter 4 - efficient medical image enhancement technique using transform HSV space and adaptive histogram equalization. In: Soft Computing Based Medical Image Analysis, pp. 51–60. Academic Press (2018). https://doi.org/10.1016/B978-0-12-813087-2.00003-8
5. de Boor, C.: A Practical Guide to Splines. Springer (2001)
6. Buck, J.R., Daniel, M.M., Singer, A.: Computer Explorations in Signals and Systems Using MATLAB. Prentice Hall (2002)
7. Caprette, D.R.: Describing colony morphology. https://bit.ly/324cqkA
8. do Carmo, M.P.: Differential Geometry of Curves and Surfaces. Prentice Hall (1976)
9. Das, B., Chakrabarty, D.: Lagrange's interpolation formula: representation of numerical data by a polynomial curve. Internat. J. Math. Trends Technol. 34, 64–72 (2016). https://doi.org/10.14445/22315373/IJMTT-V34P514
10. Dhini, A., Surjandari, I., Kusumoputro, B., Kusiak, A.: Extreme Learning Machine-Radial Basis Function (ELM-RBF) networks for diagnosing faults in a steam turbine. J. Ind. Prod. (2021). https://doi.org/10.1080/21681015.2021.1887948
11. Ferro-Flores, G., et al.: Prediction of overall survival and progression-free survival by the 18F-FDG PET/CT radiomic features in patients with primary gastric diffuse large B-Cell Lymphoma. Contrast Media Mol. Imaging (2019). https://doi.org/10.1155/2019/5963607
12. Haralick, R.M., Shanmugam, K., Dinstein, I.: Textural features for image classification. IEEE Trans. Syst. Man. SMC-**3**(6), 610–621 (1973). https://doi.org/10.1109/TSMC.1973.4309314

13. Iversen, C.: Encyclopedia of Food Microbiology: Enterobacter. Academic Press, 2nd edn. (2014)
14. Kozera, R.: Curve modeling via interpolation based on multidimensional reduced data. Studia Inform. **25**(4B), 1–140 (2004)
15. Kozera, R., Noakes, L., Wiliński, A.: Generic case of Leap-Frog Algorithm for optimal knots selection in fitting reduced data. In: ICCS 2021. LNCS, vol. 12745, pp. 337–350. Springer, Cham (2021). https://doi.org/10.1007/978-3-030-77970-2_26
16. Kozera, R., Noakes, L., Wilkołazka, M.: Parameterizations and Lagrange cubics for fitting multidimensional data. In: ICCS 2020. LNCS, vol. 12138, pp. 124–140. Springer, Cham (2020). https://doi.org/10.1007/978-3-030-50417-5_10
17. Kramer, O.: K-Nearest Neighbors, pp. 13–23. Springer, Berlin Heidelberg (2013). https://doi.org/10.1007/978-3-642-38652-7_2
18. Kruk, M., Kozera, R., Osowski, S., Trzciński, P., Paszt, L.S., Sumorok, B., Borkowski, B.: Computerized classification system for the identification of soil microorganisms. Appl. Math. Inf. Sci. **10**(1), 21–31 (2016). https://doi.org/10.18576/amis/100103
19. Kvasov, B.: Methods of Shape-Preserving Spline Approximation. World Scientific (2000)
20. Lee, C.C., Chung, P.C., Tsai, J.R., Chang, C.I.: Robust radial basis function neural networks. IEEE Trans. Syst. Man Cybern. **29**(6), 674–685 (1999). https://doi.org/10.1109/3477.809023
21. Lefakis, L., Fleuret, F.: Jointly informative feature selection made tractable by Gaussian modeling. J. Mach. Learn. Res. **17**(182), 1–39 (2016). http://jmlr.org/papers/v17/15-026.html
22. Lim, Y., et al.: Mechanically resolved imaging of bacteria using expansion microscopy. PLOS Biol. **17**(10), 1–19 (2019). https://doi.org/10.1371/journal.pbio.3000268
23. Malhi, S.S., Sahota, T.S., Gill, K.S.: Chapter 5 - potential of management practices and amendments for preventing nutrient deficiencies in field crops under organic cropping systems. In: Agricultural Sustainability, pp. 77–101. Academic Press (2013). https://doi.org/10.1016/B978-0-12-404560-6.00005-8
24. Morin, A.: Encyclopedia of Food Microbiology: Pantoea. Academic Press, 2nd edn. (2014)
25. Rao, C.R.: Generalized Inverse of a Matrix and its Applications, pp. 601–620. University of California Press (1972). https://doi.org/10.1525/9780520325883-032
26. Satoto, B.D., Utoyo, M.I., Rulaningtyas, R., Koendhori, E.B.: Classification of features shape of Gram-negative bacterial using an Extreme Learning Machine. IOP Conf. Ser. Earth Environ. Sci. **524**(1), 012005 (2020). https://doi.org/10.1088/1755-1315/524/1/012005
27. Shapiro, L.G., Stockman, G.: Computer Vision. Pearson, 1st edn. (2001)
28. Soille, P.: Morphological Image Analysis: Principles and Applications. Springer-Verlag (1999)
29. Sokolov, D.D.: Encyclopedia of Mathematics. EMS Press (2001)
30. Sorensen, J., Nybroe, O.: Pseudomonas: Volume 1 Genomics, Life Style and Molecular Architecture, chap. Pseudomonas in the Soil Environment, pp. 369–401. Springer, US (2004). https://doi.org/10.1007/978-1-4419-9086-0_12
31. Toprak, A.: Extreme Learning Machine (ELM)-based classification of benign and malignant cells in breast cancer. Med. Sci. Monit. **24**, 6537–6543 (2018). https://doi.org/10.12659/MSM.910520

32. Yang, X., et al.: Ultrasound GLCM texture analysis of radiation-induced parotid-gland injury in head-and-neck cancer radiotherapy: an in vivo study of late toxicity. Med. Phys. **39**(9), 5732–5739 (2012). https://doi.org/10.1118/1.4747526

Calibration Window Selection Based on Change-Point Detection for Forecasting Electricity Prices

Julia Nasiadka[iD], Weronika Nitka[✉][iD], and Rafał Weron[iD]

Department of Operations Research and Business Intelligence, Wrocław University
of Science and Technology, 50-370 Wrocław, Poland
weronika.nitka@pwr.edu.pl

Abstract. We employ a recently proposed change-point detection algorithm, the Narrowest-Over-Threshold (NOT) method, to select subperiods of past observations that are similar to the currently recorded values.
Then, contrarily to the traditional time series approach in which the most
recent τ observations are taken as the calibration sample, we estimate
autoregressive models only for data in these subperiods. We illustrate our
approach using a challenging dataset – day-ahead electricity prices in the
German EPEX SPOT market – and observe a significant improvement in
forecasting accuracy compared to commonly used approaches, including
the Autoregressive Hybrid Nearest Neighbors (ARHNN) method.

Keywords: Change-point detection · Narrowest-Over-Threshold
method · Electricity price forecasting · Autoregressive model ·
Calibration window

1 Introduction

Electricity price forecasting (EPF) is an extremely challenging task. A number
of methods have been developed for this purpose, ranging from linear regression
to hybrid deep learning architectures utilizing long-short term memory and/or
convolutional neural networks. While most studies focus on improving model
structures, selecting input features with more predictive power or implementing
more efficient algorithms [3,5,6], the issue of the optimal calibration window is
generally overlooked [4].

This work is inspired by a recent article [2], which utilized a relatively simple change-point detection method [9] to split the time series into segments
with the 'same' price level, and an ICCS 2021 paper [7], which employed the
k-nearest neighbors (k-NN) algorithm to select the calibration sample based on

Supported by the Ministry of Education & Science (MEiN, Poland) through Grant No.
0027/DIA/2020/49 (to WN) and the National Science Center (NCN, Poland) through
Grant No. 2018/30/A/HS4/00444 (to RW).

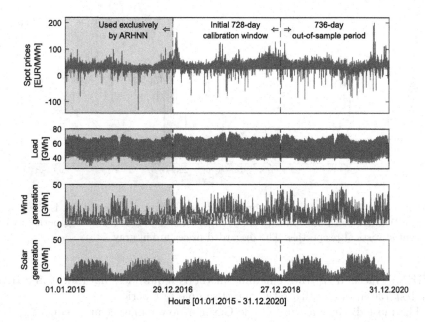

Fig. 1. Electricity spot prices and day-ahead load, wind and solar power generation forecasts in Germany. The last 736 days constitute the test period.

similarity over a subset of explanatory variables. Here, we utilize a recently proposed change-point detection algorithm – the Narrowest-Over-Threshold (NOT) method [1] – to construct an automatic method for detecting subperiods exhibiting different temporal dynamics. Once identified, those not resembling the current behavior are discarded when estimating the predictive model. In what follows, we provide empirical evidence that significant improvement in forecasting accuracy can be achieved compared to commonly used EPF approaches.

The remainder of the paper is structured as follows. In Sect. 2 we present the dataset and the transformation, which is used to standardize the data. In Sect. 3 we briefly describe the NOT method and introduce our approach to selecting subperiods for model calibration. Next, in Sect. 4 we present the forecasting models and in Sect. 5 the empirical results. Finally, in Sect. 6 we conclude and discuss future research directions.

2   The Data

For comparison purposes, we use the same dataset as in [7]. It spans six years (2015–2020) at hourly resolution and includes four series from the German EPEX SPOT market: electricity spot prices $P_{d,h}$ (more precisely: prices set in the day-ahead auction on day $d-1$ for the 24 h of day d) and day-ahead load $\hat{L}_{d,h}$, wind $\hat{W}_{d,h}$ and solar power generation $\hat{S}_{d,h}$ forecasts, see Fig. 1. The first two years are exclusively used for estimating the Autoregressive Hybrid Nearest Neighbors

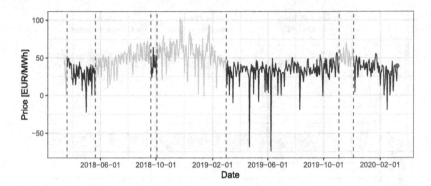

Fig. 2. A sample run of the algorithm introduced in Sect. 3. The red dot is the target day. Vertical dashed lines indicate the located change-points separating periods with different statistical properties. The discarded prices are in gray.

(ARHNN) method [7]; the remaining methods require less data for calibration. The last 736 days constitute the out-of-sample test period.

The most distinct feature of the German power market are frequent spikes and negative prices. Similarly volatile is wind energy generation, while load and solar generation are more predictable. Following [8,10], to cope with this extreme volatility, we transform the electricity prices using the area hyperbolic sine: $Y_{d,h} = \operatorname{asinh}\left(\frac{1}{b}\{P_{d,h} - a\}\right)$ with $\operatorname{asinh}(x) = \log\{x + (x^2 + 1)^{0.5}\}$, where a is the median of $P_{d,h}$ in the calibration window and b is median absolute deviation. The price forecasts are then obtained by the inverse transformation.

Note that in [7] a different transformation was used. All series, not just $P_{d,h}$, were normalized by subtracting the mean and dividing by the standard deviation in each calibration window. We denote models utilizing *asinh*-transformed data with subscript H; the remaining ones use the standard normalization, as in [7].

3   Calibration Window Selection Using NOT

The Narrowest-Over-Threshold (NOT) method [1] can detect an unknown number of change-points at unknown locations in one-dimensional time series data. The key feature is its focus on the smallest local sections of the data on which the existence of a change-point is suspected. A change-point is said to occur when the behavior of the series changes significantly [2]. See www.changepoint.info for an excellent review site and software repository on this topic. Said differently, change-points split the data into stationary subseries, see Fig. 2. This is what makes them interesting for model calibration and forecasting.

Our algorithm for calibration window selection, i.e., identifying periods with similar time series dynamics to the currently observed, is as follows:

1. Set the maximum number N_c^{max} of change-points to be identified.
2. Use the NOT method to identify $N_c \in [0, N_c^{max}]$ change-points c_i, $i = 1, \ldots, N_c$, in the initial calibration window C_0 of length τ. Additionally, denote the first observation in C_0 by c_0.

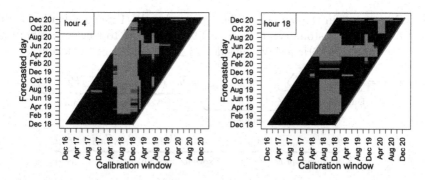

Fig. 3. Overview of NOT-selected (black) and discarded (gray) periods in the two-year calibration window (dates on the x axis) for hours 4 (*left panel*) and 18 (*right panel*). The red line indicates the forecasted day (date on the y axis).

3. If $N_c = 0$ return calibration sample $C = C_0$. Otherwise, compute the empirical quantiles q_{low} and q_{high} of the observations within the period between the most recent change-point found and the last observation in C_0.
4. For every interval between two subsequent change-points c_{i-1} and c_i compute the median m_i of its observations, $i = 1, \ldots, N_c - 1$.
5. Set $C = \bigcup_{m_i \in (q_{low}, q_{high})}^i [c_{i-1}, c_i]$.

Based on a limited simulation study, we set the maximum number of change-points $N_c^{max} = 12$ and the order of quantiles $(q_{low}, q_{high}) = (q_{0.025}, q_{0.975})$. We also use the least constraining form of NOT, i.e., we assume that the data have piecewise continuous variance and piecewise continuous mean. Any deviations from this are treated as a breach of stationarity.

A sample run of the algorithm is presented in Fig. 2. In the plot, the period closest to the forecasted day (red dot) is characterized by relatively stable, low prices with a small variance. The algorithm discards the light gray subperiods, when either the prices or their variance are significantly higher. In Fig. 3 we illustrate the results for two sample hours and the whole test window. We use a rolling scheme, i.e., once forecasts for the 24 h of the first day in the test sample are computed, the calibration window is moved forward by one day and forecasts for the 2nd day in the test sample are calculated. A clear pattern of vertical gray stripes emerges, meaning that for a range of windows the change-points are consistently detected on the same or neighboring days. Comparing these plots with the price trajectory in Fig. 1, we can observe that much fewer observations are selected by NOT when the prices tend to be more spiky, as can be seen in Spring 2020 (Apr 20 – Jun 20 on the y axis, esp. for hour 18).

4   Forecasting Models

For comparison purposes, the underlying model we use is the same as in [7]. It is an autoregressive structure with exogenous variables dubbed ARX. Since the prices $P_{d,h}$ are set in the day-ahead auction on day $d-1$ independently for the 24h of day d, it is customary in the EPF literature [4,6] to treat every hour as a separate time series. Hence, we consider 24 ARX models of the form:

$$P_{d,h} = \underbrace{\alpha_h D_d}_{\text{Dummies}} + \underbrace{\sum_{p \in \{1,2,7\}} \beta_{h,p} P_{d-p,h} + \underbrace{\theta_{h,1} P_{d-1,min} + \theta_{h,2} P_{d-1,max}}_{\text{Yesterday's price range}}}_{\text{AR component}}$$

$$+ \underbrace{\theta_{h,3} P_{d-1,24}}_{\text{Last known price}} + \underbrace{\theta_{h,4} \hat{L}_{d,h} + \theta_{h,5} \hat{W}_{d,h} + \theta_{h,6} \hat{S}_{d,h}}_{\text{Exogenous variables}} + \underbrace{\varepsilon_{d,h}}_{\text{Noise}} . \tag{1}$$

The autoregressive (AR) dynamics are captured by the lagged prices from the same hour yesterday, two and seven days ago. Following [10], yesterday's minimum $P_{d-1,min}$, maximum $P_{d-1,max}$ and the last known price $P_{d-1,24}$, as well as day-ahead predictions of the three exogenous variables are included. Finally, a 1×7 vector of dummy variables D_d is used to represent the weekly seasonality and the uncertainty is represented by white noise.

Overall, we compare seven types of approaches that all use ARX as the underlying model. The first three are the same as in [7]: (i) **Win(τ)** – the ARX model estimated using a window of τ days, with $\tau \in [56, 57, ..., 728]$, (ii) **Av(Win)** – the arithmetic average of six forecasts of the ARX model for three short ($\tau = 56, 84, 112$) and three long windows ($\tau = 714, 721, 728$), and (iii) the **ARHNN** model. The next four include: (iv) **Win$_H$(τ)** – the same as Win(τ) but calibrated to *asinh*-transformed prices, (v) **NOT$_H$(728)** – the ARX model calibrated to *asinh*-transformed prices in NOT-selected subperiods from the 728-day window, (vi) **Av(Win$_H$)** – the same as Av(Win) but calibrated to *asinh*-transformed prices, and (vii) **Av(NOT$_H$)** – the same as Av(Win$_H$) but with the forecasts for the three long windows ($\tau = 714, 721, 728$) replaced by NOT$_H$(728). The rationale behind the latter averaging scheme is that NOT$_H$(τ) performs best for long calibration windows and offers little or even no gain for $\tau < 1$ year.

5   Results

We evaluate the forecasting performance of the seven approaches presented in Sect. 4 in terms of the *root mean squared error* (RMSE; results for the mean absolute error are similar and available from the authors upon request). The RMSE values reported in Fig. 4 are aggregated (averaged) across all hours in the 736-day test sample, see Fig. 1. Additionally, to test the significance of differences in forecasting accuracy, for each pair of models we employ the multivariate variant of the Diebold-Mariano (DM) test, as proposed in [10].

Several conclusions can be drawn. Firstly, changing the preprocessing method from normalization [7] to *asinh* transformation [8] generally reduces the RMSE. Even the worst performing out of the latter approaches, Win$_H$(728), improves on ARHNN, the most accurate method in [7]. Secondly, NOT-selection yields further improvement, although not statistically significant if considered on its own. Compare NOT$_H$(728) with Win$_H$(728) and Av(NOT$_H$) with Av(Win$_H$).

Method	RMSE
Win(728)	8.2860
Av(Win)	8.0286
ARHNN	7.8605
Win$_H$(728)	7.7286
NOT$_H$(728)	7.5994
Av(Win$_H$)	7.0968
Av(NOT$_H$)	7.0831

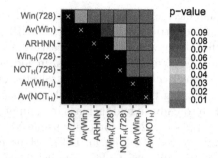

Fig. 4. RMSE errors in the out-of-sample test period (*left panel*). A heatmap of the p-values for the multivariate Diebold-Mariano test [10] for each pair of methods (*right panel*). The smaller the p-values, the more significant is the difference between the forecasts of a model on the x-axis (better) and the forecasts of a model on the y-axis (worse). Black color indicates p-values in excess of 0.1.

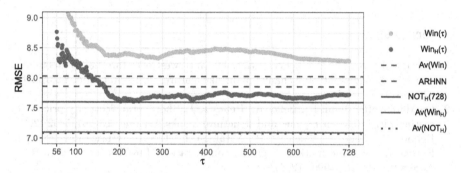

Fig. 5. RMSE values for all considered models; τ is the calibration window length.

The RMSE values for all considered approaches are presented graphically in Fig. 5. It clearly shows the significant improvement from using the *asinh* – compare between Win(τ) with Win$_H$(τ) for all τ's. While Win$_H$(728) is not the best performing of all Win$_H$(τ) models, the differences in performance are relatively minor for $\tau \geq 200$ days. Even the best ex-post known model, Win$_H$(233), is slightly worse than NOT$_H$(728). The averaged forecasts Av(Win$_H$) and Av(NOT$_H$) are further able to improve on the accuracy, although the differences between them are not significant.

6   Conclusions and Discussion

In this paper we propose a novel method for selecting calibration subperiods based on Narrowest-Over-Threshold (NOT) change-point detection [1]. Contrarily to the traditional time series approach in which the most recent observations are taken as the calibration sample, we propose to estimate the predictive models only using data in the selected subperiods. We evaluate our approach using

German electricity market data and seven variants of autoregressive models tailored for electricity price forecasting (EPF). We provide empirical evidence that significant improvement in forecasting accuracy can be achieved compared to commonly used EPF approaches, including the recently proposed ARHNN [7]. In addition to calibration sample selection, our results also emphasize the importance of using transformations like the *asinh*, in line with [8,10].

The roughly sixfold increase in computational time of the NOT-based methods – 4.97s for 24 h forecasts using $NOT_H(728)$ vs. 0.83s using $Win_H(728)$, running R ver. 3.6.3 on an i7-9750H processor – can be seen as a drawback, especially compared to less complex ways of improving forecast accuracy, like calibration window averaging [4]. However, the automation of the forecasting process may make the trade-off worthwhile. If this is the case for more complex models than the autoregressive ones considered here or the shallow neural network in [2], e.g., LASSO-estimated AR (LEAR) and deep neural networks [6], is left for future work.

References

1. Baranowski, R., Chen, Y., Fryzlewicz, P.: Narrowest-over-threshold detection of multiple change-points and change-point-like features. J. R. Stat. Soc. **81**(3), 649–672 (2019)
2. De Marcos, R., Bunn, D., Bello, A., Reneses, J.: Short-term electricity price forecasting with recurrent regimes and structural breaks. Energies **13**(20), 5452 (2020)
3. Heijden, T., Lago, J., Palensky, P., Abraham, E.: Electricity price forecasting in European day ahead markets: a greedy consideration of market integration. IEEE Access **9**, 119954–119966 (2021)
4. Hubicka, K., Marcjasz, G., Weron, R.: A note on averaging day-ahead electricity price forecasts across calibration windows. IEEE Trans. Sustain. Energy **10**(1), 321–323 (2019)
5. Jahangir, H., Tayarani, H., Baghali, S., et al.: A novel electricity price forecasting approach based on dimension reduction strategy and rough artificial neural networks. IEEE Trans. Ind. Inform. **16**(4), 2369–2381 (2020)
6. Lago, J., Marcjasz, G., Schutter, B.D., Weron, R.: Forecasting day-ahead electricity prices: a review of state-of-the-art algorithms, best practices and an open-access benchmark. Appl. Energy **293**, 116983 (2021)
7. Nitka, W., Serafin, T., Sotiros, D.: Forecasting electricity prices: autoregressive hybrid nearest neighbors (ARHNN) method. In: Paszynski, M., Kranzlmüller, D., Krzhizhanovskaya, V.V., Dongarra, J.J., Sloot, P.M.A. (eds.) ICCS 2021. LNCS, vol. 12745, pp. 312–325. Springer, Cham (2021). https://doi.org/10.1007/978-3-030-77970-2_24
8. Uniejewski, B., Weron, R., Ziel, F.: Variance stabilizing transformations for electricity spot price forecasting. IEEE Trans. Power Syst. **33**(2), 2219–2229 (2018)
9. Zeileis, A., Kleiber, C., Walter, K., Hornik, K.: Testing and dating of structural changes in practice. Comput. Stat. Data Anal. **44**, 109–123 (2003)
10. Ziel, F., Weron, R.: Day-ahead electricity price forecasting with high-dimensional structures: univariate vs. multivariate modeling frameworks. Energy Econ. **70**, 396–420 (2018)

Tackling Air Pollution in Cities with Modelling and Simulation: Remote Group Model Building as an Educational Tool Supporting System Dynamics Modelling

Ewa Duda[1]([✉]) [iD] and Agata Sawicka[2] [iD]

[1] Institute of Education, Maria Grzegorzewska University, Warsaw, Poland
eduda@aps.edu.pl
[2] Faculty of Social Sciences, Department of Geography, University of Bergen, Bergen, Norway

Abstract. The study introduces a System Dynamics Modelling (SDM) approach with a remote Group Model Building (GMB) component used as an answer to the climate problems of cities related to severe air pollution. The main objective of our research is twofold: to identify the factors and mechanisms that are key for the elimination of fossil-fuel boilers (FFBs) in Poland and to provide a system understanding of the underlying causal relationships and their implications to facilitate eradication of FFBs. The first phase of modelling process is presented: action research represented by remote GMB workshops, attended by 14 participants from Poland and Norway. The workshops' results help to identify the key variables, capture the main causal relationships between them, helping to develop the first outline of the causal structure to underlie the simulation model. Despite holding the workshops during social isolation, they proved to be good educational facilitation tool for a key element of the modelling process in the SD methodology - the collection of input data for model building, as well as raising participants' awareness of climate change issues. The novelty of the study lies in the application of a proven tool such as dynamic systems modelling to applied research for sustainable development. In particular, using modelling and simulation approaches to support development and practical implementation of the energy transition policies involving the elimination of FFBs and replacement with renewable energy sources. To the best of our knowledge, only a few such studies are being conducted.

Keywords: System dynamics modelling · Group model building · Education for sustainable development · Educational facilitation tool · Urban education

1 Introduction

The issue of caring for the quality of the environment, including air quality, has been growing in importance for many years. The search for effective solutions to improve the quality of the environment, but also for knowledge on how not to destroy what nature has given us is a challenge for many people who care about the well-being of our planet. Awareness of the negative effects of our actions plays a key role in changing old, inappropriate habits. Tools that make it easier to plan future actions, anticipating their consequences, could be a support.

Caring for the environment requires an integrated strategy, engaging the participation of diverse stakeholders, beginning from government authorities, policy makers, through non-government organizations, media, local communities, educators, ending on individual users [1]. Ensuring effective communication between them is not one of the easiest tasks. However, in order to slow down the direction of man-made climate change, we must strive for change at global level, but also for change of individual, habitual behavior. Not many are interested in acquisition of the up-to-date knowledge in the field of environmental protection and changing their individual behaviors, habits. It is not new that people learn best through co-participation [2, 3]. Involving people in the decision-making process, working out effective solutions together, learning by doing, gives the best results. However, despite good intentions, such a process can lead to wrong decisions. After all, as humans we are fallible.

Mathematics and computer science are separate fields, but they have the great advantage of being able to use the tools they provide to solve important social problems, especially complex problems. One such tool is modelling based on systems dynamics methodology, which has been proven successful in many application areas over the years [4]. In this paper we would like to introduce a Systems Dynamics Modelling (SDM) approach used with a Group Model Building (GMB) component as an answer to the climate problems of future cities related to severe air pollution. The energies transition process will be supported by providing a simulation model to allow for exploration of various scenarios and development of the common understanding of the causal relationships underlying the complex socio-technical ecosystem of the city and its residents. The following main research questions that we formulated guided us through the research process:

Q1: Can GMB support the process of modelling complex systems carried out in social isolation?

Q2: Can the GMB be an effective tool to raise awareness and increase the knowledge of residents about the energy transition?

Q3: How a SDM with GMB component can enhance a pro-environmental decision making process?

This paper presents initial results of the first phase of the project, outlining the conceptual system dynamic model of the causal structures that are likely to underlie the dynamics of the process of elimination of fossil fuel boilers in residential areas in Poland. The causal structures were identified during an online workshop. The initial causal model was developed based on the data gathered from the project experts and refined based on the review of the relevant literature. In the paper we focus on presenting the elicitation process and the resulting model with references to relevant literature where necessary. The following section reviews the literature thus far. Next, we describe the applied methodology. The following section reviews the literature thus far. Next, we describe the applied methodology. The consecutive section presents the workshop and the initial system dynamics model. Finally, we outline further steps to be taken in the successive phases of our project.

2 Related Works

System dynamics was developed by Professor Jay W. Forrester in the late 1950s at the Massachusetts Institute of Technology in the late 1950s. It is a computer-aided approach for strategy and policy design. It uses feedback systems theory to develop computer simulation models. It is an analytical approach to tackle dynamic problems arising in complex social, managerial, economic, or ecological systems [5, 6] (see also [7]). System dynamics approach is especially useful where experts from different domains try to address a complex problem with problem-owners. The system dynamics models allow to create a shared understanding of the problem, providing effective tool for common exploration of possible solutions as well as for improvement of the fragmented individual or domain-specific knowledge. As system dynamics models draw on different expert domains, they are usually developed in a process of group model building [8]. Group model building (GMB), during which stakeholders are gathered around the table to share their insights, has been seen as an important social process in the model development phase, crucial for creating a shared understanding of complex systems and providing a platform for stakeholders to exchange information and ideas [9, 10].

The specialized system dynamics literature decarbonization transitions of residential buildings revealed two main projects research, which became the starting point for further analyses and system dynamics modeling dedicated for pilot case in Poland.

The first one is the HEW project: Integrated decision-making about Housing, Energy and Wellbeing, conducted in years 2011–2016 by University College London, Energy Institute [11]. The HEW project focused on the unintended health, social and environmental harms of decarbonizing the built environment. The main conclusion stressed the need to consider cross-sectoral policy objectives in close interconnection. Despite the numerous linkages to other sectors, there was a conspicuous lack of feedback in the energy efficiency model, suggesting that emphasis should be placed on creating decarbonization policies that take into account feedbacks across sectors of the system [12].

The second project taken into account presents research conducted in Australia on an impact of individual solar power installation on the conventional electricity networks. The installation of rooftop photovoltaic panels by individual consumers leads to a reduction in the need for energy from the conventional electricity networks and thus a reduction in the amount of energy purchased. The reduction in networks revenue with fixed operating costs leads to the need to increase tariff costs to compensate for electricity networks losses. This continues to exacerbate the problem [13]. The development of renewable energy sources represented by the increasing share of private solar power plants is a progressive process, affecting the grid significantly. The author of the study emphasizes the necessity of considering the presented issue in the context of a broad transformation, forcing the steps of planning an effective long-term energy strategy.

3 Methodology

The overall objective of our research is twofold:

O1: to identify the factors and mechanisms that are key for elimination of fossil-fuel boilers (FFB) in Poland;

O2: to facilitate development of a system understanding – shared among the project partners as well as between the project team and the stakeholders – of the underlying causal relationships and their implications on the process of FFB eradication.

To achieve this overall objective, we employ the system dynamics methodology. For the first model, we intended to conduct a group model building (GMB) session with experts representing the project partners.

Due to the changes in the work and social life introduced because of the COVID-19 virus, we had to amend the plan and conducted GMB sessions using the online scripts and tools developed at the University of Bergen (UiB). In particular, we have adopted established scripts from Scriptapedia [14] to facilitate the first model building workshop, in particular:

a. Graphs Over Time;
b. Variable Elicitation;
c. Initiating and Elaborating a Causal Loop Diagram;
d. Model Review.

Graphs Over Time and the Initial Variable Elicitation were carried out in the form of survey sent out to all project partners. The survey results were then analyzed and implemented into the interactive Miro whiteboard template for GMB developed at the UiB (see Fig. 1). This whiteboard with the initial causal structure identified through the survey was presented as a point of departure for the online workshop carried out with all the project participants. During the workshop we have focused on Model Review, identifying additional variables and causal relationships. The causal map was next used to develop the first outline of the system dynamics stock-and-flow model in Vensim[1].

4   Elicitation of the General Causal Structures

The initial elicitation of the general causal structures was carried out through a pre-workshop survey and an online modeling workshop. As a first step (May 2021), project experts took part in a survey. The goal of the questionnaire was to elucidate the problem scope for the GreenHeat project. Participants completed the qualitative questionnaires at their convenience and in their preferred form (electronic or paper version sent as a scan). Input was based on participants' expert knowledge and professional experience.

---

[1] https://vensim.com/download/.

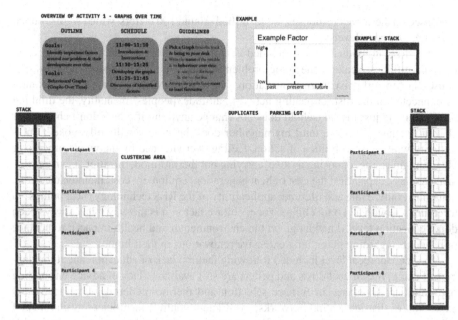

Fig. 1. Template for GMB developed at the UiB [15].

We received 11 completed questionnaires. The results of the survey were used to set the initial problem boundary, key system variables and relationships. As part of the preparation of the materials for the workshop, we analyzed the data from the questionnaires and transferred it to the Miro platform template in a structured form. Figure 2 shows a sample data from questionnaires completed by representatives of one of the partner institutions.

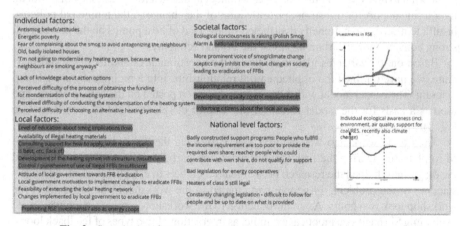

Fig. 2. Sample data from representatives of one of the partner institutions.

Based on the survey responses, we identified around 100 factors, which were grouped into three operable levels:

The first, individual level included factors like: attitudes that deny the existence/harmfulness/scale of the smog problem; poverty, including energy poverty; old, houses, lacking effective thermo-insulation – very expensive to renovate and that cannot be connected to the district heating network; attitude, peoples' mentality, the thinking that "changing just my household does not change anything if others don't change" or "I won't change the stove until my neighbor does, because it will still smoke"; unfamiliarity with the possibilities of action, feeling overwhelmed by the whole process of obtaining financing, choosing a stove, carrying out thermo-modernization; environmental awareness; capacities; the cost of heat generation; equipment cost; maintenance cost; equipment failure rate and lifetime; applicability of the RES technology (heat pumps) in a certain area with a certain climate zone; culture factor, it is the way it has always been done; education level (knowledge on the environmental and health impact); reluctance of residents regarding potentially extensive renovations in their home.

The second, local level included following factors: lack of education about the dangers of smog; illegal coal dusts and pellets are still available; lack of advice on funding opportunities, assistance in furnace selection and thermo-modernization; insufficient expansion of district heating networks; insufficient control; insufficient availability of infrastructure; conflicting interests of key institutions; insufficient availability of market offer/services; load of heat pumps on the power network; economic factor, it is expensive to move to other heating technology; fear to resistance from industry; fear to resistance from the voters.

The third, national level included following factors: challenges with the national 'Clean Air' program - poorer people cannot afford to contribute their own money, from which they will later receive a return, and those who can, are already beyond the income threshold and are not eligible for subsidies; incorrect law on the establishment of energy cooperatives; 5th class solid fuel stoves still legal; changing regulations - people do not know what to invest in, whether the rules will not change in a few years; lack of political, long-term support for transition; prioritization of the change in the municipal policies; insufficient access to nationally distributed funds; load of heat pumps on the power grid; strong conviction that carbon is part of the Poland economy.

Based on the variables identified during the questionnaires and the reviewed literature, we have outlined the implied causal relationships into a preliminary causal loop diagram. The causal model with a description has been posted on the Miro platform and used as a basis for the model review at the time of the planned online workshop (see Fig. 3).

The online workshop was held on 21/05/2021 via the Teams application and the Miro platform. During the first session, workshop participants consulted and updated the diagram. It helped the research project team members to establish a common understanding of the problem. 14 people participated in the workshop. The first part of the workshop took two hours. As an introduction to the workshop, a presentation was made explaining SD basics on practical example, as interaction of two types of feedback loops. Next the Icebreaker for participants who had not worked with Miro was provided. Then, participants were divided into two working groups, gathering representatives from each partner in each group. The model and the proposed causal relationships and loops were reviewed and discussed. The work resulted in an updated map with added structures, which we discuss briefly in turn.

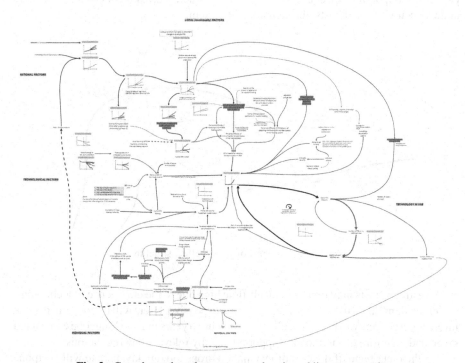

Fig. 3. Causal map based on survey and reviewed literature results.

The first group's discussion was focused on variables centered around the Word-of-mouth effect and the Expected return on investment structures.

The Word-of-mouth effect (see Fig. 4) reflects the impact of the information provided by current users of renewable energy-based systems. Very often users' knowledge can be based largely on conversations with neighbors, co-workers. If there are more satisfied users, with a pool of positive experiences, they will be a source of practical information on the process of installation replacement, system operation, will increase trust in energy innovation. The shaping of environmental awareness through social networks is shown by the phenomenon of the so-called social diffusion mechanism of innovation - the higher the density of PV panels in a neighborhood, the more likely more will appear. But if there are disgruntled users or people deliberately discouraging people from moving away from coal, their opinions will cause new investments to be abandoned, especially among those who are hesitant.

The Expected return on investment (see Fig. 5) is strongly associated with installation costs. The design, complexity, and promotion of subsidy programs affect how the projected benefits of replacement are calculated. The cost of installation and the potential payback is important especially for the poorest, reducing investment costs can be done at the expense of fixed costs, through a credit system.

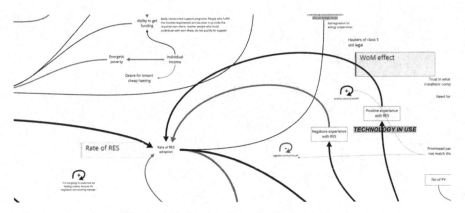

Fig. 4. Added word-of-mouth effect structure.

Investment costs may remain at an all-time high due to the emergence of technological innovations that will become more and more expensive. Users will be concerned about investing in currently expensive technology that in 10 years may no longer be considered green and therefore economically supported, e.g. by relevant law regulations.

On the other hand, if the cost of installation drops too quickly, people will postpone the decision to replace the installation because it will pay of better in the future. This complicates the calculation of return on investment. In addition, the return on investment may remain unclear because of uncertainty about the costs of various resources and because of uncertainty about long-term regulation.

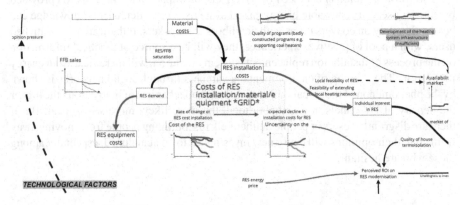

Fig. 5. Expected return on investment effect structure.

The second group's discussion was focused on variables centered around the Willingness to change the heating system effect.

The Willingness to change the heating system (see Fig. 6) is linked to understanding what renewable energy system is. Consumers do not know or acknowledge the effects of coal-fired boilers. Mainly because these effects are not direct. Users want the cheapest possible heating here and now without considering the possibility of investment. If energy policies provide people with access to tools that make it easier to replace non-environmental installations, but people's awareness is at an insufficient level, they will not want or know how to make this change. Useful information is insufficiently promoted and most recipients have no practical knowledge about transition process.

Participants of the workshop said that consumers do not know about subsidy programs. Grant programs are too complicated and confusing for individual recipients. Moreover, they usually do not provide information on who and what companies can perform the services and the recipients are forced to look for this information themselves. The programs are also poorly structured, for example, they subsidize new fossil fueled boilers, so they lead to the beginning of the problem.

On the other hand the willingness to change depends on many individual factors as for example ecological consciousness. Participants of the workshop noticed that recipients often do not have technical knowledge, do not know the principles of green technologies, do not have contact with such technologies, so they do not see their advantages. The less knowledge will lead to greater resistance to unfamiliar technological innovations. Other voices stressed that there is also a risk of polarizing public opinion by the actions of anti-climate populists who spread theses about the lie about the harmfulness of coal and smog, invented by the EU and RES producers.

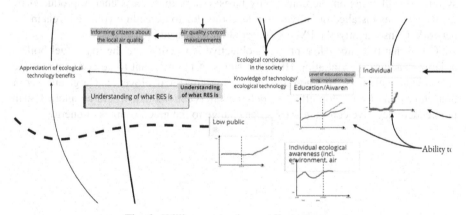

Fig. 6. Willingness to change effect structure.

The second part of the workshop took next two hours, during this time all participants worked as one group. During the discussion, the problem of creating energy storages, which on the one hand is an important element of the installation strengthening its efficiency and reliability, but on the other hand is costly, was strongly emphasized. The aim of the second part of the workshop was the further exploration of casual loop diagrams and identification of data sources needed for future analyses.

5 Outline of the Initial Causal Structures with Discussion

Based on the causal model, we have started to develop a structure for the simulation, stock-and-flow model. The overview of its initial structure is presented in Fig. 7. In this section we describe one of the parts that seems to be critical for sustainable local adoption of any renewable energy system that involves solar energy generation systems. This structure – highlighted in green – has been labeled the Death Spiral in the research report on the impact of individual solar energy systems in Australia [13].

The Death Spiral process is underlined by the reinforcing feedback mechanism triggered by the increasing number of individual households with solar energy systems (PV-systems). The more solar energy is produced by the individual households, the less demand for the traditional network energy, hence an increase in the unit network energy prices, leading in turn in even more interest and adoption of solar systems. Consequently, if uninterrupted, the process would inevitably lead to eradication of the traditional energy network providers, hence the "Death Spiral" label. Still, such an elimination of the traditional energy providers would require that individual household energy systems would be able to export or store the excess energy that is produced during daytime with much sun and relatively low energy demand.

The Death Spiral report indicates that the individual solar energy production systems need to be accompanied by the appropriate storage capacity. Otherwise, their economic value would be dramatically depleted as the excess energy will be lost, instead of being stored or exported for use in the periods of low/none energy production. The importance of energy storage systems has been fagged also by the GreenHeat project partners during the workshop discussions. In Poland there has been already cases of unsuccessful PV-systems installations on the community basis, where the excess energy produced by the PV-systems installed in individual households led to exceeding voltage levels in the network. Consequently, the PV-systems could not be used [16].

Given that the GreenHeat project's objective is to eliminate the fossil fuel boilers at large, leading to installation HP/PV systems, it is important to be conscious of the side effects likely to occur when solar energy systems are deployed in a large number of households and to develop effective policies/interventions that would counteract such unintended, negative consequences of the change to the green sources of energy.

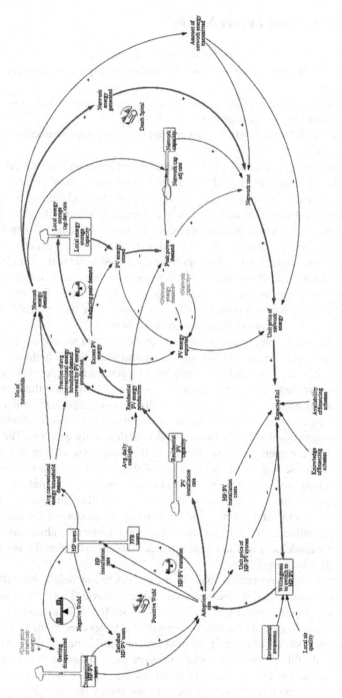

Fig. 7. The Death Spiral and the impact of local storage capacity on effectiveness of the PV systems.

6 Conclusions and Future Works

In our study, we focused on the following research questions:

Q1: Can GMB support the process of modelling complex systems carried out in social isolation?

Q2: Can the GMB be an effective tool to raise awareness and increase the knowledge of residents about the energy transition?

Q3: How a SDM with GMB component can enhance a pro-environmental decision making process?

Referring to the first research question, based on our experience we believe that GMB to support the development of models may be carried out successfully even remotely. Despite holding the workshops during social isolation, the modelling approach proved to be an effective facilitation tool for a key element of the modelling process in the SD methodology - the collection of input data for model building. The online format provided also an additional advantage of making it easier to gather experts from distant places, also from abroad, as the workshop was attended by experts from Poland and Norway, residents of different cities. On the other hand, we note that the views of the experts involved in our project have been rather similar. In case of groups with more diverse participants, with substantially different opinions or backgrounds, the remote format may show to be more challenging and less efficient.

Referring to the second research question, at the moment we can conclude that the GMB shows indeed to be an effective tool to communicate the problem understanding across the expertise domains, and in that way helps to increase awareness of other possible aspects of the problem. The causal model helps to explicitly show the problem from different interdisciplinary perspectives. Joint discussions, especially when exploring the relationships between the drivers of the energy transition process, allowed for new insights, helping to develop a more holistic understanding of the problem. The workshop also allowed for the expansion and verification of the existing knowledge of those directly involved in SD modelling. In the next phases of the project, we will use the model to facilitate the discussions with the citizens and community decision-makers.

Referring to the third research question, we can see that the GMB exercise helped the project participants to reach a better and common understanding of the multiplicity of factors likely to influence the decision making process. In the next phases of the project, we will use the model as a tool to involve people who are to be directly involved in the energy transition process.

The current, more theoretical phase of the project is concluded. We will now turn to investigate in more detail the particular case of Legionowo. These investigations will allow us to develop the model that addresses the specifics of Legionowo community. Drawing on the findings of concurrent project work packages concerning viable technical solutions, present social structure of the pilot site, or feasible financial mechanisms, we will include and calibrate the factors likely to shape the willingness to switch to HP/PV system, as well as the factors likely to fuel both the positive and negative word-of-mouth processes. Structures governing the network energy demand and supply will be elaborated. Once the structures are refined, we initial simulation-based explorations will be conducted.

The novelty and originality of the study lies in the application of a proven tool such as system dynamics modelling to research for sustainable development, in particular to support local energy transition policies involving the elimination of fossil fuel boilers and replacement with renewable energy sources. To the best of our knowledge, few such studies are being conducted. Although the tool is tested in a specific city located in Poland, its universality allows for easy transfer to any city in the world.

Acknowledgement. This research is supported by the € 1.3 million grant from Iceland, Liechtenstein and Norway through the EEA Grants. Grant agreement NOR/IdeaLab/GREENHEAT/0006/2020-00.

References

1. Maione, M., et al.: Air quality and climate change: designing new win-win policies for Europe. Environ. Sci. Policy **65**, 48–57 (2016)
2. Hodge, P., Wright, S., Barraket, J., Scott, M., Melville, R., Richardson, S.: Revisiting "how we learn" in academia: practice-based learning exchanges in three Australian universities. Stud. High. Educ. **36**(2), 167–183 (2011)
3. Carter, T.J., Adkins, B.: Situated learning communities of practice, and the social construction of knowledge. In: Wang, V.C.X. (ed.) Theory and Practice of Adult and Higher Education, pp. 113–138. Information Age Publishing, Charlotte (2017)
4. Sterman, J.: System dynamics at sixty: the path forward. Syst. Dyn. Rev. **34**(1–2), 5–47 (2018)
5. Forrester, J.W.: Industrial Dynamics. The MIT Press and Wiley, New York (1961)
6. Sterman, J.D.: Business Dynamics: Systems Thinking and Modeling for a Complex World. Irwin McGraw-Hill, Boston (2000)
7. https://systemdynamics.org/what-is-system-dynamics/. Accessed 02 Jan 2022
8. Vennix, J.A.M., Akkermans, H.A., Rouwette, E.A.J.A.: Group model-building to facilitate organizational change: an exploratory study. Syst. Dyn. Rev. **12**(1), 39–58 (1996)
9. Antunes, P., Stave, K., Videira, N., Santos, R.: Using participatory system dynamics in environmental and sustainability dialogues. In: Matthias, R. (ed.) Handbook of Research Methods and Applications in Environmental Studies, pp. 346–374. Edward Elgar, Cheltenham (2015)
10. Scott, R.J., Cavana, R.Y., Cameron, D.: Recent evidence on the effectiveness of group model building. Eur. J. Oper. Res. **249**, 908–918 (2016)
11. https://www.ucl.ac.uk/bartlett/environmental-design/research-projects/2020/nov/hew-integrated-decision-making-about-housing-energy-and-wellbeing. Accessed 10 Jan 2022
12. Macmillan, A., Davies, M., Bobrova, Y.: Integrated decision-making about housing, energy and wellbeing (HEW). Report on the Mapping Work for Stakeholders. UCL Faculty of the Built Environment, The Bartlett, London (2014)
13. Grace, W.: Exploring the death spiral: a system dynamics model of the electricity network in Western Australia. In: Sayigh, A. (ed.) Transition Towards 100% Renewable Energy. IRE, pp. 157–170. Springer, Cham (2018). https://doi.org/10.1007/978-3-319-69844-1_15
14. Hovmand, P., et al.: Scriptapedia: A Handbook of Scripts for Developing Structured Group Model Building Sessions (2011)
15. Wilkerson, B., et al.: Reflections on adapting group model building scripts into online workshops. Syst. Dyn. Rev. **36**(3), 358–372 (2020)
16. https://biznesalert.pl/magazyny-energii-transformacja-energetyka-fotowoltaika-oze-energetyka/. Accessed 30 Jan 2022

Fast Isogeometric Analysis Simulations of a Process of Air Pollution Removal by Artificially Generated Shock Waves

Krzysztof Misan, Weronika Ormaniec, Adam Kania, Maciej Kozieja,
Marcin Łoś(iD), Dominik Gryboś(iD), Jacek Leszczyński(iD),
and Maciej Paszyński[✉](iD)

AGH University of Science and Technology, Kraków, Poland
maciej.paszynski@agh.edu.pl

Abstract. Large concentrations of particulate matter in residential areas are related to the lack of vertical movements of air masses. Their disappearance is associated with the occurrence of the most common ground temperature inversion, which inhibits the natural air convection. As a result, air layers separated by a temperature inversion layer are formed, which practically do not interact with each other. Therefore, to reduce the concentration of particulate matter, mixing of air layers should be forced, or natural processes should be restored. For this purpose, it was proposed to generate shock waves of high pressure in the vertical direction to mix the polluted air and break the inversion layer locally. This paper performs fast isogeometric analysis simulations of the thermal inversion and pollution removal process. We employ a linear computational cost solver using Kronecker product-based factorization. We compare our numerical simulations to an experiment performed with an anti-hail cannon in a highly polluted city of Kraków, Poland.

Keywords: Thermal inversions · Atmoshperic pollution reduction methods · Advection-diffusion-reaction · Variational splitting · Isogeometric analysis

1 Introduction

One of the main challenges of human civilization is air pollution, which is an environmental, social, and economic problem. According to WHO data, almost 90% of people live in poor air quality, which causes over 4 million deaths annually [19]. Air pollution adversely affects human health, and causes such ailments as asthma, respiratory tract infections, and heart attacks [22]. Air pollution can be divided according to the type of harmful substances that cause this phenomenon: smog (particulate matter) and photochemical smog (ozone). Smog is a suspension of aerosol particles smaller than $10 \frac{\mu g}{m^3}$ or $2.5 \frac{\mu g}{m^3}$. Usually, it is created in the autumn and winter months and is associated with low emissions,

D. Groen et al. (Eds.): ICCS 2022, LNCS 13352, pp. 298–311, 2022.
https://doi.org/10.1007/978-3-031-08757-8_26

i.e., heating with solid fuels in households. As a result of unfavorable meteorological conditions and the formation of ground inversion, pollutants get stuck in
the living zone, reaching dangerous concentrations. The phenomenon is particularly enhanced in unfavorable terrain features, such as valleys or basins [9,15].
Due to the lack of commercial methods of removing particulate matter in open
spaces, only a change in weather conditions, i.e., the disappearance of temperature inversion, rain or snowfall or wind, causes the pollution to disperse in the
entire atmosphere and/or fall to the ground. As a result of the formation of
inversion layers, the vertical air movement between the layers of the atmosphere
above and below this layer disappears. By introducing forced mixing of the layers, it can thin out or even pierce the inversion layer. This will cause at least
a temporary and partial natural mixing of the air layers and a decrease in air
pollution. The method presented by Leszczynski et al. (2020) [14], uses a shock
wave generator which, by producing an impact, mixes the layers of air in the
atmosphere. The idea of using shock waves to mix and lift polluted air upwards
is schematically shown in Fig. 1.

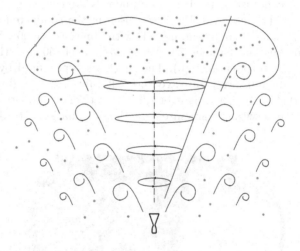

Fig. 1. The idea of using shock waves to mix and lift polluted air.

In this paper, we propose a numerical verification of the process of air pollution removal by artificially generated shock waves. We employ fast isogeometric analysis [17,18] simulations. Our solver is based on the Kronecker product
decomposition of the system of linear equations.

We employ the advection-diffusion-reaction model in a weak form. We first
discretize in space using finite elements and higher-order B-spline basis functions
employed by isogeometric analysis (IGA) [5]. Then, we employ the explicit time
marching scheme for discretization in time. This method requires factorization
of the mass matrix built from higher-order B-spline basis functions spans over
tensor product grid. The mass matrix from the explicit scheme can be factorized

in a linear computational cost resulting from the Kronecker product decomposition [7,8,16,17]. We factorize the matrix in the first step and then forward and backward substitute in the following steps. Using our solver, we first simulate the effect of thermal inversion on the formulation of clouds with the pollution particles. Next, we show how we can locally force mixing of the air layers using the artificially generated shock waves, resulting in a local decrease of the concentration of particles. We compare our numerical simulations to an experiment performed with a cannon in a highly polluted city of Kraków, Poland.

2   Experimental Verification

The method proposed by Leszczynski et al. (2020) [14], using a generation of high-pressure vertical shock waves, introduces the propagating disturbance of the air medium, which causes the air with particulate matter to be lifted out. Moreover, it generates air vortices in the adjacent volume, which causes the horizontal mixing of the air layers. The method is based on an intervention reduction of the suspended dust concentration when the standards are exceeded. The Inopower anti-hail cannon [11], shown in Fig. 2 was used in the conducted technological tests. The device consists of a container (1) with dimensions $6.00 \times 2.45 \times 2.60$ in which is the combustion chamber (4) with the volume of 150 dm^3, three fuel inlet ports (5), and a control cabinet (6). The shock wave produced by the ignition of an acetylene-air mixture is directed vertically upward through the widening conical outlet tube (2). There is a bundle of acetylene cylinders on the other side of the container with gas pressure reduction installation (3). After detecting

Fig. 2. Shock wave generator. 1 - Container, 2 - Outlet tube, 3 - Bundle of acetylene cylinders, 4 - Combustion chamber, 5 - Fuel inlet ports, 6 - Control cabinet.

unfavorable air pollution forecasts, intervention actions were taken to end the unfavorable stagnation of the atmosphere. During the tests, the acetylene gauge pressure supplied to the combustion chamber was 2.9 bar. During the explosion, the acetylene-air mixture reaches a pressure of about 1 MPa. The frequency of generating the shock wave was 0.17 Hz, and the total operating time was about 30 min. Temperature, pressure, humidity, and particulate matter concentration in the vertical profile were measured using the equipment placed on the drone DJI Matrice 200 V2. For each test, a flight was performed immediately before and after the generator was started and 20 min after its completion. Additionally, stationary Airly PM sensors measured dust concentrations at 0.1 km, 1.5 km, and 2 km from the generator. The preliminary results are shown in two papers: Jedrzejek et al. (2021) [12] and Jedrzejek et al. (2021) [13]. Figure 3 shows the measurement data of the altitude profile from 0 m to 130 m for the temperature and PM2.5 concentration for the situation before and after the generation of shock waves. The altitude profile of the temperature before the generation of the shock waves indicates a temperature inversion from 90 m to 130 m, where the temperature change was 0.38 $\frac{^\circ C}{100\,m}$. However, after using the cannon, the temperature change in a given height range decreased to about $-0.6\,\frac{^\circ C}{100\,m}$. Which is a typical humid-adiabatic gradient value for the atmosphere. Moreover, the monitored concentration of particulate matter PM2.5 decreased by up to 50% at specific points, e.g. at 120 m. The average concentration of PM2.5 from 0 to 120 m decreased from 85.1 $\frac{\mu g}{m^3}$ to 61.9 $\frac{\mu g}{m^3}$.

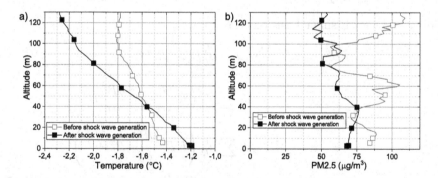

Fig. 3. Measurement data of the altitude profile: a) Temperature; b) PM2.5 concentration.

3 Numerical Simulations

We employ advection-diffusion-reaction equations to model the concentration of the water vapor forming a cloud, mixed with the pollution particles.

The equations in the strong form are

$$\frac{\partial u}{\partial t} + (b \cdot \nabla)u - \nabla \cdot (K\nabla u) + cu = f \text{ in } \Omega \times (0, T] \tag{1}$$

$$\nabla u \cdot n = 0 \text{ in } \Omega \times (0, T] \tag{2}$$

$$u = u_0 \text{ in } \Omega \times 0 \tag{3}$$

where u is the concentration scalar field, b is the assumed wind velocity vector field, $K = \begin{pmatrix} K_{11} & 0 \\ 0 & K_{22} \end{pmatrix}$ is the diffusion matrix, c is the reaction parameter, and f is the source term.

We employ the explicit method formulation

$$\frac{u^{t+1} - u^t}{dt} = \nabla \cdot (K\nabla u^t) - (b \cdot \nabla)u^t + cu^t = f^t \tag{4}$$

we derive the weak formulation, testing again with B-spline basis functions

$$\left(u^{t+1}, v\right) = \left(u^t, v\right) - dt \left(K\nabla u^t, \nabla v\right) - dt \left(b \cdot \nabla u^t, v\right) + \left(cu^t + f^t, v\right) \forall v \in V \tag{5}$$

We discretize with B-spline basis functions defined over the square domain $\Omega = [0, 1]^2$

$$u^{t+1} = \sum_{i=1,\dots,N_x; j=1,\dots,N_y} u_{ij}^{t+1} B_i^x B_j^y; \quad u^t = \sum_{i=1,\dots,N_x; j=1,\dots,N_y} u_{ij}^t B_i^x B_j^y \tag{6}$$

and we test with B-spline basis functions

$$\sum_{ij} u_{ij}^{t+1} \left(B_i^x B_j^y, B_k^x B_l^y\right) = \sum_{ij} u_{ij}^t \left(B_i^x B_j^y, B_k^x B_l^y\right)$$

$$- dt \sum_{ij} u_{ij}^t \left(K \frac{\partial B_i^x}{\partial x} B_j^y, \frac{\partial B_k^x}{\partial x} B_l^y\right)$$

$$- dt \sum_{ij} u_{ij}^t \left(K B_i^x \frac{\partial B_j^y}{\partial y}, B_k^x \frac{\partial B_l^y}{\partial y}\right)$$

$$- dt \sum_{ij} u_{ij}^t \left(b \frac{\partial B_i^x}{\partial x} B_j^y, B_k^x B_l^y\right) \tag{7}$$

$$+ dt \sum_{ij} u_{ij}^t \left(b B_i^x \frac{\partial B_j^y}{\partial y}, B_k^x B_l^y\right)$$

$$+ \sum_{ij} u_{ij}^t \left(c B_i^x B_j^y, B_k^x B_l^y\right) + \left(f^t, B_k^x B_l^y\right)$$

$$k = 1, \dots, N_x; l = 1, \dots, N_y$$

where $(u, v) = \int_\Omega uvx dy$.

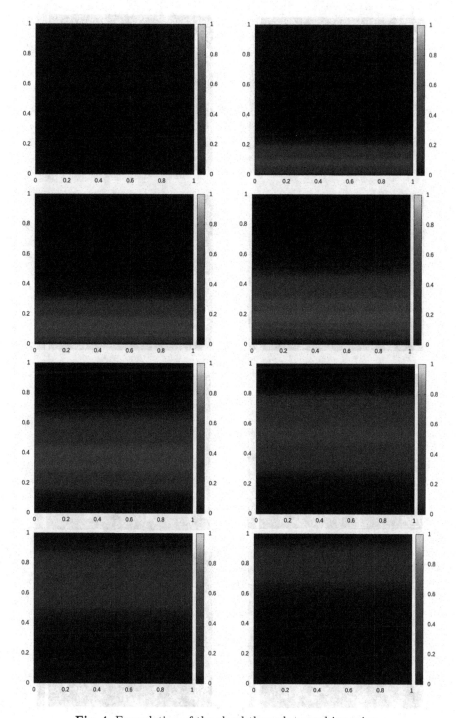

Fig. 4. Formulation of the cloud through termal inversion.

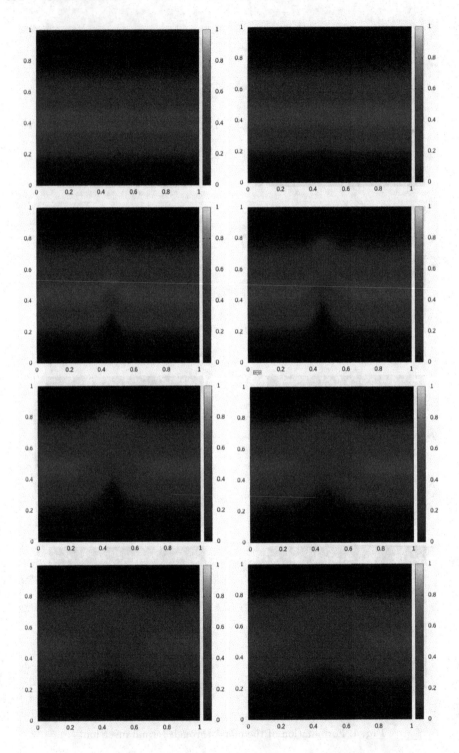

Fig. 5. Pollution reduction by generated shock waves.

We introduce

$$\mathbf{M}_x = \{(B_i^x, B_k^x)_x\}_{ik} = \{\int B_i^x B_k^x dx\}_{ik} \tag{8}$$

$$\mathbf{M}_y = \{(B_j^y, B_l^y)_y\}_{jl} = \{\int B_j^y B_l^y dy\}_{jl} \tag{9}$$

In general, Kronecker product matrix $\mathcal{M} = \mathcal{A}^x \otimes \mathcal{B}^y$ over 2D domain $\Omega = \Omega_x \times \Omega_y$ is defined as

$$\mathcal{M}_{ijkl} = \mathcal{A}_{ik}^x \mathcal{B}_{jl}^y. \tag{10}$$

Due to the fact, that one-dimensional matrices discretized with B-spline functions are banded and they have $2p + 1$ diagonals (where p stands for the order of B-splines), since

$$(\mathcal{M})^{-1} = (\mathcal{A}^x \otimes \mathcal{B}^y)^{-1} = (\mathcal{A}^x)^{-1} \otimes (\mathcal{B}^y)^{-1} \tag{11}$$

we can solve our system in a linear computational cost.

3.1 Fast Simulation of Thermal Inversion and Cloud Formation

We concentrate first on the fast simulation of cloud formation and thermal inversion. In our simulation, the scalar field u represents the water vapor forming a cloud, mixed with the pollution particles. Following [1] we introduce the linear temperature decrease from 20 C at ground level to -30 C at the height 3/4 of the domain. The thermal inversion effect is obtained by introducing the advection field as a temperature gradient.

The equations in the strong form are

$$\frac{\partial u(x, y; t)}{\partial t} + \frac{\partial T(y; t)}{\partial y} \frac{\partial u(x, y; t)}{\partial y}$$

$$-K_x \frac{\partial^2 u(x, y; t)}{\partial x^2} - K_y \frac{\partial^2 u(x, y; t)}{\partial y^2} = f(x, y; t)$$

$$(x, y; t) \text{ in } \Omega \times (0, T] \tag{12}$$

$$\nabla u(x, y; t) \cdot n(x, y) = 0, \quad (x, y; t) \text{ in } \Omega \times (0, T] \tag{13}$$

$$u(x, y; 0) = u_0 \text{ in } \Omega \times 0 \tag{14}$$

where u is the concentration scalar field, where the advection is driven by the temperature gradient in the vertical direction

$$\frac{\partial T(y; t)}{\partial y} = \{0 \text{ for } y > 30 \ -5 \text{ for } y <= 30 \tag{15}$$

$K_x = 1.0$ is the horizontal diffusion, $K_y = 0.1$ is the vertical diffusion, and the source term

$$f(x, y; t) = \{-25y \text{ for } y < 5, t < 5000 * dt \ 0 \text{ otherwise} \tag{16}$$

We employ the implementation of the linear computational cost Kronecker product structure solver as described in [18]. The explicit method formulation is implemented in the `compute_rhs` routine

```
void compute_rhs(int iter)
1 auto& rhs = u;
2 zero(rhs);
3 for(auto e:elements()) {
4   auto U = element_rhs();
5   double J = jacobian(e);
6   for (auto q : quad_points()) {
7     double w = weight(q);
8     for (auto a : dofs_on_element(e)) {
9       auto aa = dof_global_to_local(e, a);
10      value_type v = eval_basis(e, q, a);
11      value_type u = eval_fun(u_prev, e, q);
12      double gradient_prod = 1.*u.dx * v.dx + 0.1*u.dy * v.dy;
13      double val = u.val * v.val - steps.dt * gradient_prod+
                steps.dt*Tprim(e[1])*u.dy*v.val+
                steps.dt*f(e[1],iter)*v.val*50.;
14      U(aa[0], aa[1]) += val * w * J;
15    }
16  }
17  update_global_rhs(rhs, U, e);
18}
```

where the temperature gradient

```
void Tprim(int iter, int y)
1 if(y>30)
2   return 0.0;
3 return -5.;
```

The cloud formation is modeled by the right-hand side term, that is active in first 5000 steps

```
void f(int iter, int y)
1 if(y<5 && iter<5000 )
2   return (-y/5.);
3 return 0.;
```

The simulation results are summarized in Fig. 1. The water vapor formulated first at the ground level is driven by the temperature gradient.

3.2 Fast Simulation of the Pollution Removal by Artificially Generated Shock Waves

In order to simulate the atmospheric cannon, we modify the advection function in a sepparable way as

$$cannon(x,z) = const * (1-z) * sin(10 * \pi * x) * max(0, sin(\pi * t/10)) \quad (17)$$

for $t = s - 100$, where s is the time step size. In other words we run the cannon from time step 100, and we shoot for 10 time steps with a function $(1-z) * sin(10 * \pi * x)$ that runs in time like $max(0, sin(\pi * t/10))$.

```
  double cannon(double x, double y,int iter) {
1 x=x/40.; y=y/40.;
2 double t=iter;
3 if(x>0.3 && x<0.6 &&t>10000)
4    return 200.*(1.-y)*max(sin(10*PI*x),0)*
       max(0,sin(PI*(t-8000)/1000));
5 else
6    return 0.;
```

We add this cannon function to the formulation

void compute_rhs(int iter)

```
   ...
13      double val = u.val * v.val - steps.dt * gradient_prod+
           steps.dt*(delta_T(e[1])-
           cannon(e[0],e[1],iter))*u.dy*v.val+
           steps.dt*f(e[1])*v.val*50.;
   ...
```

The simulation results are summarized in Fig. 2. We can read from this simulation that shooting a cannon results in a local mixing of the layers and reduction of the water vapor mixed with the pollution particles. This local pollution and water vapor reduction maintain if the cannon creates shocking waves in a repeated period of time.

4 Comparison of Higher-Order and Continuity Isogeometric Finite Element Method with Finite Difference Method

Our simulator employs higher-order and continuity B-spline basis functions. This requires computing the integrals with higher-order B-spline basis functions on the right-hand side. We compute them element-wise:

$$\mathcal{I}_1 = \int_E B_i^x(x)B_j^y(y)B_k^x(x)B_l^y(y)dxdy$$

$$= \int_{[0,1]^2} B_i^x(x)B_j^y(y)B_k^x(x)B_l^y(y)J_E(x,y)dxdy$$

$$= \sum_q w_q B_i^x(x_q)B_j^y(y_q)B_k^x(x_q)B_l^y(y_q)J_E(x_q,y_q)$$

$$\mathcal{I}_2 = \int_E K(x,y)\frac{\partial B_i^x(x)}{\partial x}B_j^y(y)\frac{\partial B_k^x(x)}{\partial x}B_l^y(y)dxdy$$

$$= \int_{[0,1]^2} K(x,y)\frac{\partial B_i^x(x)}{\partial x}B_j^y(y)\frac{\partial B_k^x(x)}{\partial x}B_l^y(y)J_E(x,y)dxdy$$

$$= \sum_q w_q K(x_q,y_q)\frac{\partial B_i^x(x_q)}{\partial x}B_j^y(y_q)\frac{\partial B_k^x(x_q)}{\partial x}B_l^y(y_q)J_E(x_q,y_q)$$

$$\mathcal{I}_3 = \int_E K(x,y)B_i^x(x)\frac{\partial B_j^y(y)}{\partial y}B_k^x(x)\frac{\partial B_l^y(y)dxdy}{\partial y}$$

$$= \int_{[0,1]^2} K(x,y)B_i^x(x)\frac{\partial B_j^y(y)}{\partial y}B_k^x(x)\frac{\partial B_l^y(y)J_E(x,y)}{\partial y}dxdy$$

$$= \sum_q w_q K(x_q,y_q)B_i^x(x_q)\frac{\partial B_j^y(y_q)}{\partial y}B_k^x(x_q)\frac{\partial B_l^y(y_q)}{\partial y}J_E(x_q,y_q)$$

$$\mathcal{I}_4 = \int_E b_x(x,y)\frac{\partial B_i^x(x)}{\partial x}B_j^y(y)B_k^x(x)B_l^y(y)dxdy \tag{18}$$

$$= \int_{[0,1]^2} b_x(x,y)\frac{\partial B_i^x(x)}{\partial x}B_j^y(y)B_k^x(x)B_l^y(y)J_E(x,y)dxdy$$

$$= \sum_q w_q b_x(x_q,y_q)\frac{\partial B_i^x(x_q)}{\partial x}B_j^y(y_q)B_k^x(x_q)B_l^y(y_q)J_E(x_q,y_q)$$

$$\mathcal{I}_5 = \int_E b_y(x,y)B_i^x(x)\frac{\partial B_j^y(y)}{\partial y}B_k^x(x)B_l^y(y)dxdy\partial y$$

$$= \int_{[0,1]^2} b_y(x,y)B_i^x(x)\frac{\partial B_j^y(y)}{\partial y}B_k^x(x)B_l^y(y)J_E(x,y)dxdy$$

$$= \sum_q w_q b_y(x_q,y_q)B_i^x(x_q)\frac{\partial B_j^y(y_q)}{\partial y}B_k^x(x_q)B_l^y(y_q)J_E(x_q,y_q)$$

$$\mathcal{I}_6 = \int_E f(x,y)B_k^x(x)B_l^y(y)dxdy$$

$$= \int_{[0,1]^2} f(x,y)B_k^x(x)B_l^y(y)J_E(x,y)dxdy$$

$$= \sum_q w_q f(x_q,y_q)B_k^x(x_q)B_l^y(y_q)J_E(x_q,y_q)$$

We change the variables into the reference $[0,1]^2$ element and introduce the Jacobian of the element transformation $J_E(x,y)$. We employ $q = 1, ..., N_q$ quadrature points (x_q, y_q) and weights w_q to compute the integrals, where $N_q \approx \mathcal{O}(p^2)$. Finally, we collect the integrals into a global right-hand-side vector according to (7).

The mass matrix on the left-hand side is a Kronecker product $\mathcal{M} = \mathcal{M}_x \otimes \mathcal{M}_y$ of matrices with one-dimensional B-spline basis functions.

$$\{\mathcal{M}_x\}_{i,k} = \int_\Omega B_i^x(x) B_k^x(x) dx, \quad \{\mathcal{M}_y\}_{j,l} = \int_\Omega B_j^y(y) B_l^y(x) dy \qquad (19)$$

These matrices are $2p+1$ diagonal, because of the overlap of the one-dimensional B-splines. Thus, the factorization cost is $\mathcal{O}(Np^3)$.

The alternative finite-difference formulation approximates the derivates in the strong equation using the point-wise values

$$u_{i,j}^{t+1} = u_{i,j}^t - dt K_{i,j} \frac{u_{i-1,j}^t - 2u_{i,j}^t + u_{i+1,j}^t}{h} - dt K_{i,j} \frac{u_{i,j-1}^t - 2u_{i,j}^t + u_{i,j+1}^t}{h}$$
$$+ dt b_{i,j}^x \frac{u_{i,j}^t - u_{i-1,j}^t}{h} - + dt b_{i,j}^y \frac{u_{i,j}^t - u_{i,j-1}^t}{h} + dt f_{i,j}^t \qquad (20)$$

where h stands for the horizontal or vertical distance between grid points, $u_{i,j}^t$ stands for the concentrations at point (i,j), and $K_{i,j}, (b_{i,j}^x, b_{i,j}^y), f_{i,j}^t$ stand for the diffusion, advection, and forcing values estimated at the grid points. In the explicit method of the finite difference there are no matrices to factorize, there is only an $\mathcal{O}(N)$ update operation.

The advantage of the isogeometric analysis method is that we obtain a global smooth solution that is globally differentiable. The finite difference method gives us pointwise values.

The disadvantage of the isogeometric analysis is that we have to compute the integrals with higher order polynomials, which is $\mathcal{O}(p^2)$ more expensive than generating and computing pointwise values. We also have to factorize the mass matrix, which is $\mathcal{O}(p^3)$ more expensive than an update operation in the explicit method in finite differences.

5   Conclusion

The article presents a numerical verification of the experimental method of intervention reduction of particulate matter using shock waves generated by the combustion of the acetylene air mixture. The shock wave cycle restores the natural movements of air masses by temporarily acting on the inversion layer. We employed a fast isogeometric solver to model the thermal inversion, cloud formation, and pollution removal by artificially created shock waves. The numerical experiments confirm the experimental effect of mixing of the air layers and local decrease of the pollution and water vapor in the neighborhood of the created shock waves. Future work will include generalizing the developed simulational software to three dimensions. We also plan parallelization into shared memory Linux cluster nodes [4,10]. Additionally, we plan to employ inverse algorithms [2,3,6,21] to solve several related inverse problems. They are related to the following open research questions: what is the optimal location of the cannon in a given area, what is the optimal frequency and power of the cannon, and what is the best moment during the cloud and pollution formation process to start the process. A detailed comparison of the numerical accuracy, convergence, and execution times of the isogeometric analysis and alternative methods can be a subject of a future study. The future work may also include incorporation of mesh adaptation techniques [20].

Acknowledgement. The paper was partially financed by AGH University of Science and Technology Statutory Fund.

References

1. U.s. standard atmosphere vs. altitude. (2003). https://www.engineeringtoolbox. com/standard-atmosphere-d_604.html. Accessed 21 Dec 2021
2. Barabasz, B., Gajda-Zagórska, E., Migórski, S., Paszyński, M., Schaefer, R., Smołka, M.: A hybrid algorithm for solving inverse problems in elasticity. Appl. Math. Comput. Sci. **24**(4), 865–886 (2014)
3. Barabasz, B., Migórski, S., Schaefer, R., Paszyński, M.: Multi-deme, twin adaptive strategy hp-HGS. Inverse Prob. Sci. Eng. **19**(1), 3–16 (2011)
4. Calo, V., Collier, N., Pardo, D., Paszyński, M.: Computational complexity and memory usage for multi-frontal direct solvers used in p finite element analysis. Procedia Comput. Sci. **4**, 1854–1861 (2011)
5. Cottrell, J.A., Hughes, T.J.R., Bazilevs, Y.: Isogeometric Analysis: Toward Integration of CAD and FEA. John Wiley & Sons (2009)
6. Gajda-Zagórska, E., Schaefer, R., Smołka, M., Paszyński, M., Pardo, D.: A hybrid method for inversion of 3D dc resistivity logging measurements. Natl. Comput. **14**(3), 355–374 (2015)
7. Gao, L., Calo, V.M.: Fast isogeometric solvers for explicit dynamics. Comput. Meth. Appl. Mech. Eng. **274**, 19–41 (2014)
8. Gao, L., Calo, V.M.: Preconditioners based on the alternating-direction-implicit algorithm for the 2D steady-state diffusion equation with orthotropic heterogeneous coefficients. J. Comput. Appl. Math. **273**, 274–295 (2015)
9. Giemsa, E., Soentgen, J., Kusch, T., Beck, C., Munkel, C., Cyrys, J., Pitz, M.: Influence of local sources and meteorological parameters on the spatial and temporal distribution of ultrafine particles in Augsburg, Germany. Front. Environ. Sci **8** (2021)
10. Goik, D., Jopek, K., Paszyński, M., Lenharth, A., Nguyen, D., Pingali, K.: Graph grammar based multi-thread multi-frontal direct solver with Galois scheduler. Procedia Comput. Sci. **29**(29), 960–969 (2014)
11. Inopower: Hail cannon user's manual (2009)
12. Jedrzejek, F., et al.: The innovative method of purifying polluted air in the region of an inversion layer. Front. Environ. Sci. **600** (2021)
13. Jedrzejek, F., et al.: An innovative method of reducing particulate matter concentration levels in the atmosphere. Laboratorium - Przeglåd Ogólnopolski nr 1 (2021). (in Polish)

14. Leszczyński, J., et al.: The method of reducing dust accumulation in the smog layer, which is the inversion layer. Munich, Germany, European Patent Office EP20217680 (2020)

15. Li, W.W., et al.: Analysis of temporal and spatial dichotomous pm air samples in the el paso-cd. juarez air quality basin. J. Air Waste Manag. Assoc. **51**, 1551–1560 (2001)

16. Łoś, M., Paszyński, M., Kłusek, A., Dzwinel, W.: Application of fast isogeometric l2 projection solver for tumor growth simulations. Comput. Meth. Appl. Mech. Eng. **316**, 1257–1269 (2017)

17. Łoś, M., Woźniak, M., Paszyński, M., Dalcin, L., Calo, V.M.: Dynamics with matrices possessing Kronecker product structure. Procedia Comput. Sci. **51**, 286–295 (2015)

18. Łoś, M., Woźniak, M., Paszyński, M., Lenharth, A., Hassaan, M.A., Pingali, K.: IGA-ads: isogeometric analysis fem using ads solver. Computer. Phys. Commun. **217**, 99–116 (2017)

19. Observatory, W.: Mortality from Household Air Pollution. World Health Organization, Luxembourg (2018)

20. Paszyńska, A., Paszyński, M., Grabska, E.: Graph transformations for modeling *hp*-adaptive finite element method with mixed triangular and rectangular elements. In: Allen, G., Nabrzyski, J., Seidel, E., van Albada, G.D., Dongarra, J., Sloot, P.M.A. (eds.) ICCS 2009. LNCS, vol. 5545, pp. 875–884. Springer, Heidelberg (2009). https://doi.org/10.1007/978-3-642-01973-9_97

21. Paszyński, M., Barabasz, B., Schaefer, R.: Efficient adaptive strategy for solving inverse problems. Lect. Comput. Sci. **4487**, 342–354 (2007)

22. Sinha, J., Kumar, N.: Mortality and air pollution effects of air quality interventions in Delhi and Beijing. Front. Environ. Sci. **7**, 15 (2019)

Automatic Generation of Individual Fuzzy Cognitive Maps from Longitudinal Data

Maciej K. Wozniak[1] , Samvel Mkhitaryan[2] ,
and Philippe J. Giabbanelli[1(✉)]

[1] Department of Computer Science and Software Engineering, Miami University,
Oxford, OH 45056, USA
{wozniamk,giabbapj}@miamioh.edu
[2] Department of Health Promotion, CAPHRI, Maastricht University,
Maastricht, The Netherlands
s.mkhitaryan@maastrichtuniversity.nl

Abstract. Fuzzy Cognitive Maps (FCMs) are computational models that represent how factors (nodes) change over discrete steps based on causal impacts (weighted directed edges) from other factors. This approach has traditionally been used as an aggregate, similarly to System Dynamics, to depict the functioning of a system. There has been a growing interest in taking this aggregate approach at the individual-level, for example by equipping each agent of an Agent-Based Model with its own FCM to express its behavior. Although frameworks and studies have already taken this approach, an ongoing limitation has been the difficulty of creating as many FCMs as there are individuals. Indeed, current studies have been able to create agents whose traits are different, but whose decision-making modules are often identical, thus limiting the behavioral heterogeneity of the simulated population. In this paper, we address this limitation by using Genetic Algorithms to create one FCM for each agent, thus providing the means to automatically create a virtual population with heterogeneous behaviors. Our algorithm builds on prior work from Stach and colleagues by introducing additional constraints into the process and applying it over longitudinal, individual-level data. A case study from a real-world intervention on nutrition confirms that our approach can generate heterogeneous agents that closely follow the trajectories of their real-world human counterparts. Future works include technical improvements such as lowering the computational time of the approach, or case studies in computational intelligence that use our virtual populations to test new behavior change interventions.

Keywords: Genetic algorithms · Fuzzy cognitive maps · Population generation · Simulations

---

This project is supported by the Department of Computer Science & Software Engineering at Miami University.

1   Introduction

A *Fuzzy Cognitive Map* (FCM) is an aggregate computational model consisting of factors (weighted labelled nodes), which interact via causal links (directed weighted edges). This approach has been used across a broad range of domains [5], ranging from medical applications [1] to socio-environmental systems [18] and engineering [2]. Many of these applications are motivated by the need to make decisions in complex systems characterized by high uncertainty, feedback loops, and limited access to the detailed temporal datasets (e.g., delays, rates per unit of time) that would support alternative approaches such as System Dynamics (SD). Historically, a computational study would typically result in *one* FCM that depicts the functioning of a system, or *two* FCMs that serve to compare perspectives (e.g., Western views vs. Aboriginal views) about a same social system [12]. For example, a study on smart cities can yield one FCM that seeks to summarize all potential factors and every relationship [6]. A specific city would be portrayed as an instance of this overall map, based on specific node values. Similarly, a study on obesity provides a single FCM explaining how weight gain generally works (e.g., stress leads to overeating), and allowing for individuals to be represented through node values (e.g., level of stress) [11].

A growing interest in generating a multitude of FCMs to represent individuals in an Agent-Based Model (ABM) has resulted in questioning this 'one-size-fits-all' approach of creating a single map and spawning individuals only by varying node values [3,7]. Intuitively, individuals do *not only* differ on node values (e.g., some are more stressed than others) but *also* on causal links (e.g., some individuals will over-eat when stressed and some will not). In other words, agents can have different traits and also follow different rules. This was empirically confirmed by studies that created FCMs from individual participants, noting that the existence and strength of the causal edges differed widely [13,15]. Despite the realization that these virtual populations may be over-simplifying human behaviors, there has been a lack of tools to automatically generate a large and heterogeneous population. On the one hand, we can ignore the differences in causal links and focus on quickly creating a large number of agents by drawing their node values from calibrated probability density functions, while respecting relevant correlations between distributions [9]. On the other hand, an abundance of Machine Learning algorithms can automatically set the weight of causal links based on a dataset, but their focus is to create a single FCM rather than a population [14,19]. There is thus an ongoing need to address the complexity of human behaviors by accounting for heterogeneous traits and a diversity of rules [16].

Our paper contributes to addressing this need through the automatic creation of different FCMs, thus generating a population of agents that can behave differently. This overarching contribution is realized through two specific goals:

– We *propose a new method*, based on Genetic Algorithms (GAs), to generate a virtual population based on the answers of real participants to commonly administered repeated surveys. Although GAs have previously been used to create FCMs from historical multivariate time series [22], we emphasize

that our paper proposes the first method to create different FCMs through individual-level data, rather than a single FCM from aggregate data.
- We demonstrate that our method leads to a more diverse population than the 'one-size-fits-all' approach by applying it to a *real-world scenario in nutrition*. This application relies on our open-source library [17] and all scripts used in this case study are publicly accessible at https://osf.io/z543j/.

The remainder of this paper is structured as follows. In Sect. 2, we succinctly cover Fuzzy Cognitive Maps and how they can be constructed via Genetic Algorithms, while referring the reader to [22,26] for detailed technical introductions. In Sect. 3.1, we present our proposed approach, named 'one-for-each'. We describe it through a formal pseudocode, which is explained line-per-line; its implementation as a Jupyter Notebook is available at https://osf.io/z543j/. Similarly, we describe an approach that represents the current 'one-size-fits-all' paradigm in Sect. 3.2. The two approaches are experimentally compared on a case study in Sect. 4. Results are discussed in Sect. 5 together with the limitations of our work and opportunities for follow-up studies.

2   Background

2.1   Fuzzy Cognitive Maps

A Fuzzy Cognitive Map (FCM) has two components [5,7]. First, *structurally*, it is a directed, labeled, weighted graph. Each node has a value from 0 (absence) to 1 (full presence); for example, 0 on stress means no stress whatsoever, whereas 1 signifies as much stress as physically possible. Each edge has a weight from -1 to 1 representing causality. For instance, a link $A \rightarrow B$ with a positive weight (e.g., stress $\xrightarrow{0.5}$ depression) states that A causally increases B. Mathematically, the weights of the edges are stored in an adjacency matrix W and the value of the n nodes are stored in a vector whose initially value (i.e., at $t = 0$) is A.

Second, the nodes' values are *updated* over iterations by Algorithm 1. Nodes are updated *synchronously* based on (i) their current value, (ii) the value of their incoming neighbors, (iii) the weight of the incident edges, and (iv) a clipping or activation function f such as a sigmoid to keep the result within the desired target range of 0 to 1. The update is applied iteratively (Algorithm 1, lines 3–4) until one of two stopping conditions is met. The desirable condition is that stabilization has occurred, as target nodes of interest (i.e., outputs of the system) are changing between two consecutive steps by less than a user-defined threshold (Algorithm 1, lines 5–6). However, depending on conditions such as the choice of f, the system may oscillate or enter a chaotic attractor. The second condition thus sets a hard limit on the maximum number of iterations (Algorithm 1, line 2.)

2.2   Genetic Algorithms for Fuzzy Cognitive Maps

Historically, FCMs were mostly used as expert systems. A panel of subject-matter experts would be conveyed to identify the concept nodes and/or evaluate

---

Algorithm 1: SIMULATEFCM

Input: input vector A of n concepts, adjacency matrix W, clipping function f,
max. iterations t, output set O, threshold τ

1 $output^0 \leftarrow A$
2 **for** $step \in \{1, \ldots, t\}$ **do**
3 $\quad$ **for** $i \in \{1, \ldots, n\}$ **do**
4 $\quad\quad$ $output_i^{step} \leftarrow f(output_i^{step-1} + \sum_{j=1}^{n} output_j^{step-1} \times W_{j,i})$
5 $\quad$ **if** $\forall o \in O, |output_o^{step} - output_o^{step-1}| < \tau$ **then**
6 $\quad\quad$ return $output$
7 return $output$

---

their causal strength qualitatively (e.g., 'low', 'medium'). These qualifiers would then be transformed into numerical weights using fuzzy logic [21]. To improve the accuracy of an FCM, learning algorithms have emerged as an approach that uses data to either fine-tune some of the experts' weights (e.g., nonlinear Hebbian learning) or set all the weights. The latter was pioneered in 2005 when Stach and colleagues used Genetic Algorithms [26]. Their method, known as real-coded genetic algorithm (RCGA) "does not require human intervention and provides high quality models that are generated from historical data" [25]. Variations of this method have been proposed, such as the use of a density parameter to force the algorithm in generating a sparse FCM that more closely resembles the map created by experts [27]. Such sparsity mechanism and the use of evolutionary algorithms remain an active area of research to learn FCMs from data [28].

Figure 1 shows the main steps of a GA (left) and its application to FCMs in particular via the RCGA (right). The algorithm first creates 100 random matrices. *Fitness* is computed for each weight matrix to quantify the proximity between the *final results* generated through that matrix (i.e., the final node values) and the desired outputs D (Fig. 2). That is, the RCGA algorithm seeks results that are similar to the goal upon termination; intermediate results are inconsequential. *Crossover* is then applied, with a probability of 0.9, via a two steps process: (i) start with the first weight matrix and randomly choose the index of an edge (i.e., a cell in the matrix); (ii) from this cell onward, all weights until the end are swapped with the corresponding cells in the next matrix. The process repeats until all matrices have been visited. For each weight matrix, we then apply *mutations* with a probability of 0.5, which consists of randomly changing the value of selected edges into the interval $[-1, 1]$. One of two selection mechanisms is randomly chosen for each iteration: either we randomly choose the same number of edges, or we select fewer edges as iterations go. The fitness function is computed again for the mutated matrices. If any of their fitness values is sufficient (0.99), then the matrix with the best fitness is returned. Otherwise, the algorithm continues by selecting weight matrices for the next generation with a probability proportional to their fitness. Given this process, the same matrix can be repeatedly chosen, if its fitness is higher than for other matrices. The algorithm stops after at most 100000 generations, then returning the matrix with highest fitness. Several of these steps are reused (crossover, mutation, stopping conditions) in our proposed solution, hence they are formalized in Algorithm 2.

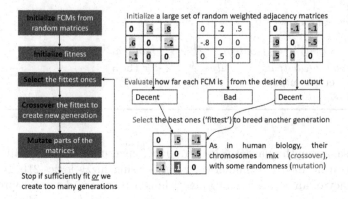

Fig. 1. Application of a Genetic Algorithm (GA) to Fuzzy Cognitive Maps.

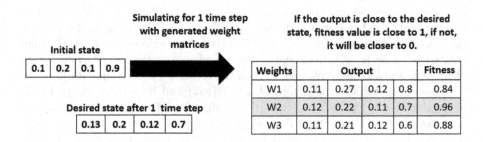

Fig. 2. Example to calculate fitness.

3   Methods

3.1   Proposed Algorithm to Generate an Individual FCM

Our 'one-for-each' approach is summarized in Algorithm 2. Parameters are listed in Table 1, while noting that they mostly originate from the use of Genetic Algorithms [26] (Sect. 2.2); practical choices for parameter values are illustrated via our case study in Sect. 4. The main steps of Algorithm 2 proceed as follows. The user starts by providing the same parameters (line 1) as in the RCGA (Sect. 2.2). Then, the initial and desired values of concepts are fed into the algorithm as $t \times n$ 2-D vectors, where t is a number of available time steps (three in our case study) and n is the number of concepts (lines 2–4). Afterwards, we calculate fitness, with one major difference from the RCGA. Instead of computing fitness only on the final step of an FCM, we compute it at each iteration (lines 5–9).

This is essential to ensure that the *trajectory* of an individual is plausible, thus setting more constraints than only replicating the final endpoint. This point is emphasized in Sect. 4. We check for the two termination conditions (lines 10–14) and, if they are not satisfied, we continue by applying crossover (lines 16–20) and mutations (lines 24–31) before calculating the fitness again (lines 24–31). Weight matrices are selected for the next generation (lines 32–35) and we repeat the process (line 16) until the termination conditions are met. If the fitness of a matrix never reaches the desired *threshold* within *max_generations*, we end by returning the matrix with the highest fitness.

3.2 Algorithms to Generate a Population from Longitudinal Data

Algorithm 2 creates an FCM that best resembles the *trajectory* of a *single* individual. In order to create a *population*, we thus need to apply that method based on the longitudinal data from *every* individual. This is achieved in Algorithm 3, where we provide the baseline data for each participant ($participant^0$) and up to T consecutive measurements. For example, if participants are seen at 3 months and 6 months, we have $T = 2$ as there are two additional measurements.

Table 1. Parameters of Algorithm 2

Variable	Notation	Default value	Meaning
Concepts	A	Vector with values between 0 and 1	Values of concepts (nodes) of FCM
Weights	W	2-D Array with values between −1 and 1	Weights of the edges connecting concepts in FCM model
Number of concepts	n	Integer	Number of concepts (nodes) in FCM model
Maximum number of generations	max_gen	100000	Number of generation (iterations) after which RCGA will stop, even if the desired fitness is not reached
Maximum fitness	$threshold$	0.99	Minimum fitness value of at least one weight matrix in the generation for the algorithm to stop
Generation size	gen_size	100	Number of samples (weight matrices) in each generation
Probability of crossover	$p\_crossover$	0.9	Probability of a pair of samples being chosen for crossover
Probability of mutation	$p\_mutation$	0.5	Probability of a sample being chosen for mutation
Number of weights chosen for mutation	$n\_mutation$	2	Number of weights chosen for mutation (the same in each time)
Probability distribution for selection	K	Vector with values between 0 and 1	Probability distribution created according to fitness values of each sample in the generation
Uniform distribution	U	n/a	Uniform distribution of mean 0

318 M. K. Wozniak et al.

Algorithm 2: MAKEINDIVIDUALFCM

---

1 **Input:** $max\_generations, gen\_size, threshold$, iterations t, number of concepts n,
    desired values of concepts D
2 $\mathbb{W} \leftarrow \{W^1_{n \times n}, \ldots, W^{gen\_size}_{n \times n}\}$ // Each matrix is initialized with random weight
    between -1 and 1
3 $\mathbb{A} \leftarrow \{[0]^1_{t \times n}, \ldots, [0]^{gen\_size}_{t \times n}\}$
4 $fitness \leftarrow [0]_{gen\_size}$
5 **for** $i \in \{1, \ldots, gen\_size\}$ **do**
6     $A^i \leftarrow simulateFCM(input\_concepts, W^i, t)$ // Simulate using each matrix
7 **for** $i \in \{1, \ldots, gen\_size\}$ **do**
8     $error \leftarrow \sum_{s=1}^{t} \sum_{m=1}^{n} |A^i_{s,m} - D_{s,m}|$ // Calculating fitness for each W
9     $fitness^i \leftarrow \frac{1}{100 \times error + 1}$
10 **if** $max(fitness_{i \in gen\_size}) \geq threshold$ **then**
11     $i \leftarrow 0$ // Checking for termination condition
12   **while** $fitness^i < threshold$ **do**
13        $i \leftarrow i + 1$
14   **return** $(W^i, fitness^i)$
15 **for** $step \in \{1, \ldots, max\_generations\}$ **do**
16   **for** $i \in [1, 3, 5, \ldots, gen\_size]$ **do**
17   **if** $p\_crossover$ **then**
18           $idr \leftarrow U(1, n)$ // Swapping edges between 2 weight matrices
19           $idc \leftarrow U(1, n)$
20           $W^i_{\substack{idr \leq i \leq n \\ idc \leq j \leq n}} \leftrightarrow W^{i+1}_{\substack{idr \leq i \leq n \\ idc \leq j \leq n}}$
21   **if** $p\_mutation$ **then**
22   **for** $i \in \{1, \ldots, n\_mutations\}$ **do**
23              $W^i_{U(1,n),U(1,n)} \leftarrow U(\{-1, -0.99, \ldots, 1\})$
24   **for** $i \in \{1, \ldots, gen\_size\}$ **do**
25        $error \leftarrow \sum_{s=1}^{t} \sum_{m=1}^{n} |A^i_{s,m} - D_{s,m}|$
26        $fitness^i \leftarrow \frac{1}{100 \times error + 1}$
27   **if** $max(fitness_{i \in gen\_size}) \geq threshold$ **then**
28        $i \leftarrow 0$
          // Checking for termination condition
29   **while** $fitness^i < threshold$ **do**
30           $i \leftarrow +1$
31   **return** $(W^i, fitness^i)$
32     $\mathbb{W}\_buffer \leftarrow \mathbb{W}$
33   **for** $i \in \{1, \ldots, gen\_size\}$ **do**
34        $idx \leftarrow K(1, gen\_size)$ // Probability distribution according to
          $\frac{fitness}{\sum fitness}$
35        $W\_buffer^i \leftarrow W^{idx}$
36     $\mathbb{W} \leftarrow \mathbb{W}\_buffer$
37 $j \leftarrow 0$ // If the $threshold$ value of fitness was not achieved before
    $max\_generation$ iterations, return W with the highest fitness value
38 $max\_value \leftarrow 0$
39 **for** $i \in \{1, \ldots, gen\_size\}$ **do**
40   **if** $fitness^i > max\_value$ **then**
41        $j \leftarrow i$
42        $max\_value \leftarrow fitness^i$
43 **return** $(W^j, fitness^j)$

Although our motivation is to create a virtual population that replicates the heterogeneity of behaviors found in real-world individuals, there are several reasons for which a simplification might be used by taking an 'average' individual profile and scaling it to an entire virtual population. For example, generic or 'one-size-fits-all' interventions are designed around the notion of an average profile [4]. In addition, health data sharing may employ privacy-preserving aggregation algorithms, or only national statistics may be available [10]; in either cases, virtual agents may have to be created based on an average archetype. Algorithm 4 captures this situation and serves to empirically establish the difference with our approach; that is, having a 'one-for-each' algorithm alongside a 'one-size-fits-all' version will let us examine the value-proposition of our approach through a case study. If the difference between the two algorithms is negligible within the application context, then the computational expense of Algorithm 3 would not be warranted. Otherwise, if the difference is noteworthy, then our tool will be able to address a practical need.

Algorithm 3: ONE-FOR-EACH

Input: $longitudinalData$, number of participants p, number of concepts n, follow-up measurements T

1 $Wouts \leftarrow [0]_{p \times n \times n}$
2 **for** $participant \in longitudinalData$ **do**
3       $maxfit \leftarrow 0$
4   **for** $i \in \{1, \ldots, 100\}$ **do**
5           $W, fitness \leftarrow rcga(1000000, 100, 0.99, participant^{0}, participant^{1,\ldots,T}, T)$
6   **if** $fitness > maxfit$ **then**
7               $Wouts^{i} \leftarrow W$
8 **return** $Wouts$

Algorithm 4: ONE-FITS-ALL

Input: $longitudinalData$, number of participants p, number of concepts n, follow-up measurements T

1 $meanValues \leftarrow \frac{\sum(longitudinalData, axis \leftarrow 1)}{p}$
2 $maxfit \leftarrow 0$
3 $Wout \leftarrow \emptyset$
4 **for** $i \in \{1, \ldots, 100\}$ **do**
5       $W, fitness \leftarrow rcga(1000000, 100, 0.99, meanValues^{T0}, meanValues^{T1,T2}, t)$
6   **if** $fitness > maxfit$ **then**
7           $Wout \leftarrow W$
8 **return** $Wout$

4   Case Study

4.1   Overview

Our case study is an intervention on promoting healthy eating among adults in the Netherlands, with a focus on fruit intake [24]. Sixteen concepts were identified as salient in this application context [24], thus we have $n = 16$ concepts in our FCMs, with one of them ('daily fruit intake') serving as outcome. Each concept is listed in Table 2, with a unique identifier (left column) used for brevity when plotting results. A total of 1149 individuals took part in the study, out of whom 722 were surveyed two more times (at one month and at four months) beyond the baseline measurement and provided complete data. Consequently, our goal is to generate 722 FCMs that follow the trajectory of each individual, from the baseline to the two ensuing measurements.

In Sect. 4.2, we empirically examine whether our proposed solution (Algorithm 2) is more suited to generate one individual than the previous works (RCGA). Then, in Sect. 4.3, we assess our ability at generating a heterogeneous population via Algorithm 3, while using Algorithm 4 for comparison.

Table 2. FCM concepts in our case study were chosen by experts in psychology and nutrition to cover relevant domains and subdomains. The unique ID (#, left column) is used to refer to each concept in Figs. 3–4.

#	Domain	Sub-domain	Question
0	Perceived Intake/Awareness		How many fruits do you think you eat?
1	Attitude	Health	I think eating two pieces of fruit per day is very... (Unhealthy–Healthy)/(Cheap–Expensive)
2		Cost	
3	Self-efficacy	Ability	Do you think you can eat more fruit per day in the next six months if you really want to?
4		Difficulty	How difficult or easy do you think it is to eat more fruit in the next six months?
5	Social norms	Normative	Most people who are important to me... (think I should eat two pieces of fruit per day) (consume two pieces of fruit per day)
6		Modeling	
7	Intention		Do you intend to eat two pieces of fruit per day?
8	Action Planning	When	I have a clear plan for... (when I am going to eat more fruit)/(which fruit I am going to eat more or less)/(how many fruit I am going to eat)
9		Which	
10		How many	
11	Coping Planning	Interferences	I have a clear plan for what I am going to do... (when something interferes with my plans to eat more fruit)/(in situations in which it is difficult to eat more fruit)
12		Difficulty	
13	Environment	Amount	How often do you have fruit products available at home?
14		Location	Where do you store the fruit products at home?
15	Daily fruit intake (pieces)		Pieces of fruit per day, including citrus fruit, other fruit and fruit juice

4.2 Comparing Algorithm 2 with the Original RCGA Procedure

One essential difference in our proposed Algorithm 2 compared to the original RCGA procedure is to further constraint the notion of fitness by requiring a fit at *each iteration* of the FCM (to follow each participant) rather than only at the end. This raises several questions. First, is there a difference between the FCMs produced? Second, how closely do they follow the ground truth data? These questions are answered by Fig. 3. We see that there is indeed a difference between the FCMs (original in green; ours in orange) across concepts. Most interestingly, we note that the difference is not limited to the intermediate steps, but also carries onto the final results. Our proposed algorithm arrives at an accurate endpoint in the last iteration *and* remains within 0.01 (on a scale of 0 to 1) of the ground truth value throughout the individual's trajectory.

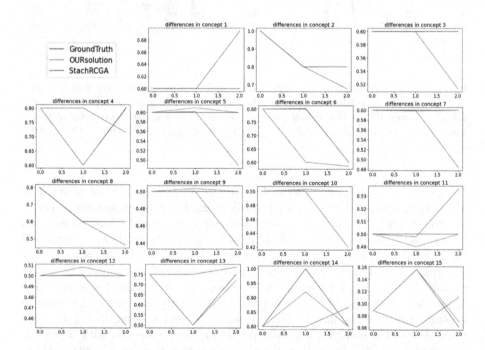

Fig. 3. Comparing ground truth data (blue), the original RCGA (green), and our proposed solution (orange) on all concepts of an individual. The two iterations are shown on the x-axis and the concept values on the y-axis. Concept #0 is not included as it is constant. When the blue line cannot be seen, it overlaps with the orange line (i.e., our solution exactly matches the ground truth). (Color figure online)

In contrast, the original RCGA procedure is always further from the endpoint *in addition to* being occasionally very far from intermediate steps. For example, on concept #15 (which is the output of the FCM hence a key concept), the RCGA starts by going down instead of going up, and then goes down instead of up, ending above the desired state as a series of erroneous directions. Wide fluctuations are similarly observed on most other concepts.

4.3   One-for-Each vs. One-Size-Fits-All

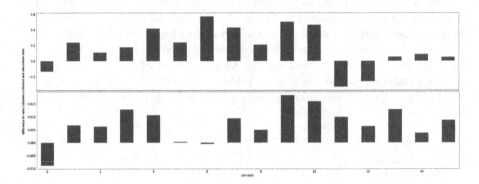

Fig. 4. Mean difference between ground truth and data generated using weight matrices obtained by Algorithm 4 ('one-size-fits-all'; top) and Algorithm 3 ('one-for-each'), calculated over 100 randomly selected individuals. Note the *different scales of the y-axis*; top errors range from −0.4 to 0.6 while bottom errors are within −0.01 to 0.02.

By applying Algorithm 3, we used the responses of each participant to generate a corresponding FCM, that is, a personalized model that describes their behavior. For analysis, 100 individuals were randomly selected and simulated over two time steps, in line with the empirical data. To assess the quality of this virtual population, we applied Algorithm 4 for comparison. The key difference is that Algorithm 4 is given the *average* participant responses, thus assuming that individuals are distributed around an average behavior. Algorithm 4 was called 100 times, thus leveraging the randomness of the underlying genetic algorithm to generate a diverse population. The two solutions are compared in Fig. 4. When using Algorithm 3, the error is within 0.01 for most concepts, with a peak at approximately 0.015 (on action planning). In contrast, using Algorithm 4, the error is an order of magnitude higher, with many concepts being erroneous by a margin of 0.2–0.4 and even a peak at almost 0.6 on social norms. In short, our approach to simulating a heterogeneous population is confirmed, as it yields accurate models. The computational costs of generating individuals are also justified, given that the computationally cheaper approach of generating an 'average' model ends up being wrong for most of the population *in our application context*.

Note that Algorithm 4 *may* be useful in settings where the population is very homogeneous and *not* exposed to an intervention. Indeed, if there are even small differences in baseline *and* behavior change taking place, then individuals may vastly different trajectories [23]. This notion of 'quantum leap for health promotion' is confirmed in our sample. Using the approach by D'Agostino and Pearson's [20], we tested the null hypothesis that our participants follow a normal distribution on each concept, at each iteration. At baseline ($t = 0$), we found that all concepts but two (#1, #15) were normally distributed. However, as participants went through the intervention, *none* of the concepts were normally distributed at later measurements ($t = 1$, $t = 2$), which justifies the creation of individual models that fit each person's trajectory instead of assuming that an initially normal distribution will remain unchanged.

5   Discussion

Fuzzy Cognitive Maps (FCMs) have long been used in fields such as environmental management or medicine [1,18], where a single map serves to define how a system functions [6]. The potential of using this aggregate-level modeling technique to capture how *individuals* function has been researched more recently [3,7]. One key challenge lies in generating enough FCMs to capture the heterogeneity of a population. As FCMs have historically served to synthesize the perspectives of experts, it would be possible to take the viewpoint of each expert and create an FCM accordingly. However, there would only be as many 'behavioral archetypes' as there are experts, which still drastically limits the heterogeneity of a population and may not fully match how individuals truly operate. Although Machine Learning (ML) has increasingly been used with simulations [8] and algorithms exist to train *one* FCM from data [14,26], there was no solution to create a large number of FCMs. Our manuscript has sought to address this gap by modifying a Genetic Algorithm [26,27] and applying it on data collected from participants over time. This type of individual-level longitudinal data is commonplace in the study of human behaviors, thus our proposed solution has broad applicability to create virtual populations with precisely calibrated behaviors. Our experimental evaluation on a case study from nutrition confirms that our algorithm can closely follow the trajectories of individuals and even outperforms previous solutions on matching their final state. The case study also illustrated the value of creating a population based on individuals instead of assuming an 'average profile', which may hold at baseline but quickly ceases to apply as participants experience a behavior change intervention.

One limitation of this approach is its computational cost. For each individual, 100 matrices are evaluated for their fitness, then selected and mutated over up to 100,000 steps. The process is repeated several times to ensure the best fitness. Even at the scale of a small population of 257 individuals (in our case study), a High Performance Computing Cluster was necessary to train the FCMs. An important follow-up study would thus consist of identifying the parameters to which the algorithm is most sensitive (Table 1) and reduce the other ones to lower the total number of combinations computed. In addition to this classic approach to simulation-based optimization, additional constraints can be placed on the algorithm to more quickly guide the generation of relevant FCMs, such as the use of sparsity to discard candidate matrices [28]. From an application standpoint, additional evaluations from other fields of human behavior would complement our case study. In particular, experiments with different trajectories can be an interesting application to assess the ability of our algorithm at closely following variations.

Ultimately, the main purpose in building a simulation is to *use it*. As we now have a procedure to generate a virtual population with a code-base that is open to the research community, we look forward to case studies in which these populations are used to test new interventions, thus continuing to use simulations to develop insight in human behaviors and find effective policies.

Acknowledgements. We thank Dr Jens Mueller for his assistance with the RedHawk high-performance computing cluster at Miami University. We also benefited from the feedback of Drs Rik Crutzen, Nanne K. de Vries, and Anke Oenema.

References

1. Amirkhani, A., Papageorgiou, E.I., Mohseni, A., Mosavi, M.R.: A review of fuzzy cognitive maps in medicine: taxonomy, methods, and applications. Comput. Methods Programs Biomed. **142**, 129–145 (2017)
2. Bakhtavar, E., Valipour, M., Yousefi, S., Sadiq, R., Hewage, K.: Fuzzy cognitive maps in systems risk analysis: a comprehensive review. Complex Intell. Syst. **7**(2), 621–637 (2020). https://doi.org/10.1007/s40747-020-00228-2
3. Davis, C.W., Giabbanelli, P.J., Jetter, A.J.: The intersection of agent based models and fuzzy cognitive maps: a review of an emerging hybrid modeling practice. In: 2019 Winter Simulation Conference (WSC), pp. 1292–1303. IEEE (2019)
4. Epstein, L.H., Myers, M.D., Raynor, H.A., Saelens, B.E.: Treatment of pediatric obesity. Pediatrics **101**(Supplement_2), 554–570 (1998)
5. Felix, G., Nápoles, G., Falcon, R., Froelich, W., Vanhoof, K., Bello, R.: A review on methods and software for fuzzy cognitive maps. Artif. Intell. Rev. **52**(3), 1707–1737 (2017). https://doi.org/10.1007/s10462-017-9575-1
6. Firmansyah, H.S., Supangkat, S.H., Arman, A.A., Giabbanelli, P.J.: Identifying the components and interrelationships of smart cities in Indonesia: supporting policymaking via fuzzy cognitive systems. IEEE Access **7**, 46136–46151 (2019)
7. Giabbanelli, P., Fattoruso, M., Norman, M.L.: CoFluences: simulating the spread of social influences via a hybrid agent-based/fuzzy cognitive maps architecture. In: Proceedings of the 2019 ACM SIGSIM Conference on Principles of Advanced Discrete Simulation, pp. 71–82 (2019)
8. Giabbanelli, P.J.: Solving challenges at the interface of simulation and big data using machine learning. In: 2019 Winter Simulation Conference (WSC), pp. 572–583. IEEE (2019)
9. Giabbanelli, P.J., Crutzen, R.: Creating groups with similar expected behavioural response in randomized controlled trials: a fuzzy cognitive map approach. BMC Med. Res. Methodol. **14**(1), 1–19 (2014)
10. Giabbanelli, P.J., Jackson, P.J., Finegood, D.T.: Modelling the joint effect of social determinants and peers on obesity among canadian adults. In: Theories and Simulations of Complex Social Systems, pp. 145–160. Springer (2014). https://doi.org/10.1007/978-3-642-39149-1_10

11. Giabbanelli, P.J., Torsney-Weir, T., Mago, V.K.: A fuzzy cognitive map of the psychosocial determinants of obesity. Appl. Soft Comput. **12**(12), 3711–3724 (2012)
12. Giles, B.G., Findlay, C.S., Haas, G., LaFrance, B., Laughing, W., Pembleton, S.: Integrating conventional science and aboriginal perspectives on diabetes using fuzzy cognitive maps. Soc. Sci. Med. **64**(3), 562–576 (2007)
13. Gray, S., Hilsberg, J., McFall, A., Arlinghaus, R.: The structure and function of angler mental models about fish population ecology: the influence of specialization and target species. J. Outdoor Recreat. Tour. **12**, 1–13 (2015)
14. Groumpos, P.P.: Intelligence and fuzzy cognitive maps: scientific issues, challenges and opportunities. Stud. Inform. Control **27**(3), 247–264 (2018)
15. Lavin, E.A., Giabbanelli, P.J., Stefanik, A.T., Gray, S.A., Arlinghaus, R.: Should we simulate mental models to assess whether they agree? In: Proceedings of the Annual Simulation Symposium, pp. 1–12 (2018)
16. Mkhitaryan, S., Giabbanelli, P.J., de Vries, N.K., Crutzen, R.: Dealing with complexity: how to use a hybrid approach to incorporate complexity in health behavior interventions. Intell.-Based Med. **3**, 100008 (2020)
17. Mkhitaryan, S., Giabbanelli, P.J., Wozniak, M.K., Napoles, G., de Vries, N.K., Crutzen, R.: FCMpy: a python module for constructing and analyzing fuzzy cognitive maps (2021)
18. Mourhir, A.: Scoping review of the potentials of fuzzy cognitive maps as a modeling approach for integrated environmental assessment and management. Environ. Modell. Softw. **135**, 104891 (2021)
19. Papageorgiou, E.I.: Learning algorithms for fuzzy cognitive maps-a review study. IEEE Trans. Syst. Man Cybern. Part C (Appl. Rev.) **42**(2), 150–163 (2011)
20. Pearson, E.S., D ''AGOSTINO, R.B., Bowman, K.O.: Tests for departure from normality: comparison of powers. Biometrika **64**(2), 231–246 (1977)
21. Pedrycz, W.: Why triangular membership functions? Fuzzy Sets Syst. **64**(1), 21–30 (1994)
22. Poczeta, K., Yastrebov, A., Papageorgiou, E.I.: Learning fuzzy cognitive maps using structure optimization genetic algorithm. In: 2015 Federated Conference on Computer Science and Information Systems (FedCSIS), pp. 547–554. IEEE (2015)
23. Resnicow, K., Vaughan, R.: A chaotic view of behavior change: a quantum leap for health promotion. Int. J. Behav. Nutr. Phys. Act. **3**(1), 1–7 (2006)
24. Springvloet, L., et al.: Short-and medium-term efficacy of a web-based computer-tailored nutrition education intervention for adults including cognitive and environmental feedback: randomized controlled trial. J. Med. Internet Res. **17**(1), e3837 (2015)
25. Stach, W., Kurgan, L., Pedrycz, W.: A divide and conquer method for learning large fuzzy cognitive maps. Fuzzy Sets Syst. **161**(19), 2515–2532 (2010)
26. Stach, W., Kurgan, L., Pedrycz, W., Reformat, M.: Genetic learning of fuzzy cognitive maps. Fuzzy Sets Syst. **153**(3), 371–401 (2005)
27. Stach, W., Pedrycz, W., Kurgan, L.A.: Learning of fuzzy cognitive maps using density estimate. IEEE Trans. Syst. Man Cybern. Part B (Cybern.) **42**(3), 900–912 (2012)
28. Wang, C., Liu, J., Wu, K., Ying, C.: Learning large-scale fuzzy cognitive maps using an evolutionary many-task algorithm. Appl. Soft Comput. **108**, 107441 (2021)

A Simulation Study of the Delayed Effect of Covid-19 Pandemic on Pensions and Welfare of the Elderly: Evidence from Poland

Bożena Mielczarek(✉) iD

Faculty of Management, Wrocław University of Science and Technology, Wrocław, Poland
bozena.mielczarek@pwr.edu.pl

Abstract. Changes in the demographic structure of the population have imposed alterations in the pension systems. In many countries, including Poland, the amount of retirement benefits is highly dependent on life expectancy, which in the case of increases in longevity leads to a decrease in accrued benefits. A dynamic Monte Carlo simulation model was developed to investigate the financial implications of the aging problem in connection with the previously unexpected demographic changes caused by the Covid-19 pandemic on future pension payments. The model uses data from Polish statistical databases. The study distinguishes different life cycle profiles, i.e. women and men with average and minimum wage earnings. Simulation experiments are conducted in two variants. The first variant takes into account the currently registered shortening of life expectancy, while the second variant assumes that life expectancy is continuously lengthening, as it was observed until the outbreak of the Covid-19 epidemic. The simulation results show that the Covid-19 pandemic has a beneficial effect for future retirees, which is reflected in the expected higher replacement rates at retirement.

Keywords: Simulation · Modeling · Retirement · Ageing

1 Introduction

Aging populations are an international problem that affects most countries in the world. We are currently observing major changes in the demographic structure of the population, the most characteristic of which are increasing life expectancy and decreasing fertility rates. Although it is true that in the last two years, as a result of the prevailing Covid 19 epidemic, life expectancy has shortened, but at the same time there has been a further decline in fertility. Slowly but unavoidably, the structure of the population is changing [1], expressed by an increasingly intense growth in the percentage of old age groups (older than 65 years) and an increasingly significant decline in the size of working age groups (20–64 years). According to [1] the number of people over 65 years of age per 100 people of working age has increased from 20 in 1980 to 31 in 2020.

Progressive demographic change is placing a significant strain on public finances. Therefore, many countries around the world are reforming their retirement policies in such a way as to maintain the stability of pension systems while covering the increased

© The Author(s), under exclusive license to Springer Nature Switzerland AG 2022
D. Groen et al. (Eds.): ICCS 2022, LNCS 13352, pp. 326–340, 2022.
https://doi.org/10.1007/978-3-031-08757-8_28

costs of ageing populations. In the past, pension systems were dominated by PAYGO DB (pay-as-you-go defined benefit) schemes, in which the amount of the pension depended on the number of years worked. However, recent years have witnessed a paradigm shift from PAYGO DB to PAYGO DC (pay-as-you-go defined contribution) schemes, in which the amount of pension benefit is closely related to the amount of earnings during the entire period of employment. The Polish pension system was also radically altered in 1999 with the establishment of a new legislative act, under which a DC system came into force.

In DC schemes, the amount of retirement benefits is highly dependent on life expectancy, which in the case of increases in longevity leads to a decrease in accrued benefits. The timing of retirement is also crucial, as the longer the pension contributions are accumulated, the higher the accumulated capital, and the shorter the estimated time of longevity. However, the impact of demographics on DC scheme benefits is even stronger, as the accumulated pension capital is annually indexed by the change of the sum of premiums accrued, which in turn depends on the level of employment and the average salary in the economy. This means that the more people in the working age group pay pension contributions and the higher the salaries, the higher the annual indexation of the accumulated capital. The currently observed changes in the demographic structure of the population, manifested by the decreasing number of people in the working-age group, add therefore another unfavorable factor, in addition to extended longevity and lowered retirement age, affecting the amount of pension benefits. The number of people in the working age group determines how intensely the accumulated capital, which defines the size of future pensions, will grow.

To calculate the financial implications of the aging problem in connection with the previously unexpected demographic changes caused by the Covid-19 pandemic, a dynamic Monte Carlo (MC) simulation model was developed. The model uses data taken from Polish statistical databases for the period 2000 to 2020, i.e., from the introduction of pension reform in Poland to the present. The simulation model is used to analyze the effects of the Covid-19 pandemic on the gross pension, depending on the decision made on the timing of retirement. One of the issues raised was how significant is the impact of changes in life expectancy on retirement. The study distinguished four different lifecycle profiles: women and men with average wage earnings and women and men with minimum wage earnings. An additional scenario was examined: a scenario in which a delayed retirement decision is made. To investigate the effect of SARS-Cov2-induced changes in life expectancy, we conducted a simulation study that accounts for the reduction in longevity observed as a result of the Covid-19 pandemic in the past year and a second study that hypothesizes that a gradual increase in life expectancy is occurring at all times.

The structure of this paper is as follows. Section 2 briefly describes the Polish pension system. Section 3 provides background information on the simulation methods used to model pension systems. Section 4 presents the MC simulation model and the data it uses. In Sect. 5 simulation results are presented and discussed. Finally, Sect. 6 contains conclusions with some remarks and perspectives.

2 The Polish Pension System

The Polish pension system is based on three pillars, of which the first two pillars are mandatory and the third is voluntary. A pension contribution of 19,52% of earnings (split between employer and employee) is recorded for personal retirement accounts and is divided into two mandatory pillars. The first mandatory pillar is in the form of a notional (nonfinancial) account. The term 'notional' means that contributions (12.22% of an employee's salary) are built into individual accounts and at retirement this capital is converted into a monthly pension using an algorithm based on life expectancy. Therefore, this capital may be treated as claims against the government [2]. The second mandatory pillar (fed by 7.3% of an employee's salary) is administered by private open pension funds and/or the state company (Social Insurance Company, ZUS), where financial assets are gathered. Each insured person can decide whether to have the entire contribution transferred to a sub-account in the ZUS or to divide it between open funds and ZUS. However, even if the insured person decides to save in the second pillar in private open fund, 10 years before retirement, the accumulated funds are systematically transferred to the sub-account in ZUS. The third pillar offers additional savings methods that are facilitated by some tax benefits.

This paper focuses only on the first two obligatory pillars.

Capital accumulated in both mandatory ZUS accounts is indexed annually. Indexation of contributions in a pension account means multiplying the sum of contributions in the insured person's account by the index. The value of the first-pillar index depends on the so-called written premiums (i.e., the increase in salaries and the number of persons insured in ZUS) from the previous year, but cannot be lower than inflation or negative. The indexation of the second pillar account is derived from the gross domestic product (GDP) growth rate of the last 5 years.

However, the assessment base for pension contributions may be limited. This happens when a person's annual income exceeds the amount equal to thirty times the projected average monthly salary in the national economy for a given calendar year.

The state pension is calculated by dividing the capital accumulated in both pillars (*Capital*1 and *Capital*2) by the average life expectancy determined for persons of the same age as a person retiring (see Eq. 1).

$$Pension_k = \frac{Capital1 + Capital2}{Life\ expectancy_k} \qquad (1)$$

where k is the retirement age.

Life expectancy is defined as the average number of months that people of a particular age could expect to live. It is determined jointly for men and women and is announced every year by the President of the Central Statistical Office. By 2045, life expectancy is expected to increase by approximately 12 months [1].

3 Simulation Methods in Retirement Planning

The existing literature on pension systems in the context of longevity risk is relatively scarce [3]. Jimeno et al. [4] identified three methodological approaches to analyze the impact of demographic change on the sustainability of pension systems. These are aggregate accounting, general equilibrium, and individual life-cycle profile models.

The first approach is based on strong assumptions about demographic and economic variables and the determination of accounting identities. In this approach, the behavior of individuals is not taken into account. Analytical methods such as discrete choice models [5], dynamic optimization models, and forecasting models [3] are used. Another class of models is general equilibrium models with overlapping generation (OLG). These models study an artificial economy populated by rational agents who make their own decisions about consumption, savings, pension claims, and labor supply. The purpose of using these models is to study the impact of different government policy options, such as those involving an increase in retirement age [6] or changes in the structure of taxes and transfers [7], on the macroeconomic variables responsible for the behavioral responses of individuals. The advantage of OLG models is, in particular, that they allow one to model the behavior of individuals [8].

The approaches discussed above have made significant contributions to understanding the impact of demographic change on various pension systems and the economy, but their main weakness is that they model the dynamics of population change at the aggregate level and that the individuals modeled behave rationally.

The third group of models, that is, the individual life cycle profile models, uses simulation methods to follow the lives of many different individuals to calculate the distribution of retirement benefits under different alternative retirement rules. Two simulation approaches are noticed here [9], i.e. Monte Carlo (MC) and dynamic microsimulation (MSM). Microsimulation operates at the level of individuals. The model simulates the life paths of a virtual population by accessing detailed demographic data and pension payment records and applying mathematical formulas to model individual behavior. The MC simulation relies on one or more typical individuals to describe the experiences of some larger group. Several input assumptions are made that relate to a typical representative of the target occupational group, and the conclusions are valid only for predefined individuals.

Van Sonsbeek [10] developed an MSM model that simulated life paths for a sample of the Dutch population to analyse the effects of ageing budgetary, redistributive and labour participation. The author formulated conclusions for policy measures intended to reduce state pension costs. Schofield et al. [11] used MSM to estimate the costs of early retirement in Australia due to back problems. They found that early retirement not only limits the retiree's income, but also reduces their long-term financial capacity. Halvorsen and Pedersen [12] studied the gender gap and its influence on the general inequality in the distribution of pension income, using an advanced MSM model.

The MC approach to solving decision-related problems in the area of retirement planning is not the predominant choice, but it has been applied a number of times. McFarland and Warshawsky [13] studied the relationship between financial market fluctuations and the degree of retirement security for retirees at age 65. The authors used historical returns and interest rates taken over a long period (95 years) and tested two investment strategies. MC simulation methods can also be used to indicate the optimal retirement age. Bieker [14], using historical data, defined three input probability distributions (inflation rate, wage growth rate, and life expectancy) and ran MC simulations for different retirement age scenarios. Mielczarek [9] built an MC simulation model to compare two variants of the Polish pension system. Simulations enabled calculation of the economic implications of the new pension system strategy from the perspective of the individual worker and comparison of the results of the previous system, assuming the same macroeconomic circumstances.

4 Methodology

4.1 Method and Output Measures

The dynamic MC simulation method was used to simulate the pension to be received in 2042 or later by a hypothetical worker from two obligatory pillars. Simulation starts in the year 2008 when a hypothetical individual is 25 years old and he/she begins earning a living. From 2008 to 2019, the growth of capital accumulated in the first and second obligatory pillars is simulated on the basis of real values of capital indexation. From 2020 to the moment of reaching the retirement age (that is, at least until 2043) the growth of the accumulated capital is simulated according to different scenarios of indexation and according to different individual life-cycle professional profiles.

Two groups of workers by income level and gender are considered: minimum- and average-wage workers. Individual earnings are assumed to grow in line with the economy's average. This means that the individual is assumed to remain at the same point of earning distribution: low-income workers are paid the minimum wage at all times, while average-income workers' wages are always at the average-wage level.

Calculations of annual changes in an individual's first and second pillar accounts were made for different numbers of working years and were replicated 500 times.

Two indicators were chosen as baseline measures in the simulation. These ratios are calculated when the decision is made to retire. The most important indicator is the gross future replacement rate (GRR). The GRR represents the level of retirement benefits from mandatory public pension systems relative to earnings while working. Following [15], the probability of achieving the target gross replacement rate (TRR) was also analyzed. Three TRR values were investigated: 40%, 45%, and 50%.

We also examined to what extent, in the face of shortened life expectancy, delayed retirement can contribute to higher pensions.

4.2 Assumptions

Several assumptions were formulated in the simulation model.

- The simulation is run separately for a hypothetical female employee and a hypothetical male employee.
- Hypothetical workers belong to two main income classes: those earning the minimum wage and those earning the average wage. The simulation is run separately for both earning classes and separately for both genders.
- The salary change occurs on the last day of the calendar year. Mid-year changes in an employee's salary are not considered.

- There are no unemployment intervals in the professional history of the individual.
- The capital accumulated in the pension account in the first pillar is increased monthly by an amount equivalent to 12.22% of the employee's gross salary.
- The capital accumulated in the pension account in the second pillar is increased monthly by an amount equivalent to 7.3% of the employee's gross salary.
- There are no delays in money transfer; therefore, no interest is added due to late payments.
- The simulation is run with a one-year step.
- There are no preretirement withdrawals.
- There is no fee contribution paid by individuals.
- Life expectancy is assumed to increase every year.

4.3 Input Distributions

The key variables in the model were described using random distributions that were fitted by analyzing historical data. All available and published values were considered (Table 1).

The indexation rates of the accumulated capital in the first pillar were forecasted based on the historical data from the period 2000–2020, published annually by the Minister of Family, Labour and Social Policy. These years cover the entire period that the current pension system has been in effect. Indexation means multiplying the sum of contributions by the index. The triangular distribution was fitted to the data.

The indexation rates of the accumulated capital in the second pillar were forecasted based on the historical data from the period 2008–2020, published annually by the Minister of Family, Labour and Social Policy. Indexation means multiplying the sum of contributions by the index. These years cover the entire period that the current pension system has been in effect. The triangular distribution was fitted to the data.

A change in the minimum wage is made once a year, taking into account the percentage rate of increase/decrease. Between 2008 and 2022, there was an increase from 1126 PLN to 3010 PLN (167.32% increase). The gamma distribution was fitted to the data. A change in the average-wage is made once a year by taking into account the percentage rate of increase/decrease. Between 2004 and 2021, there was an increase from 2748.11 PLN to 6644.39 PLN (167.32% increase). The gamma distribution was fitted to the data.

In the Polish pension system, there is a limitation of the assessment basis for pension contributions. The limit is equal to 30 times the projected average monthly salary, which is announced annually by the Minister of Family, Labour, and Social Policy and is effective for the following calendar year. Between 2008 and 2022, there was an increase from 2843 PLN to 5922 PLN (108.3% increase). The triangular distribution was fitted to the data.

4.4 Simulation Experiments

An overview of the MC model is presented in Fig. 1.

Simulation studies were conducted in two main scenarios (Table 2). Scenario 1 (Sc1) is a simulation of a retirement benefit for a person whose salary is kept at the minimum wage. Scenario 2 (Sc2) is a simulation of the retirement benefit for a person from an

Table 1. Input random distributions: fitted parameters and test statistics.

Input parameter	Random distribution	Statistic value
Indexation rates of the accumulated capital (%) – Pillar 1	Triangular: a = 101; b = 117; c = 105	Chi-square test *p-value* > 0.75 Kolmogorov-Smirnov test *p-value* > 0.15
Indexation rates of the accumulated capital (%) – Pillar 2	Triangular: a = 103; b = 105; c = 107	Kolmogorov-Smirnov test *p-value* > 0.15
Minimum wage – growth/decline in %	Gamma: alfa = 3.37; beta = 1.47; offset = 2	Kolmogorov-Smirnov test *p-value* > 0.15
Average wage – growth/decline in %	Gamma: alfa = 1.76; beta = 2.47; offset = 1	Kolmogorov-Smirnov test *p-value* > 0.15
Projected average monthly salary growth/decline in %	Triangular: a = -2; b = 5.4; c = 13	Kolmogorov-Smirnov test *p-value* > 0.15

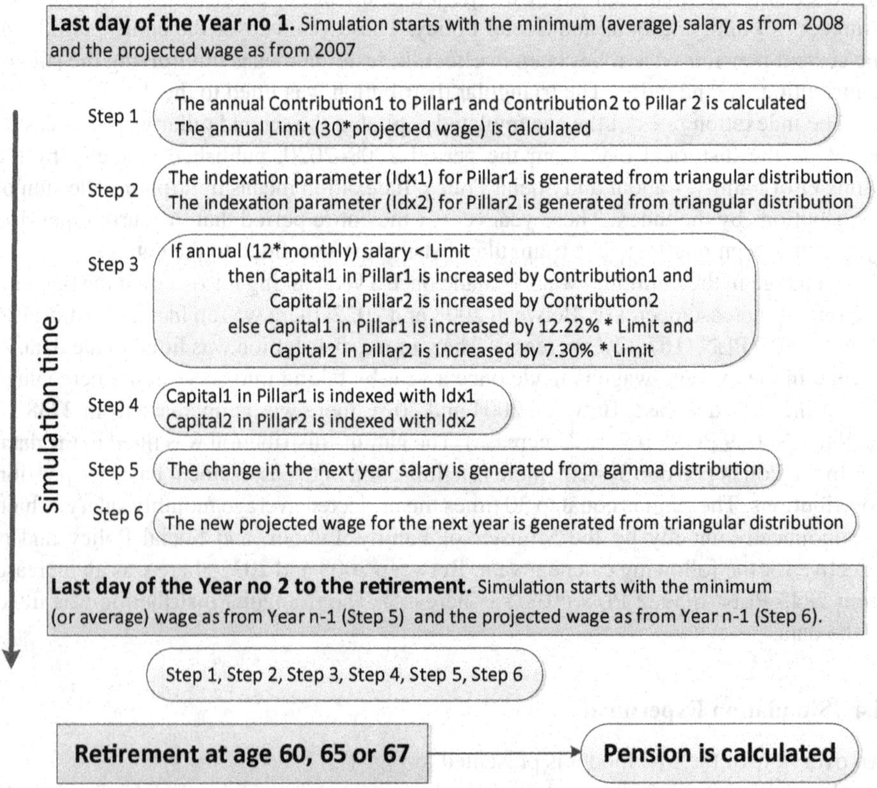

Fig. 1. An overview of the MC simulation model.

average-wage working group. Within each main scenario, simulations were performed for three different retirement ages: 60 (women), 65 (men), and 67 years. The study included an analysis for a retirement age of 67, because for a short period (from January 2013 to October 2017) this was the retirement age for both women and men in Poland.

Each of the six experiments (Sc1_60, Sc1_65, Sc1_67 and Sc2_60, Sc2_65, Sc2_67) was conducted in two variants: for shortened future life expectancy caused by the Covid 19 pandemic and for life expectancy projected before the SARS-Cov2 virus affected population longevity.

Table 2. Simulation scenarios.

Main scenario	Sub-scenario	Sub-sub scenario	Description
Sc1	Minimum-age working group		
	Sc1_60	Sc1_60Cov, Sc1_60NCov	retirement age = 60 (women)
	Sc1_65	Sc1_65Cov, Sc1_65NCov	retirement age = 65 (men)
	Sc1_67	Sc1_67Cov, Sc1_67NCov	retirement age = 67 for men and women
Sc2	Average-age working group		
	Sc2_60	Sc2_60Cov, Sc2_60NCov	retirement age = 60 (women)
	Sc2_65	Sc2_65Cov, Sc2_65NCov	retirement age = 65 (men)
	Sc2_67	Sc2_67Cov, Sc2_67NCov	retirement age = 67 for men and women

5   Simulation Results

5.1   Gross Replacement Rate (GRR)

The Covid 19 epidemic caused significant changes in the life expectancy tables. The long-observed trend of increased longevity was reversed between 2020 and 2021, and life expectancy shortened by approximately 13 months for all age groups studied (Table 3). Therefore, two variants were simulated: a variant in which life tables are significantly modified as a result of the Covid 19 outbreak and a variant in which life expectancy increases steadily, as predicted before the outbreak. Figures 2 and 3 present distributions of the GRR values for persons who earn minimum wage and retire at the age of 60 (women) and 65 (men). Figures 4 and 5 present the GRR values for people who earn an average wage and retire at the age of 60 (women) and 65 (men).

Table 3. Life expectancy values in 2020 and 2021 according to [16].

Age	Life expectancy in 2020 [months]	Life expectancy in 2021 [months]
60	201,1	188,1
65	217,6	204,3
67	261,5	247,7

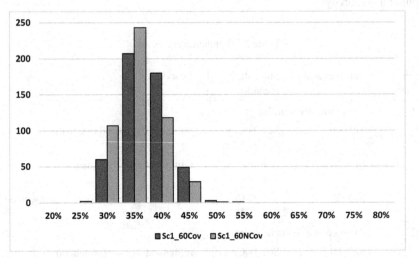

Fig. 2. Distribution of the GRR values for a person who retires at age 60 (woman) and who earns minimum wage during her professional career. Two scenarios are visualized: Life expectancy is reduced (Sc1_60Cov), or a pandemic does not affect life expectancy (Sc1_60NCov).

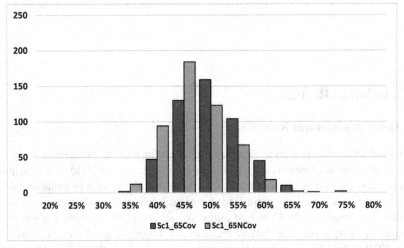

Fig. 3. Distribution of the GRR values for a person who retires at age 65 (man) and who earns minimum wage during his professional career. Two scenarios are visualized: Life expectancy is shortened (Sc1_65Cov) or a pandemic does not affect life expectancy (Sc1_65NCov).

The Figs. 2 and 3 capture the essential differences between these two variants in minimum wage scenario. The Covid-19 epidemic caused the GRR values to shift to the right in the graphs, which means that the pension has increased relative to the last salary, and we observe more instances of higher GRRs in the simulation. The average GRR for retirement age 60 in the Sc1_60NCov variant is 32.89%, while in the Sc1_60Cov variant it is 34.18%. For retirement age 65, it is 44.64% and 47.28%, respectively. In Scenario 2, in which the retirement of a person from the average-wage working group was simulated, a similar trend can be observed (Figs. 4 and 5), but the differences for both variants are much smaller. This is caused by the limitation of the contribution base for retirement benefits, which means that the capital deposited in the mandatory pension funds is constrained from above.

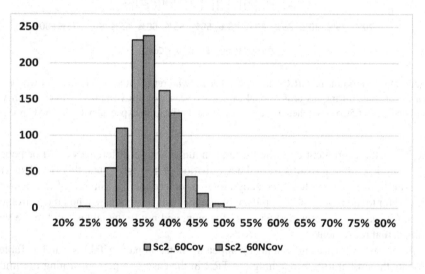

Fig. 4. Distribution of GRR values for a person who retires at age 60 (woman) and who earns an average wage during her professional career. Two scenarios are visualized: Life expectancy is shortened (Sc2_60Cov), or a pandemic does not affect life expectancy (Sc2_60NCov).

5.2 Probability of Target Replacement Rates (TRR)

The second output measure taken into account was the probability of achieving the target replacement rate (TRR). Following [15], three values were tested: 40%, 45%, and 50%. Tables 4 and 5 summarize the simulation findings. There is a large difference between the retirement ages of 60 and 65 and 67. The higher the retirement age, the higher the probability of achieving the TRR in both variants (Cov and NCov). This is a clear indication in favor of extending the period of paid work and confirms the large pension gap between men and women.

Another conclusion comes from comparing the Cov and NCov variants. The probabilities of obtaining TRR are slightly lower for the variant not affected by the pandemic.

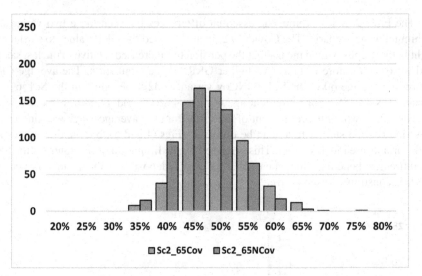

Fig. 5. Distribution of the GRR values for a person who retires at age 65 (man) and who earns an average wage during his professional career. wo scenarios are visualized: life expectancy is shortened (Sc2_65Cov) or when a pandemic does not affect life expectancy (Sc2_65NCov).

The lower the retirement age, the more pronounced these differences are. For people who retire later than it results from the acquisition of pension rights, this is a clear prerequisite to use earlier life expectancy tables to determine the amount of the pension. These refer to the tables valid on the day when the insured person reached the retirement age of 60 years for women and 65 years for men, and not to the tables valid in the year of the actual retirement.

A somewhat surprising conclusion appears when comparing Tables 4 and 5. Table 4 presents the probability of reaching the TRR in the case of a person earning minimum wage, while Table 5 refers to people in the average-wage group. It turns out that in some variants, minimum wage earners obtain a higher probability of reaching a given TRR threshold compared to average wage earners. This is the case, for instance, for a the NCov variant for retirement age equal to 60 years. The reason is the limitation of the contribution base for retirement benefits.

Table 4. Probability of three target replacement rates (TRR) for a scenario with shortened life expectancy (Cov) and with life expectancy unaffected by the Covid 19 outbreak (NCov). Applies to the minimum wage group of workers.

Variant	Cov	NCov	Cov	NCov	Cov	NCov
Retirement age	60	60	65	65	67	67
P(TRR > 0.4)	0.094	0.050	0.880	0.800	0.986	0.950
P(TRR > 0.45)	0.012	0.008	0.616	0.424	0.890	0.776
P(TRR > 0.5)	0.000	0.002	0.308	0.174	0.674	0.490

Table 5. Probability of three target replacement rates (TRR) for a scenario with shortened life expectancy (Cov) and with life expectancy unaffected by the Covid 19 outbreak (NCov). Applies to the average-wage group of workers.

Variant	Cov	NCov	Cov	NCov	Cov	NCov
Retirement age	60	60	65	65	67	67
P(TRR > 0.4)	0,098	0,042	0,908	0,782	0,982	0,976
P(TRR > 0.45)	0,012	0,002	0,610	0,448	0,898	0,794
P(TRR > 0.5)	0,000	0,000	0,284	0,170	0,642	0,512

5.3 Delayed Retirement

Figure 6 compares the GRRs for delayed treatment and with different life tables. The delayed retirement in the DC scheme allows for a higher pension due to a longer period of capital accumulation in retirement accounts and a shorter life expectancy. In the Polish pension system there is an additional possibility of increasing the pension benefits, as the insured who retires after reaching the age of retirement can choose the life table from the one currently in force and the one which was in force when he/she reached the retirement age. Therefore, the shortening of longevity due to the pandemic further contributes to the increase in retirement benefits. The differences in pension when selecting Covid nad Non-Covid tables are about 4 percentage points.

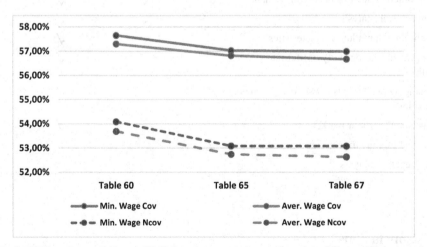

Fig. 6. GRRs when retiring at 67 (delayed treatment) and choosing different life expectancy tables. The graph shows the simulation results for minimum and average wage workers. Table 67: Life expectancy table applicable to retirement at age 67; Table 65: the life expectancy table applicable when the insured (male) has reached retirement age of 65; Table 60: the life expectancy table applicable when the insured (female) has reached retirement age of 60.

5.4 The Comparative Analysis of Simulation Techniques

Among the simulation methods used to study the impact of population aging on the sustainability of pension systems, in addition to the previously mentioned MC and MSM, agent-based simulation (ABS) is also used. Each approach has its advantages and disadvantages, and the purpose of simulation studies is usually somewhat different. Table 6 compares these three approaches in relation to the study of pension systems. The comparative analysis was carried out on the basis of selected items of the literature. All methods enable dynamic simulations useful for long-term analyses, allow for randomness and uncertainty, and are based on dynamic simulation analyses. The main differences between them concern such aspects as consideration of the full range of lifetime trajectories (MSM and ABS) versus selection of typical life trajectories (MC), and microlevel analysis (MSM and ABS) versus aggregated analysis (MC). If the goal is to study the impact of pension systems on individuals, then microsimulation and agent models allow you to explore in detail the different variants of characteristics and behaviour of employees. However, if the aim of the study is to formulate some universal relationships for typical characteristic groups of individuals, then MC simulation may be the choice. MC models do not require such detailed data about individual characteristics.

Table 6. Simulation techniques used for retirement studies: comparative analysis.

Feature	MC [9, 17]	MSM [10, 12]	ABS [18, 19]
Dynamics of lifetime trajectories	✓	✓	✓
Unlimited number of uncertain elements	✓	✓	✓
Micro-level analysis		✓	✓
Unlimited individual characteristics		✓	✓
Unlimited heterogeneity of lifetime trajectories		✓	✓
Changing preferences towards employment positions	✓		✓
Risks in the labour market	✓		✓
Typical lifetime trajectories	✓		
Aggregated analysis	✓		

6 Conclusions

The objective of the studies was to focus on the impact of the Covid-19 pandemic on projected retirement benefits. An MC simulation was successfully applied to compare two scenarios to test the impact of the Covid-19 pandemic on future pensions in two income groups and for different retirement ages. Although the presented model does not reflect all the factors that influence projected retirement benefits, it clearly shows that a shortening of life expectancy can affect future pension in various ways. Generally, this effect is beneficial for future retirees. The GRR ratios, obtained during the simulations, clearly move towards higher values, which means a higher replacement rate at retirement. An interesting conclusion can also be drawn by comparing the probabilities of obtaining TRR in the two study groups: the minimum wage and the average wage. In most cases, across retirement age classes, these probabilities, i.e. $P(TRR > 0.45)$ and $P(TRR > 0.50)$ are close to each other. This indicates a certain resilience of the pension system, unfavorable from the point of view of the retiree, to changes in the life tables.

Our study has some limitations. The most important limitation is the lack of correlation between demographic changes and capital indexation rates. The indexation parameters depend on the level of employment and the average salary in the economy. The volume of people in the working age group and the amount of contributions paid influence the annual indexation of the accumulated capital. Future research will focus on expanding the model by incorporating such dependencies.

References

1. OECD: Pensions at a Glance 2021 (2021)
2. Kurach, R., Kuśmierczyk, P., Papla, D.: The optimal portfolio for the two-pillar mandatory pension system: the case of Poland. Appl. Econ. **51**, 2552–2565 (2019)
3. Wang, H., Koo, B., O'Hare, C.: Retirement planning in the light of changing demographics. Econ. Model. **52**, 749–763 (2016)
4. Jimeno, J.F., Rojas, J.A., Puente, S.: Modelling the impact of aging on social security expenditures. Econ. Model. **25**, 201–224 (2008)
5. Etgeton, S.: The effect of pension reforms on old-age income inequality. Labour Econ. **53**, 146–161 (2018)
6. Shen, Z., Yang, J.: A simulation study of the effect of delayed retirement on welfare of the elderly: evidence from China. SAGE Open, 1–12 (2021)
7. Lin, H.-C., Tanaka, A., Wu, P.-S.: Shifting from pay-as-you-go to individual retirement accounts: a path to a sustainable pension system. J. Macroecon. **69**(C), (2021)
8. Bielecki, M., Goraus, K., Hagemejer, J., Tyrowicz, J.: Decreasing fertility vs increasing longevity: raising the retirement age in the context of ageing processes. Econ. Model. **52**, 125–143 (2016)
9. Mielczarek, B.: Simulation model to forecast the consequences of changes introduced into the 2nd pillar of the Polish pension system. Econ. Model. **30**, 706–714 (2013)
10. van Sonsbeek, J.M.: Micro simulations on the effects of ageing-related policy measures. Econ. Model. **27**, 968–979 (2010)
11. Schofield, D., Kelly, S., Shrestha, R., Callander, E., Passey, M., Percival, R.: The impact of back problems on retirement wealth. Pain **153**, 203–210 (2012)
12. Halvorsen, E., Pedersen, A.W.: Closing the gender gap in pensions: a microsimulation analysis of the Norwegian NDC pension system. J. Eur. Soc. Policy. **29**, 130–143 (2019)
13. McFarland, B., Warshawsky, M.J.: Balances and retirement income from individual accounts: U.S. historical simulations. Benefits Q. **26**, 36–40 (2010)
14. Bieker, R.F.: Using simulation as a tool in selecting a retirement age under defined benefit pension plans. J. Econ. Financ. **26**, 334–343 (2002)
15. Kurach, R., Kuśmierczyk, P., Papla, D.: Risk reduction in two-pillar mandatory pension system under regulatory constraints: simulation-based evidence from Poland. Appl. Econ. Lett. **28**, 191–195 (2021)

16. GUS: GłównyUrząd Statystyczny Homepage, https://stat.gov.pl/. Accessed 01 Apr 2022
17. Šebo, J., Danková, D., Králik, I.: Projecting a life-cycle income – a simulation model for the Slovak pension benefit statement. Olsztyn Econ. J. **15**, 271–284 (2020)
18. Murata, T., Arikawa, H.: Pension simulation with a huge number of agents. In: 2012 7th International Conference on Computing and Convergence Technology (ICCCT), pp. 1482–1487 (2012)
19. Bae, J.W., Paik, E., Kim, K., Singh, K., Sajjad, M.: Combining microsimulation and agent-based model for micro-level population dynamics. Procedia Comput. Sci. **80**, 507–517 (2016)

Non-Generic Case of Leap-Frog
for Optimal Knots Selection in Fitting
Reduced Data

Ryszard Kozera[1,2]([✉]) [iD] and Lyle Noakes[2] [iD]

[1] Warsaw University of Life Sciences - SGGW, Institute of Information Technology,
Ul. Nowoursynowska 159, 02-776 Warsaw, Poland
ryszard_kozera@sggw.edu.pl
[2] The School of Physics, Mathematics and Computing, The University of Western
Australia, 35 Stirling Highway, Crawley, W.A., 6009 Perth, Australia
lyle.noakes@uwa.edu.au

Abstract. The problem of fitting multidimensional reduced data is analyzed here . The missing interpolation knots $\mathcal{T}$ are substituted by $\hat{\mathcal{T}}$
which minimize a non-linear multivariate function $\mathcal{J}_0$. One of numerical schemes designed to compute such optimal knots relies on iterative
scheme called *Leap-Frog Algorithm*. The latter is based on merging the
respective generic and non-generic univariate overlapping optimizations
of $\mathcal{J}_0^{(k,i)}$. The discussion to follow establishes the sufficient conditions
enforcing unimodality of the non-generic case of $\mathcal{J}_0^{(k,i)}$ (for special data
set-up and its perturbation). Illustrative example supplements the analysis in question. This work complements already existing analysis on
generic case of *Leap-Frog Algorithm*.

Keywords: Data fitting · Optimization · Curve modelling

1 Introduction

Assume that n interpolation points $\mathcal{M}_n = \{x_i\}_{i=0}^n$ in arbitrary Euclidean space
$\mathbb{E}^m$ are given with the associated knots $\mathcal{T} = \{t_i\}_{i=0}^n$ unavailable. The analyzed
here class of fitting curves $\mathcal{I}$ forms piecewise C^2 functions $\gamma : [0,T] \rightarrow \mathbb{E}^m$
satisfying $\gamma(t_i) = q_i$ and $\ddot{\gamma}(t_0) = \ddot{\gamma}(T) = \mathbf{0}$. It is also assumed that $\gamma \in \mathcal{I}$ is at
least of class C^1 over $\mathcal{T}_{int} = \{t_i\}_{i=1}^{n-1}$ and extends to $C^2([t_i, t_{i+1}])$. *The unknown
internal knots $\mathcal{T}_{int}$ are called admissible if* $t_i < t_{i+1}$, for $i = 0, 1, \ldots, n-1$ (here
$t_0 = 0$ and $t_n = T$). Different choices of $\mathcal{T}_{int}$ permits *to control and model the
trajectory of* γ. A possible criterion (measuring the "average acceleration" of γ)
for a given choice of fixed knots $\mathcal{T}$ is to minimize

$$\mathcal{J}_T(\gamma) = \sum_{i=0}^{n-1} \int_{t_i}^{t_{i+1}} \|\ddot{\gamma}(t)\|^2 dt , \qquad (1)$$

D. Groen et al. (Eds.): ICCS 2022, LNCS 13352, pp. 341–354, 2022.
https://doi.org/10.1007/978-3-031-08757-8_29

(over $\mathcal{I}$) which yields a unique optimal curve $\gamma_{opt} \in \mathcal{I}$ being *a natural cubic spline* γ_{NS} - see [1,9]. Thus, varying $\mathcal{T}_{int}$ reformulates (1) into searching for an optimal natural spline γ_{NS} with $\mathcal{T}_{int}$ treated as free variables. As shown in [3] the latter reformulates into optimizing a highly non-linear function J_0 in $n-1$ variables $\mathcal{T}_{int}$ subject to $t_i < t_{i+1}$. Due to the complicated nature of J_0 most of numerical schemes used to optimize J_0 face numerical difficulties (see e.g. [3]). Equally, studying the character of critical points of J_0 forms a non-trivial task. A possible remedy is to apply *a Leap-Frog Algorithm* (see [2,3]) designed to optimize J_0 upon merging the iterative sequence of univariate overlapping optimizations of $J_0^{(k,i)}$ preserving $t_i < t_{i+1}$. Recent work [5] deals with *the generic case* of a *Leap-Frog* recomputing iteratively the knots $\{t_2, t_3, \ldots, t_{n-2}\}$.

This paper discusses[1] a *non-generic* case of *Leap-Frog* covering the recursive readjustment of knots t_1 and t_{n-1}. The latter establishes sufficient conditions for *unimodality* of $J_0^{(k,1)}$ and $J_0^{(k,n-1)}$. First a special data case (18) is considered (see Sect. 4) which in turn is subsequently extended to its perturbed version (22) preserving in practice the unimodality (once it occurs for (18)) for substantially large perturbations (see Proposition 1 and Example 1 in Sect. 5).

Numerical performance of *Leap-Frog* and comparison tests with other optimization schemes are presented in [2,3,7]. Some applications of *Leap-Frog* optimization scheme in modelling and simulation are covered in [10–12].

2   Preliminaries

A cubic spline interpolant (see [1]) $\gamma_{\mathcal{T}}^{C_i} = \gamma_{\mathcal{T}}^{C}|_{[t_i, t_{i+1}]}$, for a given admissible knots $\mathcal{T} = (t_0, t_1, \ldots, t_{n-1}, t_n)$ defined as $\gamma_{\mathcal{T}}^{C_i}(t) = c_{1,i} + c_{2,i}(t - t_i) + c_{3,i}(t - t_i)^2 + c_{4,i}(t - t_i)^3$, (for $t \in [t_i, t_{i+2}]$) satisfies (for $i = 0, 1, 2, \ldots, n-1$; $c_{j,i} \in \mathbb{R}^m$, where $j = 1, 2, 3, 4$) $\gamma_{\mathcal{T}}^{C_i}(t_{i+k}) = x_{i+k}$ and $\dot{\gamma}_{\mathcal{T}}^{C_i}(t_{i+k}) = v_{i+k}$, for $k = 0, 1$ with the velocities $v_0, v_1, \ldots, v_{n-1}, v_n \in \mathbb{R}^m$ assumed to be temporarily free parameters (*if unknown*). The coefficients $c_{j,i}$ read (with $\Delta t_i = t_{i+1} - t_i$):

$$c_{1,i} = x_i, \qquad\qquad c_{2,i} = v_i,$$

$$c_{4,i} = \frac{v_i + v_{i+1} - 2\frac{x_{i+1} - x_i}{\Delta t_i}}{(\Delta t_i)^2}, \qquad c_{3,i} = \frac{\frac{(x_{i+1} - x_i)}{\Delta t_i} - v_i}{\Delta t_i} - c_{4,i}\Delta t_i . \quad (2)$$

The latter follows from Newton's divided differences formula (see e.g. [1, Chap. 1]). Adding $n-1$ constraints $\ddot{\gamma}_{\mathcal{T}}^{C_{i-1}}(t_i) = \ddot{\gamma}_{\mathcal{T}}^{C_i}(t_i)$ for continuity of $\ddot{\gamma}_{\mathcal{T}}^{C}$ at $x_1, \ldots, x_{n-1}$ (with $i = 1, 2, \ldots, n-1$) leads by (2) (for $\gamma_{\mathcal{T}}^{C_i}$) to the m tridiagonal linear systems (strictly diagonally dominant) of $n-1$ equations in $n+1$ vector unknowns representing velocities at $\mathcal{M}$ i.e. $v_0, v_1, v_2, \ldots, v_{n-1}, v_n \in \mathbb{R}^m$:

$$v_{i-1}\Delta t_i + 2v_i(\Delta t_{i-1} + \Delta t_i) + v_{i+1}\Delta t_{i-1} = b_i ,$$

$$b_i = 3(\Delta t_i \frac{x_i - x_{i-1}}{\Delta t_{i-1}} + \Delta t_{i-1} \frac{x_{i+1} - x_i}{\Delta t_i}) . \quad (3)$$

---
[1] This work is a part of Polish National Centre of Research and Development research project POIR.01.02.00-00-0160/20.

(i) Both v_0 and v_n *(if unknown)* can be e.g. calculated from $a_0 = \ddot\gamma_{\mathcal{T}}^{C}(0) = a_n = \ddot\gamma_{\mathcal{T}}^{C}(T_c) = \mathbf{0}$ combined with (2) (this yields *a natural cubic spline interpolant $\gamma_{\mathcal{T}}^{NS}$* - a special $\gamma_{\mathcal{T}}^{C}$) which supplements (3) with two missing vector linear equations:

$$2v_0 + v_1 = 3\frac{x_1 - x_0}{\Delta t_0}, \quad v_{n-1} + 2v_n = 3\frac{x_n - x_{n-1}}{\Delta t_{n-1}}. \tag{4}$$

The resulting m linear systems, each of size $(n+1) \times (n+1)$, (based on (3) and (4)) as strictly row diagonally dominant result in one solution $v_0, v_1, \ldots, v_{n-1}, v_n$ (solved e.g. by Gauss elimination without pivoting - see [1, Chap. 4]), which when fed into (2) determines explicitly *a natural cubic spline $\gamma_{\mathcal{T}}^{NS}$* (with fixed $\mathcal{T}$). A similar approach follows for arbitrary a_0 and a_n.
(ii) If both v_0 and v_n are given then the so-called *complete spline $\gamma_{\mathcal{T}}^{CS}$* can be found with $v_1, \ldots v_{n-1}$ determined solely by (3).
(iii) If one of v_0 or v_n is unknown, this can be compensated by setting the respective terminal acceleration e.g. to $\mathbf{0}$. The above scheme relies on solving (3) with one equation from (4). Such splines are denoted here by $\gamma_{\mathcal{T}}^{v_n}$ or $\gamma_{\mathcal{T}}^{v_0}$. Two *non-generic* cases of *Leap-Frog* optimizations deal with the latter.
By (1) $\mathcal{J}_T(\gamma_{\mathcal{T}}^{NS}) = 4\sum_{i=0}^{n-1}(\|c_{3,i}\|^2\Delta t_i + 3\|c_{4,i}\|^2(\Delta t_i)^3 + 3\langle c_{3,i}|c_{4,i}\rangle(\Delta t_i)^2)$, which ultimately reformulates into (see [2]):

$$\mathcal{J}_T(\gamma_{\mathcal{T}}^{NS}) = 4\sum_{i=0}^{n-1}\left(\frac{-1}{(\Delta t_i)^3}(-3\|x_{i+1} - x_i\|^2 + 3\langle v_i + v_{i+1}|x_{i+1} - x_i\rangle\Delta t_i\right.$$
$$\left. -(\|v_i\|^2 + \|v_{i+1}\|^2 + \langle v_i|v_{i+1}\rangle)(\Delta t_i)^2\right). \tag{5}$$

As mentioned before for fixed knots $\mathcal{T}$, the natural spline $\gamma_{\mathcal{T}}^{NS}$ minimizes (1) (see [1]). Thus upon relaxing the internal knots $\mathcal{T}_{int}$ the original infinite dimensional optimization (1) reduces into finding the corresponding *optimal knots* $(t_1^{opt}, t_2^{opt}, \ldots, t_{n-1}^{opt})$ for (5) (viewed from now on as a multivariate function $J_0(t_1, t_2, \ldots, t_{n-1})$) subject to $t_0 = 0 < t_1^{opt} < t_2^{opt} < \cdots < t_{n-1}^{opt} < t_n = T$. Such reformulated non-linear optimization task (5) transformed into minimizing $J_0(\mathcal{T}_{int})$ (here $t_0 = 0$ and $t_n = T$) forms a difficult task for critical points examination as well as for the numerical computations (see e.g. [2,3,7]). One of the computationally feasible schemes handling (5) is *a Leap-Frog Algorithm*. For optimizing J_0 this scheme is based on the sequence of single variable iterative optimization which in k-th iteration minimizes $J_0^{(k,i)}(s) = \int_{t_{i-1}^k}^{t_{i+1}^{k-1}}\|\ddot\gamma_{k,i}^{CS}(s)\|^2 ds$, over $I_i^{k-1} = [t_{i-1}^k, t_{i+1}^{k-1}]$. Here t_i^k is set to be a free variable denoted as s_i. The complete spline $\gamma_{k,i}^{CS} : I_i^{k-1} \to \mathbb{E}^m$ is determined by $\{t_{i-1}^k, s_i, t_{i+1}^{k-1}\}$, both velocities $\{v_{i-1}^k, v_{i+1}^{k-1}\}$ and the interpolation points $\{x_{i-1}, x_i, x_{i+1}\}$. Once s_i^{opt} is found one updates t_i^{k-1} with $t_i^k = s_i^{opt}$ and v_i^{k-1} with the $v_i^k = \dot\gamma_{k,i}^{CS}(s_i^{opt})$. Next we pass to the shifted overlapped sub-interval $I_{i+1}^k = [t_i^k, t_{i+2}^{k-1}]$ and repeat the previous step of updating t_{i+1}^{k-1}. Note that both cases $[0, t_2^{k-1}]$ and $[t_{n-2}^{k-1}, T]$ rely on splines discussed in *(iii)*, where the vanishing acceleration replaces one of the velocities v_0^{k-1} or v_n^{k-1}. Once t_{n-1}^{k-1} is changed over the last sub-interval $I_{n-1}^{k-1} = [t_{n-2}^k, T]$

the k-th iteration is terminated and the next local optimization over $I_1^k = [0, t_2^k]$ represents the beginning of the $(k+1)$-st iteration of *Leap-Frog*. The initialization of $\mathcal{T}_{int}$ for *Leap-Frog* can follow normalized *cumulative chord parameterization* (see e.g. [9]) which sets $t_0^0 = 0, t_1^0, \ldots, t_{n-1}^0, t_n^0 = T$ according to $t_0^0 = 0$ and $t_{i+1}^0 = \|x_{i+1} - x_i\| \frac{T}{\hat{T}} + t_i^0$, for $i = 0, 1, \ldots, n-1$ and $\hat{T} = \sum_{i=0}^{n-1} \|x_{i+1} - x_i\|$.

3   Non-generic Case: First Acceleration and Last Velocity

Let for $x_0, x_1, x_2 \in \mathcal{M}$ (for $n \geq 3$) the corresponding knots (see [1]) t_0 and t_2 be somehow given together with the respective first acceleration and the last velocity $a_0, v_2 \in \mathbb{R}^m$ (without loss $t_0 = 0$). We construct now a C^2 piecewise cubic $\gamma_0^c : [t_0, t_2] \to \mathbb{E}^m$ depending on varying $t_1 \in (t_0, t_2)$ (i.e. a cubic on each $[t_0, t_1]$ and $[t_1, t_2]$) satisfying $\gamma_0^c(t_j) = x_j$ (for $j = 0, 1, 2$), $\ddot{\gamma}_0^c(t_0) = a_0$ and $\dot{\gamma}_0^c(t_2) = v_2$. With $\phi_0 : [t_0, t_2] \to [0, 1]$ (with $\phi_0(t) = (t - t_0)(t_2 - t_0)^{-1}$) the re-parameterized curve $\tilde{\gamma}_0^c = \gamma_0^c \circ \phi_0^{-1} : [0, 1] \to \mathbb{E}^m$ satisfies, for $0 < s_1 < 1$ (where $s_1 = \phi_0(t_1)$): $\tilde{\gamma}_0^c(0) = x_0$, $\tilde{\gamma}_0^c(s_1) = x_1$ and $\tilde{\gamma}_0^c(1) = x_2$, with the adjusted $\tilde{a}_0, \tilde{v}_2 \in \mathbb{R}^m$ equal to:

$$\tilde{a}_0 = \tilde{\gamma}_0^{c''}(0) = (t_2 - t_0)^2 a_0 , \quad \tilde{v}_2 = \tilde{\gamma}_0^{c'}(1) = (t_2 - t_0)v_2 . \tag{6}$$

An easy inspection shows (for each $s_1 = \phi_0(t_1)$):

$$\tilde{\mathcal{E}}_0(s_1) = \int_0^1 \|\tilde{\gamma}_0^{c''}(s)\|^2 ds = (t_2 - t_0)^3 \int_{t_0}^{t_2} \|\ddot{\gamma}_0^c(t)\|^2 dt = (t_2 - t_0)^3 \mathcal{E}_0(t_1) . \tag{7}$$

Thus critical points s_1^{crit} of $\tilde{\mathcal{E}}_0$ are mapped (and vice versa) onto the corresponding critical points $t_1^{crit} = \phi_0^{-1}(s_1^{crit}) = s_1^{crit}(t_2 - t_0) + t_0$ of $\mathcal{E}_0$. Hence optimal points of $\tilde{\mathcal{E}}_0$ and $\mathcal{E}_0$ satisfy $t_1^{opt} = \phi_0^{-1}(s_1^{opt})$. Thus by (7) to decrease $\mathcal{E}_0$ it suffices to decrease $\tilde{\mathcal{E}}_0$. To find the expression for $\tilde{\mathcal{E}}_0$ we determine $\tilde{\gamma}_0^c$ (depending on s_1)

$$\tilde{\gamma}_0^c(s) = \begin{cases} \tilde{\gamma}_0^{lc}(s) , & \text{for } s \in [0, s_1] \\ \tilde{\gamma}_0^{rc}(s) , & \text{for } s \in [s_1, 1] \end{cases} \tag{8}$$

with $c_{0j}, d_{0j} \in \mathbb{R}^m$ and $\tilde{\gamma}_0^{lc}(s) = c_{00} + c_{01}(s - s_1) + c_{02}(s - s_1)^2 + c_{03}(s - s_1)^3$ and $\tilde{\gamma}_0^{rc}(s) = d_{00} + d_{01}(s - s_1) + d_{02}(s - s_1)^2 + d_{03}(s - s_1)^3$, the following holds:

$$\tilde{\gamma}_0^{lc}(0) = x_0 , \quad \tilde{\gamma}_0^{lc}(s_1) = \tilde{\gamma}_0^{rc}(s_1) = x_1 , \quad \tilde{\gamma}_0^{rc}(1) = x_2 , \tag{9}$$

$$\tilde{\gamma}_0^{lc''}(0) = \tilde{a}_0 , \quad \tilde{\gamma}_0^{rc'}(1) = \tilde{v}_2 , \tag{10}$$

together with the smoothness (C^1 and C^2) constraints at $s = s_1$ i.e.:

$$\tilde{\gamma}_0^{lc'}(s_1) = \tilde{\gamma}_0^{rc'}(s_1) , \quad \tilde{\gamma}_0^{lc''}(s_1) = \tilde{\gamma}_0^{rc''}(s_1) . \tag{11}$$

We may assume that (upon shifting the origin of coordinate system) $\tilde{x}_0 = x_0 - x_1$, $\tilde{x}_1 = \mathbf{0}$, $\tilde{x}_2 = x_2 - x_1$ and therefore by (9) we have

$$\tilde{\gamma}_0^c(0) = \tilde{x}_0 , \quad \tilde{\gamma}_0^c(s_1) = \mathbf{0} , \quad \tilde{\gamma}_0^c(1) = \tilde{x}_2 . \tag{12}$$

In sequel, combining (11) with $\tilde{x}_1$ vanishing gives

$$\tilde{\gamma}_0^{lc}(s) = c_{01}(s - s_1) + c_{02}(s - s_1)^2 + c_{03}(s - s_1)^3 ,$$
$$\tilde{\gamma}_0^{rc}(s) = c_{01}(s - s_1) + c_{02}(s - s_1)^2 + d_{03}(s - s_1)^3 , \qquad (13)$$

with $c_{00} = d_{00} = \mathbf{0}$. The unknown vectors $c_{01}, c_{02}, c_{03}, d_{03}$ are determined by solving the system of four linear vector equations obtained from (10) and (12):

$$\tilde{x}_0 = -c_{01}s_1 + c_{02}s_1^2 - c_{03}s_1^3 ,$$
$$\tilde{x}_2 = c_{01}(1 - s_1) + c_{02}(1 - s_1)^2 + d_{03}(1 - s_1)^3 ,$$
$$\tilde{a}_0 = 2c_{02} - 6c_{03}s_1 ,$$
$$\tilde{v}_2 = c_{01} + 2c_{02}(1 - s_1) + 3d_{03}(1 - s_1)^2 . \qquad (14)$$

An inspection reveals that:

$$c_{01} = \frac{s_1^2\tilde{a}_0 + 2s_1^3\tilde{a}_0 - s_1^4\tilde{a}_0 + 4s_1^2\tilde{v}_2 - 4s_1^3\tilde{v}_2 + 6\tilde{x}_0 - 12s_1\tilde{x}_0 + 6s_1^2\tilde{x}_0 - 12s_1^2\tilde{x}_2}{2s_1(s_1^2 + 2s_1 - 3)} ,$$

$$c_{02} = -\frac{-s_1^2\tilde{a}_0 + s_1^3\tilde{a}_0 - 3s_1\tilde{v}_2 + 3s_1^2\tilde{v}_2 + 6\tilde{x}_0 - 6s_1\tilde{x}_0 + 9s_1\tilde{x}_2}{s_1(s_1 - 1)(s_1 + 3)} ,$$

$$c_{03} = -\frac{-s_1\tilde{a}_0 + s_1^3\tilde{a}_0 - 2s_1\tilde{v}_2 + 2s_1^2\tilde{v}_2 + 4\tilde{x}_0 - 4s_1\tilde{x}_0 + 6s_1\tilde{x}_2}{2s_1^2(s_1^2 + 2s_1 - 3)} ,$$

$$d_{03} = -\frac{s_1^2\tilde{a}_0 - 2s_1^3\tilde{a}_0 + s_1^4\tilde{a}_0 + 6s_1\tilde{v}_2 - 8s_1^2\tilde{v}_2 + 2s_1^3\tilde{v}_2 - 6\tilde{x}_0 + 12s_1\tilde{x}_0 - 6s_1^2\tilde{x}_0}{2s_1(s_1 + 3)(s_1 - 1)^3}$$

$$-\frac{-12s_1\tilde{x}_2 + 8s_1^2\tilde{x}_2}{2s_1(s_1 + 3)(s_1 - 1)^3} \qquad (15)$$

satisfy (14) (as functions in s_1). As $\|\tilde{\gamma}_0^{lc''}(s)\|^2 = 4\|c_{02}\|^2 + 24\langle c_{02}|c_{03}\rangle(s - s_1) + 36\|c_{03}\|^2(s-s_1)^2$, $\|\tilde{\gamma}_0^{rc''}(s)\|^2 = 4\|c_{02}\|^2 + 24\langle c_{02}|d_{03}\rangle(s - s_1) + 36\|d_{03}\|^2(s-s_1)^2$:

$$\tilde{\mathcal{E}}_0(s_1) = \int_0^{s_1} \|\tilde{\gamma}_0^{lc''}(s)\|^2 ds + \int_{s_1}^1 \|\tilde{\gamma}_0^{rc''}(s)\|^2 ds = I_1 + I_2,$$

where $I_1 = 4(\|c_{02}\|^2 s_1 - 3\langle c_{02}|c_{03}\rangle s_1^2 + 3\|c_{03}\|^2 s_1^3)$ and $I_2 = 4(\|c_{02}\|^2(1 - s_1) + 3\langle c_{02}|d_{i3}\rangle(1-s_1)^2 + 3\|d_{03}\|^2(1-s_1)^3)$. The latter combined with $a_0 = \mathbf{0}$ (and $\tilde{a}_0 = \mathbf{0}$ - see (6)), (13) and (15) yields (with *Mathematica Integrate* and *FullSimplify*):

$$\tilde{\mathcal{E}}_0(s_1) = \frac{-1}{(s_1 + 3)s_1^2(s_1 - 1)^3}(12(-\|\tilde{x}_0\|^2(s_1 - 1)^3 + s_1(\|\tilde{x}_2\|^2(3 - 2s_1)s_1$$
$$+ \|\tilde{v}_2\|^2(s_1 - 1)^2 s_1 + (s_1 - 1)^3\langle \tilde{v}_2|\tilde{x}_0\rangle - (s_1 - 3)(s_1 - 1)s_1\langle \tilde{v}_2|\tilde{x}_2\rangle$$
$$+ 3(s_1 - 1)^2\langle \tilde{x}_0|\tilde{x}_2\rangle))) . \qquad (16)$$

Note also that $\lim_{s_1 \to 0^+} \tilde{\mathcal{E}}_0(s_1) = (12\|\tilde{x}_0\|^2/0^+) = +\infty$, and $\lim_{s_1 \to 1^-} \tilde{\mathcal{E}}_0(s_1) = (12\|\tilde{x}_2\|^2/0^+) = +\infty$. Hence as $\tilde{\mathcal{E}}_0 \geq 0$ and $\tilde{\mathcal{E}}_0 \in C^1$ the global minimum $s_1^{opt} \in$

$(0, 1)$ exists (one of critical points of $\tilde{\mathcal{E}}_0$). Note that $x_i \neq x_{i+1}$ yields $\|\tilde{x}_0\| \neq 0$ and $\|\tilde{x}_2\| \neq 0$. An inspection shows that (or differentiate symbolically in *Mathematica* $\tilde{\mathcal{E}}_0$ and use *FullSimplify*):

$$\tilde{\mathcal{E}}_0'(s_1) = \frac{1}{(s_1-1)^4 s_1^3 (s_1+3)^2} (12(2\|\tilde{v}_2\|^2 (s_1-1)^2 s_1^3 (s_1+1) - 6\|\tilde{x}_2\|^2 s_1^3 (s_1^2 - 3)$$
$$- 3\|\tilde{x}_0\|^2 (s_1-1)^4 (s_1+2) + (s_1-1)^4 s_1 (2s_1+3)\langle \tilde{v}_2 | \tilde{x}_0 \rangle$$
$$- 2(s_1-1)s_1^3(-6 + (-3+s_1)s_1)\langle \tilde{v}_2 | \tilde{x}_2 \rangle$$
$$+ 3(s_1-1)^2 s_1 (-3 + s_1(4+3s_1))\langle \tilde{x}_0 | \tilde{x}_2 \rangle)) . \tag{17}$$

The numerator of (17) is the polynomial of degree 6 i.e. $N_0(s_1) = b_0^0 + b_1^0 s_1 + b_2^0 s_1^2 + b_3^0 s_1^3 + b_4^0 s_1^4 + b_5^0 s_1^5 + b_6^0 s_1^6$, with the coefficients $b_j^0 \in \mathbb{R}$ (for $j = 0, 1, \ldots, 6$) equal to (use e.g. *Mathematica* functions *Factor* and *CoefficientList*): $\frac{b_0^0}{12} = -6\|\tilde{x}_0\|^2$, $\frac{b_1^0}{12} = 21\|\tilde{x}_0\|^2 + 3\langle \tilde{v}_2 | \tilde{x}_0 \rangle - 9\langle \tilde{x}_0 | \tilde{x}_2 \rangle$, $\frac{b_2^0}{12} = -24\|\tilde{x}_0\|^2 - 10\langle \tilde{v}_2 | \tilde{x}_0 \rangle + 30\langle \tilde{x}_0 | \tilde{x}_2 \rangle$, $\frac{b_3^0}{12} = 2\|\tilde{v}_2\|^2 + 6\|\tilde{x}_0\|^2 + 18\|\tilde{x}_2\|^2 + 10\langle \tilde{v}_2 | \tilde{x}_0 \rangle - 12\langle \tilde{v}_2 | \tilde{x}_2 \rangle - 24\langle \tilde{x}_0 | \tilde{x}_2 \rangle$, $\frac{b_4^0}{12} = -2\|\tilde{v}_2\|^2 + 6\|\tilde{x}_0\|^2 + 6\langle \tilde{v}_2 | \tilde{x}_2 \rangle - 6\langle \tilde{x}_0 | \tilde{x}_2 \rangle$, $\frac{b_5^0}{12} = -2\|\tilde{v}_2\|^2 - 3\|\tilde{x}_0\|^2 - 6\|\tilde{x}_2\|^2 - 5\langle \tilde{v}_2 | \tilde{x}_0 \rangle + 8\langle \tilde{v}_2 | \tilde{x}_2 \rangle + 9\langle \tilde{x}_0 | \tilde{x}_2 \rangle$ and $\frac{b_6^0}{12} = 2\|\tilde{v}_2\|^2 + 2\langle \tilde{v}_2 | \tilde{x}_0 \rangle - 2\langle \tilde{v}_2 | \tilde{x}_2 \rangle$.

In a search for a global optimum of $\tilde{\mathcal{E}}_0$, instead of performing any iterative optimization scheme (relying on initial guess), one can invoke *Mathematica Package Solve* which easily finds all roots (real and complex) for a given low order polynomial in one variable. Upon computing the roots of $N_0(s_1)$ we select only these which are real and belong to $(0, 1)$. Next we evaluate $\tilde{\mathcal{E}}_0$ on each critical point $s_1^{crit} \in (0, 1)$ and choose s_1^{crit} with minimal energy $\tilde{\mathcal{E}}_0$ as optimal. Again this particular property of optimizing $\tilde{\mathcal{E}}_0$ is very useful for future Leap-Frog Algorithm as compared with optimizing multiple variable function (5). We analyze in the next section the character of the energy $\tilde{\mathcal{E}}_0$ for the special case (18).

The case of *first velocity and last acceleration given* (covering the last three interpolation points $x_{n-2}, x_{n-1}, x_n \in \mathcal{M}$) is symmetric and as such is omitted.

4   Special Conditions for Non-generic Case of Leap-Frog

For a global minimum of $\tilde{\mathcal{E}}_0$ the analysis of $\tilde{\mathcal{E}}_0'(s_1) = 0$ reduces into finding all real roots of the sixth order polynomial $N_0(s_1)$. Consider a special case of $\tilde{x}_0, \tilde{x}_1, \tilde{x}_2 \in \mathbb{E}^m$ and $\tilde{v}_2 \in \mathbb{R}^m$ which satisfy (for some $k \neq 0$ as $\tilde{x}_0 \neq \mathbf{0}$):

$$\tilde{x}_2 - \tilde{x}_0 = \tilde{v}_2 \qquad \tilde{x}_0 = k\tilde{x}_2 . \tag{18}$$

Remark 1. The case of $k < 0$ yields the so-called *co-linearly ordered data* $\tilde{x}_0, \tilde{x}_1, \tilde{x}_2$. Here the function $x(s) = s\tilde{x}_2 + (1-s)\tilde{x}_0$ (with $s \in (0, 1)$) satisfies required constraints (18), namely: $x(0) = \tilde{x}_0$, $x(1) = \tilde{x}_2$, $x'(1) = \tilde{x}_2 - \tilde{x}_0 = \tilde{v}_2$, and $x''(0) = \mathbf{0}$. For $k \neq 0$ the normalized cumulative chord reads

$$\hat{s}_1^{cc} = \hat{s}_1^{cc}(k) = \frac{\|\tilde{x}_0\|}{\|\tilde{x}_2\| + \|\tilde{x}_0\|} = \frac{|k|}{1 + |k|} . \tag{19}$$

Noticeably, $\|\tilde{x}_2\| \neq 0$ and $\|\tilde{x}_0\| \neq 0$ as $x_0 \neq x_1$ and $x_1 \neq x_2$. Thus since $|k| = -k$ (for $k < 0$) by (18) we have $x\left(\frac{|k|}{1+|k|}\right) = \tilde{x}_2 \frac{|k|}{1+|k|} + \tilde{x}_0 \frac{1}{1+|k|} = \tilde{x}_2\left(\frac{|k|}{1+|k|} + \frac{k}{1+|k|}\right) = 0 = \tilde{x}_1$. Hence the interpolation condition $x(s_1) = 0$ is satisfied with $s_1 = \hat{s}_1^{cc}$. In sequel since $\|x''(s)\| = 0$ over $(0,1)$, the integral (7) vanishes with $s = \hat{s}_1^{cc}$. Thus $\hat{s}_1^{cc}$ is a global minimizer of $\tilde{\mathcal{E}}_0$ for any $k < 0$. This does not hold for $k > 0$. Note that for n big the case $k > 0$ prevails. $\qquad\square$

By (18) we have $\|\tilde{x}_0\|^2 = k^2\|\tilde{x}_2\|^2$, $\|\tilde{v}_2\|^2 = (1-k)^2\|\tilde{x}_2\|^2$, $\langle \tilde{v}_2|\tilde{x}_2\rangle = (1-k)\|\tilde{x}_2\|^2$, $\langle \tilde{v}_2|\tilde{x}_0\rangle = (k-k^2)\|\tilde{x}_2\|^2$, $\langle \tilde{x}_0|\tilde{x}_2\rangle = k\|\tilde{x}_2\|^2$. Substituting the latter into (16) yields (upon using e.g. *FullSimplify* in *Mathematica*) $\tilde{\mathcal{E}}_0(s) = \tilde{\mathcal{E}}_0^d(s) = (-12(\|\tilde{x}_2\|(k+s(1-k)))^2)/((s-1)^3 s^2(3+s))$. The latter vanishes iff $s = -k/(1-k)$ which for $k < 0$ reads as $s = |k|/(1+|k|) = \hat{s}_1^{cc}(k)$ (as previously $\hat{s}_1^{cc}$ can be treated as a function in k). In a search for other critical points of $\tilde{\mathcal{E}}_0^d$, the respective derivative $\tilde{\mathcal{E}}_0^{d'}$ (e.g. use symbolic differentiation in *Mathematica* and *FullSimplify*) reads:

$$\tilde{\mathcal{E}}_0^{d'}(s) = \frac{24\|\tilde{x}_2\|^2(k+s(1-k))(-3k+6ks+s^2(4-k)+s^3(2-2k))}{(s-1)^4 s^3 (3+s)^2}. \qquad (20)$$

If $k \neq 1$ (i.e. $\tilde{x}_2 \neq \tilde{x}_0$) the first numerator's factor of (20) yields exactly one root $\hat{s}_1^L = -k/(1-k)$. Only for $k < 0$ we have $\hat{s}_1^L \in (0,1)$. Otherwise for $k > 1$ we have $\hat{s}_1^L > 1$ and for $0 < k < 1$ we have $\hat{s}_1^L < 0$. The analysis of $k < 0$ and $k > 0$ (with $k \neq 1$) for the second cubic factor in (20) is performed below. Note that with $k = 1$ the first linear factor in (20) has no roots and the second cubic factor reduces in quadratic with the roots $\hat{s}_1^{\mp} = -1 \pm \sqrt{2}$. Only $\hat{s}_1 = \hat{s}_1^+ \in (0,1)$.

(i) Assume that $k < 0$ (the data are co-linearly ordered). The latter shows that cumulative chord $\hat{s}_1^{cc}$ (see the linear factor in (20)) is a critical point of $\tilde{\mathcal{E}}_0^d$.

To find the remaining critical points of $\tilde{\mathcal{E}}_0^d$ the real roots of $M_0(s) = -3k + 6ks + s^2(4-k) + s^3(2-2k)$ over $(0,1)$ are to be examined. Note that as $M_0(0) = -3k > 0$ (as $k < 0$) and $M_0(1) = 6 > 0$ for the existence of one critical point $\hat{s}_1$ of $\tilde{\mathcal{E}}_0'$ (i.e. here with $\hat{s}_1 = \hat{s}_1^{cc}$) it suffices to show that M_0 is positive at any of its critical points $\hat{u}_0 \in (0,1)$ (i.e. at points where $M_0'(\hat{u}_0) = 0$ with $M_0(\hat{u}_0) > 0$). Indeed $M_0'(u) = 0$ iff $3k + u(4-k) + u^2(3-3k) = 0$. The discriminant (as $k < 0$) $\tilde{\Delta}(k) = 16 - 44k + 37k^2 > 0$ and thus there are two different real roots $\hat{u}_0^-$ and $\hat{u}_0^+$ (both depending on parameter $k < 0$). Since $(-k)/(k-1) < 0$ (again as $k < 0$) by Vieta's formula both roots are of opposite signs. Thus as

$$\hat{u}_0^{\pm} = \hat{u}_0^{\pm}(k) = \frac{k - 4 \pm \sqrt{\tilde{\Delta}}}{6(1-k)} \qquad (21)$$

we have $\hat{u}_0^- < 0$ (since for $k < 0$ we have $k - 4 - \sqrt{\tilde{\Delta}} < 0$ and $6(1-k) > 0$). Hence $\hat{u}_0^+ > 0$ - this can be independently shown as being equivalent to true inequality $36k(k-1) > 0$. But as $M_0'(0) = 6k < 0$ and $M_0'(1) = 14 - 2k > 0$ by Intermediate Value Theorem we have that $\hat{u}_0^+ \in (0,1)$ (and it is a unique root of $M_0'(u) = 0$ over $(0,1)$). As also $M_0'(0) < 0$ for $u \in (0, \hat{u}_0^+)$ and $M_0'(0) > 0$ for $\hat{u} \in (u_0^+, 1)$ we have minimum of M_0 at $\hat{u}_0^+$ over interval

$(0, 1)$. Evaluating $M_0(\hat{u}_0^+(k))$ (see (21)) yields a function $\bar{M}_0(k) = M_0(\hat{u}_0^+(k))$ in k (again use. e.g. *Mathematica FullSimplify* and *Factor*) which reads: $\bar{M}_0(k) = \frac{-1}{54(k-1)^2}(-64 + 426k - 606k^2 + 217k^3 + 16\sqrt{\Delta(k)} - 44k\sqrt{\Delta(k)} + 37k^2\sqrt{\Delta(k)})$. By Taylor expansion we have $f(x) = \sqrt{1+x} = 1 + (1/2)x + (-1/(2(1+\xi)^{3/2}))x^2$ (with $0 < \xi < x$ if $x > 0$ and $x < \xi < 0$ if $x < 0$). Upon applying the latter to $\sqrt{\Delta(k)} = 4\sqrt{1 + (37/16)k^2 - (11/4)k}$ with $x = (37/16)k^2 - (11/4)k$ we arrive at $\lim_{k\to 0^-} \bar{M}_0(k) = 0$. Again by Taylor expansion, the domineering factor (recall for $k \approx O^-$ we have $|k| \geq |k|^\alpha$ with $\alpha \geq 1$) in the expression for $M_0(k)$ (inside the brackets) is a linear component $426k - 88k - 176k = 162k < 0$, as $k < 0$ - (here the constant 64 is canceled). Thus $\bar{M}_0(k) > 0$, for sufficiently small $k < 0$. To show that $\lim_{k\to -\infty} \bar{M}_0(k) = -\infty$, it suffices to show $\lim_{k\to -\infty}((217k^3 + 37k^2\sqrt{\Delta(k)})/(k-1)^2) = +\infty$. The latter follows upon observation that (for $k < 0$) $217k^3 + 37k^2\sqrt{\Delta(k)} > 217k^3 + 37k^2\sqrt{36k^2} = 217k^3 + 222k^2|k| = -5k^3$. The *Mathematica* function *NSolve* applied to $\bar{M}_0(k) = 0$ yields two real roots i.e. $k_1 = 0$ (excluded as $\tilde{x}_0 = k\tilde{x}_2$) and $k_2 \approx -26.1326$. Next $\bar{M}_0'(k) = \frac{1}{54(k-1)^3}(-298 + 786k - 651k^2 + 217k^3 + 34\sqrt{\Delta(k)} - 89k\sqrt{\Delta(k)} + 37k^2\sqrt{\Delta(k)})$. Again, *Solve* applied to $\bar{M}_0'(k) = 0$ yields one root -4.61116. Combining the latter with the plotted graph of $\bar{M}_0(k)$ renders for each $k \in (k_2, 0)$ the following: $\bar{M}_0 > 0$ and thus $M_0 > 0$. Hence, for each $k \in (k_2, 0)$ there is only one critical point of $\tilde{\mathcal{E}}_0^d$ over $(0, 1)$ - i.e. cumulative chord $\hat{s}_1^{cc} = |k|/(1+|k|)$. Hence if any iterative optimization scheme for $\tilde{\mathcal{E}}_0^d$ is invoked a good initial guess can be an arbitrary number from the interval $(0, 1)$ (due to unimodality of $\tilde{\mathcal{E}}_0^d$). In case of small perturbations of (18) one expects that $\tilde{\mathcal{E}}_0^\delta$ preserves a similar pattern as its unperturbed counterpart $\tilde{\mathcal{E}}_0^0 = \tilde{\mathcal{E}}_0^d$ (see Proposition 1). Thus, for arbitrary $k_2 < k < 0$, a good initial guess to optimize $\tilde{\mathcal{E}}_0^\delta$ should be chosen in the proximity of cumulative chord $\hat{s}_1^{cc}(k)$.

Clearly for $k = k_2 \approx -26.1326$ (the second case when $k = 0$ is excluded) $M_0(\hat{u}_0^+(k_2)) = \bar{M}_0(k_2) = 0$ and $M_0'(\hat{u}_0^+(k_2)) = M_0'(k_2) = 0$ (see (21)). Thus we have one additional critical point $\hat{u}_0^+(k_2) \in (0, 1)$ of $\tilde{\mathcal{E}}_0^d$. Hence for k_2 there are exactly two critical points of $\tilde{\mathcal{E}}_0^d$ over $(0, 1)$ - one is a cumulative chord $\hat{s}_1^{cc}(k_2) \in (0, 1)$ (a global minimum) and the second one $\hat{s}_1^0 = u_0^+(k_2) \in (0, 1)$. Substituting $k_2 \approx -26.1326$ into (19) and (21) gives $\hat{u}_0^+(k_2) \approx 0.813606 < \hat{s}_1^{cc}(k_2) = (|k_2|/(1+|k_2|)) \approx 0.963144$. Of course $u_0^+(k_2)$ must be a a saddle-like point of $\tilde{\mathcal{E}}_0^d$ - recall here that $\tilde{\mathcal{E}}_0^d(\hat{s}_1^{cc}(k_2)) = 0$ (attained global minimum), which is smaller than $\tilde{\mathcal{E}}_0^d(\hat{u}_0^+(k_2)) \approx 12083.9\|\tilde{x}_2\|^2 > 0$ and that $\lim_{s\to 0^+} \tilde{\mathcal{E}}_0^d(s) = \lim_{s\to 1^-} \tilde{\mathcal{E}}_0^d(s) = +\infty$. This together with non-negativity of $\tilde{\mathcal{E}}_0^d$ and its smoothness over $(0, 1)$ implies that $\hat{u}_0^+(k_2)$ is a saddle-like point of $\tilde{\mathcal{E}}_0^d$. Thus for $k = k_2$ if any iterative optimization scheme for $\tilde{\mathcal{E}}_0^d$ is to be invoked a good initial guess can be an arbitrary number from the interval $(\hat{s}_1^{cc}(k_2), 1)$. As for small perturbations of (18) one expects the energy $\tilde{\mathcal{E}}_0^\delta$ to have as similar pattern as its unperturbed counterpart $\tilde{\mathcal{E}}_0^0 = \tilde{\mathcal{E}}_0^d$. Thus, for $k = k_2$, a good initial guess to optimize $\tilde{\mathcal{E}}_0^\delta$ should also be taken closely to cumulative chord $\hat{s}_1^{cc}(k_2)$ and preferably from $(\hat{s}_1^{cc}(k_2), 1)$.

For $k \in (-\infty, k_2)$ we have $M_0(\hat{u}_0^+(k)) = \bar{M}_0(k) < 0$. Thus for each $k \in (-\infty, k_2)$, as $M_0(0) > 0$, $M_0(1) > 0$, $M_0(\hat{u}_0^+(k)) < 0$ and M_0' vanishes over $(0,1)$ only at $\hat{u}_0^+(k) \in (0,1)$, there are exactly two different additional critical points $\hat{s}_0^1(k), \hat{s}_0^2(k) \in (0,1)$ (say $\hat{s}_0^1(k) < \hat{s}_0^2(k)$) of $\tilde{\mathcal{E}}_0^d$ (as $M_0(s_0^{1,2}(k)) = 0$). Clearly the inequalities hold $\hat{s}_0^1(k) < \hat{u}_0^+(k) < \hat{s}_0^2(k)$. Recall that the global minimum of $\tilde{\mathcal{E}}_0^d$ is attained at cumulative chord $\hat{s}_1^{cc}$, where $\tilde{\mathcal{E}}_0^d$ vanishes. Note also that here $\hat{s}_1^{cc} \approx 1$ as $\hat{s}_1^{cc}(k) = |k|/(1 + |k|) \in (|k_2|/(1 + |k_2|), 1) = (0.963144, 1)$ for $k \in (-\infty, k_2)$. To show the latter use the facts that $f(x) = (x/(1 + x))$ is increasing and $0 \le f(x) < 1$ for $0 \le x \le 1$ and $\lim_{x \to 1-} f(x) = 1$. In addition, as $M_0(\hat{s}_1^{cc}(k)) = (3 - 4k)k/(-1 + k)^2 \neq 0$ for $k \in (-\infty, k_2)$, and $M_0(\hat{s}_0^{1,2}(k)) = 0$ we have $\hat{s}_1^{cc}(k) \neq s_0^{1,2}(k)$. In Remark 2 we will show that none of $\hat{s}_0^{1,2}$ can be saddle-like points (for $k \in (-\infty, k_2)$). Thus as $\lim_{s \to 0+} \tilde{\mathcal{E}}_0^d(s) = \lim_{s \to 1-} \tilde{\mathcal{E}}_0^d(s) = +\infty$ and $\tilde{\mathcal{E}}_0^d(\hat{s}_1^{cc}(k))$ attains its global minimum both $\hat{s}_0^{1,2}$ must be on one side of $\hat{s}_1^{cc}$ (as otherwise that would imply both $\hat{s}_0^1$ and $\hat{s}_0^2$ to be saddle-like points, since no more than 3 critical points of $\tilde{\mathcal{E}}_0^d$ over $(0,1)$ exist). Thus to prove that $\hat{u}_0^{1,2}(k) < \hat{s}_1^{cc}(k)$ its suffices to show $\hat{s}_0^1(k) < \hat{s}_1^{cc}(k)$. For the latter (as $\hat{s}_0^1(k) < \hat{u}_0^+(k)$) it is sufficient to prove $\hat{u}_0^+(k) < \hat{s}_1^{cc}(k)$ (where $k \in (-\infty, k_2)$). A simple inspection shows that $\hat{u}_0^+(k) = (4 - k - \sqrt{16 - 44k + 37k^2})/(6(k - 1)) < \hat{s}_1^{cc}(k) = |k|/(1 + |k|)$ holds as this requires $4 - 7k > \sqrt{\Delta}$ which is true since $k(k - 1) > 0$ (with $k < 0$). Thus at $\hat{s}_0^1(k)$ (at $\hat{s}_0^2(k)$) we have a local minimum (maximum) of $\tilde{\mathcal{E}}_0^d$. Again, if any iterative optimization scheme for $\tilde{\mathcal{E}}_0^d$ is invoked a good initial guess can be an arbitrary number from the interval $(\hat{s}_1^{cc}, 1)$. For small perturbations of (18) one expects the energy $\tilde{\mathcal{E}}_0^\delta$ to preserve a pattern of its unperturbed counterpart $\tilde{\mathcal{E}}_0^0 = \tilde{\mathcal{E}}_0^c$ (see Proposition 1). Consequently, for any $k \in (-\infty, -26.1326)$, a good initial guess can be taken as a close to $\hat{s}_1^{cc}(k)$ and from the interval $(\hat{s}_1^{cc}(k), 1)$.

(ii) Assume now that $k > 0$ (the data are *co-linearly unordered*). As here $s_1^L = -k/(1 - k) \notin (0,1)$ (see (20)) the critical points of $\tilde{\mathcal{E}}_0^d$ coincides with the roots of a cubic $N_c(s) = -3k + 6ks + s^2(4 - k) + s^3(2 - 2k)$.

Evidently for that for $0 < k < 1$ there is only one change of signs of the coefficients (three are positive and one is negative). By Fermat sign principle there exists up to one positive root of $N_c(s)$. The analysis from Sect. 3 assures the existence of at least one critical point of $\tilde{\mathcal{E}}_0$ over $(0,1)$ ($\tilde{\mathcal{E}}_0^d$ is a special case of $\tilde{\mathcal{E}}_0$). Thus for $0 < k < 1$ there exists exactly one critical point $\hat{s}_1 \in (0,1)$ (a global minimum) of $\tilde{\mathcal{E}}_0^d$.

The case $k = 1$ (with $\tilde{x}_2 = \tilde{x}_0$) already analyzed yields one critical point $\hat{s}_1 = -1 + \sqrt{2} \in (0,1)$ of $\tilde{\mathcal{E}}_0^d$.

Finally, for $k > 1$ as $N_c(0) = -3k < 0$, $N_c(1) = 6 > 0$ and $N_c \in C^\infty$, it suffices to show that there is up to one root of $N_c'(u) = 0$, for $u \in (0,1)$. Of course no roots for $N_c'(u) = 0$ yields strictly increasing $N_c(s)$ over $(0,1)$ (as $N_c(0) < 0$ and $N_c(1) > 0$) and hence exactly one critical point $\hat{s}_1 \in (0,1)$ (a global minimum) of $\tilde{\mathcal{E}}_0^d$. The existence of exactly one root $u_1 \in (0,1)$ for $N_c'(u) = 0$, still yields (since $N_c(0) < 0$ and $N_c(1) > 0$) exactly one root $\hat{s}_1 \in (0,1)$ of $N_c(s) = 0$ (this time N_c is not monotonic with the exception of the case when $\hat{s}_1 = u_1$, i.e. when u_1 is a saddle-like point of N_c). To show that $N_c'(s) = 6k + 2s(4 - k) + 6s^2(1 - k) = 0$ has up to one root over $(0,1)$, note that as now $1 - k < 0$ we have $\lim_{s \to \pm\infty} N'_c(s) = -\infty$. The latter combined with $N'_c(0) = 6k > 0$ assures the existence of exactly one negative and one positive root $N'_c(s) = 0$. This completes the proof.

Thus for $k > 0$ the energy $\tilde{\mathcal{E}}^d_0$ has exactly one critical point $\hat{s}_1 \in (0, 1)$.

5   Perturbed Special Case

The case when $\{\tilde{x}_0, \tilde{x}_1 = \mathbf{0}, \tilde{x}_2\}$ and $\tilde{v}_2$ do not satisfy (18) is considered now. Upon introducing a perturbation vector $\delta = (\delta_1, \delta_2) \in \mathbb{R}^{2m}$ the question arises whether a unimodality of $\tilde{\mathcal{E}}^d_0 = \tilde{\mathcal{E}}^{\delta=0}_0$ (holding for special data (18)) extends to the perturbed case i.e. to $\tilde{\mathcal{E}}^\delta_0$. In doing so, assume the following holds:

$$\tilde{x}_2 - \tilde{x}_0 - \tilde{v}_2 = \delta_1 \,, \qquad \tilde{x}_0 - k\tilde{x}_2 = \delta_2 \,, \tag{22}$$

(for some $k \neq 0$) with the corresponding $\tilde{\mathcal{E}}^\delta_0$ derived as in (16) (see also (23)). For $\delta_1 = \delta_2 = \mathbf{0} \in \mathbb{R}^m$ formulas (22) reduce into (18) (i.e. $\tilde{\mathcal{E}}^0_0 = \tilde{\mathcal{E}}_0$ derived for (18)). For an explicit formula of $\tilde{\mathcal{E}}^\delta_0$ and $\tilde{\mathcal{E}}^{\delta'}_0$ we resort to (recalling (22)): $\|\tilde{x}_0\|^2 = k^2\|\tilde{x}_2\|^2 + \|\delta_2\|^2 + 2k\langle\tilde{x}_2|\delta_2\rangle$, $\|\tilde{v}_2\|^2 = (1-k)^2\|\tilde{x}_2\|^2 + \|\delta_1\|^2 + \|\delta_2\|^2 + 2\langle\delta_1|\delta_2\rangle - 2(1-k)\langle\tilde{x}_2|\delta_1\rangle - 2(1-k)\langle\tilde{x}_2|\delta_2\rangle$, $\langle\tilde{v}_2|\tilde{x}_2\rangle = (1-k)\|\tilde{x}_2\|^2 - \langle\tilde{x}_2|\delta_1\rangle - \langle\tilde{x}_2|\delta_2\rangle$, $\langle\tilde{v}_2|\tilde{x}_0\rangle = (k - k^2)\|\tilde{x}_2\|^2 + (1 - 2k)\langle\tilde{x}_2|\delta_2\rangle - k\langle\tilde{x}_2|\delta_1\rangle - \langle\delta_1|\delta_2\rangle - \|\delta_2\|^2$ and $\langle\tilde{x}_0|\tilde{x}_2\rangle = k\|\tilde{x}_2\|^2 + \langle\tilde{x}_2|\delta_2\rangle$. Substituting the latter into the formula for $\tilde{\mathcal{E}}_0$ (here $s_1 = s$) (see (16)) yields (we use here *Mathematica* function *FullSimplify*)

$$
\begin{aligned}
\tilde{\mathcal{E}}^\delta_0(s) = \frac{-1}{(s-1)^3 s^2 (s+3)} &\big(12(\|\delta_2\|^2(s-1)^2 + k^2\|\tilde{x}_2\|^2(s-1)^2 \\
&+ s(\langle\delta_1|\delta_2\rangle(s-1)^2(1+s) + s(\|\tilde{x}_2\|^2 + \|\delta_1\|^2(s-1)^2 + \langle\tilde{x}_2|\delta_1\rangle \\
&- s^2\langle\tilde{x}_2|\delta_1\rangle - 2\langle\tilde{x}_2|\delta_2\rangle) + 2\langle\tilde{x}_2|\delta_2\rangle) + k(s-1)(-2\|\tilde{x}_2\|^2 s \\
&+ (s-1)(s(1+s)\langle\tilde{x}_2|\delta_1\rangle + 2\langle\tilde{x}_2|\delta_2\rangle)))))
\end{aligned}
\tag{23}
$$

which upon simplification yields $\tilde{\mathcal{E}}^\delta_0(s) = (-12M^\delta_0(s))/((s-1)^3 s^2(s+3))$, where M^δ_0 is the 4-th order polynomial in s with the coefficients (use *Mathematica* functions *Factor* and *CoefficientList*): $a^{0,\delta}_0 = \|\delta_2\|^2 + k^2\|\tilde{x}_2\|^2 + 2k\langle\tilde{x}_2|\delta_2\rangle$, $a^{0,\delta}_1 = \langle\delta_1|\delta_2\rangle - 2\|\delta_2\|^2 + 2k\|\tilde{x}_2\|^2 - 2k^2\|\tilde{x}_2\|^2 + k\langle\tilde{x}_2|\delta_1\rangle + 2\langle\tilde{x}_2|\delta_2\rangle - 4k\langle\tilde{x}_2|\delta_2\rangle$, $a^{0,\delta}_2 = -\langle\delta_1|\delta_2\rangle + \|\delta_1\|^2 + \|\delta_2\|^2 + \|\tilde{x}_2\|^2 - 2k\|\tilde{x}_2\|^2 + k^2\|\tilde{x}_2\|^2 + \langle\tilde{x}_2|\delta_1\rangle - k\langle\tilde{x}_2|\delta_1\rangle - 2\langle\tilde{x}_2|\delta_2\rangle + 2k\langle\tilde{x}_2|\delta_2\rangle$, $a^{0,\delta}_3 = -\langle\delta_1|\delta_2\rangle - 2\|\delta_1\|^2 - k\langle\tilde{x}_2|\delta_1\rangle$ and $a^{0,\delta}_4 = \langle\delta_1|\delta_2\rangle + \|\delta_1\|^2 - \langle\tilde{x}_2|\delta_1\rangle + k\langle\tilde{x}_2|\delta_1\rangle$ with the corresponding derivative (use symbolic differentiation in *Mathematica* and apply *Factor* and *CoefficientList*) $\tilde{\mathcal{E}}^{\delta'}_0(s) = (12N^\delta_0(s))/((s-1)^4 s^3(s+3)^2)$, where N^δ_0 is the 6-th order polynomial in s with the coefficients: $b^{0,\delta}_0 = -6\|\delta_2\|^2 - 6k^2\|\tilde{x}_2\|^2 - 12k\langle\tilde{x}_2|\delta_2\rangle$, $b^{0,\delta}_1 = -3\langle\delta_1|\delta_2\rangle + 18\|\delta_2\|^2 - 6k\|\tilde{x}_2\|^2 + 18k^2\|\tilde{x}_2\|^2 - 3k\langle\tilde{x}_2|\delta_1\rangle - 6\langle\tilde{x}_2|\delta_2\rangle + 36k\langle\tilde{x}_2|\tilde{\delta}_2\rangle$, $b^{0,\delta}_2 = 10\langle\delta_1|\delta_2\rangle - 14\|\delta_2\|^2 + 20k\|\tilde{x}_2\|^2 - 14k^2\|\tilde{x}_2\|^2 + 10k\langle\tilde{x}_2|\delta_1\rangle + 20\langle\tilde{x}_2|\delta_2\rangle - 28k\langle\tilde{x}_2|\delta_2\rangle$, $b^{0,\delta}_3 = -6\langle\delta_1|\delta_2\rangle + 2\|\delta_1\|^2 - 2\|\delta_2\|^2 + 8\|\tilde{x}_2\|^2 - 6k\|\tilde{x}_2\|^2 - 2k^2\|x_2\|^2 + 8\langle\tilde{x}_2|\delta_1\rangle - 6k\langle\tilde{x}_2|\delta_1\rangle - 6\langle\tilde{x}_2|\delta_2\rangle - 4k\langle\tilde{x}_2|\delta_2\rangle$, $b^{0,\delta}_4 = -4\langle\delta_1|\delta_2\rangle - 2\|\delta_1\|^2 + 4\|\delta_2\|^2 + 4\|\tilde{x}_2\|^2 - 8k\|\tilde{x}_2\|^2 + 4k^2\|\tilde{x}_2\|^2 - 2\langle\tilde{x}_2|\delta_1\rangle - 4k\langle\tilde{x}_2|\delta_1\rangle - 8\langle\tilde{x}_2|\delta_2\rangle + 8k\langle\tilde{x}_2|\delta_2\rangle$,

$b_5^{0,\delta} = \langle \delta_1 | \delta_2 \rangle - 2\|\delta_1\|^2 - 4\langle \tilde{x}_2 | \delta_1 \rangle + k \langle \tilde{x}_2 | \delta_1 \rangle$ and $b_5^{0,\delta} = 2\langle \delta_1 | \delta_2 \rangle + 2\|\delta_1\|^2 - 2\langle \tilde{x}_2 | \delta_1 \rangle + 2k \langle \tilde{x}_2 | \delta_1 \rangle$.

The following result holds (the proof is straightforward):

Proposition 1. *Assume that for unperturbed data (18) the corresponding energy $\tilde{\mathcal{E}}_0^0$ has exactly one critical point $\hat{s}_0 \in (0,1)$ at which $\tilde{\mathcal{E}}_0^{0''}(\hat{s}_0) > 0$. Then there exists sufficiently small $\varepsilon_0 > 0$ such that for all $\|\delta\| < \varepsilon_0$ (where $\delta = (\delta_1, \delta_2) \in \mathbb{R}^{2m}$) the perturbed data (22) yield the energy $\tilde{\mathcal{E}}_0^\delta$ with exactly one critical point $\hat{s}_0^\delta \in (0,1)$ (a global minimum $\hat{s}_0^\delta$ of $\tilde{\mathcal{E}}_0^\delta$ is sufficiently close to $\hat{s}_0$).*

Remark 2. The condition $\tilde{\mathcal{E}}_0^{0'}(s) = \tilde{\mathcal{E}}_0^{0''}(s) = 0$ excludes among all possible saddle-like points of $\tilde{\mathcal{E}}_0^0 = \tilde{\mathcal{E}}_0^d$ (which as shown, e.g. happens for $k = k_2 \approx -26.1326$). In a search for other possible saddle-like points for $k \le k_2$ varying we eliminate variable s $\tilde{\mathcal{E}}_0^{0'}(s) = 0$ and $\tilde{\mathcal{E}}_0^{0''}(s) = 0$ by resorting to *Mathematica* function *Eliminate*. Indeed upon symbolic differentiation of $\tilde{\mathcal{E}}_0^{\delta'}$ (and then putting $\delta = 0$) with *Mathematica* function *FullSimplify* we obtain $\tilde{\mathcal{E}}_0^{0''}(s) = \frac{-24\|\tilde{x}_2\|^2}{(s-1)^5 s^4 (s+3)^3}(27k^2 + (18k - 102k^2)s + (-72k + 120k^2)s^2 + (96k - 12k^2)s^3 + (46+28k-53k^2)s^4 + (40-50k+10k^2)s^5 + (10-20k+10k^2)s^6$. Using *Eliminate* function in *Mathematica* applied to $\tilde{\mathcal{E}}_0^{0'}(s) = \tilde{\mathcal{E}}_0^{0''}(s) = 0$ (in fact applied to the respective numerators of the first and the second derivatives of $\tilde{\mathcal{E}}_0^0$) leads to $576k^3 - 4209k^4 + 10636k^5 - 11277k^6 + 4152k^7 + 176k^8 = 0$ which when factorized (use e.g. *Mathematica Factor*) equals to $k^3(-3+4k)^2(64-297k+276k^2+11k^3) = 0$. *Mathematica* function *NSolve* yields four non-negative roots $k = 0$, $k = 3/4$, $k = 0.300288$ and $k = 0.741423$ (excluded as $k \le k_2$) and one negative $k = -26.1326 = k_2$. The latter is not only consistent with the previous analysis but also implies that the saddle-like point can only occur for $k = k_2$. This fact was used when we analyzed the critical points of $\tilde{\mathcal{E}}_0^0$ for different $k < 0$. □

Example 1. Let $\tilde{x}_0, \tilde{x}_1 = \mathbf{0}, \tilde{x}_2 \in \mathbb{E}^m$ be *co-linearly ordered* i.e. $\tilde{x}_0 = k\tilde{x}_2$, for some $k < 0$ and $\|\tilde{x}_2\| = 1$. The energy $\tilde{\mathcal{E}}_0^d$ with $\|\tilde{x}_2\| = 1$ reads here as $\tilde{\mathcal{E}}_0^d(s) = (-12((k+s-ks)^2))/((s-1)^3 s^2 (3+s))$. The plot of $\tilde{\mathcal{E}}_0^d$ with $k = -4, -15, -25 \in (k_2 \approx -26.1326, 0)$ is shown in Fig. 1. As already proved, in this case there is only *one critical point* of $\tilde{\mathcal{E}}_0^d$ at cumulative chords $\hat{s}_1^{cc}(k) = 4/5 = 0.6$, $\hat{s}_1^{cc}(k) =$

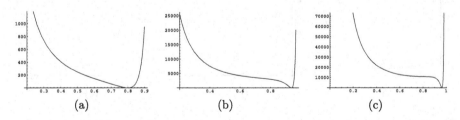

Fig. 1. The graph of $\tilde{\mathcal{E}}_0^d$ for different $\tilde{x}_0, \tilde{x}_1, \tilde{x}_2 \in \mathbb{E}^m$ co-linearly ordered with varying $k \in (k_2, 0)$ and $\|\tilde{x}_2\| = 1$: (a) $k = -4$ and a global minimum at $\hat{s}_1^{cc}(k) = 4/5 = 0.8$, (b) $k = -15$ and a global minimum at $\hat{s}_1^{cc}(k) = 15/16 \approx 0.9375$, (c) $k = -25$ and a global minimum at $\hat{s}_1^{cc}(k) = 25/26 \approx 0.961538$.

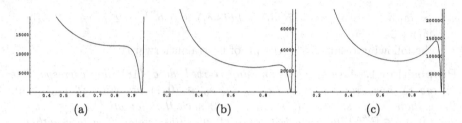

Fig. 2. The graph of $\tilde{\mathcal{E}}_0^d$ for $\tilde{x}_0, \tilde{x}_1, \tilde{x}_2 \in \mathbb{E}^m$ co-linearly ordered with varying $k \in (-\infty, k_2]$ and $\|\tilde{x}_2\| = 1$: (a) $k = k_2$ and a global minimum at $\hat{s}_1^{cc}(k_2) = (|k_2|/(1+|k_2|)) \approx 0.963144$ and saddle-like point at $\hat{s}_1^0(k_2) \approx 0.813607$, (b) $k = -35$ and a global minimum at $\hat{s}_1^{cc}(k) = 35/36 \approx 0.972222$ and two other critical points i.e. with local maximum $\hat{s}_0^2 \approx 0.897748$ and with local minimum $\hat{s}_0^1 \approx 0.743991$, (c) $k = -65$ and global minimum at $\hat{s}_1^{cc}(k) = 65/66 \approx 0.984848$ and two other critical points i.e. with local maximum $\hat{s}_0^2 \approx 0.950547$ and with local minimum $\hat{s}_0^1 \approx 0.711383$.

$15/16 \approx 0.9375$ and $\hat{s}_1^{cc}(k) = 25/26 \approx 0.961538$, respectively (where $\tilde{\mathcal{E}}_0^d(\hat{s}_1^{cc}(k)) = 0$). Similarly the plot of the corresponding energy $\tilde{\mathcal{E}}_0^d$ with $k = k_2 \approx -26.1326$ is shown in Fig. 2a). Here there are *two critical points* of $\tilde{\mathcal{E}}_0^d$ i.e. a global minimum at cumulative chord $\hat{s}_1^{cc}(k_2) \approx 26.1326/27.1326 \approx 0.963144$ (where $\tilde{\mathcal{E}}_0^d(\hat{s}_1^{cc}(k_2)) = 0$) and a saddle-like point $\hat{s}_1^0(k_2) \approx 0.813607$ (with $\tilde{\mathcal{E}}_0^d(\hat{s}_1^0(k_2)) \approx 12083.9$). Finally, the plot of $\tilde{\mathcal{E}}_0^d$ with $k = -35, -65 \in (-\infty, k_2)$ is shown Fig. 2b)–c). As established above, *a single global minimum* $\tilde{\mathcal{E}}_0^d$ is again taken at cumulative chord $\hat{s}_1^{cc}(k) = 35/36 \approx 0.972222$ (or at $\hat{s}_1^{cc}(k) = 65/66 \approx 0.984848$) with $\tilde{\mathcal{E}}_0^d(\hat{s}_1^{cc}(k)) = 0$. There are *other two critical points*: local maximum at $\hat{s}_0^2(k) \approx 0.897748$ with $\tilde{\mathcal{E}}_0^d(\hat{s}_0^2(k)) \approx 25683.8$ (or at $\hat{s}_0^2(k) \approx 0.950547$ with $\tilde{\mathcal{E}}_0^d(\hat{s}_0^2(k)) \approx 142466$) and local minimum at $\hat{s}_0^1(k) \approx 0.743991$ with $\tilde{\mathcal{E}}_0^d(\hat{s}_0^1(k)) \approx 23297$ (or at $\hat{s}_1^2(k) \approx 0.711383$ with $\tilde{\mathcal{E}}_0^d(\hat{s}_1^2(k)) \approx 86569.7$). Note that for $k = 35, 64$ as already proved the critical points and cumulative chord $\hat{s}_1^{cc}(k)$ satisfy $\hat{s}_0^{1,2}(k) < \hat{s}_1^{cc}(k)$ and $\hat{s}_1^{cc}(k) \approx 1$.

Let now $\tilde{x}_0, \tilde{x}_1 = \mathbf{0}, \tilde{x}_2 \in \mathbb{E}^m$ be *co-linearly unordered* i.e. $\tilde{x}_0 = k\tilde{x}_2$, for some $k > 0$ (here also $\|\tilde{x}_2\| = 1$). The corresponding energy $\tilde{\mathcal{E}}_0^d$ coincides with the one derived for $k < 0$. The plot of $\tilde{\mathcal{E}}_0^d$ with $k = 1/2, 1, 5$ is shown in Fig. 3. As proved for $k > 0$ there is only *one critical point (a global minimum)* of $\tilde{\mathcal{E}}_0^d$ different

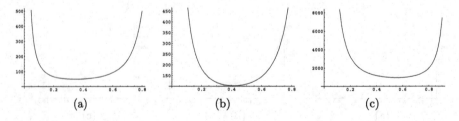

Fig. 3. The graph of $\tilde{\mathcal{E}}_0^d$ for $\tilde{x}_0, \tilde{x}_1, \tilde{x}_2 \in \mathbb{E}^m$ co-linearly unordered with varying $k \in (0, +\infty)$ and $\|\tilde{x}_2\| = 1$: (a) $k = 1/2$ and a global minimum at $\hat{s}_1 \approx 0.346272$ different than $\hat{s}_1^{cc}(k) = (|k|/(1+|k|)) = 1/3 \approx 0.333333$, (b) $k = 1$ and a global minimum at $\hat{s}_1 = 1 - \sqrt{2} \approx 0.414214$ different than $\hat{s}_1^{cc}(k) = 1/2$, (c) $k = 5$ and global minimum at $\hat{s}_1 \approx 0.556194$ different than $\hat{s}_1^{cc}(k) = 5/6 \approx 0.833333$.

than cumulative chords $\hat{s}_1^{cc}(k) = 1/3 \approx 0.333333$, $\hat{s}_1^{cc}(k) = 1/2$ and $\hat{s}_1^{cc}(k) = 5/6 \approx 0.833333$, respectively. For $k = 0.5$ (see Fig. 3a)) the global minimum $\hat{s}_1 \approx 0.346272$ yields $\tilde{\mathcal{E}}_0^d(\hat{s}_1) \approx 48.5065 < \tilde{\mathcal{E}}_0^d(\hat{s}_1^{cc}) = 48.6$. For $k = 1$ (see Fig. 3b)) the global minimum $\hat{s}_1 \approx 0.414214$ yields $\tilde{\mathcal{E}}_0^d(\hat{s}_1) \approx 101.912 < \tilde{\mathcal{E}}_0^d(\hat{s}_1^{cc}) \approx 109.714$. Note that here as proved $\hat{s}_1 = -1+\sqrt{2}$. Finally, for $k = 5$ (see Fig. 3c)) the global minimum $\hat{s}_1 \approx 0.556194$ yields $\tilde{\mathcal{E}}_0^d(\hat{s}_1) \approx 961.081 < \tilde{\mathcal{E}}_0^d(\hat{s}_1^{cc}) = 2704.7$.

Consider *the unperturbed planar data* $\tilde{x}_2 = (3/5, 4/5)$, $\tilde{x}_0 = (-12/5, -16/5)$, $\tilde{v}_2 = (3, 4)$ clearly satisfying (18) with $k = -4 \in (k_2 \approx -26.1226, 0)$. The graph of $\tilde{\mathcal{E}}_0^0$ is shown in Fig. 1a). We perturb now the co-linearity of $\tilde{x}_0, \tilde{x}_1 = \mathbf{0}, \tilde{x}_2$ by taking $\tilde{x}_0^{\delta_2} = k\tilde{x}_2 + \delta_2 = \tilde{x}_0 + \delta_2$, for some $\delta_2 \in \mathbb{R}^2$ (and any fixed $k < 0$). In this example the second interpolation point $\tilde{x}_2$ remains *fixed* with $\|\tilde{x}_2\| = 1$. Similarly we violate the first condition in (18) by choosing a new velocity $\tilde{v}_2^{\delta_1,\delta_2}$ and perturbation $\delta_1 \in \mathbb{R}^2$ such that $\tilde{x}_2 - \tilde{x}_0^{\delta_2} - \tilde{v}_2^{\delta_1,\delta_2} = \delta_1$ holds (i.e. $\tilde{v}_2^{\delta_1,\delta_2} = \tilde{v}_0 - (\delta_1 + \delta_2)$). For $k = -4$ and $\delta_1 = (-1/5, 1)$, $\delta_2 = (1, -2/5)$ (small perturbation with $(\|\delta_1\|, \|\delta_2\|) = (\sqrt{26/25}, \sqrt{29/25})$) the data $\tilde{x}_2, \tilde{x}_0^{\delta_2} = (-1\frac{2}{5}, -3\frac{3}{5})$ and $\tilde{v}_2^{\delta_1,\delta_2} = (11/5, 17/5)$ satisfy (22). Similarly, for $k = -4$ and $\delta_1 = (-4, 1)$, $\delta_2 = (7, -2)$ (big perturbation with $(\|\delta_1\|, \|\delta_2\|) = (\sqrt{17}, \sqrt{53})$) the data $\tilde{x}_2, \tilde{x}_0^{\delta_2} = (4\frac{3}{5}, -5\frac{1}{5})$ and $\tilde{v}_2^{\delta_1,\delta_2} = (0, 5)$ satisfy (22). Lastly, for $k = -4$ and $\delta_1 = (-12, 1)$, $\delta_2 = (-11, -8)$ (large perturbation with $(\|\delta_1\|, \|\delta_2\|) = (\sqrt{145}, \sqrt{185})$) the data $\tilde{x}_2, \tilde{x}_0^{\delta_2} = (-13\frac{2}{5}, -11\frac{1}{5})$ and $\tilde{v}_2^{\delta_1,\delta_2} = (26, 11)$ satisfy (22). Comparing the graph of $\tilde{\mathcal{E}}_0^d$ from Fig. 1a) with the graphs of $\tilde{\mathcal{E}}_0^\delta$ from Fig. 4 shows that unimodality of $\tilde{\mathcal{E}}_0^d$ is preserved for substantial perturbations $\delta \neq \mathbf{0}$. This trend repeats for other $k \in (k_2, 0) \cup (0, \infty)$ which indicates that in practice the perturbation $\delta = (\delta_1, \delta_2)$ from Proposition 1 can be taken as reasonably large. $\qquad\square$

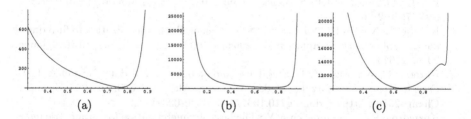

    (a)   (b)   (c)

Fig. 4. The graph of $\tilde{\mathcal{E}}_0^\delta$ for $k = -4$, $\tilde{x}_0, \tilde{x}_2^{\delta_2}, \tilde{v}_2^{\delta_1,\delta_2} \in \mathbb{R}^2$ and for (a) $\delta_1 = (-1/5, 1)$ and $\delta_2 = (1, -2/5)$ yields a global minimum at $\hat{s}_1(k) \approx 0.76615 \neq \hat{s}_1^{cc}(k) \approx 0.79435$, (b) $\delta_1 = (-4, 1)$ and $\delta_2 = (7, -2)$ yields a global minimum at $\hat{s}_1(k) \approx 0.755816 \neq \hat{s}_1^{cc}(k) \approx 0.874097$, (c) $\delta_1 = (-12, 1)$ and $\delta_2 = (-11, -8)$ yields a global minimum at $\hat{s}_1(k) \approx 0.654924 \neq \hat{s}_1^{cc}(k) \approx 0.945841$, and two other critical points i.e. a local minimum at $\hat{s}_0^1(k) \approx 0.944104$ and a local maximum at $\hat{s}_0^2(k) \approx 0.922$.

6 Conclusions

The optimization task (1) is reformulated into (5) (and (16)) to minimize a highly non-linear multivariate function $\mathcal{J}_0$ depending on knots $\mathcal{T}_{int}$. One of the numerical scheme to handle the latter is *a Leap-Frog*. The *generic case* of this algorithm is studied in [5]. Here, we complement the latter by analyzing *non-generic case* of *Leap-Frog* and formulate sufficient conditions preserving unimodality of (16). In doing so, first a special case of data (18) is addressed. Subsequently its perturbed analogue (22) is covered. Example shows that unimodality for (18) (if it occurs) is in practice preserved by large perturbations (22). The performance of *Leap-Frog* compared with *Newton's* and *Secant Methods* is reported in [2,3,7]. More applications of *Leap-Frog* are discussed in [10–12]. For related work on fitting (sparse or dense) reduced data $\mathcal{M}_n$ see e.g. [4,6,8].

References

1. de Boor, C.: A Practical Guide to Splines. 2nd edn. Springer-Verlag, New York (2001). https://www.springer.com/gp/book/9780387953663
2. Kozera, R., Noakes, L.: Optimal knots selection for sparse reduced data. In: Huang, F., Sugimoto, A. (eds.) PSIVT 2015. LNCS, vol. 9555, pp. 3–14. Springer, Cham (2016). https://doi.org/10.1007/978-3-319-30285-0_1
3. Kozera, R., Noakes, L.: Non-linearity and Non-convexity in optimal knots selection for sparse reduced data. In: Gerdt, V.P., Koepf, W., Seiler, W.M., Vorozhtsov, E.V. (eds.) CASC 2017. LNCS, vol. 10490, pp. 257–271. Springer, Cham (2017). https://doi.org/10.1007/978-3-319-66320-3_19
4. Kozera, R., Noakes, L., Wilkołazka, M.: Parameterizations and Lagrange cubics for fitting multidimensional data. In: Krzhizhanovskaya, V.V., et al. (eds.) ICCS 2020. LNCS, vol. 12138, pp. 124–140. Springer, Cham (2020). https://doi.org/10.1007/978-3-030-50417-5_10
5. Kozera, R., Noakes, L., Wiliński, A.: Generic case of Leap-Frog Algorithm for optimal knots selection in fitting reduced data. In: Paszyński, M., Kranzlmüller, D., Krzhizhanovskaya, V.V., Dongarra, J.J., Sloot, P.M.A. (eds.) ICCS 2021. LNCS, vol. 12745, pp. 337–350. Springer, Cham (2021). https://doi.org/10.1007/978-3-030-77970-2_26
6. Kozera, R., Noakes L., Wilkołazka, M.: Exponential parameterization to fit reduced data. Appl. Math. Comput. **391**, 125645 (2021). https://doi.org/10.1016/j.amc.2020.125645
7. Kozera, R., Wiliński, A.: Fitting dense and sparse reduced data. In: Pejaś, J., El Fray, I., Hyla, T., Kacprzyk, J. (eds.) ACS 2018. AISC, vol. 889, pp. 3–17. Springer, Cham (2019). https://doi.org/10.1007/978-3-030-03314-9_1
8. Kuznetsov, E.B., Yakimovich A.Y.: The best parameterization for parametric interpolation. J. Comput. Appl. Math. **191**(2), 239–245 (2006). https://core.ac.uk/download/pdf/81959885.pdf
9. Kvasov, B.I.: Methods of Shape-Preserving Spline Approximation. World Scientific Pub., Singapore (2000). https://doi.org/10.1142/4172
10. Matebese, B., Withey, D., Banda M.K.: Modified Newton's method in the Leapfrog method for mobile robot path planning. In: Dash, S.S., et al. (eds.) ICAIECES 2017, pp. 71–78. Advances in Intelligent Systems and Computing, vol. 668, Springer Nature Singapore (2018). https://doi.org/10.1007/978-981-10-7868-2$_$7
11. Noakes, L.: A global algorithm for geodesics. J. Aust. Math. Soc. Series A **65**(1), 37–50 (1998). https://doi.org/10.1017/S1446788700039380
12. Noakes, L., Kozera, R.: Nonlinearities and noise reduction in 3-source photometric stereo. J. Math. Imaging Vis. **18**(2), 119–127 (2003). https://doi.org/10.1023/A:1022104332058

Adaptive Surrogate-Assisted Optimal Sailboat Path Search Using Onboard Computers

Roman Dębski[(✉)] [iD] and Rafał Dreżewski [iD]

Institute of Computer Science, AGH University of Science and Technology,
Al. Mickiewicza 30, 30-059 Kraków, Poland
{rdebski,drezew}@agh.edu.pl

Abstract. A new surrogate-assisted dynamic programming based optimal path search algorithm – studied in the context of *high-performance
sailing* – is shown to be both effective and (energy) efficient. The key elements in achieving this – the fast and accurate physics-based surrogate
model, the integrated refinement of the solution space and simulation
model fidelity, and the OpenCL-based SPMD-parallelisation of the algorithm – are presented in detail. The included numerical results show
the high accuracy of the surrogate model (relative approximation error
medians smaller than 0.85%), its efficacy in terms of computing time
reduction (from 39.2 to 45.4 times), and the high speedup of the parallel
algorithm (from 5.5 to 54.2). Combining these effects gives (up to) 2461
times faster execution. The proposed approach can also be applied to
other domains. It can be considered as a dynamic programming based
optimal path planning framework parameterised by a problem specific
(potentially variable-fidelity) cost-function evaluator (surrogate).

Keywords: Simulation-based optimisation · Surrogate model ·
Optimal path planning · Trajectory optimisation · Heterogeneous
computing

1 Introduction

High-fidelity (HF) simulations are often computationally too expensive to be
used in a simulation-based optimisation. One obvious way to address this issue
is to parallelise the algorithm. In many instances, however, it is at most a partial
solution to the problem. This can be due to target platform constraints (e.g.,
limited number of processors/cores, energy consumption), an intrinsic strong
sequential component of the algorithm (limited parallel speedup), and/or the
cost of a single simulation, which is often by far the most computationally expensive part of the optimisation process.

Another option is to replace as many as possible high-fidelity simulations
with evaluations of an auxiliary/approximation model – *the surrogate*. This auxiliary model should be a reasonably accurate representation of the HF-model

D. Groen et al. (Eds.): ICCS 2022, LNCS 13352, pp. 355–368, 2022.
https://doi.org/10.1007/978-3-031-08757-8_30

and, at the same time (often much more importantly), remain computationally inexpensive.

The discussed issue is of particular importance in the context of optimal path search algorithms based on dynamic programming. This approach usually leads to accurate results but is computationally very expensive, mostly due to search space size[1]. In a number of cases, including (hard) real-time embedded systems but also near-real-time autonomous robot or sailboat path planners, this is not acceptable. To the best of our knowledge, surrogate-assisted optimal path search based on dynamic programming has not been studied yet (Sect. 2).

The aim of this paper, which is a significant extension of [3] and [4], is to present such an algorithm. When compared to these reference algorithms, it is equally accurate but, at the same time, is significantly faster and more energy-efficient, which is of primary importance for on-board systems (especially when at sea). The main contributions of the paper are:

1. an effective (time-constrained, fast, and accurate) physics-based surrogate model of sailboat motion (duration), defined in the spatial-domain rather than in the time domain, as the original, ODE-based model is (Sect. 4.2),
2. surrogate-assisted, SPMD-parallel, dynamic programming based optimal path search algorithm, which incorporates an integrated refinement of the solution space and simulation model fidelity (Sect. 4.3),
3. numerical results which demonstrate three important aspects of the algorithm: the accuracy of the surrogate-model, the surrogate-related speedup, and the SPMD-parallelisation capabilities (Sect. 5).

The remainder of this paper is organised as follows. The next section presents related research. Following that, the search problem under consideration is defined, and the proposed algorithm is described. After that, experimental results are presented and discussed. The last section contains the conclusion of the study.

2   Related Research

Computer simulations are now widely used to verify engineering designs and to fine-tune the parameters of designed systems. Computer simulation-based approaches are also used in the optimal search problems when the performance measures are represented as a black box. In such cases, classical optimization methods cannot be used directly, which causes that AI-based approaches are often applied [14,17]. Unfortunately, accurate (high-fidelity) simulation models for such tasks are usually computationally expensive, which makes them often hard or even impossible to apply in practice. In such cases, approaches based on the so-called surrogates are used, i.e. simplified (low-fidelity) simulation models, which however reliably represent a complex simulation model of a system or process, and are much more computationally efficient [10,11].

In general, there are two types of surrogate simulation models: *approximation-based*, in which function approximation is constructed based on

---

[1] it can be significantly larger than 10^6.

data sampled from accurate (high-fidelity) simulation models, and *physics-based*, in which surrogates are constructed based on simplified physical models of systems or processes [11].

The approximation-based surrogate models [11,19] are usually developed with the use of radial basis functions (RBF) [6], Kriging [5], polynomial response surfaces [12], artificial neural networks [7], support vector regression [16], Gaussian process regression [1], or multidimensional rational approximation [15].

Because physics-based surrogates are based on low-fidelity, simplified models of the system or processes, and thus contain some (simplified) knowledge about them, they usually require only a few accurate (high-fidelity) simulations runs to be reliably configured [11]. Due to the intrinsic knowledge about the simulated system/process, physics-based surrogates are also characterized by good generalization capabilities, so they can generate high-quality predictions of the accurate simulation model in the case of system designs or configurations not used during the training [11].

In the case of some optimization problems, evaluation of the objectives and constraints require data from physical experiments or numerical simulations, so such optimization problems are called data-driven [9]. To reduce the computation costs of data-driven problems, the surrogate-based models are used together with metaheuristic optimization algorithms, like evolutionary algorithms, particle swarm optimization and ant colony optimization. The excellent review of surrogate-assisted evolutionary algorithms can be found in [9].

The surrogate-based approach to modelling and simulation has gained a great popularity as a tool for lowering the computational costs of complex simulation models, especially in the area of engineering design problems. In the case of optimal trajectory search or path planning problems, surrogate-based methods are less frequently used [19]. The selected ones are briefly discussed below.

A surrogate based on uniform design and radial basis functions was used for mechatronic systems trajectory planning in [19]. In the proposed method, a conventional continue-time problem of optimal control was transformed into a nonlinear programming problem. The developed surrogate-based approach was applied to trajectory planning of an unmanned electric shovel. The Non-Dominated Sorting Genetic Algorithm II (NSGA-II) was used as optimization algorithm.

A surrogate-based optimization method for full space mission trajectory design was proposed in [8]. As a data sampling method, the Optimized Latin Hypercube Design was used. The surrogate model was developed with the use of Kriging approach. After replacing the accurate dynamical model of the entire trajectory by a surrogate model, the multi-objective optimization problem was solved using NSGA-II genetic algorithm.

A multi-fidelity surrogate-based framework for global trajectory optimization was proposed in [18]. In the proposed approach, the Latin hypercube sampling was used for data sampling. Trajectories with initial conditions were evaluated using GPU parallel computing. Several approaches to the development of surrogate models were applied: the quadratic response surface, artificial neural network, and Kriging. The impact of each decision variable on the output parameter was assessed with the use of variance-based global sensitivity analysis (a numerical procedure based on Sobol's variance decomposition was applied). The verification of the proposed framework was performed with the use of a mission scenario including orbital transfer from a near-rectilinear halo orbit to a low lunar orbit.

A flight path planning surrogate model based on stacking ensemble learning was introduced in [20]. The proposed surrogate-based approach allowed for accurate, real-time calculation of the flight waypoint coordinates. In the proposed approach, a path planning analytical model was first used to generate numerous samples, which were then grouped based on selected parameters from the analytical model. The samples were then used to train base-learners using radial basis function neural networks. Finally, a complete surrogate model was constructed from base-learners using a stacking method.

An adaptive surrogate model for fast optimal transfer paths planning for spacecraft formation reconfiguration on libration point orbits was proposed in [13]. The uniform design of experiments was used to obtain the initial and updated samples. To construct the surrogate model, the Kriging and radial basis functions methods were used. The ant colony optimization algorithm was used for adaptive surrogate model optimization.

All the above-mentioned surrogate-based approaches for path planning or optimal trajectory search used approximation-based surrogates. Only a few of them applied GPU-based parallel computing. *The distinctive features of the optimal path search algorithm proposed in this paper include: the use of physics-based surrogate, using dynamic programming with integrated refinement of the solution space and simulation model fidelity, and* SPMD-*parallelization.*

3   Problem Formulation

Consider a sailboat going from point $A(q_A, y_A)$ to $B(q_B, y_B)$, where (q_i, y_i) are the coordinates of the corresponding point in either the Cartesian or polar system [3,4]. The true wind vector field is given in the following way (see Fig. 1):

$$\boldsymbol{v}_t(q, y, t) = T_q(q, y, t)\, \hat{\boldsymbol{q}} + T_y(q, y, t)\, \hat{\boldsymbol{y}}, \tag{1}$$

where: $T_q(q, y, t)$, $T_y(q, y, t)$ are scalar functions, and $\hat{\boldsymbol{q}}$, $\hat{\boldsymbol{y}}$ are the unit vectors representing the axes of the corresponding coordinate system.

The problem domain consists of C^1-continuous $\overset{\frown}{AB}$ paths which cover the given sailing area S_A (see Fig. 1). We assume that the explicit formula for the objective functional is unknown. Hence, each path $(y^{(i)})$ can be evaluated only through simulation, i.e.:

$$J[y^{(i)}] = \texttt{performSimulation}\,[y^{(i)}, \text{cfg}\,(\boldsymbol{v}_t, \dots)\,], \tag{2}$$

where: J represents the given performance measure and cfg$(\boldsymbol{v}_t, \dots)$ – the simulator configuration.

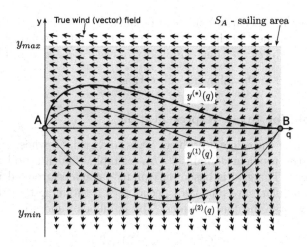

Fig. 1. Optimal sailboat path search problem: example admissible paths connecting points A and B, with $y^{(*)}(q)$ representing the optimal path [3,4].

Problem Statement. The optimal sailboat path search problem under consideration can be defined as follows ([3,4]):

- find, among all admissible paths, the one with the best value of performance measure J;
- the values of J can be found only through simulation;
- only on-board, off-line computers can be used.

In the special case, when $J[y^{(i)}] = \Delta t[y^{(i)}]$, with Δt being the sailing duration, we get the *minimum-time problem*.

4   Proposed Solution

The approach proposed in this paper is a "surrogate-accelerated" version of the one introduced in [3] and then extended in [4]. It is based on the following two main steps:

1. transformation of the continuous optimisation problem into a (discrete) search problem over a specially constructed finite graph (*multi-spline* [3]);
2. application of surrogate-assisted dynamic programming to find the approximation of the optimal path represented as a C^1-continuous cubic Hermite spline.

These two steps repeated several times form an adaptive version of the algorithm. Its key elements are:

- multi-spline based solution space and the SPMD-parallel computational topology it generates [3];

– effective (fast and accurate) surrogate model;
– integrated refinement of the solution space and simulation model fidelity that significantly reduces the time complexity of the reference algorithm.

They are discussed in the following subsections.

4.1   The Solution Space Representation

A discretisation of the original problem leads to a grid, G, whose structure can be fitted into the problem domain [3]. An example of such a grid is shown on the left of Fig. 2. The grid is based on equidistant nodes grouped in rows and columns: four regular rows plus two special ones (containing the start (A) and the end (B) points) and four columns. The number of nodes in such a grid is equal to

$$|G| = n_c (n_r - 2) + 2 \tag{3}$$

where n_c and n_r are the numbers of columns and rows (including the two special ones), respectively.

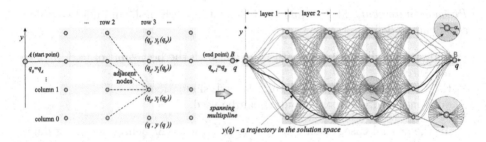

Fig. 2. Solution space representation: multi-spline spanned on regular grid G_{ex} [3]

After assigning n_{ts} additional values to every node of grid G, we obtain a new structure, G_{ex}, that can store not only the coordinates of each node but also the n_{ts} slopes (angles) of path segments which start/end in that particular node (see the right part of Fig. 2).

Joining the nodes from subsequent rows of G_{ex} by using cubic Hermite spline segments, we get a *multi-spline* which forms a discrete space of C^1-continuous functions (see Fig. 2). A detailed description of the multi-spline concept can be found in [3].

4.2   Surrogate-Based Performance Measure

The proposed performance measure approximator (*low-fidelity* model, LFM) is an extended version of the estimator introduced in [4]. It is inspired by the *work-energy principle* that states that the work done by the forces on an object, $W_{AB} = \int_A^B \vec{F} \cdot d\vec{r}$, equals the change in its kinetic energy, $0.5\, m(v_B^2 - v_A^2)$, where

m stands for the sailboat mass. This principle can be used to transform the original problem from *time domain* to *spatial domain*, i.e.,

$$m\frac{dv}{dt} = mv\frac{dv}{ds} = F \rightarrow mvdv = Fds = dW.$$

In our case, $F = F(s, v)$, or even, $F = F(t, s, v)$, but since we are building an approximation model, we can safely assume that:

$$\begin{cases} s_0 = s_A, \\ v_0 = v_A, \\ v_i^2 = v_{i-1}^2 + \frac{2}{m}F(s_{i-1}, v_{i-1})\Delta s_i \end{cases} \tag{4}$$

where: $i = 1, 2, \ldots, i_B$, $\Delta s_i = s_i - s_{i-1}$, and $s_{i_B} = s_B$. Having found the distribution of the (tangent) velocity along the sailing line, and assuming a constant value of F in each sub-interval, we can find the sailing duration. The details are shown in Algorithm 1.

Remark 1. Operating in the spatial domain is one of the key properties of the proposed approximator because the sailing duration can be calculated in a pre-defined number of steps N_s (e.g., 15) stemming from the (spatial) discretization of a path. In the original problem [3,4] – given in the time domain – the number of time-steps to be taken by the simulator (i.e., the ODE solver) to reach the final point is unknown upfront. In some cases, it can be several orders of magnitude larger than N_s.

4.3   Surrogate-Assisted Optimal Path Search Algorithm

The graph G_{ex}, on which the solution space (multi-spline) is spanned, is directed, acyclic (DAG) and has a layered structure. Since, at the beginning of the search process, the performance measure of each path segment is unknown, it has to be obtained from simulation. The *cost matrix* corresponding to the graph can be then computed using the *Principle of Optimality* [2].

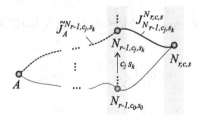

Fig. 3. Principle of Optimality in Dynamic Programming (see Eq. 5).

Algorithm 1: Surrogate model based approximation of a path segment performance measure

---

Input:
- s: the segment to be evaluated,
- v_S: the initial velocity of the sailboat (at the start point of s),
- t_S: the time of reaching the start point of s,
- m_S: the mass of the sailboat,
- t_{min}: the best (approximated) performance measure found so far,
- L_F: the "safety factor" for turning-on the (very) low-fidelity approximation,
- v_{min}: minimum non-zero velocity.

Output: the surrogate-based approximation of the performance measure of s

1 **function** v_2(v_1, dl, F, m_S):
2     $s_v \leftarrow v_1^2 + 2\frac{F\ dl}{m_S}$
3   **if** $s_v > 0$ **then return** $\sqrt{s_v}$ **else return** 0

4 **function** LFM_eval(s, v_S, t_S, t_{min}, m_S, L_F):
5     $v_1 \leftarrow v_S$
6     $v_{nz} \leftarrow \max(v_S, v_{min})$ // **last non-zero velocity**
7     $dt \leftarrow 0$
8   **foreach** sub-segment s_j of segment s **do**
9   **foreach** x_i in Gauss nodes for sub-segment s_j **do**
10        $dl_i \leftarrow$ the length of the i-th part of s_j
11        $F_i \leftarrow$ the net-force for the current position and velocity
12        $v_2 \leftarrow$ v_2 (v_1, dl_i, F_i, m_S)
13   **if** $v_2 > 0$ **then** $v_{nz} \leftarrow v_2$
14        $\bar{v} \leftarrow 0.5\,(v_1 + v_2)$
15   **if** $\bar{v} > 0$ **then** $v_{inv} \leftarrow \bar{v}^{-1}$ **else** $v_{inv} \leftarrow v_{nz}^{-1}$
16        $dt \leftarrow dt + v_{inv}\,dl_i$
17   **if** $(t_S + dt)\,t_{min}^{-1} > L_F$ **then return** (length (s) $(\sum_{k=0}^{i} dl_k)^{-1}\,dt$)
18        $v_2 \leftarrow v_1$
19   **return** $t_S + dt$

This principle can be expressed for an example path $A\text{-}N_{r,c,s}$ (see Fig. 3) in the following way [3]:

$$\widetilde{J}_A^{N_{r,c,s}} = \min_{c_j, s_k} \left(\widetilde{J}_A^{N_{r-1,c_j,s_k}} + J_{N_{r-1,c_j,s_k}}^{N_{r,c,s}} \right) \tag{5}$$

where: $c_j = (0,\ldots,n_c - 1)$, $s_k = (0,\ldots,n_{ts} - 1)$, $J_{N_s}^{N_e}$ is the cost corresponding to the path $N_s\text{-}N_e$ (N_s - start node, N_e - end node), $\widetilde{J}$ represents the optimal value of J, and $N_{r,c,s}$ is the node of G_{ex} with "graph coordinates" $\langle row, column, tangent\_slope \rangle = \langle r, c, s \rangle$.

Figure 3 (a visualization of Eq. 5) presents the computation state in which the optimal costs of reaching all nodes in row $(r - 1)$ are known – they were calculated in previous stages of this multi-stage process. The optimal cost of path A-$N_{r,c,s}$ is calculated by performing simulations for all spline segments that join node $N_{r,c,s}$, which is located in layer/row r, with nodes from the previous (i.e. $(r - 1)^{\text{th}}$) row.

The multi-spline generated computational topology is reflected in the SPMD-structure of Algorithm 2 (see annotation **@parallel**). The computation begins from point A in layer 1, taking into account the corresponding initial conditions, and is continued (layer by layer) for the nodes in subsequent rows. On the completion of the simulations for the last layer (i.e. reaching the end node B), we get the optimal path and its performance measure.

Algorithm 2: Adaptive, SPMD-parallel, surrogate-assisted optimal sailboat path search

---

Input:
- g_{AB}: initial (layered) grid with the start point, A, and the target point, B,
- $\boldsymbol{v}_t$: (true) wind vector field (see Eq. 1),
- HFM: the sailboat movement simulator (high-fidelity model),
- LFM: the surrogate-model (low-fidelity model),
- C_M: "promising" segments (according to LFM) cut-off threshold.

Output: t_{min} - the minimum-time path

1   **foreach** *refinement* $g_{AB}^{(ref)}$ of grid g_{AB} **do**
2   **foreach** *layer in* $g_{AB}^{(ref)}$ **do**
3   **@parallel foreach** *entry point* e_p of $g_{AB}^{(ref)}$ nodes **do**
4   **if** *not final refinement of* g_{AB} **then**
5           $\mathbb{S}_{ep}^{(r)} \leftarrow$ representative subset of segments ending in e_p
6           $(t_{min}^{(lfm)}, s_{best}) \leftarrow \langle \min_{s \in \mathbb{S}_{ep}^{(r)}}, \arg\min_{s \in \mathbb{S}_{ep}^{(r)}} \rangle$ (LFM_eval (s)) // **using** LFM
7           $t_{best} \leftarrow$ HFM_eval (s_{best}) // **using** HFM
8   **else**
9           $\mathbb{S}_{ep} \leftarrow$ all segments ending in e_p
10          $\mathbb{K}_b \leftarrow k$ best from $\mathbb{S}_{ep}$, according to LFM
11          $(t_{best}, s_{best}) \leftarrow \langle \min_{s \in \mathbb{K}_b}, \arg\min_{s \in \mathbb{K}_b} \rangle$ (HFM_eval (s))
12          $\Delta t_{max}^{(k)} \leftarrow \max_{s \in \mathbb{K}_b} |\frac{\text{LFM\_eval}(s)}{t_{best}} - 1|$
13          $\mathbb{R}_s \leftarrow \{ \mathbb{S}_{ep} \setminus \mathbb{K}_b \} \bigcap \{ s : | \frac{\text{LFM\_eval}(s)}{t_{best}} - 1 | < C_M \Delta t_{max}^{(k)} \}$
14          $(t'_{best}, s'_{best}) \leftarrow \langle \min_{s \in \mathbb{R}_s}, \arg\min_{s \in \mathbb{R}_s} \rangle$ (HFM_eval (s))
15   **if** $t'_{best} < t_{best}$ **then** $(t_{best}, s_{best}) \leftarrow (t'_{best}, s'_{best})$
16   save (t_{best}, s_{best}) // **save the best segm and its performance**

The next key element of the algorithm – *the integrated refinement of the solution space and simulation model fidelity* – is reflected by the conditional statement (lines 4–15). As the computation progresses, the search strategy changes from *mostly exploration* (lines 5–7) to *mostly exploitation* (lines 9–15). In the exploration phase, only the *representative subset* of segments is evaluated (line 5) and it is done with a *coarse grid* (i.e., $N_s = 15$). The exploitation phase is more complex. In the first step (line 9), *all* segments ending in a particular node are evaluated using a *fine grid* (i.e., $N_s = 30$). Following that (line 10), the k-best[2] candidate segments are evaluated using the HFM-model (i.e. the simulator). As a final step (lines 12–15), using the accuracy measure $\Delta t_{max}^{(k)}$, additional "promising" segments are selected (if there are any) and evaluated.

Complexity Analysis. Algorithm 2 average-case *time complexity* is determined by the number of solution space refinements, n_i, the average number of force evaluations[3] for a single path segment, $\bar{n}_F$, and the number of such segments, $n_c n_{ts}^2 [(n_r - 3) n_c + 2]$ (see Sect. 4.1). For the sequential version of the algorithm it can be expressed as:

$$T_s = \Theta \left(n_i \, n_r \, n_c^2 \, n_{ts}^2 \, \bar{n}_F \right). \tag{6}$$

In the SPMD-parallel version of the algorithm, the evaluations for all nodes in a given row can be performed in parallel (using p processing units), thus:

$$T_p = \Theta \left(n_i \, n_r \, n_c \, n_{ts} \left\lceil \frac{n_c \, n_{ts}}{p} \right\rceil \bar{n}_F \right). \tag{7}$$

In the same way as in reference algorithm [3], the Algorithm 2 *space complexity* formula, $\Theta (n_r \, n_c \, n_{ts})$, arises from the solution space representation.

5   Results and Discussion

To demonstrate the effectiveness of the algorithm, a series of experiments was carried out using a MacBook Pro[4] with macOS 12.2 and OpenCL 1.2. This system had two (operational) OpenCL-capable devices: Intel Core i5 @ 2.7 GHz (the CPU) and Intel Iris Graphics 6100, 1536 MB (the integrated GPU). The aim of the experiments was to investigate three important aspects of the algorithm: the accuracy of the surrogate-model, the surrogate-related computational time cost reduction, and the SPMD-parallelisation efficiency. The results are presented in the subsequent paragraphs.

The Accuracy of the Surrogate-Model. This element has a significant impact, especially in the exploration phase of the search process (detection of all potentially good segments). The results of its experimental evaluation are given in the form of a violin plot in Fig. 4.

---

[2] the value of k can be a constant or auto-adaptive variable.
[3] values of F are used both in HFM and LFM; Runge-Kutta-Fehlberg 4(5) method, used in the simulator, requires at each step six evaluations of F.
[4] Retina, 13-in, Early 2015, with 16 GB of DDR3 1867 MHz MHz RAM.

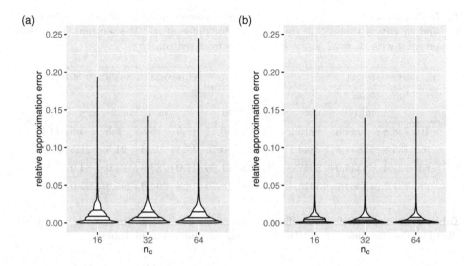

Fig. 4. Accuracy of the two surrogate models in use in the forms of their relative approximation errors (with reference to the high-fidelity model) for different numbers of multi-spline nodes (n_c): coarse model **(a)** vs. finer model **(b)**.

The plot shows the distributions of approximation relative errors, $|\frac{t_{approx} - t_{sim}}{t_{sim}}|$, for path segments from different search spaces. The highest recorded medians (0.83% for the coarse model and 0.45% for the finer model) confirm the very high accuracy of the proposed surrogate-model[5].

Efficacy of the Surrogate-Assisted Search. This element was verified using the reduction of the number of force computations, $(\bar{n}_F^{(base)} - \bar{n}_F^{(sur)})/\bar{n}_F^{(base)}$, as the measure. To test the accuracy of $\bar{n}_F$ as the measure of the algorithm time complexity (see Eqns. 6 and 7), the duration of sequential computations, t_{sim}, was also measured, and then the Pearson correlation coefficient ($\rho(\bar{n}_F, t_{sim})$) was calculated. The result was equal to 0.999999997, which means (almost) perfect linear dependence between the two variables. The corresponding experimental results are given in Table 1 and Fig. 5. The observed $\bar{n}_F$ was reduced by 98%, which gives a significant improvement (more than 2 times) when compared to the results presented in [4].

SPMD-*Parallelisation Efficiency.* Parallelisation is another way of lowering the total computation time. Contemporary mobile/on-board computers are usually equipped with more than one type of processor, typically one CPU and at least one GPU. OpenCL makes it possible to use these heterogeneous platforms effectively, since the same code can be executed on any OpenCL-capable processor. The execution times for different sizes of the solution space and the corresponding parallel-speedups are presented in Table 2 and Fig. 5.

---

[5] The samples sizes (i.e., the number of segments) used to compute the distributions of errors were large: from 85 745 for $n_c = 16$ to 1 390 047 for $n_c = 64$.

Table 1. Efficacy of surrogate-assisted search: average numbers of force evaluations, $\bar{n}_F$, and execution times, t_{sim} (in seconds), for different n_c. The solution space (two refinements) with $n_r = 32$, $n_{ts} = 8$. Statistics from ten runs.

n_c	BASE MODEL					SURROGATE-ASSISTED MODEL				
	$\bar{n}_F$	t_{sim}				$\bar{n}_F$	t_{sim}			
		min	max	avg	sd		min	max	avg	sd
16	748.8	214.6	215.2	214.9	0.17	17.9	5.5	5.5	5.5	0.01
32	735.8	841.4	844.1	842.7	0.76	17.4	21.4	21.5	21.5	0.01
64	731.4	3344.0	3347.6	3345.8	1.20	17.0	83.9	84.0	84.0	0.02
128	731.7	13364.1	13413.4	13383.8	18.79	14.8	294.8	295.4	295.0	0.21

Table 2. SPMD-parallelisation efficiency: execution times, t_{sim} (in seconds) and (parallel) speedup for different n_c. The remaining parameters as in Table 1.

n_c	t_{sim}				Speedup
	min	max	avg	sd	
16	0.973	0.985	0.980	0.003	5.6
32	1.854	1.865	1.859	0.003	11.5
64	3.360	3.379	3.368	0.006	24.9
128	5.432	5.445	5.439	0.004	54.2

Its maximum recorded value was 54.2 (see Table 1 and Fig. 5). With the reference point as the sequential search based on the full simulation, it gives in

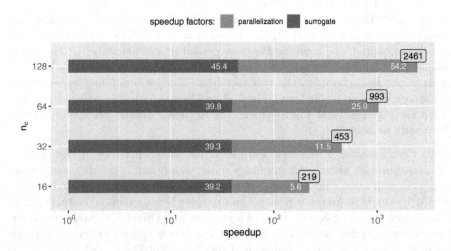

Fig. 5. Total speedup (boxed numbers at the end of each bar) and its factors: surrogate-model application (dark-grey part) and SPMD-parallelisation (light-grey part) for different n_c. The remaining parameters as in Table 1.

total 2461 times faster execution. Additionally, when compared to the results presented in [4], it gives us an execution time more than 3 times shorter.

6   Conclusions

It has been shown that the surrogate-assisted dynamic programming based optimal sailboat path planning algorithm can be both effective and (energy) efficient. The key elements in achieving this have been the fast and accurate physics-based surrogate model, the integrated refinement of the solution space (multi-spline) and simulation model fidelity, and the OpenCL-based SPMD-parallelisation of the algorithm.

The numerical results show the high accuracy of the surrogate model (the medians of relative approximation errors were smaller than 0.85%, see Fig. 4), its efficacy in terms of the reduction of computing time (from 39.2 to 45.4 times, see Table 1 and Fig. 5), and the high speedup of the parallel algorithm (its maximum observed value was 54.2, see Fig. 5). Combining these effects has given (up to) 2461 times faster execution time (see Fig. 5).

The proposed approach can also be applied to other scenarios. In fact, it can be considered as a dynamic programming based optimal path planning framework parameterised by a problem specific (potentially variable-fidelity) cost-function evaluator (surrogate). Further exploration of this idea could be the first possible future research direction. Another could be the algorithm space complexity reduction.

Acknowledgement. The research presented in this paper was partially supported by the funds of Polish Ministry of Education and Science assigned to AGH University of Science and Technology.

References

1. Angiulli, G., Cacciola, M., Versaci, M.: Microwave devices and antennas modelling by support vector regression machines. IEEE Trans. Magn. **43**(4), 1589–1592 (2007). https://doi.org/10.1109/TMAG.2007.892480
2. Bellman, R., Dreyfus, S.: Applied Dynamic Programming. Princeton University Press, Princeton, New Jersey (1962)
3. Dębski, R.: An adaptive multi-spline refinement algorithm in simulation based sailboat trajectory optimization using onboard multi-core computer systems. Int. J. Appl. Math. Comput. Sci. **26**(2), 351–365 (2016). https://doi.org/10.1515/amcs-2016-0025
4. Dębski, R., Sniezynski, B.: Pruned simulation-based optimal sailboat path search using micro HPC systems. In: International Conference on Computational Science, pp. 158–172. Springer (2021). https://doi.org/10.1007/978-3-030-77970-2_13
5. Giunta, A., Watson, L.: A comparison of approximation modeling techniques-polynomial versus interpolating models. In: 7th AIAA/USAF/NASA/ISSMO Symposium on Multidisciplinary Analysis and Optimization, p. 4758. American Institute of Aeronautics and Astronautics (1998). https://doi.org/10.2514/6.1998-4758

6. Hardy, R.L.: Multiquadric equations of topography and other irregular surfaces. J. Geophys. Res. **76**(8), 1905–1915 (1971)
7. Haykin, S.: Neural Networks: A Comprehensive Foundation. Prentice Hall, Upper Saddle River, NJ (1998)
8. He, X., Zuo, X., Li, Q., Xu, M., Li, J.: Surrogate-based entire trajectory optimization for full space mission from launch to reentry. Acta Astronaut. **190**, 83–97 (2022). https://doi.org/10.1016/j.actaastro.2021.09.030
9. Jin, Y., Wang, H., Chugh, T., Guo, D., Miettinen, K.: Data-driven evolutionary optimization: an overview and case studies. IEEE Trans. Evol. Comput. **23**(3), 442–458 (2019). https://doi.org/10.1109/TEVC.2018.2869001
10. Koziel, S., Leifsson, L. (eds.): Surrogate-Based Modeling and Optimization. Springer, New York, NY (2013). https://doi.org/10.1007/978-1-4614-7551-4
11. Koziel, S., Ogurtsov, S.: Antenna Design by Simulation-Driven Optimization. SpringerBriefs in Optimization. Springer, Cham (2014). https://doi.org/10.1007/978-3-319-04367-8
12. Myers, R.H., Montgomery, D.C., Anderson-Cook, C.M.: Response Surface Methodology: Process and Product Optimization Using Designed Experiments. John Wiley & Sons (2016)
13. Peng, H., Wang, W.: Adaptive surrogate model-based fast path planning for spacecraft formation reconfiguration on libration point orbits. Aerosp. Sci. Technol. **54**, 151–163 (2016). https://doi.org/10.1016/j.ast.2016.04.017
14. Pošík, P., Huyer, W., Pál, L.: A comparison of global search algorithms for continuous black box optimization. Evolut. Comput. **20**, 1–32 (2012)
15. Shaker, G., Bakr, M.H., Sangary, N., Safavi-Naeini, S.: Accelerated antenna design methodology exploiting parameterized Cauchy models. Progress Electromag. Res. B **18**, 279–309 (2009). https://doi.org/10.2528/PIERB09091109
16. Smola, A., Schölkopf, B.: A tutorial on support vector regression. Stat. Comput. **14**, 199–222 (2004)
17. Szłapczyński: Customized crossover in evolutionary sets of safe ship trajectories. Int. J. Appl. Math. Comput. Sci. **22**(4), 999–1009 (2012)
18. Ueda, S., Ogawa, H.: Multi-fidelity approach for global trajectory optimization using GPU-based highly parallel architecture. Aerosp. Sci. Technol. **116**, 106829 (2021). https://doi.org/10.1016/j.ast.2021.106829
19. Wang, X., Song, X., Sun, W.: Surrogate based trajectory planning method for an unmanned electric shovel. Mech. Mach. Theory **158**, 104230 (2021). https://doi.org/10.1016/j.mechmachtheory.2020.104230
20. Yang, X.Z., Cui, Z.X., Qiu, X.Y.: Flight path planning surrogate model based on stacking ensemble learning. In: IOP Conference Series: Materials Science and Engineering, vol. 751(1), 012038 (2020). https://doi.org/10.1088/1757-899x/751/1/012038

Local Search in Selected Crossover Operators

Mirosław Kordos[1]($\boxtimes$), Rafał Kulka[1], Tomasz Steblik[1], and Rafał Scherer[2]

[1] Department of Computer Science and Automatics, University of Bielsko-Biała,
Willowa 2, 43-309 Bielsko-Biała, Poland
`mkordos@ath.bielsko.pl`
[2] Department of Intelligent Computer Systems, Częstochowa University
of Technology, al. Armii Krajowej 36, 42-200 Częstochowa, Poland
`rafal.scherer@pcz.pl`

Abstract. The purpose of the paper is to analyze an incorporation of local search mechanisms into five crossover operators (KPoint, AEX, HGreX, HProX and HRndX) used in genetic algorithms, compare the results depending on various parameters and draw the conclusions. The local search is used randomly with some probability instead of the standard crossover procedure in order to generate a new individual. We analyze injecting the local search in two situations: to resolve the conflicts and also without a conflict with a certain probability. The discussed mechanisms improve the obtained results and significantly accelerate the calculations. Moreover, we show that there exists an optimal degree of the local search component, and it depends on the particular crossover operator.

Keywords: Genetic algorithms · Crossover · Local search

1 Introduction

Genetic algorithms (GA) is a highly researched topic for a long time and especially in recent years. GA have two important advantages: a fast intelligent search mechanism, which allows finding a good solution after analyzing only a small fraction of possible candidates, and a high level of universality, which allows for a broad range of practical applications, and also optimization of training data and parameters of other artificial intelligence methods [2,12,19]. GA are very good at exploring the whole solution space, however, they are not so good at exploiting the local areas of the most promising solutions. For that purpose some hybrid methods have been proposed [3,18,20].

Hybrid genetic algorithms use an additional local search method, which co-operates with the genetic algorithm in order to achieve better results, by leveraging the power of both: the global search capabilities of GA and the local search. There are two basic families of approaching the joint genetic and local search: the Lamarckian approach and the Baldwinian one [14]. In the Baldwinian approach the effects of the local search improve the fitness of the individual, but its

D. Groen et al. (Eds.): ICCS 2022, LNCS 13352, pp. 369–382, 2022.
https://doi.org/10.1007/978-3-031-08757-8_31

chromosome remains unchanged. In the Lamarckian approach, the effects of the local search are reflected in the chromosome. The advantage of the former one is better exploration of the search space, and of the second one is better exploitation. Also combinations of the both approaches are possible [8]. In the method presented in this paper, we use the Lamarckian approach, because it accelerates the search process more effectively [102]. On the other hand, by changing the chromosome of individuals, it can disrupt the building blocks created by genetic algorithms, what may lead to a fast conversion, but only to a local minimum. To prevent this, we apply the local search only with a limited frequency.

An Adaptive Hybrid Genetic Algorithm (AHGA) was proposed in [4], which contains two dynamic learning mechanisms to guide and combine the exploration and exploitation search processes adaptively. The first learning mechanism aims to assess the worthiness of conducting the local search. The second learning mechanism uses instantaneously learned probabilities to select from a set of predefined local search operators which compete against each other for selection which is the most appropriate at any particular stage of the search to take over from the evolutionary-based search process. The authors of [6] presented a hybrid genetic algorithm (HGA) that uses a sequential constructive crossover, a local search approach along with an immigration technique to find good solutions in the multiple traveling salesman problem. In [23] an inversion operation was discussed to solve this problem, which was similar to an RMS mutation, but performed only if it improves the fitness. Hybrid genetic search with dynamic population management and adaptive diversity control with a problem-tailored crossover and local search operators were analyzed in [24]. A hybrid genetic algorithm given by applying 2-opt selection strategy to two edges chosen by replacement probability, and add the new edges by 2-opt permutation algorithm was proposed in [15].

2   Crossover Operators

The purpose of the crossover operator to combine information from two or more different chromosomes (parents) into one chromosome (child) that can represent a better solution than its parents.

In this paper we consider this kind of problems, where each item can occupy only one location (one position of the chromosome) at a time and each location must be occupied by exactly one item. For example, the traveling salesman problem or product placement optimization in a warehouse. In these cases, the crossover operator must ensure that there will be no duplicate elements and that each element will be present in the newly created individual.

Several crossover operators have been developed for this purpose. This includes the well known crossover operators as order crossover (OX), partially mapped crossover (PMX), order-based crossover (OBX), and position-based crossover (PBX) and cycle crossover (CX) [1,11]. Also several newer crossover operators were developed, and we present some of them below.

Tan proposed heuristic greedy crossover (HGreX) and its variants HRndX and HProX [21], which we will explain in detail later. Other popular crossover operator is edge recombination crossover (ERX) and also several its variants were proposed [10,22], alternating edges crossover (AEX) [16] and the uniform partially matched crossover [7]. AEX also performs well after the improvements that we introduced, and for this reason we also used it in this paper.

A crossover method similar to HGreX, but with the difference that four candidates for each next position in the child were considered, was presented in [14]. These four candidates were the position placed before and after the current position in both parents, and the nearest position was selected.

Hassanat and Alkafaween [9] proposed several crossover operators, such as cut on worst gene crossover (COWGC) and collision crossover, and selection approaches, as select the best crossover (SBC). COWGC exchanges genes between parents by cutting the chromosome at the point that maximally decreases the cost. The collision crossover uses two selection strategies for the crossover operators. The first one selects this crossover operator from several examined operators, which maximally improves the fitness, and the other one randomly selects any operator. This algorithm applies multiple crossover operators at the same time on the same parents, and finally selects the best two offspring to enter the population.

In this paper we analyze incorporation of the local search in the following five crossover operators: KPoint, AEX, HRndX, HProX and HGreX. We chose them, because they perform very well and at the same time they are widely used. All the five operators work in a similar way. They use two parents to generate a child by alternatively taking the elements from one or from the other parent. They differ in the way in which the element alternation is organized.

The AEX operator can start from any position in the chromosome. In the example, we will start from the value on the first position in the first parent, this is A. Thus, A becomes the first position in the child. Then AEX looks at the second parent to find what it contains after A. This is actually D. Then again a value from Parent1 that is after D is used and so on. If following this rule would cause a conflict leading to repeating some values in the child, then randomly one from the available elements is chosen.

A sample explanation is presented in Fig. 1, where we start from two parents; P1 and P2, and perform the following steps:

1. In step 1, we start creating a child, by taking the first element from the first parent and removing that element from both parents.
2. In step 2, we must find what is after A in P2 - this is D. So the next element of the child will be D and we remove it from the parents.
3. In step 3, we look, what is after D in P1 - this is E and this will be the next element in the child.
4. In a similar way in step 4.
5. In step 5, we would normally add D to the child. However, D has already been used, so a conflict appear here, and we select randomly any of the remaining elements. Let us say G.
6. In steps 6–8, we follow the standard rule and since there are no more conflicts, we add to the child F and then H and then B.

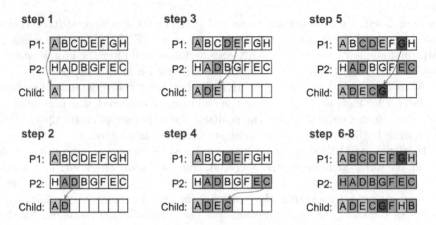

Fig. 1. Explanation of the AEX Crossover operator.

The HRndX crossover operator works similarly to AEX. The difference is, that it does not take alternatively the elements from both parents, but always randomly decides if the next element is taken from the first or from the second parent. HGreX always chooses the next element from this parent, to which the distance is closer. HProX chooses the closest element with a higher probability. The conflicts are resolved in the same way as in AEX.

The KPoint crossover operator does not take into account, which element occurs after which in the parents, but takes them sequentially alternating between the two parents. So for this example, it will generate the following child in the first three steps:

$$P1 = ABCDEFGH \qquad P2 = HADBGFEC \qquad Child = HBD\_\_\_\_\_$$

and a conflict occurs here. The conflicts are resolved in the same way as in AEX, so any unused value will be taken randomly, and then it continues with its standard scheme.

3 The Local-Search Enhancements of the Crossover Operators

3.1 Local Search Implementation in Conflict and Non-conflict Resolvers

Let us analyze the case of a conflict first. A conflict appears, if following the standard crossover rule would lead to setting the next position in the child chromosome to a value that has already been used, as discussed in the previous section. The conflicts must be resolved in order to build a valid child chromosome in which each value appears exactly once.

Random/Native Resolver is the method of conflict resolving used originally by all the five crossover operators. In case of a conflict, this resolver takes randomly any of the not used so far values and makes it the next element of the child.

Algorithm 1. The Extended Crossover Operators

1: **Input:** Two parents
2: **Output:** The child

3: **if** Both parents are equal **then**
4:    mutate one parent with RSM mutation
5: **end if**
6: **for** $k = 1$ **to** *chromocomeLength* **do**
7:   **if** No Conflict **then**
8:     **if** Radom()¿probability1 **then**
9:       set the k-th position in the child according to the crossover operator rule
10:    **else**
11:      **if** random resolver = Nearest Neighbor Resolver **then**
12:        set the k-th position in the child to the nearest from the unused positions
13:      **else if** random resolver = Tournament Resolver **then**
14:        set the k-th position in the child with the tournament selection of the unused positions
15:      **end if**
16:    **end if**
17: **else if** Conflict **then**
18:    **if** conflict resolver = Random Resolver **then**
19:      set the k-th position in the child according to the crossover rule (by selecting a random unused position)
20:    **else if** conflict resolver = Nearest Neighbor Resolver **then**
21:      set the k-th position in the child to the nearest from the unused positions
22:    **else if** conflict resolver = Tournament Resolver **then**
23:      set the k-th position in the child with the tournament selection of the unused positions
24:    **end if**
25: **end if**
26: **end for**
27: Return child

Nearest Neighbor Resolver - in this case, if a conflict occurs, the cost between the last position added to the child and all unused positions is calculated and the point with the lowest cost is chosen.

Tournament Resolver performs a tournament selection of all the unused values. A predefined number n or percentage p of the unused elements are randomly selected, and then the one with the lower cost to the recently added position in the child is chosen.

As a matter of fact, the tournament conflict resolver with $n = 1$ is equivalent to the random selection (random conflict resolver) and with n being equal to the number of remaining points is equivalent to the nearest neighbor resolver.

The Nearest Neighbor Resolver and the Tournament Resolver can also be applied with a certain probability for dealing with the situations, where there are no conflicts. In these situations, normally the next child position would be determined by the base crossover operator rule. However, using the resolver here introduces the local search component.

4   Experimental Evaluation

We performed the experiments using two sample problems: the Warehouse Optimization Problem and the Traveling Salesman Problem (TSP). The source code of our software and the experimental data can be found at *kordos.com/iccs2022*.

In this section, first we describe these two problems along with the encoding schemes and local search methods that we used, and then we present the obtained results.

The HGreX and HProX operators are only suitable for the TSP example. The KPoint operator, although being suitable for both examples, works very well for the warehouse problem, but quite poor for the TSP. For that reason we evaluate AEX, HRndX for both problems, HGreX and HProX only for TSP and KPoint only for the warehouse problem.

4.1   The Example of Warehouse Optimization Problem

For the evaluation we used 6 different warehouse structures with the size between 60 and 240 locations, which can be found at *kordos.com/iccs2022*

Problem Description. Order picking is the most time-consuming task in warehouse operations, and thus also the task generating most costs. It was discovered that about 60% of warehouse operation costs are those of picking up products when completing orders [2]. Thus, reducing the order completion time is a crucial challenge, as it gives an opportunity to significantly reduce costs and to increase sales using the same resources. The problem was studied in many literature positions [4–6, 13, 17]. A review of the scientific literature investigating order picking and planning problems can be found in [23]. With n items in the warehouse, the number of all possible their placements is $n!$ Already for 100 products it gives $100! = 9.3e157$ possible product placements. Thus, it can be easily seen that this cannot be optimized by brute force.

Thus, we consider this as a good example that can be used to analyze the performance of the crossover operators with incorporated local search.

In a paper co-authored by one of us [13] presented an optimization of product layout in the warehouse with a genetic algorithm, which used only the standard global search with the standard AEX crossover operator. The reader is referenced to that paper for more details about the problem, because due to space limitations we cannot explain them here.

In the current experiment we analyze the different ways of adding local search components to the AEX, KPoint, and HRndX crossovers to solve the warehouse optimization problem.

Objective Function. The objective function is expressed by the sum of all orders picking times within a certain time frame, e.g. one month (the smaller, the better). This objective can be measured by the length of the route that the workers must cover to complete all the orders. In practice, the length should be measured rather in time units than in distance units, as for example covering the same distance in a straight line is faster than going around a corner and especially covering the same distance in horizontal direction is faster than in vertical direction (to reach the product located on upper shelves). However, it does not matter for the functioning of the genetic algorithm.

As a byproduct of the optimization, we obtain shorter completion routes for all orders. (which is another NP-hard problem). To keep the example simple, we use the nearest neighbor algorithm followed by 2-opt to find the shortest order completion routes. This method is used only as an example to calculate the value of the objective function used by the genetic algorithm. Thus, we do not try to improve this mechanism of the order completion route generation, because it is not the purpose of the paper.

Problem Encoding. The particular locations in the warehouse (e.g. shelves) are represented by the positions in the chromosome. Each position corresponds to one location. The products are represented by particular numbers on the chromosome positions. For example, if a product number 1 is placed on the location number 3, product number 2 is placed on location number 2 and product number 3 is placed on location number 1, then the encoding of this product distribution is represented as: [3 2 1]. To optimize the objective function, the genetic algorithm needs to decide upon appropriate locations of particular products, that is, upon the appropriate order of the numbers in the chromosome.

Local Search Component. To construct the local search for the warehouse example, we use the two commonly known facts. First, the products which are most frequently purchased should be placed close to the warehouse entry. We call this Single Product Frequency Search (SPFS). Second, the products which are frequently purchased together within one order should be placed close to each other. Pairwise Product Frequency Search (PPFS). However, optimizing the two criteria together is an NP-hard problem, as the orders differ one from another and particular products appear in many orders in different combinations with other products. For that reason, we cannot use the local search only, without the genetic algorithm, which provides the global search.

FPS can be used with some probability to determine the next position in the child instead of determining this position according to the base crossover operator rule.

Before applying SPFS, two arrays are created: an array of warehouse locations A_{WL} sorted by the distance from the warehouse entry and an array of single product frequencies A_{PS} (how many times each product occurred in all the orders together). The idea of SPFS is to try placing a product in a location which is on similar position in the A_{WL} as the product position in A_{PS}. In this way, more frequently purchased product get placed closer to the entry and the less frequently get placed further. However, SPFS cannot always place the most optimal product on each location, because it can use only the products, which are not positioned yet by the crossover operator, so it searches for free products and selects this one, which is most close to the optimal one.

Before applying PPFS, a pairwise product frequency matrix M_{PF} is calculated. M_{PF} contains the frequencies with which each two products occur together in the same orders. The idea is to select such a product for the current location, which coexists frequently in the same orders with the product already placed in the closest neighbor locations in the warehouse.

SPFS and PPFS can be used as well to resolve the conflicts in the crossover operator (replacing the default method of conflict resolving, which takes randomly any available product) as to select the next product in the chromosome (instead of the given crossover rule). In the second case, it is used with a certain probability.

Also, PPFS cannot always place the most optimal product on each location, but chooses the most optimal of the still available products.

4.2 The Example of Traveling Salesman Problem

For the evaluation we used 10 travels comprising between 50 and 1600 locations, which can be found at *kordos.com/iccs2022*

Problem Description and Objective Function. In the traveling salesman problem (TSP) the task is to visit all the cities from a given list, starting from the first one and returning to the first one in such a way that the total length of the travel will be minimal. The objective function is expressed by the total length of the travel $Tdist$ (the smaller, the better). To obtain this, the cities must be visited in a proper order.

Problem Encoding. The order of visiting the cities is encoded in the chromosome. Let us say that there are six cities on the list: A, B, C, D, E, F, and A is the starting city (and the ending one). For example, the following chromosome:

$$[A\ D\ B\ C\ F\ E]$$

represents the following order of visiting the cities: A−>D−>B−>C−>F−>E −>A and this order also determines the total distance $Tdist$ that must be covered to visit the cities in this order as the $Tdist = dist(A, D) + dist(D, B) + dist(B, C) + dist(C, F) + dist(F, E) + dist(E, A)$.

Local Search Component. The common knowledge in the TSP is that cities that are close to each another should be rather visited one by one. This knowledge is implemented by the nearest neighbor algorithm, which always connects a given city with the closest city. Because of lack of the global search, it produces worse results than GA in this case. Nerveless, when used as the local search component inside GA, it allows improving the results.

4.3   Experimental Setup and Parameters

The following parameters of the genetic algorithm were used for all the experiments:

- base crossover operators: AEX, HGreX, HProX, HRndX for TSP, and KPoint, AEX, HRndX for the Warehouse Optimization Problem
- population size: 100
- number of children: 80 (80 children and the best 20 parents were promoted to the next generation)
- selection: tournament selection with 8 candidates per each parent.
- The GA was run for 1000 epochs, because no further improvement of results was observed while running the optimization for more than 1000 epochs.
- mutation: RSM mutation with the probability of 10%.
- resolver type 1: tournament selection with the number of candidates set to 25% of the available locations
- resolver type 2: nearest neighbor
- resolver type 3: random selection
- random resolver probability: we performed the experiments with 11 different probabilities: 0, 0.1, 0.2, ... 1.0.

We repeated each experiment 100 times. The average results for selected configurations are presented in Tables 1 and 2. We analyzed the fitness obtained after each number of epochs, so that the speed of achieving the results can be observed, and this is presented in Table 1, and in Fig. 1 for the warehouse optimization problem with optimal probabilities of the tournament selection resolver.

For the TSP the results were similar, so due to the limited space here we decided not to present them in this form, but instead to show a detailed snapshot of the process, to show in detail how the types and probabilities of the resolvers influence the results. We show the snapshot at the 100-th epoch, because it shows also well the speed of the convergence of the optimization with various configurations. Obviously, the differences at the end of the optimization (1000-th epoch) are smaller, but still significant - proportionally to what we can see in Table 1.

To compare the results over all the dataset in Table 2, we introduced the relative route length for each dataset as the proportion of the route length obtained with a given set of parameters to the shortest obtained route lengths. These values were calculated as the average length obtained over 100 runs of the genetic algorithms for each dataset, and then averaged over all the datasets. Thus, the lowest possible value is 1.0. In fact, 1.0 does not appear in the table, because it would mean that some set of parameters were the best for each dataset and in practice it was not the case.

4.4   Analysis of the Results Obtained for the Warehouse Problem

The first conclusion visible from Table 1, Table 2, and Fig. 1 is that the addition of local search not only improves the final results, but also drastically accelerates the genetic algorithms process.

Table 1. Fitness for the warehouse optimization problem (sum of the order picking route lengths for the warehouse-orders set) obtained with different crossover operators with and without the local search (the lower, the better). The name of the warehouse-orders set indicates the number of position in the warehouse (p), number of orders (o) and number of items in all orders (i). For example 233p25o131i is the set of 233 positions, 25 orders and 131 items.

Warehouse-orders	Epoch	Kpoint	Kpoint with local search	AEX	AEX with local search	HRndX	HrndX with local search
233p25o131i	10	7087	3745	8987	4558	8848	5153
	100	4607	3512	6541	3599	5663	3619
	1000	3905	3450	5251	3513	3691	2931
118p25o666i	10	7495	6197	12088	9218	12049	9444
	100	6435	5997	9871	8139	9469	8184
	1000	6187	5946	8476	7768	8335	7846
81p25o804i	10	33213	24146	35004	26983	34710	28235
	100	22230	20415	26938	23624	24158	24079
	1000	19600	19427	22828	22012	20892	21927
71p34o392i	10	10053	7174	11066	8037	11069	7730
	100	7330	6554	7656	6794	7469	6701
	1000	6796	6497	6836	6577	6879	6516
106p25o668i	10	30456	17578	37250	19568	34094	23196
	100	20611	15032	25888	17176	23285	17504
	1000	16351	14662	21544	16478	19722	16416
127p25o406i	10	21191	18364	28898	20897	30093	21807
	100	16089	13824	17709	20060	17864	17400
	1000	15588	12768	16939	15674	16803	14957

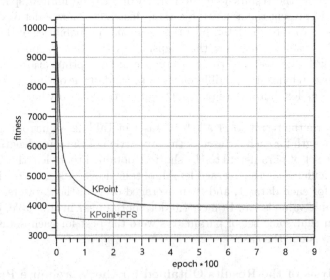

Fig. 2. Comparison of optimization progress with standard KPoint crossover (blue) and KPoint crossover with product frequency search (red) for the *100p34o434* warehouse. (Color figure online)

Table 2. Relative fitness value after 100 iterations averaged over 100 runs on 10 datasets for TSP (the lower the better).

Crossover	Conflict	Non-c.	Non-conflict resolver probability										
			p = 0	p = 0.1	p = 0.2	p = 0.3	p = 0.4	p = 0.5	p = 0.6	p = 0.7	p = 0.8	p = 0.9	p = 1
AEX	Rnd	Rnd	8.76	9.91	10.15	10.30	10.42	10.51	10.59	10.66	10.72	10.77	10.80
AEX	NN	Rnd	1.21	2.48	3.84	5.05	6.12	7.08	7.96	8.73	9.47	10.16	10.80
AEX	Trn	Rnd	1.56	3.64	4.92	6.01	6.93	7.70	8.44	9.11	9.70	10.27	10.80
AEX	Rnd	NN	8.76	5.45	4.45	3.73	3.12	2.60	2.11	1.69	1.38	1.15	1.18
AEX	NN	NN	1.21	1.17	1.14	1.11	1.10	1.09	1.09	1.08	1.09	1.09	1.18
AEX	Trn	NN	1.56	1.55	1.40	1.29	1.22	1.17	1.12	1.11	1.10	1.10	1.18
AEX	Rnd	Trn	8.76	6.20	5.42	4.70	4.13	3.59	3.06	2.52	1.97	1.43	1.17
AEX	NN	Trn	1.21	1.36	1.35	1.29	1.23	1.17	1.12	1.08	1.07	1.09	1.17
AEX	Trn	Trn	1.56	1.60	1.46	1.35	1.26	1.18	1.13	1.11	1.11	1.12	1.17
HProX	Rnd	Rnd	4.17	6.93	8.17	8.95	9.49	9.87	10.15	10.38	10.54	10.71	10.80
HProX	NN	Rnd	1.19	2.30	3.68	5.05	6.28	7.34	8.24	9.00	9.67	10.26	10.80
HProX	Trn	Rnd	1.47	3.14	4.65	5.95	7.05	7.94	8.68	9.31	9.85	10.37	10.80
HProX	Rnd	NN	4.17	2.49	2.12	1.79	1.58	1.41	1.26	1.19	1.10	1.10	1.18
HProX	NN	NN	1.19	1.11	1.09	1.09	1.08	1.08	1.08	1.08	1.08	1.09	1.18
HProX	Trn	NN	1.47	1.25	1.19	1.14	1.12	1.10	1.10	1.09	1.09	1.09	1.18
HProX	Rnd	Trn	4.17	3.79	3.52	3.17	2.77	2.39	2.00	1.62	1.36	1.20	1.16
HProX	NN	Trn	1.19	1.28	1.27	1.23	1.18	1.12	1.09	1.07	1.09	1.10	1.16
HProX	Trn	Trn	1.47	1.39	1.33	1.25	1.19	1.13	1.11	1.10	1.11	1.12	1.16
HGreX	Rnd	Rnd	2.97	5.57	7.07	8.09	8.86	9.42	9.85	10.17	10.44	10.64	10.80
HGreX	NN	Rnd	1.11	2.05	3.36	4.68	5.91	7.03	8.00	8.85	9.59	10.22	10.80
HGreX	Trn	Rnd	1.23	2.64	4.08	5.40	6.57	7.56	8.42	9.15	9.77	10.31	10.80
HGreX	Rnd	NN	2.97	1.80	1.55	1.41	1.30	1.18	1.12	1.11	1.11	1.11	1.18
HGreX	NN	NN	1.11	1.06	1.07	1.07	1.08	1.08	1.08	1.09	1.09	1.10	1.18
HGreX	Trn	NN	1.23	1.07	1.08	1.09	1.09	1.09	1.09	1.09	1.09	1.10	1.18
HGreX	Rnd	Trn	2.97	2.85	2.62	2.39	2.10	1.88	1.64	1.45	1.26	1.21	1.17
HGreX	NN	Trn	1.11	1.07	1.08	1.08	1.06	1.05	1.06	1.07	1.09	1.10	1.17
HGreX	Trn	Trn	1.23	1.11	1.11	1.10	1.07	1.08	1.09	1.10	1.11	1.12	1.17
HRndX	Rnd	Rnd	7.11	9.33	9.77	10.06	10.26	10.39	10.52	10.60	10.69	10.76	10.79
HRndX	NN	Rnd	1.47	3.47	5.07	6.19	7.07	7.81	8.51	9.17	9.77	10.34	10.79
HRndX	Trn	Rnd	2.17	4.63	6.06	6.99	7.73	8.36	8.92	9.46	9.95	10.40	10.79
HRndX	Rnd	NN	7.11	4.86	3.74	2.88	2.23	1.72	1.44	1.24	1.13	1.10	1.18
HRndX	NN	NN	1.47	1.26	1.18	1.13	1.11	1.10	1.08	1.08	1.08	1.09	1.18
HRndX	Trn	NN	2.17	1.64	1.42	1.28	1.21	1.15	1.11	1.10	1.09	1.09	1.18
HRndX	Rnd	Trn	7.11	5.70	4.79	4.04	3.42	2.88	2.35	1.86	1.43	1.20	1.16
HRndX	NN	Trn	1.47	1.58	1.50	1.40	1.30	1.21	1.14	1.10	1.09	1.10	1.16
HRndX	Trn	Trn	2.17	1.85	1.62	1.46	1.33	1.22	1.16	1.12	1.11	1.11	1.16

For the warehouse optimization problems the optimal probabilities, in a non-conflict resolvers, based on our experiments, the optimal probability of using FTS was about 0.3. In each place, where SPFS can be applied, also PPFS can be applied and the decision about which operator to use is taken randomly each time (we use SPFS with probability 0.65 and PPFS with probability 0.35).

The best performing base crossover operator was KPoint, followed by AEX, followed by HRndX. As it can be seen from Table 1 and from Fig. 1, the incorporation of the product frequency search (PFS) in the crossover operators significantly improved the results, while still preserving the order of the quality: the best one was the improved KPoint, followed by the improved AEX, followed by the improved HRndX.

The progress with the improved operators was faster, and the final results were better. Significant differences were obtained already after 10 epochs. This allows to efficiently use the crossover operators with local search for running multiple optimizations for only 100 or even only 10 epochs, and then to choose the best performing run to continue the optimization. That is because in most cases, the best run after 10 or 100 epochs is also be the best run after 1000 epochs [13]. The choice of the best run with the improved crossover operators can be done earlier than with the base crossover operators, because the improvement is much faster in the initial epochs. This is an additional benefit in terms of computational time of the improved crossover operators.

4.5   Analysis of the Results Obtained for TSP

The values that represent the basic crossover operators in Table 2 are those with random conflict resolver and with zero probability of random resolver, that is: 8.76 for AEX, 4.17 for HProX, 2.97 for HGreX, and 7.11 for HRndX.

The best results are obtained for all the four crossover operators with a combination of nearest neighbor (NN) resolver and tournament resolver (Trn) or with two nearest neighbor resolvers. For AEX, HProX and HRndX, the optimal probability of the random resolver in these cases is between 0.4 and 0.8 for two NN resolvers and between 0.6 and 0.8 for one NN and one Trn resolver. The highest optimal probability of the mixed resolver corresponds to a weaker preference for the shortest distance inside the resolvers than inside two NN resolvers. Thus, in both cases, the optimal local component is similar, but can be obtained either by the second NN resolver or by an increased random resolver probability in the mixed configuration.

HGreX, on the other hand, has already embedded some local search mechanism, as was already mentioned. This implies two results. First, in its basic form, is the best performing from the crossover operators (the relative route length 2.97). Second, it does need so strong additional local search as the other crossover operators. For that reason it obtained the best performance with random resolver probability of 0.1 for NN/NN and Trn/NN configuration and about 0.4–0.5 for NN/Trn, and also it achieved very good performance for Trn/Trn resolvers with the optimal probability of 0.4–0.5.

As it can be seen, the random resolvers used in the situation, where there is no conflict do not make any sense, as they only worsens the results, especially if they are used with high probabilities.

Although in some cases the standard crossover operators were able to find the optimal route and there was nothing to improve in terms of the results, incorporating the local search always allowed to reduce dramatically the number of epochs required to find the solution.

5   Conclusions

We analyzed the AEX, HGreX, HProX, HRndX and KPoint crossover operators with local search mechanisms and investigated the optimal balance between exploration and exploitation in the crossover operators for the genetic algorithm based optimization. In most cases, HGreX was the superior crossover operator TSP (it is worth noting that HGreX has already embedded some local search component in its basic form), while KPoint performed best for the warehouse problem. Nevertheless, by adding the resolvers to the crossover operators, the results could be improved in many cases and could be obtained faster. Especially, the great acceleration of the optimization process was observed.

Although, in the paper, we used two examples: the warehouse optimization and TSP, the approach is more universal and can be extended to to other problems, where the concept of cost between two locations can be defined. This is especially the case in many production optimization, planing and scheduling problems, which we are going to consider in our future research and practical implementations.

An important issue is to provide the optimal amount of the local search (see Table 2). It is also likely, that the results can still be improved if the local search frequency is adjusted dynamically during the optimization process, which we are going to further investigate. It is also worth including other crossover operators in the future experiments.

References

1. Abdoun, O., Abouchabaka, J.: A comparative study of adaptive crossover operators for genetic algorithms to resolve the traveling salesman problem. Int. J. Comput. Appl. **31**, 49–57 (2011)
2. Affenzeller, M., Wagner, S., Winkler, S., Beham, A.: Genetic Algorithms and Genetic Programming: Modern Concepts and Practical Applications. CRC Press (2018)
3. Al-Furhud, M.A., Ahmed, Z.H.: Experimental study of a hybrid genetic algorithm for the multiple travelling salesman problem. Math. Probl. Eng. **20**, 3431420 (2020)
4. Avdeikins, A., Savrasovs, M.: Making warehouse logistics smart by effective placement strategy based on genetic algorithms. Transp. Telecommun. **20**(4), 318–324 (2019)
5. Bartholdi, J.J., Hackman, S.T.: Warehouse and Distribution Science. https://www.warehouse-science.com/book/index.html (2019)
6. Zuñiga, J.B., et. al.: Optimization of the storage location assignment and the picker-routing problem by using mathematical programming. Appl. Sci. **10**(2), 534 (2020)
7. Cicirello, V.A.: Non-wrapping order crossover: An order preserving crossover operator that respects absolute position. In: GECCO 2006, p. 1125–1131 (2006)
8. Ha, Q.M., Deville, Y., Pham, Q.D., Hà, M.H.: A hybrid genetic algorithm for the traveling salesman problem with drone. https://arxiv.org/abs/1812.09351 (2018)
9. Hassanat, A.B.A., Alkafaween, E.: On enhancing genetic algorithms using new crossovers. Int. J. Comput. Appl. Technol. **55**, 202–212 (2017)

10. Hongxin, Z., guohui, Z., shili, C.: On directed edge recombination crossover for ATSP. In: Jiao, L., Wang, L., Gao, X., Liu, J., Wu, F. (eds.) ICNC 2006. LNCS, vol. 4221, pp. 783–791. Springer, Heidelberg (2006). https://doi.org/10.1007/11881070_104

11. Hwang, H.: An improvement model for vehicle routing problem with time constraint based on genetic algorithm. Comput. Ind. Eng. **42**, 361–369 (2002)

12. Kordos, M., Blachnik, M., Scherer, R.: Fuzzy clustering decomposition of genetic algorithm-based instance selection for regression problems. Inf. Sci. **587**, 23–40 (2021)

13. Kordos, M., Boryczko, J., Blachnik, M., Golak, S.: Optimization of warehouse operations with genetic algorithms. Appl. Sci. **10**(14), 4817 (2020)

14. Lin, B.L., Sun, X., Salous, S.: Solving travelling salesman problem with an improved hybrid genetic algorithm. J. Comput. Commun. **4**(15), 98–106 (2016)

15. Ma, M., Li, H.: A hybrid genetic algorithm for solving bi-objective Traveling Salesman Problems. J. Phys.: Conf. Ser. **887**, 012065 (2017)

16. Puljić, K., Manger, R.: Comparison of eight evolutionary crossover operators for the vehicle routing problem. Math. Commun. **18**, 359–375 (2013)

17. Rakesh, V., Kadil, G.: Layout optimization of a three dimensional order picking warehouse. IFAC-PapersOnLine **48**, 1155–1160 (2017)

18. Santos, J., Ferreira, A., Flintsch, G.: An adaptive hybrid genetic algorithm for pavement management. Int. J. Pavement Eng. **20**(3) (2019)

19. Simon, D.: Evolutionary Optimization Algorithms. Wiley (2013)

20. Singh, K., Sundar, S.: A hybrid genetic algorithm for the degree-constrained minimum spanning tree problem. Soft. Comput. **24**(3), 2169–2186 (2019). https://doi.org/10.1007/s00500-019-04051-x

21. Tan, H., Lee, L.H., Zhu, Q., Ou, K.: Heuristic methods for vehicle routing problem with time windows. Artif. Intell. Eng. **16**, 281–295 (2001)

22. Ting, C.-K.: Improving edge recombination through alternate inheritance and greedy manner. In: Gottlieb, J., Raidl, G.R. (eds.) EvoCOP 2004. LNCS, vol. 3004, pp. 210–219. Springer, Heidelberg (2004). https://doi.org/10.1007/978-3-540-24652-7_21

23. Van Gils, T., Ramaekers, K., Caris, A., De Koster, R.: Designing efficient order picking systems by combining planning problems: state-of-the-art classification and review. Eur. J. Oper. Res. **267**, 1–15 (2018)

24. Zhang, W., Zhu, J., Yuan, R.: Optimization of automated warehouse storage location assignment problem based on improved genetic algorithm. In: The 9th International Conference on Logistics, Informatics and Service Sciences, pp. 297–311 (2019)

Numerical and Statistical Probability Distribution Transformation for Modeling Traffic in Optical Networks

Alicja Poturała[1], Maria Konieczka[1], Piotr Śliwka[2]ⓘ, Sławomir Sujecki[3,4]ⓘ, and Stanisław Kozdrowski[1](✉)ⓘ

[1] Department of Computer Science, Faculty of Electronics,
Warsaw University of Technology, Nowowiejska 15/19, 00-665 Warsaw, Poland
s.kozdrowski@elka.pw.edu.pl
[2] Department of Computer Science, Cardinal Wyszynski University,
Woycickiego 1/3, 01-938 Warsaw, Poland
[3] Telecommunications and Teleinformatics Department, Wroclaw University
of Science and Technology, Wyb. Wyspianskiego 27, 50-370 Wroclaw, Poland
[4] Faculty of Electronics, Military University of Technology,
S. Kaliskiego 2, 00-908 Warsaw, Poland

Abstract. It is important for optical network operators to consider the available budget in forecasting network traffic. This is related to network expansion and equipment purchases. The underlying motivation is the constant increase in the demand for network traffic due to the development of new access technologies (5G, FTTH), which require particularly large amounts of bandwidth. The aim of this paper is to numerically calculate a transformation that allows determining probability distributions of network edge traffic based on known probability distributions of demand matrix elements. Statistical methods confirmed the proposed transformation. The study is performed for a practically relevant network within selected scenarios determined by realistic traffic demand sets.

Keywords: Optical network modeling and optimization ·
Non-parametric and parametric probability distribution · Statistical
analysis · Network congestion · Forecasting · Traffic demands

1 Introduction

Optical network operators need to continually upgrade the networks to accommodate the ever-increasing data traffic. In currently deployed optical networks, which are based on single core fibres, the data transmission rate can be increased by using either a larger per-channel bit rate or by increasing the number of available channels [8]. In order to implement such changes in an operating optical network a network operator needs to add equipment to the network nodes. This additional equipment incurs significant costs, which have to be accounted for in

ⓒ The Author(s), under exclusive license to Springer Nature Switzerland AG 2022
D. Groen et al. (Eds.): ICCS 2022, LNCS 13352, pp. 383–397, 2022.
https://doi.org/10.1007/978-3-031-08757-8_32

the company budget. It is needless to say that the minimisation of the equipment costs is critical to the commercial success of an operation network operator company. The cost minimisation is usually achieved by careful planning and optimisation of network resources at every stage of the network development. Algorithms for network optimisation have been developed for many years now [1,3,4,9]. Of specific importance to optical networks are Routing and Wavelength Assignment (RWA) and Routing and Spectrum Allocation (RSA) algorithms applied in static [2,7,11] and dynamic [16] environment that has been developed by a number of authors over the past years. More recently, the predictive capability of a network optimisation software has been improved by taking into account the physical phenomena occurring in the optical fibre [8] and developing algorithms that predict the traffic demands in optical network nodes and corresponding data transmission rates in the network edges [10].

The traffic demands at optical network nodes increase steadily due to a continued modernisation of telecommunications technologies used in access networks. A milestone in the development of modern access networks was the introduction of glass fibres, which have now practically almost completely displaced in many countries the previously used copper connections. Fibre optics dominate the access network market in many countries, as they are superior to copper cable solutions by allowing larger data transmitting rates and longer distances. The other technology, which will further increase the traffic in access networks is 5G wireless technology, supported by its backhaul infrastructure. In the context of fast technological changes in access networks the analysis of demands and more specifically an estimation of the demand matrix elements is an essential element of the optical network planning and development.

In order to analyse and predict the values of demand matrix elements for optical network it is prudent to assume that a demand can be described by a random variable. Introducing a random variable, a probability distribution for a network matrix element can be empirically determined [10]. In the next step, one can attempt to fit parametric probability distributions to the empirically determined probability distributions [5]. In this context a natural question arises regarding the transformation of the probability distributions from the known probability distributions of demand matrix elements to the probability distributions describing the traffic in the network edges, which is the main problem considered in this contribution.

It is not easy to find an analytical form of the transformation from one nonparametric probability distribution of the random variable X to another nonparametric probability distribution of the random variable Y and expressed by the analytical function f such that $Y = f(X)$, where f is not a given explicitly. In some cases, for known parametric distributions, analytical transformations can be defined. However, to the best of the authors' knowledge, this problem has not found a satisfactory solution in the literature so far. Thus, one needs to resort to the use of numerical methods, based on sampling or random number generators, and the methods of statistical data analysis.

The probability distribution transformation problem is therefore the primary objective of this contribution. It is noted that only after the traffic in the network edges is known the process of the optical network nodal equipment purchase planning can start. Also the use of statistical methods adds rigour to the way in which the problem is approached since it allows calculating formally expected values and the variance, i.e. the values and the ranges for the considered data transmission rate in a given network edge. We also provide a justification of the relationship between the known probability distributions of the demand matrix elements and the probability distributions describing traffic at the edges of the network.

The rest of the paper is organised as follows. In Sect. 2, the problem is described and the proposed methods for addressing the probability distribution transformation problem are presented. Next, in Sect. 3 the results obtained are presented together with the relevant discussion. Finally, Sect. 4, provides a summary of the main research findings.

2   Problem Formulation and Methods

A 3-node slice containing the nodes Wroclaw, Lodz and Katowice and the connections between them were selected from the Polish backbone network. The whole network together with the analysed part consisting of 3 nodes (marked using the red line) is shown on the Fig. 1.

In the following subsections the research methods are described. First, the approach used for the calculation of the demands matrix elements probability distributions is discussed. In the second subsection, the technique applied to calculate the probability distributions of the network edge data transmission rates subject to the known values of the demand matrix elements probability distributions is presented in a case of a 3 node network. In the last subsection an approach to calculating the probability distributions of the network edge data transmission rates for larger networks is discussed.

Fig. 1. Case of study, the segment of Polish core optical network.

2.1   Demands Matrix Elements Probability Distributions

When attempting to forecast the values of demands matrix elements, i.e., calculate their probability distributions in the coming years one needs historical data to build suitable stochastic models. However, such data are not generally available, as telecommunications network operators protect such information. Also SNDlib database lacks data on the historical evolution of demand matrix

388 A. Poturała et al.

elements for the Polish network considered here [12]. Consequently, the values
of the demands matrix elements had to be calculated using other methods. The
general description of the method adopted here is described in [10]. The approach
presented in [10] relies upon combining data from two statistical offices: Central Statistical Office (CSO) and European Statistical Office (Eurostat). Using
the statistical data stemming from both sources the historical values of traffic
demands for specific network nodes were estimated. Then the demands between
pairs of cities of the Polish backbone network were estimated [10].

The Table 1 shows the calculated historical values of the demands matrix elements in the years 2010–2020 between
cities of the considered network section
consisting of 3 nodes (expressed in units of
Tera bit per second [Tbps]). The demand
is modeled by the sum of the values resulting from the trend present in the histori-
cal data $(\hat{y}_t)$ and a random variable $(e_t = y_t - \hat{y}_t)$ that represents the residuals, and
is based on the relationship: $y_t = \hat{y}_t + e_t$
[15].

Table 1. Demands.

Year	Lodz–Kat.	Wro.–Kat.	Wro.–Lodz.
2010	0,0909	0,0833	0,1211
2011	0,0972	0,0919	0,1303
2012	0,1099	0,0961	0,1389
2013	0,1100	0,0959	0,1393
2014	0,1125	0,1036	0,1501
2015	0,1099	0,1036	0,1496
2016	0,1184	0,1102	0,1576
2017	0,1191	0,1124	0,1600
2018	0,1202	0,1097	0,1580
2019	0,1228	0,1215	0,1684
2020	0,1319	0,1297	0,1801

Three methods were selected, based on the following procedure:

S1. on the basis of empirical data Y, estimate the parameters $\hat{\alpha}$ of the appropriate trend function $\hat{Y} = f(\hat{\alpha}, t)$, determine residuals $e = \hat{Y} - Y$ and finally
the probability distribution function of the value of random variable Y as
one realisation of an appropriate stochastic process,
S2. generate a sequence of $\{z_1, \ldots, z_n\} \in Z$ random numbers based on
$F^{-1}(Y) := Z$, where F - distribution function of the random variable Y,
S3. based on a sequence of $\{z_1, \ldots, z_n\}$ from S2. find the "averaged" probability
distribution of the random variable Z,

These methods calculate the probability distributions of the demands matrix
elements using:

1. Extended Empirical Distribution (EED) - a stationary distribution that is
computed from data containing historical residuals (e_t) and the observations
generated from them. Limited historical data are available, so the idea is to
estimate 1000000 new ones using the inverse transform sampling. All points
are divided into k bins. Each bin represents one class with probability proportional to the number of samples;
2. Normal Distribution (ND) - in contrast to EED, describes the variables using
μ and σ parameters (computed from e). For discretisation, one million points
were generated from the $N(\mu, \sigma)$ distribution, which were divided into k bins;
3. Model with Increasing Uncertainty (MwIU) - took into account that predictions become less and less reliable with the length of the prediction horizon, which is used in fan-based methods ([5,14]). They are typically based

on the two-part normal distribution [6] represented in the Eq. (1), where $A = \frac{2}{(1/\sqrt{1-\gamma})+(1/\sqrt{1+\gamma})}$.

$$f(x; \mu, \sigma, \gamma) = \frac{A}{\sqrt{2\pi}\sigma} \begin{cases} exp\{-\frac{1-\gamma}{2\sigma^2}[(x-\mu)^2]\}, & \text{for} \quad x \leqslant \mu \\ exp\{-\frac{1+\gamma}{2\sigma^2}[(x-\mu)^2]\}, & \text{for} \quad x > \mu \end{cases} \quad (1)$$

This distribution has three parameters mode - μ, uncertainty indicator - σ and inverse skewness indicator - γ. Nonstationarity was introduced by multiplying the uncertainty coefficient values in successive years. The three-sigma rule of thumb was used for discretisation.

2.2 Calculation of Edge Data Transmission Rate Probability Distributions for a Small Network

A demand matrix is used to describe the data transmission demands across the DWDM network. Its example form for a 3-node network is shown in the Eq. (2). This matrix is a square matrix of dimension $N \times N$, where N denotes the number of nodes in the network. The elements from row l and column m contains the data transfer demand between city l and m expressed in Gbps.

$$D^{ijk} = \begin{bmatrix} 0 & d_{12}^i & d_{13}^j \\ 0 & 0 & d_{23}^k \\ 0 & 0 & 0 \end{bmatrix} \quad (2)$$

The demand matrix is a multivariate random variable. In the approach adopted here, it is assumed that each of its elements is ultimately described by a discrete probability distribution consisting of c classes (or bins), where $c \in \{3, 4, 5, 6\}$. In the Eq. (2), the indices i, j, and k are used to specify the class to which each element of the demand matrix belongs, where $i, j, k \in \{0, \cdots, c-1\}$. The label 0 corresponds to the minimum class and $c-1$ to the maximum class. To fully describe a random variable next to the values (i.e. data transfer demand between two selected cities expressed in Gbps) one needs to specify the corresponding values of the probability that the random variable assumes a specific value. The individual elements of the matrix D^{ijk} have a corresponding probability of occurrence. Thus, a matrix of the probabilities for occurrence of a specific value of the matrix D^{ijk} element are given by the corresponding elements of the matrix P^{ijk}:

$$P^{ijk} = \begin{bmatrix} 0 & p_{12}^i & p_{13}^j \\ 0 & 0 & p_{23}^k \\ 0 & 0 & 0 \end{bmatrix} \quad (3)$$

Thus, knowing the values of the matrix P^{ijk} elements and assuming that the random variables corresponding to the elements of matrix D^{ijk} are independent of each other one can calculate the probability of a specific matrix D^{ijk} realisation as the product of the corresponding matrix P^{ijk} elements. Once all the demands matrix elements are set and the corresponding probability of the specific realisation of demands matrix D^{ijk}, is known one can start calculating the values of the data traffic in the network edges by solving the optimisation problem as described in [8] subject to the known constrains. This procedure has to be repeated for each possible realisation of the demands matrix D^{ijk} to give the full probability distribution of the data traffic in the network edges, which in the considered 3 node example, and including 4 classes for matrix D^{ijk} elements discrete probability distributions gives 4^3 optimisations to calculate the full probability distribution of network edge data transmission rates.

2.3 Extending Edge Calculations to Complex Networks

The advantage of studying a small network segment is that it allows for a detailed estimation of network traffic based on demands matrix element forecasts. However, the accuracy of such approach is limited since other nodes of the network have an impact on the traffic present in the specific segment. Unfortunately, a full analysis of all realisations of the demand matrix in case of an entire backbone network using the presented approach leads to very intensive computations. First of all, it should be noted that the number of combinations of demand matrices is $c^{\frac{N(N-1)}{2}}$, where c is the number of classes used in probability distributions while N is the number of network nodes. So, even with a 12-node network and a three-class distribution, the number of combinations of demand matrices is 3^{66}. Moreover, as the complexity of the network rises, the time required to find a solution to the optimisation problem increases. For these reasons, an alternative approach has been considered giving an additional insight into the network operation. It consists in analysing the impact of changes in demands for selected city pairs. For this purpose the demand matrix element corresponding to the considered city pair is assumed to be described by a random variable whilst all other demands matrix elements are approximated by the expected value of the corresponding random variable. As an illustrative example in the next section an analysis was carried out of data traffic between selected distant city pairs: Szczecin-Rzeszow and Rzeszow-Poznan.

3 Results

The first series of experiments started by generating demand matrices for all class numbers and for all methods of forecast considered as described in Sect. 2.1. The demand matrix element values corresponding to an example pair of cities (Katowice and Lodz) with a four-class distribution are collected in Table 2. The first two distributions (EED and ND) have similar expected values, which is due to the fact that they use the same linear model and the mean values of the deviations are close to zero.

In order to assess the quality of the predictions, the Theil Index values (I^2) and relative prediction errors „ex post" ($v_{s_f} = \dfrac{\sqrt{\frac{1}{\#I_f}\sum_{t\in I_f}(y_t-y_t^f)^2}}{\bar{y}_{t\in I_f}}$, where y_t^f - forecast of y in time t, I_f - forecast verification time interval) of the distributions employed were collected in Table 3. Since there were no significant differences

Table 2. Comparison of forecast demand levels between Katowice and Lodz, 4-class distribution, (probability).

Id	Year	Levels			
		Min	Negative	Positive	Max
EED	2019	616(0.278)	638(0.278)	653(0.222)	679(0.222)
	2020	633(0.278)	655(0.278)	670(0.222)	696(0.222)
	2021	650(0.278)	672(0.278)	687(0.222)	713(0.222)
	2022	667(0.278)	689(0.278)	704(0.222)	730(0.222)
	2023	684(0.278)	706(0.278)	721(0.222)	747(0.222)
ND	2019	556(0.010)	623(0.491)	669(0.494)	743(0.005)
	2020	573(0.010)	640(0.491)	686(0.494)	760(0.005)
	2021	590(0.010)	656(0.491)	703(0.494)	777(0.005)
	2022	607(0.010)	673(0.491)	720(0.494)	794(0.005)
	2023	624(0.010)	690(0.491)	737(0.494)	811(0.005)
MwIU	2019	587(0.065)	630(0.431)	659(0.431)	702(0.065)
	2020	576(0.065)	640(0.431)	683(0.431)	748(0.065)
	2021	564(0.065)	650(0.431)	707(0.431)	794(0.065)
	2022	552(0.065)	660(0.431)	732(0.431)	839(0.065)
	2023	540(0.065)	670(0.431)	756(0.431)	885(0.065)

between the expected values, the relative prediction errors and Theil Index values are similar for the distributions considered. The least accurate predictions were obtained for the Katowice-Wroclaw demand (although most values were acceptable, as $5\% < v(s_P) < 10\%$, where a commonly acceptable level of good

Table 3. Comparison of I^2 and $v(s_P)$ values of generated forecasts for the studied distributions for 3-node network.

Bin	Demand	I^2			$v(s_P)$		
		EED	ND	MwIU	EED	ND	MwIU
3	Katowice-Lodz	0.00053	0.00053	0.00054	0.02296	0.02306	0.02332
	Katowice-Wroclaw	0.00240	0.00285	0.00286	0.04905	0.05340	0.05351
	Lodz-Wroclaw	0.00066	0.00072	0.00072	0.02563	0.02680	0.02681
4	Katowice-Lodz	0.00054	0.00056	0.00054	0.02322	0.02376	0.02324
	Katowice-Wroclaw	0.00275	0.00282	0.00282	0.05252	0.05312	0.05316
	Lodz-Wroclaw	0.00083	0.00073	0.00071	0.02877	0.02710	0.02675
5	Katowice-Lodz	0.00054	0.00054	0.00055	0.02329	0.02327	0.02357
	Katowice-Wroclaw	0.00275	0.00285	0.00285	0.05251	0.05338	0.05339
	Lodz-Wroclaw	0.00082	0.00071	0.00071	0.02865	0.02675	0.02672
6	Katowice-Lodz	0.00055	0.00055	0.00054	0.02338	0.02349	0.02332
	Katowice-Wroclaw	0.00274	0.00286	0.00286	0.05235	0.05355	0.05351
	Lodz-Wroclaw	0.00084	0.00073	0.00073	0.02904	0.02713	0.02697

fit is $v(s_P) < 10\%$ [15]). The forecasts' accuracy for the remaining demands is very good $(v(s_P) < 3\%)$. Similar conclusions can be drawn by considering the Theil coefficient.

The Kolmogorov-Smirnov goodness of fit test was used to verify the statistical concordance of the probability distributions: demand matrix elements and traffic in the network edges for the 3rd to sixth classes. Tables 4 and 5 contain the results of the KS statistics given by the Formula 4.

$$\sqrt{\frac{n \cdot n}{2n}} \cdot \sup_x |F_{ref}(x) - F_{id}(x)| \tag{4}$$

where
$F_{ref}(x)$ - the distribution function of the demand matrix elements,
$F_{id}(x)$ - the distribution function of the forecast demand level (given e.g. in Table 2).

Depending on the number of classes, the empirical KS_{emp} statistics values obtained in Tables 4 and 5, with a significance level $\alpha = 0.05$, do not exceed $KS_{theor} = 1.358$. So there are no grounds for rejecting $H_0 : F_{ref}(x) = F_{id}(x)$ against the alternative $H_1 : F_{ref}(x) \neq F_{id}(x)$. At the significance level $\alpha = 0.05$ we can assume that the respective probability distributions are consistent. As expected, an increase in the bin number generally increases the value of KS_{emp}, which in the case of high granularity of the probability distribution may lead to the relationship $KS_{emp} > KS_{theor}$, i.e. rejection of H_0 (a lack of accordance between the compared distribution functions [13]).

Table 4. Values of KS tests for stationary distributions.

Id	Bin	Demand			
		Katowice-Lodz	Katowice-Wroclaw	Lodz-Wroclaw	Max per demand
EED	3	0.0001	0.0010	0.0005	0.0010
	4	0.0789	0.0012	0.1566	0.1566
	5	0.1052	0.0691	0.2804	0.2804
	6	0.0005	0.0017	0.0009	0.0017
ND	3	0.4696	0.6177	0.3356	0.6177
	4	0.4571	0.7071	0.4578	0.7071
	5	0.4624	0.6201	0.2789	0.6201
	6	0.6750	0.4887	0.6743	0.6750

Table 5. Values of KS tests for MwIU.

Year	Bin	Demand			
		Kat.-Lodz	Kat.-Wro.	Lodz-Wro.	Max per demand
2022	3	0.3512	0.4872	0.2151	0.4872
	4	0.7071	0.7071	0.7071	0.7071
	5	0.4327	0.4457	0.4327	0.4457
	6	0.6736	0.8660	0.6736	0.8660
2023	3	0.3512	0.4872	0.2151	0.4872
	4	0.7071	0.7071	0.7071	0.7071
	5	0.4327	0.4457	0.4327	0.4457
	6	0.6736	0.8660	0.8660	0.8660

As expected, when the bin number increases, the value of KS_{emp} increases. It means that the higher number of bins, the greater probability of rejecting H_0, (which is in line with the intuition) the greater accuracy of the probability distribution, expressed with more classes. This may, from a statistical point of view, lead to a lack of accordance between the compared distribution functions.

Table 6 presents the values of the Kolmogorov-Smirnov statistics scaled by class count for the 2019 and 2023 MwIU distributions for different bins. The comparison of the remaining pairs of years had similar yields similar results and for this reason it is not included in the Table 6.

Table 6. Values of KS tests for MwIU between MwIU.

Years	Bin	Demand			
		Kat-Lodz	Kat-Wro	Lodz-Wro	Max per demand
	3	0.8384	0.8384	0.8384	0.8384
2019	4	0.6143	0.6143	0.6143	0.6143
2023	5	0.7158	0.7158	0.7158	0.7158
	6	0.8289	0.8289	0.8289	0.8289

All generated realisations of the demand matrices served as input to the optimisation process as described in Sects. 2.2 and 2.3. In total, more than 10000 experiments were performed, which were based on the results of the forecasts from all distributions for 2019–2023. The simulations were designed to calculate the probability of congestion on the network and to identify the edges that are most heavily loaded. First we consider the 3 node network.

Figure 2 shows the probability of network congestion for 3-(Fig. 2a), 4-(Fig. 2b), 5-(Fig. 2c) and 6-class (Fig. 2d) distribution, respectively. Each graph shows how the probability of network congestion has changed over the forecast years.

It can be seen that the probability of network congestion calculated using the empirical distribution (EED) ranges from 0 to 100%. This is because the differences between the lowest and highest predicted values are small - in 2019. For every realisation of the demand matrix, the algorithm found a solution to the problem in 2019 while in 2023, for every realisation of the matrix no acceptable solution could be found. Using the parametric distribution (MwIU) congestion probability never reached 100% while for the normal distribution (ND) and 2023, the congestion probability was very close to 100%.

For almost all distributions and class numbers, an increase in the congestion probability was observed in subsequent years. Only for the model with increasing uncertainty (MwIU) and the 3-, 4- and 5-class distributions, there was a decrease between 2021 and 2022. This is because in this case the range of accepted values widens within the prediction horizon.

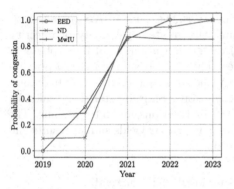

(a) 3-class demand matrix distribution.

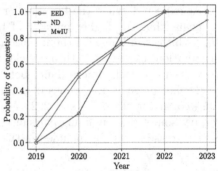

(b) 4-class demand matrix distribution.

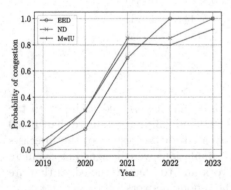

(c) 5-class demand matrix distribution.

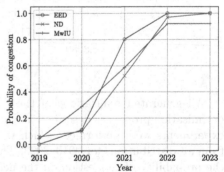

(d) 6-class demand matrix distribution.

Fig. 2. Probability of network congestion in subsequent years.

As expected, the larger the number of classes into which the data was divided, the more accurately the probability of network congestion is estimated. As the number of classes increases, the ranges associated with individual classes become narrower. Also with the number of classes the number of possible realisations of the demand matrix increases exponentially. Thus, computational effort is needed to perform simulations. On the other hand, the shapes of the cumulative distribution function, regardless of the number of classes, stay almost unchanged.

Based on the results obtained, it was decided that all three methods would also be selected for the application of the forecasts to the full network, as in the case of the 3-node network: EED, ND, and MwIU. In addition, the forecast years have been limited to 2019 and 2020. An in-depth analysis with the selected methods was carried out only for demands between distant city pairs: Szczecin-Rzeszow and Rzeszow-Poznan.

The Table 7 collects the obtained probabilities of realised classes for individual demands depending on the idea under study for 3-class distribution. The occurrence of underestimation is marked with a plus sign (+). As in the case of the forecast for the 3-node network, the EED predicted demand for 2020 is lower than in the case of ND and MwIU.

Table 7. Comparison of the probabilities of the class corresponding to the actual demand for analyzed city pairs [%].

Demand	Year	Distribution		
		EED	ND	MwIU
Rzeszow-Poznan	2019	22.21	90.33	15.69
	2020	+	5.66	15.69
Szczecin-Rzeszow	2019	22.21	88.75	15.69
	2020	+	5.53	15.69

Tables 8a and 8b show a comparison of expected values calculated using selected probability distributions between years 2019 and 2023, for the demand matrix elements corresponding to Rzeszow–Szczecin and Szczecin–Poznan network edge, respectively. As with the network slice analysis in Fig. 1, calculations were performed for all realisations of the demand matrix. The elements of the demand matrix were increased 15000 times to highlight the differences between the realisations under the adopted network parameters. Additionally, it is noted that the city pairs that were selected for in-depth analysis were characterised by small demand values.

Table 8. Comparison of expected values

(a) Rzeszow-Szczecin

Year	Distribution		
	EED	ND	MwIU
2019	1380.104	1380.373	1379.000
2020	1419.104	1419.373	1417.842
2021	1458.104	1458.373	1456.842
2022	1497.104	1497.333	1495.842
2023	1535.659	1535.430	1534.158

(b) Szczecin-Poznan

Year	Distribution		
	EED	ND	MwIU
2019	1139.433	1138.879	1139.000
2020	1175.433	1174.879	1174.842
2021	1210.765	1209.936	1210.000
2022	1246.765	1245.879	1246.000
2023	1282.433	1281.879	1282.000

Table 9 lists the I^2 and $v(s_P)$ values for the analyzed demands in the full Polish network for the 3-class distributions. Similarly to the studied network slice, the values were low enough to consider that the forecasts are either good or very good.

Table 9. Comparison of I^2 and $v(s_P)$ values of generated forecasts for the studied 3-class distributions for full Polish network.

Demand	I^2			$v(s_P)$		
	EED	ND	MwIU	EED	ND	MwIU
Rzeszow-Poznan	0.00093	0.00093	0.00096	0.03060	0.03053	0.03095
Szczecin-Rzeszow	0.00069	0.00069	0.00070	0.02627	0.02635	0.02641
Sum	0.00162	0.00162	0.00165	0.05687	0.05688	0.05736

The representative results of numerical simulations for the year 2023 using demand matrix with 3-class probability distributions are shown in Fig. 3 and 4. The top indices of the demand matrix correspond respectively to the class indices for the Rzeszow-Poznan and Szczecin-Rzeszow pairs. Maps showing the bandwidth occupancy at each edge for the extreme values of demand matrix elements and the three considered approaches (EED, ND and MwIU) are presented in Fig. 4. Figure 3 shows the channel occupancies calculated for the year 2023. The results obtained show that all distributions have similar demand expectation values.

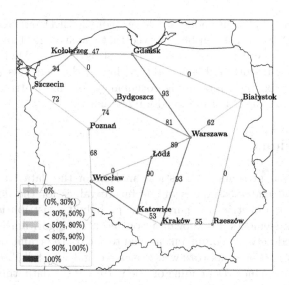

Fig. 3. Percentage edge occupancies for the 2023 forecast for D^{11} demand matrix, EED, ND and MwIU, 3-class distribution.

Also, it is noted that in all maps shown in Fig. 4 and 3 one can see that traffic tends to accumulate in the Warsaw node. Edges linked with Warsaw have the highest occupancy and the degree of this node is the only one equal to 5

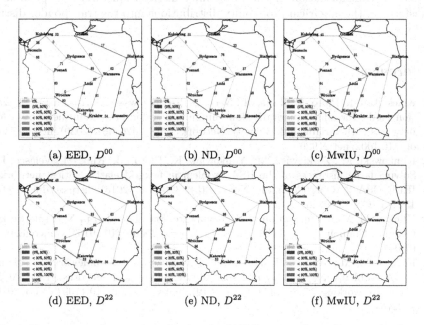

Fig. 4. Percentage edge occupancies for the 2023 forecast depending on the demand matrix and the assumed 3-class distribution.

while the degrees of all other nodes are lower. For the EED and ND distributions (Fig. 4a, 4d and 4b, 4e), it can be observed that the average channel occupancy per edge rises as the demands increase. In contrast, for the MwIU distribution this regularity is distorted (Fig. 4c, 4f). Although the demands are higher for MwIU D^{22} matrix, the average edge occupancy for this matrix was lower by about 0.6 pp than the result obtained for D^{00} matrix.

4   Conclusions

This contribution presents a statistical analysis of the data traffic in optical network and calculates estimates of the future values for the demand matrix elements with parametric/non-parametric probability distribution. Once the demand matrix elements are known, an optimisation algorithm is used to predict the required network equipment needed to satisfy the traffic demand. As an example, firstly a 3 nodes network was considered and then full 12-nodes network with selected 2 demands. Depending on the nature of the data, three approaches to determining the types of distributions have been proposed: extended empirical distribution, normal distribution and model with increasing uncertainty.

The statistical concordance distribution test confirmed that the proposed numerical methods transform the empirical probability distribution into the probability distribution of demands and can be applied in the absence of analytical methods allowing for the transformation of the considered probability distributions. The challenge is still to provide the analytical function that allows for such a transformation of the probability distributions and a step further: transforming the empirical probability distributions into the transponder distribution.

References

1. Kalesnikau, I., Pióro, M., Poss, M., Nace, D., Tomaszewski, A.: A robust optimization model for affine/quadratic flow thinning: a traffic protection mechanism for networks with variable link capacity. Networks **75**(4), 420–437 (2020). https://doi.org/10.1002/net.21929
2. Klinkowski, M., Żotkiewicz, M., Walkowiak, K., Pióro, M., Ruiz, M., Velasco, L.: Solving large instances of the RSA problem in flexgrid elastic optical networks. IEEE/OSA J. Opt. Commun. Netw. **8**(5), 320–330 (2016). https://doi.org/10.1364/JOCN.8.000320
3. Klinkowski, M.: Optimization of latency-aware flow allocation in NGFI networks. Comput. Commun. **161**, 344–359 (2020). https://doi.org/10.1016/j.comcom.2020.07.044
4. Klinkowski, M., Walkowiak, K.: An efficient optimization framework for solving RSSA problems in spectrally and spatially flexible optical networks. IEEE/ACM Trans. Networking **27**(4), 1474–1486 (2019). https://doi.org/10.1109/TNET.2019.2922761

5. Konieczka, M., Poturała, A., Śliwka, P., Sujecki, S., Kozdrowski, S.: Modeling demands forecasts with probability distributions in DWDM optical networks. In: 2021 International Conference on Software, Telecommunications and Computer Networks (SoftCOM), pp. 1–6 (2021). https://doi.org/10.23919/SoftCOM52868. 2021.9559126
6. Kotz, S., Balakrishnan, N., Johnson, N.L.: Continuous Multivariate Distributions, Volume 1: Models and Applications. John Wiley & Sons (2004)
7. Kozdrowski, S., Żotkiewicz, M., Sujecki, S.: Optimization of optical networks based on CDC-ROADM technology. Appl. Sci. **9**(3) (2019). https://doi.org/10.3390/app9030399, http://www.mdpi.com/2076-3417/9/3/399
8. Kozdrowski, S., Żotkiewicz, M., Sujecki, S.: Ultra-wideband WDM optical network optimization. Photonics **7**(1) (2020). https://doi.org/10.3390/photonics7010016
9. Kozdrowski, S., Banaszek, M., Jedrzejczak, B., Żotkiewicz, M., Kopertowski, Z.: Application of the ant colony algorithm for routing in next generation programmable networks. In: Paszynski, M., Kranzlmüller, D., Krzhizhanovskaya, V.V., Dongarra, J.J., Sloot, P.M. (eds.) Computational Science – ICCS 2021. pp. 526–539. Springer International Publishing, Cham (2021). https://doi.org/10.1007/978-3-030-77970-2_40
10. Kozdrowski, S., Sliwka, P., Sujecki, S.: Modeling traffic forecasts with probability in DWDM optical networks Springer Nature Switzerland AG, M. Paszynski et al. (eds.): ICCS 2021. LNCS 12745, pp. 1–14 (2021). https://doi.org/10.1007/978-3-030-77970-2-28
11. Kozdrowski, S., Żotkiewicz, M., Wnuk, K., Sikorski, A., Sujecki, S.: A comparative evaluation of nature inspired algorithms for telecommunication network design. Appl. Sci. **10**(19) (2020). https://doi.org/10.3390/app10196840, https://www.mdpi.com/2076-3417/10/19/6840
12. Orlowski, S., Wessäly, R., Pióro, M., Tomaszewski, A.: Sndlib 1.0-survivable network design library. Networks **55**(3), 276–286 (2010). https://doi.org/10.1002/net.20371
13. Sliwka, P.: Markov (set) chains application to predict mortality rates using extended Milevsky-Promislov generalized mortality models. J. Appl. Stat. (2021). https://doi.org/10.1080/02664763.2021.1967891
14. Sliwka, P., Socha, L.: A proposition of generalized stochastic Milevsky-Promislov mortality models. Scand. Actuar. J. **8**, 706–726 (2018). https://doi.org/10.1080/03461238.2018.1431805
15. Sliwka, P., Swistowska, A.: Economic forecasting methods with the R package. UKSW, Warszawa (2019)
16. de Sousa, A., Monteiro, P., Lopes, C.B.: Lightpath admission control and rerouting in dynamic flex-grid optical transport networks. Networks **69**(1), 151–163 (2017). https://doi.org/10.1002/net.21715

GPU Accelerated Modelling and Forecasting for Large Time Series

Christos K. Filelis - Papadopoulos[1]([📧]) [iD], John P. Morrison[1] [iD],
and Philip O'Reilly[2] [iD]

[1] Department of Computer Science, Western Gateway Building,
University College Cork, Cork, Ireland
{christos.papadopoulos-filelis,j.morrison}@cs.ucc.ie
[2] University College Cork, Cork University Business School, O'Rahilly Building,
Cork, Ireland
philip.oreilly@ucc.ie

Abstract. Modelling of large scale data series is of significant importance in fields such as astrophysics and finance. The continuous increase in available data requires new computational approaches such as the use of multicore processors and accelerators. Recently, a novel time series modelling and forecasting method was proposed, based on a recursively updated pseudoinverse matrix which enhances parsimony by enabling assessment of basis functions, before inclusion into the final model. Herewith, a novel GPU (Graphics Processing Unit) accelerated matrix based auto-regressive variant is presented, which utilizes lagged versions of a time series and interactions between them to form a model. The original approach is reviewed and a matrix multiplication based variant is proposed. The GPU accelerated and hybrid parallel versions are introduced, utilizing single and mixed precision arithmetic to increase GPU performance. Discussions around performance improvement and high order interactions are given. A block processing approach is also introduced to reduce memory requirements for the accelerator. Furthermore, the inclusion of constraints in the computation of weights, corresponding to the basis functions, with respect to the parallel implementation are discussed. The approach is assessed in a series of model problems and discussions are provided.

Keywords: Forecasting · Pseudoinverse matrix · Parallel modelling · GPU acceleration

1 Introduction

Modelling and forecasting time series has several applications is a number of scientific fields including signal processing, computational finance and astrophysics. Modelling of time series can be performed with traditional approaches [21], such as Auto-Regressive Integrated Moving Average (ARIMA) models [3] and Exponential Smoothing (ES) [4,16], or machine learning methods such as Long Short

The original version of this chapter was revised: This chapter was previously published non-open access and the acknowledgement section has been updated. The correction to this chapter is available at
https://doi.org/10.1007/978-3-031-08757-8_63

- Term Memory Networks (LSTM) [15] and Support Vector Regression (SVR) [8]. Another important family of techniques are based on orthogonalization of a set of basis functions, such as Fast Orthogonal Search [17] or Matching Pursuit [22]. These techniques construct a model from a set of basis functions (linear or non-linear) using orthogonalization procedures such as Gram-Schmidt orthogonalization. Recently, a recursive Schur complement pseudoinverse approach for modelling time series was introduced [12]. This approach avoids orthogonalization, while enabling the use of preconditioned iterative methods for reducing memory requirements in the case of a large number of basis functions [10]. These methods enable the use of arbitrary basis functions including linear, trigonometric, auto-regressive or machine learning based [11–13].

The emergence of big data led to an increase in interest in the area of parallel computing in order to reduce processing times. Extensive research has been carried out in the parallelization of machine learning methods, especially neural networks in multicore systems, distributed systems and accelerators such as GPUs [1,2,18,23]. Parallelization was used to reduce training times especially for deep neural network architectures and for very large input datasets. GPU acceleration has been also utilized to reduce training and optimization times for Support Vector Machines [19,24,25]. In the majority of these approaches, the training operations are reformed to take advantage of matrix by matrix (BLAS3) kernels that can be efficiently parallelized in GPUs and modern multicore hardware. Another important modification is the use of mixed precision arithmetic, combining half-precision, single precision and double precision arithmetic substantially improving performance and storage requirements [1,20]. In the literature, efforts have been directed also in the parallelization of Matching Pursuit type methods [6,9] for GPUs. Despite the extensive literature and software available for parallelizing machine learning methods, literature around parallelization of techniques such as Fast Orthogonal Search is limited.

Herewith, we propose a novel parallel implementation of the recently proposed recursive Schur complement pseudoinverse matrix modelling based on auto-regressive basis functions. Initially, the method is recast to take advantage of BLAS3 operations, during the basis search operation, while avoiding redundant computations which will increase computational work and memory requirements. Then, the parallel algorithm, that utilizes mixed precision arithmetic, is presented and discussed along with a pure GPU implementation and block mixed precision variants. Multiplicative interactions between auto-regressive basis functions are also discussed. Implications related to precision and memory transfers between CPU and GPU are analyzed. The proposed scheme is assessed by modelling and forecasting two time series with different characteristics and sets of candidate basis functions. Scalability results are also presented and discussed.

In Sect. 2, the recursive Schur complement based pseudoinverse matrix of basis functions is reviewed and insights on the basis functions selection, higher order basis interactions and termination criteria is given. In Sect. 3, the matrix based variant is proposed and the CPU/GPU and pure GPU implementations are presented. Moreover, a block variant is given along with discussions on memory requirements, data transfer overhead and the effects of mixed precision arithmetic. In Sect. 4, numerical results are presented depicting the applicability and accuracy of the proposed scheme along with discussions on scalability and implications of mixed precision arithmetic.

2 Recursive Schur Complement Based Time - Series Modelling

The coefficients of a model, corresponding to the time series y, with linearly independent basis functions stored in the columns of a matrix X can be computed as follows:

$$Xa = y \iff a = X^{+}y \iff a = (X^{T}X)^{-1}X^{T}y, \tag{1}$$

where a is a vector of length n retaining the coefficients corresponding to the n basis functions retained in X. The matrix $X^{T}X$ and its inverse are Symmetric Positive Definite [10]. In most cases all basis functions are not known "a priori" or their contribution to error reduction is not significant. In order to avoid such issues a recursive pseudoinverse matrix approach has been proposed [12] based on a symmetric variant of the matrix preconditioning technique introduced in [10]. Following this approach and given an additional basis F, with $X_{i+1} = [X_i \ F]$, $1 \le i \le n$, we have:

$$a_{i+1} = [X_i \ F]^{+}y = G_{i+1}D_{i+1}^{-1}G_{i+1}^{T}X_{i+1}^{T}y, \tag{2}$$

or equivalently:

$$\begin{bmatrix} a_i \\ b \end{bmatrix} = \begin{bmatrix} G_i \ -G_iD_i^{-1}G_i^{T}X_i^{T}F \\ 0 \qquad 1 \end{bmatrix} \begin{bmatrix} D_i^{-1} \ 0 \\ 0 \ s_{i+1}^{-1} \end{bmatrix} \begin{bmatrix} G_i^{T} \qquad\qquad 0 \\ -F^{T}X_iG_iD_i^{-1}G_i^{T} \ 1 \end{bmatrix} \begin{bmatrix} X_i^{T}y \\ F^{T}y \end{bmatrix}, \tag{3}$$

where $(X_i^{T}X_i)^{-1} = G_iD_i^{-1}G_i^{T}$ and $s_{i+1} = F^{T}F - F^{T}X_iG_iD_i^{-1}G_i^{T}X_i^{T}F$ denotes the Schur Complement corresponding to the addition of basis function F. The initial conditions for the recursive formulation are $G_{1,1} = 1$, $D_{1,1} = s_1^{-1} = (F^{T}F)^{-1}$ and $a_1 = s_1^{-1}F^{T}y$. The updated set of coefficients a_{i+1} can be computed alternatively using the following equations:

$$a_{i+1} = \begin{bmatrix} a_i^{*} \\ b \end{bmatrix} = \begin{bmatrix} a_i + g_{i+1}b \\ s_{i+1}^{-1}(F^{T} + g_{i+1}^{T}X_i^{T})y \end{bmatrix}, \tag{4}$$

with:

$$G_{i+1} = \begin{bmatrix} G_i \ g_{i+1} \\ 0 \quad 1 \end{bmatrix} \ and \ D_{i+1}^{-1} = \begin{bmatrix} D_i^{-1} \ 0 \\ 0 \ s_{i+1}^{-1} \end{bmatrix}, \tag{5}$$

where b denotes the coefficient corresponding to basis function F and a_i^{*} is the updated coefficients after addition of basis function F. The vector $g_{i+1} = -G_iD_i^{-1}G_i^{T}X_i^{T}F$ corresponds to the $(i+1)$ column and $s_{i+1} = F^{T}(F + X_ig_{i+1})$

is the Schur complement. The modelling error $\rho_{i+1} = \|r_{i+1}\|_2^2$ corresponding to the addition of a basis function F can be computed as:

$$\|r_{i+1}\|_2^2 = \|y - X_{i+1}a_{i+1}\|_2^2 = \|r_i\|_2^2 - \|(X_i g_{i+1} + F)b\|_2^2 = \|r_i\|_2^2 - b^T s_{i+1} b, 0 \leq i \leq n-1 \tag{6}$$

with $\|r_0\|_2^2 = \|y\|_2^2$. The quantity $e_{i+1} = b^T s_{i+1} b$ denotes the error reduction corresponding to the addition of a basis function F [12]. In order to ensure positive definiteness of the dot product matrix $X_{i+1}^T X_{i+1}$ the quantity e_{i+1} should be bounded by $0 \leq e_{i+1} \leq \|r_i\|_2^2$. It should be noted that $\rho_{i+1} = \|r_{i+1}\|_2^2 = T \cdot MSE$ is the Squared Error, T is the number of samples in time series y, and MSE denotes the Mean Squared Error. Detailed description and additional discussions regarding the method are given in [11–13].

2.1 Assessment and Selection of Basis Functions

The explicit expression of error reduction can be used to select a subset of basis functions to form a model from a candidate set U retaining N basis functions. Trigonometric, exponential and linear functions have been considered for modelling in [12,13], while a Non-Negative Adaptive Auto - Regression approach was followed in [11]. The procedure of selecting an appropriate basis requires computation of the potential error reduction for each member of the set U. This procedure is algorithmically described in Algorithm 1. The procedure, described by Algorithm 1, proceeds through all candidate basis in U sequentially, storing the respective error reductions to vector u. Then, the algorithm proceeds by selecting the index of the basis function that lead to maximum error reduction under the constraints that ensure positive definiteness.

Algorithm 1. Basis Search
$(k = bs(y, G_i, D_i^{-1}, X_i, U, \rho_i))$

1: **Let** N denote the number of candidate basis functions in U.
2: $e_i = 0, 1 \leq i \leq N$
3: **for** $i \in [1, N] \subset \mathbb{N}$ **do**
4:     $F = U_i$
5:     $g_{i+1} = -G_i D_i^{-1} G_i^T X_i^T F$
6:     $s_{i+1} = F^T (F + X_i g_{i+1})$
7:     $b = s_{i+1}^{-1}(F^T + g_{i+1}^T X_i^T)y$
8:     $e_i = b^T s_{i+1} b$
9: **end for**
10: $k = \arg\max_{i \in [1,N]} e_i$ under the constraint $0 \leq e_i \leq \rho_i$

The set U can host any type of basis functions even Machine Learning models such as Support Vector Machines [8]. In the current manuscript we focus on lagged basis function of the form:

$$U = [y_{-1} \ y_{-2} \ y_{-3} \ y_{-4} \ \cdots \ y_{-N}], \tag{7}$$

with $y_0 = y$. In practice, the number of samples in lagged time series y_{-i} is $n - N$, since the latest sample is removed (retained in the responses y), along with the first $N - 1$ samples from each candidate lagged basis to ensure that there are no missing data. In case multiplicative interactions between basis functions are allowed [14], e.g. $y_i y_j \ldots$ the number of basis functions into the candidate set are:

$$\binom{N}{k} + N\,k, \ k > 1 \tag{8}$$

where k is the order of allowed interactions. In the case of $k = 1$, then the number of candidate basis functions is equal to N.

Additional constraints can be imposed during the selection of a basis function that leads to the maximum error reduction. These constraints can be imposed during step 10 of Algorithm 1. Examples of constraints include non-negativity of the coefficients [11] or imposing a threshold on their magnitude.

After selection of an appropriate basis function or lag, addition of this basis function has to be performed and the corresponding matrices to be updated. Several basis functions can be fitted by executing the process described by Algorithm 1 followed by the process of Algorithm 2, iteratively. The fitting process is terminated based on criteria regarding the fitting error, e.g. [12]:

$$\sqrt{\rho_{i+1}} < \epsilon \sqrt{\rho_0}. \tag{9}$$

Another approach is to terminate the fitting process based on the magnitude of the coefficients:

$$|b| < \epsilon |a_1|. \tag{10}$$

where b is the coefficient corresponding to the $i + 1$ added basis function, while a_1 is the coefficient corresponding to the basis function added first. In both termination criteria ϵ is the prescribed tolerance. It should be noted that the second criterion is more appropriate in the case of lagged basis.

Algorithm 2. Add Basis Function
$([G_{i+1}, D_{i+1}^{-1}, X_{i+1}, a_{i+1}, \rho_{i+1}] = addbasis(y, G_i, D_i^{-1}, X_i, F, a_i, \rho_i))$

1: $g_{i+1} = -G_i D_i^{-1} G_i^T X_i^T F$
2: $s_{i+1} = F^T (F + X_i g_{i+1})$
3: $b = s_{i+1}^{-1} (F^T + g_{i+1}^T X_i^T) y$
4: Check termination criterion and terminate if met
5: $a_{i+1} = \begin{bmatrix} a_i + g_{i+1} b \\ b \end{bmatrix}$
6: $\rho_{i+1} = \rho_i - b^T s_{i+1} b$
7: $G_{i+1} = \begin{bmatrix} G_i & g_{i+1} \\ 0 & 1 \end{bmatrix}$; $D_{i+1} = \begin{bmatrix} D_i^{-1} & 0 \\ 0 & s_{i+1}^{-1} \end{bmatrix}$; $X_{i+1} = \begin{bmatrix} X_i & F \end{bmatrix}$

---

3   Performance Optimization and Parallelization

The procedure, described by Algorithm 1, proceeds through all candidate basis in U sequentially. It can be performed in parallel by assigning a portion of basis functions to each thread of a multicore processor. However, the most computationally intensive operations are matrix by vector and vector by vector, which are BLAS2 (Basic Linear Algebra Subroutines - Type 2) and BLAS1 operations, respectively, [5]. These operations lead to decreased performance compared to operations between matrices which are BLAS3 operations, since they require increased memory transfers and cache tiling and data re-use is limited [7]. Thus, to increase performance the most computationally intensive part, which is the basis search described by Algorithm 1, has to be redesigned in order to compute the corresponding error for all candidate basis functions.

Let us consider a set of N candidate basis functions, represented as vectors of length $N - n$, stored in the columns of matrix U ($(n - N) \times N$). The columns $g_{i+1}^{(j)}, 1 \leq j \leq N$ corresponding to each basis can be computed by the following matrix multiplication operations:

$$\mathfrak{g}_{i+1} = [g_{i+1}^{(1)} \; g_{i+1}^{(2)} \; \cdots \; g_{i+1}^{(N)}] = -G_i D_i^{-1} G_i^T X_i^T U. \tag{11}$$

The matrix $\mathfrak{g}_{i+1}$ ($i \times N$) is formed by four dense matrix multiplications, however matrix D_i^{-1} is diagonal matrix, thus it can be retained as a vector. Multiplying matrix D_i^{-1} by another matrix is equivalent to multiplying the elements of each row j with the corresponding element $d_{j,j}^{-1}$ of the vector retaining the elements of the diagonal matrix. For the remainder of the manuscript we will denote this operation as $(\star)$. This operation can be used in the process described in Algorithm 2. This reduces operations required, as well as storage requirements, since a matrix multiplication is avoided and the computation can be performed in place in memory.

Following computation of the columns stored in matrix $\mathfrak{g}_{i+1}$, the Schur Complements $s_{i+1}^{(j)}, 1 \leq j \leq N$ corresponding to the candidate basis functions are computed as follows:

$$\mathfrak{s}_{i+1} = diag(\tilde{\mathfrak{s}}_{i+1}) = diag(U^T(U + X_i \mathfrak{g}_{i+1})). \tag{12}$$

The formula $U^T(U + X_i \mathfrak{g}_{i+1})$ leads to the computation of Schur complement of the block incorporation of basis and not the individual Schur complements corresponding to the candidate basis function, which are stored in the diagonal of the result. The Schur complement $\tilde{\mathfrak{s}}_{i+1}$ is a dense matrix of dimensions $N \times N$ and requires substantial computational effort. In order to avoid unnecessary operations each diagonal element can be computed as follows:

$$(\mathfrak{s}_{i+1})_j = (U^T)_{j,:}((U)_{:,j} + X_i(\mathfrak{g}_{i+1})_{:,j}), \tag{13}$$

where $(\, . \,)_{i,j}$ denotes an element at position (i,j) of a matrix and (:) denotes all elements of a row or column of a matrix. In order to compute all elements of the diagonal concurrently, Eq. (13) can be reformed as follows:

$$\mathfrak{s}_{i+1} = (U^T \odot (U + X_i \mathfrak{g}_{i+1})^T)v, \tag{14}$$

where $\odot$ denotes the Hadamard product of two matrices and v is a vector of the form $[1 \ 1 \ \ldots \ 1]^T$. The vector $\mathfrak{s}_{i+1}$ ($N \times 1$) retains the Schur complements corresponding to the candidate basis functions in U. Dedicated (Optimized) Hadamard product is not included in the standard BLAS collection, however it is included in vendor versions or CUDA (Compute Unified Device Architecture). Following the same notation the coefficients corresponding to the basis functions in set U can be computed as:

$$\mathfrak{b} = \mathfrak{s}_{i+1}^{-1} \odot (U^T + \mathfrak{g}_{i+1}^T X_i^T)y \qquad (15)$$

and the corresponding error reductions as:

$$\mathfrak{e} = \mathfrak{b} \odot \mathfrak{s}_{i+1} \odot \mathfrak{b}. \qquad (16)$$

The most appropriate basis function is selected by finding the maximum error reduction in vector $\mathfrak{e}$. The matrix based basis selection procedure is algorithmically described in Algorithm 3.

Algorithm 3. Matrix Based Basis Search
$(k = mbbs(y, G_i, D_i^{-1}, X_i, U, \rho_i))$

1: **Let** N denote the number of candidate basis functions in U.
2: $v_i = 1, 1 \le i \le N$
3: $\mathfrak{g}_{i+1} = -G_i(D_i^{-1} \star (G_i^T X_i^T U))$
4: $\tilde{U} = U + X_i \mathfrak{g}_{i+1}$
5: $\mathfrak{s}_{i+1} = (U^T \odot \tilde{U})v$
6: $\mathfrak{b} = \mathfrak{s}_{i+1}^{-1} \odot \tilde{U}^T y$
7: $\mathfrak{e} = \mathfrak{b} \odot \mathfrak{s}_{i+1} \odot \mathfrak{b}$
8: $k = \arg\max_{i \in [1,N]} \mathfrak{e}_i$ under the constraint $0 \le \mathfrak{e}_i \le \rho_i$

A block variant of Algorithm 3 can be utilized to process batches of candidate functions. This can be performed by splitting matrix U into groups, processing them and accumulating the corresponding error reductions in vector $\mathfrak{e}$ before computing the index of the most effective basis function. Despite the advantages in terms of performance, the matrix and block based matrix approaches have increased memory requirements. The memory requirements are analogous to the number of candidate basis functions, since they have to be evaluated before assessment, while in the original approach each candidate basis is evaluated only before its assessment. Thus, the matrix approach requires $\mathcal{O}(N(n-N))$, the block approach $\mathcal{O}(max(\nu(n-N)), b_s(n-N)))$ and the original approach $\mathcal{O}(\nu(n-N))$ 64-bit words, where ν denotes the number of basis functions included in the model and b_s the number of basis in each block.

3.1   Graphics Processing Unit Acceleration

The operations involved in Matrix Based Basis Search, given in Algorithm 3, can be efficiently accelerated in a Graphics Processing Unit (GPU). However, most

GPU units suffer from substantial reduction of the double precision performance, e.g. 32× in the case of double precision arithmetic (Geforce RTX 20 series). In order to mitigate this issue, 32-bit floating point operations and 16-bit half precision floating point operations are utilized. This gives rise to mixed precision computations, where GPU related operations are performed in reduced precision, while CPU related ones are performed in double precision arithmetic. This approach enables acceleration while minimizing round off errors from reduced precision computation.

The proposed scheme utilized a similar approach in order to accelerate the most computationally intensive part of the process, which is the basis search. Computations in the GPU require data movement from the main memory to the GPU memory, which is a time consuming operation, thus should be limited. For the case of the Matrix Based Basis Search algorithm, data should be transferred in the GPU before processing. This includes the time series y, the matrix G_i and vector D_i, the matrix of included basis X_i and the previous modelling error ρ_i in order to mark basis that could hinder positive definiteness. Before copying these arrays to the GPU memory, they should be cast to single precision arithmetic to ensure increased performance during computations. The matrix U, retaining the candidate basis, and time series y can be transferred in the GPU before starting the fitting process, since they are "a priori" known, while all the other matrices should be updated after addition of new basis function to the model. However, the update process includes only a small number of values to be transferred at each iteration.

Algorithm 4. GPU accelerated modelling

1: **Let** y denote the time series to be modelled, N the maximum lag, n the number of samples in y.
2: $v_i = 1, 1 \leq i \leq n - N$
3: $\mathbf{y} = cpu2gpu(single(y))$
4: $\mathbf{U} = cpu2gpu(single([y_{-1} \ y_{-2} \ \cdots \ y_{-N}]))$
5: $\rho_0 = \|y\|_2^2; \quad \boldsymbol{\rho} = cpu2gpu(single(\rho_0))$
6: $[G_1, D_1^{-1}, X_1, a_1, \rho_1] = addbasis(y, [\,], [\,], [\,], v, [\,], \rho_0)$
7: $\mathbf{G} = cpu2gpu(single(G_1)); \quad \mathbf{D}^{-1} = cpu2gpu(single(D_1))$
8: $\mathbf{X} = cpu2gpu(single(X_1)); \quad \boldsymbol{\rho} = cpu2gpu(single(\rho_1))$
9: **for** $i \in [1, N]$ **do**
10: $\quad \mathbf{k} = mbbs(\mathbf{y}, \mathbf{G}, \mathbf{D}^{-1}, \mathbf{X}, \mathbf{U}, \boldsymbol{\rho})$ ▷ GPU
11: $\quad k = gpu2cpu(\mathbf{k}); \quad F = y_{-k}$
12: $\quad [G_{i+1}, D_{i+1}^{-1}, X_{i+1}, a_{i+1}, \rho_{i+1}] = addbasis(y, G_i, D_i^{-1}, X_i, F, a_i, \rho_i)$
13: $\quad \mathbf{G} = update(single(G_{i+1})); \quad \mathbf{D}^{-1} = update(single(D_{i+1}^{-1}))$
14: $\quad \mathbf{X} = update(single(X_{i+1})); \quad \boldsymbol{\rho} = cpu2gpu(single(\rho_{i+1}))$
15: **end for**

The process is described in Algorithm 4. The matrices and vectors stored in the GPU memory are given in bold. The process terminates if the termination criterion of Eq. (10) is met during the basis addition process or line 12 of

Algorithm 4. It should be noted that the first basis included removes the mean value of the time series y. This is performed in line 6 where the *addbasis* function is invoked. In practice, addition of the first basis function is performed using the equations: $s_1 = F^T F, a_1 = s_1^{-1} F^T y, D_1^{-1} = s_1^{-1}, G_1 = 1$, where F is substituted with a vector $((n - N) \times 1)$ with all its components set to unity.

The functions *cpu2gpu* and *gpu2cpu* are used to transfer data from CPU memory to GPU memory and from GPU memory to CPU memory, respectively. Conversion of matrices, vectors and variables from double precision to single precision arithmetic is performed with function *single* and the function *update* is utilized to update matrices and vectors involved in the Matrix Based Basis Search performed in the GPU. At each iteration $2 + i + (n - N), 1 \leq i \leq \nu$ single precision floating point numbers need to be transferred to GPU memory, with ν denoting the number of basis functions included in the model.

A full GPU version of Algorithm 4 can be also used, by forming and updating all matrices directly to the GPU memory. In this approach, the matrix of candidate matrices U and the time-series y need to be transferred to the GPU memory, before computation commences. The value of the first coefficient should also be transferred to the CPU since it is required by the termination criterion. Moreover, in every iteration the new coefficient has to be transferred to CPU in order to assess model formation through the termination criterion. This approach uses solely single precision arithmetic and is expected to yield slightly different results due to rounding errors.

4   Numerical Results

In this section the applicability, performance and accuracy of the proposed scheme is examined by applying the proposed technique to two time series. The first time series is composed of large number of samples and lagged basis functions without multiplicative interactions are used as the candidate set. The scalability of different approaches is assessed with respect to single precision, mixed precision and double precision arithmetic executed either on CPU or GPU. The second time series has a reduced number of samples, however an extended set of lagged candidate basis functions, which include second order interactions, are included. The characteristics of the time series are given in Table 1. The two time series were extracted from R studio. The error measures used to assess the forecasting error was Mean Absolute Percentage Error (MAPE), Mean Absolute Deviation (MAE) and Root Mean Squared Error (RMSE):

$$MAPE = \frac{100}{T} \sum_{i=1}^{T} \frac{|y_i - \hat{y}_i|}{|y_i|}, \quad MAE = \frac{1}{T} \sum_{i=1}^{T} |y_i - \hat{y}_i|, \quad RMSE = \sqrt{\frac{1}{T} \sum_{i=1}^{T} (y_i - \hat{y}_i)^2} \quad (17)$$

where y_i are the actual values, $\hat{y}_i$ the forecasted values and T the length of the test set. All forecasts are performed out-of-sample using a multi-step approach without retraining. All experiments were executed on a system with an Intel Core i7 9700K 3.6–4.9 GHz CPU (8 cores) with 16 GBytes of RAM memory and an NVIDIA Geforce 2070 RTX (2304 CUDA Cores) with 8 GBytes of memory. All CPU computations were carried out in parallel using Intel MKL, while the GPU computations were carried out using NVIDIA CUDA libraries.

Table 1. Model time series with description and selected splitting.

#	Name	Frequency	Train	Test	Description
1	**Call volume for a large North American bank**	5-min	22325	5391	Volume of calls, per five minute intervals, spanning 164 days starting from 3 March 2003
2	**Daily female births in California**	Daily	304	61	Daily observations in 1959

Different notation is used for the variants of the proposed scheme:

CPU-DP: Matrix based CPU implementation using double precision arithmetic. This is the baseline implementation.

CPU-SP: Matrix based CPU implementation using single precision arithmetic.

CPU-DP-GPU: Matrix based CPU/GPU implementation using mixed precision arithmetic. The basis search is performed in the GPU using single precision arithmetic, while incorporation of the basis function is performed in double precision arithmetic.

CPU-DP-GPU-block(n_b): Block matrix based CPU/GPU implementation using mixed precision arithmetic. The basis search is performed in the GPU using single precision arithmetic, while incorporation of the basis function is performed in double precision arithmetic.

GPU: Pure matrix based GPU implementation using single precision arithmetic.

The parameter n_b denotes the number of blocks. The block approach requires less GPU memory.

4.1   Time Series 1 - Scalability and Accuracy

The average value of the dataset is 192.079, the minimum value is 11 and the maximum value is 465. For this model the lagged candidate basis has been utilized, while higher degree interactions are not allowed, resulting in an additive model. The performance in seconds for all variants is given in Fig. 1, while speedups are presented in Fig. 2. The pure GPU and CPU-DP-GPU implementations have the best performance overall leading to the best speedups. The pure GPU implementation has a speedup greater than 20× for more than 50 basis functions with a maximum of approximately 27×, with respect to the baseline implementation. With respect to the CPU single precision arithmetic implementation the pure GPU approach has a speedup of approximately 10× for more than 50 basis functions. The CPU-DP-GPU has a maximum speedup of approximately 22× attained for 134 basis functions. After that point the speedup decreases because the double precision operations in the CPU do not scale with same rate, reducing the overall speedup. It should be noted that even for low number of basis functions, e.g. 6, the speedup of the pure GPU implementation is approximately 6× with respect to CPU-DP and approximately 4× with respect to CPU-SP implementation.

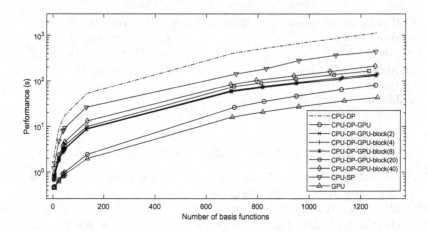

Fig. 1. Performance for all variants for different number of basis functions.

The performance of the block variants degrades when increasing the number of blocks retaining the candidate basis functions, since they require more data transfers between CPU and GPU. The speedups for the block variants range from 2.5×–8.6× over CPU-DP implementation and 1.5×–3.4× over the CPU-SP implementation.

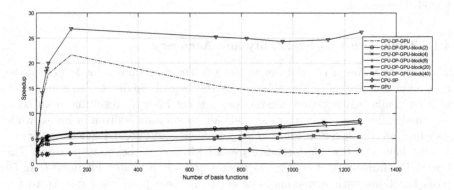

Fig. 2. Speedup for all variants for different number of basis functions.

Table 2. Minimum, maximum and average number of basis functions, RMSE, MAE and MAPE for all variants for the first time series.

	# Basis			RMSE			MAE			MAPE		
ϵ	Min	Max	Avg	Min	Max	Avg	Min	Max	Avg	Min	Max	Avg
0.1	4	4	4.0	41.96	41.96	41.96	34.25	34.25	34.25	25.80	25.80	25.80
0.07	6	6	6.0	44.86	44.86	44.86	36.87	36.87	36.87	29.29	29.29	29.29
0.04	23	23	23.0	26.32	26.32	26.32	19.61	19.61	19.61	11.76	11.76	11.76
0.02	39	39	39.0	24.07	24.07	24.07	18.07	18.07	18.07	11.23	11.23	11.23
0.01	42	42	42.0	24.13	24.14	24.14	18.12	18.12	18.12	11.22	11.22	11.22
0.009	47	47	47.0	23.86	23.86	23.86	17.96	17.96	17.96	11.45	11.45	11.45
0.006	132	136	134.1	23.31	23.64	23.47	17.65	17.96	17.81	11.66	11.92	11.83
0.005	695	716	702.8	23.16	23.64	23.22	17.50	17.96	17.56	11.49	11.92	11.55
0.004	795	833	818.7	23.15	23.64	23.21	17.49	17.96	17.55	11.48	11.92	11.53
0.003	941	960	952.3	23.16	23.17	23.17	17.50	17.52	17.51	11.50	11.52	11.51
0.002	1081	1136	1116.0	23.16	23.17	23.16	17.49	17.50	17.50	11.46	11.47	11.47
0.001	1232	1266	1257.3	23.16	23.16	23.16	17.49	17.50	17.50	11.46	11.47	11.47

The number of basis functions included in the model along with forecasting errors are given in Table 2. The number of basis functions as well as the errors are not substantially affected by the use of mixed or single precision arithmetic. More specifically, up to approximately 50 basis functions all variants produce almost identical results. However, above 50 basis functions there is a minor difference in the number of basis functions included in the model which in turn slightly affects the error measures. The difference in the number of included basis functions is caused by rounding errors in the computation of error reduction ρ. This is caused by the rounding errors in the formation of the column vectors $\mathfrak{g}_{i+1}$, involved in the computation of respective Schur complements and potential basis coefficients.

An important observation is that the error measures regarding forecasts do not significantly reduce after the incorporation of approximately 130 basis functions. Thus, additional basis functions increase the complexity of the model. In order to ensure sparsity of the underlying model, a different termination criterion can be used, since the termination criterion of Eq. (10) depends on the magnitude of the entries of the basis functions and is more susceptible to numerical rounding errors. The new termination criterion based on error reduction percentage is as follows:

$$\sqrt{\rho_i} - \sqrt{\rho_i - e_{i+1}} < \epsilon\sqrt{\rho_i}, \tag{18}$$

where e_{i+1} denotes the potential error reduction that will be caused by the incorporation of the $i+1$-th basis. $\epsilon \in [0,1] \subset \mathbb{R}$ denotes the acceptable percentage of error reduction to include a basis function. This criterion will be used to model the second time series along with higher level interactions.

4.2   Time Series 2 - Flexibility and Higher Order Interactions

The average value of the dataset is 41.9808, the minimum value is 23 and the maximum value is 73.

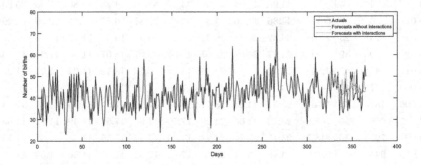

Fig. 3. Actual along with forecasted values with and without interactions.

The candidate set is composed of lagged basis functions and second order interactions of the form $y_j y_k$. The termination criterion of Eq. (18) was used with $\epsilon = 0.002$, with maximum lags equal to 50. In Fig. 3 the actuals along with the forecasted values computed with and without interactions are given. The inclusion of second order interactions results in capturing the nonlinear behavior of the time series in the forecasted values. The error measures without interactions were: $RMSE = 5.96$, $MAE = 5.11$ and $MAPE = 12.33$, while with interactions the error measures were: $RMSE = 6.18$, $MAE = 4.93$ and $MAPE = 12.20$. The inclusion of higher order interactions led to reduction of the error measures and showcases the flexibility of the approach allowing for the inclusion of arbitrary order interactions in the candidate basis functions.

The execution time for CPU-SP, CPU-DP and GPU were 1.1621, 2.2076 and 0.3932, respectively. Thus, the speedup of the pure GPU variant was approximately 3× over the CPU-SP variant and 5.6× over the CPU-DP version. The pure GPU version is efficient even for time series with small number of samples, under a sufficiently sized space of candidate basis functions.

5   Conclusion

A matrix based parallel adaptive auto-regressive modelling technique has been proposed. The technique has been parallelized in multicore CPUs and GPUs and a block variant has been also proposed, based on a matrix (BLAS3) recast of the required operations. The pure GPU variant presented speedup up to 27× over the double precision arithmetic parallel CPU version and 10× over the parallel single precision CPU version for time series with large number of samples. The use of single and mixed precision did not affect substantially the forecasting error, rendering the technique suitable for modelling and forecasting large time series. Implementation details and discussions on higher order interactions between basis functions have been also given. The applicability and effectiveness of the method were also discussed and new termination criterion based on potential error reduction of basis functions, which is invariant to scaling, has been given.

Future work is directed towards the design of an improved basis search that will reduce the search space based on a tree approach. Moreover, backfitting procedures will be considered.

Acknowledgement. This publication has emanated from research conducted with the financial support of Science Foundation Ireland under Grant number 18/SPP/3459. For the purpose of Open Access, the author has applied a CC BY public copyright licence to any Author Accepted Manuscript version arising from this submission

References

1. Abadi, M., et al.: Tensorflow: a system for large-scale machine learning. In: Proceedings of the 12th USENIX Conference on Operating Systems Design and Implementation. OSDI 2016, pp. 265–283. USENIX Association, USA (2016)
2. Ben-Nun, T., Hoefler, T.: Demystifying parallel and distributed deep learning: an in-depth concurrency analysis. ACM Comput. Surv. **52**(4) (2019). https://doi.org/10.1145/3320060
3. Box, G.E.P., Jenkins, G.M.: Time Series Analysis Forecasting and Control. Holden Day, San Francisco (1976)
4. Brown, R.G.: Smoothing, Forecasting and Prediction of Discrete Time Series. Prentice Hall, Englewood Cliffs (1963)
5. Choi, J., Dongarra, J., Ostrouchov, S., Petitet, A., Walker, D., Whaley, R.C.: A proposal for a set of parallel basic linear algebra subprograms. In: Dongarra, J., Madsen, K., Waśniewski, J. (eds.) PARA 1995. LNCS, vol. 1041, pp. 107–114. Springer, Heidelberg (1996). https://doi.org/10.1007/3-540-60902-4_13
6. Dai, Y., He, D., Fang, Y., Yang, L.: Accelerating 2D orthogonal matching pursuit algorithm on GPU. J. Supercomput. **69**(3), 1363–1381 (2014). https://doi.org/10.1007/s11227-014-1188-8
7. Dongarra, J., et al.: The Sourcebook of Parallel Computing. Morgan Kaufmann, San Francisco (2002)
8. Drucker, H., Burges, C.J.C., Kaufman, L., Smola, A., Vapnik, V.: Support vector regression machines. In: Mozer, M.C., Jordan, M., Petsche, T. (eds.) Advances in Neural Information Processing Systems, vol. 9. MIT Press, Cambridge (1997)
9. Fang, Y., Chen, L., Wu, J., Huang, B.: GPU implementation of orthogonal matching pursuit for compressive sensing. In: 2011 IEEE 17th International Conference on Parallel and Distributed Systems, pp. 1044–1047 (2011). https://doi.org/10.1109/ICPADS.2011.158
10. Filelis-Papadopoulos, C.K.: Incomplete inverse matrices. Numer. Lin. Algebra Appl. **28**(5), e2380 (2021). https://doi.org/10.1002/nla.2380
11. Filelis-Papadopoulos, C.K., Kirschner, S., O'Reilly, P.: Forecasting with limited data: predicting aircraft co2 emissions following covid-19. Submitted (2021)
12. Filelis-Papadopoulos, C.K., Kyziropoulos, P.E., Morrison, J.P., O'Reilly, P.: Modelling and forecasting based on recurrent pseudoinverse matrices. In: Paszynski, M., Kranzlmüller, D., Krzhizhanovskaya, V.V., Dongarra, J.J., Sloot, P.M.A. (eds.) ICCS 2021. LNCS, vol. 12745, pp. 229–242. Springer, Cham (2021). https://doi.org/10.1007/978-3-030-77970-2_18

412   C. K. Filelis - Papadopoulos et al.

13. Filelis-Papadopoulos, C.K., Kyziropoulos, P.E., Morrison, J.P., O'Reilly, P.: Modelling and forecasting based on recursive incomplete pseudoinverse matrices. Mathematics and Computers in Simulation, accepted (2022)
14. Friedman, J.H.: Multivariate adaptive regression splines. Ann. Stat. **19**(1), 1–67 (1991). https://doi.org/10.1214/aos/1176347963
15. Hochreiter, S., Schmidhuber, J.: Long Short-Term Memory. Neural Comput. **9**(8), 1735–1780 (1997). https://doi.org/10.1162/neco.1997.9.8.1735
16. Holt, C.C.: Forecasting seasonals and trends by exponentially weighted moving averages. Int. J. Forecast. **20**, 5–13 (2004)
17. Korenberg, M.J., Paarmann, L.D.: Orthogonal approaches to time-series analysis and system identification. IEEE Signal Process. Mag. **8**(3), 29–43 (1991)
18. Li, B., et al.: Large scale recurrent neural network on GPU. In: 2014 International Joint Conference on Neural Networks (IJCNN), pp. 4062–4069 (2014). https://doi.org/10.1109/IJCNN.2014.6889433
19. Li, Q., Salman, R., Test, E., Strack, R., Kecman, V.: GPUSVM: a comprehensive CUDA based support vector machine package. Open Comput. Sci. **1**(4), 387–405 (2011). https://doi.org/10.2478/s13537-011-0028-7
20. Lu, Y., Zhu, Y., Han, M., He, J.S., Zhang, Y.: A survey of GPU accelerated SVM. In: Proceedings of the 2014 ACM Southeast Regional Conference. ACM SE 2014, Association for Computing Machinery, New York, NY, USA (2014)
21. Makridakis, S., Spiliotis, E., Assimakopoulos, V.: The m4 competition: 100,000 time series and 61 forecasting methods. Int. J. Forecast. **36**(1), 54–74 (2020). https://doi.org/10.1016/j.ijforecast.2019.04.014. m4 Competition
22. Mallat, S., Zhang, Z.: Matching pursuits with time-frequency dictionaries. IEEE Trans. Signal Process. **41**(12), 3397–3415 (1993). https://doi.org/10.1109/78.258082
23. Oh, K.S., Jung, K.: GPU implementation of neural networks. Pattern Recogn. **37**(6), 1311–1314 (2004). https://doi.org/10.1016/j.patcog.2004.01.013
24. Paoletti, M.E., Haut, J.M., Tao, X., Miguel, J.P., Plaza, A.: A new GPU implementation of support vector machines for fast hyperspectral image classification. Remote Sens. **12**(8) (2020). https://doi.org/10.3390/rs12081257
25. Tan, K., Zhang, J., Du, Q., Wang, X.: GPU parallel implementation of support vector machines for hyperspectral image classification. IEEE J. Sel. Top. Appl. Earth Obser. Remote Sens. **8**(10), 4647–4656 (2015)

Intersection Representation of Big Data Networks and Triangle Enumeration

Wali Mohammad Abdullah, David Awosoga$^{(\boxtimes)}$, and Shahadat Hossain

University of Lethbridge, Lethbridge, AB, Canada
{w.abdullah,shahadat.hossain}@uleth.ca, odo.awosoga@gmail.com

Abstract. Triangles are an essential part of network analysis, representing metrics such as transitivity and clustering coefficient. Using the correspondence between sparse adjacency matrices and graphs, linear algebraic methods have been developed for triangle counting and enumeration, where the main computational kernel is sparse matrix-matrix multiplication. In this paper, we use an intersection representation of graph data implemented as a sparse matrix, and engineer an algorithm to compute the "k-count" distribution of the triangles of the graph. The main computational task of computing sparse matrix-vector products is carefully crafted by employing compressed vectors as accumulators. Our method avoids redundant work by counting and enumerating each triangle exactly once. We present results from extensive computational experiments on large-scale real-world and synthetic graph instances that demonstrate good scalability of our method. In terms of run-time performance, our algorithm has been found to be orders of magnitude faster than the reference implementations of the **miniTri** data analytics application [18].

Keywords: Intersection matrix · Local triangle count · Forward degree cumulative · Forward neighbours · Sparse graph · k-count

1 Introduction

The presence of triangles in network data has led to the creation of many metrics to aid in the analysis of graph characteristics. As such, the ability to count and enumerate these triangles is crucial to applying these metrics and gaining further insights into the underlying composition and distribution of these graphs. Generalizations aside, the applications of triangle counting are as ubiquitous as the triangles themselves, including transitivity ratio - the ratio between the number of triangles and the paths of length two in a graph - and clustering coefficient - the fraction of neighbours for a vertex i of a graph who are neighbours themselves. Other real-life applications of triangle counting include spam detection [4], network motifs in biological pathways [12], and community discovery [13]. However, before any network analysis can be undertaken, the underlying data structure of a graph must be critically examined and understood. An efficient

© The Author(s), under exclusive license to Springer Nature Switzerland AG 2022
D. Groen et al. (Eds.): ICCS 2022, LNCS 13352, pp. 413–424, 2022.
https://doi.org/10.1007/978-3-031-08757-8_34

representation of network data will dictate analysis capabilities and improve algorithm performance and data visualization potential [5]. Note that large real-life networks are typically sparse in nature, so efficient computations of these graphs must be able to account for their sparsity and skewed degree distribution [3]. A consistent structure makes linear algebra-based triangle counting methods appealing, and most methods use direct or modified matrix-matrix multiplication, with a notable exception being the implementation of Low et al. [11]. This paper expands upon the preliminary ideas of a poster presentation from the 2021 IEEE Big Data Conference (Big Data) [2], and here we propose an "intersection" representation of network data obtained as a list of edges [17] and based on sparse matrix data structures [8]. Our triangle enumeration algorithm derives its simplicity and efficiency by employing matrix-vector product calculations as its main computational kernel. The local triangle count and edge support information are then acquired from the enumerated triangles obtained as the result of this matrix-vector multiplication.

1.1 k-count Distribution

Application proxies provide a simple yet realistic way to assess the performance of real-life applications' architecture and design. Below, we outline the main components of the miniTri data analytics proxy [18], which we use to demonstrate the effectiveness of our intersection-based graph representation and computation.

Let $G = (V, E)$ be a connected and undirected graph without multiple edges and self loops, where V denotes the set of vertices and E denotes the set of edges. For $v \in V$ a path of length 2 through v is a sequence of vertices $u-v-w$ such that $e_1 = \{u, v\} \in E$ and $e_2 = \{v, w\} \in E$. Such a length-2 path is termed a *wedge* at vertex v. Let $d(v)$ denote the number of edges *incident* on v, also defined as the number of vertices x such that $\{v, x\} \in E$. The number of wedges in G is then given by $\sum_{v \in V} \binom{d(v)}{2}$. A wedge $u - v - w$ is a *closed wedge* or a *triangle* if $e_3 = \{v, w\} \in E$. Let $\delta(v)$ and $\delta(e)$ denote the number of triangles incident on vertex v and edge $e = \{u, v\}$ respectively. In the literature $\delta(v)$ is known as the *local triangle count* or *triangle degree* of vertex v and $\delta(e)$ is known as the *support* or *triangle degree* of edge $e = \{u, v\}$. We denote by $\Delta(G)$ the number of triangles contained in graph G. Since a triangle is counted at each of its three vertices, we have $\Delta(G) = \frac{1}{3} \sum_{v \in V} \delta(v)$. Let $H = (V', E')$ be a subgraph of G where $|V'| = k$ and each pair of vertices are connected by an edge (H is a k-clique). Then H contains $\binom{k}{3}$ triangles and $\delta(v) = \binom{(k-1)}{2}$ and $\delta(e) = (k - 2)$ for $v \in V'$ and $e \in E'$. Let t be a triangle in G and let $\delta(t_x) = \min_x \delta(x)$, where x is a vertex of t and $\delta(t_e) = \min_e \delta(e)$ where e is an edge of t. The *k-count* of triangle t is defined to be the largest k such that

1. $\delta(t_x) \geq \binom{(k-1)}{2}$ and
2. $\delta(t_e) \geq (k - 2)$

The main computational task of miniTri is to compute the k-count distribution of the triangles of an input graph. Figure 1 displays an example input graph with 7 vertices and 13 edges. Each vertex i is circled and contains a label that represents its identity. Beside each vertex i is an integer denoting its local triangle count $\delta(i)$, and there is an integer beside each edge $e = \{i, j\}$ denoting its support $\delta(e)$. The graph contains 7 triangles. The table of Fig. 1 enumerates the triangles in the graph and displays the local triangle count and support of the vertices and edges together with the k-count of the triangles. Each row of the table lists the vertex labels of a triangle followed by the local triangle count, support, and k-count. There are 4 triangles with k-count value 4 and 3 triangles with k-count value 3. Let ω be the size of the largest clique in G. Then the graph contains at least $\binom{\omega}{3}$ triangles with k-count value of at least ω. Therefore, the k-count distribution can be used to obtain a bound on the size of the largest clique of a graph.

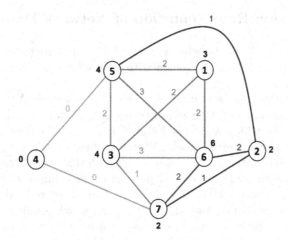

u	v	w	$\delta(u)$	$\delta(v)$	$\delta(w)$	$\delta(e_1)$	$\delta(e_2)$	$\delta(e_3)$	K-Count
1	3	5	<u>3</u>	4	4	<u>2</u>	2	2	4
1	3	6	<u>3</u>	4	6	<u>2</u>	2	3	4
1	5	6	<u>3</u>	4	6	<u>2</u>	2	3	4
2	5	6	<u>2</u>	4	6	<u>1</u>	2	3	3
2	5	7	<u>2</u>	6	2	2	<u>1</u>	2	3
3	5	6	4	4	6	<u>2</u>	3	3	4
3	6	7	4	6	<u>2</u>	3	<u>1</u>	2	3

Fig. 1. k-count table for the example input graph

The remainder of the paper is organized as follows. In Sect. 2, we introduce the notion of the intersection representation of network data and our data structure, followed by an illustrative example describing the main ideas in our intersection matrix-based triangle enumeration method. Section 3 outlines the computing environment employed to perform numerical experiments and presents triangle enumeration results on three sets of representative network data. miniTri1 [18] and its successor, which we call miniTri2, are the reference implementations by which we present comparative running times and demonstrate that our method scales very well on massive network data, and can be very flexible in its extensions to the analysis of network characteristics such as truss decomposition [7] and triangle ranking [6]. The paper is summarized in Section 4 with a discussion on future research directions.

2   Intersection Representation of Network Data

Let the vertices in V be labelled $1, 2, \ldots, |V| = n$. Using the labels on the vertices, a unique label can be assigned to each edge $e_k = \{v_i, v_j\}, i < j, k = 1, 2, \ldots, |E| = m$.

The *intersection representation* of graph G is a matrix $X \in \{0,1\}^{m \times n}$ in which for each column j of X there is a vertex $v_j \in V$ and $\{v_i, v_j\} \in E$ whenever there is a row k for which $X(k, i) = 1$ and $X(k, j) = 1$. The rows of X represent the edge list sorted by vertex labels. Therefore, matrix X can be viewed as an assignment to each vertex a subset of m labels such that there is an edge between vertices i and j if and only if the inner product of the columns i and j is 1. Since the input graph is unweighted, the edges are simply ordered pairs, and can be sorted in $O(m)$ time. Unlike the adjacency matrix which is unique (up to a fixed labelling of the vertices) for graph G, there can be more than one *intersection matrix* representation associated with graph G [1]. We exploit this flexibility to store a graph in a structured and space-efficient form.

2.1   Adjacency Matrix-Based Triangle Counting

Many existing triangle counting methods use the sparse representation of adjacency matrices in their calculations. The adjacency matrix $A(G) \equiv A \in \{0,1\}^{|V| \times |V|}$ associated with graph G is defined as,

$$A(i, j) = \begin{cases} 1 & \text{if } \{v_i, v_j\} \in E, i \neq j \\ 0 & \text{otherwise} \end{cases}$$

It is well known in the literature that the number of closed walks of length $k \geq 0$ are obtained in the diagonal entries of kth power A^k. Therefore, the total number of triangles in a graph $G, \Delta(G)$, is given by the trace of A^3,

$$\Delta(G) = \frac{1}{6} Tr(A^3).$$

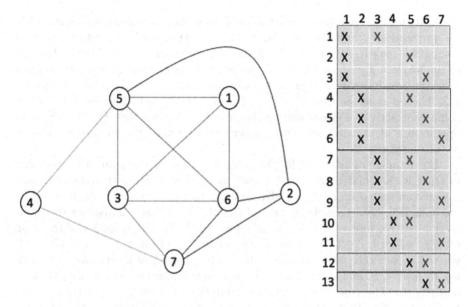

	1	2	3	4	5	6	7
1	X		X				
2	X				X		
3	X					X	
4		X			X		
5		X				X	
6		X					X
7			X	X			
8			X		X		
9			X				X
10				X	X		
11				X			X
12					X	X	
13						X	X

Fig. 2. Intersection matrix representation of the example input graph

The factor of $\frac{1}{6}$ accounts for the multiple counting of a triangle (the number of ways closed walks of length 3 can be obtained is $3! = 6$). There is a large body of literature on sparse linear algebraic triangle counting methods based on adjacency matrix representation of the data [5]. miniTri's triangle counting implementation takes the adjacency matrix A of the input graph and creates an incidence matrix B from it [18]. The enumeration and counting of the triangles occur in the overloaded matrix multiplication $C = AB$, where entries in the resultant matrix C with a value of 2 correspond to a completed triangle. This method triple-counts each triangle, once for each vertex, so the final result is divided by 3 to give the total number of triangles in the graph. Since the multiplication of two sparse matrices usually results in a dense matrix, this is a memory intensive process.

2.2 Intersection Matrix-Based Triangle Counting

Graph algorithms can be effectively expressed in terms of linear algebra operations [9], and we combine this knowledge with our proposed data representation to count the triangles in a structured three-step method. For vertex i we first find its neighbours $j > i$ such that $\{i, j\} \in E$ by multiplying the submatrix of X consisting of rows corresponding to edges incident on i (let us call them $(i - j)$−rows) by the transpose of the vector of ones of compatible length. A value of 1 in the vector-matrix product indicates that the corresponding vertex j is a neighbour of vertex i.

Next, we multiply the submatrix of X consisting of columns j identified in the previous step and the rows below the $(i - j)$–rows by a vector of ones of compatible length. A value of 2 in the matrix-vector product indicates a triangle of the form (i, j, j') where j and j' are neighbours of vertex i with $j < j'$. Let k be the row index in matrix X for which the matrix-vector product contains a 2. Then it must be that $X(k, j) = 1$ and $X(k, j') = 1$. Since each row of X contains exactly 2 nonzero entries that are 1, it follows that $\{j, j'\} \in E$. This operation is identical to performing a set intersection on the forward neighbours of vertices j and j'.

The number of triangles in the graph is given by the sum of the number of triangles associated with each vertex as described. Since the edges are represented in sorted order in our algorithm, unlike many other triangle counting methods [18], each triangle is counted exactly once. Figure 2 displays the intersection matrix representation of the input graph X. The triangles of the form $(1, j, j')$ where $j, j' \in \{3, 5, 6\}$ are obtained from the product $X(7 : 13, [3\ 5\ 6]) * \mathbf{1}$, where $\mathbf{1}$ denotes the vector of ones. The product has a 2 at locations corresponding to rows 7, 8, and 12 of X and the associated triangles are $(1, 3, 5), (1, 3, 6)$, and $(1, 5, 6)$. Therefore, there are three triangles incident on vertex 1, and it can be easily verified that the graph contains a total of 7 triangles across all of the vertices.

2.3   Data Structure

In our preliminary implementation, we use two arrays to store useful information that can be computed after we sort the edges. FDC (Forward Degree Cumulative) is an array of size n, with elements corresponding to the total number of "forward neighbours" across the vertices of a graph. Forward neighbours are defined as the neighbours of a vertex that have a higher label than the vertex of interest. With the vertices of the graph labelled, finding the forward degree of a vertex j can be calculated as fd(j) = FDC[j+1] - FDC[j]. FN is an array of size m that stores *which* vertices are the forward neighbours of a vertex j. Using FN we can find these forward neighbours of j as fn(j) = FN[k], where k ranges from FDC[j] to FDC[j+1]-1. The arrays FDC and FN thus save the vector-matrix products needed to find the forward neighbours. Figure 3 displays the arrays FDC and FN for the graph of Fig. 2.

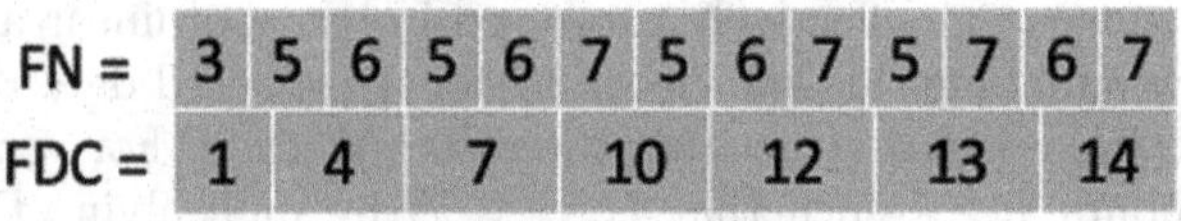

FN = | 3 | 5 | 6 | 5 | 6 | 7 | 5 | 6 | 7 | 5 | 7 | 6 | 7 |

FDC = | 1 | 4 | 7 | 10 | 12 | 13 | 14 |

Fig. 3. FN and FDC for the example graph.

2.4   Local Triangle Count and Edge Support

As discussed in Sect. 1, there are many other metrics related to triangle computation that can be found using our intersection matrix data structure. The bases for these metrics are the triangle degrees, which are the number of triangles incident on an edge (edge support) or vertex (local triangle count) of a graph. This is illustrated in Fig. 4 as **edgeDeg** and **vertDeg**, respectively, derived from Fig. 1.

edgeDeg = | 2 | 2 | 2 | 1 | 2 | 1 | 2 | 3 | 1 | 0 | 0 | 3 | 2 |

vertDeg = | 3 | 2 | 4 | 0 | 4 | 6 | 2 |

Fig. 4. vertDeg and edgeDeg for the example graph.

Let j be the column (vertex) of matrix X (graph G) currently being processed in the **fullCount** algorithm. For each pair of its forward neighbours j' and j'' there is an edge between them if and only if both of the corresponding columns contain a 1 in some row k identifying the triangle (j, j', j''). In terms of the matrix-vector multiplication in line 7 of algorithm **fullCount**, vector T will get updated as $T(k) \leftarrow 2$. Thus the triangle (j, j', j'') can be enumerated and stored instantly. The vertex triangle degrees of each triangle are dynamically updated with this same information, and stored in an array. The edge triangle degrees are stored in a separate array and updated by exploiting the structure of the FN and FDC arrays in tandem. The entries of the FDC array, while primarily used to store the forward degree of a vertex, also contain the edge number (edge id) that the forward neighbourhood of the vertex of interest begins and ends at. Since the sub-arrays in FN that correspond to the forward neighbourhood of the vertices are in the same order as the listed edges of the intersection matrix, any edge between two vertices can be identified by first finding the distance between the higher labelled vertex and the beginning of the forward neighbourhood in which it is found (using FN), and then adding this distance to the entry in FDC that corresponds to the edge of the lower numbered vertex. Finally, the k-count distribution of the triangles is used to give a bound on the maximum clique of a graph [18], and with the triangles enumerated and the edge and vertex triangle degrees computed and stored as shown in Fig. 4, the k-count calculations can be quickly computed using the method described Sect. 1. The algorithm in its entirety is given in the next section.

2.5   Algorithm

fullCount (X)
Input: Intersection matrix X
1: Calculate FDC   ▷ Forward degree cumulative
2: Calculate FN   ▷ Forward neighbour
3: $count \leftarrow 0$ ▷ Number of triangles
4: **for** $j = 1$ to $n - 1$ **do**   ▷ $j \in V$, where V is the set of vertices
5:     $fd \leftarrow FDC[j + 1] - FDC[j]$ ▷ fd is the forward degree of j
6:     **if** $fd > 1$ **then**   ▷ j has more than one forward neighbour
7:         $T = X([FDC(j + 1) : m], fn_j) * \mathbf{1}$
8:         $S = \{t \mid T[t] = 2\}$
9:         **if** $S \neq \emptyset$ **then**
10:             $count \leftarrow count + |S|$
11:             **for** $t \in S$ **do**
12:                 update edgeDeg   ▷ Array of triangle edge degrees
13:                 update vertDeg   ▷ Array of triangle vertex degrees
14:                 $Triangles \leftarrow Triangles \cup t$ ▷ Array that stores enumerated
    triangles
15: $kCountTable \leftarrow$ computeKCounts($count$, $vertDeg$, $edgeDeg$, $Triangles$)
16: **return** $count$, $vertDeg$, $edgeDeg$, $kCountTable$, and $Triangles$

3   Numerical Results

This section contains experimental results from selected test instances. The first set comprises real-world social networks from the Stanford Network Analysis Project (SNAP), obtained from the Graph Challenge website [15]. SNAP is a collection of more than 50 large network datasets containing a large number of nodes and edges, including social networks, web graphs, road networks, internet networks, citation networks, collaboration networks, and communication networks [10]. The first set of experiments were performed using a Dell Precision T1700 MT PC with a 4th Gen Intel Core I7-4770 Processor (Quad Core HT, with 3.4GHz Turbo and 8GB RAM), running Centos Linux v7.9. The implementation language was C++ and the code was compiled using $-O3$ optimization flag with a g++ version 4.4.7 compiler. Performance times are reported in seconds and were averaged over three runs where possible, using the following implementation abbreviations: $mt1$ for miniTri1, $mt2$ for miniTri2, and int for our intersection algorithm.

Figure 5 shows the speedups of our algorithm versus the two reference miniTri implementations on these real-world instances. The speedups are a unitless measurement defined as the ratio of the miniTri counting time divided by that of our algorithm. For the triangle counting algorithms, our speedups ranged from 22× to an impressive 1383× over miniTri1, and from 16× to 642× over miniTri2, with two instances ("flickrEdges" and "Cit-Patents") failing to compute with miniTri2. Instances with an "*" had speedups greater than 650× against miniTri1 and were cut off from the figure for ease of viewing.

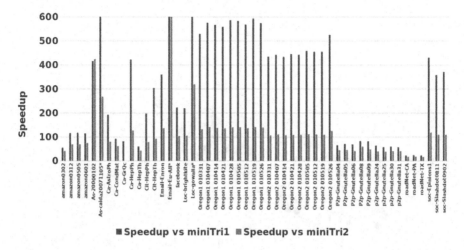

Fig. 5. Comparing our intersection algorithm with both miniTri implementations on large real world networks

Table 1. Comparing our intersection algorithm with miniTri on large synthetic networks.

Graph characteristics				Time in seconds	
Name	$|V|$	$|E|$	$\Delta(G)$	mt1	int
graph500-scale18-ef16	262144	4194304	82287285	17440	9.357
graph500-scale19-ef16	524288	8388608	186288972	49211.8	25.21
graph500-scale20-ef16	1048576	16777216	419349784	197456	72.34
graph500-scale21-ef16	2097152	33554432	935100883	N/A	171.2
graph500-scale22-ef16	4194304	67108864	2067392370	N/A	481.43
graph500-scale23-ef16	8388608	134217728	4549133002	N/A	1340.05
graph500-scale24-ef16	16777216	268435456	9936161560	N/A	3317.15
graph500-scale25-ef16	33554432	536870912	21575375802	N/A	7959.39

Table 1 compares our algorithm performance on large synthetic test instances from GraphChallenge to miniTri1 (miniTri2 was only able to compute the first instance and thus omitted). "N/A" denotes instances where miniTri1 timed out after four days of computation. Due to the large sizes of this second set of instances, they were run on the large High Performance Computing system (Graham cluster) at Compute Canada. On the first 3 instances, our method is over 1800 times faster than miniTri1, and the relative performance improves with increasing instance size, further demonstrating the scalability of our triangle counting algorithm.

Figure 6 demonstrates our algorithm's performance on relatively dense brain networks from the Network Repository [14], back in the Linux environment. These graphs have between 15 and 268 million edges and up to 42 trillion triangles, and neither miniTri implementation was able to provide results for any

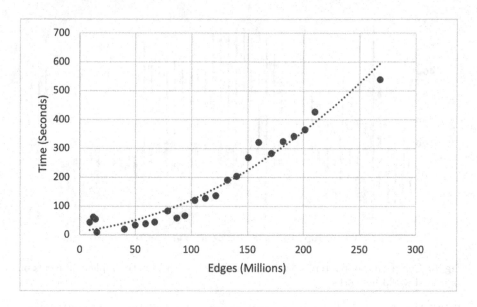

Fig. 6. Testing our intersection algorithm on networks with billions of triangles.

of the instances. The line of best fit is a polynomial of degree 2 and shows that our algorithm scales very well with graphs with massive amounts of triangles.

Our intersection-based implementation also produces competitive results when compared to the state-of-the art triangle counting algorithms [16]. Algorithms were analyzed and compared by fitting a model of graph counting times, T_{tri}, as a function of the number of edges $N_e = |E|$. This data was then used to estimate the parameters N_1 (the number of edges that can be processed in one second) and β:

$$T_{tri} = (N_e/N_1)^{\beta}$$

to compare different counting implementations. Implementations with a larger N_1 and smaller β perform the best, and the top entries from the 2019 review had N_1 values ranging from 5×10^5 to 5×10^8, and β values ranging from $\frac{1}{2}$ to $\frac{4}{3}$. For reference, our algorithm had $\beta = \frac{3}{4}$ and $N_1 = 1 \times 10^7$.

After examining the comparative performance of our triangle counting algorithm, we proceeded to expand the implementation to include the metrics described in Sect. 2.4 - triangle counting, triangle vertex degree, triangle edge degree, and k-count calculations. Similar to the basic counting experimental results, our intersection method of this "full count" was faster than miniTri1 and miniTri2 on every instance, with speedups ranging between 2× and 177× on the ten largest instances, displayed in Table 2. One noteworthy observation about these results is that due to the data structure that stored the enumerated triangles, the k-count calculation of our algorithm ran much faster than those of miniTri, even though the code implementation was nearly identical. This demonstrates the versatility of FDC and FN in their ability to perform a wide range of network analytics.

Table 2. Comparing our full count intersection algorithm with miniTri1 and miniTri2 on large real world networks.

Graph characteristics				Time in seconds			Speedup	
Name	$|V|$	$|E|$	$\Delta(G)$	mt1	mt2	int	mt1/int	mt2/int
Loc-gowalla	196591	950327	2273138	156	106.4	0.882	177	121
roadNet-PA	1090920	1541898	67150	2.067	1.792	0.076	27	24
roadNet-TX	1393383	1921660	82869	2.544	2.207	0.070	36	32
flickrEdges	105938	2316948	107987357	1112	N/A	553.1	2	∞
amazon0312	400727	2349869	3686467	26.8	22.29	0.932	29	24
amazon0505	410236	3356824	3951063	28.71	23.39	0.997	29	23
amazon0601	403394	3387388	3986507	28.24	25.09	0.998	28	25
roadNet-CA	1965206	5533214	120676	3.706	3.212	0.134	28	24
Cit-Patents	3774768	33037894	7515023	157.21	N/A	3.502	45	∞

4   Conclusion

Network data is usually input as a list of edges which can be preprocessed into a representation such as an adjacency matrix or adjacency list, suitable for algorithmic processing. We have presented a simple, yet flexible scheme based on intersecting edge labels, the intersection matrix, for the representation of and calculation with network data. A new linear algebra-based method exploits this intersection representation for triangle computation – a kernel operation in big data analytics. By using sparse matrix-vector products instead of the memory-intensive matrix-matrix multiplication, our implementation has the capacity to enumerate and extend triangle analysis in graphs so that important information such as triangle vertex and edge degree can be gleaned in a fraction of the time of reference implementation of miniTri on large benchmark instances. The computational results from a set of large-scale synthetic and real-world network instances clearly demonstrate that our basic implementation is efficient and scales well. The two arrays FDC and FN together constitute a compact representation of the sparsity pattern of network data, requiring only $n + m$ units of storage. This is incredibly useful in the exchange of network data, with the potential to allow for the computation of many additional intersection matrix-based network analytics such as rank and triangle centrality [6]. A shared memory parallel implementation of this method using OpenMP is being developed, with very optimistic preliminary results. This algorithm can still be tuned, and cache efficiency is being studied for additional optimizations, exploring temporal and spatial locality to analyze the memory footprint and provide further improvements. A natural extension of the research presented in this paper is to use the intersection representation in *graphlet* counting methods. Similar to the k-count distribution, graphlet frequency distribution (a vector of the frequency of different graphlets in a graph) provides local topological properties of graphs [17].

Acknowledgments. This research was supported in part by NSERC Discovery Grant (Individual), NSERC Undergraduate Student Research Award, and the AITF Graduate Student Scholarship. A part of our computations were performed on Compute Canada HPC system (http://www.computecanada.ca), and we gratefully acknowledge their support.

References

1. Abdullah, W.M., Hossain, S., Khan, M.A.: Covering large complex networks by cliques—a sparse matrix approach. In: Kilgour, D.M., Kunze, H., Makarov, R., Melnik, R., Wang, X. (eds.) AMMCS 2019. SPMS, vol. 343, pp. 117–127. Springer, Cham (2021). https://doi.org/10.1007/978-3-030-63591-6_11
2. Abdullah, W.M., Awosoga, D., Hossain, S.: Intersection representation of big data networks and triangle counting. In: 2021 IEEE International Conference on Big Data (Big Data), pp. 5836–5838 (2021). https://doi.org/10.1109/BigData52589.2021.9671349
3. Al Hasan, M., Dave, V.S.: Triangle counting in large networks: a review. Wiley Interdiscipl. Rev. Data Min. Knowl. Disc. **8**(2), e1226 (2018)
4. Becchetti, L., Boldi, P., Castillo, C.: Efficient algorithms for large-scale local triangle counting. ACM Trans. Knowl. Discovery Data **4**, 1–28 (2010)
5. Burkhardt, P.: Graphing trillions of triangles. Inf. Vis. **16**(3), 157–166 (2017)
6. Burkhardt, P.: Triangle centrality. arXiv:abs/2105.00110 (2021)
7. Cohen, J.: Trusses: Cohesive subgraphs for social network analysis. Natl. Secur. Agency Tech. Rep. **16**(3.1) (2008)
8. Hasan, M., Hossain, S., Khan, A.I., Mithila, N.H., Suny, A.H.: DSJM: a software toolkit for direct determination of sparse Jacobian matrices. In: Greuel, G.-M., Koch, T., Paule, P., Sommese, A. (eds.) ICMS 2016. LNCS, vol. 9725, pp. 275–283. Springer, Cham (2016). https://doi.org/10.1007/978-3-319-42432-3_34
9. Kepner, J., Gilbert, J.: Graph algorithms in the language of linear algebra. SIAM (2011)
10. Leskovec, J., Krevl, A.: SNAP Datasets: Stanford large network dataset collection, June 2014. http://snap.stanford.edu/data. Accessed 02 Oct 2019
11. Low, T.M., Rao, V.N., Lee, M., Popovici, D., Franchetti, F., McMillan, S.: First look: linear algebra-based triangle counting without matrix multiplication. In: 2017 IEEE High Performance Extreme Computing Conference (HPEC), pp. 1–6 (2017). https://doi.org/10.1109/HPEC.2017.8091046
12. Milo, R., Shen-Orr, S., Itzkovitz, S.: Network motifs: simple building blocks of complex network. Science **298**, 824–827 (2002)
13. Palla, G., Dereny, I., Frakas, I., Vicsek, T.: Uncovering the overlapping community structure of complex networks in nature and society. Nature **435**, 814–818 (2005)
14. Rossi, R.A., Ahmed, N.K.: The network data repository with interactive graph analytics and visualization. In: AAAI (2015). https://networkrepository.com
15. Samsi, S., et al.: Static graph challenge: Subgraph isomorphism (2017). http://graphchallenge.mit.edu/data-sets. Accessed 09 July 2021
16. Samsi, S., et al.: Graphchallenge.org triangle counting performance (2020)
17. Szpilrajn-Marczewski, E.: A translation of sur deux propriétés des classes d'ensembles by. Fund. Math **33**, 303–307 (1945)
18. Wolf, M.M., Berry, J.W., Stark, D.T.: A task-based linear algebra building blocks approach for scalable graph analytics. In: 2015 IEEE High Performance Extreme Computing Conference (HPEC), pp. 1–6. IEEE (2015)

Global Surrogate Modeling by Neural Network-Based Model Uncertainty

Leifur Leifsson[1]([✉])(iD), Jethro Nagawkar[2](iD), Laurel Barnet[3], Kenneth Bryden[3], Slawomir Koziel[4,5](iD), and Anna Pietrenko-Dabrowska[5](iD)

[1] School of Aeronautics and Astronautics, Purdue University,
West Lafayette, IN 47907, USA
`leifur@purdue.edu`
[2] Department of Aerospace Engineering, Iowa State University,
Ames, IA 50011, USA
`jethro@iastate.edu`
[3] Department of Mechanical Engineering, Iowa State University,
Ames, IA 50011, USA
`{labarnet,kmbryden}@iastate.edu`
[4] Engineering Optimization and Modeling Center, Department of Engineering,
Reykjavík University, Menntavegur 1, 102, Reykjavík, Iceland
`koziel@ru.is`
[5] Faculty of Electronics Telecommunications and Informatics,
Gdansk University of Technology, Narutowicza 11/12, 80-233 Gdansk, Poland
`anna.dabrowska@pg.edu.pl`

Abstract. This work proposes a novel adaptive global surrogate modeling algorithm which uses two neural networks, one for prediction and the other for the model uncertainty. Specifically, the algorithm proceeds in cycles and adaptively enhances the neural network-based surrogate model by selecting the next sampling points guided by an auxiliary neural network approximation of the spatial error. The proposed algorithm is tested numerically on the one-dimensional Forrester function and the two-dimensional Branin function. The results demonstrate that global surrogate modeling using neural network-based function prediction can be guided efficiently and adaptively using a neural network approximation of the model uncertainty.

Keywords: Global surrogate modeling · Neural networks · Model uncertainty · Error based exploration

1 Introduction

There is often a need in engineering to assess the performance of a process (e.g., through physical or computer experiments) with a limited number of evaluations. In such cases, surrogate models are often used to approximate the output response of the process over a given data [3,17,19]. The surrogates are fast to

D. Groen et al. (Eds.): ICCS 2022, LNCS 13352, pp. 425–434, 2022.
https://doi.org/10.1007/978-3-031-08757-8_35

evaluate and can be used to either explore the output response or exploit them to determine a set of parameter values that yield optimal performance.

Modern surrogate modeling strategies start by constructing a surrogate of an initial data set and then progress in cycles using prediction and uncertainty estimates (if available) to select the next sampling point [26]. Several such approaches have been proposed, including the efficient global optimization (EGO) algorithm [9] and Bayesian optimization (BO) [8,16,20,24]. The EGO algorithm [9] follows this strategy by modeling the output response as a random variable and selects the next point to be sampled by maximizing the expected improvement over the best current solution. BO follows the same idea as EGO, but the approach is formalized rigorously through Bayesian theory [16,20,24].

Gaussian process regression (GPR) (or Kriging) [3,7,11] is widely used with EGO and BO because of its unique feature of providing a prediction of the mean of the underlying data and a prediction of its uncertainty. In particular, GPR provides the mean squared error of the predictor using the same data for constructing the predictor. EGO and BO utilize the predictor and its error estimate to compute a criterion to guide the algorithm to adaptively enhance the predictor. Both EGO and BO typically use ther expected improvement as the criterion [3,27]. The major disadvantages of GPR modeling, however, are that the computational cost scales cubically with the number of observations, and does not scale well to higher dimensions [13]. This issue can be partially relieved by using graphical processing units (GPUs) and parallel computing [14].

Neural network (NN) regression modeling [6], on the other hand, scales much more efficiently for the optimization of complex and large data sets [13,22]. It should be noted that the training cost of NNs depends on various factors, such as sample size, number of epochs, and architecture complexity. A major limitation of NN regression modeling is that uncertainty estimates are, in general, not readily available for a single prediction [13]. Rather, it is necessary to make use of an ensemble of NNs with a range of predictions. Bayesian neural networks (BNNs) are an example of such class of algorithms [5,12,25]. Current BNN approaches, however, are approximation methods because exact NN-based Bayesian inference is computationally intractable. Using dropout as a Bayesian approximation to represent model uncertainty in deep NNs (DNNs) is an example of one such approach [4]. Current BNN algorithms are, however, computationally intensive.

There is recent interest in creating surrogate modeling algorithms that combine the predictive capabilities of NNs and the uncertainty estimates of GPR. Renganathan et al. [18] use DNNs in place of a polynomial to model the global trend function in GPR modeling. This approach improves the prediction capabilities while still retaining the model uncertainty of GPR. Nevertheless, that approach is still limited in the same way as the original GPR modeling approach. Zhang et al. [28] propose an algorithm that creates and adaptively enhances a multifidelity DNN by exploiting information from low-fidelity data sets. This approach is limited to exploitation only and cannot perform exploration or search a criterion that balances exploration and exploitation.

In this paper, a novel adaptive global surrogate modeling algorithm is proposed that follows the EGO strategy but uses NNs in place of GPR. Specifically, the proposed algorithm iteratively constructs two NN models, one for the prediction of a given process output and the other for the model uncertainty. The proposed algorithm uses separate data sets to construct each NN model. In each cycle, the model uncertainty is used to select the next sampling point and then update the NN prediction model. The algorithm terminates once the uncertainty measure has reached a specified tolerance or the maximum number of samples is reached. In this work, the spatial error in the prediction is used as the uncertainty measure, and it is maximized in each cycle to select the next sampling point. The proposed algorithm is tested on two low-dimensional analytical problems. The results demonstrate that global modeling using NN-based function prediction can be guided efficiently and adaptively by an NN approximation of the model uncertainty.

The next section introduces the proposed algorithm. The following section presents results of numerical experiments using one- and two-dimensional analytical functions. Finally, concluding remarks are presented.

2   Methods

The proposed approach is summarized in Algorithm 1. The algorithm requires two initial data sets that are used to fit separate neural networks. One neural network models the process output in terms of the input parameters, and the other models the spatial error in the first neural network. Let $(\mathbf{X}, \mathbf{Y})_f$ be the set of sample points used to fit the neural network to the process output, and let $(\mathbf{X}, \mathbf{Y})_u$ be the set of sample points used to fit the neural network to the spatial uncertainty. Here, $\mathbf{X}_f = \{\mathbf{x}^{(1)}, ..., \mathbf{x}^{(p)}\}^T$ is the set of the input parameter sample points and $\mathbf{Y}_f = (y^{(1)}(\mathbf{x}^{(1)}), ..., y^{(p)}(\mathbf{x}^{(p)}))^T$ the corresponding set of model outputs. Furthermore, $\mathbf{X}_u = \{\mathbf{x}^{(1)}, ..., \mathbf{x}^{(q)}\}^T$ is the set of input parameter sample points and $\mathbf{Y}_u = (y^{(1)}(\mathbf{x}^{(1)}), ..., y^{(q)}(\mathbf{x}^{(q)}))^T$ the corresponding set of model outputs. In this work, it is assumed that the data sets $(\mathbf{X}, \mathbf{Y})_f$ and $(\mathbf{X}, \mathbf{Y})_u$ are distinctly different. Both sets are created using Latin hypercube sampling (LHS) [15].

To fit the neural networks within the proposed algorithm, the mean squared error (MSE) loss function is minimized:

$$\mathcal{L} = \frac{\sum_{l=1}^{N}(\hat{y}^{(l)} - y^{(l)})^2}{N}, \tag{1}$$

where N is the number of samples in the training data. The loss function minimizes the mismatch between the training data, y, and the predicted values, $\hat{y}$, of the neural network [6,21]. To minimize the loss function, the adaptive moments (ADAM) optimization algorithm is used [10] along with the backpropagation algorithm [2] to compute the gradients. The neural network setup used in this work, in particular the number of hidden layers and the number of neurons per hidden layer, is case dependent and is described in the numerical experiments.

Algorithm 1. Adaptive global surrogate modeling algorithm with neural network-based prediction and uncertainty

Require: initial data sets $(\mathbf{X},\mathbf{Y})_f$ and $(\mathbf{X},\mathbf{Y})_u$
  repeat
    fit neural network to function with available data $(\mathbf{X},\mathbf{Y})_f$
    compute uncertainty with available data $(\mathbf{X},\mathbf{Y})_u$
    fit neural network to uncertainty with available data $(\mathbf{X},\mathbf{Y})_u$
    $\mathbf{P} \leftarrow \arg\max_{\mathbf{x}} \hat{s}^2(\mathbf{x})$
    $\mathbf{X}_f \leftarrow \mathbf{X}_f \cup \mathbf{P}$
    $\mathbf{Y}_f \leftarrow \mathbf{Y}_f \cup y(\mathbf{P})$
  until convergence

Other hyperparameters are common between the cases. Specifically, the tangent hyperbolic is used as the activation function, the learning rate is set to 0.001, and the number of epochs is fixed with a value of 3,000. The neural network algorithm is implemented using TensorFlow [1].

The neural network algorithm is used in each cycle to construct a surrogate model, $\hat{y}_f$, of the process output, y, using $(\mathbf{X},\mathbf{Y})_f$. In this work, the uncertainty measure of $\hat{y}_f$ is estimated by the square of the spatial error and is written as

$$s^2(\mathbf{x}) = (\hat{y}_f(\mathbf{x}) - y_f(\mathbf{x}))^2. \tag{2}$$

In the proposed algorithm, $s(\mathbf{x})^2$ is computed in each cycle using the data set $(\mathbf{X},\mathbf{Y})_u$, and the neural network algorithm is used to construct the surrogate model $\hat{s}(\mathbf{x})^2$.

To select the next sampling point in each cycle of the proposed algorithm, the uncertainty measure $\hat{s}(\mathbf{x})^2$ is maximized using differential evolution [23]. The algorithm is terminated if $\hat{s}(\mathbf{x})^2$ is lower than a specified tolerance or the number of cycles exceeds a specified maximum value.

3 Numerical Experiments

The results of numerical experiments with the proposed algorithm are presented in this section. Two analytical cases are considered, the first case has one input parameter and the second has two.

3.1 One-Dimensional Forrester Function

The one-dimensional analytical function developed by Forrester et al. is written as

$$y(x) = (6x - 2)^2 \sin(12x - 4), \tag{3}$$

where $x \in [0,1]$. The proposed algorithm is applied to the modeling of this function using three uniformly distributed initial samples and ten infill points. The total number of samples for modeling the function is, therefore, 13. Ten

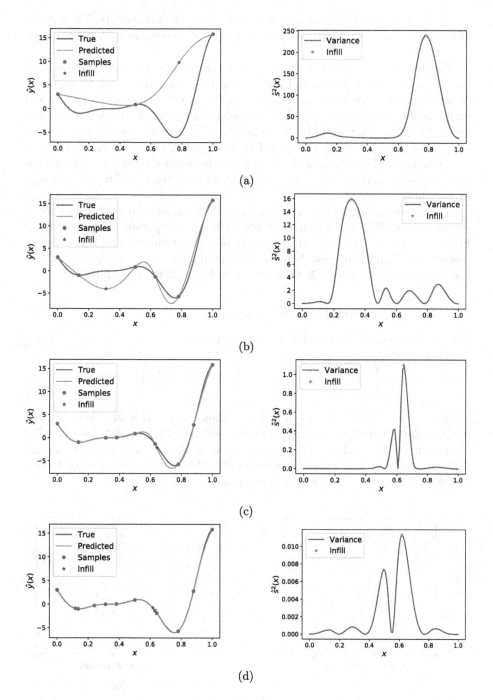

Fig. 1. Forrester function prediction (left) and uncertainty (right) at iterations: (a) 0, (b) 3, (c) 6, (d) 9.

uniformly distributed samples are used for modeling the uncertainty. The number of hidden layers is set to three and the number of neurons in each hidden layer is set to 50.

Figure 1 shows the modeling progression at iterations 0 (3 initial samples and the first infill point), 3 (3 initial and 3 prior infills and the new infill point), 6 (3 initial and 6 prior infills and the new infill), and 9 (with all the initial and infill samples), respectively. Specifically, each subfigure shows the neural network prediction $\hat{y}(x)$ along with the true function $y(x)$, sample points and the next sampling point, as well as the neural network model of the uncertainty $\hat{s}^2(x)$. The location of the maximum uncertainty in the interval from 0 to 1 guides the sampling so that the prediction quickly aligns with the true function through exploration.

The global accuracy of the surrogate models is measured using the root mean squared error (RMSE), which is evaluated using a separate testing data set. Figure 2(a) shows how the RMSE changes over the iterations and is reduced to around 0.1. Figure 2(b) shows how the maximum model uncertainty reduces over the iterations from around 250 to 0.01, or by four orders of magnitude.

3.2   Two-Dimensional Branin Function

The two-dimensional Branin function is written as

$$y(x_1, x_2) = \left(x_2 - \frac{5.1}{4\pi^2}x_1^2 + \frac{5}{\pi}x_1 - 6\right)^2 + 10\left(1 - \frac{1}{8\pi}\right)\cos(x_1) + 10, \quad (4)$$

where $x_1, x_2 \in [0, 10]$. The proposed algorithm models this function with ten initial samples selected using LHS and fifty additional infill points for a total of sixty points at the end of fifty iterations. One hundred points, selected through LHS, are used for the uncertainty model. For this case, the number of hidden layers was set to two, with fifty neurons in each hidden layer.

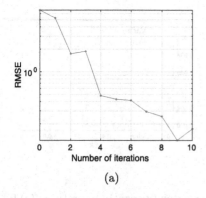
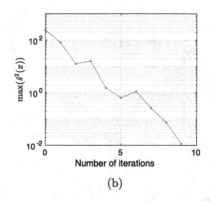

(a)   (b)

Fig. 2. Forrester function modeling error evolution: (a) root mean squared error of the prediction model, (b) maximum variance of the uncertainty model.

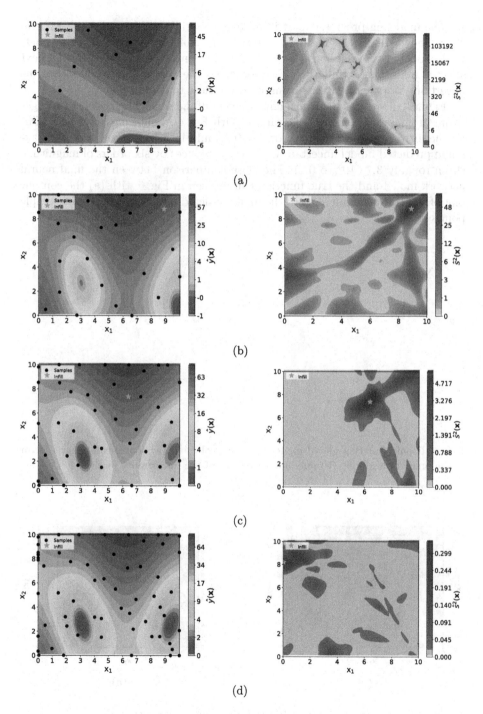

Fig. 3. Branin function prediction (left) and uncertainty (right) at iterations: (a) 0, (b) 14, (c) 31, (d) 49.

The modeling progression, at iterations 0 (initial samples only), 14, 31, and 49, is shown in Fig. 3. The left plot in each subfigure shows the neural network model of the function $\hat{y}(x_1, x_2)$ with the sample points used and the next selected infill point, while the right plot shows the neural network model of the uncertainty $\hat{s}^2(x_1, x_2)$ that is being used to select that infill point.

Figure 4 illustrates the global improvement of the surrogate model as the algorithm progresses through the iterations with Fig. 4(a) showing how the RMSE for the surrogate model reduces down to 0.2, and Fig. 4(b) showing how the maximum predicted model uncertainty $\hat{s}^2(x_1, x_2)$ reduces by six orders of magnitude (from roughly $3.7 \cdot 10^5$ to 0.1). The close comparison between the final neural network model and the true function can be seen in Fig. 5 with (a) the contour plot of the true function and (b) the final predicted model with all of the sample points indicated.

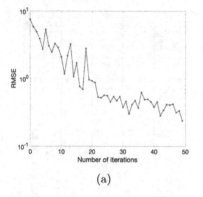
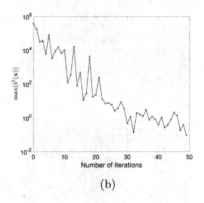

(a)   (b)

Fig. 4. Branin function global modeling error evolution: (a) root mean squared error of the prediction model, (b) maximum variance of the uncertainty model.

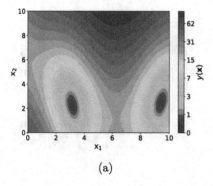
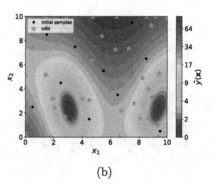

(a)   (b)

Fig. 5. Branin function: (a) true, (b) predicted.

4   Conclusion

Global modeling of large data sets is important for decision-making in experimentally and computationally-driven discoveries in engineering and science. The proposed approach of combining efficient global optimization strategies and neural network modeling directly tackles this important problem. Specifically, this paper demonstrates that global modeling using neural network-based function prediction can be guided by an auxiliary neural network approximation of the prediction spatial error that enables efficient adaptive surrogate modeling of large data sets. This capability will support scientists and engineers to make decisions on whether and where in the parameter space to do a physical experiment or computational simulation.

Future work will focus on improving the proposed algorithm to permit adaptive sampling of the uncertainty model as well as using data from multiple levels of fidelity. Furthermore, the process of updating the neural network fit in each cycle of the algorithm needs to be accelerated. Other uncertainty metrics than the prediction variance also need to be explored. An important step will be to compare the proposed approach against current state-of-the-art methods and to characterize the computational costs of each approach. Performing numerical experiments on high-dimensional problems involving physical and computational data is of current interest.

Acknowledgements. This material is based upon work supported in part by the Department of Energy under a Laboratory Directed Research and Development grant at Ames Laboratory.

References

1. Abadi, M., et al.: TensorFlow: large-scale machine learning on heterogeneous systems (2015). http://tensorflow.org/. software available from tensorflow.org
2. Chauvin, Y., Rumelhart, D.E.: Backpropagation: Theory, Architectures, and Applications. Psychology press, Hillsdale (1995)
3. Forrester, A.I.J., Keane, A.J.: Recent advances in surrogate-based optimization. Prog. Aerosp. Sci. **45**(1–3), 50–79 (2009)
4. Gal, Y., Ghahramani, Z.: Dropout as a Bayesian approximation: Representing model uncertainty in deep learning. In: Proceedings of the 33rd International Conference on Machine Learning, pp. 1050–1059 (2016)
5. Goan, E., Fookes, C.: Bayesian neural networks: an introduction and survey. In: Mengersen, K.L., Pudlo, P., Robert, C.P. (eds.) Case Studies in Applied Bayesian Data Science. LNM, vol. 2259, pp. 45–87. Springer, Cham (2020). https://doi.org/10.1007/978-3-030-42553-1_3
6. Goodfellow, I., Bengio, Y., Courville, A.: Deep Learning. The MIT Press, Cambridge (2016)
7. Huang, D., Allen, T., Notz, W., Zeng, N.: Global optimization of stochastic black-box systems via sequential kriging meta-models. J. Global Optim. **34**(3), 441–466 (2006)
8. Jones, D.R.: A non-myopic utility function for statistical global optimization. J. Global Optim. **21**, 345–383 (2001)

9. Jones, D.R., Schonlau, M., Welch, W.J.: Efficient global optimization of expensive black-box functions. J. Global Optim. **13**(4), 455–492 (1998)
10. Kingma, D.P., Ba, J.: Adam: a method for stochastic optimization. arXiv:1412.6980 (2014)
11. Krige, D.G.: Statistical approach to some basic mine valuation problems on the witwatersrand. J. Chem. Metallurgical Min. Eng. Soc. South Africa **52**(6), 119–139 (1951)
12. Lampinen, J., Vehtari, A.: Bayesian approach for neural networks - review and case studies. Neural Netw. **14**(3), 257–274 (2001)
13. Lim, Y.F., Ng, C.K., Vaitesswar, U.S., Hippalgaonkar, K.: Extrapolative Bayesian optimization with Gaussain process and neural network ensemble surrogate models. Adv. Intell. Syst. **3**, 2100101 (2021)
14. Liu, H., Ong, Y.S., Shen, X., Cai, J.: When Gaussian process meets big data: a review of scalable GPs. IEEE Trans. Neural Networks Learn. Syst. **31**(11), 4405–4423 (2020)
15. McKay, M.D., Beckman, R.J., Conover, W.J.: A comparison of three methods for selecting values of input variables in the analysis of output from a computer code. Technometrics **21**(2), 239–245 (1979)
16. Mockus, J.: Application of Bayesian approach to numerical methods of global and stochastic optimization. J. Global Optim. **4**(4), 347–365 (1994)
17. Queipo, N.V., Haftka, R.T., Shyy, W., Goel, T., Vaidyanathan, R., Tucker, P.K.: Surrogate-based analysis and optimization. Prog. Aerosp. Sci. **21**(1), 1–28 (2005)
18. Renganathan, S.A., Maulik, R., Ahuja, J.: Multi-fidelity deep neural network surrogate model for aerodynamic shape optimization. Aerosp. Sci. Technol. **111**, 106522 (2021)
19. Sacks, J., Welch, W., Michell, J.T., Wynn, P.H.: Design and analysis of computer experiments. Stat. Sci. **4**, 409–423 (1989). https://doi.org/10.1214/ss/1177012413
20. Sasena, M.J.: Flexibility and Efficiency Enhancement for Constrained Global Design Optimization with Kriging Approximations. Ph.D. thesis, University of Michigan, USA (2002)
21. Schmidhuber, J.: Deep learning in neural networks: an overview. Neural Netw. **61**, 85–117 (2015). https://doi.org/10.1016/j.neunet.2014.09.003
22. Snoek, J., et al.: Scalable Bayesian optimization using deep neural networks. In: Proceedings of the 32nd International Conference on Machine Learning, pp. 2171–2180 (2015)
23. Storn, R., Price, K.: Differential evolution - a simple and efficient heuristic for global optimization over continuous spaces. J. Global Optim. **11**, 341–359 (1997)
24. Streltsov, S., Vakili, P.: A non-myopic utility function for statistical global optimization. J. Global Optim. **14**(3), 283–298 (1999)
25. Titterington, D.M.: Bayeisan methods for neural networks and related models. Stat. Sci. **19**(1), 128–139 (2004)
26. Viana, F.A., Haftka, R.T., Watson, L.T.: Efficient global optimization algorithm assisted by multiple surrogate techniques. J. Global Optim. **56**, 669–689 (2013)
27. Zhan, D., Xing, H.: Expected improvement for expensive optimization: a review. J. Global Optim. **78**(3), 507–544 (2020). https://doi.org/10.1007/s10898-020-00923-x
28. Zhang, X., Xie, F., Ji, T., Zhu, Z., Zheng, Y.: Multi-fidelity deep neural network surrogate model for aerodynamic shape optimization. Comput. Methods Appl. Mech. Eng. **373**(1), 113485 (2021)

Analysis of Agricultural and Engineering Systems Using Simulation Decomposition

Yen-Chen Liu[1] ⓘ, Leifur Leifsson[2]([✉]) ⓘ, Anna Pietrenko-Dabrowska[3] ⓘ,
and Slawomir Koziel[3,4] ⓘ

[1] Department of Aerospace Engineering, Iowa State University, Ames, IA 50011, USA
clarkliu@iastate.edu
[2] School of Aeronautics and Astronautics, Purdue University,
West Lafayette, IN 47907, USA
leifur@purdue.edu
[3] Faculty of Electronics Telecommunications and Informatics,
Gdansk University of Technology, Narutowicza 11/12, 80-233 Gdansk, Poland
anna.dabrowska@pg.edu.pl
[4] Engineering Optimization and Modeling Center, Department of Engineering,
Reykjavík University, Menntavegur 1, 102 Reykjavík, Iceland
koziel@ru.is

Abstract. This paper focuses on the analysis of agricultural and engineering processes using simulation decomposition (SD). SD is a technique that utilizes Monte Carlo simulations and distribution decomposition to visually evaluate the source and the outcome of different portions of data. Here, SD is applied to three distinct processes: a model problem, a nondestructive evaluation testing system, and an agricultural food-water-energy system. The results demonstrate successful implementations of SD for the different systems, and the illustrate the potential of SD to support new understanding of cause and effect relationships in complex systems.

Keywords: Simulation decomposition · Food-water-energy systems ·
Nondestructive evaluation · Physics-based simulations · Parameter variability

1 Introduction

Simulation decomposition (SD) [1] is not only a great visualizing technique for engineering processes, but also an efficient application to utilize the resulting data set created by the Monte Carlo [2] sampling process. An illustration, such as a stacked bar chart, can express the decomposed variability of simulation inputs and outputs to understand the results. Furthermore, SD has the potential to understand cause and effect relationships between the input and output parameters of a given system [3].

In this work, a recent SD method [3] is applied to problems in the area of agriculture and engineering system analysis. The first problem is intended

D. Groen et al. (Eds.): ICCS 2022, LNCS 13352, pp. 435–444, 2022.
https://doi.org/10.1007/978-3-031-08757-8_36

to illustrate the SD method using is a model problem with three inputs and one output. The second problem has relevance to nondestructive evaluation and involves an ultrasonic testing (UT) system with a pillbox void. The last problem involves an agricultural model that computes the nitrogen export of the state of Iowa. The commonality of the second and the third problems is that they both involve processes that utilize physics-based computational simulations and have input parameters with variability. The SD analysis provides a graphical representation of the effect of input variability on the simulation outputs.

The next section gives the formulation of the problem and a description of the SD analysis technique. The following section presents the results of the three numerical cases. The last section summarizes this work and discusses the potential future work.

2   Methods

This section describes the general problem formulation and gives the details of the SD method.

2.1   Problem Statement

System analysis can be represented as a black box model as

$$y = f(\mathbf{x}), \tag{1}$$

where the left-side of the equation represents the model output y and the right-side of the equation represents the model f with input parameters $\mathbf{x}$. It is important to understand the effects of uncertainties of the input parameters on the output response when making design decisions. In this work, the effects of the input parameters on the output parameters are visualized using simulation decomposition (SD).

2.2   Simulation Decomposition

SD [1,3] is an approach to visualize the effects of variability on models. Furthermore, SD analysis is used to distinguish the influences of different cases of inputs affecting the model output. Figure 1 shows a flowchart of the SD workflow. The process starts by generating random samples as the input data set. First, specify the statistical distributions for each input parameter and choose the states of interest for each input parameter individually. Then, divide the distributions into sub-distributions according to the chosen states. Next, generate every possible combination of the input data from the sub-distribution using Monte Carlo sampling and categorize them into different cases. These cases would determine the model output that is to-be-decomposed. Then, conduct the simulations with the input samples and obtain the output data set. Meanwhile, register the output of each simulation according to their cases. Lastly, construct the outcome probability for each case and decompose the full outcome probability into different cases

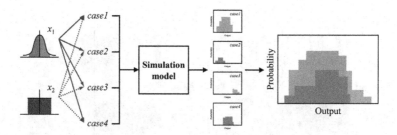

Fig. 1. A schematic of the simulation decomposition workflow.

accordingly. The importance of each case can be observed by the occurrence in the decomposed output visually, and cross-checking important cases provide a sense of importance for the input parameters.

3 Numerical Examples

This section presents the results of SD analysis of a model problem, an ultrasonic nondestructive testing system, and the Iowa food-energy-water system analyzed.

3.1 Model Problem

The simple analytical function is written as [4]

$$y(\boldsymbol{x}) = x_1 + x_2 x_3^2, \tag{2}$$

where x_1, x_2, and x_3 are the input parameters with the variabilities given in Table 1, and y is the function output parameter. A data set of a total 10^6 points is created using Latin hypercube sampling (LHS) [5].

Figure 2 shows how each input parameter is decomposed into two states that define eight cases for the parameter space. All the parameters are considered to be uniformly distributed. For x_1, 500 is set to divide the complete distribution into sub-distribution, and 50 and 5 are used to divide x_2 and x_3, respectively.

Figure 3 shows the decomposed distribution of the output y. In general, a low value of y is contributed by almost all combinations of the input parameters. However, a high value of y is mainly obtained by the cases with high values of

Table 1. Input parameters and their statistical distributions for the model problem.

Parameter	Distribution
x_1	$U(0,\ 1000)$
x_2	$U(0,\ 100)$
x_3	$U(0,\ 10)$

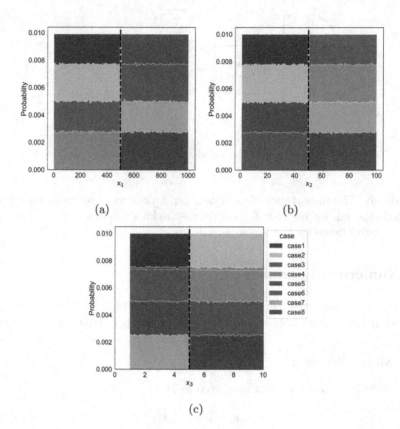

Fig. 2. Decomposed distribution of input parameters from SD for the model problem.

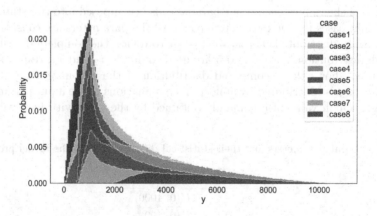

Fig. 3. Decomposed distribution of the model problem output.

x_2 and x_3. In particular, in the cases where x_1 and x_2 are both low, the simple function can still yield high y with high x_3, and vice versa. In this case, the SD analysis shows that the output of the analytical function is dominated by x_3, which is as expected when inspecting the function directly.

3.2   Ultrasonic Testing (UT) System with a Pillbox Void

Ultrasound testing (UT) is a widely used nondestructive testing (NDT) technique for flaw detection. In this problem, a pillbox-inclusion-defect under planar UT transducer is considered [6]. Figure 4 shows the setup of the problem. The planar transducer is placed in water and the probe angle (θ) and the probe coordinates (x and y) are varied. A fused quartz block with a pillbox-like void is inspected by the transducer where the distance between the transducer and the surface of the block (z_1) and the distance between the surface of the block and the defect (z_2) can vary based on the setup. The variability distributions for this problem are given in Table 2. The output response is the reflected pulse (v) received by the transducer. A data set of 10^5 data points is generated by LHS [5] for this problem.

Figure 5 shows the sub-distributions of sampled input parameters in two states. The variability of probe angle (θ) is considered normal distribution and the rest of input parameters are considered uniform distributions in this problem. For θ, the complete distribution is divided by the statistical mean of $0°$. The sub-distributions of the probe coordinates x and y are both divided by 0.5 mm. z_1 and z_2 are consider using full distribution in this work for simplicity. The total of eight cases are shown in Fig. 5.

Figure 6 shows the decomposed distribution of output response (v). A diagonal trend can be observed. In particular, the SD analysis shows that high values of the inputs yield low values of response. Furthemore, the results show that the response is dominated by θ, followed by x, and then y.

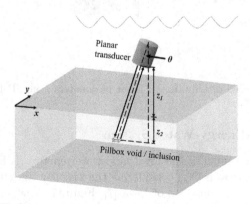

Fig. 4. A schematic showing the setup of the ultrasonic testing system.

Table 2. Ultrasonic testing system input parameters statistical distributions.

Parameter	Distribution
θ (deg)	$N(0, 0.5^2)$
x (mm)	$U(0, 1)$
y (mm)	$U(0, 1)$
z_1 (mm)	$U(24.9, 25.9)$
z_2 (mm)	$U(12.5, 13.5)$

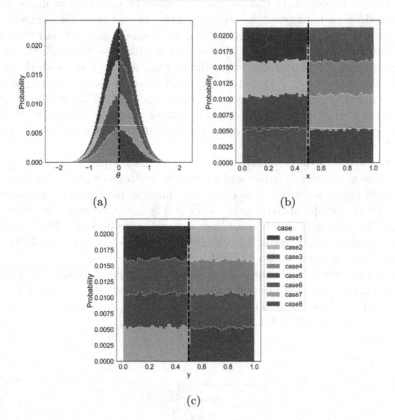

Fig. 5. Decomposed distribution of input parameters from SD for UT system.

3.3 Iowa Food-Energy-Water System

A simulation-based model of the Iowa food-water-energy (IFEW) system computes the surplus nitrogen (N_s) considering the weather, agriculture, and animal agriculture domains in the state of Iowa [7]. Figure 7 show an extended design structure matrix (XDSM) diagram of the simulation model. The input parameters, intermediate parameters and output parameters are listed in Table 3, and the details of the computation for different domains are described in [7]. This

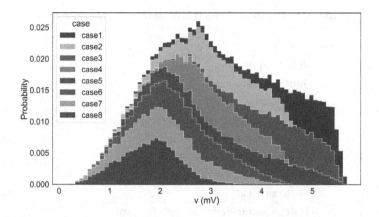

Fig. 6. Decomposed distribution of ultrasonic testing system output parameter.

work focuses on the SD analysis of the variability input parameters given in Table 4 and assumes other input parameters are fixed. A data set of 10^5 points is created for the SD analysis using LHS [5].

Figure 8 shows the distribution of sampled input parameters in two states and the four corresponding cases. The variability distributions of the July temperature (w_1) and precipitation (w_2) are considered to be normal and log-normal, respectively. A temperature of 76 °F and a precipitation 2.5 in. are used to divide the distributions into sub-distributions of w_1 and w_2, respectively. The states can be identified as regular temperature (below 76 °F) and high temperature (above 76 °F). Similarly, the states of precipitation are low precipitation (below 25 in.) and regular precipitation (above 25 in.).

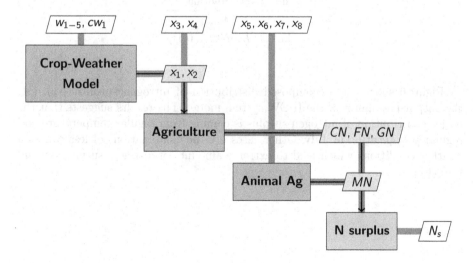

Fig. 7. Extended design structure matrix of Iowa food-water-energy system model.

Table 3. Parameters of the IFEW simulation system model.

Parameter	Description
w_1	July temperature
w_2	July precipitation
w_3	June precipitation
x_4	July-August average temperature
x_5	July-August average precipitation
cw_1	May planting progress
x_1	Corn yield
x_2	Soybean yield
x_3	Rate of commercial nitrogen for corn
x_4	Rate of commercial nitrogen for soybean
x_5	Hog/pigs population
x_6	Beef cattle population
x_7	Milk cows population
x_8	Other cattle population (heifers + slaughter cattle)
CN	Commercial nitrogen (nitrogen in commercial fertilizers)
FN	Biological fixation nitrogen of soybean crop
GN	Grain nitrogen (Nitrogen harvested in grain)
MN	Manure nitrogen (Nitrogen in animal manure)
N_s	Surplus nitrogen in soil

Table 4. Input parameters of the IFEWS simulation model with variability.

Parameter	Distribution
w_1 (°F)	$N(74, 2^2)$
w_2 (in.)	$LogN(0.4, 0.4^2)$

Figure 6 shows the decomposed distribution of nitrogen surplus (N_s), i.e., the output parameter of the IFEW system model. The results suggests that the major contribution of nitrogen surplus is coming from regular temperature and regular precipitation of July. Other cases are the combinations of less common weather conditions which lead to extreme amounts of nitrogen surplus but are rare (Fig. 9).

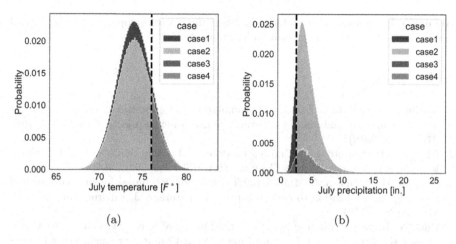

(a)   (b)

Fig. 8. Decomposed distribution of input parameters from SD for the IFEW system.

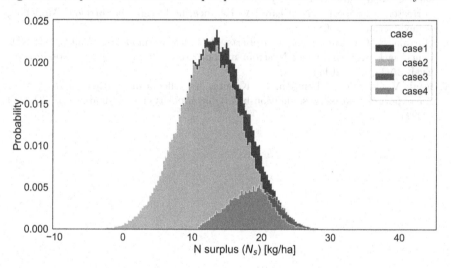

Fig. 9. Decomposed distribution of the nitrogen surplus output of the IFEW system simulation model.

4   Conclusion

This work demonstrates that simulation decomposition (SD) can support the analysis of agricultural and engineering systems involving input parameters of variability. In particular, SD can provide new insights into the effects of the model input ranges on its output. This insight can be useful in understanding cause and effect relations in complex systems. Future work will explore the potential of combining global sensitivity analysis (GSA) with the SD technique. Furthermore, the use of surrogate modeling in conjunction with those analyses will be investigated to create computationally efficient algorithms for analysis with SD and GSA.

Acknowledgements. This material is based upon work supported by the United States National Science Foundation under grant no. 1739551.

References

1. Mariia Kozlova, M.C., Luukka, P.: Simulation decomposition: new approach for better simulation analysis of multi-variable investment projects. Fuzzy Econ. Rev. **21**(1), 3–8 (2016)
2. Kroese, D.P., Brereton, T., Taimre, T., Botev, Z.I.: Why the Monte Carlo method is so important today. WIREs Comput. Stat. **6**(6), 386–392 (2014)
3. Kozlova, M., Yeomans, J.: Multi-variable simulation decomposition in environmental planning: an application to carbon capture and storage. J. Environ. Inform. Lett. **1**(1), 20–26 (2019)
4. Bilal, N.: Implementation of Sobol's method of global sensitivity analysis to a compressor simulation model. In: 22nd International Compressor Engineering Conference, p. 2385, July 2014
5. Forrester, A., Sóbester, A., Keane, A.: Engineering Design via Surrogate Modelling. Wiley, Germany (2008)
6. Gurrala, P., Chen, K., Song, J., Roberts, R.: Full wave modeling of ultrasonic NDE benchmark problems using Nystrom method. Rev. Progress Quant. Nondestruct. Eval. **36**(1), 1–8 (2017)
7. Raul, V., Liu, Y.C., Leifsson, L., Kaleita, A.: Effects of weather on Iowa nitrogen export estimated by simulation-based decomposition. Sustainability **14**(3), 1060 (2022)

Neural Network-Based Sequential Global Sensitivity Analysis Algorithm

Yen-Chen Liu[1]📛, Leifur Leifsson[2](✉)📛, Slawomir Koziel[3,4]📛,
and Anna Pietrenko-Dabrowska[4]📛

[1] Department of Aerospace Engineering, Iowa State University, Ames, IA 50011, USA
clarkliu@iastate.edu
[2] School of Aeronautics and Astronautics, Purdue University,
West Lafayette, IN 47907, USA
leifur@purdue.edu
[3] Engineering Optimization and Modeling Center, Department of Engineering,
Reykjavík University, Menntavegur 1, 102 Reykjavík, Iceland
koziel@ru.is
[4] Faculty of Electronics Telecommunications and Informatics,
Gdansk University of Technology, Narutowicza 11/12, 80-233 Gdansk, Poland
anna.dabrowska@pg.edu.pl

Abstract. Performing global sensitivity analysis (GSA) can be challenging due to the combined effect of the high computational cost, but it is also essential for engineering decision making. To reduce this cost, surrogate modeling such as neural networks (NNs) are used to replace the expensive simulation model in the GSA process, which introduces the additional challenge of finding the minimum number of training data samples required to train the NNs accurately. In this work, a recently proposed NN-based GSA algorithm to accurately quantify the sensitivities is improved. The algorithm iterates over the number of samples required to train the NNs and terminates using an outer-loop sensitivity convergence criteria. The iterative surrogate-based GSA yields converged values for the Sobol' indices and, at the same time, alleviates the specification of arbitrary accuracy metrics for the NN-based approximation model. In this paper, the algorithm is improved by enhanced NN modeling, which lead to an overall acceleration of the GSA process. The improved algorithm is tested numerically on problems involving an analytical function with three input parameters, and a simulation-based nondestructive evaluation problem with three input parameters.

Keywords: Global sensitivity analysis · Surrogate modeling · Neural networks · Sobol' indices · Termination criteria

1 Introduction

The study of sensitivity analysis (SA) [1,2] is important in many engineering and science applications. Individual effects and interactions of the input parameters

© The Author(s), under exclusive license to Springer Nature Switzerland AG 2022
D. Groen et al. (Eds.): ICCS 2022, LNCS 13352, pp. 445–454, 2022.
https://doi.org/10.1007/978-3-031-08757-8_37

on the output model response can be quantified by SA [3,4]. Engineers and scientists can use SA to understand the importance of parameters in experimental or computational investigations. There are two types of SA, local [5] and global [6] SA. Local SA usually refers to using the the local output model response to quantify the effect of small local perturbations in the inputs. In global SA, it utilizes the variance of output model response to quantify the effect due to the input variability in the entire parameter space. This work focuses on the use of global variance-based SA with Sobol' indices [3,4] for simulation-based problems.

In this paper, a recently developed algorithm for surrogate-based GSA [7] is improved and applied to new testing problems. In the NN-based sequential algorithm, the number of samples is iteratively increased with the goals of obtaining the converged Sobol' indices with the training cost as minimum as possible. This approach not only alleviates the needs to specify arbitrary surrogate modeling accuracy metrics, but also reliefs from the fact that accuracy metrics for surrogate models do not guarantee that converged Sobol' indices are obtained. In this work, the implementation of the NN training has been improved significantly, which leads to improved convergence of the GSA algorithm. The algorithm is tested numerically on two problems; an analytical function with three parameters and a simulation-based problem with five parameters.

The next section describes the problem statement and gives the details of the GSA algorithm. The following section presents results of numerical experiments. Finally, conclusions are presented and remarks on future steps are given.

2   Methods

This work proposes a sequential algorithm to quantify the global sensitivities of each input variability parameter to the simulation-based model outputs. Figure 1 shows the flowchart of the proposed algorithm. The algorithm starts from an initial sample plan, $\mathbf{x}$, which takes a small number of samples from the input parameters with their variability. Latin hypercube sampling (LHS) [8] is used in this work to randomly select sample data points from each probability distribution of the inputs. The corresponding outputs or observations, y, are then generated from the simulation model. A surrogate model, $\hat{y}(\mathbf{x})$, is constructed using these inputs and outputs as training data. The input-output behavior of the simulation model is imitated by the surrogate. Next, GSA is performed with this surrogate model using Sobol' sensitivity indices. The calculation of the Sobol' indices is a Monte Carlo process, therefore the convergence of these indices are checked within an inner-loop. In the inner-loop, the Sobol' indices are computed by sampling the current surrogate model, and the number of samples is increased during each iteration (e.g., one order of magnitude for this work) until it achieves the convergence of the inner-loop. Then, the converged inner-loop indices are checked by the outer-loop. The above process is resampled with an increasing number of sample plan from the simulation model until the outer-loop convergence criteria are met. The number of sample plan affects training an accurate surrogate model, and the outputs of the surrogate affect the precision of the GSA. The outcome of the proposed algorithm yields the corresponding surrogate model and the converged Sobol' indices of GSA.

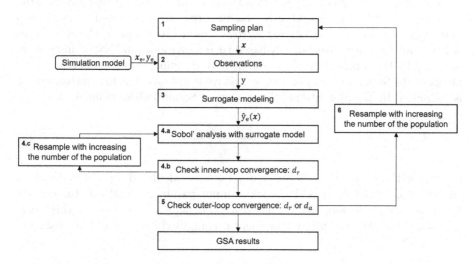

Fig. 1. A flowchart of the sequential global sensitivity analysis algorithm with neural network-based prediction.

Neural networks (NNs) are used in a variety of applications in the world. In this work, NN is used to be the surrogate and mimic the behavior of the simulation model. The general structure of NN is a hierarchy of features [9] with three parts: input layer, output layer, and hidden layers [10,11]. All the layers are composed by "neurons" which are the fundamental units of computation [9]. The number of neurons in the input and output layers are the same as the number of input and output variables of the simulation model. Hidden layers are the layers in-between the input and output layers. There could be zero or more hidden layers in a neural network. The number of hidden layers and the number of neurons in each hidden layer usually varies from case to case.

This work uses Sobol' indices [3,4] for the global sensitivity analysis. It is a variance-based method that quantifies the single effects of individual inputs and the interactions of combination of inputs on the simulation model output. The first-order Sobol' indices [4] that quantify the effect of individual inputs are given by

$$S_i = \frac{V_i}{\text{Var}(y)} = \frac{\text{Var}_{x_i}(E_{\mathbf{x}_{\sim i}}(y|x_i))}{\text{Var}(y)}, \tag{1}$$

where S_i is the contribution of individual x_i on the output variance of the simulation model. The total-order or total-effect Sobol' indices [4] that quantify the interactions of combined inputs are given by

$$S_{T,i} = 1 - \frac{\text{Var}_{\mathbf{x}_{\sim i}}(E_{x_i}(y|\mathbf{x}_{\sim i}))}{\text{Var}(y)}, \tag{2}$$

where $S_{T,i}$ measures the contribution of both individual x_i and the interactions between x_i and other input parameters on the output variance of the simulation model.

The proposed sequential algorithm includes an outer-loop that samples the simulation model to generate the NN-based surrogate models and an inner-loop that samples the trained NN-based surrogate model to computes the Sobol' sensitivity indices. The converged NN-based surrogate model and Sobol' indices are obtained by the termination of the outer- and inner-loop based on two measurements of the Sobol' indices between successive iterations. The first measurement is computed by the absolute relative change of Sobol' indices defined as

$$d_r[s_i] = \left| \frac{s_i^{(n)} - s_i^{(n-1)}}{s_i^{(1)}} \right|, \tag{3}$$

where s is the value of the Sobol' indices and is calculated separately for first- and total-order indices, i is the index of input parameter, and n is the current iteration index. The loop is terminated when $d_r[s_i] \leq \epsilon_r$ for all s_i. In this work, ϵ_r is set to 0.1. The second measurement is computed by the absolute change of Sobol' indices, given by

$$d_a[s_i] = \left| s_i^{(n)} - s_i^{(n-1)} \right|, \tag{4}$$

where s is the value of the Sobol' indices and is calculated separately for first- and total-order indices, i is the index of input parameter, and n is the current iteration index. The loop is terminated when $d_a[s_i] \leq \epsilon_a$ for all s_i. In this work, ϵ_a is set to 0.01. Both outer- and inner-loop can be terminated by either $d_r[s_i] \leq \epsilon_r$ or $d_a[s_i] \leq \epsilon_a$ being true.

3   Numerical Experiments

3.1   Case 1: Analytical Function

An analytical function with three input variables and one single output function is used to demonstrate the algorithm. The function is written as

$$f(\boldsymbol{x}) = x_1 + x_2 x_3^2, \tag{5}$$

where $x_1 \in U(0, 1000), x_2 \in U(0, 100)$, and $x_3 \in U(0, 10)$ are the input parameters and their associated PDFs, and y is the function output.

Figure 2 shows the convergence of the direct GSA of the true model. Direct GSA converged at 1000 sampling points. Figure 3 shows the inner-loop GSA convergence using the NN-based algorithm and Fig. 4 shows the outer-loop convergence. The algorithm terminates at 200 samples. The NN models are trained using two hidden layers, with twenty neurons, and the tangent hyperbolic activation function. The learning rate is set to 0.001 and the batch size is set to 16.

The maximum number of epochs and β_1 are set to 2,000 and 0.9, respectively. L2 regularization is used with $\lambda = 0.1$ while training the models.

Table 1 shows a comparison of the global sensitivities obtained from direct and NN-based numerical experiments. It can be seen that the NN-based yields global sensitivities within 1.7% of the true values while using only a fraction of the cost of the direct approach.

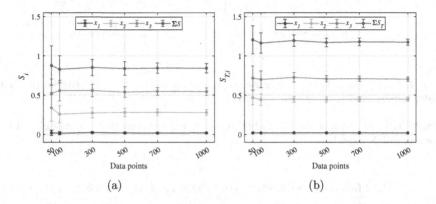

Fig. 2. Case 1 convergence of GSA directly on the true model: (a) first-order indices, and (b) total-order indices.

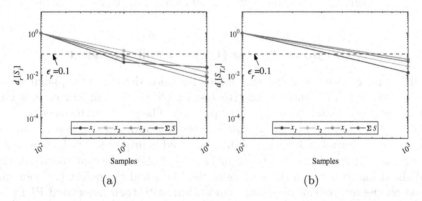

Fig. 3. Case 1 inner-loop convergence of s_i for the NN trained with 100 LHS samples: (a) first-order indices, and (b) total-order indices.

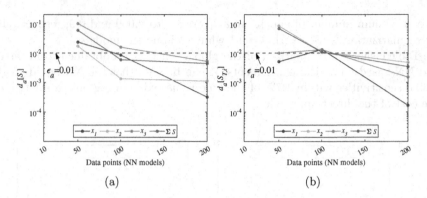

Fig. 4. Case 1 outer-loop convergence of s_i terminated on $d_a \leq \epsilon_a$ criteria: (a) first-order indices, and (b) total-order indices.

Table 1. Case 1 comparison of Sobol' index values between the true model and the converged NN model.

x	S_i			$S_{T,i}$		
	True function	Seq. GSA	% error	True function	Seq. GSA	% error
x_1	0.0199	0.0201	1%	0.0204	0.0201	1.5%
x_2	0.2793	0.2745	1.7%	0.4497	0.4453	1%
x_3	0.5456	0.5347	2%	0.7065	0.7053	0.2%

3.2   Case 2: Ultrasonic Testing (UT) of a Pillbox-Defect

In this numerical experiment the pillbox-inclusion-defect under planar transducer ultrasonic (UT) nondestructive testing (NDT) benchmark case is used [12,13]. Figure 5 shows the setup of the problem. The planar transducer is placed in water and the probe angle (θ) and the probe coordinates (x and y) are varied. A fused quartz block with a pillbox-like void is inspected by the transducer where the distance between the transducer and the surface of the block (z_1) and the distance between the surface of the block and the defect (z_2) can vary based on the setup. The variability parameters with their associated PDFs are $\theta \in N(0, 0.5^2)$ deg, $x \in U(0,1)$ mm, $y \in U(0,1)$ mm, $z_1 \in U(24.9, 25.9)$ mm, and $z_2 \in U(12.5, 13.5)$ mm. The output response is the reflected pulse (v) received by the transducer.

The simulation model for this case uses the Kirchhoff approximation (KA) to simulate the voltage wave receives by the transducer. The center frequency of the planar transducer is set to 10 MHz. The longitudinal wave speed is set to 6,200 m/s, and the shear wave speed is set to 3,180 m/s. The density of the fused quartz block is set to 4,420 kg/m^3. The NN models in this case are trained using two hidden layers, with thirty neurons. Similar to the previous case, the tangent hyperbolic activation function is used. The learning rate and β_1 of the ADAM optimizer are set to 0.001 and 0.9, respectively. The batch size is set to

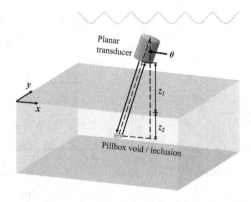

Fig. 5. A schematic of setup for the ultrasonic testing case.

16 and the maximum number of epochs is set to 2,000. L2 regularization is used with $\lambda = 0.1$ while training the models.

Figure 6 shows the direct GSA converged at 3000 sampling points. The convergence criteria of sequential GSA in this case were the same as in the previous case. The outer-loop sequentially iterated from 100 to 400 LHS samples. Figures 7(a) and 7(b) show the inner-loop GSA convergence using the NN-based algorithm for the first- and total-order indices require 10^6 and 10^3 samples, respectively. Figure 8 shows the outer-loop convergence plots for the first- and total-order indices both terminated at 400 samples. Figure 9 shows that the distance from the transducer to the defect has a negligible effect on the output response, while the probe angle has the highest effect follows by y coordinate then x coordinate. Table 2 compares the Sobol' indices values from the proposed method to those from the true function. It shows a good match of the the Sobol' indices values while the cost of sequential GSA is an order of magnitude less than the direct GSA.

Table 2. Case 2 comparison of Sobol' index values between the true model and the converged NN model.

x	S_i			$S_{T,i}$		
	True model	Seq. GSA	% error	true model	Seq. GSA	% error
θ	0.5599	0.56	0.02%	0.7424	0.7456	0.4%
x	0.0623	0.0571	8.3%	0.2086	0.2082	0.2%
y	0.1841	0.1919	4.2%	0.2449	0.2467	0.7%
z_1	5.2×10^{-5}	6.5×10^{-5}	–	5.3×10^{-5}	1.0×10^{-4}	–
z_2	6.6×10^{-4}	5.2×10^{-4}	–	8.4×10^{-4}	7.6×10^{-4}	–

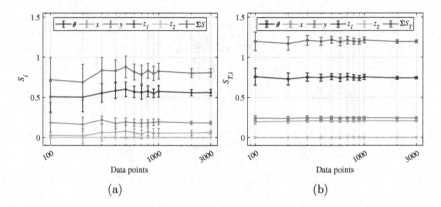

Fig. 6. Case 2 convergence of GSA directly on the physics model: (a) first-order indices, and (b) total-order indices.

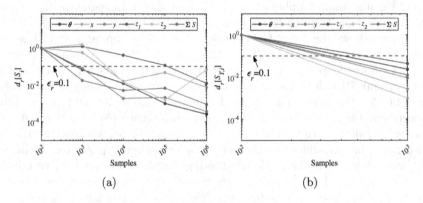

Fig. 7. Case 2 inner-loop convergence of s_i for the NN trained with 400 LHS samples: (a) first-order indices, and (b) total-order indices.

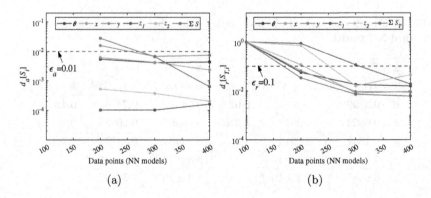

Fig. 8. Case 2 outer-loop convergence of s_i terminated on (a) $d_a \leq \epsilon_a$ criteria: first-order indices, and on (b) $d_r \leq \epsilon_r$ criteria: total-order indices.

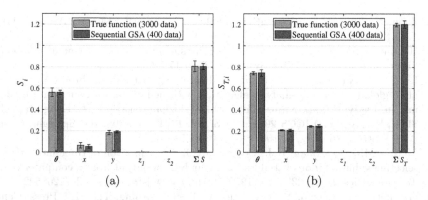

Fig. 9. Case 2 Sobol' index values of input parameters computed by the converged NN model: (a) first-order indices, and (b) total-order indices.

4 Conclusion

Global sensitivity analysis (GSA) of large-scale data sets is an important problem in engineering and science decision-making. The algorithm presented in this paper directly tackles this important task in the context of simulation-based problems. In particular, this work demonstrates that simulation-based GSA using neural network-based function prediction can be iteratively improved and terminated once accurate sensitivities are obtained, thereby enabling efficient adaptive GSA for large-scale problems.

Future steps in this work will focus on how to adaptively sample the NN model using model uncertainty, which will alleviate the resampling stage of the algorithm. An important step in this work will be to characterize the computational cost and benchmark it against current state-of-the-art methods, as well as to perform numerical experiments on high-dimensional problems involving physical and computational data.

Acknowledgements. This material is based upon work supported by the United States National Science Foundation under grant no. 1739551.

References

1. Ferretti, F., Saltelli, A., Tarantola, S.: Trends in sensitivity analysis practice in the last decades. Sci. Total Environ. **568**, 666–670 (2016). https://doi.org/10.1016/j.scitotenv.2016.02.133
2. Iooss, B., Saltelli, A.: Introduction to Sensitivity Analysis. Springer International Publishing, Switzerland (2015). https://doi.org/10.1007/978-3-319-12385-1_31
3. Sobol', I., Kucherekoand, S.: Sensitivity estimates for nonlinear mathematical models. Math. Model. Comput. Exp. **1**, 407–414 (1993)
4. Sobol', I.: Global sensitivity indices for nonlinear mathematical models and their monte carlo estimates. Math. Comput. Simul. **55**, 271–280 (2001)

5. Zhou, X., Lin, H.: Local sensitivity analysis. In: Encyclopedia of GIS, pp. 1116–1119 (2017)

6. Homma, T., Saltelli, A.: Importance measures in global sensitivity analysis of nonlinear models. Reliabil. Eng. Syst. Saf. **52**, 1–17 (1996)

7. Liu, Y.-C., Nagawkar, J., Leifsson, L., Koziel, S., Pietrenko-Dabrowska, A.: Iterative global sensitivity analysis algorithm with neural network surrogate modeling. In: Paszynski, M., Kranzlmüller, D., Krzhizhanovskaya, V.V., Dongarra, J.J., Sloot, P.M.A. (eds.) ICCS 2021. LNCS, vol. 12745, pp. 298–311. Springer, Cham (2021). https://doi.org/10.1007/978-3-030-77970-2_23

8. McKay, M.D., Beckman, R.J., Conover, W.J.: A comparison of three methods for selecting values of input variables in the analysis of output from a computer code. Technometrics **21**(2), 239–245 (1979). http://www.jstor.org/stable/1268522

9. Goodfellow, I., Bengio, Y., Courville, A.: Deep Learning. The MIT Press, Cambridge (2016)

10. Haykin, S.S.: Neural Networks and Learning Machines, 3rd edn. Pearson Education, Upper Saddle River (2009)

11. Schmidhuber, J.: Deep learning in neural networks: an overview. Neural Netw. **61**, 85–117 (2015). https://doi.org/10.1016/j.neunet.2014.09.003

12. Gurrala, P., Chen, K., Song, J., Roberts, R.: Full wave modeling of ultrasonic NDE benchmark problems using Nystrom method. Rev. Progr. Quant. Nondestruct. Eval. **36**(1), 1–8 (2017)

13. Du, X., Leifsson, L., Meeker, W., Gurrala, P., Song, J., Roberts, R.: Efficient model-assisted probability of detection and sensitivity analysis for ultrasonic testing simulations using stochastic metamodeling. J. Nondestruct. Eval. Diagnost. Prognost. Eng. Syst. **2**(4), 041002(4) (2019). https://doi.org/10.1115/1.4044446

Development of an Event-Driven System Architecture for Smart Manufacturing

Maksymilian Piechota[1], Mikołaj Nowak[1], and Dariusz Król[2]([✉])

[1] Faculty of Information and Communication Technology, Wrocław University
of Science and Technology,
Wrocław, Poland
[2] Department of Applied Informatics, Wrocław University of Science and Technology,
Wrocław, Poland
dariusz.krol@pwr.edu.pl

Abstract. This paper describes the automated production data acquisition and integration process in the architectural pattern Tweeting Factory. This concept allows the use of existing production equipment with PLCs and the use of industrial IoT prepared for Industry 4.0. The main purpose of the work is to propose an event-driven system architecture and to prove its correctness and efficiency. The proposed architecture is able to perform transformation operations on the collected data. The simulation tests were carried out using real data from the factory shop-floor, services prepared for production monitoring, allowing the calculation of KPIs. The correctness of the solution is confirmed on 20 production units by comparing its results with the blackboard architecture using SQL queries. Finally, the response time for calculating ISO 22400 performance indicators is examined and it was verified that the presented solution can be considered as a real-time system.

Keywords: Application case studies · Data integration · Engineering optimization and design · Event-driven system architecture · Tweeting factory

1 Introduction

The paper focuses on a concept called Tweeting Factory [3,8]. Currently, in the manufacturing industry for data acquisition, the PLC controllers are mostly used [5]. However together with the development of the Industry 4.0 concept, intelligent sensors that are capable to communicate with various other systems [10] are getting more recognition. Additionally they can do some initial data processing. In the case of intelligent sensors, the whole process can be done by the sensor itself and the output values can be sent further over the protocols like AMQP or MQTT. Those new features open a variety of new capabilities to the production line systems. The Tweeting Factory is an architecture proposal for utilizing new features and providing the capabilities. It is an architecture pattern

D. Groen et al. (Eds.): ICCS 2022, LNCS 13352, pp. 455–468, 2022.
https://doi.org/10.1007/978-3-031-08757-8_38

providing a communication between production units, data processing services, and output data subscribers [12]. The machines are responsible for messages dispatched that are then processed by the services and the services send new messages. All messages in the system can be utilized by the subscribers. Every tenant in the system has access to all messages - the architecture implements the Publish-Subscribe pattern [6]. There are no constraints for the messages exchanged in the system. Only a few recommendations are made for the messages' metadata, such as the origin, subject, and publishing time. The services can read or send new messages, communicate with external systems and other data sources in order to leverage external information, data processing outside the system, or data recording.

The paper aims to validate the described architecture. Furthermore, the experiment designed for the verification purpose is described: based on the data acquisition process, the production environment project was designed that was the basis for the simulation environment. The simulation experiment was performed with the real production data provided by the cooperating company. The simulation results were compared with the results calculated by the blackboard style data acquisition process. Additionally, the Tweeting Factory performance was measured and analyzed. Finally, in the last section, the conclusions and possibilities for further research are discussed.

Before describing the methods in detail, we introduce the related work that influenced our research. Tweeting Factory term was used for the initial literature review. However, only four papers [3,8,11,12] were found using the term. Additionally, an alternative name for the architectural model was discovered - LISA, i.e., Line Information System Architecture [11]. Taking the alternative name into account and the low number of papers found in initial survey, there was a reasonable doubt that the Tweeting Factory concept exists, but the different name is used. Another term that was found during the analysis is called Digital Twin. It seems that Digital Twin is the higher abstraction concept and Tweeting Factory might be one of its implementations.

During the analysis, no comparison with the blackboard style SQL data acquisition method was found. The authors find this subject very important to eventually prove the advantages of the Tweeting Factory over the existing solution. Therefore, following research gaps were determined, that aims to be the authors contribution:

- RQ1: Is the Tweeting Factory concept correct?
- RQ2: Does the Tweeting Factory concept fulfill the real-time system constraints?
- RQ3: How the Tweeting Factory concept can improve the data acquisition process?
- RQ4: Is the Tweeting Factory a correct replacement of the SQL method?

2   Overview of the Tweeting Factory

2.1   Data Acquisition

Data acquisition is a process of measuring the physical values and storing them in a digital form [7]. The following phases of the process can be highlighted:

1. Physical value extraction
2. The value measurement by a sensor
3. The measured values transmission to a registration device
4. The values conversion to a digital form by ADC
5. Storing of the digital values

Data can serve different purposes, for example, KPIs computation, historical data analysis. event reporting such as machines' condition monitoring, environment parameters control, failure reporting, and material fatigue. Evolution of the Industry 4.0 concept is strictly coupled to intelligent sensors. In comparison with the previous generation sensors, intelligent sensors are capable to communicate with the other environment's participants, follow precise production, leverage production ontology, and use machine learning algorithms for independent conversation [10]. Industry 4.0 compatible sensors are capable to address all phases of data acquisition, besides the last one - the data storage. For this purpose, we use the cloud computing infrastructure. The paper's authors consider Tweeting Factory concept as a good candidate solution to address this requirement.

2.2   From Tweets to Decisions

Tweeting Factory concept provides additional benefits for the manufacturing environments. Additional features of the architecture were found in the papers [8,12]: KPI indexes calculation, hybrid systems optimization, energy consumption optimization, production process planning, quality assurance, production machines management, and monitoring. It is possible to use Tweeting Factory concept for data storing only [3], however additional services provided by the concept create its value. For example, for KPIs calculation, the stored data from a requested period can be used instead of single data coming directly from PLCs. All above-mentioned data are calculated in real-time, so it is possible to immediately generate a notification event when the specific index reaches a predefined threshold. A significant advantage of Tweeting Factory allowing to add services operating on available data is the ability to connect to external services. Based on that, the systems using the architecture are capable to easily connect and take part in the Shared Industry concept [14] or to be used in other applications outside the manufacturing industry [16]. Considering the Tweeting Factory as a data stream, it is also possible to apply machine learning algorithms.

2.3   Tweeting Factory Architecture

The base sources of the data are production units. Temperature or humidity are examples of the working environment measured values. Machine's work parameters such as state (idle/active), specific state time, or smaller component state

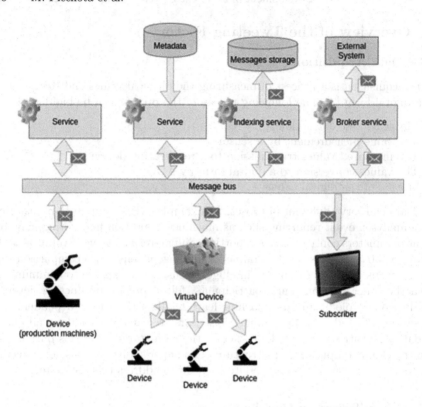

Fig. 1. The Tweeting Factory architecture model

are also measured. The machines can send that information themselves or to the connected PLCs. Devices other than production units can be found on the production line, for instance, a barcode reader. They can also send data themselves or to the PLC controllers. Another category of data sources are services. Services can also subscribe by receiving various types of messages available on Tweeting Factory's bus. In most cases, the data published by the service is just a transformation of the data received. Tweeting Factory provides support to connect and leverage external systems. Metadata models proposed in [11] allow to decorate the messages with the context of the external systems. However, the metadata is not the only advantage. Tweeting Factory provides a way to emit events registered in external systems. It allows to gather context data, not limited to data generated by production machines, for example, information entered manually by an employee or imported from HR systems, such as work hours, holidays, employee availability, or workplace assignments.

The Tweeting Factory architecture is depicted in Fig. 1. Production units, including virtual ones, can measure and convert the physical values to a digital representation itself. However, they are not capable to store data. This responsibility belongs to indexing services and subscribers.

When dealing with messages from different sources, it is highly possible to get incompatible messages' formats. The incompatibility might be a reason for incorrect data handling. This problem is called the interoperability issue and is considered on a semantic level [13]. Production machines and external systems produce information with incompatible labels, what leads the subscriber service's data parsing issues. A well-known solution of the interoperability issue is an application of a domain ontology or an application of an existing information exchange standard, such as ISO 10303 [4]. Standardization of the system's message layout solves the interoperability issue on the semantic level [13]. The literature provides different solutions [1], however, considering a variety of manufacturing industry applications, it is impossible to establish one unified standard. Different ontologies might be leveraged in providing a unified standard for the specific application. For example, in the paper, the authors decided to replace labels created by specific machines with URL labels. The interoperability issue resolved can be seen in Fig. 2. The external system's messages need to be translated by the broker service. For the production machines, if it is not possible to configure their messages' labels, translation services [15] should be introduced.

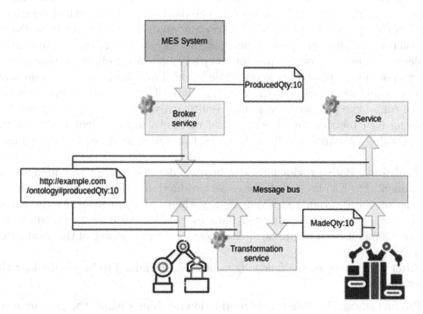

Fig. 2. Ontology based solution of the interoperability issue in the Tweeting Factory environment

3 Experiments

In this section, an experiment validating the Tweeting Factory concept and its results are presented. In the experiment, KPI metrics were calculated by Tweeting Factory simulation environment and by the blackboard style method. The analysis of the experiment's results allowed the authors to answer the presented paper's research questions. To prove the Tweeting Factory correctness (RQ1), the calculated metrics were compared with the metrics calculated by standard methods. The comparison proved also if the Tweeting Factory is a good replacement for the legacy SQL method (RQ4). Additionally, the performance of the system is measured and analyzed (RQ2).

The experiment was divided into the following steps:

1. Analyze source data.
2. Calculate KPI indexes by the standard method.
3. Design the new data acquisition method.
4. Complete the implementation.
5. Conduct a simulation run.
6. Analyze the results.

Source data analysis (1) aims to check if the provided production data is correct for the experiment, what quality it is, and what time range and units should be chosen as an input to the simulation. The initial KPI calculation (2) was done by the legacy SQL method. The OEE-based metrics were chosen for the experiment, considering their universality [2]. The data acquisition method design (3) requires the specific production infrastructure analysis and leads to the Tweeting Factory architecture preparation for the specific environment. Simulation implementation based on the previous step's results. A set of programming tools was prepared that played the service role in the Tweeting Factory environment. Another tools were verifying the resulting data. The simulation run (5) was a simple run of the prepared tools. The results analysis (6) aimed to compare the resulting KPI metrics with the standard method's initial calculation from the previous step (2). Additionally, calculation times were analyzed for performance metrics.

Real-world data provided by the heating devices factory was used in the experiment. It provides such features as: production order reporting, machines' production data automated acquisition, gathered data analysis. In the experiment, the automatically collected data as well as the data provided by the employees was used. The provided database served as a source of the production data.

After data analysis, the following data was determined to be required for the experiment:

1. Products data: the data source production machine's name, the product completion time, the amount of the successfully produced and rejected products, estimated reference unit time of work. Described data was available in various tables in provided database.
2. Production machine's work time: the machine's code, the event's start timestamp, the event's end timestamp, the event's type, the event's value.
   The event's value corresponding to work time is either Work or PW. This type of event can take value 0 (stopped working) or 1 (started working). These records were emitted every time a machine started or stopped to work and also at the top of every hour.

Based on determined event types it is possible to determine KPI metrics to calculate:

- PQ - Produced Quantity - based on products' completion events,
- RQ - Rejected Quantity - based on products' completion events,
- PRI - Planned Run Time - based on products' completion events,
- APT - Actual Production Time - based on a machine's work time events,
- PBT - Planned Busy Time - based on a machine's work time events.

In order to calculate the OEE index [2], the following component metrics are required:

$$A = \frac{APT}{PBT}, \quad E = \frac{PRI * PQ}{APT}, \quad Q = \frac{PQ - RQ}{PQ} \tag{1}$$

None of the above indexes can't reach a value greater than 1, however it can happen as a result of false PRI estimation. Then the performance index is limited to 1. The OEE index is a product of three indexes:

$$OEE = A * E * Q = \frac{PRI * (PQ - RQ)}{BPT} \tag{2}$$

Assuming the available experiment's data is sufficient for the OEE index calculation.

The last step of the source data analysis is to determine the event data subset for the simulation. Every event data contains the code of a machine and its timestamp. It allows to connect the machine's event data with a product's events. The number of products' completion events versus production unit is presented in Fig. 3(a). To achieve better readability, the machines with the number of such events less than 20 were filtered out. From the plot, it can be noticed that four machines emit more events: 0005, 5309-1, 5309-2, 8105.

The number of a production machine's work time events is presented in Fig. 3(b). It can be noticed that the machines highlighted above, except for 0005, have also a lot of work time events.

(a) *Products' completion events*   (b) *Work time events*

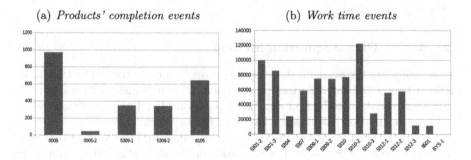

Fig. 3. Distribution versus production unit

Assuming the three machines highlighted above (5309-1, 5309-2, 8105) were chosen for the simulation, as those were the only ones that had a high number of events of both types. To determine the time range of events for the simulation, the number of both types of events generated by the chosen machines was analyzed versus months. The results are presented in Fig. 4(a) and Fig. 4(b).

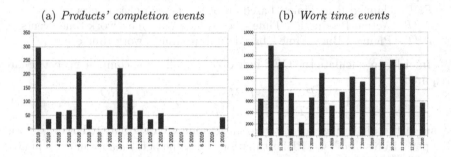

(a) *Products' completion events*   (b) *Work time events*

Fig. 4. Distribution versus months

To use the largest unit of data, the following time range was selected from 2018-09-17 to 2019-09-16. It can be noticed that there are months when no products' completions were registered. It may be because employees refrain from registering the events in the system terminals. In such cases, it is impossible to calculate the quality, performance and the OEE indexes, but the availability index is still possible to calculate. Additionally, considering a low number of products' completion events in general, the indexes will be calculated versus days.

3.1   KPI Indexes Calculation by the Legacy Method

To verify the Tweeting Factory concept, the KPI metrics were calculated offline using SQL query. The well-known considerations and limitations were applied to the legacy method.

3.2   A New Data Acquisition Method

Considering that the available data set does already contain the production information of units and entered by employees, only OEE index acquisition is considered. The available production data is further considered as production events (employee or machine). Following Tweeting Factory architecture, the events are transmitted by the message bus between the machines and the services that subscribe and publish the events. For the needs of the simulation, the assumption was made that the system is publishing the data on the message bus instead of storing it. The following services were introduced to the designed environment:

– a service transforming the machines' events,
– a service calculating the KPI indexes in real time,
– a service storing the KPI indexes into a database.

The designed architecture is presented in Fig. 5.

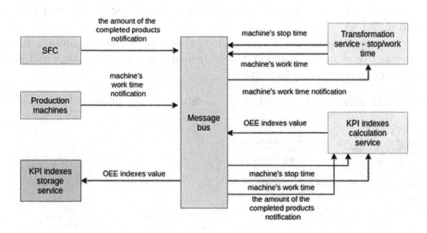

Fig. 5. The Tweeting Factory architecture proposal

Having the Tweeting Factory not specifying the message formats [12], JSON format is applied in the architecture. The base message format includes: start, end, machine, e_type, and payload. The e_type and payload field values are specific to the registered event. The e_type field is the key value of the registered event. The payload field contains the data specific for the event:

– the number of completed product notifications: QtyAll - an amount of produced products, QtyRejected - an amount of rejected products, RefTime - an estimated reference unit time for a product,
– work time notification: work - 0 for stopped, 1 for running
– work time: time - work time in seconds
– stop time: time - stop time in seconds
– OEE indexes value: pq - an amount of all produced products, rq - an amount of all rejected products, pri_pq - the product of of manufactured products and planned working time per unit, in seconds, apt - machine's work time in seconds, pbt - machine's estimated work time in seconds, quality - quality index value, - efficiency - efficiency index value, availability - availability index value, oee - OEE index value.

3.3 Simulation Implementation

To adjust the design to the simulation conditions, the following extensions were made:

– the source events generated originally by the system and the machines are emitted by the simulation program,

- in order to shorten the simulation time (365 days of the chosen time range), the simulation clock event is introduced,
- start and end events indicating the whole simulation start and end time are introduced,
- in order to measure a KPI calculation time, the message's metadata were extended by a new parameter trace_id, being an unique message's id and a new service calculating the time.

The extensions implemented on the architecture are presented in Fig. 6.

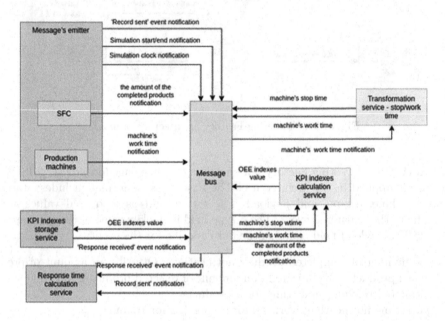

Fig. 6. Simulation environment's extended architecture

The Tweeting Factory concept does not specify any software for the message bus [12] implementation. The authors decided to use RabbitMQ[1]. The main advantage of the software over the Apache ActiveMQ[2] is ease of use and an extensive documentation. All services are implemented using Python3 and the Pika library[3] that enables a communication with the RabbitMQ server over the AMQP 0.9.1 protocol. Additionally, a CSV file with the input data for the simulation messages emitter was prepared. The CSV contains an array of source events, sorted by the publication time. A new type of messages were added in the extended simulation architecture:

---

[1] https://www.rabbitmq.com/.
[2] https://activemq.apache.org/.
[3] https://pypi.org/project/pika/.

- 'record sent' event notification: real_time - the event's timestamp, id - id for time measurement,
- 'response received' event notification: real_time - the event's timestamp, id - id for time measurement,
- simulation clock notification: clock - simulation timestamp,
- simulation start notification,
- simulation end notification.

3.4 Simulation Run

The run was performed on a PC equipped with AMD Ryzen 5 3600 processor unit and 16GB 3200MHz RAM memory. The message bus component was implemented in the Podman container[4] and docker.io/library/rabbitmq:3.8.4-management container image was used.

The average KPI indexes for the machines are presented in Table 1.

Table 1. The average KPI indexes for the production units

Unit	Quality	Efficiency	Availability	OEE
(a)	(b)	(c)	(d)	(e)
5309-1	0.9988	0.4625	0.4644	0.2657
5309-1	0.9949	0.5431	0.4087	0.2543
8105	0.9884	0.7952	0.6920	0.5857

Column (a) shows the symbolic codes of the selected production units. In column (b) the quality value is placed, while in the next two columns there are efficiency (c) and availability values (d). The last column (e) lists the total value of OEE.

The correctness verification was performed by the comparison of the Tweeting Factory approach results with the results of the legacy method. The following metrics were compared: PQ, RQ, PRI*PQ, APT, PBT, Quality, Efficiency, Availability, OEE. The comparisons have occurred 1095 times (365 days, 3 machines). No significant differences between the output results were found. The performance verification was done by the analysis of the data generated by the simulation indicating the waiting time for the KPI index calculation. The time was measured 113542 times. Average time was 0.012520 s, i.e., 12.52 ms. In the case of about 99% of the results, the calculation time was less than 25 ms. Figure 7(a) presents the box plot of the data. Large number of outliers can be generated by the temporary high load of the KPI calculation service or the temporary spike of the processor demand of the operating system.

Anderson-Darling test [9] was performed to determine whether a normal distribution adequately describes a set of data. Significance level 0.05 was assumed. The following hypotheses were tested:

---

[4] https://podman.io/.

- H0: the results are normally distributed,
- H1: the results are not normally distributed.

The output p-value was 3.7×10^{-24}, so the alternative hypothesis was accepted. Shapiro-Wilk test [17] was also performed to confirm the previous test output. The input data for the test was limited to 2000 records to keep the test's power. Significance level 0.05 was assumed. The following hypotheses were tested:

- H0: the results are normally distributed.
- H1: the results are not normally distributed.

The output p-value was 6.923531×10^{-57}, so the alternative hypothesis was accepted. Quantile distribution of the KPI indexed calculation time is presented in Fig. 7(b). The distribution is not linear, which means the distribution is not normal.

(a) *Box plot*   (b) *Quantile normal distribution*

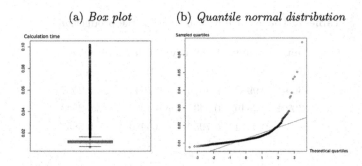

Fig. 7. KPI indexes calculation

To check if the calculated average value is a good representation of the random variable, the Wilcoxon test [37] was performed for one attempt. Significance level 0.05 was assumed. The following hypotheses were tested: - H0 - random variable's value's distribution is symmetric around = 0.012520. - H1 - random variable's value's distribution is not symmetric around = 0.012520. The output p-value was 0, so the alternative hypothesis was accepted. The median and percentile values are considered in further analysis. Statistical analysis showed no coincidence with the normal distribution of the computation time KPI indicators and the asymmetry of the time distribution with respect to the mean.

4   Conclusions

No differences between KPI index values calculated by the Tweeting Factory and the blackboard style method prove the Tweeting Factory concept correctness (RQ1). The response time of the proposed system that was shorter than 17ms in more than 95% cases puts the system among the real-time systems (RQ2).

Considering the above two conclusions, the Tweeting Factory becomes a great candidate for a replacement of the SQL legacy method (RQ4), providing a lot of advantages (RQ3):

- ease of new production units and service integration,
- capability to add and remove new components without a need for reconfiguration,
- control over the messages flow,
- ESB controls data exchange security,
- complex metrics calculation in real-time,
- support to integrate the system with external systems.

There are also cons that need to be taken into account when considering the concept:

- data transformations required for the components using unsupported protocols,
- single point of failure - ESB,
- security of data exchange under the sole control of the message broker.

The literature analysis demonstrated a low number of concept applications in the industry. Taking the experiment's promising results into account the authors state that the lack of real life application studies is a major gap that may be a subject for the further research. Considering the literature survey conclusion that the Tweeting Factory concept can be used as a Digital Twin core component, the Digital Twin with Tweeting Factory application in Shared Industry [14]. Leveraging the real-time feature of the concept, another interesting research subject could be managed and optimized using machine learning algorithms.

Acknowledgment. Part of the work presented in this paper was received financial support from the statutory funds at the Wrocław University of Science and Technology and DSR Ltd.

References

1. Çetiner, G., Ismail, A., Hassan, A.: Ontology of manufacturing engineering. In: 5th International Advanced Technologies Symposium, p. 6 (2009)
2. De Ron, A., Rooda, J.: Equipment effectiveness: OEE revisited. IEEE Trans. Semicond. Manuf. **18**(1), 190–196 (2005)
3. Dressler, N.: Towards The Tweeting Factory. Master's thesis, KTH Industrial Engineering and Management, SE-100 44 Stockholm (2015)
4. Feeney, A.: The step modular architecture. J. Comput. Inf. Sci. Eng. **2**(2), 132–135 (2002)
5. Gittler, T., Gontarz, A., Weiss, L., Wegener, K.: A fundamental approach for data acquisition on machine tools as enabler for analytical industrie 4.0 applications. Procedia CIRP **79**, 586–591 (2019)
6. Hoffmann, M.: Smart Agents for the Industry 4.0, 1st edn. Springer Vieweg, Heidelberg (2019). https://doi.org/10.1007/978-3-658-27742-0

7. Kos, T., Kosar, T., Mernik, M.: Development of data acquisition systems by using a domain-specific modeling language. Comput. Ind. **63**(3), 181–192 (2012)
8. Lennartson, B., Bengtsson, K., Wigström, O., Riazi, S.: Modeling and optimization of hybrid systems for the tweeting factory. IEEE Trans. Autom. Sci. Eng. **13**(1), 191–205 (2016)
9. Nelson, L.: The Anderson-Darling test for normality. J. Qual. Technol. **30**(3), 298–299 (1998)
10. Schütze, A., Helwig, N., Schneider, T.: Sensors 4.0 - smart sensors and measurement technology enable industry 4.0. J. Sensors Sensor Syst. **7**(1), 359–371 (2018)
11. Theorin, A., et al.: An event-driven manufacturing information system architecture. IFAC-PapersOnLine **48**(3), 547–554 (2015)
12. Theorin, A., et al.: An event-driven manufacturing information system architecture for industry 4.0. Int. J. Prod. Res. **55**(5), 1297–1311 (2017)
13. Tursi, A.: Ontology-Based approach for Product-Driven interoperability of enterprise production systems. Phd thesis, Université Henri Poincaré - Nancy 1, Politecnico di Bari (2009)
14. Wang, G., Zhang, G., Guo, X., Zhang, Y.: Digital twin-driven service model and optimal allocation of manufacturing resources in shared manufacturing. J. Manuf. Syst. **59**, 165–179 (2021)
15. Woolf, B.: Enterprise Integration Patterns: Designing, Building, and Deploying Messaging, 1st edn. Addison-Wesley, Boston (2003)
16. Yan, X.: Knowledge Acquisition from Streaming Data through a Novel Dynamic Clustering Algorithm. Phd thesis, North Carolina Agricultural and Technical State University (2018)
17. Zhao, L., Chuang, Z., Ke-Fu, X., Meng-Meng, C.: A computing model for real-time stream processing. In: 2014 International Conference on Cloud Computing and Big Data, pp. 134–137 (2014)

Boundary Geometry Fitting with Bézier Curves in PIES Based on Automatic Differentiation

Krzysztof Szerszeń$^{(\boxtimes)}$ and Eugeniusz Zieniuk

Institute of Computer Science, University of Bialystok, Konstantego Ciołkowskiego 1M,
15-245 Białystok, Poland
{kszerszen,ezieniuk}@ii.uwb.edu.pl

Abstract. This paper presents an algorithm for fitting the boundary geometry with Bézier curves in the parametric integral equation system (PIES). The algorithm determines the coordinates of control points by minimizing the distance between the constructed curves and contour points on the boundary. The minimization is done with the Adam optimizer that uses the gradient of the objective function calculated by automatic differentiation (AD). Automatic differentiation eliminates error-prone manual routines to evaluate symbolic derivatives. The algorithm automatically adjusts to the actual number of curves and their degrees. The presented tests show high accuracy and scalability of the proposed approach. Finally, we demonstrate that the resulting boundary may be directly used by the PIES to solve the boundary value problem in 2D governed by the Laplace equation.

Keywords: Automatic differentiation · Parametric curve fitting · Bézier curves · Parametric integral equation system (PIES) · Boundary value problems

1 Introduction

One of the main difficulties during computer simulation of boundary value problems (BVP) is the appropriate definition of the computational domain. Typically it is done by dividing the problem domain into finite elements (FEM) [1] or only the boundary of that domain into boundary elements (BEM) [2]. However, such discretization is extremely laborious as it requires hundreds or thousands of elements and even more nodes are necessary to declare them. The alternative is to introduce the mathematical and geometric tools used in computer graphics and CAD/CAM systems. This is used in the parametric integral equation system (PIES) where the boundary of the computational domain is bounded by parametric curves [3] and surfaces [4]. Moreover, due to the analytical integration of the boundary geometry directly in the PIES formula, there is a lot of freedom in choosing the appropriate structure of the boundary representation. Our previous studies have shown that it is particularly effective to declare such boundary geometries by employing Bézier parametric curves defined by a small set of control points. This allows for a continuous representation of the boundary and the number of the curves is significantly smaller than finite or boundary elements required to solve the same problem.

© The Author(s), under exclusive license to Springer Nature Switzerland AG 2022
D. Groen et al. (Eds.): ICCS 2022, LNCS 13352, pp. 469–475, 2022.
https://doi.org/10.1007/978-3-031-08757-8_39

Despite these advantages, parametric curves also have some implementation difficulties since its shape is determined by control points that generally do not lie on the
curve. Curve fitting is a fundamental problem in computer graphics and has become a
very active scientific area. Mathematically, it can be formulated as optimization problem to minimize the distance between the points describing the approximated shape
and the points on the parametric curve. There is a rich literature to solve such optimization problem with several linear [5] and non-linear [6] least-squares techniques.
Another approaches are based on biologically inspired solutions: genetic algorithms [7],
simulated annealing [8], particle swarm optimization [9], evolutionary algorithms [10],
artificial immune systems [11]. In [12] neural network-based curve fitting technique is
presented. However, most existing algorithms are based on gradient descent minimization [13]. On the other hand, analytical evaluation of gradients is a labor-intensive task
and sensitive to human errors, especially in the case of real problems defined by many
design variables related to the coordinates of control points.

This paper attempts a new look at the parametric curve fitting applied to the boundary
approximation in PIES. This is done by minimizing the distance between the constructed
curves and the contour points on the boundary with the Adam optimizer [14], where the
required derivatives of the objective function are computed by automatic differentiation
(AD). The idea of AD is based on a decomposition of an input function into elementary
operations, whose local derivatives are easy to compute. While AD has a long history
and is supported by many publications [15, 16], the recent revolution in deep learning
and artificial intelligence has brought a new stage in the development of advanced AD
tools and new optimization algorithms. Popular libraries such as TensorFlow [17] and
PyTorch [18] provide efficient AD and optimization algorithms for large models with
thousands or even millions of parameters. The results of the presented tests illustrate
good fitting properties of the proposed algorithm for various shapes of the boundary and
high scalability. The fitted boundaries are directly used in PIES to solve BVP governed
by the Laplace equation.

2 Defining the Boundary by Bézier Curves in PIES

We consider a boundary value problem governed by the Laplace equation defined in a
domain Ω with a boundary Γ, as shown in Fig. 1a.

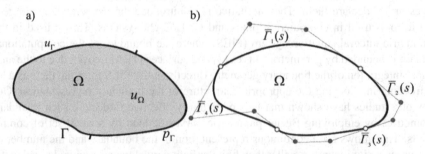

Fig. 1. BVP defined in the domain Ω with the boundary Γ(a), boundary description in PIES with
4 Bézier curves and 12 control points (b).

The boundary of that domain is described in PIES by a set of Bézier curves. A single Bézier curve of degree m is defined by the location of the $m + 1$ control points P_i and is mathematically described as

$$\overline{\Gamma}(s) = \sum_{i=0}^{m} \binom{m}{i} s^i (1 - s)^{m-i} P_i, \tag{1}$$

where s is a parametric coordinate along the curve ($0 \leq s \leq 1$). Figure 1b shows a practical definition of the boundary formed by joining 4 cubic Bézier curves. The formula of PIES for 2D BVP governed by the Laplace equation is presented below

$$0.5 u_l(\overline{s}) = \sum_{j=1}^{n} \int_{s_{j-1}}^{s_j} \left\{ \overline{U}_{lj}^*(\overline{s}, s) p_j(s) - \overline{P}_{lj}^*(\overline{s}, s) u_j(s) \right\} J_j(s) ds, \tag{2}$$

where $s_{j-1} \leq s \leq s_j, s_{l-1} \leq \overline{s} \leq s_l, l = 1, 2, ..., n$.

The Bézier curves $\overline{\Gamma}(s)$ are included analytically in the kernels $\overline{U}_{lj}^*(\overline{s}, s)$ and $\overline{P}_{lj}^*(\overline{s}, s)$ written as

$$\overline{U}_{lj}^*(\overline{s}, s) = \ln \frac{1}{[\eta_1^2 + \eta_2^2]^{0.5}}, \tag{3}$$

$$\overline{P}_{lj}^*(\overline{s}, s) = \frac{\eta_1 n_1^{(j)}(s) + \eta_2 n_2^{(j)}(s)}{\eta_1^2 + \eta_2^2}, \tag{4}$$

$$\text{where } \eta_1 = \overline{\Gamma}_l^{(1)}(\overline{s}) - \overline{\Gamma}_j^{(1)}(s), \eta_2 = \overline{\Gamma}_l^{(2)}(\overline{s}) - \overline{\Gamma}_j^{(2)}(s). \tag{5}$$

Moreover, $n_1^{(j)}(s), n_2^{(j)}(s)$ denote the components of the normal vector to the boundary and $J_j(s)$ is the Jacobian. The reader can find details of the PIES mathematical formulation in several previous papers, for example in [3].

3 Proposed Algorithm

Let G_Γ be a set of data points sampled along the boundary and $G_{\overline{\Gamma}}$ is a set of points lying on Bézier curves. The objective function $loss_{L2}(G_\Gamma, G_{\overline{\Gamma}})$ of our optimization problem is the L_2 distance between G_Γ and $G_{\overline{\Gamma}}$. The gradient of the objective function with respect to the corresponding control points P of the curves is written as

$$\frac{\partial loss(P)}{\partial P} = \frac{\partial loss_{L2}(G_\Gamma, G_{\overline{\Gamma}})}{\partial \overline{\Gamma}} \frac{\partial \overline{\Gamma}}{\partial P}. \tag{6}$$

The gradient (6) can be derived analytically, however, is laborious and prone to human errors. Therefore, we compute it employing AD which decomposes the derivative of the objective function into those of elementary operations based on the chain rule. This decomposition can be described by a computational graph that shows the relations between individual differentials. For demonstrating the procedure, an elementary case with one cubic Bézier curve defined by 4 control points is analyzed below. In this case, the formula (6) reduces to the following form

$$\frac{\partial loss(P)}{\partial P} = \frac{\partial}{\partial P}(G_\Gamma - CAP)^2, \tag{7}$$

where C, A, P are referred to a matrix representation for the cubic Bézier curve

$$C = \begin{bmatrix} s^3 & s^2 & s & 1 \end{bmatrix}, A = \begin{bmatrix} -1 & 3 & -3 & 1 \\ 3 & -6 & 3 & 0 \\ -3 & 3 & 0 & 0 \\ 1 & 0 & 0 & 0 \end{bmatrix}, P = \begin{bmatrix} P_1 \\ P_2 \\ P_3 \\ P_4 \end{bmatrix}. \tag{8}$$

Finally, the computational graph for formula (7) is presented below.

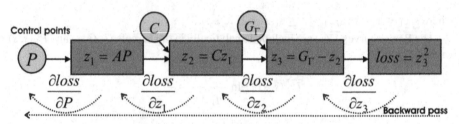

Fig. 2. Structure of the computational graph for a single cubic Bézier curve.

Table 1 contains the vertices z_i corresponding to the intermediate computed variables for elemental operations and their derivatives for forward and backward data flow.

Table 1. Forward and backward propagation of derived values for the graph shown in Fig. 2.

	Forward pass	Backward pass
$z_1 = AP$	$\frac{\partial z_1}{\partial P} = A$	$\frac{\partial loss}{\partial P} = \frac{\partial loss}{\partial z_1}\frac{\partial z_1}{\partial P} = -2CA$
$z_2 = Cz_1$	$\frac{\partial z_2}{\partial P} = \frac{\partial z_2}{\partial z_1}\frac{\partial z_1}{\partial P} = CA$	$\frac{\partial loss}{\partial z_1} = \frac{\partial loss}{\partial z_2}\frac{\partial z_2}{\partial z_1} = -2C$
$z_3 = G_\Gamma - z_2$	$\frac{\partial z_3}{\partial P} = \frac{\partial z_3}{\partial z_2}\frac{\partial z_2}{\partial P} = -CA$	$\frac{\partial loss}{\partial z_2} = \frac{\partial loss}{\partial z_3}\frac{\partial z_3}{\partial z_2} = -2$
$loss = z_3^2$	$\frac{\partial loss}{\partial P} = \frac{\partial loss}{\partial z_3}\frac{\partial z_3}{\partial P} = -2CA$	$\frac{\partial loss}{\partial z_3} = 2$

The computational graph can dynamically update its structure for more Bézier curves and control points. The full diagram of the procedure with backward mode AD as best suited for our problem is shown in Fig. 3.

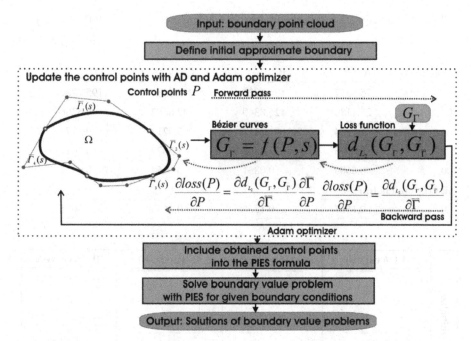

Fig. 3. Schematic flowchart of the proposed curve fitting combined with PIES.

To determine the coordinates of control points, the ADAM [14] optimization algorithm is used characterized by fast convergence, and resistance to local minima.

4 Results and Evaluation

The scheme given in Fig. 3 is implemented in the PyTorch framework providing access to AD and Adam optimizer modules. Figure 4 shows the results of our experimental evaluation for 4 identified boundaries generated from 16 to 40 cubic Bézier curves.

The curves require 48 and 120 control points to be defined. The data set G_Γ consists of evenly sampled points along the boundary. The first three shapes are sampled into 160 points and the last one into 400 points.

As it can be seen from the graphic results, all tested shapes are fitted successfully. The related numerical results are listed in Table 2. The final boundaries are directly used in PIES to simulate BVP governed by the Laplace equation for the following Dirichlet boundary conditions

$$\phi(x_1, x_2) = \cos(x_1)\cosh(x_2) + \sin(x_1)\sinh(x_1). \tag{9}$$

This is possible since the control points are included analytically in the kernels (3, 4). In order to solve the problem on the boundary, the collocation method [3] is adopted with 6 colocation points per each Bézier curve. The last row in Table 2 shows the L_2 error norm of the PIES solutions for all shapes from the 400th iteration of the fitting.

Table 2. Convergence of the objective functions during iterations and the PIES accuracy.

$loss_{L2}(G_\Gamma, G_{\overline{\Gamma}})$

Iteration	A	B	C	D
1	75.3377	186.8021	212.6234	195.2782
50	25.1174	125.8349	147.8667	98.5266
100	1.0304	81.2112	87.9212	20.3172
200	0.0935	17.3265	1.5679	0.1621
400	0.0685	0.0831	0.0657	0.0954

PIES (L_2 error norm)

	A	B	C	D
400	0.1567	0.1235	0.3643	0.3025

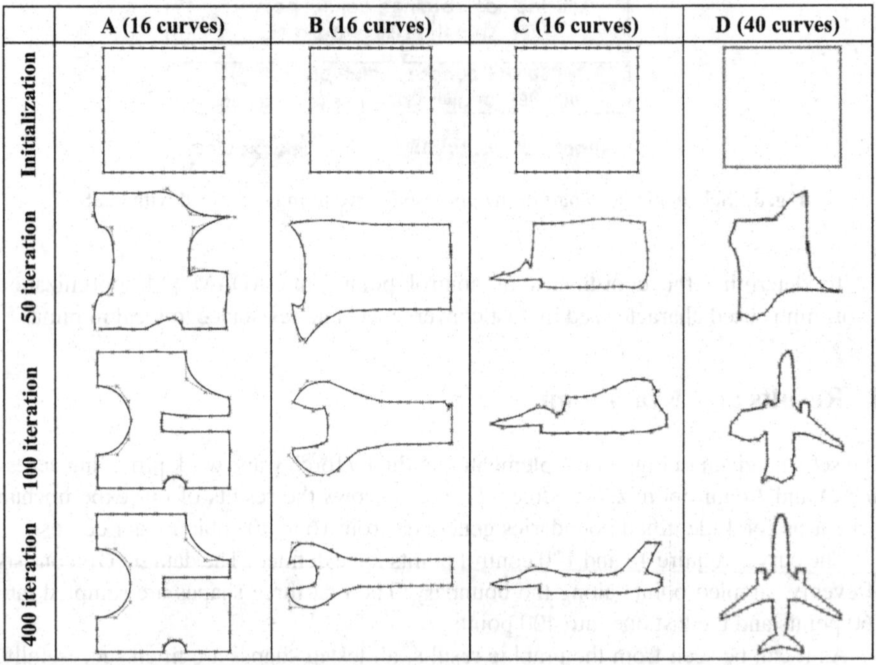

Fig. 4. Iterations 1, 50, 100, and 400 in the fitting process.

5 Conclusions

The results show that the proposed approach can be applied not only to academic but also to real-life problems with hundreds of design variables considered as the positions of control points. The boundary defined by the Bézier curves is integrated analytically with the PIES computational method so that it can be directly used to solve BVP, as shown in the example. The approach can use different types of objective functions and optimization algorithms. We also hope to extend the research to study the boundary reconstruction from an unstructured point cloud as well to 3D problems.

References

1. Zienkiewicz, O.C., Taylor, R.L.: The finite element method for solid and structural mechanics. Elsevier (2005)
2. Brebbia, C.A., Telles, J.C.F., Wrobel, L.C.: Boundary Element Techniques: Theory and Applications in Engineering. Springer, Heidelberg (2012). https://doi.org/10.1007/978-3-642-488 60-3
3. Zieniuk, E., Szerszeń, K.: NURBS curves in direct definition of the shapes of the boundary for 2D Stokes flow problems in modified classical BIE. Appl. Numer. Math. **132**, 111–126 (2018)
4. Zieniuk, E., Szerszeń, K.: Nonelement boundary representation with Bézier surface patches for 3D linear elasticity problems in parametric integral equation system (PIES) and its solving using Lagrange polynomials. Numer. Methods Part. Differ. Eqn. **34**(1), 51–79 (2018)
5. Piegl, L., Tiller, W.: The NURBS Book. Springer, Heidelberg (1996). https://doi.org/10.1007/978-3-642-97385-7
6. Borges, C.F., Pastva, T.: Total least squares fitting of Bézier and B-spline curves to ordered data. Comput. Aided Geom. Des. **19**(4), 275–289 (2002)
7. Zhou, M., Wang, G.: Genetic algorithm-based least square fitting of B-Spline and Bézier curves. J. Comput. Res. Dev. **42**(1), 134–143 (2005)
8. Zhang, J., Wang, H.: B-Spline curve fitting based on genetic algorithms and the simulated annealing algorithm. Comput. Eng. Sci. **33**(3), 191–193 (2011)
9. Gálvez, A., Iglesias, A.: Efficient particle swarm optimization approach for data fitting with free knot B-splines. Comput. Aided Des. **43**(12), 1683–1692 (2011)
10. Pandunata, P., Shamsuddin, S.M.H.: Differential evolution optimization for Bézier curve fitting. In: 2010 Seventh International Conference on Computer Graphics, Imaging and Visualization, pp. 68–72 (2010)
11. Iglesias, A., Gálvez, A., Avila, A.: Discrete Bézier curve fitting with artificial immune systems. Stud. Comput. Intell. **441**, 59–75 (2013)
12. Bishop, C.M., Roach, C.M.: Fast curve fitting using neural networks. Rev. Sci. Instrum. **63**(10), 4450–4456 (1992)
13. Samir C., Absil, P.A., Srivastava, A., Klassen, E.: A gradient-descent method for curve fitting on Riemannian manifolds. Found. Comput. Math. **12**(1), 49–73 (2012)
14. Kingma, D.P., Ba, J.: Adam: A method for stochastic optimization. arXiv preprint arXiv: 1412.6980 (2014)
15. Margossian, C.C.: A review of automatic differentiation and its efficient implementation. Wiley Interdisc. Rev. Data Mining Knowl. Disc. **9**, 1305e (2019)
16. Neidinger, R.D.: Introduction to automatic differentiation and MATLAB object-oriented programming. SIAM Rev. **52**(3), 545–563 (2010)
17. Abadi, M., et al.: {TensorFlow}: A System for {Large-Scale} Machine Learning. In: 12th USENIX symposium on operating systems design and implementation (OSDI 16), pp. 265–283 (2016)
18. Paszke, A., et al.: Pytorch: An imperative style, high-performance deep learning library. Adv. Neural Inf. Process. Syst. **32** (2019)

Partitioning Dense Graphs
with Hardware Accelerators

Xiaoyuan Liu[1,2]($\boxtimes$), Hayato Ushijima-Mwesigwa[2], Indradeep Ghosh[2],
and Ilya Safro[1]

[1] University of Delaware, Newark, DE, USA
{joeyxliu,isafro}@udel.edu
[2] Fujitsu Research of America, Inc., Sunnyvale, CA, USA
{hayato,ighosh}@fujitsu.com

Abstract. Graph partitioning is a fundamental combinatorial optimization problem that attracts a lot of attention from theoreticians and practitioners due to its broad applications.In this work, we experiment with solving the graph partitioning on the Fujitsu Digital Annealer (a special-purpose hardware designed for solving combinatorial optimization problems) and compare it with the existing top solvers. We demonstrate limitations of existing solvers on many dense graphs as well as those of the Digital Annealer on sparse graphs which opens an avenue to hybridize these approaches.

Keywords: Graph partitioning · Dense graphs · Digital annealer · Quantum-inspired

1 Introduction

There are several reasons to be optimistic about the future of quantum-inspired and quantum devices. However, despite their great potential, we also need to acknowledge that state-of-art classical methods are extremely powerful after years of relentless research and development. In classical computing, the development of algorithms, the rich mathematical framework behind them, and sophisticated data structures are relatively mature, whereas the area of quantum computing is still at its nascent stage. Many existing classical algorithms do not have provable or good enough bounds on the performance (e.g., they might not have ideal performance in the worst case), but in many applications, the worst-case scenarios are rather rarely seen. As a result, such algorithms, many of which heuristics, can achieve excellent results in terms of the solution quality or speed. Therefore, when utilizing emerging technologies such as quantum-inspired hardware accelerators and quantum computers to tackle certain problems, it is important to compare them not only with possibly slow but provably strong algorithms but also with the heuristic algorithms that exhibit reasonably good results on the instances of interest.

© The Author(s), under exclusive license to Springer Nature Switzerland AG 2022
D. Groen et al. (Eds.): ICCS 2022, LNCS 13352, pp. 476–483, 2022.
https://doi.org/10.1007/978-3-031-08757-8_40

The graph partitioning [2] is one of the combinatorial optimization problems for which there exists a big gap between rigorous theoretical approaches that ensure best known worst-case scenarios, and heuristics that are designed to cope with application instances exhibiting a reasonable quality-speed trade-off. Instances that arise in practical applications often contain special structures on which heuristics are engineered and tuned. Because of its practical importance, a huge amount of work has been done for a big class of graphs that arise in such areas as combinatorial scientific computing, machine learning, bioinformatics, and social science, namely, *sparse graphs*. Over the years, there were several benchmarks on which the graph partitioning algorithms have been tested and compared with each other to mention just a few [1,3,21]. However, *dense graphs* can be rarely found in them. As a result, most existing excellent graph partitioning heuristics do not perform well in practice on dense graphs, while provable algorithms with complexity that depends on the number of edges (or non-zeros in the corresponding matrix) are extremely slow. As we also show in computational results, a graph sparsification does not necessarily practically help to achieve high-quality solutions.

Multilevel Algorithms. This class of heuristics is one of the most successful for a variety of cut-based graph problems such as the minimum linear arrangement [15], and vertex separator [7]. Specifically for a whole variety of (hyper)graph partitioning versions [10,11,16,18] these heuristics exhibit best quality/speed trade-off [2]. In multilevel graph partitioning frameworks, a hierarchy of coarse graph representations is constructed in such a way that each next coarser graph is smaller than the previous finer one, and a solution of the partitioning for the coarse graph can approximate that of the fine graph and be further improved using fast local refinement. Multilevel algorithms are ideally suited for sparse graphs and suffer from the same problems as the algebraic multigrid (which generalizes, to the best of our knowledge, all known multilevel coarsening for partitioning) on dense matrices. In addition, a real scalability of the existing refinement for partitioning is achieved only for sparse local problems. Typically, if the density is increasing throughout the hierarchy construction, various ad-hoc tricks are used to accelerate optimization sacrificing the solution quality. When such things happen at the coarse levels, an error is quickly accumulated. Here we compare our results with KaHIP [17] which produced the best results among several multilevel solvers [2].

Hardware Accelerators for Combinatorial Problems. Hardware accelerators such as GPU have been pivotal in the recent advancements of fields such as machine learning. Due to the computing challenges arising as a result of the physical scaling limits of Moore's law, scientists have started to develop special-purpose hardware for solving combinatorial optimization problems. These novel technologies are all unified by an ability to solve the Ising model or, equivalently, the quadratic unconstrained binary optimization (QUBO) problem. The general QUBO is NP-hard and many problems can be formulated as QUBO [14]. It is also often used as a subroutine to model large neighborhood local search [13].

The Fujitsu Digital Annealer (DA) [4], used in this work, utilizes application-specific integrated circuit hardware for solving fully connected QUBO problems. Internally the hardware runs a modified version of the Metropolis-Hastings algorithm for simulated annealing. The hardware utilizes massive parallelization and a novel sampling technique. The novel sampling technique speeds up the traditional Markov Chain Monte Carlo by almost always moving to a new state instead of being stuck in a local minimum. Here, we use the third generation DA, which is a hybrid software-hardware configuration that supports up to 100,000 binary variables. DA also supports users to specify inequality constraints and special equality constraints such as 1-hot and 2-way 1-hot constraints.

Our Contribution. The goal of this paper is twofold. First, we demonstrate that existing scalable graph partitioning dedicated solvers are struggling with the dense graphs not only in comparison to the special-purpose hardware accelerators but even sometimes if compared to generic global optimization solvers that are not converged. At the same time, we demonstrate a clear superiority of classical dedicated graph partitioning solvers on sparse instances. Second, this work is a step towards investigating what kind of problems we can solve using combinatorial hardware accelerators. Can we find problems that are hard for existing methods, but can be solved more efficiently with novel hardware and specialized algorithms? As an example, we explore the performance of Fujitsu Digital Annealer (DA) on graph partitioning and compare it with general-purpose solver Gurobi, and also graph partitioning solver KaHIP.

We do not attempt to achieve an advantage for every single instance, especially since at the current stage, the devices we have right now are still facing many issues on scalability, noise, and so on. However, we advocate that hybridization of classical algorithms and specialized hardware (e.g., future quantum and existing quantum-inspired hardware) is a good candidate to break the barriers of the existing quality/speed trade-off.

2   Graph Partitioning Formulations

Let $G = (V, E)$ be an undirected, unweighted graph, where V denotes the set of n vertices, and E denotes the set of m edges. The goal of perfect balanced k-way graph partitioning (GP), is to partition V into k parts, $V_1, V_2, \cdots, V_k$, such that the k parts are disjoint and have equal size, while minimizing the total number of *cut edges*. A *cut edge* is an edge that has two end vertices assigned to different parts. Sometimes, the quality of the partition can be improved if we allow some imbalance between different parts. In this case, we allow some imbalance factor $\epsilon > 0$, and each part can have at most $(1 + \epsilon)\lceil n/k \rceil$ vertices.

Binary Quadratic Programming Formulation of GP. We first review the integer quadratic programming formulation for k-way GP [8,20]. When $k = 2$, we introduce binary variables $x_i \in \{0, 1\}$ for each vertex $i \in V$, where $x_i = 1$ if vertex

i is assigned to one part, and 0 otherwise. We denote by $\mathbf{x}$ the column vector $\mathbf{x} = (x_1, x_2, \cdots, x_n)^T$. The quadratic programming is then given by

$$\min_{\mathbf{x}} \mathbf{x}^T L \mathbf{x} \quad \text{such that } x_i \in \{0, 1\}, \; \forall i \in V, \tag{1}$$

where L is the Laplacian matrix of graph G. For perfect balance GP, we have the following equality constraint:

$$\mathbf{x}^T \mathbb{1} = \left\lceil \frac{n}{2} \right\rceil, \tag{2}$$

where $\mathbb{1}$ is the column vector with ones. For the imbalanced case, we have the following inequality constraint $\mathbf{x}^T \mathbb{1} \leq (1 + \epsilon) \left\lceil \frac{n}{2} \right\rceil$.

When $k > 2$, we introduce binary variables $x_{i,j} \in \{0, 1\}$ for each vertex $i \in V$ and part j, where $x_{i,j} = 1$ if vertex i is assigned to part j, and 0 otherwise. Let $\mathbf{x}_j$ denote the column vector $\mathbf{x}_j = (x_{1,j}, x_{2,j}, \cdots, x_{n,j})^T$ for $1 \leq j \leq k$. The quadratic programming formulation is then given by

$$\min_{\mathbf{x}} \; \frac{1}{2} \sum_{j=1}^{k} \mathbf{x}_j^T L \mathbf{x}_j$$

$$\text{s.t.} \; \sum_{j=1}^{k} x_{i,j} = 1, \quad \forall i \in V,$$

$$x_{i,j} \in \{0, 1\}, \quad \forall i \in V, \; 1 \leq j \leq k.$$

Again, for perfect balance GP, we have another set of equality constraints:

$$\mathbf{x}_j^T \mathbb{1} = \left\lceil \frac{n}{k} \right\rceil, \quad 1 \leq j \leq k.$$

For the imbalance case, we have the following inequality constraints:

$$(1 - \epsilon) \left\lceil \frac{n}{k} \right\rceil \leq \mathbf{x}_j^T \mathbb{1} \leq (1 + \epsilon) \left\lceil \frac{n}{k} \right\rceil, \quad 1 \leq j \leq k.$$

QUBO Formulation. To convert the problem into QUBO model, we will need to remove the constraints and add them as penalty terms to the objective function [14]. For example, in the quadratic programming (1) with the equality constraint (2), we obtain the QUBO model as follows:

$$\min_{\mathbf{x}} \; \mathbf{x}^T L \mathbf{x} + P \left(\mathbf{x}^T \mathbb{1} - \left\lceil \frac{n}{2} \right\rceil \right)^2$$

$$\text{s.t.} \; x_i \in \{0, 1\}, \quad \forall i \in V,$$

where $P > 0$ is a postive parameter to penalize the violation of constraint (2). For inequality constraints, we will introduce additional slack variables to first convert the inequality to equality constraints, and then add them as penalty terms to the objective function.

3   Computational Experiments

The goal of the experiments was to identify the class of instances that is more suitable to be solved using the QUBO framework and the current hardware. We compare the performance of DA with exact solver Gurobi [5], and the state-of-the-art multilevel graph partitioning solver KaHIP [17]. We set the time limit for DA and Gurobi to be 15 min. For KaHIP, we use KaFFPaE, a combination of distributed evolutionary algorithm and multilevel algorithm for GP. KaFFPaE computes partitions of very high quality when the imbalance factor $\epsilon > 0$, but does not perform very well for the perfectly balanced case when $\epsilon = 0$. Therefore we also enable a recommended by the developers KaBaPE ran with 24 parallel processes, and the time limit of 30 min.

To evaluate the quality of the solution, we compare the approximation ratio, which is computed using the GP cut found by each solver divided by the best-known value. For some graphs, we have the best-known provided from the benchmark [21], otherwise we use the best results found by the three solvers as the best known. Since this is a minimization problem, the minimum possible value of the approximation ratio is 1, the smaller the better. For each graph and each solver used, we also provide the objective function value, i.e., the number of cut edges. Due to space limitation, we present only the summary of the results. Detailed results are available in [12].

Main Conclusion: We have focused on demonstrating practical advantage of software and hardware approaches for GP. We found that dense graphs exhibit limitations of the existing algorithms which can be improved by the hardware accelerators.

Graph Partitioning on Sparse Graphs. We first test the three solvers on instances from the Walshaw graph partitioning archive [21]. We present the summary of the results with box plots in Fig. 1 (a), (d). We observe that in Fig. 1 (d), where we compare DA and Gurobi, DA can find the best-known partition for most instances, and perform better compared to Gurobi. However, for several sparse graphs, i.e., $d_{avg} < 3$, for example, uk, add32 and 4elt, DA can not find the best-known solutions. For these sparse graphs, multilevel graph partitioning solvers such as KaHIP can usually perform an effective coarsening and uncoarsening procedure based on local structures of the graph and therefore find good solutions quickly. As shown in Fig. 1 (a), KaHIP performs better than DA. Based on the numerical results, we conclude that for the sparse graphs, generic and hardware QUBO solvers do not lead to many practical advantages. However, graphs with more complex structures, that bring practical challenges to the current solvers might benefit from using the QUBO and hardware accelerators.

Graph Partitioning on Dense Graphs. To validate our conjecture, in the next set of experiments, we examine dense graphs from the SuiteSparse Matrix Collection [3] The experimental results are presented in Fig. 1 (b), (e). We observe that for these dense graphs, in general, DA is able to find solutions that are usually at

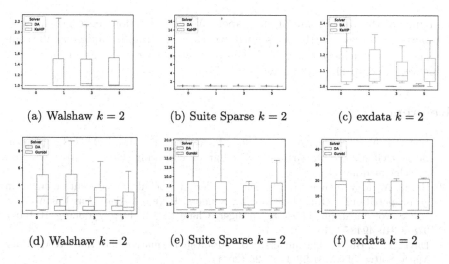

(a) Walshaw $k = 2$ (b) Suite Sparse $k = 2$ (c) exdata $k = 2$

(d) Walshaw $k = 2$ (e) Suite Sparse $k = 2$ (f) exdata $k = 2$

Fig. 1. Comparison of DA with KaHIP (dedicated GP solver), and Gurobi (general-purpose solver) for sparse and dense graphs respectively. The y-axis represents the approximation ratio (solution to best-solution ratio), the minimum possible value of the approximation ratio is 1, the smaller the better. The x-axis represents the imbalance factor as percentage

least as good as those produced by KaHIP and Gurobi. In particular, we find that for one instance, `exdata_1`, KaHIP fails significantly. We therefore use a graph generator MUSKETEER [6] to generate similar instances[1]. The parameters used to generate the instances can be found in the appendix of the full version. In short, MUSKETEER applies perturbation to the original graph with a multilevel approach, the local editing preserves many network properties including different centralities measures, modularity, and clustering. The experiment results are presented in Fig. 1 (c), (f). We find that in most instances, DA outperforms KaHIP and Gurobi, demonstrating that in this class of problems, specialized hardware such as DA is having an advantage.

Currently, to tackle GP on dense graphs, the main practical solution is to first sparsify the graphs (hoping that the sparsified graph still preserves the structure of the original dense graph), solve GP on the sparsified graph, and finally project the obtained solution back to the original graph. We have applied the Forest Fire sparsification [9] available in Networkit [19]. This sparsification is based on random walks. The vertices are burned starting from a random vertex, and fire may spread to the neighbors of a burning vertex. The intuition is that the edges that are visited more often during the random walk are more important in the graph. In our experiments, we eliminate 30% of the edges. Then we solve GP using KaHIP (KaffpaE version) and project the obtained solution back to the original dense graph. Results and details of the experiments can

---

[1] The `exdata` graph files are available here: https://github.com/JoeyXLiu/dense-graph-exdata.

be found in the full version of the paper. We find that for dense graphs with complex structures, KaHIP does not outperform DA, and graph sparsification does not help to achieve this goal. In this case, we advocate the use of the QUBO framework and specialized hardware.

References

1. Bader, D.A., Meyerhenke, H., Sanders, P., Wagner, D.: 10th DIMACS implementation challenge-graph partitioning and graph clustering (2011). https://www.cc.gatech.edu/dimacs10/
2. Buluç, A., Meyerhenke, H., Safro, I., Sanders, P., Schulz, C.: Recent advances in graph partitioning. In: Kliemann, L., Sanders, P. (eds.) Algorithm Engineering. LNCS, vol. 9220, pp. 117–158. Springer, Cham (2016). https://doi.org/10.1007/978-3-319-49487-6_4
3. Davis, T.A., Hu, Y.: The university of Florida sparse matrix collection. ACM Trans. Math. Softw. (TOMS) 38(1), 1–25 (2011)
4. Fujitsu: Fujitsu Digital Annealer (2022). https://www.fujitsu.com/global/services/business-services/digital-annealer/
5. Gurobi Optimization, I.: Gurobi optimizer reference manual (2018). https://www.gurobi.com/
6. Gutfraind, A., Safro, I., Meyers, L.A.: Multiscale network generation. In: 2015 18th International Conference on Information Fusion, pp. 158–165. IEEE (2015)
7. Hager, W.W., Hungerford, J.T., Safro, I.: A multilevel bilinear programming algorithm for the vertex separator problem. Comput. Optim. Appl. 69(1), 189–223 (2017). https://doi.org/10.1007/s10589-017-9945-2
8. Hager, W.W., Krylyuk, Y.: Graph partitioning and continuous quadratic programming. SIAM J. Discret. Math. 12(4), 500–523 (1999)
9. Hamann, M., Lindner, G., Meyerhenke, H., Staudt, C.L., Wagner, D.: Structure-preserving sparsification methods for social networks. Soc. Netw. Anal. Min. 6(1), 1–22 (2016). https://doi.org/10.1007/s13278-016-0332-2
10. Karypis, G., Kumar, V.: A fast and high quality multilevel scheme for partitioning irregular graphs. SIAM J. Sci. Comput. 20(1), 359–392 (1999)
11. Karypis, G., Kumar, V.: Multilevel algorithms for multi-constraint graph partitioning. In: SC 1998: Proceedings of the 1998 ACM/IEEE Conference on Supercomputing, p. 28. IEEE (1998)
12. Liu, X., Ushijima-Mwesigwa, H., Ghosh, I., Safro, I.: Partitioning dense graphs with hardware accelerators. arXiv preprint arXiv:2202.09420 (2022)
13. Liu, X., Ushijima-Mwesigwa, H., Mandal, A., Upadhyay, S., Safro, I., Roy, A.: Leveraging special-purpose hardware for local search heuristics. Computational Optimization and Applications (2022, to appear)
14. Lucas, A.: Ising formulations of many np problems. Front. Phys. 2, 5 (2014)
15. Safro, I., Ron, D., Brandt, A.: Graph minimum linear arrangement by multilevel weighted edge contractions. J. Algorithms 60(1), 24–41 (2006)
16. Safro, I., Sanders, P., Schulz, C.: Advanced coarsening schemes for graph partitioning. ACM J. Exp. Algorithm. (JEA) 19, 2 (2015)
17. Sanders, P., Schulz, C.: Think locally, act globally: highly balanced graph partitioning. In: Bonifaci, V., Demetrescu, C., Marchetti-Spaccamela, A. (eds.) SEA 2013. LNCS, vol. 7933, pp. 164–175. Springer, Heidelberg (2013). https://doi.org/10.1007/978-3-642-38527-8_16

18. Shaydulin, R., Chen, J., Safro, I.: Relaxation-based coarsening for multilevel hypergraph partitioning. SIAM Multiscale Model. Simul. **17**, 482–506 (2019)
19. Staudt, C.L., Sazonovs, A., Meyerhenke, H.: NetworKit: a tool suite for large-scale complex network analysis. Netw. Sci. **4**(4), 508–530 (2016)
20. Ushijima-Mwesigwa, H., Negre, C.F., Mniszewski, S.M.: Graph partitioning using quantum annealing on the D-Wave system. In: Proceedings of the Second International Workshop on Post Moores Era Supercomputing, pp. 22–29 (2017)
21. Walshaw, C.: The graph partitioning archive (2009). https://chriswalshaw.co.uk/partition/

A Taxonomy Guided Method to Identify Metaheuristic Components

Thimershen Achary$^{(\boxtimes)}$ and Anban W. Pillay

University of KwaZulu-Natal, Durban, South Africa
thimershenzn@gmail.com

Abstract. A component-based view of metaheuristics has recently been promoted to deal with several problems in the field of metaheuristic research. These problems include inconsistent metaphor usage, non-standard terminology and a proliferation of metaheuristics that are often insignificant variations on a theme. These problems make the identification of novel metaheuristics, performance-based comparisons, and selection of metaheuristics difficult. The central problem for the component-based view is the identification of components of a metaheuristic. This paper proposes the use of taxonomies to guide the identification of metaheuristic components. We developed a general and rigorous method, TAXONOG-IMC, that takes as input an appropriate taxonomy and guides the user to identify components. The method is described in detail, an example application of the method is given, and an analysis of its usefulness is provided. The analysis shows that the method is effective and provides insights that are not possible without the proper identification of the components.

Keywords: Metaheuristic · General metaheuristic · Taxonomy

1 Introduction

The metaheuristic research field has been criticized for inconsistent metaphor usage, non-standard terminology [1, 2], and use of poor experimental setups, validation, and comparisons [1–3]. These factors have contributed to challenges in the field such as a proliferation of novel metaheuristics and 'novel' approaches being very similar to existing approaches [1, 2, 4]. Several researchers have thus proposed that a component-based view of metaheuristics that explicitly lists metaheuristic components, will assist in identifying novel components [1, 5], promote component-based performance comparison and analyses, and facilitate component-wise selection of metaheuristics for comparative studies [1, 2, 6, 7].

A component-based view is especially important for general metaheuristics, which has enjoyed increasing popularity in recent literature. General metaheuristics, also known as general metaheuristic frameworks [8], unified metaheuristic frameworks [9], and generalized metaheuristic models [10] are used for tasks such as metaheuristic generation [10], performance analysis [11, 12], metaheuristic-similarity analysis [13], and classification of metaheuristics [7]. General metaheuristics are an abstraction of a set of

D. Groen et al. (Eds.): ICCS 2022, LNCS 13352, pp. 484–496, 2022.
https://doi.org/10.1007/978-3-031-08757-8_41

metaheuristics, i.e., they are generalizations of the components, structure, and information utilized by a set of metaheuristics [6, 12]. They thus also take a component-based view. General metaheuristics make use of a set of component-types, also referred to as general metaheuristics structures [12], component-categories [6], main ingredients [14], or key components [15].

However, general metaheuristics still suffer the challenges outlined above viz. inconsistent metaphor usage and non-standard terminology. They also suffer from similar problems if components are not properly identified. Thus, the identification of components takes on special importance.

This work promotes the systematic use of taxonomies to guide the identification of components. Our proposed method uses formal taxonomy theory, which appears to be absent in several recent metaheuristic studies that involve the creation or incorporation of taxonomies such as [7, 16–19]. Taxonomies, ideally, are built using a rigorous taxonomy building-method e.g. [20, 21]. Taxonomies are intrinsic prerequisites to understanding a given domain, differentiating between objects, and facilitating discussion on the state and direction of research in a domain [22]. Taxonomies may thus help solve the issues affecting metaheuristic research, such as non-standard terminology and nomenclature.

This work proposes the use of taxonomies to guide the identification of metaheuristic components. We developed a general and rigorous method, TAXONOG-IMC, that takes as input an appropriate taxonomy and guides the user to identify components. TAXONOG-IMC promotes the use of taxonomies to guide component identification for any metaheuristic subset, and provides guidance for the proper use of taxonomies to perform component identification.

This paper presents the method, provides an example of its application, and gives an analysis of its usefulness. The rest of the paper is structured as follows: Sect. 2 provides a literature review, Sect. 3 comprehensively describes TAXONOG-IMC, Sect. 4 demonstrates the use of the method by applying it to two taxonomies to showcase its effectiveness, Sect. 5 provides an analysis of the method by showing its effectiveness in analysing nature-inspired, population-based metaheuristics. Section 6 concludes the study.

2   Literature Review

The need for a component-based view is best appreciated in general metaheuristics. However, many general metaheuristics lack a rigorous method for identifying components. Many studies proposing a general metaheuristic provide guidance through examples of their usage. Several broad-scoped general metaheuristics follow this trend, such as general metaheuristics for population-based metaheuristics [9] and metaheuristics in general [10, 11, 13]. The general metaheuristics proposed by [6, 9, 10, 13] use mathematical formulations for their component-types. Since these mathematical formulations are sometimes in-part derived from text, the researcher can choose how to formulate a component based on their judgement and interpretation. However, this process can be negatively impacted by inconsistent metaphor usage and non-standard terminology. Components that are essentially the same can be regarded as different. Using examples for guidance may not account for all contingencies.

A general metaheuristic built on the assumption that differentiating the components in detail and using relatable terminology may help resolve challenges in component identification, is presented in [12]. However, most of their component-types of the general metaheuristic were a renaming of the components in [13] and may consequently face the same challenges. Some component-categories in literature were listed, but using them for the general metaheuristic may be difficult; if they consist of combinations of components, then they themselves need to be decomposed, which requires expert knowledge.

Several studies used taxonomies and/or classification-schemes to support the design of general metaheuristics. The advantage of using a taxonomy for this purpose is that it declares a convention by which the components will be identified. It provides a list of possible components that a component-type encompasses. If an issue is taken with the convention, then it can be argued at the taxonomy level. There are studies, such as [23, 24], that propose general metaheuristics whose components make use of a presented taxonomy, and there are studies that make use of existing taxonomies for a proposed general metaheuristic, such as [7, 15]. The studies that proposed both a general metaheuristic and a taxonomy are likely to work well, as the taxonomy is built for the general metaheuristic; however, taxonomies are not necessarily built with general metaheuristics in mind.

Works that use existing taxonomies lack guidance on how to use taxonomies effectively. Existing taxonomies and viewpoints were used in [15] to create a new taxonomy to guide the usage of a proposed general metaheuristic. The taxonomy presented used examples at the lowest level of its hierarchy to illustrate its usage. However, examples do not account for every contingency. The essence of the multi-level classification method proposed in [7] is meritorious; however, a misuse of the behaviour taxonomy presented in [5], led to a classification that is questionable in terms of the taxonomy used, i.e., tabu search is depicted as possessing the differential vector movement behaviour. Some studies consider tabu search as population-based but viewing tabu search as being single-solution based has a stronger consensus [25] and appears to be followed by [5], i.e., the behaviour taxonomy presented by [5] is not applicable to tabu search in its canonical sense.

The study in [14] presents a taxonomy for evolutionary algorithms based on their main components. The same study uses the taxonomy to facilitate the expression of evolutionary algorithms in terms of their main components, and the distinguishing between various evolutionary algorithm classes. This study is notable for its use of a vector representation for its components. Our work uses a similar representation.

3   Taxonomy Guided Identification of Metaheuristic Components: TAXONOG-IMC

This section proposes TAXONOG-IMC (see Fig. 1), a general, rigorous method that guides the identification of metaheuristic components using taxonomies.

We use the definition of a taxonomy provided in [20] that lends itself to a flat representation of the metaheuristics or metaheuristic component-types, which facilitates tabular analysis. A taxonomy T is formally defined in [20] as:

$$T = \{D_i, (i = 1, \ldots, n) | D_i = \{C_{ij}, (j = 1, \ldots, k_i); k_i \geq 2\}\} \tag{1}$$

where T is an arbitrary taxonomy, D_i is an arbitrary dimension of T, $k_i \geq 2$ is the number of possible characteristics for dimension D_i, C_{ij} an arbitrary characteristic for dimension D_i. Characteristics for every dimension are mutually exclusive and collectively exhaustive, i.e., each object under consideration must have one and only one C_{ij} for every D_i. This organization, using dimensions and characteristics, is likely to be relevant in all cases since they are fundamental to understanding the properties of objects in a domain; hence the definition (1) is used.

Some important terms concerning taxonomies are explained below:

1. Dimensions: A dimension represents some attribute of an object and can be thought of as a variable that has a set of possible values.
2. Characteristics: The characteristics of a given dimension are the possible values that can be assigned to a particular dimension.
3. Taxonomy dimension: A taxonomy dimension refers to a dimension that is part of the taxonomy under consideration. The method has steps where dimensions are proposed – these are not part of the taxonomy but are under consideration to be included. We refer to these as candidate dimensions that may then become part of the taxonomy.
4. Specialized dimension: A specialized dimension is a characteristic of a taxonomy that is promoted to dimension status; specialized dimensions are candidate dimensions.
5. Generalized dimension: A generalized dimension is created by partitioning characteristics of a taxonomy dimension or partitioning the combination of characteristics from multiple taxonomy dimensions. A generalized dimension is a candidate dimension.

To illustrate each term, consider the following dimensions of some metaheuristic: initializer, search operator, and selection. Characteristics of search operator may be, e.g., genetic crossover, swarm dynamic, differential mutation. A taxonomy for evolutionary algorithms in [14] has population, structured population, information sources etc., as its dimensions. Then population would be a taxonomy dimension. Using the behaviour taxonomy presented in [5], solution creation can be thought of as a generalized dimension of the combination and stigmergy dimensions. If we use solution-creation as a taxonomy dimension, then combination would be a specialized dimension.

3.1 Comprehensive Description of Method Process

A good start for step 1 (select or create a taxonomy), is to conduct a literature search for relevant taxonomies using keywords, key-phrases, publication titles, etc. However, if no appropriate taxonomy is found, then an appropriate taxonomy building method should be used to create a taxonomy.

Expressing a taxonomy using definition (1), ensures the taxonomy is in a standard format for subsequent steps. The dimensions, and the dimensions' characteristics must be clearly stated to avoid ambiguity.

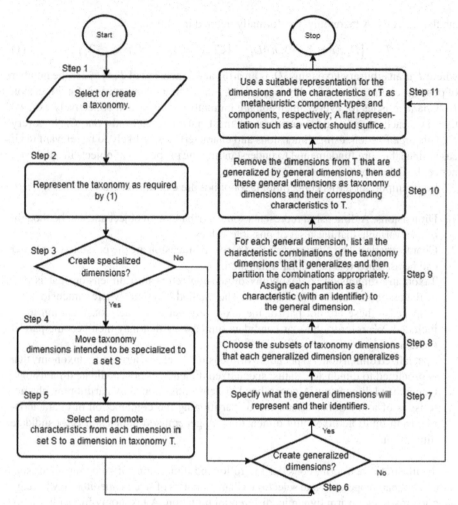

Fig. 1. Flowchart depicting the processes of TAXONOG-IMC

Steps 3 to 5 guides the creation of specialized dimensions. Using specialized dimensions will allow for focusing on specific components. The role of set S, introduced in step 4, is to store a collection of dimensions that are to be replaced by one of their characteristics in taxonomy T. In the metaheuristic context, a dimension may be replaced by more than one of its characteristics; this decision accommodates for hybrid-metaheuristics that have more than one characteristic for a dimension. When characteristics become dimensions, they will each need a set of possible characteristics of their own that will be derived from literature or the expertise of the researcher.

The addition of specialized dimensions to the Taxonomy may result in an overwhelmingly large number of taxonomy dimensions. Generalizing an appropriate number of taxonomy dimensions may help with this challenge.

Creating generalized dimensions is guided by steps 7 to 10. It is essential to name the general dimensions clearly and their characteristics to ensure no ambiguities nor confusion arises as to which dimension or characteristic a trait falls under. It is important to note that each subset of taxonomy dimensions, chosen in step 8, must be disjoint. Note that not every taxonomy dimension needs to be integrated into a general dimension.

As an example of when and how general dimensions can be used, consider a chosen set of metaheuristics that have a large diversity on certain taxonomy dimensions. They may be grouped by their characteristic combinations on these dimensions. A generalized dimension could then have two possible values, 1 representing a metaheuristic having a required combination of characteristics for those dimensions, and 0 representing a metaheuristic not having such a combination of characteristics for those dimensions.

4 Application of Method

To demonstrate the method, we use it to generate binary component vectors to represent nature-inspired, population-based metaheuristics in terms of their inspiration and behaviour components. We use the behaviour and natural-inspiration taxonomies provided in [5]. In this study, we consider the metaphor/inspiration of a metaheuristic to be a component, but more specifically, a non-functional component. The nature-inspiration taxonomy was created to ascertain the natural-inspiration category of a metaheuristic without ambiguity. The behavioural taxonomy is based on the metaheuristic behaviour, i.e., focusing on the means by which new candidate solutions are obtained, and disregarding its natural inspiration. See Sect. 4.3 for descriptions of all dimensions used by the behaviour and natural-inspiration taxonomies.

4.1 Behavior Taxonomy

- Step 1: We use the behavior taxonomy from [5].
- Step 2: We express the taxonomy using the definition given in (1) as follows. A characteristic of 1 means that it is present and 0 means it is not.

  - b_1 - Combination (characteristics are $\{0, 1\}$)
  - b_2 - Stigmergy (characteristics are $\{0; 1\}$)
  - b_3 - All population Differential Vector Movement (DVM) (characteristics are $\{0; 1\}$)
  - b_4 - Groups-based (DVM) (characteristics are $\{0; 1\}$)
  - b_5 - Representative based (DVM) (characteristics are $\{0; 1\}$)

- Step 3: We create specialized dimensions.
- Step 4: S = {Groups-based (DVM)}, The step at this phase dictates that we only select one characteristic to promote to dimension status, but with regards to metaheuristics, which can be hybridized and still be metaheuristics, an exception can be made such that numerous characteristics can be promoted during specialization (this depends on the characteristics, if the characteristics are single-solution and population-based then these can't both be used as component-types for a metaheuristic at the same time, since

there is a possibility that both can be set to 1, which does not make intuitive sense). Therefore, we promote both Sub-population (DVM) and Neighborhood (DVM) to dimensions with their characteristics being binary $\{0; 1\}$. b_4 is set to Sub-population (DVM) and b_5 is set to Neighborhood (DVM), b_6 is set to Representative based (DVM).

- Step 5: Groups-based (DVM) is not referenced by any dimension and can thus be discarded. $T = \{b_1; b_2; b_3; b_4; b_5; b_6 \mid b_i = \{0; 1\}; (i = 1, 2, 3, 4, 5, 6)\}$
- Step 6: We do not create generalized dimension.
- Step 11: The vector representation derived from the behavoiur taxonomy is:

$$\begin{bmatrix} b_1 & b_2 & b_3 & b_4 & b_5 & b_6 \end{bmatrix} \qquad (2)$$

4.2 Natural-Inspiration Taxonomy

- Step 1: We use the natural-inspiration taxonomy from [5].
- Step 2: We express the taxonomy using the definition given in (1) as follows:

  - n_1 - Breeding-based evolution (characteristics are $\{0; 1\}$)
  - n_2 - Aquatic animals (characteristics are $\{0; 1\}$)
  - n_3 - Terrestrial animals (characteristics are $\{0; 1\}$)
  - n_4 - Flying animals (characteristics are $\{0; 1\}$)
  - n_5 - Microorganisms (characteristics are $\{0; 1\}$)
  - n_6 - Others (characteristics are $\{0; 1\}$)
  - n_7 - Physics-based (characteristics are $\{0; 1\}$)
  - n_8 - Chemistry-based (characteristics are $\{0; 1\}$)
  - n_9 - Social human behaviour algorithms (characteristics are $\{0; 1\}$)
  - n_{10} - Plants based (characteristics are $\{0; 1\}$)
  - n_{11} - Miscellaneous (characteristics are $\{0; 1\}$)

- Step 3: We do not create specialized dimensions.
- Step 6: We create general dimensions.
- Step 7: We create two general dimensions that will be identified as Swarm-intelligence and Physics and Chemistry Based. (This is already done in the taxonomy, but we are redoing it in this process for demonstration).
- Step 8: Aquatic animals, Terrestrial animals, Flying animals, Microorganisms, Others are allocated to the Swarm-intelligence general dimension. Physics-based, Chemistry-based are allocated to the Physics and Chemistry Based general dimension.
- Step 9: The characteristics of Swarm-intelligence are $\{0; 1\}$. 1 indicating that either Aquatic animals, Terrestrial animals, Flying animals, Microorganisms, or Others are present, 0 indicating that Aquatic animals, Terrestrial animals, Flying animals, Microorganisms, and Others are absent. The characteristics of Physics and Chemistry Based are $\{0; 1\}$. 1 indicating that either Physics-based or Chemistry-based is 1, 0 indicating that Physics-based and Chemistry-based are absent.
- Step 10: Since n_2 to n_8 are removed, n_2 will be the dimension for Swarm-intelligence, n_3 will be the dimension for Physics and Chemistry Based, n_4 will be the dimension for Social human behavior algorithms, n_5 will be the dimension for Plants based, n_6

will be the dimension for Miscellaneous; n_7 to n_{11} do not refer to any dimensions so they can be discarded. $T = \{n_1; n_2; n_3; n_4; n_5; n_6 \mid n_i = \{0; 1\}, (i = 1, 2, 3, 4, 5, 6)\}$
- Step 11: The vector representation definition derived from the selected taxonomy is:

$$\begin{bmatrix} n_1 & n_2 & n_3 & n_4 & n_5 & n_6 \end{bmatrix}$$ (3)

4.3 Dimension Descriptions

In this sub-section, the nodes of each hierarchal taxonomy presented in [5] are unambiguously defined as dimensions using the descriptions of each node provided in the same study; from these definitions, we can define the dimensions in the initial steps and proceed to modify them in subsequent steps by adding and/or dropping these dimensions due to using generalized or specialized dimensions.

Behaviour Dimensions

- Differential vector movement: New solution is obtained by movement relative to an existing solution
- All population Differential Vector Movement (DVM): All individuals in the population are used to generate the movement of each solution.
- Representative-based (DVM): The movements of each solution are only influenced by a small group of representative solutions, e.g., the best solutions found
- Group-based (DVM): Sub-populations or subsets of the populations are considered, without representative solutions.
- Sub-population (DVM): The movements of each solution are influenced by a subset or group of solutions in the population, and no representative solutions are determined and used in the trajectory calculation at hand.
- Neighborhood (DVM): Each solution is only influenced by solutions in its local neighborhood.
- Combination: New solutions are selected and combined via some method to create new solutions.
- Stigmergy: An indirect communication and coordination strategy is used between different solutions to create new solutions.
- Creation: Exploration of search domain by generating new solution, differential vector movement not present.

Natural-Inspiration Dimensions

- Breeding-based evolution: Inspired by the principle of natural evolution and references to producing offspring, successive generations.
- Swarm Intelligence: Inspired by the collective behavior of animal societies.
- Flying animals: Agent movements inspired by flying movements.
- Terrestrial animals: Agent movements inspired by foraging or movements of terrestrial animals.

- Aquatic animals: Agent movements inspired by animals living in aquatic ecosystems.
- Microorganisms: Agent movements inspired by food search by bacteria or how viruses spread infection.
- Others: Very low popularity inspiration sources from the collective behavior of animals.
- Physics and Chemistry Based: Imitate the behavior of physical/chemical phenomena (field of physics and chemistry).
- Social Human Behavior Algorithms: Inspired by human social concepts.
- Plants Based: Inspired by plants, where there is no communication between agents.
- Miscellaneous: Not inspired by any identified category.

5 Analysis and Discussion

We now demonstrate the use of the method. Information showing the application frequency of different nature-inspired metaheuristics to feature selection in disease diagnosis is depicted in Table 10 taken from the study in [26]. It is stated that data for the table was obtained by executing various search queries on google scholar. RA is not population-based, and thus is ignored since it is out of scope for the vector derived in the current paper. In this section, the amount of information extracted from Table 10 in [26] is extended using the derived vector. The aim is to reconfigure the table to attribute the frequencies to the component-types of the derived vector. This task is accomplished via the following steps:

1. List all metaheuristic abbreviations and ascertain their full name.
2. Represent each of the nature-inspired, population-based metaheuristics using the vector formats derived, i.e., (2) and (3), as shown in Table 1. If the metaheuristics were not present in the tables, the descriptions of the dimensions of the taxonomies presented in [5] would have to be used to derive their vector representation.
3. Let B be a matrix representing the data of Table 1, i.e., $B[p][q]$ will indicate whether the component-type at column index q is present in the metaheuristic at row index p. Let D be a matrix where each intersection of row i and column j is the frequency of application of metaheuristic at row index i to the disease at column index j (D holds the data of Table 10 in [26]). Let F be the matrix that holds the component-type to disease diagnosis application frequencies (Table 2), i.e., where j is index number of the disease in the columns of Table 10 presented in [26] and q is the index number of the component-type in the vector:

$$F[j][q] = \sum_{x=0}^{N} B[x][q] \times D[x][j] \qquad (4)$$

4. Matrix F contains the data of Table 2 that depicts the table of frequency of appli cation of a component-type to disease diagnosis. From this table, further analysis can be done.

It can be observed from Table 2 that b_6 (Representative-based (DVM)) is the dominant behaviour and n_2 (Swarm intelligence) is the dominant natural-inspiration. It is

Table 1. Representation of nature-inspired, population-based metaheuristics in terms of derived vector formats

KEY: Harmony search (HS), Artificial bee colony (ABC), Glow-worm swarm optimization (GSO), Ant colony optimization (ACO), Firefly algorithm (FA), Monkey algorithm (MA), Cuckoo search (CS), Bat algorithm (BA), Dolphin echolocation (DE), Flower pollination algorithm (FPA), Grey wolf optimizer (GWO), Dragonfly algorithm (DA), Krill herd algorithm (KHA), Elephant search algorithm (ESA), Ant lion optimizer (ALO), Moth-flame optimization (MFO), Multi-verse optimizer (MVO), Runner-root algorithm (RRA), Laying chicken algorithm (LCA), Killer whale algorithm (KWA), Butterfly optimization algorithm (BOA).

PMBH	b1	b2	b3	b4	b5	b6	n1	n2	n3	n4	n5	n6
HS	1	0	0	0	0	0	0	0	1	0	0	0
ABC	0	0	0	0	0	1	0	1	0	0	0	0
GSO	0	0	0	0	0	1	0	1	0	0	0	0
ACO	0	1	0	0	0	0	0	1	0	0	0	0
FA	0	0	1	0	0	0	0	1	0	0	0	0
MA	0	0	0	0	0	1	0	1	0	0	0	0
CS	1	0	0	0	0	0	0	1	0	0	0	0
BA	0	0	0	0	0	1	0	1	0	0	0	0
DE	1	0	0	0	0	0	0	1	0	0	0	0
FPA	0	0	0	0	0	1	0	0	0	0	1	0
GWO	0	0	0	0	0	1	0	1	0	0	0	0
DA	0	0	0	0	0	1	0	1	0	0	0	0
KHA	0	0	0	0	0	1	0	1	0	0	0	0
ESA	0	0	0	0	0	1	0	1	0	0	0	0
ALO	0	0	0	0	0	1	0	1	0	0	0	0
MFO	0	0	0	0	0	1	0	1	0	0	0	0
MVO	0	0	0	0	0	1	0	0	1	0	0	0
RRA	0	0	0	0	0	1	0	0	0	0	1	0
LCA	1	0	0	0	0	0	0	1	0	0	0	0
KWA	0	0	0	0	0	1	0	1	0	0	0	0
BOA	0	0	0	0	0	1	0	1	0	0	0	0

interesting to note that in [26], it is stated that ACO is dominant in the use of diagnosis of different human disorders. However, the behaviour associated with ACO is Stigmergy (b_2) is not the dominant behaviour; instead, representative-based differential movement (b_6) is the dominant behaviour for this application domain.

Literature such as [1] has shown that the names and metaphors of metaheuristics sometimes mask the substantial similarities between the metaheuristics and their differences are so minute that they can be considered marginal variants. ACO is popular, but the problem could lie with many metaheuristics, which have behavioural component-type b_6, being diverse in names as this trend is either diluting the core algorithm's popularity or is misguiding users to believe that different metaheuristic names entail that they have nearly orthogonal behaviours.

From Table 2, it can be ascertained that scope for future research lies in applying metaheuristics with behavioural component-types: sub-population (DVM), neighbourhood

Table 2. Frequencies of component-type usage, in literature, in various disease diagnosis applications

Disease diagnosis	b_1	b_2	b_3	b_4	b_5	b_6	n_1	n_2	n_3	n_4	n_5	n_6
Breast cancer	413	619	216	0	0	893	0	1859	236	0	46	0
Prostate cancer	35	73	9	0	0	68	0	161	21	0	3	0
Lung cancer	105	157	41	0	0	154	0	400	51	0	6	0
Oral cancer	4	3	2	0	0	6	0	12	3	0	0	0
Neck cancer	4	4	0	0	0	9	0	13	3	0	1	0
Skin cancer	19	4	15	0	0	53	0	81	8	0	2	0
HIV	40	114	24	0	0	80	0	237	18	0	3	0
Stroke	116	120	36	0	0	129	0	330	60	0	11	0
Schizophrenia	8	44	9	0	0	16	0	72	4	0	1	0
Parkinson	91	144	52	0	0	233	0	434	62	0	24	0
Heart disease	129	34	58	0	0	234	0	390	55	0	10	0
Anxiety	17	65	9	0	0	50	0	135	5	0	1	0
Insomnia	1	6	0	0	0	2	0	9	0	0	0	0
Sum	982	1387	471	0	0	1927	0	4133	526	0	108	0

(DVM), breeding-based evolution, social-human behaviour algorithms, and miscellaneous to disease diagnosis. Even though the three latter component-types are natural-inspirations, and literature has motivated that this category of component-types has little contribution to performance. Applying them increases their presence in a population, from which data can be sampled, i.e., a diverse population is good.

The taxonomies in [5] organized the metaheuristics using their canonical versions. This study relies on the assumption that if two or more metaheuristic-algorithms are associated with the same metaheuristic, then they should possess the behaviour of that metaheuristic. The proposed method can be used to select components for metaheuristic frameworks, classification schemes, representations, and comparative analysis.

6 Conclusion

This study proposes TAXONOG-IMC, a structured method that provides guidance for metaheuristic component identification using taxonomies. An example application is provided to showcase how TAXONOG-IMC can aid in metaheuristic analysis.

Identification of metaheuristic components is an important task for the effective use of general metaheuristics, and the metaheuristic component-based view by and large. General metaheuristic publications use strategies such as providing examples, using finer-grain component-types, relying on existing taxonomies or creating new ones to assist in component identification. However, examples don't account for all contingencies that a researcher may encounter, and finer-grain components can also be affected by non-standard terminology and inconsistent metaphor usage. There are general meta-heuristic publications that use taxonomies to assist in component identification; some propose their own taxonomy, and others use an existing taxonomy. The ones that propose their own taxonomy are likely to be compatible with the general metaheuristic since they are created for that purpose; however, some of the publications that use existing taxonomies made questionable decisions during the demonstration of general metaheuristic use – indicating a lack of proper use of taxonomy.

Future research lies in using taxonomies for component-identification for many other metaheuristic subsets, metaheuristics analysis, and use in general metaheuristics.

References

1. Sörensen, K.: Metaheuristics-the metaphor exposed. Int. Trans. Oper. Res. **22**, 3–18 (2015). https://doi.org/10.1111/itor.12001
2. Aranha, C., et al.: Metaphor-based metaheuristics, a call for action: the elephant in the room. Swarm Intell. (2021). https://doi.org/10.1007/s11721-021-00202-9
3. García-Martínez, C., Gutiérrez, P.D., Molina, D., Lozano, M., Herrera, F.: Since CEC 2005 competition on real-parameter optimisation: a decade of research, progress and comparative analysis's weakness. Soft. Comput. **21**(19), 5573–5583 (2017). https://doi.org/10.1007/s00 500-016-2471-9
4. Tzanetos, A., Dounias, G.: Nature inspired optimization algorithms or simply variations of metaheuristics? Artif. Intell. Rev. **54**(3), 1841–1862 (2020). https://doi.org/10.1007/s10462-020-09893-8
5. Molina, D., Poyatos, J., Ser, J.D., García, S., Hussain, A., Herrera, F.: Comprehensive tax-onomies of nature- and bio-inspired optimization: inspiration versus algorithmic behavior, critical analysis recommendations. Cogn. Comput. **12**(5), 897–939 (2020). https://doi.org/ 10.1007/s12559-020-09730-8
6. Peres, F., Castelli, M.: Combinatorial optimization problems and metaheuristics: review, chal-lenges, design, and development. Appl. Sci. **11**, 6449 (2021). https://doi.org/10.3390/app111 46449
7. Stegherr, H., Heider, M., Hähner, J.: Classifying Metaheuristics: towards a unified multi-level classification system. Natural Comput. (2020). https://doi.org/10.1007/s11047-020-09824-0.
8. Birattari, M., Paquete, L., Stützle, T.: Classification of metaheuristics and design of experiments for the analysis of components (2003)
9. Liu, B., Wang, L., Liu, Y., Wang, S.: A unified framework for population-based metaheuristics. Ann. Oper. Res. **186**, 231–262 (2011). https://doi.org/10.1007/s10479-011-0894-3
10. Cruz-Duarte, J.M., Ortiz-Bayliss, J.C., Amaya, I., Shi, Y., Terashima-Marín, H., Pillay, N.: Towards a generalised metaheuristic model for continuous optimisation problems. Mathematics **8**, 2046 (2020). https://doi.org/10.3390/math8112046
11. De Araujo Pessoa, L.F., Wagner, C., Hellingrath, B., Buarque De Lima Neto, F.: Component analysis based approach to support the design of meta-heuristics for MLCLSP providing guidelines. In: 2015 IEEE Symposium Series on Computational Intelligence, pp. 1029–1038. IEEE, Cape Town (2015). https://doi.org/10.1109/SSCI.2015.149
12. Stegherr, H., Heider, M., Luley, L., Hähner, J.: Design of large-scale metaheuristic com-ponent studies. In: Proceedings of the Genetic and Evolutionary Computation Conference Companion, pp. 1217–1226. ACM, Lille France (2021). https://doi.org/10.1145/3449726. 3463168

13. de Armas, J., Lalla-Ruiz, E., Tilahun, S.L., Voß, S.: Similarity in metaheuristics: a gentle step towards a comparison methodology. Natural Comput. (2021). https://doi.org/10.1007/s11047-020-09837-9
14. Calégari, P., Coray, G., Hertz, A., Kobler, D., Kuonen, P.: A taxonomy of evolutionary algorithms in combinatorial optimization. J. Heuristics 5, 145–158 (1999). https://doi.org/10.1023/A:1009625526657
15. Raidl, G.R.: A unified view on hybrid metaheuristics. In: Almeida, F., et al. (eds.) HM 2006. LNCS, vol. 4030, pp. 1–12. Springer, Heidelberg (2006). https://doi.org/10.1007/11890584_1
16. Kaviarasan, R., Amuthan, A.: Survey on analysis of meta-heuristic optimization methodologies for node network environment. In: 2019 International Conference on Computer Communication and Informatics (ICCCI), pp. 1–4. IEEE, Coimbatore, Tamil Nadu, India (2019). https://doi.org/10.1109/ICCCI.2019.8821838
17. Fister, I., Perc, M., Kamal, S.M., Fister, I.: A review of chaos-based firefly algorithms: perspectives and research challenges. Appl. Math. Comput. 252, 155–165 (2015). https://doi.org/10.1016/j.amc.2014.12.006
18. Diao, R., Shen, Q.: Nature inspired feature selection meta-heuristics. Artif. Intell. Rev. 44(3), 311–340 (2015). https://doi.org/10.1007/s10462-015-9428-8
19. Donyagard Vahed, N., Ghobaei-Arani, M., Souri, A.: Multiobjective virtual machine placement mechanisms using nature-inspired metaheuristic algorithms in cloud environments: a comprehensive review. Int J Commun Syst. 32, e4068 (2019). https://doi.org/10.1002/dac.4068
20. Nickerson, R.C., Varshney, U., Muntermann, J.: A method for taxonomy development and its application in information systems. Eur. J. Inf. Syst. 22, 336–359 (2013). https://doi.org/10.1057/ejis.2012.26
21. Usman, M., Britto, R., Börstler, J., Mendes, E.: Taxonomies in software engineering: a systematic mapping study and a revised taxonomy development method. Inf. Softw. Technol. 85, 43–59 (2017). https://doi.org/10.1016/j.infsof.2017.01.006
22. Szopinski, D., Schoormann, T., Kundisch, D.: because your taxonomy is worth it: towards a framework for taxonomy evaluation. Research Papers (2019)
23. Krasnogor, N., Smith, J.: A tutorial for competent memetic algorithms: model, taxonomy, and design issues. IEEE Trans. Evol. Comput. 9, 474–488 (2005). https://doi.org/10.1109/TEVC.2005.850260
24. Stork, J., Eiben, A.E., Bartz-Beielstein, T.: A new taxonomy of global optimization algorithms. Natural Comput. (2020). https://doi.org/10.1007/s11047-020-09820-4
25. Glover, F., Laguna, M.: Tabu search background. In: Tabu Search, pp. 1–24. Springer US, Boston, MA (1997). https://doi.org/10.1007/978-1-4615-6089-0_1
26. Sharma, M., Kaur, P.: A comprehensive analysis of nature-inspired meta-heuristic techniques for feature selection problem. Archives Comput. Methods Eng. 28(3), 1103–1127 (2020). https://doi.org/10.1007/s11831-020-09412-6

Camp Location Selection in Humanitarian Logistics: A Multiobjective Simulation Optimization Approach

Yani Xue[1], Miqing Li[2], Hamid Arabnejad[1], Diana Suleimenova[1], Alireza Jahani[1], Bernhard C. Geiger[3], Zidong Wang[1], Xiaohui Liu[1], and Derek Groen[1(✉)]

[1] Department of Computer Science, Brunel University London, London, UK
Derek.Groen@brunel.ac.uk
[2] School of Computer Science, University of Birmingham, Birmingham, UK
[3] Know-Center GmbH, Graz, Austria

Abstract. In the context of humanitarian support for forcibly displaced persons, camps play an important role in protecting people and ensuring their survival and health. A challenge in this regard is to find optimal locations for establishing a new asylum-seeker/unrecognized refugee or IDPs (internally displaced persons) camp. In this paper we formulate this problem as an instantiation of the well-known facility location problem (FLP) with three objectives to be optimized. In particular, we show that AI techniques and migration simulations can be used to provide decision support on camp placement.

Keywords: Facility location problem · Multiobjective optimization · Simulation · Evolutionary algorithms

1 Introduction

Forced displacement is a complex global phenomenon, which refers to the movement of people away from their home or origin countries due to many factors, such as conflict, violence, persecution, etc. In 2020, almost 26.4 million people had fled their countries according to the UNHCR (https://www.unhcr.org/uk/figures-at-a-glance.html). In this situation, relocating asylum-seekers/unrecognized refugees to camps becomes an urgent issue to humanitarian organizations or governments. Camps, as important infrastructures, provide protection and allocate available humanitarian resources to thousands of forcibly displaced people. As resources are commonly limited, it is critical to make optimal decisions in seeking the best location for establishing a new camp. Camp placement can be formulated as the well-known facility location problem (FLP) [6]. The FLP can be considered as a multiobjective optimization problem (MOP), which includes two or more objectives to be optimized simultaneously. The objectives of the FLP can include minimizing the total travel distance and maximizing the demand coverage, meanwhile satisfying some constraints [8].

© The Author(s), under exclusive license to Springer Nature Switzerland AG 2022
D. Groen et al. (Eds.): ICCS 2022, LNCS 13352, pp. 497–504, 2022.
https://doi.org/10.1007/978-3-031-08757-8_42

Several MOP-FLP approaches have been proposed, including traditional goal programming, ϵ-constraint approaches and, more recently, metaheuristic optimization algorithms [13] such as particle swarm optimization (PSO) and evolutionary algorithm (EA). As a population-based metaheuristic optimization approach, EA may effectively handle MOPs as it can generate a set of trade-off solutions in a single run. It has specifically been applied to tackle the FLP in disaster emergency management [14], making it natural to employ EA in the context of camp placement. The main challenge here is to have exact number of forcibly displaced persons arriving in destination countries. Due to the ongoing conflicts in origin countries, the number of asylum-seekers/unrecognized refugees or IDPs continuously changes over time.

Here we aim to assist humanitarian organizations and governments in their decision-making on camp placement, and the paper has the following contributions: (1) we present an MOP for camp placement with three objectives regarding travel distance, demand coverage, and idle camp capacity; (2) we use an agent-based simulation to capture the demand uncertainty (i.e., the number of camp arrivals), which is crucial for camp placement but has not been considered in most existing literature; (3) we present a new multiobjective simulation optimization approach for our MOP, which consists of EA and an agent-based forced migration simulation; and (4) we successfully apply the proposed approach to a case study of the South Sudan conflict, and identify a group of optimal solutions for decision-makers.

1.1   Related Work

The camp location selection problem is a complex task for the humanitarian organizations to deploy aid. The research areas related to this problem can be generally divided into the modelling the movements of people [11], and the FLP in humanitarian logistics [1,4,7,9]. Here we attempt to address the optimization problem of how to find the optimal locations for establishing a new camp. This problem can be formulated as an MOP. Current approaches for multiobjective FLPs can be classified into two categories. The first is concerned with the traditional single-objective optimization approach, such as the goal programming approach [1], the weighted sum approach [9] and the ϵ-constraint [4]. The second is the multiobjective optimization approach searching for the whole Pareto front, from which the decision makers choose their preferred solution. For example, the classic NSGA-II and a multiobjective variant of the PSO algorithm were applied in the earthquake evacuation planning problem [7]. The reason we consider the second category is that optimization approaches in the first may require prior knowledge, such as the relative importance of the objectives in the weighted sum approach. Such knowledge may not be easy to access, and even if it is available it has been shown that the search aiming for the whole Pareto front may be more promising since it can help the search escape the local optima [3]. Another strand of research is multiobjective optimization under uncertainty. Recently, some studies have proposed a number of robust or stochastic models for FLPs

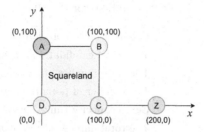

Fig. 1. An illustration of the route network for a basic camp placement, where 1) a source country is represented by a square region with one conflict zone (i.e., point A), three towns (i.e., points B, C, and D) and all possible links among these points, and 2) one camp (i.e., point Z) is connected to the nearest location in the source country.

under uncertainty [2]. However, there is a lack of studies on FLPs under uncertainty that take the preferences of people into account. As popular simulation approaches, different agent-based modelling frameworks have been developed to model the movements of displaced persons (or the preferences of those people).

2   A Multiobjective Camp Location Selection Model

Our multiobjective model aims to determine the optimal location of a new camp and is constructed according to four main steps. First, we create a source country with conflict zones and towns, and interconnecting links. Second, we add a camp at given coordinates in a destination country. Third, we create a link between the camp and its nearest location in the source country, and lastly we run the Flee simulation [11] and calculate the objectives. Figure 1 illustrates the route network for a basic camp placement problem with one conflict zone, three towns and one camp, and interconnecting roads (lines). The coordinates (x, y) associated with each conflict zone, town or camp are used to indicate their positions.

We have the following model assumptions: the locations of conflict zones and towns, the number of asylum-seekers/unrecognized refugees or IDPs (i.e., agents in Flee simulation), and the conflict period are given, agents are spawned in conflict zones, destination countries are represented by a continuous region, camps have limited capacities, agents move during each time step based on predefined rules in [11], and agents stop moving once they reach the camp. With the notation in Table 1, the MOP can be formulated as follows:

$$\text{minimize}: f_1(j) = \frac{\sum_i^{n_{sim,t,j}} d_{sim,t,i,j}}{n_{sim,t,j}}, \quad t = T \tag{1}$$

$$\text{maximize}: f_2(j) = n_{sim,t,j}, \quad t = T \tag{2}$$

$$\text{minimize}: f_3(j) = \frac{\sum_t |c - n_{sim,t,j}|}{T}, \quad t = 1, 2, \ldots, T \tag{3}$$

Table 1. Notations for the MOP.

Notations	Type	Explanation of notations
J	Set	The set of candidates sites indexed by j
a	Parameter	The total number of agents in all conflict zones
n	Parameter	The number of potential camp sites
c	Parameter	Camp capacity (unit: agent)
k	Parameter	The total number of new camps that will be placed and open
T	Parameter	The simulation period or the conflict period (unit: day)
j	Decision variable	The index of a candidate site
$d_{sim,t,i,j}$	Dependent variable	The distance travelled by an agent $i \in I_{sim,t,j}$ in the new camp at candidate site j at time t based on the simulation predictions
$n_{sim,t,j}$	Dependent variable	The number of agents served by the new camp at candidate site j at time t based on simulation predictions, indexed by i

subject to

$$1 \leq j \leq n \tag{4}$$

The objective function Eq. (1) minimizes the average distance travelled by each arriving agent in a destination camp at the end of the simulation. This objective focuses on the efficiency (i.e., distance) of allocating people to facilities. The objective function Eq. (2) maximizes the number of people in the camp at the end of the simulation. This objective function can be easily changed to a minimization problem by calculating the negative value of successful arrivals (i.e., $-n_{sim,t,j}, t = T$). The objective function Eq. (3) minimizes the average idle camp capacity over simulation days for the new camp. Note that the new camp can be overpopulated, and if the idle capacity is a negative value, we simply take the absolute value. Constraint (4) restricts the search space of the MOP (i.e., a set of n possible sites), from which we select the optimal camp site. In our MOP, the decision variable j is known as a solution to the problem. Different from the single-objective optimization problem, the MOP has a set of trade-off solutions, called Pareto front, rather than a single optimal solution. In this paper, only one camp will be established (i.e., $k = 1$) and we aim to find the Pareto front of the MOP. This MOP can be further extended to jointly solve the MOP for multiple camps by replacing the current single decision variable with a set of decision variables, expressed as a k-dimensional decision vector $\vec{j} = (j_1, j_2, \ldots, j_k)$, and considering all people who arrived at these new camps.

2.1   A Simulation-Optimization Approach

We develop a simulation-optimization approach, which combines a (Flee) simulation with a multiobjective optimization algorithm. For the optimization algorithm, we adopt a representative multiobjective evolutionary algorithm, called NSGA-II [5]. Our algorithm works as follows: for each generation of NSGA-II, a group of candidate solutions (each solution is a sequence of k selected sites) are generated, followed by the Flee simulation taking the coordinates corresponding to each solution as input parameters, and assessing and outputting the objective values for the optimization stage. To implement NSGA-II, a candidate solution is represented as a chromosome using a grid-based spatial representation strategy. Each grid cell has longitude and latitude coordinates corresponding to its centroid. The chromosome is then sequentially encoded by the indexes of k selected site(s), where k is the number of camps that will be placed and opened. Note that in this paper we only consider one new camp (i.e., $k = 1$). To automate the simulation process, we utilize FabFlee [12], which is a plugin of FabSim3 (https://github.com/djgroen/FabSim3). Due to data complexity, simulation runs for a group of solutions (i.e., candidate camp locations) are computationally expensive. To reduce the runtime, we employ QCG-PilotJob (http://github.com/vecma-project/QCG-PilotJob) to schedule submitted ensemble runs for different camp locations.

3   Test Setup and Results

To demonstrate the application of our MOP, we conducted a case study for the South Sudan conflict in 2013. The geographic coordinates of examined region are $N0° - N16°$ and $E20° - E40°$, and the region was divided into 26842 $0.1° \times 0.1°$ (around 11 km $\times$11 km) grids. Our simulation instances (*ssudan_c1* and *ssudan_c2*) are constructed based on the South Sudan simulation instance presented in [12], which involves almost 2 million fleeing people in a simulation period of 604 days starting from the 15th December 2013, 25 conflict zones and 16 towns in South Sudan, as well as ten camps in neighboring countries Sudan, Uganda and Ethiopia. The *ssudan_c1* has no camp in place yet and aims to establish one new camp with a capacity of $80,000$ (i.e., $c = 80,000$), while the *ssudan_c2* involves all ten existing established camps and aims to add one new camp with a capacity of $12,000$ (i.e., $c = 12,000$). For both simulation instances, the distance between camp and its nearest location in South Sudan was estimated by using the route planning method in [10]. Furthermore, to shorten the execution time, we reduced the number of agents from all conflict zones by a factor 100 (i.e., $a = a/100$), and accordingly, the camp capacity for *ssudan_c2* and *ssudan_c2* are reduced to 800 and 120, respectively. Figure 2 plots the optimal camp locations for the two conflict instances. The objective values of optimal solutions obtained by NSGA-II are summarized in Table 2. For each conflict instance, NSGA-II can find a set of optimal solutions, which are incomparable based on the concept of Pareto optimality. In other words, each solution is a trade-off among average travel distance, the number of camp arrivals, and the average idle camp capacity.

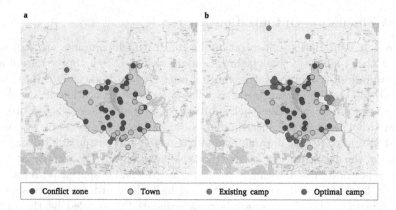

Fig. 2. Optimal camp locations (blue circles) obtained by NSGA-II on the (a) ssu-
dan_c1 and (b) ssudan_c2 conflict instances, respectively. (Color figure online)

Table 2. The objective values of the optimal solutions obtained by the NSGA-II on
the ssudan_c1 and ssudan_c2 conflict instances.

Conflict instance	Camp location		Objectives		
	Longitude	Latitude	Travel distance	No. camp arrivals	Idle capacity
ssudan_c1	30.55	3.75	1380.2211	801	77.0182
	25.25	11.25	6785.469	809	173.0762
	31.55	3.65	1354.2624	803	82.1556
	30.25	3.35	1995.5878	804	91.2666
ssudan_c2	30.35	3.85	558.905	166	49.7136
	29.85	3.85	651.9379	124	11.6589
	29.95	3.65	598.6553	120	7.6788
	28.25	10.35	440.0152	120	8.2483
	28.85	9.65	226.7078	143	29.096
	28.35	9.45	283.3134	150	35.351
	28.55	9.55	313.1518	160	44.2268
	28.45	9.55	281.1019	156	40.6904
	28.65	9.55	433.7734	147	32.5613
	28.05	10.05	507.1222	121	8.9636
	28.45	9.65	336.701	140	26.048
	28.55	9.85	262.416	132	19.0679
	30.75	3.45	580.9609	126	13.6225
	28.55	9.75	364.0978	129	16.0894
	28.45	9.45	397.1481	158	42.5331
	28.35	10.05	539.0705	131	18.0646
	29.75	4.15	634.4269	123	10.6474
	28.55	9.65	322.0341	138	24.2169
	28.05	9.45	371.0897	135	21.9901
	28.55	9.45	388.0439	144	29.7268

4 Conclusion

In this paper, a multiobjective model for the FLP in the context of humanitarian support for forcibly displaced people has been proposed, and the model has been solved by using a simulation-optimization approach. The proposed model has been employed in a case study of South Sudan conflict with a simulation period of 604 days from 15th December 2013. The results obtained by our simulation-optimization approach have demonstrated its ability to provide decision makers with diverse solutions, which strike a balance among the individual travel distance, the number of camp arrivals, and the average idle camp capacity. In the future, other algorithms in multiobjective optimization will be explored. In addition, it would be interesting to consider other factors in the context of forced migration, e.g., construction and transportation costs.

Acknowledgements. This work is supported by the ITFLOWS and HiDALGO projects, which have received funding from the European Union Horizon 2020 research and innovation programme under grant agreement nos 882986 and 824115. The authors are grateful to Prof. Simon J E Taylor and Dr. Anastasia Anagnostou for their constructive discussions on this work.

References

1. Barzinpour, F., Esmaeili, V.: A multi-objective relief chain location distribution model for urban disaster management. Int. J. Adv. Manuf. Technol. **70**(5), 1291–1302 (2014)
2. Boonmee, C., Arimura, M., Asada, T.: Facility location optimization model for emergency humanitarian logistics. Int. J. Disaster Risk Reduct. **24**, 485–498 (2017)
3. Chen, T., Li, M.: The weights can be harmful: Pareto search versus weighted search in multi-objective search-based software engineering. ACM Trans. Softw. Eng. Methodol. **25**(2), 17 (2022)
4. Cilali, B., Barker, K., González, A.D.: A location optimization approach to refugee resettlement decision-making. Sustain. Urban Areas **74**, 103153 (2021)
5. Deb, K., Pratap, A., Agarwal, S., Meyarivan, T.A.M.T.: A fast and elitist multiobjective genetic algorithm: NSGA-II. IEEE Trans. Evol. Comput. **6**(2), 182–197 (2002)
6. Estrada, L.E.P., Groen, D., Ramirez-Marquez, J.E.: A serious video game to support decision making on refugee aid deployment policy. Procedia Comput. Sci. **108**, 205–214 (2017)
7. Ghasemi, P., Khalili-Damghani, K., Hafezalkotob, A., Raissi, S.: Uncertain multiobjective multi-commodity multi-period multi-vehicle location-allocation model for earthquake evacuation planning. Appl. Math. Comput. **350**, 105–132 (2019)
8. Ma, Y., Xu, W., Qin, L., Zhao, X.: Site selection models in natural disaster shelters: a review. Sustainability **11**(2), 399 (2019)
9. Manopiniwes, W., Irohara, T.: Stochastic optimisation model for integrated decisions on relief supply chains: preparedness for disaster response. Int. J. Prod. Res. **55**(4), 979–996 (2017)
10. Schweimer, C., et al.: A route pruning algorithm for an automated geographic location graph construction. Sci. Rep. **11**(1), 1–11 (2021)

11. Suleimenova, D., Bell, D., Groen, D.: A generalized simulation development app-roach for predicting refugee destinations. Sci. Rep. **7**(1), 1–13 (2017)
12. Suleimenova, D., Groen, D.: How policy decisions affect refugee journeys in south Sudan: a study using automated ensemble simulations. J. Artif. Soc. Soc. Simul. **23**(1) (2020)
13. Xu, W., Zhao, X., Ma, Y., Li, Y., Qin, L., Wang, Y., Du, J.: A multi-objective optimization based method for evaluating earthquake shelter location-allocation. Geomat. Nat. Haz. Risk **9**(1), 662–677 (2018)
14. Zhao, M., Chen, Q.W., Ma, J., Cai, D.: Optimizing temporary rescue facility loca-tions for large-scale urban environmental emergencies to improve public safety. J. Environ. Inform. **29**(1) (2017)

A Sparse Matrix Approach for Covering Large Complex Networks by Cliques

Wali Mohammad Abdullah$^{(\boxtimes)}$ and Shahadat Hossain

University of Lethbridge, Lethbridge, AB, Canada
{w.abdullah,shahadat.hossain}@uleth.ca

Abstract. A classical NP-hard problem is the *Minimum Edge Clique Cover (minECC)* problem, which is concerned with covering the edges of a network (graph) with the minimum number of cliques. There are many real-life applications of this problem, such as in food science, computational biology, efficient representation of pairwise information, and so on. Borrowing ideas from [8], we propose using a compact representation, the intersection representation, of network data and design an efficient and scalable algorithm for minECC. Edges are considered for inclusion in cliques in degree-based orders during the clique construction step. The intersection representation of the input graph enabled efficient computer implementation of the algorithm by utilizing an existing sparse matrix package [11]. We present results from numerical experiments on a representative set of real-world and synthetically constructed benchmark graph instances. Our algorithm significantly outperforms the current state-of-the-art heuristic algorithm of [4] in terms of the quality of the edge clique covers returned and running time performance on the benchmark test instances. On some of the largest graph instances whilst existing heuristics failed to terminate, our algorithm could finish the computation within a reasonable amount of time.

Keywords: Adjacency matrix · Clique cover · Intersection matrix · Ordering · Sparse graph

1 Introduction

The graph kernel operations, such as identification of and computation with dense subgraphs, frequently arise in areas as diverse as sparse matrix determination and complex network analysis [13,14]. In social networks, identification of special interest groups or characterization of information propagation are examples of frequently performed network analytics tasks [23]. The *Edge Clique Cover problem (ECC)* considered in this paper is concerned with finding a collection of complete subgraphs or cliques such that every edge and every vertex of the input graph is included in some clique. The computational challenge is to find an ECC with the smallest number of cliques (*minECC*). The minECC problem is computationally intractable or NP-hard [16].

© The Author(s), under exclusive license to Springer Nature Switzerland AG 2022
D. Groen et al. (Eds.): ICCS 2022, LNCS 13352, pp. 505–517, 2022.
https://doi.org/10.1007/978-3-031-08757-8_43

Effective representation of network data is critical to meeting algorithmic challenges for exactly or approximately solving intractable problems, especially when the instance sizes are large and sparse. In this paper, we use sparse matrix data structures to enable compact representation of sparse network data based on an existing sparse matrix framework [11] to design efficient algorithms for the minECC problem.

Let $G = (V, E)$ be an undirected connected graph, where V is the set of vertices, and E is the set of edges. A clique is a subset of vertices such that every pair of distinct vertices are connected by an edge in the induced subgraph. In graph G, an edge clique cover of size k is a decomposition of set V into k subsets $C_1, C_2, \ldots, C_k$ such that $C_i, i = 1, 2, \ldots, k$ induces a clique in G and each edge $\{u, v\} \in E$ is included in some C_i. A trivial clique cover with $k = m, |E| = m$ can be specified by the set of edges E with each edge being a clique. Finding a clique cover with the minimum number of cliques (and many variants thereof) is known to be an NP-hard problem [16].

In 1973, Bron and Kerbosch [2] proposed an algorithm to find all maximal cliques of a given graph. That algorithm uses a branch-and-bound technique. The algorithm is made more efficient by cutting off branches of the search tree that will not lead to new cliques at a very early stage. Etsuji Tomita et al. [22] presented a depth-first search algorithm for generating all maximal cliques of an undirected graph, in which pruning methods are employed as in the Bron-Kerbosch algorithm.

Many algorithms have been proposed in the literature to solve the ECC problem approximately. At the same time, there are only a few exact methods that are usually limited to solving small instance sizes. A recent heuristics approach is described by Conte et al. [4] to find an edge clique cover in $O(m\Delta)$ time, where m is the number of edges and Δ is the highest degree of any vertex in the graph.

In this paper, we use a compact representation of network data based on sparse matrix data structures [11] and provide an improved algorithm motivated by the works of Bron et al. [2], and E. Tomita et al. [22] for finding clique covers. In [1], we used a similar compact representation of network data. In that paper, we employ a "vertex-centric" approach where a vertex, in some judiciously chosen order, together with its edges incident on a partially constructed clique cover, is considered for inclusion in an existing clique. The preliminary implementation produced smaller-sized clique covers when compared with the method of [9] on a set of test instances. While the vertex-centric ECC algorithm frequently produced smaller clique covers compared with other methods, the high memory footprint of the method made it less scalable on very large problem instances. In this paper, we propose an "edge-centric" minECC method. Our method is characterized by a significantly reduced memory footprint and exhibits very good scalability when applied to extremely large synthetic and real-life network instances.

Our approach is based on the simple but critical observation that for a sparse matrix $A \in \mathbb{R}^{m \times n}$, the row intersection graph of A is isomorphic to the adjacency graph of $AA^\top$, and that the column intersection graph of A is isomorphic to the adjacency graph of $A^\top A$ [11]. Therefore, the subset of rows corresponding to nonzero entries in column j induces a clique in the adjacency graph of $AA^\top$, and the subset of columns corresponding to nonzero entries in row i induces a clique in the adjacency graph of $A^\top A$. Note that matrices $A^\top A$ and $AA^\top$ are most likely dense even if matrix A is sparse. We exploit the close connection between sparse matrices and graphs in the reverse direction. We show that given a graph (or network), we can define a sparse matrix, *intersection matrix*, such that graph algorithms of interest can be expressed in terms of the associated intersection matrix. This structural reduction enables us to use the existing sparse matrix computational framework to solve graph problems [11]. This duality between graphs and sparse matrices has also been exploited where the graph algorithms are expressed in the language of sparse linear algebra [14,15]. However, they use adjacency matrix representation which is different from our intersection matrix representation.

The paper is organized as follows. In Sect. 2, we consider representations of sparse graph data and introduce the notion of intersection representation and cast the minECC problem as a matrix compression problem. Section 3 presents the new edge-centric minECC algorithm. An important ingredient of our algorithm is to select edges incident on the vertex being processed in specific orders. The details of the implementation steps are described, followed by the presentation of the ECC algorithm. The section ends with a discussion on the computational complexity of the algorithm. Section 4 contains results from elaborate numerical experiments. We choose 5 different sets of network data consisting of real-world network and synthetic instances. Finally, the paper is concluded in Sect. 5.

2   Compact Representation and Edge Clique Cover

For efficient computer implementation of many important graph operations, representing graphs using adjacency matrix or adjacency lists is inefficient. Adjacency matrix stored as a two-dimensional array is costly for sparse graphs, and typical adjacency list implementations employ pointers where indirect access leads to poor cache utilization [19]. The intersection matrix representation that we propose below enables an efficient representation of pairwise information and allows us to utilize the computational framework DSJM to implement the new ECC algorithm.

2.1   Intersection Representation

We require some preliminary definitions. The *adjacency graph* associated with a symmetric matrix $A \in \mathbb{R}^{n \times n}$ is an undirected graph $G = (V, E)$ in which for each column or row k of A there is a vertex $v_k \in V$ and $A(i,j) \neq 0, i \neq j$ if and only if $\{v_i, v_j\} \in E$.

Let $G = (V, E)$ be an undirected and connected graph without self-loops or multiple edges between a pair of vertices. The adjacency matrix $A(G) \equiv A \in \{0,1\}^{|V| \times |V|}$ associated with graph G is defined as,

$$A(i,j) = \begin{cases} 1 & \text{if } \{v_i, v_j\} \in E, i \neq j \\ 0 & \text{otherwise} \end{cases}$$

We now introduce the intersection representation, an enabling and efficient representation of pairwise information. The *intersection representation* of graph G is a matrix $X \in \{0,1\}^{k \times n}$ in which for each vertex v_j of G there is a column j in X and $\{v_i, v_j\} \in E$ if and only if there is a row l for which $X(l,i) = 1$ and $X(l,j) = 1$. A special case is obtained for $k = m$. Then, the rows of X can be uniquely labeled by the edge list sorted by vertex labels. Therefore, matrix $X \in \{0,1\}^{m \times n}$ can be viewed as an assignment to each vertex a subset of m labels such that there is an edge between vertices i and j if and only if the inner product of the columns i and j is 1. Since the input graph is unweighted, the edges are simply ordered pairs and can be sorted in $O(m)$ time. Unlike the adjacency matrix, which is unique (up to fixed labeling of the vertices) for graph G, there can be more than one *intersection matrix* representation associated with graph G [1]. We exploit this flexibility to store a graph in a structured and space-efficient form.

Let $X \in \{0,1\}^{m \times n}$ be the intersection matrix as defined above associated with a graph $G = (V, E)$. Consider the product $B = X^{\top} X$.

Theorem 1. *The adjacency graph of matrix B is isomorphic to graph G.* [1]

Theorem 1 establishes the desired connection between a graph and its sparse matrix representation. The following result follows directly from Theorem 1.

Corollary 1. *The diagonal entry $B(i,i)$ where $B = X^{\top} X$ and X is the intersection matrix of graph G, is the degree $d(v_i)$ of vertex $v_i \in V, i = 1, \ldots, n$ of graph $G = (V, E)$.* [1]

Intersection matrix X defined above represents an edge clique cover of cardinality m for graph G. Each edge $\{v_i, v_j\}$ constitutes a clique of size 2. In the intersection matrix X, edge $e_l = \{v_i, v_j\}$ is represented by row l with $X(l,i) = X(l,j) = 1$ and other entries in the row being zero. In general, column indices j' in row l where $X(l,j') = 1$ constitutes a clique on vertices $v_{j'}$ of graph G. Thus the minECC problem can be cast as a *matrix compression* problem.

minECC Matrix Problem. Given $X \in \{0,1\}^{m \times n}$ determine $X' \in \{0,1\}^{k \times n}$ with k minimized such that the intersection graphs of X and X' are isomorphic.

3   An Edge-Centric MinECC Algorithm

The algorithm that we propose for the ECC problem is motivated by the maximal clique algorithm due to Bron et al. [2], and E. Tomita et al. [22]. For ease of presentation, we discuss the algorithm in graph-theoretic terms. However, our computer implementation uses a sparse matrix framework of DSJM [11], and all computations are expressed in terms of intersection matrices.

3.1 Selection of Uncovered Edges

An edge $\{u,v\} \in E$ is said to be *covered* if both of its incident vertices have been included in some clique; otherwise the edge is *uncovered*. In our algorithm, we select an uncovered edge $\{u,v\}$ and try to construct a maximal clique, C, containing the edge. The algorithm selects vertices and edges in a prespecified order during the clique construction process. Note that it may or may not be possible to include additional uncovered edges while building a clique after selecting an uncovered edge. This subsection will give details on how the algorithm selects an uncovered edge.

Vertex Ordering. We recall that $d(v)$ denotes the degree of vertex v in graph $G = (V,E)$. Let $Vertex\_Order$ be a list of vertices of graph G using one of the ordering schemes below.

- **Largest-Degree Order (LDO) (see [12]):** Order the vertices such that $\{d(v_i), i = 1, \ldots, n\}$ is nonincreasing.
- **Degeneracy Order (DGO) (see [7,21]):** Let $V' \subseteq V$ be a subset of vertices of G. The subgraph induced by V' is denoted by $G[V']$. Assume the vertices $V' = \{v_n, v_{n-1}, \ldots, v_{i+1}\}$ have already been ordered. The ith vertex in DGO is an unordered vertex u such that $d(u)$ is minimum in $G[V \setminus V']$ where, $G[V \setminus V']$ is the graph obtained from G by removing the vertices of set V' from V.
- **Incidence-Degree Order (IDO) (see [3]):** Assume that the first $k-1$ vertices $\{v_1 \ldots, v_{k-1}\}$ in incidence-degree order have been determined. Choose v_k from among the unordered vertices that has maximum degree in the subgraph induced by $\{v_1, \ldots, v_k\}$.

Edge Ordering. After the vertices have been ordered using one of the above schemes, the algorithm proceeds to choose a vertex in that specific order, which has at least one uncovered incident edge. If there is more than one uncovered edge incident on the vertex being processed, the order in which the edges are processed (i.e., to include in a clique) is as follows. Place all the edges $\{u,v\}$ before $\{p,q\}$ in an ordered edge list, $Edge\_Order$, such that vertex u or vertex v is ordered before vertices p and q in $Vertex\_Order$ list.

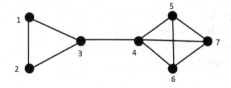

Fig. 1. An example of an undirected graph.

Figure 1 shows an undirected graph. $\{4, 3, 5, 6, 7, 1, 2\}$ would be the list with LDO. The edge list induced by the $Vertex\_Order$ will have the following form.

$$Edge\_Order = \{\{4,3\}, \{4,5\}, \{4,6\}, \{4,7\}, \{3,1\}, \{3,2\}, \{5,6\}, \{5,7\}, \{6,7\}, \{1,2\}\}$$

Edge Selection. We select an edge to $\{u, v\} \in E$ to include in a new clique if $\{u, v\}$ is uncovered and ordered before all uncovered edges in $Edge\_Order$. The clique that gets constructed with edge $\{u, v\}$ may cover other uncovered edges that are further down the list.

We consider three variants of edge selection for our algorithm, denoted by L, D, and I.

- **L**: In this variant, the set of vertices are ordered using the Largest-Degree Ordering (LDO) scheme. We select a vertex u in that order and then return all the uncovered edges of the form $\{u, v\}$.
- **D**: All the vertices are ordered using Degeneracy Ordering (DGO) scheme. Select a vertex u in that order, and then return all the uncovered edges of the form $\{u, v\}$.
- **I**: Finally, this variant orders the set of vertices using the Incidence-Degree Ordering (IDO) scheme. We select a vertex u in that order and return all the uncovered edges $\{u, v\}$.

3.2   The Algorithm

Let $E_{\mathcal{P}} = \{e_1, \ldots, e_{i-1}\}$ be the set of edges that have been assigned to one or more cliques $\{C_1, \ldots, C_{k-1}\}$ and let $e_i = \{v_j, v_{j'}\}$ be the edge currently being processed according to the ordered edge list. Denote by

$$W = \{v_l \mid \{v_j, v_l\}, \{v_{j'}, v_l\} \in E\}$$

the set of common neighbors of v_j and $v_{j'}$.

The complete algorithm is presented below.

EO-ECC ($Edge\_Order$)

Input: $Edge\_Order$, set of edges in a predefined order using schemes L, D, or I

```
1: k ← 0                                          ▷ Number of cliques
2: for index = 1 to m do                          ▷ m is the number of edges
3:     {u, v} ← Edge_Order[index]
4:     if {u, v} is uncovered then
5:         W ← FindCommonNeighbors(u, v)
6:         if W = ∅ then
7:             k + +
8:             Ck ← {u, v}
9:             Mark {u, v} as covered
10:        else
```

11:          $k++$
12:          $C_k \leftarrow \{u, v\}$
13:          Mark $\{u, v\}$ as covered
14:          **while** $W \neq \emptyset$ **do**
15:                  let t be a vertex in W
16:                  $W \leftarrow W \setminus t$
17:                  **if** $\{t, s\} \in E$ for each $s \in C_k$ **then**
18:                          Mark $\{t, s\}$ as covered
19:                          $C_k \leftarrow C_k \cup \{t\}$
20:                          $FindCommonNeighbors(W, FindNeighbors(t))$
21: **return** $C_1, C_2, ..., C_k$

3.3 Discussion

In this subsection, we analyze algorithm EO-ECC to derive it's asymptotic running time. The two kernel operations used in the algorithm are "FindCommon-Neighbors" and "FindNeighbors." The *FindCommonNeighbors* operation merges two sorted lists (of integers) and computes the intersection of the lists. The list (of vertices) that this operation returns after each call has at least one fewer vertices. Thus, to construct a clique C_i, the total cost would be $(\frac{\rho_i(\rho_i-1)}{2})$, where $|C_i| = \rho_i$. Let, $C = \{C_1, C_2, \ldots, C_k\}$ be a clique cover returned by the algorithm EO-ECC. Then the total cost of calling *FindCommonNeighbors* for the algorithm would be $O(\sum_{i=1}^{k} \frac{\rho_i(\rho_i-1)}{2})$. The operation *FindNeighbors* in algorithm EO-ECC computes the neighbors set of vertex $v \in V$ [12]. In line 20, *FindNeighbors* operation is used to compute the neighbors of a vertex. Since an uncovered edge gets covered only once, the total cost of *FindNeighbors* operation is at most $O(m)$. Thus, the overall running time of algorithm EO-ECC is $O(m + \sum_{i=1}^{k} \frac{\rho_i(\rho_i-1)}{2})$. The following result follows immediately from the above running time expression.

Theorem 2. *If the input graph G is triangle-free, then the algorithm EO-ECC runs in $O(m)$ time.*

4 Numerical Testing

In this section, we provide results from numerical experiments on selected test instances. 10th Discrete Mathematics and Theoretical Computer Science (DIMACS10) instances and Stanford Network Analysis Platform (SNAP) instances are obtained from the University of Florida Sparse Matrix Collection [5]. (SNAP) is a collection of more than 50 large network datasets containing large number of nodes and edges including social networks, web graphs, road networks, internet networks, citation networks, collaboration networks, and communication networks [17]. We also experiment with synthetic graph instances. We generated 182 Erdös-Rényi and Small-World instances using the Stanford

Network Analysis Project (SNAP) [18] instance generator. The number of edges
of these generated graphs is varied from 800 to 72 million.

The experiments were performed using a PC with 3.4 GHz Intel Xeon CPU,
with 8 GB RAM running Linux. The implementation language was C++ and the
code was compiled using $-O2$ optimization flag with a g++ version 4.4.7 compiler.
We employed the High-Performance Computing system (Graham cluster) at
Compute Canada for large instances that could not be handled by the PC.

In what follows, we refer to the vertex-centric ECC algorithm from [1] as Vertex Ordered Edge Clique Cover (VO-ECC). We also refer to the ECC algorithm
due to Conte et al. as (Conte-Method). Finally, the edge-centric minECC algorithm of this paper is identified as Edge Ordered Edge Clique Cover (EO-ECC).
EO-ECC has three variants associated with the three different edge ordering
schemes D, L, and I. They are: EO-ECC-D, EO-ECC-L, and EO-ECC-I respectively.
In these results, m denotes the number of edges, n denotes the number of vertices
of the graph; $|C|$ denotes the number of cliques in the cover, and t is the time in
seconds to get the cover. In the presented tables, the smallest cardinality clique
cover is marked in **bold**.

Table 1. Test Results (Number of cliques) for SNAP instances.

Graph			$|C|$		
Name	m	n	VO-ECC using [1]	Conte-Method using [4]	EO-ECC
p2p-Gnutella04	39994	10878	38474	38491	**38449**
p2p-Gnutella24	65369	26518	63726	63725	**63689**
p2p-Gnutella25	54705	22687	53368	53367	**53347**
p2p-Gnutella30	88328	36682	85823	85822	**85717**
ca-GrQc	14496	5242	3777	3753	**3717**
as-735	13895	7716	8985	**8938**	10130
Wiki-Vote	103689	8297	42914	**39393**	51145
Oregon-1	23409	11492	15631	**15491**	15527
ca-HepTh	25998	9877	9663	9270	**9162**

Table 1 displays the size of clique covers returned by three algorithms: the
edge-centric algorithm (EO-ECC), the vertex-centric algorithm (VO-ECC) discussed in [1] and algorithm (Conte-Method) discussed in [4]. Conte-Method
randomly selects an edge and attempts to build a clique around the selected
edge. As the table illustrates, EO-ECC produces smaller cardinality edge clique
cover than VO-ECC except for two instances. On the other hand, it outperforms
Conte-Method on six out of nine instances.

Table 2. Test Results (number of cliques) for DIMACS10 matrices.

Graph			Number of cliques			
Name	m	n	Conte-Method	EO-ECC-D	EO-ECC-L	EO-ECC-I
chesapeake	170	39	**75**	76	**75**	76
delaunay_n10	3056	1024	1250	**1233**	1275	1241
delaunay_n11	6127	2048	2485	**2449**	2544	2481
delaunay_n12	12264	4096	4993	**4906**	5095	4939
delaunay_n13	24547	8192	9989	**9881**	10211	9920
delaunay_n14	49122	16384	19974	**19672**	20435	19855
delaunay_n15	98274	32768	39923	**39501**	40876	39782
delaunay_n16	196575	65536	79933	**78792**	81528	79445
delaunay_n17	393176	131072	159900	**157792**	163321	158851
delaunay_n18	786396	262144	319776	**315684**	326741	317987
com-DBLP	1049866	317080	238854	237713	**237685**	**237685**
belgium_osm	1549970	1441295	1545183	1545183	1545183	1545183
delaunay_n19	1572823	524288	639349	**631354**	653383	635877
delaunay_n20	3145686	1048576	1279101	**1262843**	1307080	1271229
delaunay_n21	6291408	2097152	2557828	**2525301**	2613106	2542333

Test results for the selected test instances from group DIMACS10 are reported in Table 2. For comparison, we show the results of Conte-Method, EO-ECC-D, EO-ECC-L, and EO-ECC-I. On twelve out of fifteen instances, EO-ECC-D gives the least number of cliques to cover all the edges of the given graph. On the graph named com-DBLP EO-ECC-L and EO-ECC-I produce smaller cardinality covers. Overall, EO-ECC emerges as the clear winner over Conte-Method in terms of the size of the clique covers.

Besides DIMACS10 selected instances, we compare these algorithms on 182 generated instances where the number of edges is varied from 800 to 7.2×10^7. Using SNAP tool [18], we generated 72 "Small-world" and 110 "Erdös-Rényi" graphs. EO-ECC produces smaller (on 47.3%instances) or equal (on 52.7% instances) cardinality clique covers compared with Conte-Method.

Rodrigues [20] used different graph instances to evaluate their edge clique cover algorithms. The well-known instances to evaluate edge clique cover problem are from the application "compact letter display" [10]. On thirteen out of fourteen instances, Conte-Method [4] gives optimum results. Both Rodrigues's algorithm and our EO-ECC give optimum results for all the instances.

The performance comparison between Conte-Method and EO-ECC is shown in Fig. 2. We compare the time required to find edge clique cover for the given graph.

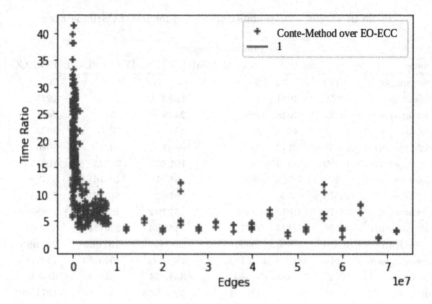

Fig. 2. Ratio between the time used by `Conte-Method` and `EO-ECC` for each graph, as a function of the number of the edges (y-axis is in log-scale).

We use fifteen DIMACS10 instances and 182 Erdös-Rényi and Small-World instances. In the figure, a cross mark represents the ratio between the time needed by and `EO-ECC`, as a function of the number of the edges. The green line at height 10^0 means that `Conte-Method` took the same time as `EO-ECC` to process the corresponding graph, and a cross mark at height 10^1 means that `Conte-Method` was ten times slower. As the figure clearly demonstrates, `EO-ECC` is always faster than `Conte-Method`, and more than 40 times faster on some of the test instances.

Table 3. Graph processing rate (number of edges processed per sec).

Group	Total instances	Largest rate	Smallest rate	Average rate
DIMACS10	15	2.7E6	3.0E5	1.7E6
SNAP	9	2.5E6	6.2E4	1.5E6
Erdös-Rényi	110	2.0E6	1.2E5	8.9E5
Small World	72	1.7E6	4.3E5	1.1E6

The graph processing rate is one of the quality assessment metrics for an algorithm. We report the processing rate of our algorithm for a selection of real-world (DIMACS10, SNAP) and synthetically generated (Erdös-Rényi, Small World) graphs in Table 3. Table 3 shows the largest rate, the smallest rate, and the average rate for each set of graph instances. On DIMACS10 instances, the algorithm performs the best, while on Erdös-Rényi instances, the algorithm is not as efficient. This can be explained by the structural properties of graphs. Real-life and Small World synthetic instances display a power-law degree distribution resulting in a large proportion of vertices with very small degrees. Thus, the set intersection operation in our algorithm can be very efficient on those types of graphs.

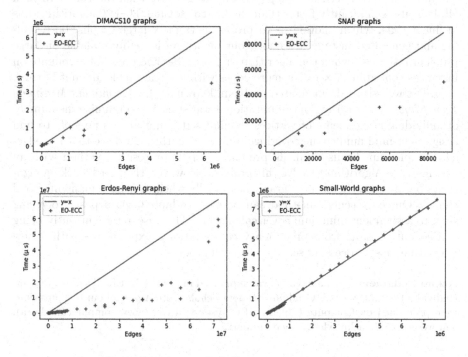

Fig. 3. Runtime to find clique cover using EO-ECC.

Finally, in Fig. 3, we demonstrate the superior scalability of our algorithm. The figure plots the time used to compute clique covers by EO-ECC, where the time is a function of the number of edges in the graph. We report the time in microseconds. A dot (x, y) states that the graph has x edges, and the algorithm spent y microseconds to finish the computation. The figure also displays the line $y = x$ for comparison with the actual running time. On each of the four sets of test instances, the running time shows a linear relationship with the number of edges, demonstrating that the running time of EO-ECC is linear in practice.

5   Conclusion

In this work, we have proposed a compact representation of network data. The edge clique cover problem is recast as a sparse matrix determination problem.

The notion of *intersection matrix* provides a unified framework that facilitates the compact representation of graph data and efficient implementation of graph algorithms. The adjacency matrix representation of a graph can potentially have many nonzero entries since it is the product of an intersection matrix with its transpose. We have compared our results concerning the clique cover size and runtime with the current state-of-the-art algorithm for minECC [4]. Our algorithm achieves significantly smaller clique covers on the vast majority of the test instances and never returns a clique cover that is larger than the `Conte-Method` [4]. It is also significantly faster than the `Conte-Method`. `EO-ECC` algorithm runs in linear time, which allowed us to process extremely large graphs, both real-life and generated instances. Finally, our algorithm is highly scalable on large problem instances, while the algorithm of `Conte-Method` does not terminate on instances containing 7×10^7 or more edges within a reasonable amount of time.

A less well-studied but related problem, known as the *Assignment Minimum Edge Clique Cover* arising in computational statistics, is to minimize the number of individual assignments of vertices to cliques It is not always possible to find assignment-minimum clique coverings by searching through those that are edge-clique-minimum. Ennis et al. [6] presented a post-processing method with an existing ECC algorithm to solve this problem. However, their backtracking algorithm becomes costly for large graphs, especially when they have many maximal cliques. Our edge-centric method can be easily adapted, via a post-processing step, to assignment minimum cover calculation. This research is currently being carried out. Results from preliminary computational experiments with a new linear-time post-processing scheme are favourable.

Acknowledgments. This research was supported in part by NSERC Discovery Grant (Individual), and the AITF Graduate Student Scholarship. A part of our computations were performed on Compute Canada HPC system (http://www.computecanada.ca), and we gratefully acknowledge their support.

References

1. Abdullah, W.M., Hossain, S., Khan, M.A.: Covering large complex networks by cliques–a sparse matrix approach. In: Kilgour, D.M., Kunze, H., Makarov, R., Melnik, R., Wang, X. (eds.) Recent Developments in Mathematical, Statistical and Computational Sciences. pp. 117–127. Springer International Publishing, Cham (2021)
2. Bron, C., Kerbosch, J.: Algorithm 457: finding all cliques of an undirected graph. Commun. ACM **16**(9), 575–577 (1973). https://doi.org/10.1145/362342.362367
3. Coleman, T.F., Moré, J.J.: Estimation of sparse jacobian matrices and graph coloring blems. SIAM J. Numerical Anal. **20**(1), 187–209 (1983)
4. Conte, A., Grossi, R., Marino, A.: Large-scale clique cover of real-world networks. Inf. Comput. **270**, 104464 (2020)
5. Davis, T., Hu, Y.: Suitesparse matrix collection. https://sparse.tamu.edu/. Accessed 10 Feb 2019
6. Ennis, J., Ennis, D.: Efficient Representation of Pairwise Sensory Information. IFPress **15**(3), 3–4 (2012)

7. Eppstein, David, Strash, Darren: Listing all maximal cliques in large sparse real-world graphs. In: Pardalos, Panos M.., Rebennack, Steffen (eds.) SEA 2011. LNCS, vol. 6630, pp. 364–375. Springer, Heidelberg (2011). https://doi.org/10.1007/978-3-642-20662-7_31

8. Erdös, P., Goodman, A.W., Pósa, L.: The representation of a graph by set intersections. Canadian J. Math. **18**, 106–112 (1966)

9. Gramm, J., Guo, J., Huffner, F., Niedermeier, R.: Data reduction, exact and heuristic algorithms for clique cover. In: Proceedings of the Eighth Workshop on Algorithm Engineering and Experiments (ALENEX), SIAM, pp. 86–94 (2006)

10. Gramm, J., Guo, J., Huffner, F., Niedermeier, R., Piepho, H., Schmid, R.: Algorithms for compact letter displays: comparison and evaluation. Comput. Stat. Data Anal. **52**, 725–736, 104464 (2007)

11. Hasan, M., Hossain, S., Khan, A.I., Mithila, N.H., Suny, A.H.: DSJM: a software toolkit for direct determination of sparse Jacobian matrices. In: Greuel, G.-M., Koch, T., Paule, P., Sommese, A. (eds.) ICMS 2016. LNCS, vol. 9725, pp. 275–283. Springer, Cham (2016). https://doi.org/10.1007/978-3-319-42432-3_34

12. Hossain, S., Khan, A.I.: Exact Coloring of Sparse Matrices. In: Kilgour, D.M., Kunze, H., Makarov, R., Melnik, R., Wang, X. (eds.) AMMCS 2017. SPMS, vol. 259, pp. 23–36. Springer, Cham (2018). https://doi.org/10.1007/978-3-319-99719-3_3

13. Hossain, S., Suny, A.H.: Determination of large sparse derivative matrices: structural: orthogonality and structural degeneracy. In: B. Randerath, H., Roglin, B., Peis, O., Schaudt, R., Schrader, F., Vallentin and V. Weil. 15th Cologne-Twente Workshop on Graphs & Combinatorial Optimization, Cologne, Germany, pp. 83–87 (2017)

14. Kepner, J., Gilbert, J.: Graph Algorithms in the Language of Linear Algebra, Society for Industrial and Applied Mathematics. Philadelphia, PA, USA (2011)

15. Kepner, J., Jananthan, H.: Mathematics of Big Data: Spreadsheets, Databases, Matrices, and Graphs. MIT Press (2018)

16. Kou, L., Stockmeyer, L., Wong, C.: Covering edges by cliques with regard to keyword conflicts and intersection graphs. Commun. ACM **21**(2), 135–139 (1978)

17. Leskovec, J., Krevl, A.: SNAP Datasets: Stanford large network dataset collection, June 2014. http://snap.stanford.edu/data. Accessed 10 Feb 2019

18. Leskovec, J., Sosič, R.: Snap: A general-purpose network analysis and graph-mining library. ACM Trans. Intell. Syst. Technol. (TIST) **8**(1), 1 (2016)

19. Park, J.S., Penner, M., Prasanna, V.K.: Optimizing graph algorithms for improved cache performance. IEEE Trans. Parallel Distrib. Syst. **15**(9), 769–782 (2004)

20. Rodrigues, M.O.: Fast constructive and improvement heuristics for edge clique covering. Discrete Opt. **39**, 100628 (2021)

21. Rossi, R.A., Gleich, D.F., Gebremedhin, A.H., Patwary, M.M.A.: Fast maximum clique algorithms for large graphs. In: Proceedings of the 23rd International Conference on World Wide Web, pp. 365–366 (2014)

22. Tomita, E., Tanaka, A., Takahashi, H.: The worst-case time complexity for generating all maximal cliques and computational experiments. Theor. Comput. Sci. **363**(1), 28–42 (2006)

23. Wasserman, S., Faust, K.: Social Network Analysis: Methods and Applications. Cambridge University Press (1994)

DSCAN for Geo-social Team Formation

Maryam MahdavyRad, Kalyani Selvarajah$^{(\boxtimes)}$, and Ziad Kobti

School of Computer Science, University of Windsor, Windsor, ON, Canada
{mahdavy,kalyanis,kobti}@uwindsor.ca

Abstract. Nowadays, geo-based social group activities have become popular because of the availability of geo-location information. In this paper, we propose a novel Geo-Social Team Formation framework using DSCAN, named DSCAN-GSTF, for impromptu activities, aim to find a group of individuals closest to a location where service requires quickly. The group should be socially cohesive for better collaboration and spatially close to minimize the preparation time. To imitate the real-world scenario, the DSCAN-GSTF framework considers various criteria which can provide effective Geo-Social groups, including a required list of skills, the minimum number of each skill, contribution capacity, and the weight of the user's skills. The existing geo-social models ignore the expertise level of individuals and fail to process a large geo-social network efficiently, which is highly important for an urgent service request. In addition to considering expertise level in our model, we also utilize the DSCAN method to create clusters in parallel machines, which makes the searching process very fast in large networks. Also, we propose a polynomial parametric network flow algorithm to check the skills criteria, which boosts the searching speed of our model. Finally, extensive experiments were conducted on real datasets to determine a competitive solution compared to other existing state-of-the-art methods.

Keywords: Geo-social networks · Geo-social groups · DSCAN

1 Introduction

Nowadays, geo-based social group activities have become popular because of the availability of geo-location information. In this paper, we propose a novel Geo-Social Team Formation framework using DSCAN, named DSCAN-GSTF, for impromptu activities, aim to find a group of Geo-Social Networks (GeoSNs) is online social networks that allow geo-located information to be shared in real-time. The availability of location acquisition technologies such as GPS and WiFi enables people to easily share their position and preferences to existing online social networks. Here, the preferences can be common interests, behavior, social relationships, and activities. This information is usually derived from a history of an individual's locations and Geo-tagged data, such as location-tagged photos and the place of the current event [27]. Thus, we have several popular GeoSNs such as Facebook, Twitter, Flickr, Foursquare, Yelp, Meetup,

© The Author(s), under exclusive license to Springer Nature Switzerland AG 2022
D. Groen et al. (Eds.): ICCS 2022, LNCS 13352, pp. 518–533, 2022.
https://doi.org/10.1007/978-3-031-08757-8_44

Gowalla, and Loopt. Consequently, GeoSNs have drawn significant attention in recent years by researchers on many applications, including finding friends in the vicinity [13,23], group-based activity planning [4], and marketing [6].

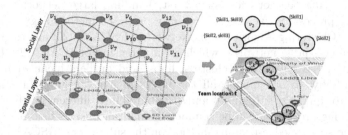

Fig. 1. Identifying individuals for impromptu social activity from a GeoSNs.

An impromptu activity planning is one of the popular motivating applications in GeoSNs search. For example, the COVID19 outbreak is affecting every part of human lives. At the initial stage of COVID19, essential services such as health, finance, food, and safety suffered a lot due to unexpected lockdown because they did not have the required human resources as expected. At the same time, fulfilling societies' requirements are also equally important. Therefore, bringing skilled people to the location where the services with diverse demands were crucial and had become a very challenging process. The location of the individual should be close to the place where service is required, and the individual with the required skills needs to be suitable for services. In this example, a service might require several skills. Additionally, each individual may contribute to as many skills as possible in various activities and might have a specific capacity to be involved in multiple activities.

To support this situation, we highly believe that forming groups from GeoSN is an effective solution. In general, this is called the Geo-social group search problem [18,19], which aim to identify groups of individual who are socially cohesive and spatially closest to a location [4]. In other words, the Geo-social group should satisfy that the participants are socially close within the group to confirm good communication between each member and spatially closest to the location of the service to bring them as soon as possible. Figure 1 represents a general Geo-Social Network, where the social layer is to show the social connections between individuals, and the spatial layer is to show the locations of the individuals.

Many existing Geo-social group models considered social cohesiveness and spatial closeness to find successful groups. In addition to these two requirements, recently, Chen et al. [4] incorporated a few essential parameters such as the collective capabilities and capacity of each member. However, to the best of our knowledge, none of the existing Geo-social group models considered the weights of the user's skills which assist in choosing the exact qualified individual. Moreover, efficiently searching required keywords that have high expertise, capacity constraints, social constraint, and spatial closeness altogether have not been explored in the existing studies. Forming a search framework that can quickly narrow the search space while preserving the correct result is an NP-Hard. It is an open problem and essential to solve in polynomial time [4].

This paper proposes a novel framework to search efficiently on large GeoSNs while preserving the correct result. First, handling large networks is a time-consuming process. So we adopt a recently proposed methodology, Distributed Structural Clustering Algorithm for Networks (DSCAN) [20] algorithm to efficiently manage large networks, which is an extension of SCAN [24]. The basic idea of SCAN is to discover clusters, hubs, and outliers included in a given graph. Initially, in our model, all the nodes of a given graph are randomly divided into equal size sub-graphs and distributed into different machines so that the remaining processing can be conducted simultaneously. Then, by employing the skewness-aware edge-pruning method on the sub-graphs, DSCAN eliminates unwanted edges and moves missing neighbors of nodes from one sub-graph to another. Second, producing socially cohesive groups from these sub-graphs is another essential process. So, DSCAN collects the Core nodes with higher structural similarity and creates a set of clusters from these sub-graphs. Parallelly, the set of skills is collected from each cluster and stored on a map. The third requirement is to choose a spatially close group to the location (∇) where the service is required. We pick a node randomly from each cluster and evaluate the geographical distance from ∇. The clusters are ranked based on the distance in ascending order. Then a cluster with the lowest rank is selected and tested to see whether it satisfies the requirements of the query or not. If it does not satisfy, move to the following cluster and test the requirement. This process will be repeated till we find the right cluster. Finally, we propose a new polynomial algorithm based on the parametric flow network [8], which checks the skills requirements of the query and contribution capacity of each individual in the selected cluster while considering the user's skills weights.

Our Contribution: The followings are the summary of our contributions:

1. We propose a Geo-Social Team Formation (GSTF) model by considering the group's collective capabilities as required for the activities while considering the capacity of contribution from each member and expertise level.
2. We utilize the benefits of Distributed Structural Graph Clustering (DSCAN) to manage the large GeoSNs efficiently.
3. We propose a new polynomial searching algorithm based on the parametric flow network, which satisfies Minimum keywords, expertise level, and capacity constraints.

The rest of the paper is organized as follows. Section 2 discusses related existing work. Section 3 defines the problem definitions of our proposed model. Our framework is presented in Sect. 4. Following that, Sect. 5 illustrates the experimental setup and the corresponding results. Finally, Sect. 6 concludes the research idea of this paper with directions for future work.

2 Related Work

Forming a group of individuals for various purposes has been tackled in many different ways. The team formation problem in Social networks was first introduced by Lappas et al. [12]. Later many studies [10,16] have been conducted by incorporating various parameters which influence the successful formation of teams in several applications, including academic collaborations, healthcare [17], and human resource management. However, many of these studies focused highly on minimizing or maximizing some social constraints such as communication cost between members in a team based on their past relationship, profit, and productivity cost.

The concept of GeoSNs services was first introduced by Huang et al. [9]. Many studies have focused on querying geo-social data in order to derive valuable information from both the users' social interactions and current locations [1,21]. Among these, forming Geo-social groups has taken considerable attention of researchers recently since this aims to identify a set of most suitable individuals for various activities which can be planned or unplanned. The unplanned activities such as groups for various purposes during unexpected events, for example, Wildfire and flooding, are relatively complex than the planned activities such as a group for a party or a game. Much existing research proposed various models for both situations while satisfying social constraints and optimizing spatial proximity [4,21]. Many of these focused on forming a group that satisfies a single social constraint while optimizing the spatial proximity. But for impromptu activities, in reality, we require individuals who have diverse demands of skills for multiple tasks or services to serve in a specific location. Recently, Chen et al. [4] introduced a novel framework to discover a set of groups that is socially cohesive while spatially closest to a location for diverse demands. Here, the groups of individuals do not necessarily know each other in the past. However, When there is a tie between two members, their model gives higher priority to the individual who is highly cohesive to the team. The concept of multiple social constraints for various activities has already been studied in [15,17]. However, they considered the frameworks on social networks with known individuals.

Searching cohesive subgroups from a large network is another challenging process in the team formation problem. Structural Clustering Algorithm for Networks (SCAN) algorithm was proposed to detect cohesive subgroups from a network [24]. However, SCAN is a computationally expensive method for a large network because it requires iterative calculation for all nodes and edges. Later, to overcome the limitation with SCAN, many clustering methods have been proposed, such as PSCAN [26], and DSCAN [20]. Since DSCAN is efficient, scalable, and exact, we employ this methodology in our model. To the best of our knowledge, we, for the first time, applied DSCAN in the Geo-Social group search problem.

3   Problem Definition

Given an undirected graph $G = (V, E)$, where V is a set of vertices and E is a set of edges. The Graph G incorporates network structures, spatial information, and textual information. In real networks, vertices are users or people, and edges between them may be friendship or previous collaboration. Additionally, each vertex $v \in V$ includes location information, which can be represented as $\nabla = (v.x, v.y)$, where $v.x$ is latitude and $v.y$ is longitude, and a set of keyword attributes which can be represented as $v.A$. The textual information can be a set of skills $S = \{s_1, s_2, \ldots, s_k\}$ of a vertex $v \in V$, where k is the number of skills that a person is expert in. Along with the skills, a vertex has a set weight $W = \{w_1, w_2, \ldots w_k\}$ to represent how much a person expert in each skill.

Definition 1. *Query (Q): The query defines the requirements of skills and number of people in each skills. This includes a Geo-location ∇ (latitude (x) and longitude (y)), a set of required skills $S = \{s_1, s_2, \ldots, s_r\}$, a set of required expertness in each skills $P = \{p_1, p_2, \ldots, p_r\}$ and a contribution capacity of an expert c for every query keyword needs to be assigned.*

Definition 2. *Geo-Social Team (B): For a given location ∇, a set of the required number of experts who satisfies social cohesiveness and spatial closeness is selected from a Geo-Social network G while considering contribution capacity c and person's skill weight in each skill.*

In our model, we exploit the DSCAN to handle larger data efficiently. To understand the concept of DSCAN, the following definition are necessary.

Definition 3. *Structural Neighborhood (N_v): The structural neighborhood N_v of vertex v can be defined as,*

$$N_v = \{w \in V | (v, w) \in E\} \cup \{v\} \tag{1}$$

Definition 4. *Structural similarity: The structural similarity $\sigma(v, w)$ between v and w can be defined as,*

$$\sigma(v, w) = |N_v \cap N_w| / \sqrt{|N_v||N_w|} \tag{2}$$

If $\sigma(v, w) \geq \epsilon$, vertex v shares similarity with w and $\epsilon \in \mathbb{R}$ is a density threshold which we assigned.

When a vertex has enough structurally identical neighbors, it becomes the seed of a cluster, named core node. Core nodes have at least μ number of neighbors with a structural similarity ($\sigma(v, w)$) that exceeds the threshold ϵ.

Definition 5. *Core: For a given ϵ and μ, A vertex $w \in V$ is called a core, iff $N_{w,\epsilon} \geq \mu$. Where $N_{w,\epsilon}$ is the set of neighbor nodes of core node w, and structural similarity of $N_{w,\epsilon}$ is greater than ϵ.*

Definition 6. *Cluster ($\mathcal{C}_w$): Assume node w be a core node. SCAN collects all nodes in $N_{w,\epsilon}$ into the same cluster ($\mathcal{C}_w$) of node w, initially $\mathcal{C}_w = \{w\}$. SCAN outputs a cluster $\mathcal{C}_w = \{v \in N_{u,\epsilon} | u \in \mathcal{C}_w\}$.*

DSCAN Algorithm: When DSCAN [20] receives a graph, it first deploys disjointed subgraphs of the given graph G to distributed memories on multiple machines $M = \{M_1, M_2, \ldots, M_n\}$ for a given a density threshold $\epsilon \in \mathbb{R}$ and a minimum size of a cluster $\mu \in \mathbb{N}$, where n is number of machines. Initially DSCAN randomly moves a set of vertices V_i in subgraph $G_i = (V_i, E_i)$ for each machine M_i. The subgraphs are then processed in a parallel and distributed fashion. Additionally, DSCAN uses edge-pruning based on skewness to improve efficiency further.

Skewness-Aware Edge-Pruning: DSCAN applies $\omega-$skewness edge-pruning to remove unwanted edges and move missing neighbors of nodes from one subgraph to another. Given graph $G = (V, E)$, consider an edge (u, v) be in E and

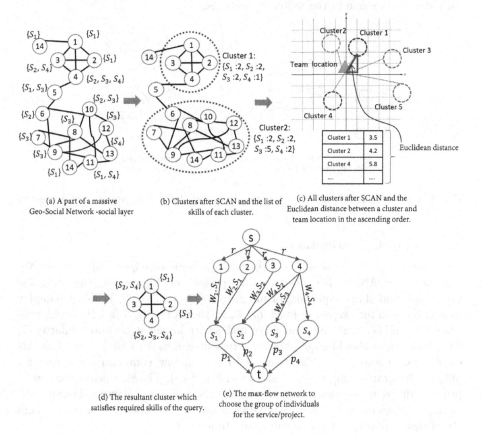

(a) A part of a massive Geo-Social Network -social layer

(b) Clusters after SCAN and the list of skills of each cluster.

(c) All clusters after SCAN and the Euclidean distance between a cluster and team location in the ascending order.

(d) The resultant cluster which satisfies required skills of the query.

(e) The max-flow network to choose the group of individuals for the service/project.

Fig. 2. The DSCAN-GSTF framework: (a) a part of a massive geo-social network - social layer, (b) the clusters and the list of skills that each cluster satisfies. (c) The ordered distance between a cluster and the team location (d) selected cluster which satisfies spatial constraint and skill constraint. (e) The max-flow network to choose the successful individuals for the service.

the structural neighborhood of node u is $N_u = \{v \in V | ((u,v) \in E)\} \cup \{u\}$. And $\omega-$skewness for each edge can be defined as,

$$\omega(u,v) = min\left\{\frac{d_u}{d_v}, \frac{d_v}{d_u}\right\} \qquad (3)$$

where $d_i = |N_i|$. If $\omega(u,v) < \epsilon^2$, then the edge (u,v) is considered dissimilar and prune from the graph [20].

4   Our Framework

The Geo-Social team formation framework, DSCAN-GSTF consists of three primary processes. 1) Distribute G into multiple machines and perform local clustering. 2) Choose the cluster proximate to the location. 3) Apply parametric flow algorithm to select a competent Geo-Social team of experts. We describe each step one by one in the following sections.

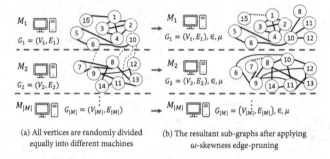

(a) All vertices are randomly divided   (b) The resultant sub-graphs after applying
equally into different machines ω-skewness edge-pruning

Fig. 3. The overview of parallel processing by using DSCAN to replace the process of Fig. 2 (b).

4.1   Network Distribution

The GeoSNs are very large networks with millions of edges and vertices. We replicated DSCAN [20] framework in our application. A given large network G is randomly divided into equal size of sub-graphs $\{G_1, G_2, \ldots, G_n\}$. We then deploy each sub-graph into separate machine M_i as shown in Fig. 3 (a). However, sub-graphs G_i and G_j might have neighbor nodes with higher structural similarity ($\geq \mu$). Those nodes should communicate across machines M_i and M_j. So, DSCAN employs skewness-aware edge-pruning to keep a low communication cost for billion-edge graphs [20], which is shown in Fig. 3 (b). The skewness-aware edge-pruning drops unnecessary edges to avoid the unwanted communication cost among the machines and moves missing neighbors of nodes which have a high structural similarity, from one sub-graph to another.

The sub-graphs which are placed in each machine, are again clustered based on the structural similarity [26]. Here, DSCAN finds all core nodes in each sub-graph and constructs clusters with the nodes which have high structural similarities. Additionally, we store the list of skills of each cluster. The example of resultant clusters and the list of skills are displayed in Fig. 2 (b).

4.2   Suitable Cluster Selection

From the list of clusters, we then select the clusters that satisfy the skill constraints of the given query. To ensure the spatial proximity, we evaluate the Euclidean distance between each cluster and the location ∇ where the service is required (Fig. 2 (c)). We order these distances in ascending order and choose the nearest one that is satisfy the required list of skills as shown in Fig. 2 (d). The selected one is then sent to searching algorithms to find a competent geo-social team. We discuss this process in the following section.

4.3   Geo-social Team Formation

We propose a polynomial searching algorithm based on the parametric flow network [8] to find a competent Geo-Social team (B). We describe the preliminaries of the parametric flow network one by one.

Flow Network: A flow network $N_G = (N_V, N_E)$ is a directed graph that contains a source node s, a target node t, a set of middle nodes N_V, and directed edge set N_E. Additionally, each edge has a weight and receive a flow. An edge's weight cannot be exceeded by the amount of flow that passes through it.

Max Flow and min $s - t$ cut: Let's say f is a flow of N_G, the flow across an $s - t$ cut (S, T) divide its nodes into S and T parts so that the sum of the capacities across S and T is minimized. So, the maximum amount of flow moving from an $s - t$ cut in N_G, say $f(S, T)$ is equal to the total weight of the edges in a minimum cut, $\sum_{u \in S, v \in T} f(u, v)$.

Preflow: A PreFlow f on N_G is a real-valued function that satisfies the capacity and anti-symmetric constraints on node pairs. The relaxation of the conservation constraint can be defined as, $\sum_{u \in V(D)} f(u, v) \geq 0, \forall u \in V \backslash \{s\}$

Valid Labelling: A valid labelling h for a preflow f is a function which is attached to the vertices and has positive integers, such that $h(t) = 0$, $h(s) = |N_V(N_G)|$, where $|N_V(N_G)|$ is the number of vertices in network N_G [3]. For every directed edge from node v to u, the relabeling of $h(v) \leq h(u)+1$ should be created to have a valid flow. In other words, for any node v is a valid labelling if $h(v) \leq min\{h_f(v,t), h_f(v,s) + |V(D)|\}$. The purpose of such labelling $h(v)$ is to estimate the shortest distance from the vertex v to s or t [7].

Calculation of min s-t cut: After running the max-flow algorithm, a minimum cut can be found as follows. For each node $v \in V$, replace $h(v)$ by $min\{h_f(s,v), h_f(t,v) + |N_V(N_G)|\}$. Now the $s - t$ cut is equal to $S = \{v | h(v) \geq |N_V(N_G)|\}$ where the sink partition T is of the minimum size.

Parametric Network Flow: The maximum or minimum value of the flow is determined using a max-flow algorithm based on some criteria. In a parametric-flow network N_R, the capacities on arcs out of s and into t are functions of a real-valued parameter λ, and edges possess the following characteristics [8]. For all $v \neq t$ the cost of the edges from source node to v nodes $C_{(s,v)}(\lambda)$ is a non-decreasing function of λ. Also, for all $v \neq s$ the cost of the edges from v nodes to target node t, $C_{(v,t)}(\lambda)$ is a non-increasing function of λ. And finally, for all $v \neq s$ and $v \neq t$ the cost of the edges from node v to node u, and $C_{(u,v)}(\lambda)$ is a constant. Parametric networks measure maximum flow or minimum cut based on a particular parameter value λ.

Triangle in Graphs: A triangle in G is a cycle of length 3. A triangle generated on vertices $u, v, w \in V(G)$ is denoted as $Tri(uvw)$.

Context Weighted Density (CW): In the selected subgraph $H \subset G$ which satisfies the requirement of query Q, vertices that are related to the query Q may be loosely or densely connected. To balance both these situation, we decided to evaluate context weighted density, (CW) so that we can have cohesive group [22]. The context weighted density, (CW) can be calculated with the use of both weighted triangle density and weighted edge density. For a given edge $(u, v) \in E(H)$, $(u, v, w) \in Tri(H)$, the context scores can be defined as below,

$$\text{Edge context score:} w(e(u,v)) = |Q \cap A(u)| + |Q \cap A(v)| \quad (4)$$

$$\text{Triangle context score:} W(T(u,v,w)) = \sum_{e \in \{(u,v),(u,w),(v,w)\}} w(e) \quad (5)$$

where $w(e(u,v))$ is the weight of edge (u, v) and $A(u)$ and $A(u)$ are the set of attributes of vertex u and v respectively.

$$\text{context weighted density: } CW(H) = \frac{\sum_{\Delta \in Tri(\Delta)} w(\Delta) + \sum_{e \in E(H)} w(e)}{|V(H)|} \quad (6)$$

Algorithm 1 shows how to find required skills using a tailored parametric preflow algorithm. It starts by considering the whole input subgraph H as a candidate team. The candidate team is the group of members who satisfies the query criteria. In the line 2, We construct a parametric flow network based on the steps explained in the part below. Then, we use the stop condition in line 3 to check whether the subgraph H itself is a candidate team or not. If not we generate a better solution by solving sub problem $l(ad_0, H_0 \leftarrow H_0')$, is defined as below [3],

$$l(ad_0, H_0 \leftarrow H_0') = \sum_{\Delta \in Tri(H_0')} w(\Delta) + \sum_{e \in E(H_0')} w(e) - ad_0 \times (|V(H_0')|) \quad (7)$$

Algorithm 1 considers the progressively modified $ad(H_0)$ as a parameterized capacity in N_R. The overall structure of the algorithm is similar to optimization

---

Algorithm 1. Skills Query Search Algorithm

Input: cluster $H \in G$, Query Q

1: $H_0 \leftarrow H$, $ad_0 \leftarrow AD(H)$
2: Construct an adapted parametric flow network N_R and $\lambda = AD(H_0)$
3: obtain H_0' from min s-t cut in N_R
4: **while** $l(ad_0, H_0 \leftarrow H_0')) \neq 0$ **do**
5:    $ad_0 \leftarrow AD(H_0), \lambda = ad_0$
6:    obtain H_0' from min s-t cut in N_R
7: **end while**
8: **return** H_0

---

algorithm, i.e., it continuously generates H_0 with higher context weighted density until reaching the stop condition. During each iteration, internally the algorithm maintains preflow labels via updating the labels computed from the previous iteration. In order to compute H_0', preflow value and some edge capacities are updated according to H_0 generated in the previous iteration. The improved solution gets generated repeatedly until the stop condition is met, i.e., a candidate team is found.

4.4   Complexity Analysis

Assume $|V| = n$ and $|E| = m$. In first step of Geo-social team Formation, it takes $O(m^{1.5})$ time to compute structural similarity. As a result, on each machine M_i extracting all the core nodes from G_i takes $O(m^{1.5})$ time. Consequently finding all dissimilar edges of E_i requires $O(\frac{m}{|M|})$ time. The skills checking complexity can be bounded by $O(|V(cluster)^3|)$, making use of the maximum-flow algorithm. However, providing parametric-network flow help us to solve this in a time complexity of solving one min-cut problem.

5   Experimental Analysis

We conducted experiments to demonstrate the efficiency and effectiveness of our framework. From the efficiency point of view, we show that the Geo-Social team formation model is faster than the state-of-the-art algorithms on large graphs. The proposed framework finds a resulting team with specified features in a Geo-social network having 1.5 billion edges within $8s$. Furthermore, for demonstrating the effectiveness of DSCAN-GSTF, two illustrative queries are analyzed on various real datasets and various metrics based on spatial and social cohesiveness.

Dataset: Table 1 describes the statistical information of five real-world GeoSNs with ground-truth clusters use to evaluate our framework.

Table 1. Statistics of real-world datasets.

Dataset	# of Nodes	# of Edges	Ave-Deg
Gowalla [5]	196,591	950,327	9.177
Dianping [2]	2,673,970	1,922,977	12.184
Orkut-2007 [25]	3,072,626	34,370,166	76.277
Ljournal-2008 [14]	5,363,260	79,023,142	14.734
Twitter-2010 [11]	41,652,098	1,468,365,182	35.253

5.1 Experiment Setup

We compare our framework, DSCAN-GSTF, with state-of-the-art models MKC-SSG [4] and geo-social group queries model (GSGQ) [28]. The MKCSSG model satisfies minimum keyword, contribution capacity, as well as social and spatial constraints. The GSGQ did not consider the required number of experts for each skill. Therefore, we change the GSGQ and add the skill constraint to the team's required skills query such that the skills attributes of the members in the resulting team should cover all required skills.

All the above models are implemented in python using NetworkX, Panda, Tensorflow, Numpy, and pyWebGraph libraries. For the distributed and multiple processing in DSCAN-GSTF, we used MPI. All the experiments are performed on a computer cluster of 16 machines with an interconnecting speed of 9.6 GB/s running GNU/Ubuntu Linux 64-bit. Furthermore, each machine's specifications were Intel Xeon E5-2665 64-bit CPU and 256 GB of RAM (8 GB/core). Moreover, MKCSSG and GSGQ are implemented on one machine since they are not distributed algorithms. Each model is executed 20 times, and the average score is recorded.

5.2 Effectiveness Evaluation

To show the effectiveness of the Geo-Social team formation framework, we analyzed two representative queries on the Gowalla dataset.

Query 1: Parameters are set as follow: location $\nabla = (36.11, -115.13)$, Skills $S = \{salad, chicken, beef, BBQ\}$, Number of each skills $E = \{10, 10, 10, 10\}$, Contribution capacity $c = 4$. This query can be used to find fans of BBQ party around Las Vegas. We assume that the tweets of each user is their favorite dish. We set $\epsilon = 0.5$ and $\mu = 10$ for the first query on Geo-Social team formation framework.

Query 2: intends to create a music band around Las Vegas. Query 2 parameters are set as follows: location $\nabla = (36.11, -115.13)$, Skills $S = \{guitar, piano, violin\}$, Number of each skills $E = \{2, 1, 2\}$, Contribution capacity $c = 1$. We set $\epsilon = 0.5$ and $\mu = 10$ for the second query on Geo-Social team formation model.

Table 2. Effectiveness evaluation

Model	SC	GD	ED	Query
GSGQ	0.14	0.71	0.51	Query 1: food
MKCSSG	0.18	0.41	0.67	
DSCAN-GSTF	0.21	0.23	0.74	
GSGQ	0.56	0.54	0.38	Query 2: music band
MKCSSG	0.67	0.28	0.63	
DSCAN-GSTF	0.56	0.23	0.81	
GSGQ	0.27	0.62	0.43	Query 3: board game
MKCSSG	0.43	0.32	0.52	
DSCAN-GSTF	0.42	0.25	0.68	

Query 3: intends to create a board game groups. Query 3 parameters are set as follows: location ∇ = $(36.11, -115.13)$, Skills S = $\{chess, backgammon, monopoly\}$, Number of each skills E = $\{2, 8, 10\}$, Contribution capacity $c = 2$. We set $\epsilon = 0.4$ and $\mu = 9$ for the second query on Geo-Social team formation model.

Evaluation Metrics: Here we define the evaluation metrics use to compare the performance of state-of-the-art methods with DSCAN-GSTF.

Spatial Closeness (SC): The spatial cohesiveness is to show how closely the team members are located to ∇. Our framework uses the spatial distance of one random member of the result team B to the query location ∇. $SC_{\nabla,B} = \{(Euclidean\_Distance(\nabla, u)), u \in V(B)\}$

Graph Diameter (GD): It calculates the topological length or extent of a graph by counting the number of edges in the shortest path between the most distant vertices. In other words, graph diameter indicates the **social closeness** of the team. $GD_B = max\{ShortestPath(v, w))|(v, w) \in V(B)\}$

Edge Density (ED): We consider another parameter ED to show the social cohesiveness. $ED_B = |E(B)|/|V(B)|$, where $E(B)$ is the number of edges in team B and $V(B)$ is the number of vertices in team B.

The comparison results of analyzed metrics are presented in Table 2. The results are normalized to a value between 0 and 1. Results with a lower score are better except for Edge Density (ED). Overall speaking, we can see DSCAN-GSTF has outperformed in spatial and social cohesiveness in both queries. However, the spatial distance is not significantly better, but the social cohesiveness shows excellent improvement in both queries. Furthermore, applying graph structural communities in DSCAN-GSTF framework improves the social Cohesion

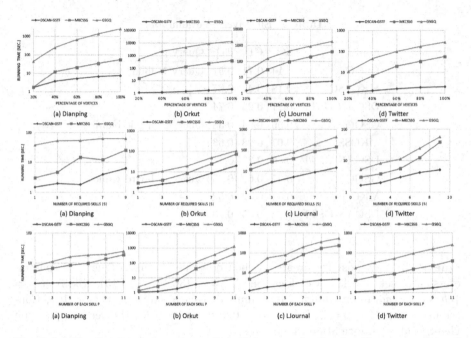

Fig. 4. (a) The first raw is to compare the efficiency evaluation based on percentage of vertices. (b) The middle raw is to compare the efficiency evaluation on various number of required skills. (c) The last raw is to compare the efficiency evaluation on various requirement on expert set R.

and indicates teams with much less graph diameter than GSGQ and MKCSSG, which utilize the minimum degree and c-truss constraint, respectively.

5.3 Efficiency Evaluation

To compare the performance of various models, we use the running time of the queries. We compare the efficiency of GSGQ and MKCSSG with our proposed model, DSCAN-GSTF. Our experiments uses various parameter settings for a query: percentage of vertices, sets of required skills $|S| = \{1, 3, 5, 7, 9\}$, and the minimum number of each skill $E = \{1, 3, 5, 7, 9\}$. We set the default value of both $|S|$ and E to 3. The location for each query is created randomly. We select reasonable values for ϵ and μ based on each dataset. In Dianping dataset $\epsilon = 0.3$ and $\mu = 2$, in the Orkut dataset $\epsilon = 0.5$ and $\mu = 5$, in the LJournal dataset $\epsilon = 0.6$ and $\mu = 5$, and finally in the Twitter dataset $\epsilon = 0.5$ and $\mu = 6$. When a parameter is changing for evaluation, other parameter values are set to their default value.

We divide each dataset into various percentages to evaluate the scalability of proposed model. The result is presented in Fig. 4 (a) for different datasets while comparing various methods. In general, our DSCAN-GSTF is much more scalable compared to other methods on different datasets. That is because of the methodology and substantially the properties of DSCAN-GSTF which can limit the search space in quicker time while preserving optimum results.

Figure 4 (b) shows the running time when the number of required skills increases for different datasets; as the required skills increase, the running time for all methods increases. However, this increase is slower in DSCAN-GSTF because checking existing skills on each cluster using the attached summation list of skills has constant time complexity. When $|S|$ is small, the GSGQ requires comparatively high running time to find optimum results. However, when the $|S|$ grows, it provides a result in half a minute. In Fig. 4 (c), changing the number of required skills E on each dataset using various models is presented. Again, for all the datasets, the running time increase as the required number of skills is increased. Nevertheless, because of distributed environment in DSCAN-GSTF, the increasing running time is on a slow increasing slope.

6   Conclusions

This paper has explored the Geo-Social team Formation framework and proposed a new model DSCAN-GSTF. In this, we incorporated various criteria to replicate the real-world scenario and exploited DSCAN for the efficient process of large networks. The DSCAN-GSTF introduced a novel polynomial algorithm based on a parametric flow algorithm to identify the successful team members for impromptu activities from GeoSNs. We compared our proposed DSCAN-GSTF model with the state-of-art methods, MKCSSG and GSGQ. Extensive experiments were conducted to examine the efficiency and effectiveness of the proposed model on four real-world datasets and recorded the running time under various system settings. Overall, our proposed model generated the output faster than the state-of-the-art methods. As for future work, we plan to extend DSCAN-GSTF to incorporate more sophisticated queries.

References

1. Armenatzoglou, N., Papadopoulos, S., Papadias, D.: A general framework for geo-social query processing. Proc. VLDB Endow. 913–924 (2013)
2. Bu, J., et al.:: ASAP: a Chinese review dataset towards aspect category sentiment analysis and rating prediction (2021)
3. Chen, L.: Efficient cohesive subgraph search in big attributed graph data (2018)
4. Chen, L., Liu, C., Zhou, R., Xu, J., Yu, J.X., Li, J.: Finding effective geo-social group for impromptu activities with diverse demands. In: ACM SIGKDD International Conference on Knowledge Discovery and Data Mining (2020)
5. Cho, E., Myers, S.A., Leskovec, J.: Friendship and mobility: user movement in location-based social networks. In: Proceedings of the 17th ACM SIGKDD International Conference on Knowledge Discovery and Data Mining (2011)
6. Cliquet, G., Baray, J.: Location-Based Marketing: Geomarketing and Geolocation. Wiley, Hoboken (2020)
7. Cormen, T.H., Leiserson, C.E., Rivest, R.L., Stein, C.: Introduction to Algorithms. MIT Press, Cambridge (2009)

8. Gallo, G., Grigoriadis, M.D., Tarjan, R.E.: A fast parametric maximum flow algorithm and applications. SIAM J. Comput. **18**(1), 30–55 (1989)
9. Huang, Q., Liu, Y.: On geo-social network services. In: 2009 17th International Conference on Geoinformatics, pp. 1–6. IEEE (2009)
10. Kargar, M., An, A., Zihayat, M.: Efficient bi-objective team formation in social networks. In: Flach, P.A., De Bie, T., Cristianini, N. (eds.) ECML PKDD 2012. LNCS (LNAI), vol. 7524, pp. 483–498. Springer, Heidelberg (2012). https://doi.org/10.1007/978-3-642-33486-3_31
11. Kwak, H., Lee, C., Park, H., Moon, S: What is Twitter, a social network or a news media? In: Proceedings of the 19th International Conference on World Wide Web (2010)
12. Lappas, T., Liu, K., Terzi, E.: Finding a team of experts in social networks. In: Proceedings of the 15th ACM SIGKDD International Conference on Knowledge Discovery and Data Mining, pp. 467–476 (2009)
13. Liu, W., Sun, W., Chen, C., Huang, Y., Jing, Y., Chen, K.: Circle of friend query in geo-social networks. In: Lee, S., Peng, Z., Zhou, X., Moon, Y.-S., Unland, R., Yoo, J. (eds.) DASFAA 2012. LNCS, vol. 7239, pp. 126–137. Springer, Heidelberg (2012). https://doi.org/10.1007/978-3-642-29035-0_9
14. Rossi, R., Ahmed, N.: The network data repository with interactive graph analytics and visualization. In: Proceedings of the AAAI Conference on Artificial Intelligence, vol. 29 (2015)
15. Selvarajah, K.: Investigation of team formation in dynamic social networks (2020)
16. Selvarajah, K., Zadeh, P.M., Kargar, M., Kobti, Z.: Identifying a team of experts in social networks using a cultural algorithm. Procedia Comput. Sci. (2019)
17. Selvarajah, K., Zadeh, P.M., Kobti, Z., Kargar, M., Ishraque, M.T., Pfaff, K.: Team formation in community-based palliative care. In: Innovations in Intelligent Systems and Applications (2018)
18. Shen, C.Y., Yang, D.N., Huang, L.H., Lee, W.C., Chen, M.S.: Socio-spatial group queries for impromptu activity planning. IEEE Trans. Knowl. Data Eng. **28**(1), 196–210 (2015)
19. Shen, C.Y., Yang, D.N., Lee, W.C., Chen, M.S.: Spatial-proximity optimization for rapid task group deployment. ACM Trans. Knowl. Discov. Data (TKDD) **10**(4), 1–36 (2016)
20. Shiokawa, H., Takahashi, T.: DSCAN: distributed structural graph clustering for billion-edge graphs. In: Hartmann, S., Küng, J., Kotsis, G., Tjoa, A.M., Khalil, I. (eds.) DEXA 2020. LNCS, vol. 12391, pp. 38–54. Springer, Cham (2020). https://doi.org/10.1007/978-3-030-59003-1_3
21. Sohail, A., Cheema, M.A., Taniar, D.: Geo-social temporal top-k queries in location-based social networks. In: Borovica-Gajic, R., Qi, J., Wang, W. (eds.) ADC 2020. LNCS, vol. 12008, pp. 147–160. Springer, Cham (2020). https://doi.org/10.1007/978-3-030-39469-1_12
22. Tsourakakis, C.: The k-clique densest subgraph problem. In: Proceedings of the 24th International Conference on World Wide Web, pp. 1122–1132 (2015)
23. Valverde-Rebaza, J.C., Roche, M., Poncelet, P., de Andrade Lopes, A.: The role of location and social strength for friendship prediction in location-based social networks. Inf. Process. Manag. **54**(4), 475–489 (2018)
24. Xu, X., Yuruk, N., Feng, Z., Schweiger, T.A.: Scan: a structural clustering algorithm for networks. In: Proceedings of the 13th ACM SIGKDD International Conference on Knowledge Discovery and Data Mining, pp. 824–833 (2007)

25. Yang, J., Leskovec, J.: Defining and evaluating network communities based on ground-truth. Knowl. Inf. Syst. **42**(1), 181–213 (2013). https://doi.org/10.1007/s10115-013-0693-z
26. Zhao, W., Martha, V., Xu, X.: PSCAN: a parallel structural clustering algorithm for big networks in MapReduce. In: International Conference on Advanced Information Networking and Applications (AINA) (2013)
27. Zheng, Y.: Location-based social networks: users. In: Zheng, Y., Zhou, X. (eds.) Computing with Spatial Trajectories, pp. 243–276. Springer, New York (2011). https://doi.org/10.1007/978-1-4614-1629-6_8
28. Zhu, Q., Hu, H., Xu, C., Xu, J., Lee, W.-C.: Geo-social group queries with minimum acquaintance constraints. VLDB J. **26**(5), 709–727 (2017). https://doi.org/10.1007/s00778-017-0473-6

Data Allocation with Neural Similarity Estimation for Data-Intensive Computing

Ralf Vamosi and Erich Schikuta[✉]

Faculty of Computer Science, University of Vienna, Vienna, Austria
ralf.vamosi@cs.univie.ac.at, erich.schikuta@univie.ac.at

Abstract. Science collaborations such as ATLAS at the high-energy particle accelerator at CERN use a computer grid to run expensive computational tasks on massive, distributed data sets.

Dealing with big data on a grid demands workload management and data allocation to maintain a continuous workflow. Data allocation in a computer grid necessitates some data placement policy that is conditioned on the resources of the system and the usage of data.

In part, automatic and manual data policies shall achieve a short time-to-result. There are efforts to improve data policies. Data placement/allocation is vital to coping with the increasing amount of data processing in different data centers. A data allocation/placement policy decides which locations sub-sets of data are to be placed.

In this paper, a novel approach copes with the bottleneck related to wide-area file transfers between data centers and large distributed data sets with high dimensionality. The model estimates similar data with a neural network on sparse and uncertain observations and then proceeds with the allocation process. The allocation process comprises evolutionary data allocation for finding near-optimal solutions and improves over 5% on network transfers for the given data centers.

Keywords: Data allocation · Data placement · Similarity estimation · Parallel computing · Wide-area file transfers

1 Introduction

Physics collaborations such as the ATLAS collaboration [10] at CERN store their data as files in a worldwide computing grid. The LHC Computing Grid [3] provides distributed storage and processing for physics data from extensive experiments. Experts from all over the world submit tasks with the data stored as files across many data centers. The data centers can be considered geographically distant sites that spawn a fully connected network, as shown in Fig. 1 as an example. Many data-intensive, high-performance applications use large numbers of files that are in some way labeled or categorized. In the case of ATLAS, data sets label a various number of files that jobs can collectively handle. A computational job is a process that reads a data set and eventually outputs another

© The Author(s), under exclusive license to Springer Nature Switzerland AG 2022
D. Groen et al. (Eds.): ICCS 2022, LNCS 13352, pp. 534–546, 2022.
https://doi.org/10.1007/978-3-031-08757-8_45

data set. Data sets can be used none or several times. The population of data sets is changing. Some may be removed after their lifespan, and data sets are created from time to time to label files. Because of this changing population of data sets, they cannot be the basic unit of data placement/allocation, but the constituent files are.

Regardless of the particular content, data sets are managed by users. A user defines a class on its own, using particular data sets. Their work interest and funding correlate with data sets. Data sets in a more or less narrow range will be processed to fulfill their research. Those have some commonalities, so some aspects are equal or similar. A metric based on machine learning will be introduced later to find similarities within data sets.

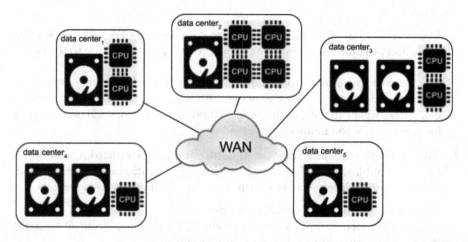

Fig. 1. Example of five data centers inter-connected with wide area network (WAN), over which each data center can read files on an external data center, which incurs costs. Each HPC site facilitates different amount of resources.

In our case, there are two types of jobs to be executed in large quantities:

– Producing final data with parameters and calibration data from the experiment. Calibration data is updated from time to time and therefore production must be repeated, for example.
– Analysis to find events in the productive data. Examples around the well-known Higgs boson H^0 are $t + \bar{t}$ into H^0, or $\gamma + \gamma$ into H^0. An H^0 boson is formed in both processes, but they are different processes with different probability amplitudes.

Jobs for production and analysis are submitted onto data sets. Computational resources are facilitated in the form of job queues at data centers whose task is to run jobs. A job requires the entire data set to be at the local data center. Missing files of the data set must be transferred to the job, that is, the location of the compute job.

The network latency causes a significant delay in the workflow. The *wide area network (WAN)*, the network between data centers, is a scarce resource and represents a bottleneck in a data-intensive workflow [14]. The way to improve the situation related to the WAN is to leave wide-area transfers out, i.e., local processing for each data set. However, jobs cannot be placed arbitrarily to the appropriate data sets: Jobs are allocated to proper sites. First, they must be compatible; the site must support the requested computation. Say, a high-memory worker is needed. Load balancing takes place according to provided computing volumes in data centers. The one with the majority of files of proper sites is preferred as the target. This process continues, and sites are filled with jobs.

On the other hand, data cannot be freely allocated as there are natural storage limitations. Our focus is to tackle data allocation under the following condition:

- Every time a job kicks in, missing input data must be transferred over the bottleneck, the inter-data center network/WAN, from the source to the target site.
- The network consists of non-uniform sites that provide resources with considerable differences. Some possess more storage capacity, and others provide more processing capabilities due to local funding constraints and even support for various user communities.

In order to deal with a large amount of data, clustering is utilized. Clustering is a broad field applied for different tasks such as classification and segmentation, matching commonalities such as identifying normal samples versus outliers.

The *novel contributions* of this paper are:

A *similarity metric*, a distance, is introduced for data sets to be able to put clustering into action and to decide on proper sub-sets.

Combined with this similarity metric, a *loss function* related to the data allocation task is induced. Based on this loss, allocation can be performed regarding spatial patterns of data set use, that is, locations at data centers. Following patterns of data use is necessary for improving data allocation because data sets will be differently used depending on jobs/tasks on distant sites. The jobs will be distributed in some way across data centers. Some of these data centers are capable of more than others. Say, some store more data sets or provision a processing capability for some job class such as large memory. Large memory jobs require larger memory on the CPU.

2   State of the Art

The data allocation problem is synonymously referred to hereto as file allocation. It was investigated when distributed databases were studied, and parallelization had to be utilized.

One of the first data placement/allocation papers was [5]. The work models the task on different abstraction levels. It is further proven that the data placement problem is very difficult to solve and generally NP-Complete.

The file placement problem is investigated under concurrency conditions to build a model with storage cost and communication cost in [17]. Constraints are the multiplicity of databases, variable routing of data, and available capacity of the network nodes.

Some work attempts to cluster data sets according to their inter-dependency [11]. Subsequently to clustering, clusters are stored on separate machines. A different clustering algorithm is described in [25] that uses k-means algorithm for finding locations for the clustered data, resulting in task allocation to the data centers with most of the input data set. This is comparable to the ATLAS workflow. This paper utilizes data sets as small clusters and imposes several conditions such as uncertainty and sparsity on the data sets.

Evolutionary algorithms have been applied to almost any kind of problem, and it is no surprise that these were applied to this kind of problem as well. In [12], data allocation strategies have been investigated to reduce transaction costs. A genetic algorithm was used here to limit communication effort between data centers by balancing the load.

Placement of files and replication across different nodes may improve on metrics such as job execution, makespan, and bandwidth [8].

Further efforts have been undertaken in previous studies for database opti-mizations. The authors of [22] discuss database allocation optimization and pro-pose a mathematical model concerning average waiting time. Other database approaches attempt to arrange data effectively over the network nodes, such as in [1,2,6]. However, these studies investigate idealized database cases. For example, they focus on a single query type or do not consider any constraints on communication characteristics. Room for improvement would comprise user behavior and workflow characteristics. Analysis of access patterns can be bene-ficial for network utilization.

A well-known approach is ranking data according to the number of accesses per time unit. This characteristic is referred to, for example, as data *popular-ity* [7]. In [7], a successful popularity model is established which uses historical data. A popularity model is implemented as an autonomous service for finding obsolete data and used in the cleaning process [15,20].

Due to complexity and variety, simplified models and local optimizations were studied and applied. A thorough analysis of data usage and data optimization for data grids is necessary. Storage resources and the use of data has to be appropriately treated in the process of file placement [18,19].

Summing up, the actual research on data partitioning and allocation tech-niques is specific, and often the models aim at a narrow case. Especially in data-intensive high-performance computing (HPC), large volumes of data are pro-cessed, and patterns for data usage have to be observed and taken into account to utilize the performance and improve the workflow fully. At this point, we would like to present our research. In general, data management strongly impacts usability, performance, and cost.

3   Methods

3.1   Background

Data sets collect files in various numbers by giving them a common name (data set name). To certain data sets, jobs will be submitted by a magnitude of 1000 users around the globe. The data set name is a high-dimensional pattern on an abstract level. It consists of a text string in variable lengths, with different sub-strings such as numbers and terms. A data set name (label) looks like

$$mc15.13TeV.362233.Sherpa.CT10.Zee.Pt70$$
$$.140.BFilter.ckkw30.evgen.EVNT.e4558.tid07027172.00$$

There are many identifiers composed into this name to describe the data set. mc15 stands for Monte Carlo; 13 TeV stands for $13 * 10^{12}$ electron volt, which implies the energy of the run; and so on. With this notation, it is challenging to find interrelations between data set names. At least, users manage jobs processing and data sets. The decision is a user-dependent process that was introduced in Sect. 1. User's co-variates are background and funding, and they do usually not select randomly at their whim. Creating data sets and submitting jobs is a confusing process from the outside. However, the common factor is the user between task and data.

Two important points must be addressed in a data allocation policy:

- How likely is it, that data sets will be read by a job of a certain type. It is certain that data sets that are not processed, do not incur any costs from a network's point of view. The opposite case is when a data set is processed at least once or several times.
- What is the impact from the cost perspective, data sets residing on specific nodes.

In the model context, the term node is used for data centers. A node represents a black box, which is not decomposed in its parts, such as network storage, different computing nodes, etc., but rather is described on a top-level with the provision of resources from a job's perspective.

With data sets also comes a lack of information. The use of data sets is uncertain two-folds:

- Data sets change over time. Some of them will overlap.
- Furthermore, there is no information on how likely they will be used for most data sets. Only a tiny minority of data sets have a popularity value that indicates how likely they will be read. There are some investigations on popularity, which are not covered in this work [4].

Our model consists of two sub-models joined together to perform the two steps depicted in Fig. 2:

1. Transform the variable space of data sets into a metric space. This model is introduced to generate variables of data sets that allow better handling. The first model in Sect. 3.2 covers this part.

2. Generate disjoint file sets from data sets and allocate them to nodes. This can be pictured in two dimensions as on the right side in Fig. 2. The allocation is the process that divides files, comprised in data sets, into the number of nodes and maps them to the nodes, which are data centers in grid. The necessary steps are done in the second model in Sect. 3.3.

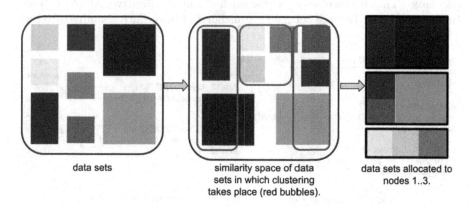

data sets | similarity space of data sets in which clustering takes place (red bubbles). | data sets allocated to nodes 1..3.

Fig. 2. Two models for two transformations of the data sets. First, the data sets are mapped into a similarity space. Second, the allocation model disjoints the data sets into proper sub-sets to allocate to nodes. (Color figure online)

3.2 Neural Similarity Estimator

The relationship between users and data sets, and thereby data sets to others, has been discussed. The similarity model proposed in this work relates data sets with others. The neural similarity model is placed first to extend the variables-space for the allocation.

With the similarity model, given a data set as input, another similar data set may be found in the population of data sets. Even new data sets that have not yet been seen can be assigned to other data sets. This is important because new data sets are constantly being created during operation. On the one hand, similarity plays a role in the allocation in order to select a specific kind of data sets. On the other hand, data sets also depend on each other because data sets can overlap several others. They will be more or less mixed, which makes them not separable.

The virtue of such a similarity model is that it can handle complex and inter-related (text) strings. Triplet learning sometimes referred to as triplet loss or triplet comparison, comes in helpful for this task. This technique makes inferences about observations based on commonalities and non-commonalities within the variables [24]. A specific loss function is applied to learn a similarity or distance metric by recognizing similarities (and dissimilarities). It was successfully applied for large learning applications in image similarities [9,23] as well as a variety of digital image processing tasks such as face recognition [16].

540   R. Vamosi and E. Schikuta

In the case of face recognition and person re-identification, similar samples are images of the same person. Figure 3 illustrates the core idea. The so-called anchor a and the positive sample + are from the same user, and the other example - is from another user. The triplet comparison operates on three different example inputs. On average, the a and + are more similar to each other. During network training, the cost function applied to a triplet of samples is gradually decreased. The distance between the anchor and the positive sample becomes smaller, while the distance between the anchor and the negative sample increases.

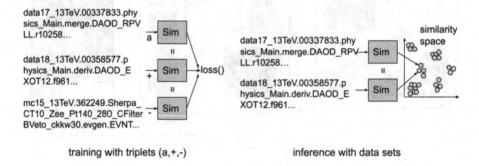

training with triplets (a,+,-)   inference with data sets

Fig. 3. Training of the model is shown on the left, in which triplets of data sets (a, +, −) are fed from the training set while the model tries to lower the loss(). The inference is shown on the right. Data sets are mapped into the latent similarity space, where next data sets have higher similarity.

In our case, data sets with their names are the input from users. As data set names are depicted in Sect. 3.1, the first step is text pre-processing to obtain a tuple of word vectors that are the representations for the data set names. Terms and numbers occurring only once in the training set of data set names are replaced by a default value since they are not informative. For example, 933 and aabb are replaced as default value.

The model consists of several steps:

1. The data set names are tokenized. This process makes a discrete set with unique numbers from each alpha-numeric subtext. In a standard text case, this would be the word level, and each word would convert into a number in the discrete co-domain.
2. Each token is converted into a word vector.
3. In training, a latent metric space is spawned under the loss function from Eq. 1. Data sets are just passed through the trained model in the inference phase. Data sets become labeled with the vector of the low-dimensional neural co-domain of the similarity model Fig. 3.

The model's network architecture is strongly inspired by image recognition projects with a triplet loss, like [13], we do not need a particular network type such as CNN for the short input vectors at hand. Even though a CNN would be possible, we go with a fully connected multi-layer feed-forward network: a multi-layer perceptron. As for other recognition tasks, as mentioned, a triplet loss is used in the network:

$$loss = max[d(\ a,\ +) - d(\ a,\ -) + margin, 0] \tag{1}$$

where a triplet is $(a, +, -)$ with some data set, a, a positive example data set from the same user $+$, and a negative example data set from some other user, $-$.

The training phase takes a magnitude of one day on a single GPU. In the prediction phase, the model maps each piece of input data set into its output space, a metric space in which similar inputs are placed closer to each other. This is depicted on the right in Fig. 3. Inference on one data set is just a mapping with the fixed weights in the model and takes a magnitude of 1 ms.

3.3 Data Allocation Model

For the allocation process, an evolutionary clustering method has been developed. The goal is to find improved data allocations over the course of the run time in terms of a loss according to the following loss function decreases:

$$cost_{tx}(\boldsymbol{S}) = \sum_{j \in \{jobs\}} cost_{tx}(S, j) \tag{2}$$

where $cost_{tx}(S, j)$ denotes an affine function of the number of non-local files for job j given $\boldsymbol{S}$, that is, the number of triggered file transfers over the network (WAN) by j.

The data allocation model generates an allocation matrix for all files, $\boldsymbol{S}$. As explained, network transfers between nodes are expensive, which is why the total number of these transfers is regarded in the total loss denoted as $cost_{tx}(S)$. Data center-internal transfers are much faster and do not need to be considered (zero cost). Given the amount of work, i.e., a set of computational jobs, the problem can be formalized as minimization problem with parameter S specifying the allocation, and a loss function $cost_{tx}(S)$ depending on system variables and the set of jobs:

$$\underset{\boldsymbol{S}}{argmin}\ cost_{tx}(\boldsymbol{S})$$

$$so\ that \tag{3}$$

$$\boldsymbol{S}^T \times \mathbf{w}_{file} \leq \mathbf{w}_{store}$$

where $cost_{tx}(\boldsymbol{S})$ is the loss function introduced before, $\mathbf{w}_{file}$ is a column vector which represents the file sizes for $file_1, ..., file_M$ and $\mathbf{w}_{store}$ is a column vector which represents the storage capacities of data centers 1, 2, ..., N. $cost_{tx}(\boldsymbol{S})$ penalizes external heavy transfers. $\mathbf{w}_{store}$ is needed by the model to control the cluster sizes. Each cluster becomes parameterized for the sake of collecting files for sub-sets in the allocation process. No other variables are needed except relative costs, i.e., $cost(\boldsymbol{S}_{new}) - cost(\boldsymbol{S})$ while the model probes the nodes iteratively with variations of data allocations and descent on the plane of the loss function.

The allocation process can be outlined in the following steps:

- Parameterize and prioritize clusters. Set a seed according to popularity and type variable for each cluster.
- Disjoint the data sets into the number of clusters into subsets to be allocated. This produces the allocation matrix S (Eq. 3). Boundary conditions $\mathbf{w}_{store}$ are needed to control the cluster sizes.
- Calculate costs by Eq. 2 and repeat the process.

The presented model here relies on work management and solves the data allocation problem. The model does not see the actual workload, but the cost that comes with the workload related to the data allocation. The consideration by only costs allows for a more flexible way of not having to expose the underlying grid and jobs.

The evolution of the model comes into the described allocation process. After finding a S which betters the situation, it is placed into an internal pool with the label of the cost that indicates the quality or fitness:

- Populate the pool and delete the bottom solutions
- Vary single solutions
- Combine parameters of two solutions into one novel solution

The model runs on a single PC with parallelization. In-memory data on files and data sets are fixed and can be shared. Each data allocation iteration is done on a worker process, which just communicates the parameters of a solution. Other workers take the work on combining solutions.

4   Justification and Evaluation

Data centers/HPC sites (in some jargon, just sites) are referred to here as nodes. The number of nodes in the optimization has been reduced from over 100 to a low 2-digit number, see Fig. 4. Details are listed below. This is a feasible number, and the set can cover the most important ones, say, nodes with the highest capacities. There is an exponential drop-off from the largest to the smallest data centers. At the peak, there are large ones working on the majority of data sets, and at the tail, there are many small ones. Large ones covered in the optimization process imply a high potential for improvement.

For each iteration of the experiment, the following parameters are defined for nodes in such a way that the values are sampled in a uniformly random way:

- the provisioned work performance in terms of average jobs
- the provisioned queues for jobs for two types (A and B)
- the provisioned storage capacity for files

Data is defined in the following way:

- 10 thousand data sets from 100 users are taken. This is used for the validation of the model.

- Each data set possesses 1 to 100 files.
- For training of the similarity model, a subset of the data set user records are taken, a sample size of 2 GB plain text. This record indicates the data set, their files, and the associated user to each data set. This sub-set covers 100 users from the full user records comprising about 25 GB per month. The selected time window of one month is short for picking up short correlations for the demonstration.
- For these data sets, a popularity value is chosen between 0 and 1. Highly active users possess data sets with higher popularity on average, and more frequent data sets are represented with higher popularity. The model can only see 10% of the popularity values.

With the data and the provision on the computer side, the work of the model can be outlined:

- Similarity estimation is done once, and the final similarity values are kept in memory beside the original variables.
- The data allocation part produces a data distribution while observing the total costs according to Eq. 2. Over the course of evolution, the allocation model permutates the output to find a fitter solution. In other words, a solution with a smaller cost.

The improvement by the data allocation can be depicted in the Fig. 4. The model runs three configurations, and in the smallest one, it performs better. This is probably due to the fact that the solution space is smaller. A smaller cardinality makes the model performs iteration quicker, and more vectors in the solution space can be probed. The improvement is shown to be at least 5.5% despite uncertainties in the system that the allocation model cannot control, on the input side, where sparse information comes from the data sets, and second, the work on each node that depends on the jobs and the nodes.

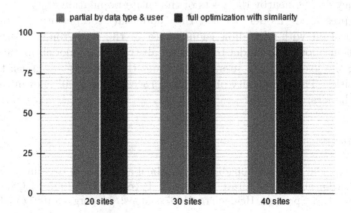

Fig. 4. On the left, data sets are allocated by clustering by type and user. The full optimization, on the right, is done with similarity model and the exploitation of popularity. It saves 6%, ranging from almost 6.3% for with 20 sites, and 5.5% points for 40 sites.

5   Conclusion and Outlook

Six percent improvement in the most important data centers is a perceivable saving on time and cost in terms of money. This amount corresponds to the provisioned traffic of one to two large data centers to get an idea. Provisioned bandwidth of data centers is often $10\,\text{Gb/s}$. Our experiment shows that savings are possible without making a structural change or manual investigation. In order to contribute to a better solution, more detailed exploitation of data centers, e.g., including different types of storage, compute nodes, etc., for example, is possible in the following way: A second optimization level can be added, and both algorithms will go hand in hand. Our model works on the top level regarding inter-data center traffic. Another model, with the proper cost metric also a variation of this model, will act on the second level, an entity for the internal costs per data center.

The first part of the composed model, the similarity model from Sect. 3.2, can be updated from time to time. This update process would run in the background to update to the next version of this sub-model.

The core model is the allocation model from Sect. 3.3. It can also be set up in such a way that it operates on a subset of the data sets. Say, new data sets shall be placed according to its policy. Free resources must be observed by the model. A flag would be added to distinguish between data sets to be moved and fixed data sets, i.e., data sets allocated/placed recently [21].

The volume of data sets is small compared to the population in the real grid. Anyhow, the outcome of the data allocation can be used in a top-down approach, in which missing data sets of the entire population would be clustered to the initial data sets. So, for example, 1 data set becomes 10, and 10 becomes 100, until all data sets are included. The solution contains clusters with contiguous elements, as with the solution of the '1' data sets with the heuristic variables' similarity and popularity. In other words, the sparse '1' data sets have a high dependency on the nearby data sets of the entire population.

The presented model shows the potential to perform data placement/allocation with high-dimensional data in the background. Further, the loss function allows the model to be more abstract. The model operates solely regarding the loss and only sees the data and the relevant resources; these are the capacities. The complexity of the data processing with all relevant resources hides in the loss.

References

1. Abdel-Ghaffar, K.A.S., Abbadi, A.E.: Optimal allocation of two-dimensional data (extended abstract). In: Afrati, F., Kolaitis, P. (eds.) ICDT 1997. LNCS, vol. 1186, pp. 409–418. Springer, Heidelberg (1997). https://doi.org/10.1007/3-540-62222-5_60
2. Atallah, M.J., Prabhakar, S.: (almost) Optimal parallel block access to range queries. In: Proceedings of the Nineteenth ACM SIGMOD-SIGACT-SIGART Symposium on Principles of Database Systems, pp. 205–215. ACM (2000)

3. Atlas, C., et al.: Atlas computing: technical design report (2005)
4. Beermann, T., et al.: Methods of data popularity evaluation in the atlas experiment at the LHC. In: EPJ Web of Conferences (2021)
5. Bell, D.A.: Difficult data placement problems. Comput. J. **27**(4), 315–320 (1984)
6. Berchtold, S., Böhm, C., Braunmüller, B., Keim, D.A., Kriegel, H.P.: Fast parallel similarity search in multimedia databases. In: Proceedings of the 1997 ACM SIG-MOD International Conference on Management of Data. SIGMOD 1997, pp. 1–12. ACM, New York (1997). https://doi.org/10.1145/253260.253263
7. Bonacorsi, D., et al.: Exploiting CMS data popularity to model the evolution of data management for run-2 and beyond. J. Phys. Conf. Ser. **664**, 032003 (2015). IOP Publishing
8. Chang, R.S., Chang, H.P.: A dynamic data replication strategy using access-weights in data grids. J. Supercomput. **45**(3), 277–295 (2008)
9. Chechik, G., Sharma, V., Shalit, U., Bengio, S.: Large scale online learning of image similarity through ranking. J. Mach. Learn. Res. **11**(3) (2010)
10. Collaboration, A., et al.: The atlas experiment at the cern large hadron collider (2008)
11. Foster, I., Kesselman, C., Tuecke, S.: The anatomy of the grid: enabling scalable virtual organizations. Int. J. High Perform. Comput. Appl. **15**(3), 200–222 (2001)
12. Guo, W., Wang, X.: A data placement strategy based on genetic algorithm in cloud computing platform. In: 2013 10th Web Information System and Application Conference (WISA), pp. 369–372. IEEE (2013)
13. Hoffer, E., Ailon, N.: Deep metric learning using triplet network. In: Feragen, A., Pelillo, M., Loog, M. (eds.) SIMBAD 2015. LNCS, vol. 9370, pp. 84–92. Springer, Cham (2015). https://doi.org/10.1007/978-3-319-24261-3_7
14. Liu, Y., Liu, Z., Kettimuthu, R., Rao, N., Chen, Z., Foster, I.: Data transfer between scientific facilities-bottleneck analysis, insights and optimizations. In: 2019 19th IEEE/ACM International Symposium on Cluster, Cloud and Grid Computing (CCGRID), pp. 122–131. IEEE (2019)
15. Megino, F.B., et al.: Implementing data placement strategies for the CMS experiment based on a popularity model. J. Phys. Conf. Ser. **396**, 032047 (2012). IOP Publishing
16. Parkhi, O.M., Vedaldi, A., Zisserman, A.: Deep face recognition (2015)
17. Ram, S., Marsten, R.E.: A model for database allocation incorporating a concurrency control mechanism. IEEE Trans. Knowl. Data Eng. **3**(3), 389–395 (1991)
18. Sato, H., Matsuoka, S., Endo, T.: File clustering based replication algorithm in a grid environment. In: Proceedings of the 2009 9th IEEE/ACM International Symposium on Cluster Computing and the Grid, pp. 204–211. IEEE Computer Society (2009)
19. Sato, H., Matsuoka, S., Endo, T., Maruyama, N.: Access-pattern and bandwidth aware file replication algorithm in a grid environment. In: Proceedings of the 2008 9th IEEE/ACM International Conference on Grid Computing, pp. 250–257. IEEE Computer Society (2008)
20. Spiga, D., Giordano, D., Barreiro Megino, F.H.: Optimizing the usage of multi-petabyte storage resources for LHC experiments. In: Proceedings of the EGI Community Forum 2012/EMI Second Technical Conference (EGICF12-EMITC2), 26–30 March 2012. Munich, Germany (2012). https://pos.sissa.it/162/107/
21. Vamosi, R., Lassnig, M., Schikuta, E.: Data allocation service ADAS for the data rebalancing of atlas. In: EPJ Web of Conferences, vol. 214, p. 06012. EDP Sciences (2019)

22. Wang, J.Y., Jea, K.F.: A near-optimal database allocation for reducing the average waiting time in the grid computing environment. Inf. Sci. **179**(21), 3772–3790 (2009)
23. Wang, J., et al.: Learning fine-grained image similarity with deep ranking. In: Proceedings of the IEEE Conference on Computer Vision and Pattern Recognition, pp. 1386–1393 (2014)
24. Weinberger, K.Q., Sha, F., Saul, L.K.: Convex optimizations for distance metric learning and pattern classification [applications corner]. IEEE Sig. Process. Mag. **27**(3), 146–158 (2010)
25. Yuan, D., Yang, Y., Liu, X., Chen, J.: A data placement strategy in scientific cloud workflows. Futur. Gener. Comput. Syst. **26**(8), 1200–1214 (2010)

Computer Graphics, Image Processing and Artificial Intelligence

Your Social Circle Affects Your Interests: Social Influence Enhanced Session-Based Recommendation

Yan Chen[1,2], Wanhui Qian[1,2], Dongqin Liu[1,2], Mengdi Zhou[1,2], Yipeng Su[2(✉)], Jizhong Han[2], and Ruixuan Li[2]

[1] School of Cyber Security, University of Chinese Academy of Sciences, Beijing, China
{chenyan1,qianwanhui,liudongqin,zhoumengdi,suyipeng}@iie.ac.cn
[2] Institute of Information Engineering, Chinese Academy of Sciences, Beijing, China
{hanjizhong,liruixuan}@iie.ac.cn

Abstract. Session-based recommendation aims at predicting the next item given a series of historical items a user interacts with in a session. Many works try to make use of social network to achieve a better recommendation performance. However, existing works treat the weights of user edges as the same and thus neglect the differences of social influences among users in a social network, for each user's social circle differs widely. In this work, we try to utilize an explicit way to describe the impact of social influence in recommender system. Specially, we build a heterogeneous graph, which is composed of users and items nodes. We argue that the fewer neighbors users have, the more likely users may be influenced by neighbors, and different neighbors may have various influences on users. Hence weights of user edges are computed to characterize different influences of social circles on users in a recommendation simulation. Moreover, based on the number of followers and PageRank score of each user, we introduce various computing methods for weights of user edges from a comprehensive perspective. Extensive experiments performed on three public datasets demonstrate the effectiveness of our proposed approach.

Keywords: Session-based recommendation · Social recommendation · Social influence

1 Introduction

In an era of information explosion, the rapid growth of online shopping and services makes it difficult for users to choose what they prefer from innumerable goods and services. As is known to all, different people's interests and preferences are completely different, and in some scenarios, individuals are not very certain about their needs. Driven by the above backgrounds, recommendation system emerges as the times require [4,25,38].

© The Author(s), under exclusive license to Springer Nature Switzerland AG 2022
D. Groen et al. (Eds.): ICCS 2022, LNCS 13352, pp. 549–562, 2022.
https://doi.org/10.1007/978-3-031-08757-8_46

Recommendation system is applied to capture users' interests based on their personal information and historical interactions with the items [12,13]. Further, it predicts the next item to users that they may interact with according to user preferences [3,14,30].

Session-based recommendation (SR) is first introduced to tackle the case that users' personal information (even the user IDs) can not be acquired [8]. A session is defined as a continuous interaction sequence of items in close proximity [23]. User preferences are usually captured by mining sequential transition patterns in a session [18,21]. Recurrent Neural Network (RNN) is proposed to model the sequence dependencies to learn users' preferences in [27]. Domain-Aware Graph Convolutional Network (DA-GCN) [5] builds a graph and applies a graph neural network to gain users' interests.

Fortunately, user IDs can be obtained in many cases, but we can still make use of SR to conduct a recommendation [1]. Even given the same session historical items, various people may interact with different items out of their personalized interests. So customized session-based recommendation can be made for users [6,22,34].

When user IDs are available, users' social network can be acquired as well. It is obvious that user's interest are easily influenced by their friends [17,24,35]. As a result, a better recommendation can be made when considering the preferences of users' friends. Lots of works have made efforts to take advantage of users' social network to get a more accurate recommendation [11,26,37]. However, when it comes to SR, brand new methods should be considered due to the above-mentioned special features (modeling sequential dependencies for recommendation). In social SR, recent work [2] builds a heterogeneous graph, which consists of social network and all historical user behaviors. It learns social-aware user and item representations, and gets a state-of-the-art (SOTA) performance. Although exiting works take the social influence into account, they set the weights of user-user edges as the same value, and do not detail the differences of social influences among users in social network, instead using a model to capture the impact of social influence on user preferences.

In this situation, we argue that the probabilities of users being affected are different among various users in SR and diverse users may have various influences on their followers. Therefore, we analyze social influence for SR in an explicit manner. Specially, we argue that the fewer people one follows, the more likely he/she may be affected by his/her social circle. Moreover, the more influential users are, the larger impact they will impose on their followers. To verify our thought, we propose a social influence enhanced model, which uses Social-aware Efficient Recommender (SERec) [2] as a backbone. The in-degree and PageRank [19] score for each user node in the social network are primarily computed, which reflect the social influences. Then we take the results as weights among connected user nodes in the graph. Finally, the social network obtains the knowledge of social influences, and we can utilize it to get a more accurate preference for each user.

Our contributions are summarized as follows:

1. We incorporate influence degree of each user affecting and being affected into the social network to capture a more accurate user preference in an explicit way.
2. We come up with some simple but effective methods to obtain the influence weights of social network to make our approach practical.
3. Extensive experiments are performed to verify the effectiveness of our proposed method.

In the rest of this paper, related work is introduced in Sect. 2. We detail our method in the following Sect. 3. Last but not least, we conduct extensive experiments in Sect. 4 and sum up our work in Sect. 5.

2 Related Works

Since we focus on social session-based recommendation (SR), we first discuss the relative works of SR and then talk about social recommendation.

Session-Based Recommendation: Session-based recommendation can be regarded as a sequential modeling task for the reason that a session is composed of a series of user historical interactions with the items. Naturally, RNNs are preferred to model item transition patterns [7,9]. In [8], Hidasi et al. first give a formal definition of session-based recommendation and come up with a multi-layered Gated Recurrent Unit (GRU) model, which is a variant of RNN to capture sequential dependencies in a session. This work is generally treated as a pioneer attempt for SR. Following Hidasi's work, an improved RNN [27] creatively points out a data augmentation technique to improve the performance of RNN. Subsequent works take Convolutional Neural Networks (CNNs) into consideration to model sequential dependencies [28,29]. A Dilated CNN [36] proposes a stack of holed convolutional layers to learn high-level representation from both short- and long-range item dependencies. Furthermore, attentive mechanism is introduced into recommendation to reduce noise item impact and focus on users' main purposes, i.e. Neural Attentive Recommendation Machine (NARM) [15], and Short-Term Attention/Memory Priority model (STAMP) [16]. Recently, Graph Neural Networks (GNNs) have been widely used in a large number of tasks on account of their superior performances besides session-based recommendation [10,20,39]. SR-GNN [33] is a typical GNN work for SR. It builds all session data into graphs, in which items are regarded as nodes and an edge is added if there is a transit between two items. SR-GNN employs a gated neural network to capture complex transitions of items. These methods try to model item dependencies in sessions, but do not take social influence into account.

Social Recommendation: As a widely studied research field, social network has been applied for recommendation in depth [31,32]. However, when social session-based recommendation is mentioned, there is not yet much work for the reason that SR is a relatively new topic, and previous social recommendation methods are not suitable for SR due to their lack of modeling sequence dependencies. Recently Dynamic Graph Recommendation (DGRec) [26] models dynamic user behaviors with a recurrent neural network, and captures context-dependent social influence with a graph-attention neural network. DGRec gets a better recommendation performance, but its inefficiency can not be ignored, because it has to deal with lots of extra sessions to make a recommendation. To solve the efficiency issue that DGRec meets, SERec [2] implements an efficient framework for session-based social recommendation. In detail, SERec precomputes user and item representations from a heterogeneous graph neural network that integrates the knowledge from social network. As a result, it reduces computations during predicting stage. Efficient as it is, SERec just adds users' social network into the interaction graph and sets weights among user nodes as the same value without considering influence differences among various users.

3   Modeling Methods

3.1   Problem Definition

In session-based recommendation, we define $U = \{u_1, u_2..., u_N\}$ as the set of users, $I = \{i_1, i_2, ...i_M\}$ as the set of items. Each user $u \in U$ generates a set of sessions, $S^u = \{s_1^u, s_2^u, ...s_T^u\}$. For each session, there are a series of items that a user interacts with sorted by timestamp. For example, $s_1^u = \{i_1^u, i_2^u, ...i_L^u\}$, and L is the length of session s_1^u. All sessions of users constitute users' historical behaviors dataset B. The goal of session-based recommendation is as follows: given a new session of user u, $S = \{i_1^u, i_2^u, ...i_n^u\}$, predict i_{n+1}^u for the user u by recommending top-K items $(1 \leq K \leq M)$ from all items I that might be interesting to the user u.

In addition, in social session-based recommendation, besides users' historical behaviors B, a social network can be utilized to improve recommendation. Let $SN = (U, E)$ denote the social network, where U is the nodes of users, E is the set of edges. There is an edge if a social link exits between two users. For example, an edge (u, v) from u to v means that u is followed by v, in other words, v follows u.

3.2   Social Influence Modeling

Social Session-Based Behaviors Graph Building. We first construct a heterogeneous graph from all users' behaviors B and social network SN. Then we apply a GNN (Graph Neural Network) [37] to learn representations of users and items. The user representations can capture user preferences and more accurate social influences. Item representations can learn useful information from user-item interactions and cross-session item transition dependencies.

In the heterogeneous graph, all the users and items in B and SN make up the graph nodes, and the set of edges consists of four kinds of edges. A user-user edge

(u, v) exists if v follows u (in other words, u is followed by v). It is worth noting that we make such a design because in our model a user node representation is learned by the incoming edges and users are more influenced by those they follow than those following them. If a user u has ever clicked an item i in any session, there will be two edges, namely (u, i) and (i, u). Lastly, there is an edge (i, j) if item i transmits directly to item j.

Now, we take edge weights into consideration. The weight of user-item (u, i) and item-user (i, u) is the times of user u clicked item i. And the weight of item-item (i, j) is the times of item i transmitted to j.

When considering weights of user-user (u, v), SERec [2] defines all the weights of user-user as an identical number 1. However, we argue that different weights should be designed among users in the social network, for various influences may have on different users in the social network. As a result, we compute every weight between user-user edge to explicitly represent the social influence.

Social Influence Computing. After the heterogeneous graph is constructed, we compute the weights among user nodes in the following method inspired by [40].

Given a user node v, the node in-degree d denotes the number of users that v follows. We view all the incoming edges' weights W as the user v's degree of being influenced. In other words, the weight of edge (u, v) is calculated in the following equation:

$$W(u, v) = C/d, \tag{1}$$

where C is a positive constant to control the range of weight.

For example, if a node v follows three users u_1, u_2, u_3, then there are three edges $(u_1, v), (u_2, v), (u_3, v)$. So the in-degree of v is 3, and the weights of the three edges are all set to $W = C/3$.

We further make a research on the ability of users to influence their followers. We apply PageRank [19] to calculate the importance of a user in the social network, which is a way of measuring the importance of website pages. Intuitively, the larger of the importance value users get, the more influence users may have on their followers. In detail, the influence of node u is computed as follows:

$$F_u = (1 - A)/N + A * (F_{v_1}/out(v_1) + F_{v_2}/out(v_2) + ... + F_{v_n}/out(v_n)), \tag{2}$$

where $out(v_i)$ denotes the out-degree of node v_i. $v_1...v_n$ is the followers of user u, A is a coefficient, N is the number of user nodes.

For example, if a node u is followed by three users v_1, v_2, v_3, then there are three edges $(u, v_1), (u, v_2), (u, v_3)$. Then weights of the three edges are all set to be F_u.

3.3 Model Architecture and Training

In this section, we briefly illustrate how to capture user preferences by modeling user sequential patterns. Our model selects SERec as a backbone, and utilizes

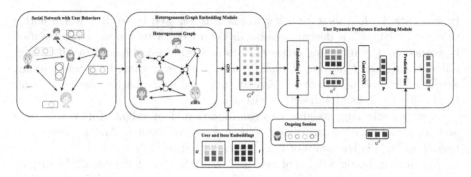

Fig. 1. An overview of model architecture. Heterogeneous graph embedding module is applied to learn representations of users and items. User dynamic preference embedding module applies a gated GNN to obtain a user preference embedding in the ongoing session.

a Graph Neural Network (GNN) [37] and gated GNN [33] to capture user preferences. As is demonstrated in Fig. 1, our model is composed of two modules: heterogeneous graph embedding module and user dynamic preference embedding module.

Heterogeneous Graph Embedding Module. We apply GNN (graph neural network) to model user and item embedding, which has fused social influence. Supposing GNN is comprised of F layers, let $G^f[x]$ denote the representation of node x at layer f, where x may be a user or an item. The new node representation $G^f[x]$ is computed as follows:

$$G^f[x] = ReLU(W_1^f(G^{f-1}[x]||\hat{G}^f[x]) + b_1^f), \qquad (3)$$

where $G^{f-1}[x]$ is the old node representation. $\hat{G}^f[x]$ is the aggregated information from node x's neighbors $N(x)$, and y belongs to $N(x)$ if there is an edge (y, x) pointing to x. W_1^f and b_1^f are learnable parameters.

The aggregation information from node x's neighbors is calculated by the equation below:

$$\hat{G}^f[x] = \sum_{y \in N(x)} Attention(y, x) * (W_2^f G^{f-1}[y] + b_2^f), \qquad (4)$$

where W_2^f and b_2^f are learnable parameters. *Attention* function is detailed in [2].

User Dynamic Preference Embedding Module. Given an ongoing session S of user x, a graph $G = (V, E)$ is constructed in the same way mentioned in Sect. 3.2, where V denotes items in S, E denotes item transitions, and the weight of an edge is the times of item transitions. Since we build a heterogeneous graph on all historical sessions of users and social network, global user and item representations are obtained. Then for the ongoing session S, we first retrieve the relative user and item representations to initialize node representations Z. We utilize a gated GNN [33] to model the session-specialized item representation:

$$g_i = v_i \odot tanh(W_c(\hat{N}[z_i]||z_i) + b_c) + (1 - v_i) \odot W_z z_i, \tag{5}$$

$$v_i = sigmod(W_i(\hat{N}[z_i]||z_i) + b_i), \tag{6}$$

where z_i is the node vector of an item in session S, $\hat{N}[z_i]$ is the aggregated information from z_i's neighboring nodes, g_i represents the vector of node i for specialized session S. W_c, b_c, W_z, W_i and b_i are learnable parameters. Based on g_i, we get a user preference embedding in the ongoing session S:

$$P = \sum_{1 \leq i \leq |S|} b_i * g_i, \tag{7}$$

$$b_i = softmax(r^T sigmod(W_v g_i + W_{last} g_{last} + W_u u^S)), \tag{8}$$

where g_{last} is the embedding of the last item in session S to capture user's recent interest, u^S is the user embedding to capture user u's general preference. r, W_v, W_{last} and W_u are learnable parameters.

Finally, we generate the score q for every item in I via multiplying its embedding e_i by user preference embedding in the ongoing session S:

$$q = softmax(P^T e_i). \tag{9}$$

We apply a cross-entropy of the prediction and the ground truth as the loss function in the following form:

$$L = - \sum_{m=1}^{M} q_m log(\hat{q}) + (1 - q_m)log(1 - \hat{q}), \tag{10}$$

where $\hat{q}$ denotes the one-hot encoding vector of the ground truth item.

4 Experiments

4.1 Experimental Setup

Datasets. Following the existing classic social recommendation works [2,26], we evaluate our proposed method on three public real-world benchmark datasets:

(1) **Foursquare:** The Foursquare[1] dataset is a publicly online available dataset which consists of users' check-in records on different venues in a period of time. Records are regarded as the same session if the check-in time interval is shorter than a day, and records are viewed as different sessions if interval is longer than a day.

[1] https://sites.google.com/site/yangdingqi/home/foursquare-dataset.

Table 1. Datasets statistics

Datasets	All users	All items	All clicks	Social links	All sessions
Foursquare	39,302	45,595	3,627,093	304,030	888,798
Gowalla	33,661	41,229	1,218,599	283,778	258,732
Delicious	1313	5793	266,190	9130	60,397

(2) **Gowalla:** Gowalla[2] is another check-in data, and the social network is based on location social network website. Sessions are extracted by the same way as Foursquare.

(3) **Delicious:** Delicious[3] is an online bookmarking system. We consider a sequence of tags a user has marked to a book as a session (marking time is recorded).

We split each dataset into training/validation/test sets, following the settings in [2]. And we conduct our experiments on augmented datasets. To be specific, for a session $S = (v_1, v_2, v_3, ..., v_n)$, we generate a series of interaction sequences and labels $([v_1], v_2), ([v_1, v_2], v_3), ..., ([v_1, v_2, ..., v_{n-1}], v_n)$, where $[v_1, v_2, ..., v_{n-1}]$ is user's historical sequence, v_n is the next-clicked item, namely the label.

The statistics of datasets are summarized in Table 1.

Evaluation Metrics. We evaluate all models with three widely used ranking-based metrics:

(1) **REC@K:** It measures the recall of the top-K ranked items in the recommendation list over all the testing instances. In our experiments, only one item is set as the label, so REC@K is used to measure whether the label item is contained in the top-K ranked items according to the scores.

(2) **MRR@K:** It measures the mean reciprocal rank of the predictive position of the true target item on the top-K ranked items in the recommendation list. The target item is expected to rank ahead in terms of ranking scores.

(3) **NDCG@K:** NDCG is a standard ranking metric. In the context of session-based recommendation, it also measures the position of target item in the recommendation list.

K is set to 10 and 20 in our experiments.

Comparison Methods. SERec [2] and DGRec [26] are regarded as two typical works related to social session-based recommendation, and SERec as the SOTA social SR model has proved it enjoys a more effective performance than DGRec [2]. Consequently, we only compare our method with SERec. Moreover, our work is realized on the basis of graph structure by taking social influence

---

[2] https://snap.stanford.edu/data/loc-gowalla.html.
[3] https://grouplens.org/datasets/hetrec-2011/.

explicitly into consideration, we also want to verify if it is effective compared with the existing none-social SR methods like SR-GNN [33], which applies a gated graph convolutional layer to learn item transitions. Last but not least, as mentioned above in Sect. 3.2, we put forward two ways (in-degree and PageRank) on how to compute the social influences, and we test the two different types and combination of both on SERec and SR-GNN to prove our proposed method.

Implementation Details. Following the backbone method, we set the model hyper-parameters as mentioned in [2]. Generally, user and item IDs are made an embedding into low dimensional latent spaces with the same dimensionality 128. Adam optimizer was used to train the models and the batch size for mini-batch is 128. The above models' performance are reported under their optimal hyper-parameter settings [2]. Specially, we find that our model gets an optimal performance when C is set to 100 and A to 0.85, so we report our model performance under this optimal settings. To compute PageRank scores of user nodes, we first initialize PageRank score of each user node to $1/N$ (N is the number of user nodes in the social network), then update all user nodes' PageRank scores using Eq. (2) for five times iteratively. Since we filter out user nodes without followers or followees, each user node can obtain a PageRank score in the social network.

4.2 Experimental Evaluations

Weights Calculating for Social Network. We first compute $W(u, v)$ and F_v for each user in the social network using Eqs. (1) and (2), and set the weights of user-user edges in the following three ways to verify our proposed method:

(1) Using $W(u, v)$ only as the weight for edge (u, v) to demonstrate our idea that the fewer people that users follow, the more influence their social circles may have on their interests;
(2) Using F_u only as the weight of edge (u, v) to prove that the more influential that users are (the PageRank scores are high), the more impact they may have on their followers' behaviors;
(3) Using the sum of $W(u, v)$ and F_u as the weight of edge (u, v) to test the relation between the above two weight computing ways.

Results Analysis. As is mentioned above, the explicit social influence is evaluated in three manners, and experiments are implemented based on SERec and SR-GNN to check if our method can achieve a high performance. The experimental results of overall performance are reported in Tables 2, 3 and 4. The optimal and suboptimal results of each column are highlighted in boldface and underline for SERec and SSR-GNN respectively. We denote SSR-GNN for social-aware SR-GNN and denote model with the three weight computing ways by adding a postfix 'W, F, C' respectively. And 'R, M, N' is short for 'REC, MRR, NDCG'. The following observations can be drawn from the results.

Table 2. Performance on foursquare

Model	Foursquare					
	R@10	M@10	N@10	R@20	M@20	N@20
SERec	61.67	34.11	40.69	70.07	34.71	42.84
SERecF	63.15	36.83	43.11	70.86	37.37	45.11
SERecC	63.21	36.82	43.12	70.89	37.37	45.11
SERecW	**63.34**	**36.91**	**43.27**	**70.94**	**37.44**	**45.20**
SSRGNN	60.94	33.62	40.15	69.28	34.21	42.29
SSRGNNF	61.02	33.56	40.13	69.53	34.16	42.29
SSRGNNC	61.04	33.55	40.12	**69.65**	34.15	42.31
SSRGNNW	**61.21**	**33.83**	**40.39**	69.60	**34.42**	**42.52**

Table 3. Performance on Gowalla

Model	Gowalla					
	R@10	M@10	N@10	R@20	M@20	N@20
SERec	45.88	25.26	30.06	53.48	25.78	31.93
SERecF	46.76	**26.54**	31.34	53.85	**27.05**	33.17
SERecC	46.75	26.14	31.00	**54.07**	26.65	32.85
SERecW	**46.87**	26.53	**31.49**	53.92	27.03	**33.27**
SSRGNN	45.32	24.81	29.68	52.94	25.33	31.59
SSRGNNF	**45.50**	24.85	29.74	**53.01**	25.37	31.62
SSRGNNC	45.45	**25.01**	**29.88**	52.87	**25.53**	**31.76**
SSRGNNW	45.29	24.76	29.65	52.79	25.28	31.53

First of all, let us focus on the performance of our method on the SOTA model SERec. In general, our proposed method outperforms SERec. It is proved that the model can learn more accurate user preferences by explicitly adding social influence as the weights of user-user edges. Furthermore, considering users' influence to their followers has a similar improvement on the model with thinking about the degree of users being influenced. To be more specific, SERecW may win a little bit than SERecF. However, to our surprise, simply summing up the first two social influence computing results as the weights does not gain a best performance. One possible reason may be that the two weights computing methods represent different views of social influence, and can not be added on the same dimension.

Secondly, when it comes to SSR-GNN, after applying our proposed method to original SSR-GNN, it can make a progress on recommendation results. Other conclusions are nearly the same as SERec, except that SSR-GNNC achieves the best results on some metrics. This further illustrates the uncertainty of summing up the two different weights for their various practical significance.

Table 4. Performance on delicious

Model	Delicious					
	R@10	M@10	N@10	R@20	M@20	N@20
SERec	**40.22**	21.22	25.70	**49.50**	21.87	28.05
SERecF	39.79	**21.39**	**25.80**	<u>49.33</u>	**22.05**	**28.19**
SERecC	40.12	21.11	**25.80**	48.96	21.75	<u>28.15</u>
SERecW	<u>40.15</u>	<u>21.31</u>	25.64	49.18	<u>21.98</u>	28.09
SSRGNN	39.73	**21.55**	**25.85**	48.78	**22.18**	**28.13**
SSRGNNF	<u>39.92</u>	<u>21.53</u>	<u>25.83</u>	49.23	22.10	28.06
SSRGNNC	**40.09**	21.48	<u>25.83</u>	**49.62**	<u>22.11</u>	**28.13**
SSRGNNW	39.77	21.35	25.68	<u>49.33</u>	22.01	28.10

We have to mention that among the three datasets, our method performs better on Foursquare and gowalla, but is not stable on Delicious. Through deep research, we find that the data characters may lead to such a result. Let us review the statistics of the three datasets in Table 1, Foursquare and Gowalla have rich social links, which can help models learn user preferences by considering the social influence in a more explicit way. On the contrary, Delicious has far less social links than the other two, so models learn little extra information from social network by adding social influence among users, even performs worse possibly due to large data variance. All in all, our proposed method can gain a significant performance promotion on large amounts of datasets, but may meet unstableness on less amounts of datasets.

5   Conclusion

In this paper, we propose an explicit view to discuss how users' social network influences their behaviors. Based on the sense that the smaller one person's social circle is, the more influence his/her social network may have on his/her interests, and the more influential users are, the more likely that they may affect their followers. To verify our idea, we build a heterogeneous social graph, and explicitly compute the influences as weights in social network graph according to in-degrees and PageRank scores of user nodes. Finally, extensive experiments are conducted on three public datasets. It is demonstrated that modeling social influences in an explicit way can outperform the SOTA model on large datasets while get a degradation on a small dataset. In the future, we will explore the influence of social network formation to session-based recommendation. We consider how a user's social network is formed. For example, they share the same topic. This kind of social network may lead to a more stable influence on users' interest, we would study such influence to session-based recommendation.

Acknowledgements. We gratefully thank the reviewers for their insightful comments. This research is supported in part by the National Key Research and Development Program of China under Grant 2018YFC0806900.

References

1. Chen, T., Wong, R.C.W.: Handling information loss of graph neural networks for session-based recommendation. In: Proceedings of the 26th ACM SIGKDD International Conference on Knowledge Discovery and Data Mining, pp. 1172–1180 (2020)
2. Chen, T., Wong, R.C.W.: An efficient and effective framework for session-based social recommendation. In: Proceedings of the 14th ACM International Conference on Web Search and Data Mining, pp. 400–408 (2021)
3. Chipchagov, M., Kublik, E.: Model of the cold-start recommender system based on the Petri-Markov nets. In: Paszynski, M., Kranzlmüller, D., Krzhizhanovskaya, V.V., Dongarra, J.J., Sloot, P.M.A. (eds.) ICCS 2021. LNCS, vol. 12743, pp. 87–91. Springer, Cham (2021). https://doi.org/10.1007/978-3-030-77964-1_7
4. Choi, M., Kim, J., Lee, J., Shim, H., Lee, J.: Session-aware linear item-item models for session-based recommendation. In: Proceedings of the Web Conference 2021, pp. 2186–2197 (2021)
5. Guo, L., Tang, L., Chen, T., Zhu, L., Nguyen, Q.V.H., Yin, H.: DA-GCN: a domain-aware attentive graph convolution network for shared-account cross-domain sequential recommendation. In: Proceedings of the Thirtieth International Joint Conference on Artificial Intelligence. IJCAI, pp. 2483–2489 (2021)
6. Guo, L., Yin, H., Wang, Q., Chen, T., Zhou, A., Quoc Viet Hung, N.: Streaming session-based recommendation. In: Proceedings of the 25th ACM SIGKDD International Conference on Knowledge Discovery and Data Mining, pp. 1569–1577 (2019)
7. Hidasi, B., Karatzoglou, A.: Recurrent neural networks with top-k gains for session-based recommendations. In: Proceedings of the 27th ACM International Conference on Information and Knowledge Management, pp. 843–852 (2018)
8. Hidasi, B., Karatzoglou, A., Baltrunas, L., Tikk, D.: Session-based recommendations with recurrent neural networks. In: 4th International Conference on Learning Representations, ICLR (2016)
9. Hidasi, B., Quadrana, M., Karatzoglou, A., Tikk, D.: Parallel recurrent neural network architectures for feature-rich session-based recommendations. In: Proceedings of the 10th ACM Conference on Recommender Systems, pp. 241–248 (2016)
10. Hsu, C., Li, C.T.: RetaGNN: relational temporal attentive graph neural networks for holistic sequential recommendation. In: Proceedings of the Web Conference, pp. 2968–2979 (2021)
11. Jiang, B., Lu, Z., Liu, Y., Li, N., Cui, Z.: Social recommendation in heterogeneous evolving relation network. In: Krzhizhanovskaya, V.V., et al. (eds.) ICCS 2020. LNCS, vol. 12137, pp. 554–567. Springer, Cham (2020). https://doi.org/10.1007/978-3-030-50371-0_41
12. Kużelewska, U.: Effect of dataset size on efficiency of collaborative filtering recommender systems with multi-clustering as a neighbourhood identification strategy. In: Krzhizhanovskaya, V.V., et al. (eds.) ICCS 2020. LNCS, vol. 12139, pp. 342–354. Springer, Cham (2020). https://doi.org/10.1007/978-3-030-50420-5_25

13. Kuzelewska, U.: Quality of recommendations and cold-start problem in recommender systems based on multi-clusters. In: Computational Science - ICCS 2021-21st International Conference, pp. 72–86 (2021)
14. Landin, A., Parapar, J., Barreiro, Á.: Novel and diverse recommendations by leveraging linear models with user and item embeddings. In: Jose, J.M., et al. (eds.) ECIR 2020. LNCS, vol. 12036, pp. 215–222. Springer, Cham (2020). https://doi.org/10.1007/978-3-030-45442-5_27
15. Li, J., Ren, P., Chen, Z., Ren, Z., Lian, T., Ma, J.: Neural attentive session-based recommendation. In: Proceedings of the 2017 ACM on Conference on Information and Knowledge Management, pp. 1419–1428 (2017)
16. Liu, Q., Zeng, Y., Mokhosi, R., Zhang, H.: Stamp: short-term attention/memory priority model for session-based recommendation. In: Proceedings of the 24th ACM SIGKDD International Conference on Knowledge Discovery and Data Mining, pp. 1831–1839 (2018)
17. Ma, H., Zhou, D., Liu, C., Lyu, M.R., King, I.: Recommender systems with social regularization. In: Proceedings of the fourth ACM International Conference on Web Search and Data Mining, pp. 287–296 (2011)
18. Meng, W., Yang, D., Xiao, Y.: Incorporating user micro-behaviors and item knowledge into multi-task learning for session-based recommendation. In: Proceedings of the 43rd International ACM SIGIR Conference on Research and Development in Information Retrieval, pp. 1091–1100 (2020)
19. Page, L., Brin, S., Motwani, R., Winograd, T.: The PageRank citation ranking: bringing order to the web. Technical report, Stanford InfoLab (1999)
20. Pan, Z., Cai, F., Chen, W., Chen, H., de Rijke, M.: Star graph neural networks for session-based recommendation. In: Proceedings of the 29th ACM International Conference on Information and Knowledge Management, pp. 1195–1204 (2020)
21. Pan, Z., Cai, F., Ling, Y., de Rijke, M.: Rethinking item importance in session-based recommendation. In: Proceedings of the 43rd International ACM SIGIR Conference on Research and Development in Information Retrieval, pp. 1837–1840 (2020)
22. Quadrana, M., Karatzoglou, A., Hidasi, B., Cremonesi, P.: Personalizing session-based recommendations with hierarchical recurrent neural networks. In: Proceedings of the Eleventh ACM Conference on Recommender Systems, pp. 130–137 (2017)
23. Ren, P., Chen, Z., Li, J., Ren, Z., Ma, J., De Rijke, M.: RepeatNet: a repeat aware neural recommendation machine for session-based recommendation. In: Proceedings of the AAAI Conference on Artificial Intelligence, pp. 4806–4813 (2019)
24. Sanz-Cruzado, J., Macdonald, C., Ounis, I., Castells, P.: Axiomatic analysis of contact recommendation methods in social networks: an IR perspective. In: Jose, J.M., et al. (eds.) ECIR 2020. LNCS, vol. 12035, pp. 175–190. Springer, Cham (2020). https://doi.org/10.1007/978-3-030-45439-5_12
25. Sato, M., Singh, J., Takemori, S., Zhang, Q.: Causality-aware neighborhood methods for recommender systems. In: Hiemstra, D., Moens, M.-F., Mothe, J., Perego, R., Potthast, M., Sebastiani, F. (eds.) ECIR 2021. LNCS, vol. 12656, pp. 603–618. Springer, Cham (2021). https://doi.org/10.1007/978-3-030-72113-8_40
26. Song, W., Xiao, Z., Wang, Y., Charlin, L., Zhang, M., Tang, J.: Session-based social recommendation via dynamic graph attention networks. In: Proceedings of the Twelfth ACM International Conference on Web Search and Data Mining, pp. 555–563 (2019)

27. Tan, Y.K., Xu, X., Liu, Y.: Improved recurrent neural networks for session-based recommendations. In: Proceedings of the 1st Workshop on Deep Learning for Recommender Systems, pp. 17–22 (2016)

28. Tuan, T.X., Phuong, T.M.: 3D convolutional networks for session-based recommendation with content features. In: Proceedings of the Eleventh ACM Conference on Recommender Systems, pp. 138–146 (2017)

29. Twardowski, B., Zawistowski, P., Zaborowski, S.: Metric learning for session-based recommendations. In: Hiemstra, D., Moens, M.-F., Mothe, J., Perego, R., Potthast, M., Sebastiani, F. (eds.) ECIR 2021. LNCS, vol. 12656, pp. 650–665. Springer, Cham (2021). https://doi.org/10.1007/978-3-030-72113-8_43

30. Wang, S., Cao, L., Wang, Y., Sheng, Q.Z., Orgun, M.A., Lian, D.: A survey on session-based recommender systems. ACM Comput. Surv. (CSUR) **54**(7), 1–38 (2021)

31. Wang, W., Yin, H., Du, X., Hua, W., Li, Y., Nguyen, Q.V.H.: Online user representation learning across heterogeneous social networks. In: Proceedings of the 42nd International ACM SIGIR Conference on Research and Development in Information Retrieval, pp. 545–554 (2019)

32. Wang, X., Hoi, S.C., Ester, M., Bu, J., Chen, C.: Learning personalized preference of strong and weak ties for social recommendation. In: Proceedings of the 26th International Conference on World Wide Web, pp. 1601–1610 (2017)

33. Wu, S., Tang, Y., Zhu, Y., Wang, L., Xie, X., Tan, T.: Session-based recommendation with graph neural networks. In: Proceedings of the AAAI Conference on Artificial Intelligence, pp. 346–353 (2019)

34. Wu, S., Zhang, M., Jiang, X., Xu, K., Wang, L.: Personalizing graph neural networks with attention mechanism for session-based recommendation. CoRR abs/1910.08887 (2019)

35. Xiao, L., Min, Z., Yongfeng, Z., Yiqun, L., Shaoping, M.: Learning and transferring social and item visibilities for personalized recommendation. In: Proceedings of the 2017 ACM on Conference on Information and Knowledge Management, pp. 337–346 (2017)

36. Yuan, F., Karatzoglou, A., Arapakis, I., Jose, J.M., He, X.: A simple convolutional generative network for next item recommendation. In: Proceedings of the Twelfth ACM International Conference on Web Search and Data Mining, pp. 582–590 (2019)

37. Zhang, C., Song, D., Huang, C., Swami, A., Chawla, N.V.: Heterogeneous graph neural network. In: Proceedings of the 25th ACM SIGKDD International Conference on Knowledge Discovery and Data Mining, pp. 793–803 (2019)

38. Zhong, J., Ma, C., Zhou, J., Wang, W.: PDPNN: modeling user personal dynamic preference for next point-of-interest recommendation. In: Krzhizhanovskaya, V.V., et al. (eds.) ICCS 2020. LNCS, vol. 12142, pp. 45–57. Springer, Cham (2020). https://doi.org/10.1007/978-3-030-50433-5_4

39. Zhou, H., Tan, Q., Huang, X., Zhou, K., Wang, X.: Temporal augmented graph neural networks for session-based recommendations. In: SIGIR 2021: The 44th International ACM SIGIR Conference on Research and Development in Information Retrieval, pp. 1798–1802 (2021)

40. Zhu, Y., Xu, Y., Yu, F., Liu, Q., Wu, S., Wang, L.: Graph contrastive learning with adaptive augmentation. In: Proceedings of the Web Conference 2021, pp. 2069–2080 (2021)

Action Recognition in Australian Rules Football Through Deep Learning

Stephen Kong Luan$^{(\boxtimes)}$, Hongwei Yin, and Richard Sinnott

School of Computing and Information Systems, University of Melbourne,
Melbourne, Australia
rsinnott@unimelb.edu.au

Abstract. Understanding player's actions and activities in sports is crucial to analyze player and team performance. Within Australian Rules football, such data is typically captured manually by multiple (paid) spectators working for sports data analytics companies. This data is augmented with data from GPS tracking devices in player clothing. This paper focuses on exploring the feasibility of action recognition in Australian rules football through deep learning and use of 3-dimensional Convolutional Neural Networks (3D CNNs). We identify several key actions that players perform: kick, pass, mark and contested mark, as well as non-action events such as images of the crowd or players running with the ball. We explore various state-of-the-art deep learning architectures and developed a custom data set containing over 500 video clips targeted specifically to Australian rules football. We fine-tune a variety of models and achieve a top-1 accuracy of 77.45% using R2+1D ResNet-152. We also consider team and player identification and tracking using You Only Look Once (YOLO) and Simple Online and Realtime Tracking with a deep association metric (DeepSORT) algorithms. To the best of our knowledge, this is the first paper to address the topic of action recognition in Australian rules football.

Keywords: Action recognition · 3D CNN · Australian rules football

1 Introduction

Action recognition has been explored by many researchers over the past decade. The typical objective is to detect and recognize human actions in a range of environments and scenarios. Action recognition, unlike object detection, needs to consider both spatial and temporal information in order to make classifications. In this paper we focus on using 3-dimensional Convolutional Neural Networks (3D CNNs) to achieve action recognition for players in Australian rules football.

Australian rules football, commonly referred to as "footy" in Australia, is a popular contact sport played between two 18-player teams on a large oval. The premier league is the Australian Football League (AFL). The ultimate aim is to kick the ball between 4 goal posts for a score (6 points if the ball goes through

D. Groen et al. (Eds.): ICCS 2022, LNCS 13352, pp. 563–576, 2022.
https://doi.org/10.1007/978-3-031-08757-8_47

the middle two posts) or a minor score (1 point if the ball goes through the one of the inner/outer posts). This is achieved by players doing a range of actions to move the ball across the pitch. These include kicking, passing (punching the ball), catching, running (up to 15 m whilst carrying the ball) and tackling.

The understanding of player actions and player movements in sports are crucial to analyse player and team performances. Counting the number of effective actions that take place during a match is key to this. This paper focuses on development of a machine learning application that is able to detect and recognize player actions through the use of deep artificial neural networks.

2   Literature Review

Prior to deep learning, approaches based on hand-engineered features for computer vision tasks were the primary method used for action recognition. Improved Dense Trajectories (IDT) [26] is representative of such approaches. This achieved good accuracy and robustness, however hand engineering features is limited. Deep learning architectures based on CNNs have achieved unparalleled performance in the field of computer vision. Deep Video developed by Karpathy et al. [17] was one of the first approaches to apply 2D CNNs for action recognition tasks. This used pre-trained 2D CNNs applied to every frame of the video and fusion techniques to learn spatio-temporal relationships. However, its performance on the UCF-101 data set [20] was worse than IDT, indicating that 2D CNNs alone are sub-optimal for action recognition tasks since they do not adequately capture spatio-temporal information.

Two-stream networks such as [19] add a stream of optical flow information [11] as a representation of motion besides the conventional RGB stream. The approach used two parallel streams that were combined with fusion based techniques. This approach was based on 2D CNNs and achieved similar results to IDT. This approach sparked a series of research efforts focused on improving two-stream networks. This included works focused on improvement in fusion [6], and use of recurrent neural networks including Long Short-Term Memory (LSTM) [4,15]. Other methods include Temporal Segment Networks (TSN) [27] capable of understanding long range video content by splitting a video into consecutive temporal segments, and multi-stream networks that consider other contextual information such as human poses, objects and audio in video. The framework of two-stream networks was widely adopted by many researchers, however, a major limitation of two-stream networks was that optical flows require pre-processing and hence require considerable hand-engineering of features. Generating optical flows for videos can be both computationally and storage demanding. This also affected the scale of training data sets required.

3D CNNs can be thought of as a natural way to understand video content. Since video is a series of consecutive frames of images, a 3-dimensional convolutional filter can be applied to both the spatial and temporal domain. Initial research was explored by [13] in 2012, then in 2015 by Tran et al. [22] who proposed a 3D neural network architecture called C3D using $3 \times 3 \times 3$ convolutional kernels. They demonstrated that 3D CNNs were better at learning spatio-temporal features than 2D CNNs. The introduction of C3D marked the start of a new chapter in action recognition. 3D CNNs were shown to be suited to extracting and learning spatio-temporal features from video - a core demand for real-time action recognition. However C3D were difficult to train with the training process usually taking weeks on large data sets, due to the cost incurred in training with an overwhelming number of parameters in the full 3D architecture.

In 2017, Carreira et al. [2] proposed Inflated 3D ConvNets (I3D), which utilized transfer learning and outperformed all other models using the UCF-101 data set. I3D avoided the necessity for training from scratch by using some well-developed 2D CNN architectures that were pre-trained on large scale data sets such as the ImageNet [3]. I3D added an additional temporal dimension, where the model weights were used. The proposed I3D model was implemented for both the two-stream and single stream approach. Weights from an Inception-V1 model [12] pre-trained on ImageNet were used and trained on the Kinetics-400 data set [18]. This was subsequently fine-tuned on the UCF-101 data set to achieve a top-1 accuracy of 95.1% with RGB stream only. I3D demonstrated that 3D CNNs could benefit from the weights of 2D CNNs pre-trained on large scale data. This has since become a popular strategy adopted by many that has sparked a model benchmark standard based on the Kinetics-400 data.

Tran et al. [24] proposed the R2+1D architecture in 2018. This focused on factorizing spatio-temporal 3D convolutions into 2D spatial convolutional blocks and 1D temporal convolutional blocks. This decomposition provided simplicity for model optimization and improved the efficiency of training, while also enhancing the model's ability to represent complex functions by increasing the number of non-linearities through adding Rectified Linear Unit activation functions (ReLU) between the 2D and 1D blocks. The R2+1D model used the Deep Residual Network (ResNet) [10] architecture as the backbone and achieved similar performances to I3D on data sets such as Kinetics-400 and UCF-101.

Non-local blocks proposed by Wang et al. in 2018 [28] introduced a new form of operational building block that was able to capture long range temporal features similar to the self attention mechanism [25]. This was compatible to most architectures with minimal effort. The authors implemented their model by adding non-local blocks into the I3D architecture and achieved consistent improvement of performance over the original model using several data sets. In 2019 Tran et al. [23] proposed Channel Separated Networks (CSN) which focused on factorizing 3D CNNs by separating channel-wide interactions and spatio-temporal interactions by introducing regularization measures into the architecture to improve the overall accuracy. CSN are regarded as an efficient and lightweight architecture, where the model interaction-reduced channel-separated network (ir-CSN) using a ResNet-152 backbone reported a top-1 accuracy of 79.2% on the Kinetics-400 data set.

Feichtenhofer et al. [5] proposed the SlowFast networks framework. This consisted of a fast and a slow stream. The fast stream was used for extracting temporal motion features at a high frame rate, whilst the slow stream was used for extracting spatial features at a low frame rate. These two streams were later fused by lateral connections, commonly seen in two-stream network models. However, the architecture of SlowFast networks was fundamentally different to two-stream networks since it was based on streams of different temporal frame rates and not two separate streams of spatial and temporal features. The SlowFast network provided a generic and efficient framework that could be applied to various spatio-temporal architectures. Furthermore, the fast stream was lightweight as the channel capacity was greatly reduced by only focusing on temporal features. The proposed network used ResNet architecture as the backbone and achieved a better performance than I3D and R2+1D on the Kinetics-400 data set.

Another similar framework was the Temporal Pyramid Network (TPN) proposed by Yang et al. [30]. This used a pyramid structure for processing frames at multiple feature levels to capture the variation in speed for different actions - so called visual tempos. TPN had the ability to use various 3D or 2D architectures as the backbone, where the set of hierarchical features extracted by the backbone undergoes down-sampling with a spatial module and a temporal rate module for processing features rich in both visual tempos and spatial information. These could then be aggregated by an information flow process. TPN used a ResNet-101 backbone and achieved better performance in Kinetics-400 over the SlowFast network.

3   Australian Rules Football Data Set

A well-defined and high-quality data set is crucial for action recognition tasks. This should contain enough samples for deep neural networks to extract motion patterns, and offer enough variance for different scenarios and camera positions for performance analysis. No such data set exists for AFL, hence we construct our own action recognition data set for AFL games. In this process, we referred to some well-known data sets for video content understanding including Youtube-8M [1], UCF 101 [20], Kinetics-400 [18], SoccerNet [8] and others. All the training and testing videos used here were retrieved from YouTube.

As AFL games are popular in Australia, there are more than enough videos on YouTube, including real match recordings, training session recordings, tutorial guides etc. However, manually creating and labelling data from video content (individual frames) is a challenging and time-consuming task. In order to feed enough frames and information for temporal feature extraction into deep learning models, we set the standard that each video clip should be at least 16 frames in length and it should be not a long-distance shot with low resolution of action tasks.

Players in an AFL match are highly mobile hence actions only exist for a very limited amount of time and are often interfered with by other players through tackles. As a result, actions sometime may end up in failure. This brings significant challenges to the construction process of the data set, e.g. judging the actual completeness of actions. This work focuses on recognizing the patterns and features of attempted actions, and pays less attention to whether the action has been completed or not. All action clips within the data set have a high level of observable features, where the actual completeness of those actions was less of a concern.

In AFL games, some actions like marks (catching the ball kicked by a player on the same team) have a specific condition that needs to be met. According to AFL rules, a mark is only valid when a player takes control of the ball for a sufficient amount of time, in which the ball has been kicked from at least 15 m away and does not touch the ground and has not been touched by another player. We aim to identify specific action patterns based only on the camera images and as such we do not consider the precision of whether the kicker was 15 m away. Marks can be separated into marks and contested marks, where the latter is when multiple players attempt to catch (or knock the ball away) at the same time.

The videos from YouTube comprise many meaningless frames. We clip videos from longer videos and label them into five different classes:

(1) Kick: This class refers to the action whereby a player kicks the ball. The ball could come from various sources: the player himself holding the ball in front and dropping/kicking it, or kicking it directly off the ground.

(2) Mark: A player catches a kicked ball for sufficient time to be judged to be in control of the ball and without the ball being touched/interfered with by another player.

(3) Contested mark: Contested mark, is a special form of mark. This refers to the action that one player is trying to catch the ball and one or more opponents are either also trying to catch the ball at the same time or they are trying to punch the ball away.

(4) Pass: A player passes (punches) the ball to another player in the same team.

(5) Non-Action: This class includes players running, crowds cheering etc. This class is used to control the model performance as during the match there are many non-action frames. Without this class, the model would always try to classify video content into the previous four classes.

The details of each class in the data set are shown in Table 1, and example of each action class is shown in Fig. 1. Compared to other classes, the non-action class has a relatively low number of instances in the data set. The reason is that this class spans many different scenes, and too many instances in this class would drive the attention of the model away from key features of the four key action classes.

There are several challenges when using a data set for action recognition. Some actions share the same proportion of representations. One example is marking and passing the ball. In a video clip of relatively long distance passing, if the camera does not capture the whole passing process, e.g. it starts from somewhere in the middle, the representing features of this action might be similar to a mark action, i.e. someone catches the ball. The data set could also be modified by combing two classes of mark and contested mark, as sometimes it is hard to identify a mark compared to a contested mark. If a player is trying to catch the ball, and in the background an opponent is also trying to catch the ball, but

Table 1. Number of instances of each class

Class	# of instances		
	Training	Testing	Total
Kick	158	20	178
Contested mark	94	20	114
Mark	61	20	81
Pass	83	21	104
Non-action	66	21	87
Total	**462**	**102**	**564**

Fig. 1. Kick, contested mark, mark, completed pass

they do have not any physical contact at any time from one angle it may be considered as a mark. From a different camera angle, where there appears to be some degree of physical contact, it might seem more like a contested mark.

4   Implementation and Discussion of Results

Given the complexity and diversity of the architectures mentioned above, we use the Gluon CV toolkit [9]. This provides a Pytorch model implementation, and importantly, the ability to train custom data sets. In order to fully utilize the benefit of transfer learning and to compensate for the limited amount of data, we used models pre-trained on largely scaled action recognition data sets such as the Kinetics-400, and then fine-tune those models using the custom AFL data set. The final implementation involves a slightly modified version of Gluon CV which includes a few algorithmic alterations and some minor bug fixes. The architectures and pre-trained models we used along with their specifications and top-1 accuracy on Kinetics-400 are listed below in Table 2 [31]. Here R2+1D ResNet-50 model was calculated using a $112 \times 112 \times 3 \times 16$ input data size, R2+1D ResNet-152 model was calculated using a $112 \times 112 \times 3 \times 32$ input data size, and all other models were calculated based on a $224 \times 224 \times 3 \times 32$ input data size.

Table 2. Model specifications

Model	Pre-trained	#Mil parameters	GFLOPS	Accuracy (%)
I3D ResNet-50	ImageNet	28.863	33.275	74.87
I3D ResNet-101 Non-Local	ImageNet	61.780	66.326	75.81
I3D SlowFast ResNet-101	ImageNet	60.359	342.696	78.57
R2+1D ResNet-50	–	53.950	65.543	74.92
SlowFast-8x8 ResNet-101	–	62.827	96.794	76.95
TPN ResNet-101	–	99.705	374.048	79.70
R2+1D ResNet-152* [7]	IG65M	118.227	252.900	81.34
irCSN ResNet-152* [7]	IG65M	29.704	74.758	**83.18**

All model architectures are in 3D. I3D and I3D SlowFast models were based on inflated 2D ResNet pre-trained on ImageNet. irCSN and R2+1D ResNet-152 were pre-trained on IG-65M, and all other models were trained from scratch. All models used the Kinetics-400 data set for training [9].

The final training dataset was randomly split into training and validation data sets in the ratio of 70% and 30% respectively. A sub-clip of 16 frames was evenly sampled from each video clip at a regular interval depending on the clip's length. The number of input frames was selected as most actions happen in a short time period. If the sampled frames were less than 16, replacements would be randomly selected from the rest of the frames. The sampled frames would then be processed by standard data augmentation techniques, where it would be first resized to a resolution of 340×256, while R2+1D resized the frames to 171×128. The frames were then subject to a random resize with bi-linear interpolation and a random crop size 224×224. The crop size for R2+1D was 112×112. Following this, the frames were randomly flipped along the horizontal axis with a probability of 0.5, and normalized with means of $(0.485, 0.456, 0406)$ and standard deviations of $(0.229, 0.224, 0.225)$ with respect to each channel.

The training process used stochastic gradient descent (SGD) as the optimizer, with custom values of learning rate, momentum and weight decay, which were specific to each model. The value of learning rate plays a very important role in the model training process, where the correct learning rates will allow the algorithms to converge, whereas the wrong learning rates will result in the model not generalizing at all. Since we fine-tune pre-trained models, the initial learning rate was set much lower than the original model. The common values of the learning rate were 0.01 and 0.001, with a momentum of 0.9, a weight decay of $1e^{-5}$, and learning rate policy set to either step or cosine, depending on each model's architecture and level of complexity. Cross entropy loss was used for the model criterion with class weights taken into consideration since the training data set was imbalanced between the different classes. The number of epochs was set at 30 with an early stopping technique used to prevent over-fitting. The epoch with the lowest validation loss was saved as the best weight.

The top-1 accuracy on the testing data set for the fine-tuned models is shown in Table 3.

Table 3. Top-1 accuracy on the AFL test data set

Model	Accuracy (%)
I3D ResNet-50	56.86
I3D ResNet-101 Non-local	61.77
SlowFast-8x8 ResNet-101	69.61
TPN ResNet-101	70.59
I3D SlowFast ResNet-101	71.57
R2+1D ResNet-50	72.55
irCSN ResNet-152	74.51
R2+1D ResNet-152	**77.45**

As seen, the best performing model was the R2+1D ResNet-152 model pre-trained on the (very large) IG65M dataset. This achieved a top-1 accuracy of 77.45%. The final classification of action recognition results are shown in Table 4. As seen, the classification for marks had the lowest recall of 0.55, while contested marks had a recall of 0.85. This is possibly due to marks and contested marks being difficult to distinguish in some circumstances due to the presence of other players in the background. The classification for non-action has the lowest precision of 0.57 and the lowest f1 score of 0.65. The reason for this is that the non-action class is very broad and contains many sub-classes, such as scenes of audiences and players running and cheering. Splitting the class into multiple distinct classes in the future may improve the non-action accuracy. Among all classes, the classification of kicks has the highest f1-score at 0.89, since a kick has arguably the most distinct and recognizable features.

Table 4. Final classification results

Action	Precision	Recall	F1-score
Kick	1.00	0.80	0.89
Contested mark	0.74	0.85	0.79
Mark	0.85	0.55	0.67
Pass	0.86	0.90	0.88
Non-action	0.57	0.76	0.65

The results for the top-1 accuracy of the AFL testing data set are generally consistent with the model performance using the Kinetics-400 dataset, however the R2+1D ResNet-50 model achieved some noteworthy improvements. The model I3D ResNet-50 performed poorly with a top-1 accuracy of 56.86%, whilst the model I3D ResNet-101 Non-Local only achieved an accuracy of 61.77%. It might be inferred that the inflated 2D ResNets (I3D) are limited in their ability to capture spatio-temporal features, while R2+1D is more capable in this regard as it utilizes the factorization of the 3D ResNet architecture. It was also found that non-local blocks may not be suitable for Australian rules football, as they are designed to capture long range temporal features. Actions in AFL are relatively fast and diverse which results in the model under-performing.

It was found that the performance of models generally depends on their backbone architecture. The complexity of the ResNet architecture is closely related to the prediction accuracy, hence it could be argued that the more complex the architecture is, the more likely the model will generalize and make the right predictions. Comparing ResNet-50 with ResNet-152, there is a significant difference in complexity and number of parameters, which could be one reason for the relatively large performance difference. Another major factor to consider is that both R2+1D ResNet-152 and irCSN used IG65M for model pre-training and hence benefit from the very largely scaled data set. It is also interesting to note that R2+1D uses a 112 × 112 resolution input after data augmentation, whilst the rest of the models use a 224 × 224 input. Despite this R2+1D is still able to produce some of the best results overall.

SlowFast and TPN networks both model visual tempos in video clips. When incorporating I3D into SlowFast network, the model I3D SlowFast ResNet-101 performed evidently better than the other I3D models, indicating that the Slow-Fast networks are capable at better extracting spatio-temporal features and that modelling visual tempos improves the overall model performance. However, Slow-Fast is a more strict framework that limits the number of frames of different streams, whilst TPN is more flexible due to its pyramid structure. As a result, TPN ResNet-101 performed slightly better than SlowFast ResNet-101.

There are several important limitations to the presented models. Firstly, incomplete actions will likely be classified as actions. As shown in Fig. 2(a), an incomplete contested mark has been classified as a contested mark. This is due to the incomplete action sharing a lot of similar features to a completed action. The model does not always possess the ability to recognise whether the ball has been cleanly caught (or not). Secondly, the model tends to perform poorly in complex scenes and environments. From Fig. 2(b), it can be seen that there are many players present in the background and a player is tackling another player who has the ball. In this case, the model mis-classifies the scenario into a pass as it is similar to the scenarios of pass in the training data set.

(a) Incomplete contested mark   (b) Realistic complex environment

Fig. 2. Mis-classified actions

5   Team Identification and Associated Limitations

Many action events depend on distinguishing teams, e.g. a completed pass requires the ball to be passed by a player within the same team. Team identification is thus important to any Australian football model. In this work, we utilize the You-Only-Look-Once (YOLO) v5 [14] framework and the DeepSORT algorithm [29] to identify and track multiple objects at the same time. The implementation of this module inputs raw frames to be classified, filters and then keeps player location information in each frame. The DeepSORT algorithm is capable of tracking object movement across different frames, and assign unique IDs to team players.

As with many team sports, AFL players wear team jerseys with colors representing their team. In this way, audiences are able to identify (distinguish) players from the two teams. We apply color distribution extractors to images to extract the differences in player jersey colors. The distribution can then be used as an input to construct a high-dimensional features such as KMeans clustering [16] to cluster players into groups. A screenshot of the results of the application is shown in Fig. 3.

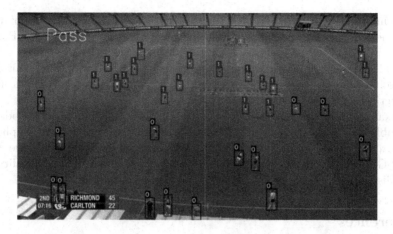

Fig. 3. Team identification classification example

The performance of this module is limited by the resolution of the video. With a higher resolution, the jersey color of players in the foreground is clear but those in the far background is less clear. Another challenge faced is rapid camera movement and viewpoint changes. In real matches, sudden viewpoint changes from long-distance views to close-up views (and vice versa) happens continually. Ideally (from the model perspective) there would be a single camera angle - akin to what a spectator sees in a game, but this never happens in reality when games are shown on television. These continuous viewpoint changes make it challenging to track a specific player's movement. Nevertheless, the team identification is able to distinguish the teams within a few milliseconds. The performance of the system also greatly depends on teams wearing clearly identifiable jerseys. This is always the case however so does not limit the model. If players get especially muddy for example this might be an issue, but this is a rarity in Australia.

6 Conclusions and Future Work

This paper explored the feasibility of action recognition for Australian rules football using 3D CNN architectures. Various action recognition models including state-of-the-art models pre-trained on large-scale data sets were utilised. We fine-tune those models on a newly developed AFL data set, and reported a 77.45% top-1 accuracy for the best performing model R2+1D ResNet-152. A smoothing strategy allowed the algorithm to localize the frame range for actions in long video segments. We also developed a team identification solution and an action recognition application that showed both the potential and viability of applying real time end-to-end action recognition to AFL matches.

There are many future extensions to the work. The team identification framework opens up further improvements on action recognition in AFL matches for specific teams. Actions such as pass and contested mark require additional team information in order to be classified correctly. Moreover, the use of attention mechanisms in machine learning and use of transformers such as Bidirectional Encoder Representations from Transformers (BERT) [21] has the ability to model contextual information with mechanisms for self attention. This could be useful in scenes that contain multiple players and allow to achieve a higher prediction accuracy.

Examples of the application of the models and the source code are available at: https://youtu.be/I7490fyuiK8 and https://github.com/stephenkl/Research-project respectively. This research was undertaken using the LIEF HPC-GPGPU Facility hosted at the University of Melbourne. This Facility was established with the assistance of LIEF Grant LE170100200.

References

1. Abu-El-Haija, S., et al.: YouTube-8M: a large-scale video classification benchmark. CoRR abs/1609.08675. arXiv: 1609.08675 (2016)
2. Carreira, J., Zisserman, A.: Quo Vadis, action recognition? A new model and the kinetics dataset. In: 2017 IEEE Conference on Computer Vision and Pattern Recognition (CVPR), pp. 4724–4733. IEEE, Honolulu (July 2017). https://doi.org/10.1109/CVPR.2017.502, http://ieeexplore.ieee.org/document/8099985/
3. Deng, J., Dong, W., Socher, R., Li, L.J., Kai Li, Li Fei-Fei: ImageNet: a large-scale hierarchical image database. In: 2009 IEEE Conference on Computer Vision and Pattern Recognition, pp. 248–255. IEEE, Miami (June 2009). https://doi.org/10.1109/CVPR.2009.5206848, https://ieeexplore.ieee.org/document/5206848/
4. Donahue, J., et al.: Long-term recurrent convolutional networks for visual recognition and description. In: 2015 IEEE Conference on Computer Vision and Pattern Recognition (CVPR), pp. 2625–2634. IEEE, Boston (June 2015). https://doi.org/10.1109/CVPR.2015.7298878, http://ieeexplore.ieee.org/document/7298878/
5. Feichtenhofer, C., Fan, H., Malik, J., He, K.: SlowFast networks for video recognition. In: 2019 IEEE/CVF International Conference on Computer Vision (ICCV), pp. 6201–6210. IEEE, Seoul (October 2019). https://doi.org/10.1109/ICCV.2019.00630, https://ieeexplore.ieee.org/document/9008780/
6. Feichtenhofer, C., Pinz, A., Zisserman, A.: Convolutional two-stream network fusion for video action recognition. In: 2016 IEEE Conference on Computer Vision and Pattern Recognition (CVPR), pp. 1933–1941. IEEE, Las Vegas (June 2016). https://doi.org/10.1109/CVPR.2016.213, http://ieeexplore.ieee.org/document/7780582/
7. Ghadiyaram, D., Tran, D., Mahajan, D.: Large-scale weakly-supervised pretraining for video action recognition. In: 2019 IEEE/CVF Conference on Computer Vision and Pattern Recognition (CVPR), pp. 12038–12047. IEEE, Long Beach (June 2019). https://doi.org/10.1109/CVPR.2019.01232, https://ieeexplore.ieee.org/document/8953267/
8. Giancola, S., Amine, M., Dghaily, T., Ghanem, B.: SoccerNet: a scalable dataset for action spotting in soccer videos. CoRR abs/1804.04527. arXiv: 1804.04527 (2018)
9. Guo, J., et al.: GluonCV and GluonNLP: deep learning in computer vision and natural language processing. arXiv:1907.04433 (February 2020)

10. He, K., Zhang, X., Ren, S., Sun, J.: Deep residual learning for image recognition. In: 2016 IEEE Conference on Computer Vision and Pattern Recognition (CVPR), pp. 770–778. IEEE, Las Vegas (June 2016). https://doi.org/10.1109/CVPR.2016.90, http://ieeexplore.ieee.org/document/7780459/

11. Horn, B.K., Schunck, B.G.: Determining optical flow. Artif. Intell. **17**(1–3), 185–203 (1981). https://doi.org/10.1016/0004-3702(81)90024-2, https://linkinghub.elsevier.com/retrieve/pii/0004370281900242

12. Ioffe, S., Szegedy, C.: Batch normalization: accelerating deep network training by reducing internal covariate shift. In: Proceedings of the 32nd International Conference on International Conference on Machine Learning, ICML 2015, JMLR.org, Lille, France, vol. 37, pp. 448–456 (July 2015)

13. Ji, S., Xu, W., Yang, M., Yu, K.: 3D convolutional neural networks for human action recognition. IEEE Trans. Pattern Anal. Mach. Intell. **35**(1), 221–231 (2013). https://doi.org/10.1109/TPAMI.2012.59

14. Jocher, G., Stoken, A., Chaurasia, A., Borovec, J.: NanoCode012, Taoxie, Kwon, Y., Michael, K., Changyu, L., Fang, J., V, A., Laughing, tkianai, yxNONG, Skalski, P., Hogan, A., Nadar, J., imyhxy, Mammana, L., AlexWang1900, Fati, C., Montes, D., Hajek, J., Diaconu, L., Minh, M.T., Marc, albinxavi, fatih, oleg, wanghaoyang0106: ultralytics/yolov5: v6.0 - YOLOv5n 'Nano' models, Roboflow integration, TensorFlow export, OpenCV DNN support (October 2021). https://doi.org/10.5281/zenodo.5563715

15. Yue-Hei Ng, J., Hausknecht, M., Vijayanarasimhan, S., Vinyals, O., Monga, R., Toderici, G.: Beyond short snippets: deep networks for video classification. In: 2015 IEEE Conference on Computer Vision and Pattern Recognition (CVPR), pp. 4694–4702. IEEE, Boston (June 2015). https://doi.org/10.1109/CVPR.2015.7299101, http://ieeexplore.ieee.org/document/7299101/

16. Kanungo, T., Mount, D., Netanyahu, N., Piatko, C., Silverman, R., Wu, A.: An efficient k-means clustering algorithm: analysis and implementation. IEEE Trans. Pattern Anal. Mach. Intell. **24**(7), 881–892 (2002). https://doi.org/10.1109/TPAMI.2002.1017616

17. Karpathy, A., Toderici, G., Shetty, S., Leung, T., Sukthankar, R., Fei-Fei, L.: Large-scale video classification with convolutional neural networks. In: 2014 IEEE Conference on Computer Vision and Pattern Recognition, pp. 1725–1732. IEEE, Columbus (June 2014). https://doi.org/10.1109/CVPR.2014.223, https://ieeexplore.ieee.org/document/6909619

18. Kay, W., et al.: The Kinetics Human Action Video Dataset. arXiv:1705.06950 (May 2017)

19. Simonyan, K., Zisserman, A.: Two-stream convolutional networks for action recognition in videos. In: Proceedings of the 27th International Conference on Neural Information Processing Systems, NIPS 2014, vol. 1, pp. 568–576. MIT Press, Montreal (December 2014)

20. Soomro, K., Zamir, A.R., Shah, M.: UCF101: a dataset of 101 human actions classes from videos in the wild. arXiv:1212.0402 (December 2012)

21. Su, W., et al.: VL-BERT: pre-training of generic visual-linguistic representations. CoRR abs/1908.08530. arXiv: 1908.08530 (2019)

22. Tran, D., Bourdev, L., Fergus, R., Torresani, L., Paluri, M.: Learning spatiotemporal features with 3D convolutional networks. In: 2015 IEEE International Conference on Computer Vision (ICCV), pp. 4489–4497. IEEE, Santiago (December 2015). https://doi.org/10.1109/ICCV.2015.510, http://ieeexplore.ieee.org/document/7410867/

23. Tran, D., Wang, H., Feiszli, M., Torresani, L.: video classification with channel-separated convolutional networks. In: 2019 IEEE/CVF International Conference on Computer Vision (ICCV), pp. 5551–5560. IEEE, Seoul (October 2019). https://doi.org/10.1109/ICCV.2019.00565, https://ieeexplore.ieee.org/document/9008828/

24. Tran, D., Wang, H., Torresani, L., Ray, J., LeCun, Y., Paluri, M.: A closer look at spatiotemporal convolutions for action recognition. In: 2018 IEEE/CVF Conference on Computer Vision and Pattern Recognition, pp. 6450–6459 (June 2018). https://doi.org/10.1109/CVPR.2018.00675, iSSN: 2575-7075

25. Vaswani, A., et al.: Attention is all you need. In: Proceedings of the 31st International Conference on Neural Information Processing Systems, NIPS 2017, pp. 6000–6010. Curran Associates Inc., Long Beach (December 2017)

26. Wang, H., Schmid, C.: Action recognition with improved trajectories. In: 2013 IEEE International Conference on Computer Vision, pp. 3551–3558 (December 2013). https://doi.org/10.1109/ICCV.2013.441, iSSN: 2380-7504

27. Wang, L., et al.: Temporal segment networks: towards good practices for deep action recognition. In: Leibe, B., Matas, J., Sebe, N., Welling, M. (eds.) ECCV 2016, Part VIII. LNCS, vol. 9912, pp. 20–36. Springer, Cham (2016). https://doi.org/10.1007/978-3-319-46484-8_2

28. Wang, X., Girshick, R., Gupta, A., He, K.: Non-local neural networks. In: 2018 IEEE/CVF Conference on Computer Vision and Pattern Recognition, pp. 7794–7803. IEEE, Salt Lake City (June 2018). https://doi.org/10.1109/CVPR.2018.00813, https://ieeexplore.ieee.org/document/8578911/

29. Wojke, N., Bewley, A., Paulus, D.: Simple online and realtime tracking with a deep association metric. In: 2017 IEEE International Conference on Image Processing (ICIP), pp. 3645–3649. IEEE, Beijing (September 2017). https://doi.org/10.1109/ICIP.2017.8296962, http://ieeexplore.ieee.org/document/8296962/

30. Yang, C., Xu, Y., Shi, J., Dai, B., Zhou, B.: Temporal pyramid network for action recognition. In: 2020 IEEE/CVF Conference on Computer Vision and Pattern Recognition (CVPR), pp. 588–597. IEEE, Seattle (June 2020). https://doi.org/10.1109/CVPR42600.2020.00067, https://ieeexplore.ieee.org/document/9157586/

31. Zhu, Y., et al.: A comprehensive study of deep video action recognition. arXiv:2012.06567 (December 2020)

SEGP: Stance-Emotion Joint Data Augmentation with Gradual Prompt-Tuning for Stance Detection

Junlin Wang[1,2], Yan Zhou[1(✉)], Yaxin Liu[1,2], Weibo Zhang[1,2], and Songlin Hu[1,2]

[1] Institute of Information Engineering, Chinese Academy of Sciences, Beijing, China
{wangjunlin,zhouyan,liuyaxin,zhangweibo,husonglin}@iie.ac.cn
[2] School of Cyber Security, University of Chinese Academy of Sciences, Beijing, China

Abstract. Stance detection is an important task in opinion mining, which aims to determine whether the author of a text is in favor of, against, or neutral towards a specific target. By now, the scarcity of annotations is one of the remaining problems in stance detection. In this paper, we propose a Stance-Emotion joint Data Augmentation with Gradual Prompt-tuning (SEGP) model to address this problem. In order to generate more training samples, we propose an auxiliary sentence based Stance-Emotion joint Data Augmentation (SEDA) method, formulate data augmentation as a conditional masked language modeling task. We leverage different relations between stance and emotion to construct auxiliary sentences. SEDA generates augmented samples by predicting the masked words conditioned on both their context and auxiliary sentences. Furthermore, we propose a Gradual Prompt-tuning method to make better use of the augmented samples, which is a combination of prompt-tuning and curriculum learning. Specifically, the model starts by training on only original samples, then adds augmented samples as training progresses. Experimental results show that SEGP significantly outperforms the state-of-the-art approaches.

Keywords: Stance detection · Data augmentation · Curriculum learning

1 Introduction

The goal of stance detection is to classify a piece of text as either being in support, opposition, or neutrality towards a given target, the target may not be directly contained in the text. With the rapid development of social media, more and more people post online to express their support or opposition towards various targets. Stance detection is known to have several practical application areas such as polling, public health surveillance, fake news detection, and so on. These conditions motivate a large number of studies to focus on inferring

Table 1. Examples of stance detection task.

Text	Stance
Wearing a mask is common sense and kind to your fellow human. We all have to do our part to slow the spread of COVID-19	Favor
Spend the day outside, get some sun and fresh air. Without a face mask. Best way to keep up your immune system	Against
Any skincare suggestions for breakouts because of face masks?	Neutral

the stances of users from their posts. Table 1 shows some examples on target "Wearing a Face Mask", annotated with the stance labels.

One of the biggest challenges in stance detection task is the scarcity of annotated samples. Data augmentation is commonly used to address data scarcity, which aims to generate augmented samples based on limited annotations. Zhang et al. [37] replace words with WordNet [19] synonyms to get augmented sentences. Wei et al. [33] propose EDA, which is a combination of token-level augmentation approaches. These methods are effective, but the replacement strategies are simple, thus can only generate limited diversified patterns. To enhance the consistency between augmented samples and labels, Wu et al. [35] propose CBERT, the segmentation embeddings of BERT [11] are replaced with the annotated labels during augmentation. However, these methods fail to take targets into consideration. To solve this problem, Li et al. [16] propose ASDA, which uses the conditional masked language modeling (C-MLM) task to generate augmented samples under target and stance conditions.

Although ASDA [16] achieves highly competitive performance, there still exist two limitations. First, they neglect the emotional information during augmentation. It should be noted that emotion can affect the judgment of stance. There exists a number of studies that use emotional information to assist stance detection and achieve good results [6,14,20]. Thus, we posit that in addition to stance and target information, the introduction of emotional information through auxiliary sentences can further improve the label consistency of augmented samples. Second, they neglect the linguistic adversity problem [17,31] during training. This problem is introduced by data augmentation method and therefore can be seen as a form of noising, where noised data is harder to learn from than unmodified original data.

In this paper, we propose a Stance-Emotion joint Data Augmentation with Gradual Prompt-tuning (SEGP) model to address the above limitations. Specifically, we present an auxiliary sentence based Stance-Emotion joint Data Augmentation (SEDA) method that generates target-relevant and stance-emotion-consistent samples based on C-MLM task. We suppose that there are "Consistency", "Discrepancy" and "None" relations between stance and emotion. The auxiliary sentences are constructed on the premise of these relations as well as the target. With the help of C-MLM task, SEDA augment the dataset by predicting the masked words conditioned on both their context and the auxiliary sentences. Furthermore, to address the linguistic adversity problem in augmented samples, we propose a Gradual Prompt-tuning method, which combines prompt-tuning with curriculum learning to train our model. We design a template that contains target and stance information. After that, we create an artificial curriculum in the training samples according to the disturbance degree in data augmentation. Starting by training on original samples, we feed augmented samples with a higher level of noising into the model as training progresses. The model learns to explicitly capture stance relations between sentence and target by predicting masked words. Our main contributions can be summarized as follows:

- We propose a Stance-Emotion joint Data Augmentation (SEDA) method, which introduces emotional information in the conditional data augmentation of stance detection.
- We further propose a Gradual Prompt-tuning method to overcome the linguistic adversity problem in augmented samples, which combines prompt-tuning with curriculum learning.
- Experimental results show that our methods significantly outperform the state-of-the-art methods.

2   Related Work

2.1   Stance Detection

Stance detection aims to automatically infer the stance of a text towards specific targets [1,13], which is related to argument mining, fact-checking, and aspect-level sentiment analysis. Early stance detection tasks concentrate on online forums and debates [27,29]. Later, a series of studies on different types of targets emerge. The targets become political figures [15,26], controversial topics [7], and so on. At present, the research tasks are mainly divided into three types, in-target stance detection [36], cross-target stance detection [3], and zero-shot stance detection [2]. In this paper, we focus on in-target stance detection, which means the test target can always be seen in the training stage.

2.2   Data Augmentation

Lexical substitution is a commonly used augmentation strategy, which attempts to substitute words without changing the meaning of the entire text.

The first commonly used approach is the thesaurus-based substitution, which means taking a random word from the sentence and replacing it with its synonym using a thesaurus. Zhang et al. [37] apply this and search synonyms in WordNet [19] database. Mueller et al. [21] use this idea to generate additional training samples for their sentence similarity model. This approach is also used by Wei et al. [33] as one of the four random augmentations in EDA.

The second approach is the word-embedding substitution, which replaces some words in a sentence with their nearest neighbor words in the embedding space. Jiao et al. [10] apply this with GloVe embeddings [23] to improve the generalization of their model on downstream tasks, while Wang et al. [30] use it to augment tweets needed to learn a topic model.

The third approach is based on the masked language model, which has to predict the masked words based on their context. Therefore, the model can generate variations of a text using the mask predictions. Compared to previous approaches, the generated text is more grammatically coherent as the model takes context into account when making predictions. Grag et al. [8] use this idea to generate adversarial samples for text classification. Wu et al. [35] formulate the data augmentation as a C-MLM task. Li et al. [16] propose an Auxiliary Sentence based Data Augmentation (ASDA) method that generates samples based on C-MLM task. Inspired by ASDA, we investigate how to introduce more information via auxiliary sentences.

2.3   Curriculum Learning

Curriculum learning is proposed by Bengio et al. [4], which is a training strategy that imitates the meaningful learning order in human curricula. It posits that models train better when training samples are organized in a meaningful order. In the beginning, researchers assume that there exists a range of difficulties in the training samples [28, 34]. They leverage various heuristics to sort samples by difficulty and train models on progressively harder samples. Korbar et al. [12] propose instead of discovering a curriculum in existing samples, samples can be intentionally modified to dictate an artificial range of difficulty. Wei et al. [32] combine this idea with data augmentation and propose a curriculum learning strategy, but the performance is still constricted by the gap of objective forms between pre-training and fine-tuning.

2.4   Prompt-Tuning

Pre-trained language models like GPT [5] and BERT [11] capture rich knowledge from massive corpora. To make better use of the knowledge, prompt-tuning is proposed. In prompt-tuning, downstream tasks are also formalized as some objectives of language modeling by leveraging language prompts. The results of language modeling can correspond to the solutions of downstream tasks. With specially constructed prompts and tuning objectives [18, 24], we can further inject and stimulate the task-related knowledge in pre-trained models, thus boosting the performance. To our knowledge, there is currently a lack of research on applying prompt-tuning to the stance detection task.

3   Method

In this section, we first introduce the variables and definitions that appear in this paper. Then provide the overall architecture of SEGP and explain it in detail.

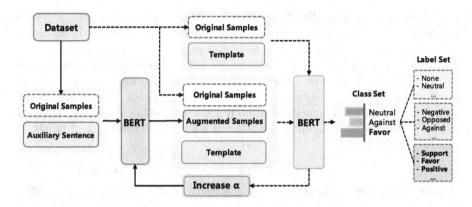

Fig. 1. The overall architecture of SEGP, where α represents the degree of disturbance in the augmentation stage. Solid arrows indicate Stance-Emotion joint Data Augmentation stage and dashed arrows indicate Gradual Prompt-tuning stage.

3.1   Preliminaries

We first give some essential preliminaries. Suppose a given training dataset of size n is $D_{\text{train}} = \{X, S, T, E\}$, where $X = \{x_1, x_2, \ldots, x_n\}$ is the set of input samples. For each $x_i \in X$, it consists of a sequence of l words $x_i = [w_i^1, w_i^2, \ldots, w_i^l]$. We define a stance label set $S = \{s_1, s_2, \ldots, s_{|M|}\}$, a target set $T = \{t_1, t_2, \ldots, t_{|C|}\}$ and an emotional label set $E = \{e_1, e_2, \ldots, e_{|N|}\}$, where the values of $|M|$, $|C|$ and $|N|$ depend on the dataset settings.

3.2   Overall Architecture

In this paper, we propose a Stance-Emotion joint Data Augmentation with Gradual Prompt-tuning (SEGP) model, and the overall architecture is shown in Fig. 1. SEGP consists of two stages, as we can see from Fig. 1, they are indicated by solid arrows and dashed arrows respectively. The first stage is to get more training samples using the SEDA method. The second is the training stage, which uses the Gradual Prompt-tuning method to overcome the linguistic adversity problem in augmented samples.

3.3   Stance-Emotion Joint Data Augmentation

The objective of a data augmentation method is to generate training samples based on the existing limited annotations. In this paper, we propose a novel conditional data augmentation method called SEDA, which is based on C-MLM task. We leverage stance, emotion, and target information to construct auxiliary sentences. SEDA generates target-relevant and stance-emotion-consistent augmented samples by predicting masked words conditioned on context and auxiliary sentences.

Construction of Auxiliary Sentences. Many approaches achieve better results by taking emotional information as auxiliary information. It should be noted that stance could be inferred independently from the emotional state, the emotions contained in a text may be positive but expresses an opposition stance to a given target. This is due to the complexity of interpreting a stance because it is not always directly consistent with the emotional polarity. We analyze the distribution of stance and emotional labels in COVID-19-Stance dataset. As shown in Fig. 2, there is a large gap in the distribution of these two types of labels.

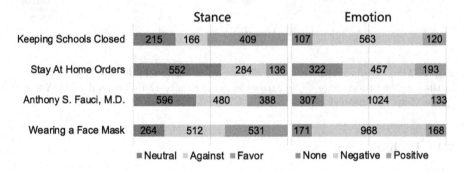

Fig. 2. Stance and emotion distribution in COVID-19-Stance dataset.

Our research is based on stance label set S = { "Neutral", "Against", "Favor"} and emotional label set E = { "None", "Negative", "Positive"}. In order to integrate these two types of information, we define a cross label set $C = \{S - E\}$, which is generated by stance label s and emotional label e. For example, given $s=$ "Favor" and $e=$ "Negative", we can obtain the cross label $c=$ "Favor-Negative". Before constructing auxiliary sentences, we put forward the following relations between stance and emotion:

- Consistency: When cross label c is in { "Favor-Positive", "Against-Negative", "Neutral-Positive", "Neutral-Negative"}, we suppose that the stance is consistent with emotion, so emotional information can be directly introduced into the auxiliary sentence.
- Discrepancy: When cross label c is in { "Favor-Negative", "Against-Positive"}, we suppose that there is a difference between stance and emotion, so we need to consider this contradiction when constructing auxiliary sentences.
- None: When the emotional label $e =$ "None", we suppose that the emotional information is not helpful. In this case, the auxiliary sentence only needs to introduce stance information.

Therefore, we leverage the above mentioned relations to construct three kinds of auxiliary sentences regarding target, stance, and emotion. We also place slots in the auxiliary sentences, $\{a_i\}$ is used to fill target words, $\{s_i\}$ is used to fill stance label, and $\{e_i\}$ is used to fill emotional label. Experiments show that grammar correctness is not important. Table 2 shows how to select the corresponding auxiliary sentence according to a cross label. After obtaining the auxiliary sentence, we prepend both another training sample x_j that has the same target and cross label with x_i. The complete input form for each training sample x_i is: Auxiliary sentence$+x_j+$"the text is"$:+x_i$.

Table 2. Correspondence between relations, cross labels, and auxiliary sentences.

Relations	Cross labels	Auxiliary sentences
Consistency	Favor-Positive Against-Negative Neutral-Positive Neutral-Negative	The following texts have $\{s_i\}$ stance and $\{e_i\}$ emotion to $\{a_i\}$.
Discrepancy	Favor-Negative Against-Positive	Although the emotion is $\{e_i\}$, the following texts are both $\{s_i\}$ to $\{a_i\}$.
None	Favor-None Neutral-None Against-None	The following texts have $\{s_i\}$ stance to $\{a_i\}$.

For example, given the input x_i: *I don't need to wear a mask to live a healthy life.* with the stance label $s =$ "Against" and emotional label $e =$ "Positive". The corresponding target is "Wearing a face mask". First, we get its cross label $c =$ "Against-Positive" and choose the discrepancy auxiliary sentence. Second, we find another training sample x_j: *The death rate is falling so fast, we don't need to wear masks at all.* So the complete input is: *Although the emotion is {positive}, the following texts are both {against} to {wearing a face mask}. The death rate is falling so fast, we don't need to wear masks at all. The text is: I don't need to wear a mask to live a healthy life.* The introduction of the auxiliary sentence and x_j not only helps to generate more diversified samples, but also provides a strong guideline to help generate target-relevant and label-compatible samples.

Data Generation. We fine-tune the pre-trained model via C-MLM task. For a training sample x_i from X, we specify that the model can only randomly mask words in the input sample x_i and the mask radio is α. Because we want to preserve all of the target, stance, and emotional information. After prepending the corresponding auxiliary sentence and x_j to obtain the masked sentence, a pre-trained language model like BERT is used to predict the masked words. The prediction of masked words depends not only on the context of x_i, but also on their target, stance, and emotion.

After fine-tuning the model on the training dataset for a few epochs, we use the well-trained model for augmentation. Similar to the fine-tuning procedure, the model randomly masks words of the training sample, then prepend the auxiliary sentence and another training sample. The model is used to predict the masked words, we repeat these steps over training samples to get augmented samples.

3.4   Gradual Prompt-Tuning

In this paper, we apply the training strategy of curriculum learning to prompt-tuning. We aim to solve the linguistic adversity problem [17,31] in augmented samples as well as make better use of the knowledge contained in pre-trained language models.

Prompt-Tuning. In order to bridge the gap of objective forms between pre-training and fine-tuning, prompt-tuning is proposed. By tuning a pre-trained language model with the cloze-style task, prompt-tuning can manipulate the model behavior to fit various downstream tasks that more fully utilize task-related knowledge in pre-trained language models. Formally, prompt is consists of a template $P(\cdot)$ and a set of stance labels S. For stance detection task, a pre-trained language model uses input sentences and prompt to predict the stance label for a given target. In order to provide more information, we place two slots into the template, $\{t_i\}$ is used to fill target words, and $[MASK]$ is for the model to fill a label word. We set the template $P(\cdot) = $ "The stance to $\{t_i\}$ is $[MASK]$", and map x to the prompt input $x_{prompt} = x+$"The stance to $\{t_i\}$ is $[MASK]$". After that, x_{prompt} is fed into a pre-trained model.

The model first converts the input $x_{prompt} = \{w_i^1, w_i^2, \ldots, [MASK], \ldots, w_i^l\}$ to sequence $\{[CLS], w_i^1, w_i^2, \ldots, [MASK], \ldots, w_i^l, [SEP]\}$, then compute the hidden vector $h_{[MASK]}$ of $[MASK]$. Given $s \in S$, the model calculates the probability for s can fill the masked position, where $\mathbf{s}$ is the embedding of s in a pre-trained language model. The probability is calculated as follows:

$$p\left([MASK] = s \mid x_{\text{prompt}}\right) = \frac{\exp\left(s \cdot h_{[MASK]}\right)}{\sum_{s \in S} \exp\left(\mathbf{s} \cdot h_{[MASK]}\right)} \tag{1}$$

There also exists an injective mapping function φ that bridges the set of classes Y and the set of label words S, we define $\varphi = Y \rightarrow S$. With the verbalizer φ, we can formalize the probability distribution over Y with the probability distribution over S at the masked position. i.e., $p(y \mid x) = p([MASK] = \phi(y) \mid x_{prompt})$. We map the supporting stance to "Favor", the opposing stance to "Against" and other stances to "Neutral". According to model fills the masked position of x_{prompt} with "Favor", "Against" or "Neutral", we can get the stance of x. For prompt-tuning, with a template $P(\cdot)$, a label set S and verbalizer φ, the learning objective is to maximize $\frac{1}{|X|} \sum_{x \in X} \log p\left([MASK] = \phi\left(y_x\right) \mid P(x)\right)$.

Curriculum Learning. The data augmentation method might introduce linguistic adversity and can be seen as a form of noising, where noised data is harder to learn from than unmodified original data. Curriculum learning posits that the model train better when training samples are organized in a meaningful order that gradually shows more concepts and complexity. Therefore, we apply the training strategy of curriculum learning to prompt-tuning. We define the mask radio $0.0 \leq \alpha \leq 0.15$ as disturbance degree for SEDA stage, create an artificial curriculum in training samples according to the disturbance degree of the augmented samples. A larger mask ratio α represents a larger variation in the training samples, thus harder to learn from than unmodified original samples. During training, we begin with a disturbance degree of $\alpha = 0.0$ (equivalent to no augmentation), then linearly increase α by 0.05 every time validation loss plateaus, up to a final of $\alpha = 0.15$.

4 Experiment

In this section, we first present the dataset used for evaluation and several baseline methods. Then introduce experimental details and analyze the results.

4.1 Dataset and Baseline Methods

We carry out experiments on the stance detection dataset COVID-19-Stance [9], which is collected by crawling Twitter, using Twitter Streaming API. It contains the tweets of four targets (i.e., "Stay At Home Orders", "Wearing a Face Mask", "Keeping Schools Closed" and "Anthony S. Fauci, M.D"), and the stance label of each tweet is either "Favor" or "Against" or "Neutral".

We compare SEGP with the following baseline methods:

- BiLSTM [25]: Bi-Directional Long Short Term Memory Network takes tweets as input and is trained to predict the stance towards a target, without explicitly using the target information.
- CT-BERT [22]: A pre-trained language model that predicts the stance by appending a linear classification layer to the hidden representation of $[CLS]$ token, pre-trained on a corpus of messages from Twitter about COVID-19.
- CT-BERT-v2 [22]: It is identical to CT-BERT, but trained on more data, resulting in higher downstream performance.
- EDA [33]: A simple data augmentation method that consists of four operations: synonym replacement, random deletion, random swap, and random insertion.
- ASDA [16]: A data augmentation method that generates target-relevant and label-consistent data samples based on C-MLM task.

4.2 Experimental Results

SEGP is implemented based on CT-BERT-v2 [22], using a batch size of 8. The learning rate of Adam optimizer is 1e−5 and the maximum sequence length is 256. Experimental results are shown in Table 3, the best model configuration is selected according to the highest performance on the development set.

We first compare SEGP with BiLSTM [25], CT-BERT [22] and CT-BERT-v2 [22]. It can be seen that SEGP is superior to all baselines in accuracy and F1 score, which demonstrates the validity of our model in stance detection tasks. Besides, we compare SEGP with different data augmentation methods, i.e., EDA and ASDA. We can observe that SEGP performs the best, while EDA and ASDA methods have limited improvement in performance. Furthermore, when target = "Anthony S. Fauci, M.D.", the result is even worse than CT-BERT that only trained on original samples.

SEGP has better performance on all targets, which proves it can not only generate more diversified samples but also have the ability to overcome the linguistic adversity problem and better utilize task-related knowledge in pre-trained language models.

Table 3. Performance of SEGP and different baseline methods for stance detection on four targets in the COVID-19-Stance dataset. The performance is reported in terms of accuracy(Acc), precision(P), recall(R), and F1 score(F1). We highlight the best results in bold.

Model	Wearing a face mask				Stay at home orders			
	Acc	P	R	F1	Acc	P	R	F1
BiLSTM	57.80	56.90	58.00	56.70	73.50	67.90	64.00	64.50
CT-BERT	81.00	81.80	80.30	80.30	84.30	81.60	78.80	80.00
CT-BERT-v2	81.25	80.49	81.99	80.13	86.00	82.56	88.00	84.78
EDA	81.50	79.77	78.61	79.07	85.50	81.96	84.50	83.09
ASDA	82.50	80.96	80.24	80.53	87.00	83.04	85.09	83.99
SEGP	**84.50**	**83.20**	**83.49**	**83.34**	**89.00**	**86.33**	**89.37**	**87.71**
Model	Anthony S. Fauci, M.D.				Keeping schools closed			
	Acc	P	R	F1	Acc	P	R	F1
BiLSTM	63.80	63.90	63.10	63.00	62.70	57.00	54.50	54.80
CT-BERT	81.70	81.60	83.00	81.80	77.20	76.50	76.10	75.50
CT-BERT-v2	80.25	80.16	81.36	80.42	81.00	78.81	79.14	78.85
EDA	80.50	80.82	81.01	80.55	83.00	80.92	81.66	80.98
ASDA	81.00	81.49	81.04	81.06	83.50	81.29	80.95	81.01
SEGP	**82.50**	**82.60**	**82.57**	**82.57**	**86.00**	**84.04**	**84.45**	**84.23**

4.3   Analysis of Stance-Emotion Joint Data Augmentation

We conduct experiments to prove the following two points: (1) the effectiveness of introducing emotional information into data augmentation; (2) the effectiveness of introducing emotional information through different types of auxiliary sentences.

In order to prove the first point, we compare the results of Stance-Emotion joint Data Augmentation (SEDA) with ASDA, which does not take emotional information into account. We present several augmented samples generated by these two methods in Table 4. It can be observed that the generated words of SEDA are more consistent with the label information. Furthermore, according to the experimental results in Table 5, SEDA outperforms ASDA on all targets, which further demonstrates the validity of emotional information.

Table 4. Examples generated by ASDA and SEDA. Italicized texts represent generated words.

Target	Wearing a face mask
Source	In the USA, Walmart will now serve mask-less customers. Hopefully the same will happen in the UK
ASDA	In the USA, Walmart will *today* serve mask-less customers. Hopefully the *fight* will *spread* in the UK
SEDA	In the USA, Walmart will now serve mask-less customers. Hopefully the same will happen *sooner to* the *globe*

In order to prove the second point, we compare the results of using different auxiliary sentences. The auxiliary sentences are constructed based on the relations between stance and emotion. "Consistency only" means we only use the "Consistency" relation between stance and emotion to introduce emotional information, thus SEDA(Consistency only) only contains the auxiliary sentence: The following texts have $\{s_i\}$ stance and $\{e_i\}$ emotion to $\{a_i\}$. "Discrepancy only" means we only use the "Discrepancy" relation, thus SEDA(Discrepancy only) only contains: Although the emotion is $\{e_i\}$, the following texts are both $\{s_i\}$ to $\{a_i\}$. SEDA is what we propose in this paper, which introduces emotional information based on "Consistency", "Discrepancy" and "None" relations. Therefore, as shown in Table 2, SEDA contains three types of auxiliary sentences. The experimental results in Table 5 show the performance impact of different auxiliary sentences, we can see that SEDA performs the best, indicating the effectiveness of the way we introduce emotional information.

4.4 Analysis of Gradual Prompt-tuning

We further explore the effectiveness of curriculum learning by comparing SEGP with SEP, which does not use the training strategy of curriculum learning. Curriculum learning requires a series of training samples with different disturbance degrees. In our method, the disturbance degree is determined by the mask ratio α in augmentation stage. Therefore, the artificial curriculums in the training samples are created according to α. Experimental results are shown in Table 6, which indicates that we can further improve performance by combining prompt-tuning with curriculum learning.

Table 5. Performance comparison of introducing emotional information in different ways. We highlight the best results in bold.

Model	Wearing a face mask				Stay at home orders			
	Acc	P	R	F1	Acc	P	R	F1
ASDA	82.50	80.96	80.24	80.53	87.00	83.04	85.09	83.99
SEDA(Consistency only)	81.50	79.57	80.18	79.83	86.50	83.05	86.40	84.51
SEDA(Discrepancy only)	80.50	78.98	77.74	78.25	86.00	82.33	86.68	84.17
SEDA	**83.50**	**82.50**	**82.26**	**82.36**	**87.50**	**84.51**	**86.32**	**85.36**
Model	Anthony S. Fauci, M.D.				Keeping schools closed			
	Acc	P	R	F1	Acc	P	R	F1
ASDA	81.00	81.49	81.04	81.06	83.50	81.29	80.95	81.01
SEDA(Consistency only)	80.00	79.93	81.17	80.32	83.50	80.89	81.51	81.14
SEDA(Discrepancy only)	80.50	80.31	81.66	80.79	82.00	80.38	81.77	80.85
SEDA	**82.00**	**82.11**	**82.20**	**82.09**	**85.50**	**83.79**	**83.09**	**83.40**

Table 6. Performance comparison of applying different training strategies. We highlight the best results in bold.

Model	Wearing a face mask				Stay at home orders			
	Acc	P	R	F1	Acc	P	R	F1
SEP	83.50	82.50	82.26	82.36	87.50	84.51	86.32	85.36
SEGP	**84.50**	**83.20**	**83.49**	**83.34**	**89.00**	**86.33**	**89.37**	**87.71**
Model	Anthony S. Fauci, M.D.				Keeping schools closed			
	Acc	P	R	F1	Acc	P	R	F1
SEP	82.00	82.11	82.20	82.09	85.50	83.79	83.09	83.40
SEGP	**82.50**	**82.60**	**82.57**	**82.57**	**86.00**	**84.04**	**84.45**	**84.23**

5   Conclusion

In this paper, we propose SEGP to address the scarcity of annotations problem in stance detection. SEGP is mainly composed of two stages, i.e., Stance-Emotion joint Data Augmentation (SEDA) and Gradual Prompt-tuning. With the help of C-MLM task, SEDA generates target-relevant and label-compatible samples by predicting the masked word conditioned on both their context and the auxiliary sentences. Gradual Prompt-tuning can make better use of the augmented samples as well as the knowledge contained in pre-trained models. The experimental results show that SEGP obtains superior performance over all baseline methods. Since our methods are not designed for a certain model, we will investigate how to extend them to other tasks in the future.

References

1. AlDayel, A., Magdy, W.: Stance detection on social media: state of the art and trends. Inf. Process. Manage. **58**(4), 102597 (2021)
2. Allaway, E., McKeown, K.: Zero-shot stance detection: a dataset and model using generalized topic representations. arXiv preprint arXiv:2010.03640 (2020)

3. Augenstein, I., Rocktäschel, T., Vlachos, A., Bontcheva, K.: Stance detection with bidirectional conditional encoding. arXiv preprint arXiv:1606.05464 (2016)
4. Bengio, Y., Louradour, J., Collobert, R., Weston, J.: Curriculum learning. In: Proceedings of the 26th Annual International Conference on Machine Learning, pp. 41–48 (2009)
5. Brown, T.B., et al.: Language models are few-shot learners. arXiv preprint arXiv:2005.14165 (2020)
6. Chauhan, D.S., Kumar, R., Ekbal, A.: Attention based shared representation for multi-task stance detection and sentiment analysis. In: Gedeon, T., Wong, K.W., Lee, M. (eds.) ICONIP 2019. CCIS, vol. 1143, pp. 661–669. Springer, Cham (2019). https://doi.org/10.1007/978-3-030-36802-9_70
7. Du, J., Xu, R., He, Y., Gui, L.: Stance classification with target-specific neural attention networks. In: International Joint Conferences on Artificial Intelligence (2017)
8. Garg, S., Ramakrishnan, G.: Bae: Bert-based adversarial examples for text classification. arXiv preprint arXiv:2004.01970 (2020)
9. Glandt, K., Khanal, S., Li, Y., Caragea, D., Caragea, C.: Stance detection in covid-19 tweets. In: Proceedings of the 59th Annual Meeting of the Association for Computational Linguistics and the 11th International Joint Conference on Natural Language Processing, vol. 1 (2021)
10. Jiao, X., et al.: Tinybert: distilling bert for natural language understanding. arXiv preprint arXiv:1909.10351 (2019)
11. Kenton, J.D.M.W.C., Toutanova, L.K.: Bert: pre-training of deep bidirectional transformers for language understanding. In: Proceedings of NAACL-HLT, pp. 4171–4186 (2019)
12. Korbar, B., Tran, D., Torresani, L.: Cooperative learning of audio and video models from self-supervised synchronization. In: Advances in Neural Information Processing Systems 31 (2018)
13. Küçük, D., Can, F.: Stance detection: a survey. ACM Comput. Surv. (CSUR) 53(1), 1–37 (2020)
14. Li, Y., Caragea, C.: Multi-task stance detection with sentiment and stance lexicons. In: Proceedings of the 2019 Conference on Empirical Methods in Natural Language Processing and the 9th International Joint Conference on Natural Language Processing (EMNLP-IJCNLP), pp. 6299–6305 (2019)
15. Li, Y., Caragea, C.: A multi-task learning framework for multi-target stance detection. In: Findings of the Association for Computational Linguistics: ACL-IJCNLP 2021, pp. 2320–2326 (2021)
16. Li, Y., Caragea, C.: Target-aware data augmentation for stance detection. In: Proceedings of the 2021 Conference of the North American Chapter of the Association for Computational Linguistics: Human Language Technologies, pp. 1850–1860 (2021)
17. Li, Y., Cohn, T., Baldwin, T.: Robust training under linguistic adversity. In: Proceedings of the 15th Conference of the European Chapter of the Association for Computational Linguistics: Volume 2, Short Papers, pp. 21–27 (2017)
18. Liu, P., Yuan, W., Fu, J., Jiang, Z., Hayashi, H., Neubig, G.: Pre-train, prompt, and predict: a systematic survey of prompting methods in natural language processing. arXiv preprint arXiv:2107.13586 (2021)
19. Miller, G.A.: Wordnet: a lexical database for English. Commun. ACM 38(11), 39–41 (1995)
20. Mohammad, S.M., Sobhani, P., Kiritchenko, S.: Stance and sentiment in tweets. ACM Trans. Internet Technol. (TOIT) 17(3), 1–23 (2017)

21. Mueller, J., Thyagarajan, A.: Siamese recurrent architectures for learning sentence similarity. In: Proceedings of the AAAI Conference on Artificial Intelligence, vol. 30 (2016)
22. Müller, M., Salathé, M., Kummervold, P.E.: Covid-twitter-bert: a natural language processing model to analyse covid-19 content on twitter. arXiv preprint arXiv:2005.07503 (2020)
23. Pennington, J., Socher, R., Manning, C.D.: Glove: Global vectors for word representation. In: Proceedings of the 2014 Conference on Empirical Methods in Natural Language Processing (EMNLP), pp. 1532–1543 (2014)
24. Reynolds, L., McDonell, K.: Prompt programming for large language models: Beyond the few-shot paradigm. In: Extended Abstracts of the 2021 CHI Conference on Human Factors in Computing Systems, pp. 1–7 (2021)
25. Schuster, M., Paliwal, K.K.: Bidirectional recurrent neural networks. IEEE Trans. Signal Process. **45**(11), 2673–2681 (1997)
26. Sobhani, P., Inkpen, D., Zhu, X.: A dataset for multi-target stance detection. In: Proceedings of the 15th Conference of the European Chapter of the Association for Computational Linguistics: Volume 2, Short Papers, pp. 551–557 (2017)
27. Somasundaran, S., Wiebe, J.: Recognizing stances in ideological on-line debates. In: Proceedings of the NAACL HLT 2010 Workshop on Computational Approaches to Analysis and Generation of Emotion in Text, pp. 116–124 (2010)
28. Tsvetkov, Y., Faruqui, M., Ling, W., MacWhinney, B., Dyer, C.: Learning the curriculum with bayesian optimization for task-specific word representation learning. arXiv preprint arXiv:1605.03852 (2016)
29. Walker, M., Anand, P., Abbott, R., Grant, R.: Stance classification using dialogic properties of persuasion. In: Proceedings of the 2012 Conference of the North American Chapter of the Association for Computational Linguistics: Human Language Technologies, pp. 592–596 (2012)
30. Wang, W.Y., Yang, D.: That's so annoying!!!: a lexical and frame-semantic embedding based data augmentation approach to automatic categorization of annoying behaviors using# petpeeve tweets. In: Proceedings of the 2015 Conference on Empirical Methods in Natural Language Processing, pp. 2557–2563 (2015)
31. Wang, X., Pham, H., Dai, Z., Neubig, G.: Switchout: an efficient data augmentation algorithm for neural machine translation. arXiv preprint arXiv:1808.07512 (2018)
32. Wei, J., Huang, C., Vosoughi, S., Cheng, Y., Xu, S.: Few-shot text classification with triplet networks, data augmentation, and curriculum learning. arXiv preprint arXiv:2103.07552 (2021)
33. Wei, J., Zou, K.: Eda: easy data augmentation techniques for boosting performance on text classification tasks. arXiv preprint arXiv:1901.11196 (2019)
34. Weinshall, D., Cohen, G., Amir, D.: Curriculum learning by transfer learning: Theory and experiments with deep networks. In: International Conference on Machine Learning, pp. 5238–5246. PMLR (2018)
35. Wu, X., Lv, S., Zang, L., Han, J., Hu, S.: Conditional bert contextual augmentation. In: International Conference on Computational Science, pp. 84–95. Springer (2019)
36. Zhang, B., Yang, M., Li, X., Ye, Y., Xu, X., Dai, K.: Enhancing cross-target stance detection with transferable semantic-emotion knowledge. In: Proceedings of the 58th Annual Meeting of the Association for Computational Linguistics, pp. 3188–3197 (2020)
37. Zhang, X., Zhao, J., LeCun, Y.: Character-level convolutional networks for text classification. Adv. Neural. Inf. Process. Syst. **28**, 649–657 (2015)

Image Features Correlation with the Impression Curve for Automatic Evaluation of the Computer Game Level Design

Jarosław Andrzejczak$^{(\boxtimes)}$ ⓘ, Olgierd Jaros, Rafał Szrajber ⓘ,
and Adam Wojciechowski ⓘ

Institute of Information Technology, Lodz University of Technology,
215 Wólczańska Street, 90-924 Lodz, Poland
{jaroslaw.andrzejczak,rafal.szrajber}@p.lodz.pl
http://it.p.lodz.pl

Abstract. In this study, we present the confirmation of existence of the correlation of the image features with the computer game level Impression Curve. Even a single image feature can describe the impression value with good precision (significant strong relationship, Pearson $r > 0{,}5$). Best results were obtained using by combining several image features using multiple regression (significant very strong positive relationship, Pearson $r = 0{,}82$ at best). We also analyze the different set of image features at different level design stages (from blockout to final design) where significant correlation (strong to very strong) was observed regardless of the level design variant. Thanks to the study results, the user impression of virtual 3D space, can be estimated with a high degree of certainty by automatic evaluation using image analysis.

Keywords: Image analysis · Virtual reality · Impression curve · Level design · Automatic evaluation

1 Introduction

In [1] study, we have shown that Virtual Reality space affects different users in a similar way. That sense can be stored and described as Impression Curve[1] for this space. Therefore, Impression Curve can be used in 3D VR space evaluation such as 3D level design. Still, it requires tests with many users to gather proper data. It would be a great improvement if designers could estimate the sense of 3D space during the development process and then verify it at the end with the users. Especially with the growing popularity of level designs generated by

---

[1] Impression Curve is a measure of the visual diversity and attractiveness of a game level. It assesses subjective attraction of a given space. For the detailed information about the Impression Curve, its acquisition method, its strengths and weaknesses in the domain of the 3D space evaluation, please refer to [1].

D. Groen et al. (Eds.): ICCS 2022, LNCS 13352, pp. 591–604, 2022.
https://doi.org/10.1007/978-3-031-08757-8_49

algorithms [16]. We focused our efforts to provide such computationally low cost tool for 3D space evaluation in the context of estimating user experience.

The purpose of the research was to verify the existence of the correlation of the image features (gathered using automatic image analysis) with the Impression Curve (obtained during previous studies conducted on 112 people). The study described in this article involves examining the impact of various image features such as mean brightness and contrast, features based on saliency and movement maps (such as complexity or density), as well as descriptive statistics like entropy, skewness and kurtosis.

The contributions to research concerning automatic evaluation of the immersive Virtual Reality space, especially in case of the Impression Curve estimation presented in this article, are:

- Confirmation of the existence of a correlation between data gathered using image analysis and user-generated Impression Curve.
- Tests verifying the correlation between individual image feature and the Impression Curve for the VR space.
- Tests verifying the correlation between combined image features and the Impression Curve for the VR space.
- Analysis of usability of each image features depending on the level design stages and changing factors of the 3D space.
- Proposition of the best image features (with the highest correlation values with Impression Curve) for evaluation of individual level design stages.

We start with a related work overview in the domain of image analysis for feature extraction in the next section. Then we describe hypotheses and an evaluation method. Next, both test results and their discussion will be presented, as well as observations about data gathered. Finally, ideas for further development and final conclusions will be given.

2   Image Features

There are many image features available to consider in terms of image analysis for automatic feature extraction and image description. Our goal was to test as diverse set of features as possible. The three groups of features were used: color and luminance-based (such as mean brightness, mean color contrast) [6], features based on saliency and motion maps (such as balance and density) [5], as well as descriptive statistics (entropy, skewness and kurtosis) [7]. Therefore, a total of thirteen features were selected for this study:

- **Color and luminance group:** Average Contrast, Average Luminance and Average Saturation.
- **Saliency and motion maps group:** Alignment Complexity, Balance Complexity, Density Complexity, Grouping Complexity, Size Complexity and Total Complexity.
- **Descriptive statistics group:** Entropy, Kurtosis, Skewness and Fractal Complexity.

For the calculations of the image **Average Contrast**, **Average Luminance** and **Average Saturation**, the definitions for the HSL color palette were used. Average Contrast was calculated using the mean square of the Luminance of individual pixels [6].

Image features from the second group are based on classification and analysis of areas indicated in saliency maps [2] and motion maps [8]. Those maps are combined (with a weight of 50% of each, as we considered them equally important) and classified to be used with the metrics described in [5]. This stage requires the greatest number of computations. We start with the creation of the saliency map using the fast background detection algorithm [2], which is then denoised using the method described in [3]. The result is a black and white image, with white pixels representing the relevant ones. A motion map is created as a difference of the pixels of two subsequent video frames converted to grayscale with a Gaussian blur applied to them (which allows limiting the influence of details and noise on motion detection) [4]. The resulting image is denoised and thresholded to obtain a black and white image and combined with the saliency map to obtain the final visual attention saliency map [14]. Then the classification of regions, objects and their contours as well as shape recognition is made.

Regions of attention (representing grouped objects) and their centroids are calculated using K-Means with 30 starting points (pixels) picked randomly on visual attention saliency map white pixels. For each iteration, the closest region centroid for each point is calculated and the region centroid weights are updated. The algorithm runs for 1000 epochs or until each region centroid remains unchanged in two subsequent epochs. During this process, centroids, which for two ages were not the closest one for any point, are permanently removed from the set to optimize the calculations. Centroids calculated for one frame become the starting points for the next frame, with one new random starting point added (to allow the new area recognition).

Objects of attention are found by applying erosion filter and OpenCV shape detection [10] on the final visual attention saliency map. Next, the object's contour is calculated using the contour approximation method [10]. For each of the identified object, a centroid is calculated. Please note that object's centroid is usually different from region centroid, as one region can contain many objects.

Localized object's contours are therefore used for shape recognition [10]. Only simple geometric shapes are taken into account, and every object with a number of vertices greater than or equal to five is classified as a circle (for the purpose of further analysis).

All of the above final visual attention saliency map characteristic is then used with the metrics for UI complexity analysis described in [5]. Each metric gives a final score in the range [0,1] where a score closer to zero means less complexity. The **Alignment Complexity** determines the complexity of the interface in terms of the position of the found shapes relative to each other. The evaluation consists of the calculation of the local and global alignment coefficients for grouped and ungrouped objects. The **Density Complexity** determines the comparison of the visual attention object size to the entire image frame size.

The **Balance Complexity** describes the distribution of visual attention objects on the quarters of the screen. It is calculated as the arithmetic mean of two mean values: the proportion of the number of objects between pairs of quarters and the proportion of the size of objects between pairs of quarters. The **Size Complexity** is calculated due to the grouping of objects on the screen in terms of shape. For each shape type, the number of occurrences of the size of objects is checked. Then The sum of the occurrences of unique object regions is divided by the number of objects in the particular group of shapes. The **Grouping Complexity** determines how many of the objects are grouped into shape type groups. It is the sum of the ratio of ungrouped objects to all occurring and the number of groups of shapes occurring in the region of objects from all possible shapes types. The **Total Complexity** is a combined metric of all previous with weights as proposed in [5]:

$$TotalComplexity = \frac{0,84 \times Alignment + 0,76 \times Balance+}{0,8 \times Density + 0,72 \times Size + 0,88 \times Grouping} \quad (1)$$

The third group of image features is based on statistical descriptors of a data set's distribution. The **Skewness** is a measure of the asymmetry of a distribution of the mean. The higher the Skewness, the more asymmetric data distribution. The **Kurtosis** is a measure of how results are concentrated around the mean. The high Kurtosis value would suggest outliers in the data set and low Kurtosis value the lack of outliers [11]. The **Entropy** of an image is used as a measure of the amount of information it contains [7]. The more detailed the image, the higher the value of the Entropy will be. Entropy, Kurtosis and Skewness were counted separately for Hue, Saturation and Luminosity as they operate on the single variable (grayscale image as input). The **Fractal Complexity** is a measure of self-similarity. It determines how much it is possible to break an image or fractal into parts that are (approximately) a reduced copy of the whole. This parameter was used to assess the complexity of the image [9][2].

3   Evaluation

The goal of the evaluation was to verify the existence of the correlation of the image features (gathered using automatic image analysis) with the Impression Curve. For this purpose, the Pearson and Spearman correlation were used [13]. All the level design stages as well as the influential factors on the 3D space impression (such as lightening condition changes, geometrical and material changes) described in [1] were used (Fig. 1).

---

[2] At this stage of the Impression Curve automatic evaluation study, we have used the controlled 3D space designs to minimize the influence of the such factors as action, gameplay rules and restrictions, story and lore present in commercial game designs. After confirmation of existence of the correlation of the image features with the computer game level Impression Curve described in this article, we moved to testing level design from popular games. The results of this study will be published in the future, as it is in development at the time of writing this article.

The study was divided into two parts. First, the correlation of the individual image features with the Impression Curve was analyzed. After that, image features with the highest correlation value were combined into sets and once again tested for correlation with Impression Curve to see if there is any gain in the strength of the correlation.

The hypotheses in individual parts were as follows:

1. First part: there is a significant correlation (positive or negative) between an individual image feature and the Impression Curve for the same VR space.
2. Second part: the correlation (positive or negative) with the Impression Curve is higher for the combined image features than for the individual image features.
3. Additional observation: different set of image features presents the highest correlation values for different level design stages.

What is more, different level design stages and changing factors of the 3D space (for example: lightening condition, geometrical detail or material changes) of the same game level allow us to observe if there is any difference in correlation between data gathered using image analysis and user-generated Impression Curve. Thanks to this, we were able to point out the best automatic evaluation measures in the form of selected image features, to use at each design stage (blockout, models without materials, textured models as well as lightning and atmospheric effects such as rain).

The twelve level variants showing successive design stages were used according to our previous research, described in details in [1]. There were as follows: simple blockout (A), advanced blockout (B), main models without materials (C), main models with monochromatic materials (D) and final materials (E) as well as with extra fine detailed models (called final level version) (F), main models with geometrical changes (G) and final level with changes of visual factors as lightening condition (L), weather condition (W), different materials (M), added expression (X) as well as with extra models and objects in the environment (O). Existence of correlation between image features and Impression Curve values would allow creation of a tool to automatically estimate Impression Curve for a VR space with a high degree of probability. And as a result, to automatically evaluate expected user impression even on an early Virtual Reality space design stage.

4   Results and Analysis

During the study, hundreds of correlation plots were gathered and analyzed. We assumed that per frame comparison will be sensitive to rapid image changes, effecting low or no correlation at all. That is why, the mean and median of an image feature for a few consecutive frames were calculated. A small range of 4–5 frames allow us to eliminate minor fluctuations, where a larger range of 20–30 frames softened the charts quite significantly. However, a larger range considerably reduces the number of data samples, which had an impact on the

Fig. 1. The twelve level variants showing successive design stages used in this study for image analysis and correlation with Impression Curve. A - simple blockout; B - advanced blockout; C - models without materials; D - models with monochromatic materials; E - models with final materials; F - final level version; G - geometrical changes; L - lightening condition changes; W - weather changes; M - material changes; X - expression added; O - extra models added.

significance value p. Thus, we started from a range of four frames and increased this interval by four from that point. As a result, four to twenty frames, we observed increased correlation value for most of the image features while preserving low value of $p < 0,05$. For frame range greater than twenty, results were not significant anymore ($p > 0,05$). Also, above this point, the correlation value for many image features dropped below the value of 0,3. Thus, we choose a range of twenty frames for our study, as it shows the highest correlation values with significance $p < 0,05$ (in many cases $p < 0,01$). In the other hand, we gathered image data more often (thirty times per second - video recorded with a 30 FPS frame rate) than during study with users. Thus, the Impression Curve data had to be interpolated between measure points (as we assumed linear change). This way we were able to compare this data even per frame.

The experiment stages were as follows: first, for each of the video game level variants the Impression Curve data (gathered with users) was interpolated between the measure points to match the frequency of data calculated using image analysis for this level variant walkthrough video; next, the image features were calculated and refined using respectively mean and median for 20 subsequent frame intervals; finally, the Pearson and Spearman correlation between those data were calculated.

The recordings of twelve variants of the video game level variants (used in [1]), including twenty-nine thousand three hundred and thirty-nine frames in total, were analyzed. As a result, thirty-six data sets were obtained and used to generate two hundred and ninety-nine correlation plots.

4.1   Individual Image Features Correlation

The first part of the study involved testing each of the thirteen image features individually for correlation with an interpolated Impression Curve for each of twelve variants of the video game level described earlier. The result of a single feature-variant pair was stored in numerical way and also as a correlation plot for easier analysis (Fig. 2). Each data point in the graph shows respectively the mean or median (depending on which one was used) over an interval of 20 frames of the video data. The feature values are marked in red, while the values of the Impression Curve are marked in green. The charts contain the calculated Pearson correlation for a whole Impression Curve. When this value is below 0,5 the Spearman correlation is calculated as well to compensate possible outliers and check for nonlinear relation. Two numbers are presented for each correlation. The first is the mean correlation value, the second is the calculated p value of this correlation.

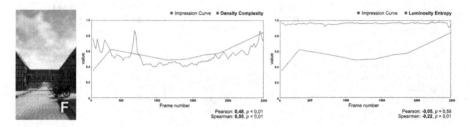

Fig. 2. Correlation plot examples for final level design variant (F variant, on the left). Two image feature correlation plots are presented: one with significant strong positive relationship - Density Complexity (Pearson r = 0,48 with $p < 0,01$, center) other with no significant linear relationship and weak non-linear relationship - Luminosity Entropy (Pearson r = -0,05 with $p = 0,59$, right). A linear relationship can be observed for Density Complexity. (Color figure online)

   Then the correlation values of every image feature tested for a single level design variant were juxtaposed with each other (Table 1 shows the results for only one variant as an example - the same was done for each of twelve level design variants).

   We observed many significant correlation values (positive and negative) between image features and Impression Curve value. Observation varied from a few weak relationships (r value between 0,20 and 0,29) to moderate relationship in most cases (r value between 0,30 and 0,39) and even over a dozen strong relationship (r value between 0,40 and 0,69). There was not a single variant without at least one significantly related image feature, and in most cases there were several moderate relationships. What is more, some image features tend to correlate more often than others, where others given at least weak relationship only once or twice (Table 2). We did not observe a significant difference between

598   J. Andrzejczak et al.

Table 1. Pearson's correlation values for individual image features of the final version of the level (F). Feature values were calculated respectively as the mean and median for the intervals of twenty frames. The highest correlation results are marked with a gray background color and bold text. The significant p values are marked with a gray background color. We can observe that the same image features show the highest correlation and similar values for both the mean and the median, with only one feature (Grouping Complexity) presenting lower correlation using the median. r - Pearson correlation coefficient value; p - significance value.

Image feature	Mean		Median	
	r	p	r	p
Alignment Complexity	0,07	0,463	0,06	0,516
Balance Complexity	**0,33**	<0,001	**0,32**	p <0,001
Density Complexity	**0,48**	<0,001	**0,52**	p <0,001
Grouping Complexity	**0,28**	0,002	0,16	0,087
Size Complexity	**0,43**	<0,001	**0,46**	p <0,001
Total Complexity	**0,53**	<0,001	**0,48**	p <0,001
Average Contrast	**-0,28**	0,002	**-0,28**	0,002
Average Luminance	0,01	0,951	-0,01	0,908
Average Saturation	**-0,36**	<0,001	**-0,35**	0,000
Fractal Complexity	-0,05	0,551	-0,06	0,544
Hue Entropy	**0,35**	<0,001	**0,35**	p <0,001
Hue Kurtosis	-0,04	0,651	-0,03	0,737
Hue Skewness	0,05	0,582	0,07	0,457
Saturation Entropy	-0,16	0,083	-0,15	0,103
Saturation Kurtosis	-0,03	0,719	-0,07	0,477
Saturation Skewness	-0,04	0,701	-0,04	0,652
Luminosity Entropy	0,03	0,719	0,05	0,587
Luminosity Kurtosis	**0,33**	<0,001	**0,33**	p <0,001
Luminosity Skewness	0,06	0,508	0,07	0,456

mean and median values ($t - test\ p = 0,52$) thus only median will be used in further analysis as less valuable for outliers.

For all but one image features, we observe no significant difference between Pearson and Spearman correlation coefficient values, which suggest a linear nature of the relationship. Thus, in further combined image features we focused on Pearson correlation coefficient as linear relationship is more desired for the future video game level design automatic evaluation system. Only for Density Complexity feature, we observed significant difference ($t-test\ p = 0,05$) between Pearson and Spearman results with Spearman correlation coefficient values being higher most of the time giving moderate to high positive relationship (also with much lower p value). This indicates the existence of a non-linear relationship between Density Complexity feature and the Impression Curve.

The results are dominated by a positive correlation, with six image features tending to present a negative relationship more often than positive. Those are: Grouping Complexity, Fractal Complexity, Average Contrast, Average Saturation and Entropy (for Saturation and Luminosity). Most of them present many moderate to strong relationships (also variants with low correlation value results were not significant with $p > 0,05$). The highest single image features correlation value observed was 0,57 (strong positive relationship, $p < 0,01$) for a Size Complexity feature in variant of models with the final materials (E).

Table 2. Pearson's correlation values for individual image features for all twelve level design variants. Feature values were calculated as median for the intervals of twenty frames. The significant correlation results (with $p =< 0,01$) are marked with a grayscale background color (the darker the color, the higher the correlation value) and bold text. Strong relationship (r value between 0,40 and 0,69) was outlined with a white text color. We can observe that some image features as Size Complexity or Grouping Complexity tend to present high correlation value in many variants. A - simple blockout; B - advanced blockout; C - models without materials; D - models with monochromatic materials; E - models with final materials; F - final level version; G - geometrical changes; L - lightening condition changes; W - weather changes; M - material changes; X - expression added; O - extra models added.

Image Feature	Level Design Variant											
	A	B	C	D	E	F	G	L	W	M	X	O
Alignment Complexity	-0,15	-0,01	-0,04	0,13	0,14	0,06	0,10	0,09	-0,25	0,02	-0,07	-0,10
Balance Complexity	0,04	0,06	0,33	0,25	0,18	0,32	0,08	0,14	0,19	0,47	0,33	0,31
Density Complexity	0,23	0,17	0,20	0,33	0,33	0,52	0,18	0,20	0,39	0,20	-0,09	0,15
Grouping Complexity	-0,14	-0,25	-0,40	-0,33	0,06	0,16	-0,27	-0,36	-0,24	-0,02	0,18	0,06
Size Complexity	-0,21	-0,19	-0,36	-0,32	0,57	0,46	-0,16	-0,42	0,38	0,24	0,34	0,39
Total Complexity	-0,21	0,05	0,00	0,23	0,38	0,48	-0,06	-0,20	0,19	0,28	0,04	0,14
Average Contrast	-0,55	-0,28	-0,14	0,10	-0,21	-0,28	-0,30	-0,37	-0,27	-0,06	0,12	0,02
Average Luminance	-0,10	-0,15	0,42	-0,24	0,25	-0,01	0,03	-0,03	-0,03	0,19	0,03	0,08
Average Saturation	-0,41	-0,10	-0,20	0,47	-0,34	-0,35	-0,40	0,26	0,29	-0,31	-0,07	-0,16
Fractal Complexity	0,22	0,01	-0,26	-0,25	-0,02	-0,06	0,00	-0,06	-0,28	-0,05	-0,21	-0,19
Hue Entropy	-0,08	-0,18	0,11	0,32	0,27	0,35	-0,12	0,21	0,06	0,30	0,32	0,30
Hue Kurtosis	0,28	0,29	0,01	0,37	0,37	-0,03	0,17	0,06	0,43	0,05	-0,11	-0,04
Hue Skewness	-0,25	-0,32	0,03	0,47	0,38	0,07	-0,04	-0,08	-0,56	0,12	-0,05	-0,09
Saturation Entropy	-0,31	-0,07	-0,14	0,05	-0,38	-0,15	-0,25	0,19	0,13	-0,22	0,21	0,00
Saturation Kurtosis	0,04	0,06	0,23	-0,55	0,21	-0,07	0,17	-0,05	0,28	0,34	-0,09	0,07
Saturation Skewness	0,06	-0,01	0,29	-0,49	0,20	-0,04	0,22	0,29	0,35	0,37	0,11	0,21
Luminosity Entropy	-0,36	-0,32	0,07	0,16	0,05	0,05	-0,25	-0,22	-0,43	0,16	0,07	0,16
Luminosity Kurtosis	0,27	0,18	0,35	0,12	0,55	0,33	0,15	0,53	0,22	0,15	0,12	0,09
Luminosity Skewness	0,17	0,13	-0,41	0,13	-0,33	0,07	0,00	0,45	0,10	-0,12	-0,09	-0,11

We also observed that the earlier the level creation stage, the lower the correlation values of most image features (Table 2). The materials used in the virtual space design has a great influence on the correlation value. In the case of variant C (3D models without materials), a significant strong relationship weak relationship with the Average Luminance can be noticed. This correlation decreases after adding materials to the models (variants D with monochromatic materials and E with final materials) effecting with no significant relation in final level variant (F with lightning). Similar observation can be made with Saturation Kurtosis and Saturation Skewness giving the highest correlation values for variant with monochromatic materials (D) and also no significant relation in the final level variant. Another interesting observation can be made in first design stage (simple blockout - variant A). In such a simple block design, the color-based image features gave the highest correlation values with significant strong negative relationship for Average Contrast (Pearson r = -0,55, $p < 0,01$). This relation weakens with the addition of final models and textures. It is also worth paying attention to the fact that with the appearance of the final materials, the sign of the correlation for the Size Complexity image feature changes from negative to positive relationship.

It must be remembered that the value at a given point for the correlating images feature shows the general tendency of the Impression Curve (increase or decrease of it) - not the exact values of it. To reproduce the value of the curve, it is necessary to know its value at one point at least. At the same time, the change in perception of virtual space (increase or decrease) is a feature shared by users (as shown in the research presented in [1]), while the assignment of a numerical value to the Impression Curve may depend on the user and the definition of the rating scale. Therefore, the use of the Immersion Curve value change in the automatic evaluation system of the game level is not only a more reliable, but also more universal (less dependent on the user).

4.2   Combined Image Features Correlation

Among the image features tested, the most common correlation between them and Impression Curve can be observed in seven cases (Table 2). They were divided into two groups:

- The most promising that gives the highest correlation values, especially in final level design variant (F). Those are: **Density Complexity, Size Complexity, Total Complexity and Balance Complexity**. This group formed a base set for all the combined set (and will be referred to as DTSBC hereinafter).
- The second most promising with a little lower correlation value than the first group or high relationship with variants other than final level design (F). Those are: **Grouping Complexity, Average Contrast, Average Saturation**. They were added, in every possible combination, to the first group and checked for improvement in relationship strength.

In addition to the above, color-based image features of Entropy, Kurtosis and Skewness for Hue, Saturation and Luminosity were also included in described sets as they presented significant correlation values in different stages of design (especially in early stages A to E). Image features in those sets were combined using multiple regression. From all the combined sets, those with the best Pearson's correlation values were selected (Table 3).

There was significant strong or very strong positive relationship in all cases. The best results overall were achieved for the sets *DSTBC + Average Contrast + Average Saturation + Hue Entropy* and *DTSBC + Average Contrast + Average Saturation + Hue Entropy + Luminosity Kurtosis* where the latter works for a larger number of variants (thus it is more universal). In almost all cases, the combined feature sets correlated significantly better than the single ones included in them (Fig. 3). These isolated opposite cases arise when one feature in a combination did not correlate individually. It can be observed in variant G (geometrical changes) where combined result of DSTBC + Average Contrast is

Table 3. The best Pearson's correlation values for combined image features for all twelve level design variants. Features were combined using multiple regression. Three best correlated image features: Density Complexity + Size Complexity + Total Complexity + Balance Complexity (called DTSBC for short) were bases for all four combined sets. The correlation results are color-coded as a heatmap with a grayscale background color (the darker the color, the higher the correlation value). Very strong relationship (r value higher than 0,70) was outlined with a white text color and bold text. All the results were significant ($p =< 0,01$). There was significant correlation in all cases, where the best results were achieved for sets DTSBC + Average Contrast + Average Saturation + Hue Entropy + Luminosity Kurtosis. We can observe that the more advanced level design stage (B to F) the stronger the correlation. Also, the combination of HSL Entropy, Kurtosis and Skewness can be useful for variants with lightning and weather changes. DTSBC - image features: Density Complexity + Size Complexity + Total Complexity + Balance Complexity; A - simple blockout; B - advanced blockout; C - models without materials; D - models with monochromatic materials; E - models with final materials; F - final level version; G - geometrical changes; L - lightening condition changes; W - weather changes; M - material changes; X - expression added; O - extra models added.

	Level Design Variant											
	A	B	C	D	E	F	G	L	W	M	X	O
DSTBC + Average Contrast	0,66	0,42	0,61	0,55	0,63	**0,73**	0,40	0,64	0,60	0,65	0,52	0,55
DSTBC + Average Contrast + Average Saturation + Hue Entropy	0,67	0,43	0,63	0,60	**0,72**	**0,81**	0,53	0,68	0,62	**0,75**	0,66	0,63
DSTBC + Average Contrast + Average Saturation + Hue Entropy + Luminosity Kurtosis	0,68	0,43	0,65	0,61	**0,82**	**0,82**	0,54	**0,70**	0,62	**0,75**	**0,71**	0,63
DSTBC + Hue Kurtosis + Hue Skewness + Saturation Entropy	0,63	0,47	0,62	0,59	**0,71**	**0,74**	0,46	0,55	**0,77**	0,68	0,52	0,57
DSTBC + Hue Kurtosis + Hue Skewness + Saturation Entropy + Luminosity Entropy	0,63	0,48	0,63	0,62	**0,71**	**0,77**	0,50	**0,77**	**0,80**	0,69	0,52	0,66

equal to single Average Saturation correlation value (but with negative sign). On the other hand, the combined sets presented strong and very strong relationship for those level variants that for a single feature had only a few weak or moderate relationships: M (material changes), X (added expression) and O (extra models added). We can also observe that the more advanced level design stage (B to F) the stronger the correlation (Table 2). Even the worst level design variant for single feature - geometrical changes (G) - now shows significant strong positive relationship (Pearson $r = 0,54$ with $p < 0,01$ at best).

There is significant difference in correlation values for color-based features (color, luminance as well as descriptive statistics for HSL) between single feature correlation (Table 2) and combined value using those image features (Table 3). The single feature correlation values are rather small or even not significant on later design variants (G to O). However, when they are combined with other image features, they have shown the highest or the second-highest correlation value. This happens even if, for a given variant of the level design, a single color-based feature did not show a correlation with the Impression Curve (mostly due to the high values of p).

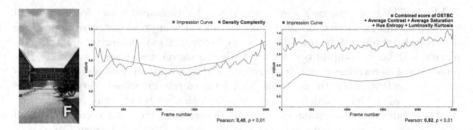

Fig. 3. Correlation plot examples for final level design variant (F variant, on the left). Two image feature correlation plots are presented: single image feature - Density Complexity (Pearson r = 0,48 with $p < 0,01$, center) and combined score of Density Complexity, Size Complexity, Total Complexity, Balance Complexity (DSTBC for short), Average Contrast, Average Saturation, Hue Entropy and Luminosity Kurtosis (Pearson r = 0,82 with $p < 0,01$, right). The combined score presents much higher correlation value than the component features separately with significant very strong positive relationship.

4.3 Best Features for Different Level Design Stages

Another aspect of the evaluation of the results was the changes of individual feature correlation at the subsequent stages of the game level design. Thanks to such approach, it was possible to assess the usefulness of the automatic evaluation method at different stages of the level design (from simple blockout with gray objects, trough materials and textures, to final design with lightning and atmospheric effects). The best image features or their combination to be used in such evaluation system will be the ones correlating regardless of the variant we are dealing with. The results for similar versions (such as first stage simple design been analyzed together) of the level were also compared. The image features with similar correlation values for each variant were considered the most promising. Such approach allowed us to eliminate those image features that correlated only in a single case.

For early stages of design that use blockout (A and B) we observed a significant correlation with the color-based image features, where at later stages (C to F) features from saliency and motion maps group showed better results (Table 2). What is more, most of the color-based image features tends to not show significant correlation at later stages, especially at the final level design. The exception here are the values of Hue Kurtosis, Hue Skewness and Luminosity Entropy for the weather change variant (W) with strong relationship. This showed that those image features could be added to the combined set to help verify how atmospheric effects affects users' impression of virtual space.

It is worth noticing that combination of HSL Entropy, Kurtosis and Skewness showed high or very high significant correlation results for variant the most visually different from the rest - L - where lightning conditions are changed (day to night). For example, set combined of DSTBC + Hue Kurtosis + Hue Skewness + Saturation Entropy + Luminosity Entropy resulted for weather changes level design variant in very strong relationship (Pearson's $r = 0,80$, $p < 0,01$).

Even this set is not universal for the whole process (other combined sets have given higher correlation results), it could improve Impression Curve estimation at design with rapid lightning or weather changes. At the same time, changes in lighting, weather or materials did not affect the shape of the Impression Curve, but did have a significant effect on the image features correlation.

5   Conclusion

The aim of this study was to investigate the existence of the correlation of the image features with the Impression Curve for game level design. The study shows that even a single image feature can describe the impression value with good precision (strong relationship, Pearson $r > 0.5$) for final level design. Best results were obtained by combining several image features using multiple regression (for image features: Density Complexity, Size Complexity, Total Complexity, Balance Complexity, Average Contrast, Average Saturation, Hue Entropy and Luminosity Kurtosis combined using multiple regression). Such set produced very strong positive relationship with Impression Curve values (Pearson $r = 0.82$ with $p < 0.01$ at best). What is more, significant correlation (strong to very strong) was observed regardless of level design variant, which makes it possible to apply image analysis at every stage of the level design process, making such solution more universal. The study also analyzed the possibility to use a different set of image features at different level design stages to get the highest results. The color-based image features were the best in this regard to be used at blockout stage of design (A and B, moderate to strong relationship) and HSL Entropy, Kurtosis and Skewness at stages with lightning and weather changes (L and W, moderate to strong relationship).

We saw many development opportunities for the idea of the automatic evaluation of game level design. The tests can be performed on production versions of game levels (taken from popular games), data about Impression Curve as well as image analysis could be obtained in real time or the study could be to extend with an Eye-tracker (to verify if there is a relation between the eye movement and Impression Curve). Also, the joined signals of EEG and Eye-tracker data can be analyzed as in [15]. Those four research ideas are carried out by us at the time of writing this article and the results will be published in the future. Improvements can be made in terms of calculation time as well, with a goal of real time analysis. For example, by applying faster classified like the one used in [12] for HUD detection, to obtain saliency maps in short time.

To sum up, the study has shown that Impression Curve value, and hence, the user impression of virtual 3D space, can be estimated with a high degree of certainty by automatic evaluation using image analysis of such level walkthrough. We propose usage of the combined image feature set for better estimation of Impression Curve. For early stages of design (blockout and models without textures) different set can be used to increase the relationship strength.

References

1. Andrzejczak, J., Osowicz, M., Szrajber, R.: Impression curve as a new tool in the study of visual diversity of computer game levels for individual phases of the design process. In: Krzhizhanovskaya, V.V., et al. (eds.) ICCS 2020. LNCS, vol. 12141, pp. 524–537. Springer, Cham (2020). https://doi.org/10.1007/978-3-030-50426-7_39
2. Dudek, P., Wang, B.: A fast self-tuning background subtraction algorithm. In: Proceedings of the IEEE Conference on Computer Vision and Pattern Recognition Workshops, pp. 401–404 (2014)
3. Jamil, N., Sembok, T.M.T., Bakar, Z.A.: Noise removal and enhancement of binary images using morphological operations. In: Proceedings of the International Symposium on Information Technology, pp. 1–6 (2008)
4. Milanfar, P.: A tour of modern image filtering: new insights and methods, both practical and theoretical. IEEE Signal Process. Mag. **30**(1), 106–128 (2013)
5. Magel, K., Alemerien, K.: GUIEvaluator: A metric-tool for evaluating the complexity of graphical user interfaces. In: Proceedings of the International Conference on Software Engineering and Knowledge Engineering, SEKE (2014)
6. Peli, E.: Contrast in complex images. J. Opt. Soc. Am. A Opt. Image Sci. Vis. **7**, 2032–2040 (1990)
7. Kumari, S., Vijay, R.: Image quality estimation by entropy and redundancy calculation for various wavelet families. Int. J. Comput. Inf. Syst. Ind. Manag. Appl. **4**, 027–034 (2012)
8. Simons, D.J., Franconeri, S.L.: Moving and looming stimuli capture attention. Percept. Psychophys. **65**, 999–1010 (2003)
9. Della-Bosca, D., Patterson, D., Roberts, S.: An analysis of game environments as measured by fractal complexity. In: Proceedings of the Australasian Computer Science Week Multiconference (ACSW 2017), Association for Computing Machinery, USA, Article 63, pp. 1–6 (2017)
10. Kaehler, A., Bradski, G.: Learning OpenCV 3: Computer Vision in C++ with the OpenCV Library. O'Reilly Media, Sebastopol (2016). ISBN 9781491937969
11. Engineering Statistics Handbook: NIST/SEMATECH e-Handbook of Statistical Methods. http://www.itl.nist.gov/div898/handbook/. Accessed 02 Feb 2022
12. Kozłowski, K., Korytkowski, M., Szajerman, D.: Visual analysis of computer game output video stream for gameplay metrics. In: Krzhizhanovskaya, V.V., et al. (eds.) ICCS 2020. LNCS, vol. 12141, pp. 538–552. Springer, Cham (2020). https://doi.org/10.1007/978-3-030-50426-7_40
13. Lazar, J., et al.: Research Methods in Human-Computer Interaction, 2nd edn. Wiley, Hoboken (2017). ISBN 9780128053904
14. Rogalska, A., Napieralski, P.: The visual attention saliency map for movie retrospection. Open Phys. **16**(1), 188–192 (2018)
15. Szajerman, D., Napieralski, P., Lecointe, J.-P.: Joint analysis of simultaneous EEG and eye tracking data for video images. COMPEL Int. J. Comput. Math. Electr. Electron. Eng. **37**(5), 1870–1884 (2018)
16. Ølsted, P.T., Ma, B., Risi, S.: Interactive evolution of levels for a competitive multiplayer FPS, evolutionary computation (CEC). In: 2015 IEEE Congress on IEEE, pp. 1527–1534 (2015)

ACCirO: A System for Analyzing and Digitizing Images of Charts with Circular Objects

Siri Chandana Daggubati and Jaya Sreevalsan-Nair[✉] [ID]

Graphics-Visualization-Computing Lab (GVCL), International Institute
of Information Technology, Bangalore (IIITB), 26/C, Electronics City,
Bengaluru 560100, Karnataka, India
{daggubati.sirichandana,jnair}@iiitb.ac.in
http://www.iiitb.ac.in/gvcl

Abstract. Automated interpretation of digital images of charts in documents and the internet helps to improve the accessibility of visual representation of data. One of the approaches for automation involves extraction of graphical objects in the charts, *e.g.,* pie segments, scatter points, etc., along with its semantics encoded in the textual content of the chart. The scatter plots and pie charts are amongst the widely used infographics for data analysis, and commonly have circle objects. Here, we propose a chart interpretation system, ACCirO (Analyzer of Charts with Circular Objects), that exploits the color and geometry of circular objects in scatter plots, its variants, and pie charts to extract the data from its images. ACCirO uses deep learning-based chart-type classification and OCR for text recognition to add semantics, and templatized sentence generation from the extracted data table for chart summarization. We show that image processing and deep learning approaches in ACCirO have improved the accuracy compared to the state-of-the-art.

Keywords: Image processing · Scatter plots · Pie charts · Dot plots ·
Bubble plots · Circle geometry · Circle Hough Transform (CHT) ·
Spectral clustering · Chart data extraction · Text recognition

1 Introduction

Given the ubiquity of charts for visualizations, there is recent interest in automating its interpretation. The motivating applications include filtering significant charts from image databases, generating visual question-answering (QA) systems, etc., given the raster format of the charts. Though many image processing techniques have been used, there are still gaps in existing technology for automated chart interpretation owing to the diversity and complexity of chart content

---

This work has been generously supported by the Machine Intelligence and Robotics (MINRO) Grant, Government of Karnataka. This paper has benefitted from discussions with GVCL members and peers at IIITB.

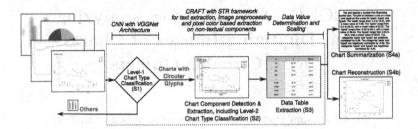

Fig. 1. Our proposed workflow for ACCirO is an automated system for data extraction from images of charts with circular objects and color, and for which geometric and color-based information extraction techniques are used.

and the requirement of human-in-the-loop in most cases. Here, we consider pie charts, scatter plots, and its variants, *dot plots*, and *bubble plots*, with circular objects. The source images are from documents in portable document format (.pdf), websites, outputs from plotting tools, and curated image databases [7]. The data extracted from images comes from its non-textual and textual content in charts. The non-textual content implies the geometric objects as per chart type, *e.g.*, pie segments and scatter points. The text content is from chart, legend, and axes titles, which are localized using annotation and extracted using image processing and text recognition.

When using the second-order gradient tensor field-based approach for object extraction from charts [4,5], we observe that only regions with high color gradients (edges and corners) are extracted, *e.g.*, boundaries of bar objects. But in pie charts, the gradients are concentrated in the corners of the largest rectangle enclosed within the pie owing to the high curvature gradient. At the same time, the pie chart has a circle geometry that can be exploited for sector extraction. Thus, we propose using color and geometry information in pie charts for automated annotation and data extraction. Given that circle geometry is predominantly used for scatter points in scatter, dot, and bubble plots, our proposed method is generalized for our selected four chart types.

Thus, our contribution is integrating an end-to-end system, ACCirO (Analyzer for Chart images with *Ci*rcular Objects), generalized for four chart types. ACCirO has a four-step workflow (Fig. 1): S_1 the chart classification, S_2 a novel color-based annotation along with text extraction, S_3 a novel color-based data extraction, S_{4a} text summarization and S_{4b} chart reconstruction. ACCirO specifically works for charts where color encodes class information and improves on S_2 and S_3 in BarChartAnalyzer [4], and ScatterPlotAnalyzer [5]. *Alpha blending* leads to the blended colors in bubble plots, which are different from the colors given in the chart legend. Our algorithm in S_3 addresses the challenge of computing the color, radius, and center of constituent circles in overlap regions.

Related Work: We generalize data extraction for different chart types, as is the current focus [2,4,5,8], but by using color information exclusively. CHT [6] has been used for object extraction from charts [8]. For the chart-types with circular objects in the foreground with non-textured background, CHT suffices.

2   The Workflow of ACCirO

We propose a fully automated workflow for ACCiro for pie charts and dot, bubble, and scatter plots. We consider bubble and dot plots as variants of scatter plots, owing to the similarity in using circular objects as graphical objects with positional information. The four key components of our workflow (Fig. 1) are: (S_1) chart type classification, S_2 chart component detection and extraction, and S_3 data table extraction, optionally followed by S_{4a} chart summarization or S_{4b} chart reconstruction. We use the implementation from our previous work [4,5] for S_1 to pick scatter plots and pie charts, S_{4a}, and S_{4b}. Scatter plots are further subclassified to its variants based on position and size variations of scatter points.

S_2: Chart Component Detection and Extraction

A chart is structurally composed of specific elements, referred to as *chart components*, whose characteristic properties in its raster format are exploited for their extraction. The seven components are: *canvas, legend, chart title, XY-axis titles*, and *XY-axis labels*. The region of the chart that *contains* the graphical objects, *e.g.*, pie sectors, scatter points, and bounded by axes, is the *canvas*. The process of localizing and retrieving them from the chart images is called *chart component extraction*. A separate component-wise analysis is more effective for stepwise chart interpretation than joint extraction using the entire image.

We first extract textual content by using a DL-based Optical Character Recognition (OCR), namely, Character Region Awareness for Text Detection (CRAFT), followed by a scene text recognition framework (STR), as used in ScatterPlotAnalyzer. This text is now removed to extract the canvas and graphical markers/objects in the legend. But, the filtered image still contains "noise" such as axis lines, gridlines, ticks, and small text fragments missed by the OCR.

In pie charts, we use CHT to extract the entire pie from the image, and the unique pixel colors in the pie pixels are used to locate objects in the legend. In scatter plots and their variants, object extraction using geometry is not reliable owing to the variety in marker styles and the presence of overlapping scatter points. So we initially remove axes and gridlines based on the property of the periodic arrangement of their straight pixels lines. We then locate the color-based clusters of pixels and tag them as "objects." The color histogram of the image gives colors of high frequency needed for the localization of pixel clusters. The bounding box of pie and axes in scatter plots gives the canvas.

Legend extraction follows after the canvas extraction in our workflow to accommodate cases of legend being placed in the canvas. The color-based clusters with relatively smaller pixel coverage and placed adjacent to text are identified as legend markers with corresponding labels. The final step is to semantically classify the textboxes outside the canvas region based on their role, *i.e.*, chart title, axes titles, and labels. The axes are usually found at the bottom, and the left of the canvas region, and the chart title at its top. In legend-free pie charts, text boxes in the proximity of arc centers of the pie sectors give the class information.

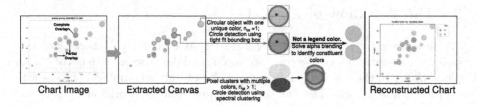

Fig. 2. Our proposed method for circular object extraction from bubble chart images.

S_3: Data Table Extraction from Graphical Objects

We extract information from the pixel clusters in the image space and convert them to data space using the cues from the extracted text.

Pixel-Based Data Extraction from Pie Charts: Percentage data is obtained from the sector area, which is determined using the fraction of pixel counts in the sector [2]. We implement this on the "donut" with one-third of the radius of the pie removed from the center. The donut is used instead of the entire pie to reduce the discrepancy from the missing pixels owing to text removal in the pie region. The sectors are then mapped to their labels based on colors. In legend-free charts, sectors are mapped to the closest text labels.

Pixel-Based Data Extraction from Scatter, Dot, and Bubble Plots: Here, the 2D data is encoded in the positional vectors of the scatter object. The variants use additional visual encodings for more attributes. Such as, the height of stacks of dots in dot plots represents bar height in an equivalent bar chart, and bubble plots use the size and color of objects to encode additional attributes. Contours are extracted for color pixel clusters from]step2. The positional information of scatter points is determined using the contour centroids of the clusters. In the case of overlapping points, the number of points involved is computed based on the ratio of contour area with the smallest contour observed in the chart. We then use k-means clustering to get centroids in the overlapping region. In the case of a dot plot, we determine the count of objects stacked with the same x-coordinate value of contour centroids. This count is given a class label using the legend color or the x-tick mark label.

Bubble plots have the unique challenge of overlapping circular objects with varying sizes and transparencies. The resultant alpha blending of overlapped scatter object regions poses a challenge in identifying the number and parameters of the constituent scatter objects. The resultant color from alpha blending is given by: $C = \alpha.F+(1-\alpha).B$, where F and B are the foreground and background colors respectively; and $0 \leq \alpha \leq 1$. We resolve the challenge using contour colors (Fig. 2). To estimate the number of overlapping points in a contour, we get n_{cs} segments based on the unique colors in contours. Using the value of n_{cs}, we extract the center, radius, and class label of the scatter point. If $n_{cs} = 1$, the center and radius of the scatter point are given by a tight-fitting bounding box of the contour. When $n_{cs} > 1$, we use spectral clustering in the pixel cluster of contour and determine circle parameters with best-fit circle regression of n_{cs}

clusters. In all cases, we use the class label corresponding to its contour color, as given in the legend.

For those circles which do not correspond to legend colors in the above two cases, we use the blending equation to solve for the legend colors and corresponding transparency, α, values that give the resultant blended colors C. We use an *exhaustive* search of the solution space to find the closest solution. We assume that the color C of the spectral cluster is a blend of two or more legend colors. Hence, we solve for $2 \leq p \leq m = |C_L|$, for m legend colors and the set of legend colors C_L. We use mC_p combinations of colors, where each experimental run uses a subset of legend colors $\{C_1, C_2, \ldots, C_p\}$, where $C_i \in C_L, \forall i \in [1, p]$. The blending equation is now modified as a linear combination of C_i:

$$C' = \sum_{i=1}^{p} \alpha_i.C_i, \text{ where } \sum_{i=1}^{p} \alpha_i = 1 \text{ and } 0 < \alpha_i < 1, \forall i \in [1, p].$$

We can reduce the search space by limiting the value of α_1 to be a value in the interval $[0.5, 0.95]$, with a step size of 0.05. We also implement a greedy algorithm terminating the search when the closest color is obtained.

Data Transformation for Scatter Plots: Finally, to transform data in pixel space to the numerical space, we use the scaling factor computed from the semantics of the text, as done in ScatterPlotAnalyzer. However, this does not work for circle size encoding in bubble charts, as the factor for circle sizes can be obtained from either the size legend (as present in some charts) or the parameter setting used in different plotting tools *e.g.,* DPI, for the chart generation. Thus, the exact data mapping for bubble charts will be explored in future.

3   Experiments and Results

Qualitative Assessment: Through the visualizations, we observe that our method performs superior to the tensor field-based method in ScatterPlotAnalyzer [5] (Fig. 3). The color-based data extraction technique is an improvement over tensor fields [5] in the case of cluttered scatter points. Figure 4 shows outputs at different stages of ACCirO for a sample of each chart type, demonstrating similarities between the reconstructed and sources chart images. Visual analysis of the results of ACCiro on pie charts and scatter plots images in the FigureQA [7] gives reconstruction accuracy of ~90.11% and ~90.5%, respectively. To improve the circle detection using CHT [6] in a pie chart, which has an average of 96% accuracy, advanced circle detection methods such as RANSAC may be used.

Quantitative Assessment: For quantitative assessment, we use synthetically generated chart images from publicly available data sources, *e.g., Kaggle*. We have generated a set of 15 images each for pie charts, dot and bubble charts, and 24 for scatter plots. In total, we use a test set of 69 images here.

We use the F1 Score and MAPE (Mean Absolute Percentage Error) metrics to measure the success and failure rates of the performance of ACCirO (Table 1). We

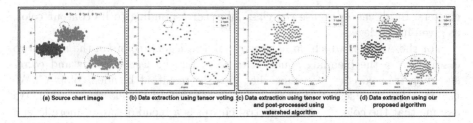

Fig. 3. Comparison of (a) source image and charts reconstructed using similar methods: (b) ScatterPlotAnalyzer [5], (c) modified ScatterPlotAnalyzer, and (d) our color-based method, for a multiclass scatter plot with a high degree of overlap of scatter points. The cluster of scatter points (red dotted ellipses) extracted is highlighted. (Color figure online)

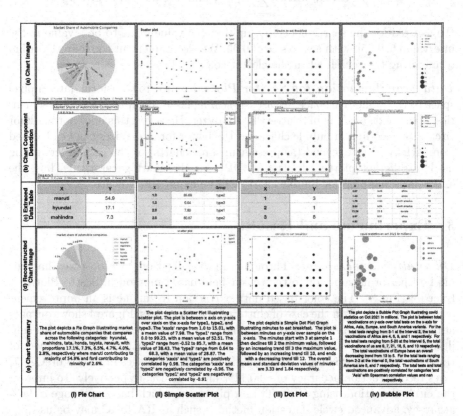

Fig. 4. Different stages of chart data extraction followed by chart reconstruction and summarization from the sample source images of (i) pie chart, (ii) scatter plot, (iii) dot plot, and (iv) bubble plot. The text summary is best visible at 220+% zoom level.

Table 1. Accuracy of data table extraction using ACCirO

Chart type → Accuracy measure ↓	Pie chart	Dot plot	Bubble plot	Scatter plot	Overall measures
Average precision	0.94	1.00	1.00	0.97	**0.96**
Average recall	1.00	1.00	0.99	0.95	**0.95**
Success rate: F1 score > 0.8	93%	100%	100%	96%	**96.4%**
Success rate: MAPE < 0.2	100%	100%	93%	84%	**88.7%**

consider the data extraction as a *success* with F1 Score > 0.8 as in [3]. Despite the smaller test dataset, the data extraction accuracy of ACCirO from scatter plots surpasses the state-of-the-art methods with F1-Score 97%, compared to 90.5% and 88% for MECDG [1] and Scatteract [3], respectively. Our data extraction for dot plots is 100% accurate, owing to the structured point layout. Even for bubble plots, despite the complex challenges due to transparency and overlapping points, we get an F1-Score of 100%. It must be noted that, since the bubble/object size is in pixel measure, we exclude it from the F1 score computation. We observe that the normalized radius values are closer to raw data.

To determine the numerical precision errors in S_3, we compare the difference between the source and extracted data values using the Mean Absolute Percentage Error (MAPE) as in [4]. Here, our alternative definition for the *success* of data extraction is when the error rate MAPE < 0.2. MAPE is augmented in the case of omission and precision errors owing to cluster centroids overlapping with scatter points and pixel space to data space transformations, respectively. Pie charts have been an exception for error-free data extraction of percentage values.

4 Conclusions

ACCirO has two known limitations which are to be resolved in future work. Firstly, owing to ACCiro being a color-based method, it fails in cases where the shape and texture of scatter points encode class or type information. Secondly, the STR text recognition model fails to interpret superscript symbols, and recognition of 'o', '0', and '−.' In summary, our proposed color-based end-to-end chart image interpretation system, ACCirO, has been generalized for the chart with circular objects, such as pie charts, and scatter, dot and bubble plots.

References

1. Chen, L., Zhao, K.: An approach for chart description generation in cyber-physical-social system. Symmetry **13**(9), 1552 (2021)
2. Choi, J., Jung, S., Park, D.G., Choo, J., Elmqvist, N.: Visualizing for the non-visual: enabling the visually impaired to use visualization. In: Computer Graphics Forum, vol. 38, pp. 249–260. Wiley Online Library (2019)

3. Cliche, M., Rosenberg, D., Madeka, D., Yee, C.: Scatteract: automated extraction of data from scatter plots. In: Ceci, M., Hollmén, J., Todorovski, L., Vens, C., Džeroski, S. (eds.) ECML PKDD 2017. LNCS (LNAI), vol. 10534, pp. 135–150. Springer, Cham (2017). https://doi.org/10.1007/978-3-319-71249-9_9
4. Dadhich, K., Daggubati, S.C., Sreevalsan-Nair, J.: BarChartAnalyzer: digitizing images of bar charts. In: International Conference on Image Processing and Vision Engineering (IMPROVE), pp. 17–28. INSTICC, SciTePress (2021)
5. Dadhich, K., Daggubati, S.C., Sreevalsan-Nair, J.: ScatterPlotAnalyzer: digitizing images of charts using tensor-based computational model. In: Paszynski, M., Kranzlmüller, D., Krzhizhanovskaya, V.V., Dongarra, J.J., Sloot, P.M.A. (eds.) ICCS 2021. LNCS, vol. 12746, pp. 70–83. Springer, Cham (2021). https://doi.org/10.1007/978-3-030-77977-1_6
6. Duda, R.O., Hart, P.E.: Use of the Hough transformation to detect lines and curves in pictures. Commun. ACM **15**(1), 11–15 (1972)
7. Kahou, S.E., Atkinson, A., Michalski, V., Kádár, Á., Trischler, A., Bengio, Y.: FigureQA: an annotated figure dataset for visual reasoning. CoRR abs/1710.07300 (2017)
8. Savva, M., Kong, N., Chhajta, A., Fei-Fei, L., Agrawala, M., Heer, J.: Revision: automated classification, analysis and redesign of chart images. In: Proceedings of the 24th Annual ACM Symposium on User Interface Software & Technology, pp. 393–402 (2011)

Learning Scale-Invariant Object Representations with a Single-Shot Convolutional Generative Model

Piotr Zieliński[✉][ID] and Tomasz Kajdanowicz[ID]

Wrocław University of Science and Technology, Wybrzeże Wyspiańskiego 27,
50-370 Wrocław, Poland
p.zielinski@pwr.edu.pl

Abstract. Contemporary machine learning literature highlights learning object-centric image representations' benefits, i.e. interpretability, and the improved generalization performance. In the current work, we develop a neural network architecture that effectively addresses the task of multi-object representation learning in scenes containing multiple objects of varying types and sizes. In particular, we combine SPAIR and SPACE ideas, which do not scale well to such complex images, and blend them with recent developments in single-shot object detection. The method overcomes the limitations of fixed-scale glimpses' processing by learning representations using a feature pyramid-based approach, allowing more feasible parallelization than all other state-of-the-art methods. Moreover, the method can focus on learning representations of only a selected subset of types of objects coexisting in scenes. Through a series of experiments, we demonstrate the superior performance of our architecture over SPAIR and SPACE, especially in terms of latent representation and inferring on images with objects of varying sizes.

Keywords: Deep autoencoders · Representation learning · Generative models · Scene analysis

1 Introduction

The ability to discriminate and reason about individual objects in an image is one of the important tasks of computer vision, which is why object detection and instance segmentation tasks have drawn vast attention from researchers throughout the years. The latest advances in artificial intelligence require a more insightful analysis of the image to provide more profound reasoning about its contents. It can be achieved through representation learning, which facilitates extracting useful information about objects, allowing transferring more general knowledge to other tasks [2]. One can see multi-object representation learning as a natural extension to the aforementioned computer vision tasks. Here, the objective is to produce a valuable abstract feature vector of each of the inferred

© The Author(s), under exclusive license to Springer Nature Switzerland AG 2022
D. Groen et al. (Eds.): ICCS 2022, LNCS 13352, pp. 613–626, 2022.
https://doi.org/10.1007/978-3-031-08757-8_51

objects and hence produce a structured representation of the image, allowing for its more insightful understanding.

Recently, the most successful methods are based on the variational autoen-coder (VAE) framework [16,21], with structured latent space, which includes individual objects' representations. The original approach consists in extract-ing object latent vectors with a recurrent network [1,3,7–9]. Alternatively, each object's representation can be produced with a single forward pass through the network by employing a convolution-based single-shot approach [4,18]. However, these methods are limited by a single feature map utilized to create objects' latent vectors and hence cannot be used when object sizes vary.

In this paper, we propose a single-shot method for learning multiple objects' representations, called *Single-Shot Detect, Infer, Repeat* (SSDIR[1]). It is a con-volutional generative model applying the single-shot approach with a feature pyramid for learning valuable, scale-invariant object representations. By pro-cessing multi-scale feature maps, SSDIR can attend to objects of highly varying sizes and produce high-quality latent representations directly, without the need of extracting objects' glimpses and processing them with an additional encoder network. The ability to focus on individual objects in the image is improved by leveraging knowledge learned in an SSD [19] object detection model. In exper-iments, we compare the SSDIR model on multi-scale scattered MNIST digits, CLEVR [15] and WIDER FACE [23] datasets with other single-shot approaches, proving the ability to focus on individual objects of varying sizes in complicated scenes, as well as the improved quality of objects' latent representations, which can be successfully used in other downstream problems, despite the use of an uncomplicated convolutional backbone.

We summarize our contributions as follows. We present a model that enhances multi-object representation learning with a single-shot, feature pyramid-based approach, retaining probabilistic modeling of objects. We provide a framework for generating object representations directly from feature maps without extracting and processing glimpses, allowing easier scaling to larger images. We compare the method with other single-shot multi-object represen-tation learning models and show its ability to attend to objects, the improved latent space quality, and applicability in various benchmark problems.

2   Related Works

Multi-object representation learning has recently been tackled using unsuper-vised, VAE-based models. Two main approaches include sequential models, attending to a single object or part of the image at a time, and single-shot methods, which generate all representations in a single forward pass through the network.

The original approach to this problem was presented by Ali Eslami *et al.* in [1]. The *Attend, Infer, Repeat* (AIR) model assumes a scene to consist of objects,

---

[1] Code available at: https://github.com/piotlinski/ssdir.

represented with *what* vector, describing the object's appearance, *where* vector indicating its position on the image and *present* vector, describing if it is present in the image, controlling termination of the recurrent image processing. The model attends to a single object at a time, generating representations sequentially with a recurrent network until a non-present object is processed. Other studies, including [10] and [22] proposed a different approach, where objects representations are learned using Neural Expectation-Maximization, without structuring the latent representations explicitly. These methods suffer from scaling issues, not being able to deal with complex scenes with multiple objects.

Alternatively, an image might be described with a scene-mixture approach, as in MONet [3], IODINE [9] and GENESIS [7,8]. Here, the model does not explicitly divide the image into objects but instead generates masks, splitting the scene into components, which the model encodes. In the case of MONet and GENESIS, each component is attended and encoded sequentially, while IODINE uses amortized iterative refinement of the output image. However, these methods are not a good fit for learning object representations in an image, as scene components usually consist of multiple objects. Furthermore, masks that indicate particular objects limit the model's scalability due to this representation requiring more memory than bounding box coordinates.

GENESIS belongs to a group of methods, which focus on the ability to generate novel, coherent and realistic scenes. Among them, one should notice recent advances with methods leveraging generative adversarial networks (GANs), such as RELATE [6] or GIRAFFE [20]. Compared to VAE-based methods, they can produce sharp and natural images, which are more similar to original datasets. However, these models do not include an explicit image encoder, and therefore cannot be applied for multi-object representation learning directly. What is more, the process of training GANs tends to be longer and more complicated than in the case of VAEs.

Recently, methods such as GMAIR [24] postulate that acquiring valuable *what* object representations is crucial for the ability to use objects encodings in other tasks, such as clustering. Here, researchers enhanced the original *what* encoder with Gaussian Mixture Model-based prior, inspired by the GMVAE framework [11]. In our work, we also emphasize the importance of the *what* object representation and evaluate its applicability in downstream tasks.

One of the promising methods of improving model scalability of VAE-based multi-object representation learning models was presented in SPAIR [4], where the recurrent attention of the original AIR was replaced with a local feature maps-based approach. In analogy to single-shot object detection models like SSD [19], the SPAIR first processes image with a convolutional backbone, which returns a feature map with dimensions corresponding to a fixed-sized grid. Each cell in the grid is then used to generate the locations of objects. Objects representations' are inferred by processing these cells sequentially, generating *what*, *depth* and *present* latent variables, describing its appearance, depth in the scene, and the fact of presence. This approach has recently been extended in SPACE [18], which fixes still existing scalability issues in SPAIR by employing parallel latent components inference. Additionally, the authors used the scene-mixture approach to model the image background, proving to be applicable for learning objects' representations in more complex scenes. However, both methods rely on a single grid of fixed size, which makes it difficult for this class of models to attend to objects of highly varying sizes. What is more, both of them employ glimpse extraction: each attended object is cut out of the input image and processed by an additional encoder network to generate objects' latent representations; this increases the computational expense of these methods.

Latest advances in the field of multi-object representation learning try to apply the aforementioned approaches for inferring representations of objects in videos. SQAIR [17] extends the recurrent approach proposed in AIR for sequences of images by proposing a propagation mechanism, which allows reusing representations in subsequent steps. A similar approach was applied to single-shot methods by extending them with a recurrent network in SILOT [5] and SCALOR [14]; here, the representations were used in the object tracking task. An interesting approach was proposed by Henderson and Lambert [12]. Authors choose to treat each instance within the scene as a 3D object; the image is then generated by rendering each object and merging their 2D views into an image. This allows for a better understanding of objects' representations, at the cost of significantly higher computational complexity.

3   Method

SSDIR (**S**ingle-**S**hot **D**etect, **I**nfer, **R**epeat) is a neural network model based on a variational autoencoder architecture [16,21] as shown in Fig. 1; its latent space consists of structured objects' representations z, enhanced by leveraging knowledge learned in a single-shot object detection model SSD [19], both sharing the same convolutional backbone.

3.1   The Proposed Model: SSDIR

Our model extends the idea of single-shot object detection. Let x be the image representing all relevant (i.e. detected by the SSD) objects present in the image. SSDIR is a probabilistic generative model, which assumes that this image is generated from a latent representation z according to a likelihood distribution. This representation consists of a set of latent vectors assigned to each grid cell in the feature pyramid of SSD's convolutional backbone and is sampled from a prior distribution $p(z)$. Since the likelihood distribution is unknown, we approximate it using the decoder network θ, which parametrizes the likelihood $p_\theta(x|z)$. Then, the generative model can be described as a standard VAE decoder (1).

$$p(x) = \int p_\theta(x|z) p(z) dz \tag{1}$$

To do inference in this model, SSDIR applies variational method and approximates the intractable true posterior with a function $q_\phi(z|x) \approx p(z|x)$,

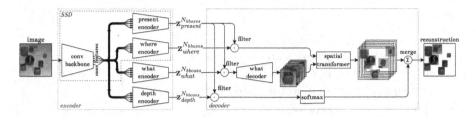

Fig. 1. Illustration of the SSDIR model. It consists of two fully-convolutional neural networks: an *encoder* and a *decoder*. The *encoder* uses a convolutional backbone as a feature extractor, which builds a pyramid of multi-scale features processed by each latent component encoder. Each object's position z_{where} and presence $z_{present}$ latent vectors are computed using a trained object detection model *SSD*, indicating grid cells, which refer to detected objects; z_{what} and z_{depth} are computed with additional convolutional encoders, which process the feature maps from the pyramid in a similar manner to *SSD*. In the decoder, all latents are filtered to include only present objects for reconstructions. *What* decoder reconstructs appearances of each present object, which are then put in their original place with an affine transformation in the *spatial transformer* module. Finally, object reconstructions are merged using weighted sum, created by applying *softmax* on objects' *depth* latents.

parametrized by ϕ (encoder parameters). This allows us to use ELBO (Evidence Lower Bound) as the loss function (2):

$$\mathcal{L}(\theta, \phi) := \mathbb{E}_{z \sim q_\phi(z|x)}\left[\log p_\theta(x|z)\right] - D_{KL}\left(q_\phi(z|x) \| p(z)\right) \tag{2}$$

where D_{KL} is the KL divergence.

Object Representation. SSDIR extends the grid-based approach with a feature pyramid for object detection proposed in SSD to produce objects' latent representations. We assume each object can be described by four latent variables:

- $z_{where} \in \mathbb{R}^4$ – the object's bounding box position and size,
- $z_{present} \in \{0, 1\}$ – a binary value indicating if given cell contains any object,
- $z_{what} \in \mathbb{R}^D$ – D-sized vector describing the object appearance,
- $z_{depth} \in \mathbb{R}$ – a real number indicating how deep in the scene the given object was observed (we assume, that objects with a bigger value of z_{depth} appear in front of those with a lower value).

To simplify the process of objects discovery, we reuse a trained SSD model to get bounding box position and size, as well as the detected object class. SSDIR utilizes detections to produce z_{where} and $z_{present}$ as shown in (3) and (4).

$$z_{where}^i = \begin{bmatrix} cx_i & cy_i & w_i & h_i \end{bmatrix} \tag{3}$$

$$z_{present}^i \sim Bernoulli\left(\beta^i\right) \tag{4}$$

P. Zieliński and T. Kajdanowicz

where:

i refers to the cell in the feature pyramid,
cx, cy are the bounding box' center coordinates,
w, h are the bounding box' width and height dimensions,

$$\beta^i = \begin{cases} \arg\max_k c_i & \text{if an object detected in the cell,} \\ 0 & \text{otherwise,} \end{cases}$$

c are the object's predicted class confidences.

The two remaining latent components: z_{what} and z_{depth} are modeled with Gaussian distributions, as shown in (5) and (6).

$$z^i_{what} \sim \mathcal{N}\left(\mu^i_{what}, \sigma^i_{what}\right) \tag{5}$$

$$z^i_{depth} \sim \mathcal{N}\left(\mu^i_{depth}, \sigma^i_{depth}\right) \tag{6}$$

where:

μ_{what}, μ_{depth} are means, encoded with *what* and *depth* encoders,
σ_{what}, σ_{depth} are standard deviations, which are treated as model's hyperparameters.

SSDIR Encoder Network. To generate the latent representation of objects contained in an image, we apply the feature pyramid-based object detection approach. The function of the encoder $q_\phi(z|x)$ is implemented with a convolutional backbone (VGG11) accepting images of size $300 \times 300 \times 3$, extended with a feature pyramid, and processed by additional convolutional encoders, as shown in Fig. 1. Specifically, *where*, *present* and *depth* encoders contain single convolution layer with 3×3 kernels (1 in case of *present* and *depth* and 4 for *where* encoder) per each feature map in the pyramid, whereas *what* encoder may include sequences of convolution layers with ReLU activations, finally returning D-sized vector for each cell in each feature pyramid grid. The outputs of these encoders are used to generate latent vectors z_{where}, $z_{present}$, z_{what} and z_{depth}.

The backbone's, as well as *where* and *present* encoders' weights are transferred from an SSD model trained with supervision for detection of objects of interest in a given task and frozen for training; *what* and *depth* encoders, which share the same pretrained backbone, are trained with the decoder network. Such architecture allows parallel inference, since neither latent component depends on any other, without the need of extracting glimpses of objects and processing them with a separate encoder network – in SSDIR latent representations are contained within feature maps directly, improving its scalability.

SSDIR Decoder Network. Latent representations of objects in the picture are forwarded to the decoder network to generate reconstructions of areas in the input image that contain objects of interest, i.e. those detected by the SSD network. First, the latent variables are filtered according to $z_{present}$, leaving only those objects, which were found present in the image by the SSD network.

Next, per-object reconstructions are generated by passing filtered z_{what} vectors through a convolutional *what* decoder, producing M images of size $64 \times 64 \times 3$, representing each detected object's appearance. These images are then translated and scaled according to the tight bounding box location z_{where} in the *spatial transformer* module [13]. The resulting M $300 \times 300 \times 3$ images are merged using a weighted sum, with softmaxed, filtered z_{depth} as the weights. The output of the model might then be normalized with respect to the maximum intensity of pixels in the reconstruction to improve the fidelity of the reconstruction.

SSDIR does not require special preprocessing of the image, apart from the standard normalization used widely in convolutional neural networks. Originally, the background is not included in the reconstruction phase, since its representation is not crucial in the task of multi-object representation learning; we assume that this way SSDIR learns to extract the key information about all objects from the image. The background might however be reconstructed as well by including an additional z_{what} encoder and treating the background as an extra object, which is transformed to fill the entire image and put behind all other objects.

The parallel nature of the model is preserved in the decoder. The operations of filtering, transforming, and merging are implemented as matrix operations, allowing good performance and scalability.

Training. The SSDIR model is trained with a modified ELBO loss function. We extend the original form (2), which intuitively includes reconstruction error of an entire image and KL divergence for latent and prior distributions with a normalized sum of each detected object's reconstruction error. This allows the model to reach high quality of reconstructions (and as a result – high quality of z_{what} latent representations) and correct order of objects' z_{depth}, preserving transformation function continuity thanks to KL divergence-based regularization. The final form of the loss function is shown in (7).

$$
\begin{aligned}
\mathcal{L}\left(\boldsymbol{x}, \theta, \phi\right) = {} & \alpha_{obj} \mathbb{E}_{\boldsymbol{z}}\left[\log p_{\theta}\left(\boldsymbol{x}|\boldsymbol{z}\right)\right] + \alpha_{rec} \frac{1}{M} \sum_{i}^{M} \mathbb{E}_{z_i}\left[\log p_{\theta}\left(x_i|z_i\right)\right] \\
& - \alpha_{what} D_{KL}\left(q_{\phi}\left(\boldsymbol{z}_{what}|\boldsymbol{x}\right) \| p\left(\boldsymbol{z}_{what}\right)\right) \\
& - \alpha_{depth} D_{KL}\left(q_{\phi}\left(\boldsymbol{z}_{depth}|\boldsymbol{x}\right) \| p\left(\boldsymbol{z}_{depth}\right)\right)
\end{aligned}
\tag{7}
$$

where:

$\mathbb{E}_{\boldsymbol{z}}\left[\log p_{\theta}\left(\boldsymbol{x}|\boldsymbol{z}\right)\right]$ is the likelihood of the reconstruction generated by the decoder,
$\mathbb{E}_{z_i}\left[\log p_{\theta}\left(x_i|z_i\right)\right]$ is the likelihood of an i-th detected object reconstruction,
α_{obj}, α_{rec}, α_{what}, α_{depth} are loss components coefficients, modifying the impact of each one on the learning of the model,
M is the number of objects detected by the SSD model in a given image.

In case of both z_{what} and z_{depth} we assume the prior to be a standard normal distribution $\mathcal{N}\left(\boldsymbol{0}, \boldsymbol{I}\right)$. The training objective is described by (8) for each image x_i in the training dataset. The model is trained jointly with gradient ascent using Adam as the optimizer, utilizing the reparametrization trick for back-propagating gradients through the sampling process. The process of learning representations is unsupervised, although the backbone's and *where* and *present* encoders' weights are transferred from a pretrained SSD model.

$$\theta^*, \phi^* = \arg\max_{\theta, \phi} \sum_i \mathcal{L}(x_i, \theta, \phi) \qquad (8)$$

Table 1. Differences between **SSDIR** and baseline methods. "*semi-*" indicates that the object detection model is trained with supervision, while the representation learning procedure is unsupervised. "*glimpses*" refers to the process of learning object's z_{what} by extracting a sub-image containing the object (based on its z_{where} latent vector) and encoding it with a separate VAE; "*single-shot*" is the approach adopted in SSDIR.

Criterion	Basic VAE	SPAIR [4]	SPACE [18]	**SSDIR**
Unsupervised	Semi-	✓	✓	Semi-
Inferring representations	Glimpses	Glimpses	Glimpses	Single-shot
Varying sizes	✓	✗	✗	✓
Particular objects type	✓	✗	✗	✓
Parallel encoding	✗	✗	✗	✓

4   Experiments

In this section, we evaluate the performance of SSDIR and compare it with two baseline methods: SPAIR [4] and SPACE [18]. We focus on verifying the ability to learn valuable representations of objects, which sizes vary; this is conducted by analyzing the quality of reconstructions produced by the decoder of each method and applying the produced representations in a downstream task. Besides, we conduct an ablation study to analyze the influence of the dataset characteristics on SSDIR performance.

Our implementation of SPAIR is enhanced with a convolutional encoder instead of the original, fully-connected network, which should improve its performance on more complicated datasets. Since in this work we focus on learning objects' representation, we consider models without background: SPAIR does not explicitly model it, whereas in SPACE we analyze the foreground module outputs, which tries to reconstruct individual objects in the image. In Table 1 we included a comparison between the analyzed methods, together with an approach employing an object detector, a spatial transformer for extracting glimpses, and a VAE for learning their representations (denominated as *SSD+STN+VAE*).

The datasets used in the research were chosen to resemble common choices among recent multi-object representation learning methods. Among them, we decided to include datasets of various complexity, providing the ability to validate the model on simple images and prove its performance on complex, realistic images. Therefore, we conducted our experiments using three datasets: 1) multi-scale, scattered MNIST digits (with configured minimum and maximum digit size, as well as grids for scattering digits), 2) CLEVR dataset [15] (containing artificially generated scenes with multiple objects of different shape, material, and size, used widely in the field of scene generation and multi-object representation learning), 3) WIDER FACE [23] (face detection benchmark dataset, with images containing multiple people; the dataset was used to demonstrate the ability of SSDIR to focus on objects of a particular type).

4.1 Per-object Reconstructions

In this section, we present a comparison of images' and objects' reconstructions for the proposed model and the baseline methods. In Fig. 2 we show inputs and reconstructions of representative images from each dataset (test subset, i.e. images not used for training), as well as some individual object reconstructions. Note, that due to the number of objects presented in the image and the nature of the models, it would not be possible to show all reconstructed objects.

Both SPAIR and SPACE can reconstruct the scattered MNIST dataset's image correctly. However, looking at the *where* boxes inferred by these models it is visible, that due to their limited object scale variability they are unable to attend to individual objects with a single latent representation, often reconstructing one digit with multiple objects. This is confirmed by the analysis of object reconstructions: SPAIR builds object reconstructions by combining reconstructed parts of digits, whereas SPACE can reconstruct digits of sizes similar to its preset, but divides bigger ones into parts. SSDIR is able to detect and reconstruct the MNIST image accurately: the use of a multi-scale feature pyramid allows for attending to entire objects, creating scale-invariant reconstructions, which are then mapped to the reconstruction according to tight *where* box coordinates.

SPAIR did not manage to learn object representations in the other two datasets. Instead, it models the image with rectangular boxes, containing a bigger part of an image. The aberrations visible in CLEVR dataset with SPAIR are caused by a transparency mask applied in this model and the fact, that these objects are heavily transformed when merging into the reconstruction. The tendency to model the image with rectangles is even more visible in the WIDER FACE dataset, where SPAIR divides the image in almost equal rectangles, aligned with the reconstruction grid. This effect allows for a fair quality of overall image reconstructions but does not yield valuable object representations.

In the case of SPACE, the model was not able to learn objects' representation in the CLEVR dataset, despite an extensive grid search of the hyperparameters relevant to the foreground module (especially the object's size). Instead, it models them using the background module, which cannot be treated as object representations since they gather multiple objects in one segment (this lies in line with problems reported in the GitHub repository[2]). Hence, objects reconstructions visible in Fig. 2 for this dataset contain noise. When applied to the WIDER FACE dataset, SPACE tends to approach image reconstruction in the

---

[2] https://github.com/zhixuan-lin/SPACE/issues/1.

622   P. Zieliński and T. Kajdanowicz

same way as SPAIR, dividing the image into rectangular parts, reconstructed as foreground objects. Similarly, this leads to an acceptable reconstruction quality but does not provide a good latent representation of the image's objects.

SSDIR shows good performance on the CLEVR dataset: it can detect individual objects and produce their latent representations, which results in good quality reconstructions. Similarly, in the case of the WIDER FACE dataset, the model is able to reconstruct individual faces. However, due to the simple backbone design and low resolution of object images, the quality of reconstructed faces is low. Additionally, as a result of using a multi-scale feature pyramid, SSDIR returns multiple image reconstructions for individual objects.

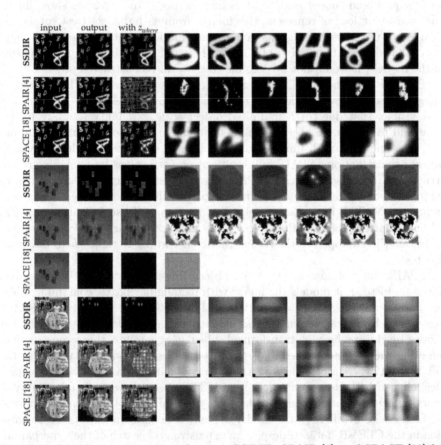

Fig. 2. Model inference comparison between **SSDIR**, SPAIR [4], and SPACE [18] for three typical images from each dataset. The first column presents the input image, the second and third contain image reconstruction without and with inferred bounding boxes; the remaining columns include some of the reconstructed individual objects. The number of images is limited due to the number of objects reconstructed by each model; for SSDIR, objects are meaningful and visually sound, while SPAIR and SPACE tend to divide bigger objects into smaller ones, or, in case of more complicated scenes, reconstruct them by dividing into rectangles, returning a redundant number of latents.

4.2   Latent Space

In this section, we present the analysis of the SSDIR model's latent space and compare it with the latent space of SPAIR and SPACE. Figure 3 visualizes latent spaces for the scattered multi-scale MNIST dataset. For each model, we process the test subset to generate latent vectors of each image. Then, individual objects' z_{where} vectors were compared with ground truth bounding boxes, and labels were assigned to latent representations by choosing the maximum intersection over union between predicted and true boxes. Each z_{what} vector was then embedded into two-dimensional space using t-SNE.

Table 2. Comparison of metrics for digit classification task using latent objects' representations and logistic regression. Results are averaged over 3 random seeds.

Method	Accuracy	Precision	Recall	F1-Score
SSDIR	**0.9789 ± 0.0016**	**0.9787 ± 0.0017**	**0.9786 ± 0.0016**	**0.9786 ± 0.0016**
SPAIR [4]	0.1919 ± 0.0073	0.1825 ± 0.0087	0.2019 ± 0.0092	0.1803 ± 0.0102
SPACE [18]	0.2121 ± 0.0432	0.2020 ± 0.0431	0.2158 ± 0.0435	0.1992 ± 0.0462

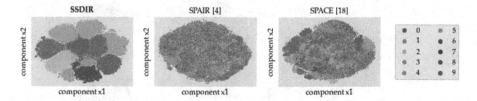

Fig. 3. Visualization of z_{what} latent space for scattered MNIST test dataset. Each object representation was converted using t-SNE to a two-dimensional space and plotted; the labels were inferred by choosing maximum intersection over union of predicted z_{where} and the ground truth bounding box and label. SSDIR shows a structured latent space, allowing easier distinguishing between digits.

Comparing the latent spaces, it is visible that SSDIR embeds the objects in a latent space, where digits can be easily distinguished. What is more, the manifold is continuous, without visible aberrations. The baseline methods' latent spaces are continuous as well, but they do not allow easy discrimination between each object class. The main reason is probably the fact, that both SPAIR and SPACE tend to divide large objects into smaller parts, according to the preset object size, as shown in Sect. 4.1.

Next, we tried to use the latent representations of objects in images for a downstream task of digit classification. For each of the methods, we trained models on the scattered MNIST dataset using three random seeds and produced latent representations for both train and test subset, assigning labels to each object's z_{what} based on intersection over union between z_{where} and ground truth boxes. Then, for each model and seed, we trained a logistic regression model to classify the digits based on their latent representations. Test subset classification metrics are gathered in Table 2. SSDIR latent space proves to be more valuable than the baseline methods', reaching high values of each metric.

4.3 Ablation Study

To test the influence of the dataset's characteristics on the model performance, we performed an ablation study. The scattered MNIST dataset is generated by drawing random cells in a preset grid and inserting a random-sized MNIST digit inside it with a random offset. The number and size of grids, as well as the minimum and maximum size of a digit, are the hyperparameters of the dataset generation researched in the ablation study.

An SSDIR model was trained on each of the generated datasets and evaluated on a test subset with regard to the mean square error of reconstructions. The results of the study are shown in Fig. 4.

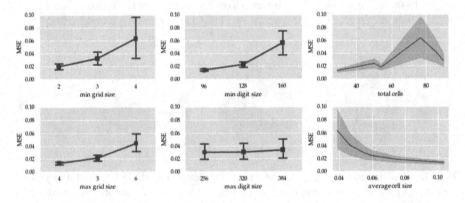

Fig. 4. Influence of the dataset generation parameters on the model performance. Parameters generating a dataset with larger or more occluded digits causes the model's performance to mitigate. SSDIR works best for non-occluded, small digits.

It is visible, that the model is sensitive to the size of objects in images. Bigger objects cause the mean square error to rise, mainly due to the transformation of small-sized reconstructions to the output image. Another factor that causes the error to increase is the number of digits in the image, which usually leads more occlusions to appear in the final image. The upturn is visible with increasing the minimum and maximum grid size, as well as the total number of cells.

5 Conclusions

In this paper, we proposed SSDIR, a single-shot convolutional generative model for learning scale-invariant object representations, which enhances existing solutions with a multi-scale feature pyramid-based approach and knowledge learned in an object detection model. We showed the improved quality of latent space inferred by SSDIR by applying it in a downstream task and proved its ability to learn scale-invariant representations of objects in simple and complex images.

Among the method's drawbacks, one should mention limited input image size, which makes it struggle with very complicated scenes, especially in case of occlusions. What is more, learning representations of objects in complex scenes could be improved by more advanced modeling of objects' interactions. These issues will be addressed in future works, which include applying a more advanced convolutional backbone and larger input images for improving the ability to detect objects and the quality of their representations. The latent vectors inferred by SSDIR could potentially be used in other advanced tasks, such as object tracking or re-identification. In such a case, the model could benefit from the increased sophistication of the model architecture. Additionally, SSDIR could be extended for processing videos by utilizing a recurrent network to consider temporal dependencies between subsequent frames.

References

1. Ali Eslami, S.M., et al.: Attend, infer, repeat: fast scene understanding with generative models. In: Advances in Neural Information Processing Systems (Nips), pp. 3233–3241 (2016)
2. Bengio, Y., Courville, A., Vincent, P.: Representation learning: a review and new perspectives. IEEE Trans. Pattern Anal. Mach. Intell. **35**(8), 1798–1828 (2013). https://doi.org/10.1109/TPAMI.2013.50
3. Burgess, C.P., et al.: MONet: Unsupervised Scene Decomposition and Representation, pp. 1–22 (2019). http://arxiv.org/abs/1901.11390
4. Crawford, E., Pineau, J.: Spatially invariant unsupervised object detection with convolutional neural networks. In: 33rd AAAI Conference on Artificial Intelligence, AAAI 2019, 31st Innovative Applications of Artificial Intelligence Conference, IAAI 2019 and the 9th AAAI Symposium on Educational Advances in Artificial Intelligence, EAAI 2019, pp. 3412–3420 (2019)
5. Crawford, E., Pineau, J.: Exploiting spatial invariance for scalable unsupervised object tracking. In: Proceedings of the AAAI Conference on Artificial Intelligence 34(04), pp. 3684–3692, April 2020. https://doi.org/10.1609/aaai.v34i04.5777, https://ojs.aaai.org/index.php/AAAI/article/view/5777
6. Ehrhardt, S., et al.: RELATE: physically plausible multi-object scene synthesis using structured latent spaces. NeurIPS (2020)
7. Engelcke, M., Kosiorek, A.R., Jones, O.P., Posner, I.: Genesis: generative scene inference and sampling with object-centric latent representations. In: International Conference on Learning Representations (2020). https://openreview.net/forum?id=BkxfaTVFwH
8. Engelcke, M., Parker Jones, O., Posner, I.: GENESIS-V2: inferring unordered object representations without iterative refinement. arXiv preprint arXiv:2104.09958 (2021)
9. Greff, K., et al.: Multi-object representation learning with iterative variational inference. In: 36th International Conference on Machine Learning, ICML 2019 2019-June, pp. 4317–4343 (2019)

10. Greff, K., Van Steenkiste, S., Schmidhuber, J.: Neural expectation maximization. Adv. Neural Inf. Process. Syst. **2017**-Decem(Nips), 6692–6702 (2017)
11. Gu, C., Xie, H., Lu, X., Zhang, C.: CGMVAE: Coupling GMM prior and GMM estimator for unsupervised clustering and disentanglement. IEEE Access **9**, 65140–65149 (2021). https://doi.org/10.1109/ACCESS.2021.3076073
12. Henderson, P., Lampert, C.H.: Unsupervised object-centric video generation and decomposition in 3D. In: Advances in Neural Information Processing Systems (NeurIPS) 33 (2020)
13. Jaderberg, M., Simonyan, K., Zisserman, A., Kavukcuoglu, K.: Spatial transformer networks. In: Proceedings of the 28th International Conference on Neural Information Processing Systems - Volume 2, NIPS 2015, pp. 2017–2025. MIT Press, Cambridge (2015)
14. Jiang*, J., Janghorbani*, S., Melo, G.D., Ahn, S.: Scalor: generative world models with scalable object representations. In: International Conference on Learning Representations (2020). https://openreview.net/forum?id=SJxrKgStDH
15. Johnson, J., Hariharan, B., van der Maaten, L., Fei-Fei, L., Zitnick, C.L., Girshick, R.: Clevr: A diagnostic dataset for compositional language and elementary visual reasoning. In: 2017 IEEE Conference on Computer Vision and Pattern Recognition (CVPR), pp. 1988–1997 (2017). DOI: https://doi.org/10.1109/CVPR.2017.215
16. Kingma, D.P., Welling, M.: Auto-encoding variational bayes. In: 2nd International Conference on Learning Representations, ICLR 2014, Banff, AB, Canada, April 14–16, 2014, Conference Track Proceedings (2014)
17. Kosiorek, A.R., Kim, H., Posner, I., Teh, Y.W.: Sequential attend, infer, repeat: Generative modelling of moving objects. Advances in Neural Information Processing Systems 2018-Decem (NeurIPS), pp. 8606–8616 (2018)
18. Lin, Z., et al.: Space: unsupervised object-oriented scene representation via spatial attention and decomposition. In: International Conference on Learning Representations (2020). https://openreview.net/forum?id=rkl03ySYDH
19. Liu, W., Anguelov, D., Erhan, D., Szegedy, C., Reed, S., Fu, C.-Y., Berg, A.C.: SSD: single shot MultiBox detector. In: Leibe, B., Matas, J., Sebe, N., Welling, M. (eds.) ECCV 2016. LNCS, vol. 9905, pp. 21–37. Springer, Cham (2016). https://doi.org/10.1007/978-3-319-46448-0_2
20. Niemeyer, M., Geiger, A.: Giraffe: representing scenes as compositional generative neural feature fields. In: Proceedings of IEEE Conference on Computer Vision and Pattern Recognition (CVPR) (2021)
21. Rezende, D.J., Mohamed, S., Wierstra, D.: Stochastic backpropagation and approximate inference in deep generative models. In: Proceedings of the 31st International Conference on International Conference on Machine Learning - Volume 32, ICML 2014, pp. II-1278-II-1286. JMLR.org (2014)
22. Van Steenkiste, S., Greff, K., Chang, M., Schmidhuber, J.: Relational neural expectation maximization: Unsupervised discovery of objects and their interactions. In: 6th International Conference on Learning Representations, ICLR 2018 - Conference Track Proceedings, pp. 1–15 (2018)
23. Yang, S., Luo, P., Loy, C.C., Tang, X.: Wider face: a face detection benchmark. In: 2016 IEEE Conference on Computer Vision and Pattern Recognition (CVPR), pp. 5525–5533 (2016). https://doi.org/10.1109/CVPR.2016.596
24. Zhu, W., Shen, Y., Yu, L., Sanchez, L.P.A.: Gmair: unsupervised object detection based on spatial attention and gaussian mixture (2021)

How to Sort Them? A Network for LEGO Bricks Classification

Tomasz Boiński$^{(\boxtimes)}$ ⓘ, Konrad Zawora, and Julian Szymański ⓘ

Faculty of Electronics, Telecommunication and Informatics, Gdańsk University
of Technology, 11/12 Narutowicza Street, 80-233 Gdańsk, Poland
{tomboins,julszyma}@pg.edu.pl

Abstract. LEGO bricks are highly popular due to the ability to build
almost any type of creation. This is possible thanks to availability of
multiple shapes and colors of the bricks. For the smooth build process
the bricks need to properly sorted and arranged. In our work we aim
at creating an automated LEGO bricks sorter. With over 3700 different LEGO parts bricks classification has to be done with deep neural
networks. The question arises which model of the available should we
use? In this paper we try to answer this question. The paper presents a
comparison of 28 models used for image classification trained to classify
objects to high number of classes with potentially high level of similarity. For that purpose a dataset consisting of 447 classes was prepared.
The paper presents brief description of analyzed models, the training
and comparison process and discusses the results obtained. Finally the
paper proposes an answer what network architecture should be used for
the problem of LEGO bricks classification and other similar problems.

Keywords: Image classification · LEGO · Neural networks

1 Introduction

LEGO bricks are highly popular among kids and adults. They can be used to
build vast array of, both very simple and very complex, constructions. This is
achieved by availability of multiple, sometimes very different, yet compatible
brick shapes. For the smooth build process the bricks need to properly sorted
and arranged - constant searching for proper bricks in a big pile of LEGO is
discouraging and limits creativity. Usually the sorting is done by shape. The
colors and decals can be easily distinguished even in a big pail of bricks [2].
Still, with over 3700 different LEGO parts [24] (and the number is constantly
growing) even disregarding the color makes the problem complex.

No solution for this problem was proposed so far. LEGO Group provides only
a simple sorting mechanism, based on the brick size, in form of the 2011 released,
now discontinued, LEGO Sort and Store item. Fan offered solutions usually rely
on optimization of the manual sorting process (e.g. [1]). Some fans tried to build
AI powered sorting machines [10,38] with some success. Independently from the
way of building the sorting machine, it requires a well-trained neural network

© The Author(s), under exclusive license to Springer Nature Switzerland AG 2022
D. Groen et al. (Eds.): ICCS 2022, LNCS 13352, pp. 627–640, 2022.
https://doi.org/10.1007/978-3-031-08757-8_52

able to distinguish between different, often very similar bricks. The solution should at least divide them into smaller number of categories aggregating bricks similar in shape and usage, allowing further manual selection of proper bricks. Thus LEGO oriented object classification solution is needed.

Problems like object detection, image segmentation, content-based image retrieval, or most commonly, object classification lie in domain of computer vision. In the last case the given, previously detected object, is assigned a one or more labels. The objects can have either one label assigned (multi-class classification) or many labels assigned (multi-label classification).

Computer vision is an actively research sub-domain of machine learning. It originated as far as in late 60ties of the 20-th century [27]. What was at the beginning portrayed as a simple task, assigned to students in summer school, currently remains a complex and not yet fully solved problem.

Across the recent years multiple deep neural network architectures emerged. For their comparison a standardised approach was established - *ImageNet Large Scale Visual Recognition Challenge* (ILSVRC) competition [29]. During the competition the models should classify objects to one of the 1000 classes based on 1.2 million of training images. The model accuracy is tested on 150000 images. Two metrics are calculated – Top1 (the percentage of directly correctly classified images) and Top5 (the percentage of images that were classified among the 5 with the highest probability). There are other commonly used datasets like CIFAR-10 and CIFAR-100 [20], SIFT10M [8], Open Images Dataset [19,21], Microsoft Common Objects in Context (COCO) [23]. As each dataset contains photos from different categories, with different size etc., good standing with one of the datasets does not guarantee the same results with the other. Furthermore the datasets try to be very general whereas in some cases the images contain similar objects. That is why further evaluation is still required.

In our research we undertook construction of AI-powered sorting machine [6] treating LEGO recognition as multi-class classification. To search for the best architecture that matches our scenario we decided to base our dataset in that prepared for ILSVRC. This way we could speed up training process thanks to *transfer learning* approach. As candidate architectures we selected the ones that achieved the best results in the aforementioned competition.

The structure of this paper is as follows. In Sect. 2 a description of compared network topologies is given. Later on, in Sect. 3, the used dataset is presented. Further in Sect. 4 details how the training was done and the testing methodology are presented. Section 5 discusses results obtained during the tests. Finally, some conclusions are given.

2   Network Topologies

In this paper we tested 28 network topologies from 7 families:

- EfficientNet – EfficientNetB0, EfficientNetB1, EfficientNetB2, Efficient-NetB3, EfficientNetB4, EfficientNetB5, EfficientNetB6 and EfficientNetB7 variants,

- NASNet – NASNetMobile and NASNetLarge variants,
- ResNet – ResNet50, ResNet50V2, ResNet101, ResNet101V2, ResNet152 and ResNet152V2 variants,
- MobileNet – MobileNet, MobileNetV2, MobileNetV3Large and MobileNetV3-Small variants,
- Inception – InceptionV3, InceptionResNetV2 and Xception variants,
- DenseNet – DenseNet121, DenseNet169 and DenseNet201 variants,
- VGG – VGG16 and VGG19 variants.

In this section a brief introduction to each family and variant is given, portraying its strengths and rationale behind the used architecture.

EfficientNet architecture was defined in 2019 [37]. The model aims at efficient scaling of convolutional deep neural networks. The authors distinguished three dimensions of scaling: depth scaling, width scaling and resolution scaling. Depth scaling is the most commonly used approach, as it allows increase in number and complexity of detected features by increasing the number of convolutions. However, with increasing network depth, the training process gets longer and a problem of vanishing gradient can be observed [13]. Width scaling relies on increase of number of channels in each convolution. It is commonly used in shallow networks, where width scaling increased both training speed and classification quality [39]. Resolution scaling allows potential extraction of additional features. With all three scaling approaches there is a point of diminishing returns, beyond which additional computational overhead is not being compensated by better accuracy. EfficientNet uses so-called compound scaling, where all three parameters are equally scaled using ϕ parameter.

The base model here is similar to MnasNet [36] and MobileNetV2 [30]. Each model in this family differs by the ϕ parameter value (starting with $\phi = 0$).

Care needs to be taken when using the model in TensorFlow framework [31], as zero-padding is used for convolutions with resolutions that cannot be divided by 8. The number of channels also needs to be divisible by 8. The real compound scaling parameters applied when using TensorFlow are thus different.

ResNet50 was proposed in 2015 [11], as a solution to vanishing and exploding gradient problems. Thanks to so-called residual connections, it allows training of very deep networks (over 1000 convolutional layers). Residual connections perform elementwise addition of identity function between convolution blocks. This improves gradient flow, by skipping non-linear activation functions usually placed in convolutional blocks.

In 2016 a revision of the original model was proposed (called ResNet V2) [12]. The whole family of this model (in both ResNet and ResNet V2 revisions) achieves very high results in ILSVR competition reaching 74.9%–78% accuracy in Top1 and 92.1%–94.2% accuracy in Top5 categories.

DenseNet was defined in 2016 [16]. Similarly as in ResNet, the aim is to solve the vanishing gradient by shortening its flow path. DenseNet uses so-called *dense blocks* to achieve it. The dense block consists of 1×1 and 3×3 blocks and output of every block within it is connected with input of every next block. Each layer within a dense block has thus direct access to its output which limits the flow path. DenseNet has also low width of the convolutional layers. Each variant of DenseNet architecture differs in terms of size of the last two dense blocks.

In 2018 DenseNet achieved the highest score in ILSVR competition Top1 category reaching accuracy of 75% for DenseNet-121, 76.2% for DenseNet-169, 77.42% for DenseNet-201 and 77.85% for DenseNet-264 77.85%.

Inception architecture was defined in 2014 [34] with Inception v1/GoogLeNet. The aim was to reduce the risk of overfitting and eliminate the problems with gradient flow. A special Inception block was proposed - it is composed of three layers with different filters (1×1, 3×3 and 5×5). This led to high calculation complexity so a reduction was introduced that limited the number of entry channels. 9 Inception v1 blocks were combined as GoogLeNet architecture.

Inception v2 and v3 were defined in 2015 [35]. They increased performance, limiting information loss and computational complexity. Inception v3 achieves 77.9% accuracy in ILSVR competition Top1 category and 93.7% in Top5.

In 2016 Inception v4, InceptionResNetV1 and InceptionResNetV2 architectures were proposed [33]. The main goal was simplification and unification of the Inception models. ResNet residual connections were also included in the model. The best results were obtained by InceptionResNetV2 model. In Top1 category of the ILSVRC competition it achieved accuracy of 80.3% and in Top5 95.3%.

In 2017 an extension to Inception V3, by replacing the inception block with so-called extreme inception, was defined [7]. The original block was modified so that for each 1×1 convolution output corresponds one 3×3 convolution. This architecture, called Xception, proved to be easier to define and modify in software frameworks than the original Inception model.

Xception achieved better results in ILSVR competition than the original Inception v3 model. For Top1 category the accuracy was 79% and for Top5 94.5%. It also had less parameters (22.86 million vs 23.63 million).

NASNet model was defined in 2017 [41]. It was created thanks to Google AI's *AutoML* [28] and Neural Architecture Search [40]. The creation of optimal network architecture is treated here as reinforcement learning problem, with the final network accuracy as a reward. This induced a very high computational cost, so the search space had to be narrowed considerably. Based on the analysis of other models the authors first defined a general architecture, which composed of only 2 blocks - *normal cell* and *reduction cell*.

This significantly reduced the time needed to find the optimal model. Still, the training time remained very long. However, the model achieved good results. For ILSVRC Top 1 category it reached accuracy of 74.4% and 82.5% for smaller NASNetMobile variant, and larger NASNetLarge variant respectively.

MobileNet model was defined in 2017 [15]. It was designed to allow fast inference on mobile and embedded devices. The authors of this solution point out that after a certain level of network complexity, the increase in inference time is much bigger than the increase in accuracy, making the potential gain computationally unprofitable. To further increase the performance of inference, authors defined a special convolution, called *depthwise separable convolution*. It separates the operation into two phases - filtering and combination. This approach allowed up to 9 times lower computational complexity with only a 1% lower accuracy [15] (for ILSVRC Top 1 category).

Few versions of MobileNet architecture were proposed, each introducing usage of different approaches (like residual connections) or different numbers of channels. The original MobileNet model achieved for Top1 category of ILSVRC competition the accuracy equal to 70.6%. MobileNetV2 [30] achieved 72.0% with around 20% lower number of parameters and 47% lower computational cost. MobileNetV3 [14] introduced 2 versions - Small with 2.5 million parameters and Large with 5.4 million parameters. The accuracy for Top 1 category of ILSVRC competition was 75.2% for MobileNetV3Large and 67.4% for MobileNetV3Small.

VGG is one of the oldest architectures, was defined in 2014 [32]. Different variants of this model vary by the number of trainable layers. For Top 1 category of ILSVRC competition, VGG16 and VGG19 reach accuracy of 71.3%. For Top5 category, VGG16 reaches accuracy of 90.1%, whereas VGG19 of 90%.

As we can see, all of the aforementioned models achieved very good result in the ILSVRC competition. At the time of their publication they gained the highest score and usually became the state of the art. As mentioned in Sect. 1 it doesn't always translate to the same results for other datasets.

3   The Dataset

During the training we used custom dataset containing both real photos and renders of LEGO bricks, belonging to 447 classes. The bricks were taken from authors personal collection of over 150 LEGO sets and represents the most commonly available brick shapes. The whole dataset consists of 620082 images, where 52601 were real photos and 567481 were life-like renders. The renders were created using Blender tool [9] based on 3D models from LDraw library [17].

The renders were used to speed up data gathering. We created a script that randomly selected a brick type, color and alignment simulating its move on a conveyor belt below a fixed positioned camera. Thanks to Blender and its extension called ImportLDraw [26] we managed to generate realistic images of LEGO bricks. Sample renders, after being cropped, can be seen in Fig. 1.

Real photos were created to increase the representativeness of the training set. For that we created a dedicated Android app allowing quick tagging and automatic cropping of LEGO bricks on pictures taken with phone camera. Sample real photos can be seen in Fig. 2.

The full set of rendered images (before cropping) and real photos are publicly available – [5] and [3] respectively. The complete dataset is also available [4].

Before the training the dataset was prepared so that all networks would be trained on the same images. The images need to be standardised in terms of size and proportions. As some of the bricks are long and narrow (e.g. brick 3002), we decided to scale the longer edge to the desired size, and the shorter edge proportionally (otherwise we could loose some information). Then, the image canvas was extended to form a square and was filled with white background. Next, all images were augmented using imgaug library [18]. The transformation included the following operations applied with 50% probability:

Fig. 1. Sample renders for brick number 3003

Fig. 2. Sample real photos of brick number 3003

- scaling to randomly selected size (80%–120% of the original size),
- random rotation between −45° and 45°,
- random shift by up to 20%,
- random transformation into a trapezoid with an angle of up to 16°.

Next, 5 randomly selected operations were applied, from the following list:

- Gaussian, median or averaged blur with a random intensity,
- sharpening filter with random blending factor and brightness,
- emboss filter with random blend factor and brightness,
- superimpose the contours detected by the edge detection filter, with a probability of 50%,
- Gaussian noise of random intensity,
- dropout of random pixels or a group of pixels,
- inversion of every image channel, with probability of 5%
- addition of a random value to each pixel,
- random brightness change of the image,
- random contrast change of the image,
- generation of a grayscale image and overlaying it with random transparency over the original photo.

The augmentations were done once, so that the results will be comparable. During the training process, to reduce the risk of overfitting, we performed additional augmentation before each epoch - the images were rotated by random angle up to 15° and the contrast was changed by random value (up to 10%).

The data gathered were divided into training and validation sets. The training set contains 447000 images (1000 images each class, 650 renders and 350 real photos). The validation set contained 44700 images (100 images each class, 50 renders, 50 real photos). The numbers were obtained experimentally.

The test set consisted of real photos created independently. It contains 4000 images of bricks belonging to 20 classes (200 images each). The set was created in separate session using bricks from other set (Lego Creative Box Classic – 10698). We used 2 variants of the set - easy and hard. Both have the same number of photos, however the hard set contained images that are hard or even impossible to distinguish but belongs to different classes (e.g. bricks 3001 and 3010).

4 Training Process

All models presented in Sect. 2 were trained using transfer learning approach. It consisted of 2 phases:

- pre-training – done with the base model locked, only the newly added top layers are trained,
- fine tuning – the base model was partially or completely unlocked, all unlocked layers could be trained.

Pre-training is characterised by a high learning rate (we've used 0.01) with relatively low computation cost, as the backward pass needs to be calculated only for the newly added layers. After this stage, we could observe Top1 accuracy for the 447 LEGO classes at around 50–70%. During the fine-tuning stage, some of the layers are unlocked and the training is repeated for those layers. The problem here is how many layers should be unlocked. If the number will be low, then the training process will be faster, but we might not get to the desired accuracy. The number of unlocked layers also depends on the initial size of the base model.

We aimed at comparing different architectures so we designed adaptive fine tuning algorithm. It goes as follows:

1. `N := 0`
2. `N := N + unfreeze_interval`
3. `top1_history := []`
4. Unlock `N` top layers and recompile the model
5. Perform 1 training epoch
6. Perform 1 validation epoch
7. Add the Top1 accuracy on the validation set to `top1_history` list
8. If `top1_history` contains no less than `patience` elements and the Top1 accuracy on validation set did not increase by at least `min_delta` during last `patience` epochs, go to step 2
9. If `top1_history` contains `max_epochs_per_fit` elements, go to step 2.
10. Go to step 5.

where:

- `unfreeze_interval` (15 by default) – the number of layers to be unlocked within on fine-tuning iteration,
- `max_epochs_per_fit` (50 by default) – max number of training epochs in one fine-tuning iteration,
- `patience` (5 by default) – the number of epochs in one iteration, after which the model quality is evaluated,
- `min_delta` (0.01 by default) – minimal requested Top1 accuracy improvement reached in `patience` epochs .

The aforementioned algorithm was run to train each model for limited time. To increase the training speed and limit memory footprint we used so called *mixed precision training* [25] and XLA [22] compiler. Both approaches allowed us to train the networks with larger batch sizes.

For fine tuning we've used `learning_rate` = 0.0001. Both phases were trained using categorical cross-entropy loss function and Adam optimizer. All networks were trained with `batch_size` = 128, except for EfficientNetB5, EfficientNetB6 and EfficientNetB7, which used 64, 32 and 32 respectively.

5   The Results

In total 28 network topologies were tested. The comparison process was divided into two stages. First, all models were trained for four hours using adaptive fine tuning approach described in Sect. 4. The second stage lasted twelve hours. It was done with the same approach as stage 1, but only 5 best models and the best out of each family was trained. All tests were done on dual Intel Xeon Gold 6130 server with 256 GiB RAM and dual NVIDIA GeForce RTX 2080 (8 GiB GDDR6 RAM each) GPU cards. Each training was done on single GPU (two models were trained at once). The default batch size was 128. Due to the memory constraints some models used smaller batch size, namely: EfficientNetB5 (64), EfficientNetB6 (32) and EfficientNetB7 (32). In both stages we used transfer learning, where for the first stage we used a model trained on ImageNet data.

5.1   Stage I - The Four-Hour Training

Summary of obtained results (ranked from best to worst) are presented in Table 1. The best model in each family is marked with bold font.

EfficientNet models achieved varied results. The best variant was EfficientNetB1. It reached 84.4% Top1 accuracy and 95.85% Top5 accuracy, giving it the 8th place. EfficientNetB3 and EfficientNetB0 got slightly worse results, whereas EfficientNetB7 and EfficientNetB6 were one of the worst models. The reason for such outcome was the compound scaling which caused small number of frames (images) processed in the give time frame. This led to relatively small number of epochs and thus lower accuracy. The differences between EfficientNet variant are sustainable. For the next stage only EfficientNetB1 variant was selected.

Table 1. Results after the first stage (measured on the validation set)

Model	Top1 accuracy	Top5 accuracy	Epochs run	Training time
VGG16	**92.99%**	**99.00%**	**11**	**04:01:07**
VGG19	91.70%	98.64%	11	04:05:07
ResNet50	**87.40%**	**96.96%**	**14**	**04:04:07**
ResNet101V2	87.20%	96.94%	14	04:10:42
ResNet152V2	86.92%	96.74%	15	04:11:10
ResNet50V2	86.57%	96.72%	14	04:05:15
ResNet152	86.00%	96.28%	13	04:09:27
EfficientNetB1	**84.40%**	**95.85%**	**15**	**04:12:06**
ResNet101	84.09%	95.44%	14	04:16:57
EfficientNetB3	83.58%	95.63%	10	04:14:06
EfficientNetB0	82.87%	95.03%	15	04:15:20
MobileNetV3Large	**82.32%**	**94.80%**	**17**	**04:09:28**
Xception	**82.31%**	**94.95%**	**9**	**04:27:31**
EfficientNetB2	81.33%	94.50%	11	04:04:16
InceptionResNetV2	79.43%	93.90%	7	04:01:10
MobileNet	78.56%	94.04%	14	04:09:26
DenseNet201	**77.41%**	**92.18%**	**14**	**04:01:57**
MobileNetV2	75.75%	92.29%	16	04:06:31
DenseNet169	75.44%	91.38%	13	04:14:47
DenseNet121	74.01%	90.49%	14	04:06:56
MobileNetV3Small	73.07%	89.97%	17	04:13:45
InceptionV3	71.19%	89.22%	10	04:16:48
EfficientNetB5	66.72%	87.58%	3	04:14:28
EfficientNetB4	65.60%	86.67%	6	04:25:51
NASNetMobile	**59.60%**	**82.08%**	**16**	**04:11:54**
EfficientNetB6	58.34%	82.15%	2	04:35:13
EfficientNetB7	54.38%	78.60%	1	04:03:32
NASNetLarge	53.39%	77.89%	4	04:26:13

ResNet models achieved very good results. The best variant was ResNet50 reaching 87.40% Top1 and 96.96% Top5 accuracy. The other variants achieved similar results. 3 models were selected: ResNet50, ResNet101V2 and ResNet152V2.

Inception models reached mediocre results. The best one was Xception reaching 82.31% Top1 and 94.95% Top5 accuracy. All models finished pre-training stage and reached the fine-tuning phase. However, we observed very low performance in terms of processed images per second, which might have been the cause of mediocre accuracy. Thus only the Xception model was selected.

DenseNet models did not perform too well. The best results were obtained by DenseNet201 variant (77.41% Top1 and 92.18% Top5 accuracy). Contrary to other models, the poor quality did not come from performance problems. DenseNet training showed one of the highest images per second rate. The problem lies in low increase of accuracy between epochs. We suspect it is caused by the design of DenseNet, specifically the concatenation operation. Unlike other tested architectures, in DenseNet, last convolutional layers are just a small part of the final feature map. By tuning a small amount of top convolutional layers, we're potentially leaving a big part of the feature map intact. This could be fixed by changing the training methodology and training all convolutional layers, but it has not been attempted in this phase.

NASNet models also got poor results. The best one, NASNetMobile, reached 59.60% Top1 and 82.08% Top5 accuracy placing 25 out of 28 tested models. Once again performance was the reason for the results. For the second stage only NASNetMobile was selected.

MobileNet scored averagely, the best variant being MobileNetV3Large reaching 82.32% Top1 and 94.8% Top5 accuracy. This variant, despite being targeted for mobile devices, outperforms deeper models like Xception or DenseNet201 thanks to the highest images per second rate and thus the highest training performance. For the second stage MobileNetV3Large was selected.

The best results were obtained by the VGG network variants - VGG16 placed first (with 92.99% Top1 and 99% Top5 accuracy) and VGG19 placed second (with 91.70% Top1 and 98.64% Top5 accuracy). The results came unexpected, as this is the oldest tested architecture. During the ILSVRC competition it was outperformed over the years by all other tested models, with exception of some MobileNet variants. The VGG are relatively shallow, but very wide. This allows fast unlocking of many layers in the fine-tuning approach and thus leads to very fast learning times. For the second stage both VGG16 and VGG19 were selected.

5.2   Stage II - The Twelve-Hour Training

During this stage 10 models were further trained. The aggregated results (ranked from best to worst) can be seen in Table 2.

All models managed to get better results. In most cases (except NASNetMobile and DenseNet201) twelve-hour limit was sufficient to achieve convergence.

During this stage, we observed the similar results as in the previous one. Once again, VGG16 and ResNet50 proved to be the best. However, the quality difference, both in Top1 and Top5 accuracy, between models that reached convergence is not big - the biggest difference is only 2.18% points. This is true even for mobile models, like MobileNetV3Large. This network required however more epochs to reach convergence.

What came as a surprise is that, once again, VGG16 model achieved the best results. In ILSVRC competition this model is outperformed by every other non-mobile approach presented in this paper. In the problem presented here

Table 2. Result after the second stage (measured on the validation set)

Model	Top1 accuracy	Top5 accuracy	Epochs run	Training time	Parameters
VGG16	94.56%	99.21%	31	12:13:49	138.3M
ResNet50	93.81%	99.10%	41	12:13:23	25.6M
ResNet101V2	93.19%	98.77%	45	12:08:03	44.6M
MobileNetV3Large	92.65%	98.68%	48	12:02:37	5.4M
VGG19	92.62%	98.79%	29	12:26:21	143.6M
Xception	92.49%	98.69%	27	12:06:12	22.9M
ResNet152V2	92.45%	98.54%	40	12:02:22	60.3M
EfficientNetB1	92.38%	98.51%	37	12:19:43	7.8M
DenseNet201	85.26%	95.85%	41	12:07:23	20.2M
NASNetMobile	78.59%	93.46%	41	12:09:57	5.3M

(distinguishing LEGO bricks), VGG16 model trains very fast and reaches superb accuracy. This model is, however, characterised by high number of parameters and thus costly in terms of calculation time both at the time of training and inference. For practical application, the second model, ResNet50, might be thus a better choice, as it has Top1 accuracy lower only by 0.75% point, while 5.4 times lower the number of parameters. This model might also be a better choice after extending the training set with images representing other LEGO bricks, that currently are not taken into consideration (and thus extending the number of classes almost tenfold).

Very good results were also obtained by a mobile-oriented models, especially MobileNetV3Large, which had only 1.91% point lower Top1 accuracy than VGG16 model. Furthermore, it contains only 5.4 million parameters (in contrast to 138.3 million for VGG16). Thus in applications where computing performance is scarce, MobileNetV3Large should be used over any more complicated model. Despite its size, it outperforms in terms of accuracy other, more complicated models (VGG19, Xception, ResNet152V2 and EfficientNetB1).

DenseNet201 and NASNetMobile did not reach convergence in the twelve-hour time limit and thus did not achieve good results. DenseNet201 suffered from overfitting and NASNetMobile had very slow accuracy increase and would require much longer training time.

5.3   Final Tests

We performed some final tests on the two best models. The results for the easy and hard sets are presented in Table 3. As can be seen, both models reach similar accuracy.

To test the models in real life application we implemented a mobile app which took photos of LEGO bricks laying on a white background and combined it with the pre-trained models. VGG16 correctly recognized 39 out of 40 bricks. Wrongly labeled 822931 brick was classified as 3003 due to their similarity from the camera

Table 3. VGG16 and ResNet50 accuracy for easy and hard tests

	Easy set		Hard set	
	Top1 accuracy	Top5 accuracy	Top1 accuracy	Top5 accuracy
VGG16	92.37%	99.02%	86.08%	98.22%
ResNet50	90.40%	99.05%	86.30%	98.60%

perspective. ResNet50 correctly classified all bricks. The networks were tested in different conditions. We used an intensive pink light to illuminate the test environment. This made the background pink and most of the bricks appeared as having different, not seen before, colors. VGG16 correctly recognized 37 bricks out of 40. ResNet50 model once again correctly classified all of the bricks.

6   Conclusions

The paper presents extensive analysis of deep neural network architectures in order to verify their suitability for classification of LEGO bricks. The problem is characterized with the need to distinguish objects between multiple, often similar classes, as there are over 3700 different LEGO brick shapes. For this purpose, a new dataset was created containing 447 classes and a set of tools automating the analysis process were implemented. In total, 28 network architectures, belonging to 7 families, were analyzed and compared. For the comparison, we used our proposed training algorithm with adaptive fine-tuning approach.

Results showed that VGG16 model proved to be the best with its Top1 accuracy of 94.56% and Top5 accuracy of 99.21%). Surprisingly, in ILSVRC competition this model was outperformed by other solutions. The model is characterized, however, with very big number of parameters (138.3 million) and high number of floating point operations during training and inference process (15.3 GFLOPs). Not falling far behind was ResNet50 model (Top1 93.81%, Top5 99.10%) which had lower parameter count (25.6 million) and required far lower system performance (3.87 GFLOPs). In many cases, this might be the best choice for similar problems, where there are a lot of similar objects to classify. Surprisingly, also the smaller, mobile models proved to be worthwhile. MobileNetV3Large achieved very good accuracy (Top1 92.65%, Top5 98.68%), with very low parameter count (5.4 million) and low performance requirements (0.21 GFLOPs).

The two best models also were tested in real life application. They proved to be very accurate in both synthetic test on predefined test sets and during live classification of LEGO bricks.

In the near future we plan on extending the dataset with additional classes to cover as much LEGO brick shapes as possible to provide a deep neural network able to classify any type of LEGO bricks. Such network could be used in LEGO sorting machines, software recommending constructions based on the bricks available, automatic brick database creation and many more.

Acknowledgment. The authors would like to thank Bartosz Śledź and Sławomir Zaraziński for help with part of the implementation and dataset creation.

References

1. Adam: LEGO sorting chart (2019). https://go.gliffy.com/go/publish/12232322. Accessed 24 Mar 2022
2. Aplhin, T.: The LEGO storage guide (2020). https://brickarchitect.com/guide/. Accessed 34 Mar 2022
3. Boiński, T.: Images of LEGO bricks (2021). https://doi.org/10.34808/xz76-ez11. Accessed 24 Mar 2022
4. Boiński, T., Zaraziński, S., Śledź, B.: LEGO bricks for training classification network (2021). https://doi.org/10.34808/3qfs-rt94. Accessed 24 Mar 2022
5. Boiński, T., Zawora, K., Zaraziński, S., Śledź, B., Łobacz, B.: LDRAW based renders of LEGO bricks moving on a conveyor belt (2020). https://doi.org/10.34808/jykr-8d71. Accessed 24 Mar 2022
6. Boiński, T.M.: Hierarchical 2-step neural-based LEGO bricks detection and labeling. In: Proceedings of 37th Business Information Management Association Conference, pp. 1344–1350 (2021)
7. Chollet, F.: Xception: Deep learning with depthwise separable convolutions (2017)
8. Dua, D., Graff, C.: UCI machine learning repository (2017). http://archive.ics.uci.edu/ml. Accessed 22 Nov 2021
9. Foundation, T.B.: Blender (2002). https://www.blender.org/. Accessed 22 Nov 2021
10. Garcia, P.: LEGO Sorter using TensorFlow on Raspberry Pi (2018). https://medium.com/@pacogarcia3/tensorflow-on-raspbery-pi-lego-sorter-ab60019dcf32. Accessed 22 Nov 2021
11. He, K., Zhang, X., Ren, S., Sun, J.: Deep Residual Learning for Image Recognition (2015)
12. He, K., Zhang, X., Ren, S., Sun, J.: Identity mappings in deep residual networks (2016)
13. Hochreiter, S.: The vanishing gradient problem during learning recurrent neural nets and problem solutions. Internat. J. Uncertain. Fuzziness Knowl. Based Syst. **6**(02), 107–116 (1998)
14. Howard, A., et al.: Searching for mobilenetv3 (2019)
15. Howard, A.G., et al.: MobileNets: efficient convolutional neural networks for mobile vision applications (2017)
16. Huang, G., Liu, Z., van der Maaten, L., Weinberger, K.Q.: Densely Connected Convolutional Networks (2018)
17. Jessiman, J.: LDraw. http://www.ldraw.org (1995). Accessed 22 Nov 2021
18. Jung, A.: imgaug source code. https://github.com/aleju/imgaug. Accessed 22 Nov 2021
19. Krasin, I., et al.: Openimages: a public dataset for large-scale multi-label and multi-class image classification. Dataset available from https://storage.googleapis.com/openimages/web/index.html (2017)
20. Krizhevsky, A., Nair, V., Hinton, G.: Cifar-10 and cifar-100 datasets. https://www.cs.toronto.edu/~kriz/cifar.html. Accessed 24 Mar 2022
21. Kuznetsova, A., et al.: The open images dataset v4: Unified image classification, object detection, and visual relationship detection at scale. IJCV (2020)

22. Li, M., et al.: The deep learning compiler: a comprehensive survey. IEEE Trans. Parallel Distrib. Syst. **32**(3), 708–727 (2021). https://doi.org/10.1109/tpds.2020.3030548
23. Lin, T., et al.: Microsoft COCO: common objects in context. CoRR abs/1405.0312 (2014). http://arxiv.org/abs/1405.0312
24. Maren, T.: 60 fun LEGO facts every LEGO fan needs to know (2018). https://mamainthenow.com/fun-lego-facts/
25. Micikevicius, P., et al.: Mixed precision training (2018)
26. Nelson, T.: ImportLDRaw. https://github.com/TobyLobster/ImportLDraw. Accessed 16 June 2021
27. Papert, S.A.: The summer vision project (1966). https://dspace.mit.edu/handle/1721.1/6125. Accessed 22 Nov 2021
28. Le, Q., Zoph, B., Research Scientists, G.B.t.: Using machine learning to explore neural network architecture (2017). https://ai.googleblog.com/2017/05/using-machine-learning-to-explore.html. Accessed 22 Nov 2021
29. Russakovsky, O., et al.: ImageNet large scale visual recognition challenge. Int. J. Comput. Vision **115**(3), 211–252 (2015). https://doi.org/10.1007/s11263-015-0816-y
30. Sandler, M., Howard, A., Zhu, M., Zhmoginov, A., Chen, L.C.: Mobilenetv 2: Inverted residuals and linear bottlenecks (2019)
31. Shukla, N., Fricklas, K.: Machine learning with TensorFlow. Manning Shelter Island, Ny (2018)
32. Simonyan, K., Zisserman, A.: Very Deep Convolutional Networks for Large-Scale Image Recognition (2015)
33. Szegedy, C., Ioffe, S., Vanhoucke, V., Alemi, A.: Inception-v4, Inception-ResNet and the Impact of Residual Connections on Learning (2016)
34. Szegedy, C., et al.: Going Deeper with Convolutions (2014)
35. Szegedy, C., Vanhoucke, V., Ioffe, S., Shlens, J., Wojna, Z.: Rethinking the Inception Architecture for Computer Vision (2015)
36. Tan, M., Chen, B., Pang, R., Vasudevan, V., Sandler, M., Howard, A., Le, Q.V.: Mnasnet: platform-aware neural architecture search for mobile. In: Proceedings of the IEEE/CVF Conference on Computer Vision and Pattern Recognition, pp. 2820–2828 (2019)
37. Tan, M., Le, Q.V.: EfficientNet: Rethinking Model Scaling for Convolutional Neural Networks (2020)
38. West, D.: LEGO sorting machine (2019). https://twitter.com/JustASquid/status/1201959889943154688. Accessed 08 Feb 2021
39. Zagoruyko, S., Komodakis, N.: Wide residual networks (2017)
40. Zoph, B., Le, Q.V.: Neural architecture search with reinforcement learning (2017)
41. Zoph, B., Vasudevan, V., Shlens, J., Le, Q.V.: Learning Transferable Architectures for Scalable Image Recognition (2018)

Novel Photoplethysmographic Signal Analysis via Wavelet Scattering Transform

Agnieszka Szczęsna[1]([✉]) [ID], Dariusz Augustyn[2] [ID], Henryk Josiński[1] [ID],
Adam Świtoński[1] [ID], Paweł Kasprowski[2] [ID], and Katarzyna Harężlak[2] [ID]

[1] Department of Computer Graphics, Vision and Digital Systems,
Faculty of Automatic Control, Electronics and Computer Science,
Silesian University of Technology, Akademicka 16, 44-100 Gliwice, Poland
agnieszka.szczesna@polsl.pl
[2] Department of Applied Informatics, Faculty of Automatic Control, Electronics and
Computer Science, Silesian University of Technology, Akademicka 16,
44-100 Gliwice, Poland

Abstract. Photoplethysmography (PPG) is a non-invasive optical technique, applied in clinical settings to measure arterial oxygen saturation. Using modern technology, PPG signals can be measured by wearable devices. This paper presents a novel procedure to study the dynamics of biomedical signals. The procedure uses features of a wavelet scattering transform to classify signal segments as either chaotic or non-chaotic. To this end, the paper also defines a chaos measure. Classification is made using a model trained on a dataset consisting of signals generated by systems with known characteristics. Using an example PPG signal, this paper demonstrates the usefulness of the wavelet scattering transform for the analysis of biomedical signals, and shows the importance of correctly preparing the training set.

Keywords: Wavelet scattering transform · Chaos · PPG · Classification · Biomedical signals

1 Introduction

Photoplethysmography (PPG) is an non-invasive optical measurement technique. By using a light source to illuminate skin tissue, either the transmitted or reflected light intensity is collected by a photodetector to record the photoplethysmogram. Traditionally, PPG signals have been recorded using red or near infra-red light. In recent years, green light has been used for wearable devices, such as wristbands and smartwatches, to provide highly usable and accessible

This publication was supported by the Department of Graphics, Computer Vision and Digital Systems, under statue research project (Rau6, 2022), Silesian University of Technology (Gliwice, Poland).

D. Groen et al. (Eds.): ICCS 2022, LNCS 13352, pp. 641–653, 2022.
https://doi.org/10.1007/978-3-031-08757-8_53

daily health monitoring [8,10,12,16,31]. As such, a proper understanding of green light PPG is of critical importance. Recorded light intensity variations have traditionally been associated with blood volume pulsations in the microvascular bed of the tissue. The tissue penetration depth of green light is approximately 530 nm. The source of the chaotic properties of PPG is unclear. Such properties could originate in the upper layers of the skin, due to changes in capillary density caused by arterial transmural pressure, or in deeper layers, due to changes in vessel blood volume [27].

Despite uncertainty concerning the mechanisms of PPG, the technique is generally accepted to provide valuable clinical information about the cardiovascular system. PPG signals are used to monitor pulse rate, heart rate, oxygen saturation, blood pressure, and blood vessel stiffness [2,7,13,15,17,21–23,28]. Unfortunately, such signals are often corrupted by noise, motion artifacts, and missing data.

Biological signals contain deterministic and stochastic components, both of which contribute to the underlying dynamics of the physiological system. All biological signals contribute information on the underlying physiological processes. Therefore, by studying such signals, the physiological systems that generate them can be better understood.

In early studies, PPG as well as ECG (electrocardiogram) and HRV (heart rate variability) were claimed to be chaotic mostly based on the results of time-delay reconstructed trajectory, correlation dimension and largest Lyapunov exponent [29]. Subsequently, with the development of nonlinear time series analysis methods for real-world data, further evidence of the chaotic nature of such biological signals has emerged. However, many tools that were previously thought to provide clear evidence of chaotic motion have been found to be sensitive to noise and prone to producing misleading results. Thus, controversy remains concerning the topic of chaos in biological signals [11,26,27].

Sviridova and Sakai [26] applied nonlinear time series analysis methods to PPG signals to identify the unique characteristics of the underlying dynamical system. Such methods included time delay embedding, largest Lyapunov exponent, deterministic nonlinear prediction, Poincaré section, the Wayland test, and the method of surrogate data. Results demonstrated that PPG dynamics are consistent with the definition of chaotic motion, and the chaotic properties were somewhat similar to Rössler's single band chaos with induced dynamical noise.

A more recent approach to signal analysis is the use of machine learning or deep learning methods. Such methods can generalize knowledge acquired from a training dataset, and apply it to the analysis of a testing dataset. Boullé et al. [3] used a deep neural network to classify univariate time series' generated by discrete and continuous dynamical systems based on the presence of chaotic behavior. The study suggests that deep learning techniques can be used to classify time series' obtained by real-life applications into chaotic or non-chaotic.

De Pedro-Carracedo et al. [9] found that the dynamics of a PPG signal were predominantly quasi-periodic over a small timescale (5000 data points 250 Hz).

Over a longer timescale (600000 data points 250 Hz), more diverse and complex dynamics were observed, but the signal did not display chaotic behavior. This analysis used a deep neural network to classify the PPG signals. The following dynamics classes were defined: periodic, quasi-periodic, non-periodic, chaotic, and random. Unfortunately, the dataset used to train the network contained only one system for each class. Given that chaotic systems are difficult to generalize [6], this is not sufficient to accurately classify the dynamics of the real-life signal.

De Pedro-Carracedo et al. [20] applied a modified 0–1 test to the same PPG time series' as the above study. They also found that the majority of PPG signals displayed quasi-periodic behavior across a small timescale, and that as the timescale increased the dynamics became more complex, due to the introduction of additional cardiac rhythm modulation factors. Under specific physiological conditions, such as stress, illness, or physical activity, a transition from quasi-periodicity to chaos can be possible. This phenomenon provides the motivation for measuring the presence of chaos within PPG signals under various conditions.

The objective of this study is to analyze the dynamics of PPG signals during different everyday activities. We propose a novel approach to classify signals using features of a wavelet scattering transform (WST) and a support vector machine (SVM) classifier. This approach was simplified by defining only two classes of signals: chaotic and non-chaotic. Compared to previous research, the training data was prepared in greater detail, and included noise, which was omitted in previous works.

Wavelet analysis provides a unifying framework for the description of many time series phenomena [25]. Introduced by Mallat [18], WST has a similar architecture to convolutional neural network. Despite requiring no parameter learning, WST performs strongly, particularly in constrained classification tasks. WST is a cascade of complex wavelet transforms and modulus non-linearities. At a chosen scale, averaging filters provide invariance to shifts and deformations within signals [1]. Hence, WST can be applied accurately and efficiently to small datasets, whereas convolutional neural network require a large amount of training data. Consequently, WST features possess translation invariance, deformation, stability, and high-frequency information [4]. As such, WST is highly suitable feature extractors for non-linear and non-stationary signals, and has been widely used in audio, music, and image classification.

Moreover, WST is often used to analyze time series', including biomedical signals. By inputting WST features to an SVM classifier, electroencephalography signals were correctly classified as belonging to alcoholic or non-alcoholic patients [5]. In addition, a WST was used to classify heart beats based on ECG signals, with an accuracy of 98.8–99.6%. Jean Effil and Rajeswari [14] used a WST and a deep learning long short-term memory algorithm to accurately estimate blood pressure from PPG signals.

2 Materials and Methods

2.1 The Wavelet Scattering Transform

The wavelet transform is convolutions with dilated wavelets. For 2D transformations, the wavelets are also rotated. Being localized waveforms, wavelets are stable to deformations, unlike Fourier sinusoidal waves. A scattering transform creates nonlinear invariants using wavelet coefficients with modulus and averaging pooling functions. Such transforms yield representations that are time-shift invariant, robust to noise, and stable to time-warping deformations. These attributes are highly useful for many classification tasks, and wavelet transforms are the most common method applied to limited datasets. Andén and Mallat [1] provide a brief overview of the key properties of scattering transforms, including stability to time-warping deformation and energy conservation, and describe a fast computational algorithm.

The WST consists of three cascading stages. In the first stage, the signal x undergoes decomposition and convolution with a dilated mother wavelet ψ of center frequency λ, giving $x*\psi_\lambda$. Following this, the convolved signal is subjected to a nonlinear modulus operator, which typically increases the signal frequency and can compensate for the loss of information due to down sampling. Finally, a time-average/low-pass filter in the form of a scaling function ϕ is applied to the absolute convolved signal, giving $|x*\psi_\lambda|*\phi$.

The zero-order scattering coefficients S_0 describe the local translation invariance of the signal:

$$S_0 = x * \phi.$$

At each level, the averaging operation causes the high-frequency parts of the convolved signal to be lost. These parts can be recovered via the convolution of the signal with the wavelet in the following level.

The first-order scattering coefficients S_1 are therefore defined as the average absolute amplitudes of wavelet coefficients for any scale $1 \leq j \leq J$, over a half-overlapping time window of size 2^j:

$$S_1 = |x * \psi_{\lambda_1}| * \phi.$$

The second-order scattering coefficients S_2 are calculated by repeating the above steps:

$$S_2 = ||x * \psi_{\lambda_1}| * \psi_{\lambda_2}| * \phi.$$

The higher order wavelet scattering coefficients can be calculated by iterating the above process.

The scattering coefficients for each level of the wavelet scattering transform are obtained by processing the defined constant-Q filter bank, where Q is the number of wavelets per octave. Each level can have a filter bank with different Q parameters.

During implementation we used the MATLAB (version R2021b) *waveletScattering* function. The two-layer WST was obtained using Gabor wavelet. For the first and second levels $Q_1 = 8$ and $Q_2 = 1$ respectively. The transform is invariant to translations up to the invariance scale, which is set to half of the signal length in the default implementation. The scaling function determines the duration of the invariant in time. Moreover, the invariance scale affects the spacing of the wavelet center frequencies in the filter banks. The output $R^{paths \times windows \times signals}$ is a feature tensor. This tensor was reshaped into a matrix which is compatible with the SVM classifier. The columns and rows of the matrix correspond to scattering paths and scattering time windows respectively. This results in a feature matrix of $signals \cdot windows$ rows and $paths - 1$ columns.

The zero-order scattering coefficients are not used. Given that multiple scattering windows are obtained for each signal, repeated labels were created that corresponded to the labels (0, 1). Following this, normalization was applied. Scattering coefficients of order greater than 0 were normalized by their parents along the scattering path. Using the defined parameters for the N input signals (runs), each composed of 1000 samples, this procedure produced a $102 \times 8 \times N$ WST feature tensor, which was then transformed into a $N \cdot 8 \times 101$ matrix.

2.2 Classification Model

The testing and training datasets were created using 13 dynamical systems (five chaotic and eight non-chaotic) of first, second, or third order. Table 1 shows the training set characteristics. Each system was provided with 1000 created test files, each of which contained 1000 samples. Augmentation was applied by randomizing the initial conditions, defined by the $\mathbf{x}_0$ vector, according to the formula $[2 \cdot rand() - 1] \cdot \mathbf{x}_0$, where $rand()$ generates pseudorandom numbers that are uniformly distributed in the interval $(0, 1)$. The chaotic systems are represented by driven or autonomous dissipative flows. Previously described as A.4.5, A.5.1, A.5.2, A.5.13, and A.5.15 [24], we describe these flows as CHA_1, CHA_2, CHA_3, CHA_4, and CHA_5, respectively. The non-chaotic systems were divided into the following classes: i) periodic, including the OSC_1, OSC_2, DOSC_1, and IOSC systems; ii) quasi-periodic, including QPS_1 and QPS_2; and iii) non-periodic, including DS_1 and DS_2. The quasi-periodic systems are described by the general function

$$x = f(t) = A_1 \cdot sin(\omega_1 \cdot t + \varphi_1) + A_2 \cdot sin(\omega_2 \cdot t + \varphi_2),$$

where the ratio ω_1/ω_2 is irrational.

Based on previous PPG signal analysis, we made the following experimental design choices:

- A signal with a length of 1000 samples was obtained from each system using 1000 runs with different initial parameters. Each dimension of the multidimensional systems was treated separately. This corresponded to analysis using windows with a short time horizon of 31.2 s for 32 Hz PPG signal.
- We used SVM classification with a radial basis kernel similar to that proposed by Buriro et al. [5]. The classification is made based on WST features.
- Two classes were defined for the classification task: chaotic (class 1) and non-chaotic (class 0). The decision to use just two classes, and therefore fold

periodic, quasi-periodic, and non-periodic behavior into the same class, was made to test the thesis that PPG signals are never chaotic [9]. In further work, a larger number of more distinct classes will be used.

- Sviridova and Sakai [26] show that PPG signals display some similarity to Rössler's chaos with induced dynamical noise. As such, we used Rössler's system as one of the signals with chaotic behavior, as shown in Fig. 1. Furthermore, all signals with additive white Gaussian noise were added to the whole set.

The accuracy of the trained models was checked by 10-fold cross-validation.

To investigate the properties of the training set, the following models were trained:

- *Model01* was trained without output signals from Rössler's system (CHA_3). Using 10-fold cross-validation, accuracy was validated as 100%. Testing using Rössler's system signal showed 32% accuracy. Based on the model, it is impossible to effectively classify the signals produced by chaotic systems. It is therefore important to include the signals from CHA_3 within the training set.
- *Model02* was trained without the quasi-periodic systems QPS_1 and QPS_2. Using 10-fold cross-validation, accuracy was validated as 100%. Testing using QPS_1 and QPS_2 showed 78.5% accuracy. On this basis, we determine that quasi-periodic systems are easier to correctly classify.
- *Model03* was trained on signals without additional noise. Using 10-fold cross-validation, accuracy was validated as 100%. Testing using signals with additive Gaussian noise with a signal to noise ratio of 7dB showed 94.03% accuracy. Although the analytical methods for the assessment of chaotic behavior are highly sensitive to noise, the prepared model is not, and even noisy signals can be classified with high accuracy.
- *Model01N* and *Model02N* are variants of models *Model01* and *Model02* respectively, trained additionally with noisy signals.
- *ModelAll* was trained using all signals, both with and without additive Gaussian noise, with a signal to noise ratio of 7 dB. Using 10-fold cross-validation, accuracy was validated as 99.88%.

2.3 PPG Dataset

The dataset used in this work is the public available PPG dataset for motion compensation and heart rate estimation in daily life activities (PPG-DaLiA[1]) [21]. Given that the database contains a reference ECG measurement, it is often used to test heart rate estimation algorithms [31]. The dataset contains a total of 36 h of recording for 15 study participants undertaking eight different types of physical everyday life activities: working, sitting, walking, eating lunch, driving, cycling, playing football, and climbing stairs. The sensor data was obtained from commercially available devices. In our case, 64 Hz PPG signals which we used for testing the trained models were recorded by the wrist-worn Empatica E4 device.

---

[1] https://ubicomp.eti.uni-siegen.de/home/datasets/sensors19/, accessed July 2021.

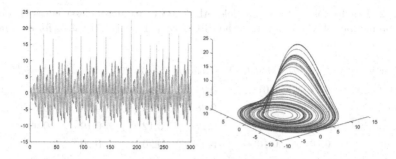

Fig. 1. 3D signals and a phase portrait of the Rössler system (CHA_3).

Table 1. The training set characteristics.

Name and symbol	Class	Dimension	Short description
Ueda oscillator CHA_1	Chaotic (class 1)	2	Driven dissipative flow
Lorenz attractor CHA_2	Chaotic (class 1)	3	Autonomous dissipative flow
Rössler attractor CHA_3	Chaotic (class 1)	3	Autonomous dissipative flow
Halvorsen's cyclically symmetric attractor CHA_4	Chaotic (class 1)	3	Autonomous dissipative flow
Rucklidge attractor CHA_5	Chaotic (class 1)	3	Autonomous dissipative flow
Undamped oscillator 1 OSC_1	Periodic (class 0)	2	Slow oscillations with constant amplitude
Undamped oscillator 2 OSC_2	Periodic (class 0)	2	Fast oscillations with constant amplitude
Damped oscillator 1 DOSC_1	Periodic (class 0)	2	Fast oscillations with decreasing amplitude
Oscillator with increasing amplitude of oscillations IOSC	Periodic (class 0)	2	Oscillations with growing amplitude
Damped system 1 DS_1	Non-periodic (class 0)	3	Slow fading signals
Damped system 2 DS_2	Non-periodic (class 0)	3	Fast fading signals
Quasi-periodic system 1 QPS_1	Quasi-periodic (class 0)	1	Irrational ratio: $\omega_1/\omega_2 = \pi$
Quasi-periodic system 2 QPS_2	Quasi-periodic (class 0)	1	Irrational ratio: $\omega_1/\omega_2 = (1 + \sqrt{5})/2$
16 000 000 samples	0		All samples of signals with non-chaotic behavior
16 000 000 samples	1		All samples of signals with chaotic behavior

The base frequency of the PPG signals was adjusted to match the training set. We counted the number of signal zero crossings within a given time interval when using the CHA_5 system. We were required to increase the frequency of zero crossings by a factor of two, giving a final PPG signal frequency 32 Hz.

3   Results

The PPG signal was split into 1000 samples (31.2 s) segments, following fitting and resampling. For each segment, the WST features were obtained. As part

Table 2. Results for all tested models—the ratio of windows classified as chaotic to the total number of windows within the PPG signal. The highest values for each model have been marked.

Participant	Model01	Model01N	Model02	Model02N	Model03	ModelAll
S01	0.27	0.07	0.25	0.14	0.26	0.11
S02	0.29	0.12	0.28	0.15	0.29	0.15
S03	0.22	0.07	0.17	0.12	0.19	0.09
S04	0.30	0.16	0.28	0.20	0.29	0.19
S05	**0.56**	**0.33**	**0.53**	**0.34**	**0.54**	**0.33**
S06	0.26	0.11	0.21	0.14	0.23	0.13
S07	0.42	0.13	0.36	0.20	0.40	0.17
S08	0.23	0.11	0.22	0.14	0.23	0.13
S09	0.27	0.14	0.26	0.18	0.26	0.18
S10	0.23	0.10	0.22	0.12	0.22	0.13
S11	0.47	0.25	0.44	0.28	0.46	0.26
S12	0.20	0.07	0.16	0.11	0.17	0.09
S13	0.50	0.22	0.44	0.26	0.47	0.24
S14	0.22	0.06	0.16	0.11	0.17	0.08
S15	0.18	0.06	0.14	0.10	0.16	0.08

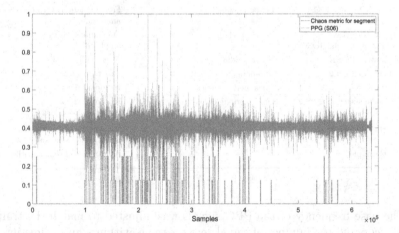

Fig. 2. Chaos measure values when using *ModelAll*, for each 1000 sample segment for the *S06* participant.

of the analysis, the signal was split into $W = 8$ scattering windows of 125 samples each. A classification result was generated for each scattering window. The overall classification of a segment is the ratio of the number of windows classified as having chaotic behavior (class 1), to the total number of windows.

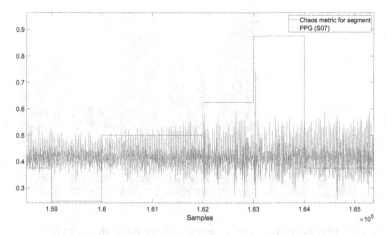

Fig. 3. Chaos measure values when using *ModelAll*, for each 1000 sample segment for a fragment of the PPG signal of the *S07* participant.

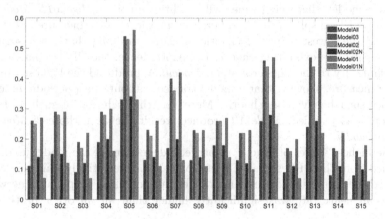

Fig. 4. Chaos measure values for each participant.

Hence, a segment that contains eight chaotic windows represents a fully chaotic segment of signal. Within the overall signal, we define the chaotic measure to be the ratio of chaotic windows to total number of windows. Tests were conducted for all models, as shown in Table 2. Figures 2 and 3 present the PPG signal and the value of the chaotic measure for each segment.

The results show that those models trained without additional noise—*Model01*, *Model02*, and *Model03*—display high chaotic measure values (see Fig. 4). The differences between each of these models are not significant. This confirms that noise is present within the data, and is highly relevant to its evaluation.

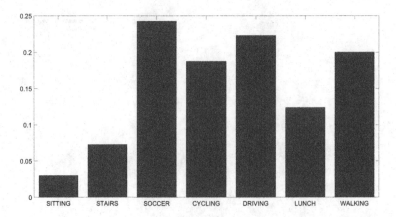

Fig. 5. Average chaos measure values for each activity.

Those models that were trained with additional noise—*Model01N*, *Model02N*, and *ModelAll*—display greater differences in the chaotic measure. *Model01N* produced the lowest values of chaotic measure, meaning that fewer segments were classified as exhibiting chaotic behavior. Hence, the PPG signals display some similarity to the Rössler system. *Model02N* produced the highest values of chaotic measure, showing that a model trained without quasi-periodic functions classifies such behavior as chaotic. Moreover, the inclusion of such a class of functions was justified. *ModelAll* produces results between those of *Model01N* and *Model02N*.

The highest values of chaotic measure, independent of the model used, were obtained from the PPG signal of participant *S05*. Interestingly, this participant reported the greatest errors in heart rate estimation using deep neural networks. The mean heart rate was significantly higher than for all other participants [21,31].

Figure 5 shows that the greatest values of chaotic measure are obtained during the cycling, soccer, driving, and walking activities. The values differ between activities, indicating that the measure is influenced by movement or changes in heart rate as a result of physical activity, stress, or the general condition and health of the participant. The relationship is not well defined, as the participants had the highest heart rate when climbing stairs and cycling. Further analysis of this phenomenon is required.

The calculations were performed on the computer with the following parameters: Windows 10, Intel(R) Core(TM) i7-7700HQ CPU 2.80 GHz, 32 GB, Matlab R2021b. The biggest differences in training and classification times are between the models based signals without noise and with additional noisy signals. The average time of determining WST features for models *Model_01*, *Model_02*, *Model_03* is 14.55956667 s (std 0.996971916), the training time is 19.45903333 s (std 2.239389959), and the average classification time of one window is 0.000109 s (std 0.000061). Taking into account the models *Model_01N*,

Model_02N, *Model_All* the average times are as follows: WST features calculations - 42.526825 s (std 4.489965561), training - 289.576025 s (std 21.05237103), classifications - 0.000435 s (std 0.000044).

4   Conclusions

The results show that signal classification based on system features requires careful preparation of the training set. Chaotic systems create signals that are difficult to predict. Therefore, insufficient data within the training set may cause misclassification. The noise within real measured signals must also be accounted for.

Furthermore, a WST can be used to successfully determine signal features for the purpose of classification. Given that such wavelet analysis is well understood, parameters can be chosen straightforwardly. Moreover, the training model is supposed to be much faster than the use of deep learning methods, which is worth future investigation.

The analysis showed that PPG signals display chaotic features over short time spans. The measure of chaos is dependent upon the activity performed. Given that wearable devices are easily available and are increasingly used for medical diagnosis [19,30], understanding this phenomenon is highly important and the topic requires further detailed analysis.

The aim of this study was to demonstrate the usefulness of WST for the analysis of biomedical signals, and to show the importance of correctly preparing the training set. Interest in the classification of real signals by deep learning methods is increasing; such methods may lead to erroneous conclusions if the training sets are inadequately prepared.

References

1. Andén, J., Mallat, S.: Deep scattering spectrum. IEEE Trans. Signal Process. **62**(16), 4114–4128 (2014)
2. Biswas, D., Simões-Capela, N., Van Hoof, C., Van Helleputte, N.: Heart rate estimation from wrist-worn photoplethysmography: a review. IEEE Sens. J. **19**(16), 6560–6570 (2019)
3. Boullé, N., Dallas, V., Nakatsukasa, Y., Samaddar, D.: Classification of chaotic time series with deep learning. Physica D **403**, 132261 (2020)
4. Bruna, J., Mallat, S.: Invariant scattering convolution networks. IEEE Trans. Pattern Anal. Mach. Intell. **35**(8), 1872–1886 (2013)
5. Buriro, A.B., et al.: Classification of alcoholic EEG signals using wavelet scattering transform-based features. Comput. Biol. Med. **139**, 104969 (2021)
6. Chen, Z., Xiu, D.: On generalized residual network for deep learning of unknown dynamical systems. J. Comput. Phys. **438**, 110362 (2021)
7. Chung, H., Ko, H., Lee, H., Lee, J.: Feasibility study of deep neural network for heart rate estimation from wearable photoplethysmography and acceleration signals. In: 2019 41st Annual International Conference of the IEEE Engineering in Medicine and Biology Society (EMBC), pp. 3633–3636. IEEE (2019)

8. Coughlin, S.S., Stewart, J.: Use of consumer wearable devices to promote physical activity: a review of health intervention studies. J. Environ. Health Sci. **2**(6) (2016)
9. De Pedro-Carracedo, J., Fuentes-Jimenez, D., Ugena, A.M., Gonzalez-Marcos, A.P.: Is the PPG signal chaotic? IEEE Access **8**, 107700–107715 (2020)
10. Friel, C.P., Garber, C.E.: An examination of the relationship between motivation, physical activity, and wearable activity monitor use. J. Sport Exerc. Psychol. **42**(2), 153–160 (2020)
11. Glass, L.: Introduction to controversial topics in nonlinear science: is the normal heart rate chaotic? Chaos: Interdisc. J. Nonlinear Sci. **19**(2), 028501 (2009)
12. Hannan, A.L., Harders, M.P., Hing, W., Climstein, M., Coombes, J.S., Furness, J.: Impact of wearable physical activity monitoring devices with exercise prescription or advice in the maintenance phase of cardiac rehabilitation: systematic review and meta-analysis. BMC Sports Sci. Med. Rehabil. **11**(1), 1–21 (2019)
13. Ismail, S., Akram, U., Siddiqi, I.: Heart rate tracking in photoplethysmography signals affected by motion artifacts: a review. EURASIP J. Adv. Sig. Process. **2021**(1), 1–27 (2021). https://doi.org/10.1186/s13634-020-00714-2
14. Jean Effil, N., Rajeswari, R.: Wavelet scattering transform and long short-term memory network-based noninvasive blood pressure estimation from photoplethys-mograph signals. SIViP **16**(1), 1–9 (2021). https://doi.org/10.1007/s11760-021-01952-z
15. Kamshilin, A.A., et al.: A new look at the essence of the imaging photoplethys-mography. Sci. Rep. **5**(1), 1–9 (2015)
16. Kossi, O., Lacroix, J., Ferry, B., Batcho, C.S., Julien-Vergonjanne, A., Mandigout, S.: Reliability of ActiGraph GT3X+ placement location in the estimation of energy expenditure during moderate and high-intensity physical activities in young and older adults. J. Sports Sci., 1–8 (2021)
17. Kumar, A., Komaragiri, R., Kumar, M., et al.: A review on computation methods used in photoplethysmography signal analysis for heart rate estimation. Arch. Comput. Methods Eng., 1–20 (2021)
18. Mallat, S.: Understanding deep convolutional networks. Philos. Trans. Roy. Soc. A Math. Phys. Eng. Sci. **374**(2065), 20150203 (2016)
19. Manninger, M., et al.: Role of wearable rhythm recordings in clinical decision making-the wEHRAbles project. Clin. Cardiol. **43**(9), 1032–1039 (2020)
20. de Pedro-Carracedo, J., Ugena, A.M., Gonzalez-Marcos, A.P.: Dynamical analysis of biological signals with the 0–1 test: a case study of the photoplethysmographic (PPG) signal. Appl. Sci. **11**(14), 6508 (2021)
21. Reiss, A., Indlekofer, I., Schmidt, P., Van Laerhoven, K.: Deep PPG: large-scale heart rate estimation with convolutional neural networks. Sensors **19**(14), 3079 (2019)
22. Salehizadeh, S., Dao, D., Bolkhovsky, J., Cho, C., Mendelson, Y., Chon, K.H.: A novel time-varying spectral filtering algorithm for reconstruction of motion artifact corrupted heart rate signals during intense physical activities using a wearable photoplethysmogram sensor. Sensors **16**(1), 10 (2016)
23. Schäck, T., Muma, M., Zoubir, A.M.: Computationally efficient heart rate estimation during physical exercise using photoplethysmographic signals. In: 2017 25th European Signal Processing Conference (EUSIPCO), pp. 2478–2481. IEEE (2017)
24. Sprott, J.C.: Chaos and Time-Series Analysis, vol. 69. Oxford University Press (2003)
25. Staszewski, W., Worden, K.: Wavelet analysis of time-series: coherent structures, chaos and noise. Int. J. Bifurcat. Chaos **9**(03), 455–471 (1999)

26. Sviridova, N., Sakai, K.: Human photoplethysmogram: new insight into chaotic characteristics. Chaos Solitons Fractals **77**, 53–63 (2015)
27. Sviridova, N., Zhao, T., Aihara, K., Nakamura, K., Nakano, A.: Photoplethysmogram at green light: where does chaos arise from? Chaos Solitons Fractals **116**, 157–165 (2018)
28. Tamura, T.: Current progress of photoplethysmography and SPO 2 for health monitoring. Biomed. Eng. Lett. **9**(1), 21–36 (2019)
29. Tsuda, I., Tahara, T., Iwanaga, H.: Chaotic pulsation in human capillary vessels and its dependence on the mental and physical conditions. In: Proceedings of the Annual Meeting of Biomedical Fuzzy Systems Association: BMFSA 4, pp. 1–40. Biomedical Fuzzy Systems Association (1992)
30. Vandecasteele, K., et al.: Automated epileptic seizure detection based on wearable ECG and PPG in a hospital environment. Sensors **17**(10), 2338 (2017)
31. Wilkosz, M., Szczęsna, A.: Multi-headed Conv-LSTM network for heart rate estimation during daily living activities. Sensors **21**(15), 5212 (2021)

Convolutional Neural Network Compression via Tensor-Train Decomposition on Permuted Weight Tensor with Automatic Rank Determination

Mateusz Gabor[✉][iD] and Rafał Zdunek[iD]

Faculty of Electronics, Photonics, and Microsystems,
Wroclaw University of Science and Technology, Wybrzeze Wyspianskiego 27,
50-370 Wroclaw, Poland
{mateusz.gabor,rafal.zdunek}@pwr.edu.pl

Abstract. Convolutional neural networks (CNNs) are among the most commonly investigated models in computer vision. Deep CNNs yield high computational performance, but their common issue is a large size. For solving this problem, it is necessary to find effective compression methods which can effectively reduce the size of the network, keeping the accuracy on a similar level. This study provides important insights into the field of CNNs compression, introducing a novel low-rank compression method based on tensor-train decomposition on a permuted kernel weight tensor with automatic rank determination. The proposed method is easy to implement, and it allows us to fine-tune neural networks from decomposed factors instead of learning them from scratch. The results of this study examined on various CNN architectures and two datasets demonstrated that the proposed method outperforms other CNNs compression methods with respect to parameter and FLOPS compression at a low drop in the classification accuracy.

Keywords: Neural network compression · Convolutional neural network · Tensor decomposition · Tensor train decomposition

1 Introduction

The area of convolutional neural networks (CNNs) has attracted growing attention in the field of computer vision for achieving one of the best results in tasks such as image classification [11], segmentation [27] or object detection [26].

However, achieving better results of CNNs is mostly done by designing deeper neural networks, which translates into larger architectures requiring more space and more computing power. Because most of the deep neural networks are over-parametrized [5], there exists a possibility of compressing them without reducing

D. Groen et al. (Eds.): ICCS 2022, LNCS 13352, pp. 654–667, 2022.
https://doi.org/10.1007/978-3-031-08757-8_54

the quality of the network significantly. The neural network compression methods can be classified into weight sharing, pruning, knowledge distillation, quantization and low-rank approximations [1,16,22]. The weight sharing method is the simplest form of compressing a neural network size, in which the weights of the neural network are shared between layers. From this approach, clustering-based weight sharing can be distinguished, in which the clustering is performed on weights, and at the end clustered weights are merged into new compressed weights. In the pruning approach, the redundant connections between neurons are removed, which results in a lower number of parameters and FLOPs. In most cases, the fine-tuning is necessary to recover the original accuracy of the network and often pruning/fine-tuning is alternately repeated in loop to gain larger compression. Quantization is another approach to compress neural network weights. In this method, the neural network weights are represented in a lower-precision format, the most popular is INT8, but the most extreme quantization is based on binary weights. On the other hand, knowledge distillation methods learn a small (student) network from a large one (teacher) using supervision. In short, a student network mimics a teacher network and leverages the knowledge of the teacher, achieving a similar or higher accuracy.

Besides the aforementioned methods, it is possible to compress the neural network using dimensionality reduction techniques such as matrix/tensor decompositions [24] in which the neural network weights are represented in a low-rank format. The low-rank compression methods can be divided into direct decomposition and tensorization. Direct decomposition methods use the factors obtained from the decomposition as new approximated weights, perform all operations on them, and are simple in implementation because they use basic convolutional neural network blocks from deep learning frameworks. The most popular two approaches of using the direct tensor decomposition to compress convolutional layers are the Tucker-2 [15] and CP [18] decomposition. The CP decomposition transforms the original weight tensor into a pipeline of two 1×1 convolutions and two depthwise separable convolutions, and the Tucker-2 into two 1×1 convolutions and one standard convolution, which is the same as the Bottleneck block in ResNet networks. Recently, Hameed *et al.* [9] proposed a new direct tensor decomposition method in which the Kronecker product decomposition is generalized to be applied to compress CNN weights. On the other hand, in the tensorization approach, the original weight tensor is tensorized into a higher-order tensor format and new weights are initialized randomly. In this approach, the decomposition algorithm is not used, and therefore the pretrained information from the baseline network is lost. By using tensorization, the achieved compression is relatively high, but the quality of the compressed network is significantly worse than the baseline model. The first tensorization approach to CNN compression was proposed by Garipov *et al.* [8], in which the tensor-train (TT) format was used to matricized weight tensor. The input feature maps tensor was reshaped into a matrix, and the convolution operation was performed as a sequence of tensor contractions. Garipov *et al.* also proposed a *naive* direct TT compression method in which the weight tensor was directly decomposed. All the decomposed cores were kept in memory, but during the convolution operation, the TT cores were reshaped into the original weight tensor, and the initialization was performed randomly. Among other methods of tensorization, one can mention the tensor ring format [21] or hierarchical Tucker format [31].

In this study, we propose a novel direct low-rank neural network compression method using direct tensor-train decomposition on the permuted kernel weight tensor with automatic rank determination. This method will be referred to as TTPWT. In our approach, each original convolutional layer is replaced and initialized with a sequence of four layers obtained from the decomposed factors, and the original convolution is approximated with four smaller convolutions, which is profitable both with respect to computational and storage complexity. The proposed compression method was applied to four neural networks: TT-conv-CNN [8], VGGnet [28], ResNet-56 [11] and ResNet-110 [11]. The experiments run on the CIFAR-10 and CIFAR-100 datasets showed that the TTPWT considerably outperforms many state-of-the-art compression methods with respect to parameter and FLOPS compression at a low drop in the classification accuracy.

The remainder of this paper is organized as follows. Section 2 presents the notation and the preliminaries to fundamental mathematical operations on tensors. It also contains a short description of the TT decomposition method. The proposed TT-based compression model is presented in Sect. 3. Numerical experiments performed using various CNN architectures tested on the CIFAR-10 and CIFAR-100 datasets are presented and discussed in Sect. 4. The final section provides concluding statements.

2 Preliminary

Notation: Multi-way arrays, matrices, vectors, and scalars are denoted by calligraphic uppercase letters (e.g., $\mathcal{X}$), boldface uppercase letters (e.g., $\boldsymbol{X}$), lowercase boldface letters (e.g., $\boldsymbol{X}$), and unbolded letters (e.g., x), respectively. Multi-way arrays will be equivalently referred to as tensors. We used Kolda's notation [17] for standard mathematical operations on tensors.

Mode-$n$unfolding: The mode-n unfolding of the N-order tensor $\mathcal{X} \in \mathbb{R}^{I_1 \times \cdots \times I_N}$ rearranges its entries by placing its mode-n fibers as the columns of matrix $\boldsymbol{X}_{(n)} = [x_{i_n,j}] \in \mathbb{R}^{I_n \times \Pi_{p \neq n} I_p}$ for $n \in \{1, \ldots, N\}$, where $j = 1 + \sum_{k=1 k \neq n}^{N}(i_k - 1)j_k$ with $j_k = \prod_{m=1 m \neq n}^{k-1} I_m$, and $i_n = 1, \ldots, I_n$.

Mode-$\{n\}$ Canonical Matricization: This matricization reshapes tensor $\mathcal{X}$ into matrix $\boldsymbol{X}_{<n>} \in \mathbb{R}^{\Pi_{p=1}^{n} I_p \times \Pi_{r=n+1}^{N} I_r}$ by mapping tensor element $x_{i_1,\ldots,i_N}$ to matrix element $x_{i,j}$, where $i = 1 + \sum_{p=1}^{n}(i_p - 1)\prod_{m=1}^{p-1} I_m$ and $j = 1 + \sum_{r>n}^{N}(i_r - 1)\prod_{m=n+1}^{r-1} I_m$. The mode-$n$ unfolding is a particular case of the mode-$\{n\}$ canonical matricization.

Mode-$n$product (also known as the tensor-matrix product): The mode-n product of tensor $\mathcal{X} \in \mathbb{R}^{I_1 \times \cdots \times I_N}$ with matrix $\boldsymbol{U} \in \mathbb{R}^{J \times I_n}$ is defined by

$$\mathcal{Z} = \mathcal{X} \times_n \boldsymbol{U}, \tag{1}$$

where $\mathcal{Z} = [z_{i_1,\ldots,i_{n-1},j,i_{n+1},\ldots,i_N}] \in \mathbb{R}^{I_1 \times \ldots \times I_{n-1} \times J \times I_{n+1} \times \ldots \times I_N}$, and

$$z_{i_1,\ldots,i_{n-1},j,i_{n+1},\ldots,i_N} = \sum_{i_n=1}^{I_n} x_{i_1,i_2,\ldots,i_N} u_{j,i_n}.$$

Tensor Contraction: The tensor contraction of tensor $\mathcal{X} = [x_{i_1,\ldots,i_N}] \in \mathbb{R}^{I_1 \times \ldots \times I_N}$ across its n-th mode with tensor $\mathcal{Y} = [y_{j_1,\ldots,j_M}] \in \mathbb{R}^{J_1 \times \ldots \times J_M}$ across its m-th mode, provided that $I_n = J_m$, gives tensor $\mathcal{Z} = \mathcal{X} \times_n^m \mathcal{Y}$ whose entries are given by:

$$z_{i_1,\ldots,i_{n-1},i_{n+1},\ldots,i_N,j_1,\ldots,j_{m-1},j_{m+1},\ldots,j_M}$$
$$= \sum_{i_n=1}^{I_n} x_{i_1,\ldots,i_{n-1},i_n,i_{n+1},\ldots,i_N} y_{j_1,\ldots,j_{m-1},i_n,j_{m+1},\ldots,j_M}. \tag{2}$$

For the matrices: $\boldsymbol{A} \times_2^1 \boldsymbol{B} = \boldsymbol{AB}$. The contraction: $\times_N^1$ will be denoted by the symbol $\bullet$. Thus: $\mathcal{X} \bullet \mathcal{Y} = \mathcal{X} \times_N^1 \mathcal{Y}$.

Kruskal Convolution: Let $\mathcal{X} = [x_{i_1,i_2,c}] \in \mathbb{R}^{I_1 \times I_2 \times C}$ be any activation tensor in any convolutional layer with C input channels, $\mathcal{W} = [w_{t,c,d_1,d_2}] \in \mathbb{R}^{T \times C \times D_1 \times D_2}$ be the kernel weight tensor, Δ be the stride, and P be the zero-padding size. The Kruskal convolution maps input tensor $\mathcal{X}$ to output tensor $\mathcal{Y} = [y_{\tilde{i}_1,\tilde{i}_2,t}] \in \mathbb{R}^{\tilde{I}_1 \times \tilde{I}_2 \times T}$ by the following linear mapping:

$$y_{\tilde{i}_1,\tilde{i}_2,t} = x_{i_1,i_2,c} \star w_{t,c,d_1,d_2} = \sum_{c=1}^{C} \sum_{d_1=1}^{D} \sum_{d_2=1}^{D} w_{t,c,d_1,d_2} x_{i_1(d_1),i_2(d_2),c}, \tag{3}$$

where $i_1(d_1) = (\tilde{i}_1 - 1)\Delta + i_1 - P$ and $i_2(d_2) = (\tilde{i}_2 - 1)\Delta + i_2 - P$.

1×1 **Convolution**: If $D = 1$, $\Delta = 1$, and $P = 0$, then $\mathcal{W} \in \mathbb{R}^{T \times C \times 1 \times 1}$, and the Kruskal convolution comes down to the 1×1 convolution: $y_{i_1,i_2,t} = \sum_{c=1}^{C} w_{t,c} x_{i_1,i_2,c}$. Using the notation of the mode-n product in (1), the 1×1 convolution takes the form:

$$\mathcal{Y} = \mathcal{X} \times_3 \boldsymbol{W}, \tag{4}$$

where $\boldsymbol{W} = [w_{tc}] \in \mathbb{R}^{T \times C}$.

Tensor Train (TT) Decomposition: The TT model [23] decomposes tensor $\mathcal{X} = [x_{i_1,\ldots,i_N}] \in \mathbb{R}^{I_1 \times \ldots \times I_N}$ to a chain of smaller (3-way) core tensors that are connected by the tensor contraction with operator $\bullet$. It can be formulated as follows:

$$\mathcal{X} = \mathcal{X}^{(1)} \bullet \mathcal{X}^{(2)} \bullet \ldots \bullet \mathcal{X}^{(N)}, \tag{5}$$

where $\mathcal{X}^{(n)}$ is the n-th core tensor of size $R_{n-1} \times I_n \times R_n$ for $n = 1,\ldots,N$. The number $\{R_0,\ldots,R_N\}$ determine the TT ranks. Assuming $R_0 = R_N = 1$, we have $\mathcal{X}^{(1)} = \boldsymbol{X}^{(1)} \in \mathbb{R}^{I_1 \times R_1}$ and $\mathcal{X}^{(N)} = \boldsymbol{X}^{(N)} \in \mathbb{R}^{R_{N-1} \times I_N}$, i.e. the first and the last core tensors become matrices. Model (5) can be expressed equivalently as:

$$x_{i_1,\ldots,i_N} = \sum_{r_1=1}^{R_1} \sum_{r_2=1}^{R_2} \cdots \sum_{r_{N-1}=1}^{R_{N-1}} x_{i_1,r_1}^{(1)} x_{r_1,i_2,r_2}^{(2)} \cdots x_{r_{N-2},i_{N-1},r_{N-1}}^{(N-1)} x_{r_{N-1},i_N}^{(N)}, \quad (6)$$

where $\forall n : \mathcal{X}^{(n)} = [x_{r_{n-1},i_n,r_n}^{(n)}] \in \mathbb{R}^{R_{N-1} \times I_n \times R_n}$.

Assuming $I_1 = \ldots = I_N = I$ and $R_1 = \ldots = R_N = R$, the storage complexities of the CANDECOM/PARAFAC (CP) [2,10], Tucker [29], and TT decomposition models can be approximated by $\mathcal{O}(NIR)$, $\mathcal{O}(NIR + R^N)$, and $\mathcal{O}(NIR^2)$. It is thus obvious that the CP model has the lowest storage complexity, but its flexibility in adapting to the observed data is very low, especially for tensors that have strongly unbalanced modes. Unfortunately, this is the case in the discussed problem because two modes of the decomposed tensor have small dimensions, but the other modes are large. Hence, it is difficult to select the optimal rank. The Tucker decomposition relaxes these problems considerably, but its storage complexity grows up exponentially with the size of the core tensor, which is also not favorable in our case because the ranks for large modes are usually pretty large. The TT model assures the best trade-off between the CP and Tucker decompositions, alleviating the curse of dimensionality and yielding a flexible decomposition with multiple TT ranks. Hence, these advantages of the TT model motivate this study.

3 Proposed Method

We assume that each convolutional layer has C input and T output channels, and the size of the filter is $D \times D$. Hence, it can be represented by the kernel weight tensor $\mathcal{W} = [w_{t,c,d_1,d_2}] \in \mathbb{R}^{T \times C \times D \times D}$. The input data is represented by activation tensor $\mathcal{X} = [x_{i_1,i_2,c}] \in \mathbb{R}^{I_1 \times I_2 \times C}$ that consists of C activation maps – each has the resolution of $I_1 \times I_2$ pixels. Each layer performs a linear mapping of tensor $\mathcal{X}$ to output activation tensor $\mathcal{Y} = [y_{\tilde{i}_1,\tilde{i}_2,t}] \in \mathbb{R}^{\tilde{I}_1 \times \tilde{I}_2 \times T}$, where the mapping is determined by the Kruskal convolution in (3). Each output activation map has the resolution of $\tilde{I}_1 \times \tilde{I}_2$ pixels, and there are T output channels.

3.1 Model

To reduce the number of parameters and FLOPS in each convolutional layer, the kernel weight tensor $\mathcal{W}$ is decomposed with the TT model.

Remark 1. Note that if $\mathcal{W} \in \mathbb{R}^{T \times C \times D \times D}$ is decomposed according to (5), ranks R_2 and R_3 cannot be greater than D^2 and D, respectively. This restriction limits the flexibility of compression only to rank R_1. Furthermore, the 3D core tensor capturing the second mode of $\mathcal{W}$ could not be processed with a simple 1×1

convolution. Thus, we propose to apply the circular permutation to $\mathcal{W}$ with one left shift lag. Thus:

$$\tilde{\mathcal{W}} = \texttt{circular\_permutation}\,(\mathcal{W}, -1) \in \mathbb{R}^{C \times D \times D \times T}. \tag{7}$$

Applying the TT decomposition to $\tilde{\mathcal{W}} = [\tilde{w}_{c,d_1,d_2,t}]$, we have:

$$\tilde{w}_{c,d_1,d_2,t} = \sum_{r_1=1}^{R_1} \sum_{r_2=1}^{R_2} \sum_{r_3=1}^{R_3} \tilde{w}_{c,r_1}^{(1)} \tilde{w}_{r_1,d_1,r_2}^{(2)} \tilde{w}_{r_2,d_2,r_3}^{(3)} \tilde{w}_{r_3,t}^{(4)}. \tag{8}$$

Inserting model (8) to mapping (3) and rearranging the summands, we get:

$$
\begin{aligned}
y_{\tilde{i}_1,\tilde{i}_2,t} &= \sum_{c=1}^{C} \sum_{d_1=1}^{D} \sum_{d_2=1}^{D} \sum_{r_1=1}^{R_1} \sum_{r_2=1}^{R_2} \sum_{r_3=1}^{R_3} \tilde{w}_{c,r_1}^{(1)} \tilde{w}_{r_1,d_1,r_2}^{(2)} \tilde{w}_{r_2,d_2,r_3}^{(3)} \tilde{w}_{r_3,t}^{(4)} x_{i_1(d_1),i_2(d_2),c} \\
&= \sum_{r_3=1}^{R_3} \tilde{w}_{r_3,t}^{(4)} \left[\sum_{d_2=1}^{D} \sum_{r_2=1}^{R_2} \tilde{w}_{r_2,d_2,r_3}^{(3)} \sum_{d_1=1}^{D} \sum_{r_1=1}^{R_1} \tilde{w}_{r_1,d_1,r_2}^{(2)} \right. \\
&\qquad \times \left. \left(\underbrace{\sum_{c=1}^{C} \tilde{w}_{c,r_1}^{(1)} x_{i_1(d_1),i_2(d_2),c}}_{1 \times 1 \text{ conv.}} \right) \right]
\end{aligned}
$$

$$
= \sum_{r_3=1}^{R_3} \tilde{w}_{r_3,t}^{(4)} \left[\sum_{d_2=1}^{D} \sum_{r_2=1}^{R_2} \tilde{w}_{r_2,d_2,r_3}^{(3)} \left(\underbrace{\sum_{d_1=1}^{D} \sum_{r_1=1}^{R_1} \tilde{w}_{r_1,d_1,r_2}^{(2)} z_{i_1(d_1),i_2(d_2),r_1}}_{D_1 \times 1 \text{ conv.}} \right) \right]
$$

$$
= \sum_{r_3=1}^{R_3} \tilde{w}_{r_3,t}^{(4)} \left[\underbrace{\sum_{d_2=1}^{D} \sum_{r_2=1}^{R_2} \tilde{w}_{r_2,d_2,r_3}^{(3)} z_{i_1,i_2(d_2),r_2}^{(V)}}_{1 \times D_2 \text{ conv.}} \right] = \underbrace{\sum_{r_3=1}^{R_3} \tilde{w}_{r_3,t}^{(4)} z_{i_1,i_2,r_3}^{(V,H)}}_{1 \times 1 \text{ conv.}} \tag{9}
$$

It can be easy to note that z_{i_1,i_2,r_1} in (9) can be computed with the 1×1 convolution. According to (4), we have:

$$\mathcal{Z} = \mathcal{X} \times_3 \tilde{\boldsymbol{W}}^{(1)T} \in \mathbb{R}^{I_1 \times I_2 \times R_1}, \tag{10}$$

where $\tilde{\boldsymbol{W}}^{(1)} = [\tilde{w}_{c,r_1}^{(1)}] \in \mathbb{R}^{C \times R_1}$. Physically, to perform operation (10), the first sublayer with the 1×1 convolutions in the analyzed convolutional layer is created. The activation tensor $\mathcal{Z}$ computed in the first sub-layer is then provided to the second convolutional sublayer represented by $\tilde{\mathcal{W}} = [\tilde{w}_{r_1,d_1,r_2}^{(2)}] \in \mathbb{R}^{R_1 \times D \times R_2}$, which is much smaller than $\mathcal{W}$, and this sublayer computes the 1D convolutions along the 1-st mode (vertically):

$$z_{i_1,i_2(d_2),r_2}^{(V)} = \sum_{d_1=1}^{D} \sum_{r_1=1}^{R_1} \tilde{w}_{r_1,d_1,r_2}^{(2)} z_{i_1(d_1),i_2(d_2),r_1} \tag{11}$$

As a result, we get the second-sublayer output activation tensor

$$\mathcal{Z}^{(V)} = [z_{i_1,i_2(d_2),r_2}^{(V)}] \in \mathbb{R}^{\tilde{I}_1 \times I_2 \times R_2}.$$

Next, the third 1D convolutional sublayer is created to compute the 1D convolutions along the horizontal direction. The output activation tensor obtained from this sublayer has the form: $\mathcal{Z}^{(V,H)} = [z_{i_1,i_2,r_3}^{(V,H)}] \in \mathbb{R}^{\tilde{I}_1 \times \tilde{I}_2 \times R_3}$. Finally, the fourth sublayer is created, which performs 1×1 convolutions according to the model:

$$\mathcal{Y} = \mathcal{Z}^{(V,H)} \times_3 \tilde{\boldsymbol{W}}^{(4)T} \in \mathbb{R}^{\tilde{I}_1 \times \tilde{I}_2 \times T}, \tag{12}$$

where $\tilde{\boldsymbol{W}}^{(4)} = [\tilde{w}_{r_3,t}^{(4)}] \in \mathbb{R}^{R_3 \times T}$.

3.2 TT Decomposition Algorithm

The TT decomposition of $\tilde{\mathcal{W}}$ in (7) can be obtained by using sequential SVD-based projections. In the first step, TSVD with a given precision δ_1 is applied to $\tilde{\mathcal{W}}$ unfolded with respect to its first-mode. Thus:

$$\tilde{\boldsymbol{W}}_{(1)} = \boldsymbol{U}\boldsymbol{\Sigma}\boldsymbol{V}^T + \boldsymbol{E}_1, \tag{13}$$

under the assumption the truncation error satisfies the condition $||\boldsymbol{E}_1||_F \leq \delta_1$. Matrix $\tilde{\boldsymbol{W}}^{(1)} \in \mathbb{R}^{C \times R_1}$ is created from $\boldsymbol{U}$ that contains the first R_1 left singular vectors (associated with the most significant singular values) of $\tilde{\boldsymbol{W}}_{(1)}$. Note that rank R_1 is determined by a given threshold δ_1 for the truncation error. In the second step, $\tilde{\boldsymbol{W}}^{(2)} \in \mathbb{R}^{R_1 I_2 \times R_2}$ is created from the first R_2 left singular vectors of the matrix obtained by reshaping matrix $\boldsymbol{\Sigma}\boldsymbol{V}^T$ using the mode-2 canonical matricization. In this step, $||\boldsymbol{E}_2||_F \leq \delta_2$ and the core tensor is obtained by reshaping $\tilde{\boldsymbol{W}}^{(2)}$ accordingly. The similar procedure is applied in the third step, where $\tilde{\mathcal{W}}^{(3)} \in \mathbb{R}^{R_2 \times I_3 \times R_3}$ is created from the first R_3 singular vectors, and $\tilde{\boldsymbol{W}}^{(4)} \in \mathbb{R}^{R_3 \times T}$ is created from the scaled right singular vectors. Oseledets [23] showed that $||\tilde{\mathcal{W}} - \tilde{\mathcal{W}}^{(1)} \bullet \ldots \bullet \tilde{\mathcal{W}}^{(N)}||_F \leq \sqrt{\sum_{n=1}^{N-1} \delta_n^2}$. Assuming $\delta = \delta_1 = \ldots = \delta_{N-1}$, then the truncation threshold can be set to $\delta = \frac{\epsilon}{\sqrt{N-1}}||\tilde{\mathcal{W}}||_F$, where $\epsilon > 0$ is a prescribed relative error.

In our approach, the optimal rank of TSVD for matrix $\boldsymbol{M} \in \mathbb{R}^{P \times R}$ in each step was computed by using the energy-threshold criterion. Thus:

$$R_* = \arg\min_j \left\{ \frac{\sum_{i=1}^{j} \sigma_i^2}{\sum_{i=1}^{I} \sigma_i^2} > \tau \right\}, \tag{14}$$

Algorithm 1. TT-SVD

Input : $\mathcal{W} \in \mathbb{R}^{T \times C \times D \times D}$ – input kernel weight tensor, τ - threshold
Output: $\{\mathcal{W}^{(1)}, ..., \mathcal{W}^{(4)}\}$ - estimated core tensors

Compute $\tilde{\mathcal{W}} \in \mathbb{R}^{C \times D \times D \times T}$ with (7) and set $R_0 = 1$ and $N = 4$,
$M = \tilde{\boldsymbol{W}}_{(1)} = \texttt{unfolding}(\tilde{\mathcal{W}}, 1)$;   // Unfolding
for $n = 1, \ldots, N - 1$ **do**

$\quad$ Compute: $\left[\tilde{\boldsymbol{U}}, \tilde{\boldsymbol{S}}, \tilde{\boldsymbol{V}}, R_n\right] = \texttt{TSVD}_\delta(M, \tau)$;   // TSVD
$\quad$ $\tilde{\mathcal{W}}^{(n)} = \texttt{reshape}(\tilde{\boldsymbol{U}}, [R_{n-1}, I_n, R_n])$
$\quad$ $M = \texttt{reshape}(\tilde{\boldsymbol{S}}\tilde{\boldsymbol{V}}^T, [R_n I_{n+1}, \prod_{p=n+2}^{N} I_p])$;   // Canonical matricization
end
$\tilde{\mathcal{W}}^{(4)} = \texttt{reshape}(M, [R_{N-1}, I_N, 1])$

---

where $Q = \min\{P, Q\}$, σ_i is the i-th singular value of M, and $\tau = \frac{\epsilon}{\sqrt{N-1}}$ is a given threshold. The energy captured by i components (singular vectors) is expressed in the nominator of (14), the total energy is presented in the denominator.

Due to the low-rank approximation, the TT model always assures the compression [25], i.e.

$$R_n \leq \min \left\{ \prod_{i=1}^{n} I_i, \prod_{j=n+1}^{N} I_j \right\}, \quad \text{for} \quad n = 1, \ldots, N-1. \tag{15}$$

The complete sequential routine is presented in Algorithm 1. Function $\texttt{TSVD}_\delta$ performs the δ-truncated SVD at a given threshold δ, where the optimal rank R_* is computed by the energy-based criterion (14).

3.3 Implementation

The procedure for training/fine-tuning networks was implemented in the deep learning framework *PyTorch* and the *tensor-train* decomposition in *Matlab*. The convolutional kernel is the main component of the convolutional layer, which is represented as the 4-th order tensor (top block, Fig. 1). After using permutation, the weight tensor can be decomposed into four factors, including two matrices and two 3-rd order tensors. All the factors are used as new weights in a sequence of four sublayers (Tensor-Train model, Fig. 1). Because the basic class of convolutional layer in *PyTorch* accepts only 4-th order tensor as weights, it is necessary to add extra two dimensions to matrices: $\tilde{\boldsymbol{W}}^{(1)} \in \mathbb{R}^{C \times R_1} \rightarrow \tilde{\mathcal{W}}^{(1)} \in \mathbb{R}^{R_1 \times C \times 1 \times 1}$ (Fig. 1, sublayer Conv2D.1), $\tilde{\boldsymbol{W}}^{(4)} \in \mathbb{R}^{R_3 \times T} \rightarrow \tilde{\mathcal{W}}^{(4)} \in \mathbb{R}^{T \times R_3 \times 1 \times 1}$ (Fig. 1, sublayer Conv2D.4) and extra one dimension to 3-rd order tensors $\tilde{\mathcal{W}}^{(2)} \in \mathbb{R}^{R_1 \times D \times R_2} \rightarrow \tilde{\mathcal{W}}^{(2)} \in \mathbb{R}^{R_2 \times R_1 \times D \times 1}$ (Fig. 1, sublayer Conv2D.2), $\tilde{\mathcal{W}}^{(3)} \in \mathbb{R}^{R_3 \times D \times R_2} \rightarrow \tilde{\mathcal{W}}^{(3)} \in \mathbb{R}^{R_3 \times R_2 \times 1 \times D}$ (Fig. 1, sublayer Conv2D.3) and permute the modes accordingly.

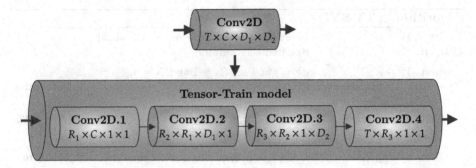

Fig. 1. Visual representation of how the decomposed factors are used as new weights in *PyTorch* framework (output channels × input channels × filter height × filter width) for the compressed convolutional layer in the TT model.

3.4 Computational Complexity

The space and time complexity of the convolution is defined as $\mathcal{O}(CTD^2)$ and $\mathcal{O}(CTD^2I_1I_2)$. By applying the tensor-train decomposition, the time and space complexity is bounded by $\mathcal{O}(CR_1 + R_2D(R_1 + R_3) + R_3T)$ and $\mathcal{O}(R_1I_1I_2 + R_2D(R_1I_1I_2 + R_3\tilde{I}_1I_2) + R_3T\tilde{I}_1\tilde{I}_2)$ respectively, where I_1 and I_2 define the height and width of the input image, respectively, and $\tilde{I}_1$ and $\tilde{I}_2$ define the reduced height and width after convolution.

4 Results

We evaluated our method on two datasets (CIFAR-10 and CIFAR-100). Each consists of 60,000 examples, including 50,000 in the training dataset, and 10,000 in the validation dataset with 10 and 100 classes respectively. To evaluate effectiveness of our method on networks of various sizes, we selected the following networks: TT-conv-CNN [8], VGGnet [28], ResNet-56 [11] and ResNet-110 [11]. The networks cover the range of models with a medium to a large number of parameters and FLOPS. The total number of FLOPS and the parameters of the mentioned networks are listed in Table 1. The compression experiments were performed with the following scheme:

energy threshold selection $\longrightarrow$ baseline CNN compression $\longrightarrow$ fine-tuning.

All the convolutional layers were compressed in each neural network except for the first one whose size is small. All the baseline networks were trained according to the source guidelines. TT-conv-CNN was trained for 100 epochs using stochastic gradient descent (SGD) with a momentum of 0.9, the weight decay was set to 0, the initial learning rate was set to 0.1, and it was decreased by a factor of 0.1 after every 20 epochs. For the fine-tuning process, all the hyperparameters remained the same. VGGnet, ResNet-56, and ResNet-110 were trained for 200 epochs using the SGD with a momentum of 0.9, the weight decay was set to 10^{-4}, the initial learning rate was set to 0.1, and it was decreased by a factor of 0.1 after 80 and 120 epochs for VGGnet, and after 100 and 150 for ResNets. In the fine-tuning step, the learning rate was lowered to 0.01 and the weight decay was increased to 10^{-3} for the VGGnet, and the hyperparameters were unchanged for ResNet-56 and ResNet-110.

To evaluate the network compression and performance, we used two metrics, such as the parameter compression ratio (PCR) that is defined as $\Downarrow Param = \frac{\text{Param(baseline network)}}{\text{Param(compressed network)}}$, and the FLOPS compression ratio (FCR) defined as $\Downarrow FLOPS = \frac{\text{FLOPS(baseline network)}}{\text{FLOPS(compressed network)}}$. The quality of the network was evaluated with the drop in the classification accuracy of the compressed network with respect to the baseline network, i.e. $\Delta Acc = Acc_{compressed} - Acc_{baseline}$. The error rate is often shown alongside with the accuracy in the literature. However, the error rate may be misleading since we fine-tune the neural networks from decomposed factors. Hence, the accuracy is sufficient to be shown for better interpretability of results. The values of PCR or FCR are not provided in all the papers, which we refer to as the reference results. Hence, the unavailable data are marked with the "–" sign in the tables.

Table 1. Total number of FLOPs and parameters for baseline networks.

Network	Params	FLOPS
TT-conv-CNN	558 K	105 M
ResNet-56	853 K	125 M
ResNet-110	1.73 M	255 M
VGGnet	20 M	399 M

4.1   CIFAR-10

TT-conv-CNN: Table 2 shows the results obtained for the TT-conv-CNN compression. We compared our method with the tensorized tensor-train version of the matricized weight tensor (TT-conv), direct tensorized tensor-train weight tensor (TT-conv (naive)), and the weight sharing method – Deep k-Means [32]. As we can see, our method outperforms both TT-based methods proposed by Garipov *et al.* in terms of PCR, FCR, and the drop in accuracy is at a much lower level. Compared with Deep k-Means, our method achieved better accuracy with higher compression.

VGGnet: The VGGnet network is a modified VGG-19 neural network adopted for CIFAR datasets. It is the largest neural network analyzed in this study, with 20M of parameters and 399M of FLOPS. Compression of VGGnet using our method was compared with the following pruning approaches: DCP [35], Random-DCP [35], WM+ [35], CP [14] and PFEC [19]. The results given in Table 3 demonstrate that our method outperforms all the compared approaches in terms of PCR and FCR. Our compressed network achieved a positive drop (gain) in the accuracy compared to the baseline network, and only DCP obtained a higher gain but with worse parameter compression. Moreover, TTPWT reduces FLOPS nearly 2.35 times more than DCP.

Table 2. Results of the TT-conv-CNN [8] compression on the CIFAR-10 validation dataset. Different rows of the TT-conv and TT-conv (naive) mean different ranks. The value in parentheses denotes the energy threshold.

Method	ΔAcc	$\Downarrow$Param	$\Downarrow$FLOPS
TT-conv-1 [8]	−0.80	2.02	–
TT-conv-2 [8]	−1.50	2.53	–
TT-conv-3 [8]	−1.40	3.23	–
TT-conv-4 [8]	−2.00	4.02	–
TT-conv-1 (naive) [8]	−2.40	2.02	–
TT-conv-2 (naive) [8]	−3.10	2.90	–
Deep k-Means [32]	+0.05	2.00	–
TTPWT (0.6)	**+0.14**	3.06	2.95
TTPWT (0.5)	−0.25	**5.03**	**4.73**

Table 3. Results of VGGnet compression on the CIFAR-10 validation dataset. The value in parentheses denotes the energy threshold.

Method	ΔAcc	$\Downarrow$Param	$\Downarrow$FLOPS
DCP [35]	+0.31	1.93	2.00
Random-DCP [35]	+0.03	1.93	2.00
WM+ [35]	−0.10	1.93	2.00
CP [14,35]	−0.32	1.93	2.00
PFEC [19,35]	+0.15	2.78	1.52
TTPWT (0.6)	+0.15	**3.03**	**4.71**

4.2   CIFAR-100

ResNet-56: ResNet-56 was the first network evaluated by us on the CIFAR-100 dataset. We compared the obtained results of our method with pruning approaches. As pruning competitors, we chose the following methods: SFP [12], FPGM [13], DMPP [20], CCPrune [4], FPC [3], and FPDC [36]. As can be seen in Table 4 our method achieved the largest FCR and PCR, and the lowest accuracy drop. It is interesting that TTPWT reduces FLOPS twice as much as SFP, FPGM and CCPrune.

Table 4. Results of ResNet-56 compression on the CIFAR-100 validation dataset. The value in parentheses denotes the energy threshold.

Method	ΔAcc	⇓Param	⇓FLOPS
SFP [12,20]	−2.61	3.20	2.11
FPGM [13,20]	−1.75	3.30	2.11
CCPrune [4]	−0.63	1.69	2.94
FPDC [36]	−1.43	1.93	1.99
TTPWT (0.55)	**−0.50**	**3.93**	**4.19**

ResNet-110: As the second network, we selected ResNet-110 that is one of the largest ResNet networks developed for CIFAR datasets. Similar to the previous results, ResNet-110 was compared with different pruning approaches [6,7,12,13, 30,33,34,36]. As shown in Table 5, it is clear that TTPWT achieved the lowest drop in accuracy and the largest PCR and FCR over all the compared methods.

Table 5. Results of ResNet-110 compression on the CIFAR-110 validation dataset. The value in parentheses denotes the energy threshold.

Method	ΔAcc	⇓Param	⇓FLOPS
OED [30]	−3.83	2.31	3.23
FPDC [36]	−0.61	1.93	3.24
PKPSMIO [33]	−0.14	3.40	3.24
PKP [34]	−0.61	2.42	2.37
TAS [7]	−1.90	–	2.11
FPGM [7,13]	−1.59	–	2.10
SFP [7,12]	−2.86	–	2.10
LCCL [6,7]	−2.01	–	1.46
TTPWT (0.55)	**−0.03**	**3.96**	**4.21**

5   Conclusions

This study proposes a new approach to low-rank compression of CNNs. The proposed method is based the tensor train decomposition of a permuted weight tensor with automatic rank determination. The original convolution is approximated with a pipeline of four smaller convolutions, which allows us to significantly reduce a number of parameters and FLOPS at the cost of a low drop in accuracy. The results obtained on two datasets using four networks of different sizes confirm that our method outperforms the other neural network compression methods presented in this study. Further research is needed to investigate the compression of larger CNNs on the ImageNet dataset and to extend the current approach for higher order convolutional neural networks, including 3D CNNs.

References

1. Alqahtani, A., Xie, X., Jones, M.W.: Literature review of deep network compression. In: Informatics, vol. 8, p. 77. Multidisciplinary Digital Publishing Institute (2021)
2. Carroll, J.D., Chang, J.J.: Analysis of individual differences in multidimensional scaling via an n-way generalization of Eckart-Young decomposition. Psychometrika **35**, 283–319 (1970)
3. Chen, Y., Wen, X., Zhang, Y., He, Q.: FPC: filter pruning via the contribution of output feature map for deep convolutional neural networks acceleration. Knowl.-Based Syst. **238**, 107876 (2022)
4. Chen, Y., Wen, X., Zhang, Y., Shi, W.: CCPrune: collaborative channel pruning for learning compact convolutional networks. Neurocomputing **451**, 35–45 (2021)
5. Denil, M., Shakibi, B., Dinh, L., Ranzato, M., De Freitas, N.: Predicting parameters in deep learning. In: Advances in Neural Information Processing Systems (NIPS), pp. 2148–2156 (2013)
6. Dong, X., Huang, J., Yang, Y., Yan, S.: More is less: a more complicated network with less inference complexity. In: Proceedings of the IEEE Conference on Computer Vision and Pattern Recognition (CVPR), pp. 5840–5848 (2017)
7. Dong, X., Yang, Y.: Network pruning via transformable architecture search. Adv. Neural Inf. Process. Syst. **32**, 760–771 (2019)
8. Garipov, T., Podoprikhin, D., Novikov, A., Vetrov, D.: Ultimate tensorization: compressing convolutional and fc layers alike. [Online] arXiv preprint arXiv:1611.03214 (2016)
9. Hameed, M.G.A., Tahaei, M.S., Mosleh, A., Nia, V.P.: Convolutional neural network compression through generalized Kronecker product decomposition. arXiv preprint arXiv:2109.14710 (2021)
10. Harshman, R.A.: PARAFAC2: mathematical and technical notes. UCLA Work. Papers Phonet. **22**, 30–44 (1972)
11. He, K., Zhang, X., Ren, S., Sun, J.: Deep residual learning for image recognition. In: Proceedings of the IEEE Conference on Computer Vision and Pattern Recognition (CVPR), pp. 770–778 (2016)
12. He, Y., Kang, G., Dong, X., Fu, Y., Yang, Y.: Soft filter pruning for accelerating deep convolutional neural networks. In: International Joint Conference on Artificial Intelligence (IJCAI), pp. 2234–2240 (2018)
13. He, Y., Liu, P., Wang, Z., Hu, Z., Yang, Y.: Filter pruning via geometric median for deep convolutional neural networks acceleration. In: Proceedings of the IEEE/CVF Conference on Computer Vision and Pattern Recognition (CVPR), pp. 4340–4349 (2019)
14. He, Y., Zhang, X., Sun, J.: Channel pruning for accelerating very deep neural networks. In: Proceedings of the IEEE International Conference on Computer Vision (CVPR), pp. 1389–1397 (2017)
15. Kim, Y.D., Park, E., Yoo, S., Choi, T., Yang, L., Shin, D.: Compression of deep convolutional neural networks for fast and low power mobile applications. In: International Conference on Learning Representations (ICLR) (2015)
16. Kirchhoffer, H., et al.: Overview of the neural network compression and representation (NNR) standard. IEEE Trans. Circ. Syst. Video Technol. 1 (2021). https://doi.org/10.1109/TCSVT.2021.3095970
17. Kolda, T.G., Bader, B.W.: Tensor decompositions and applications. SIAM Rev. **51**(3), 455–500 (2009)

18. Lebedev, V., Ganin, Y., Rakhuba, M., Oseledets, I., Lempitsky, V.: Speeding-up convolutional neural networks using fine-tuned CP-decomposition. In: International Conference on Learning Representations (ICLR) (2014)
19. Li, H., Kadav, A., Durdanovic, I., Samet, H., Graf, H.P.: Pruning filters for efficient convnets. In: International Conference on Learning Representations (ICLR) (2016)
20. Li, J., Zhao, B., Liu, D.: DMPP: differentiable multi-pruner and predictor for neural network pruning. Neural Netw. **147**, 103–112 (2022)
21. Li, N., Pan, Y., Chen, Y., Ding, Z., Zhao, D., Xu, Z.: Heuristic rank selection with progressively searching tensor ring network. Complex Intell. Syst. 1–15 (2021)
22. Neill, J.O.: An overview of neural network compression. arXiv preprint arXiv:2006.03669 (2020)
23. Oseledets, I.V.: Tensor-train decomposition. SIAM J. Sci. Comput. **33**(5), 2295–2317 (2011)
24. Panagakis, Y., et al.: Tensor methods in computer vision and deep learning. Proc. IEEE **109**(5), 863–890 (2021)
25. Phan, A.H., Cichocki, A., Uschmajew, A., Tichavský, P., Luta, G., Mandic, D.P.: Tensor networks for latent variable analysis: novel algorithms for tensor train approximation. IEEE Trans. Neural Netw. Learn. Syst. **31**(11), 4622–4636 (2020)
26. Redmon, J., Divvala, S., Girshick, R., Farhadi, A.: You only look once: unified, real-time object detection. In: Proceedings of the IEEE Conference on Computer Vision and Pattern Recognition (CVPR), pp. 779–788 (2016)
27. Ronneberger, O., Fischer, P., Brox, T.: U-Net: convolutional networks for biomedical image segmentation. In: Navab, N., Hornegger, J., Wells, W.M., Frangi, A.F. (eds.) MICCAI 2015. LNCS, vol. 9351, pp. 234–241. Springer, Cham (2015). https://doi.org/10.1007/978-3-319-24574-4_28
28. Simonyan, K., Zisserman, A.: Very deep convolutional networks for large-scale image recognition. In: International Conference Learning Representations (ICLR) (2015)
29. Tucker, L.R.: The extension of factor analysis to three-dimensional matrices. In: Gulliksen, H., Frederiksen, N. (eds.) Contributions to mathematical psychology, pp. 110–127. Holt, Rinehart and Winston, New York (1964)
30. Wang, Z., Lin, S., Xie, J., Lin, Y.: Pruning blocks for CNN compression and acceleration via online ensemble distillation. IEEE Access **7**, 175703–175716 (2019)
31. Wu, B., Wang, D., Zhao, G., Deng, L., Li, G.: Hybrid tensor decomposition in neural network compression. Neural Netw. **132**, 309–320 (2020)
32. Wu, J., Wang, Y., Wu, Z., Wang, Z., Veeraraghavan, A., Lin, Y.: Deep k-means: re-training and parameter sharing with harder cluster assignments for compressing deep convolutions. In: International Conference on Machine Learning (ICML), pp. 5363–5372. PMLR (2018)
33. Zhu, J., Pei, J.: Progressive Kernel pruning with saliency mapping of input-output channels. Neurocomputing **467**, 360–378 (2022)
34. Zhu, J., Zhao, Y., Pei, J.: Progressive kernel pruning based on the information mapping sparse index for CNN compression. IEEE Access **9**, 10974–10987 (2021)
35. Zhuang, Z., et al.: Discrimination-aware channel pruning for deep neural networks. In: Bengio, S., Wallach, H., Larochelle, H., Grauman, K., Cesa-Bianchi, N., Garnett, R. (eds.) Advances in Neural Information Processing Systems 31 (NeurIPS), pp. 881–892. Curran Associates, Inc. (2018)
36. Zuo, Y., Chen, B., Shi, T., Sun, M.: Filter pruning without damaging networks capacity. IEEE Access **8**, 90924–90930 (2020)

Comparing Explanations from Glass-Box and Black-Box Machine-Learning Models

Michał Kuk[1]([✉])[iD], Szymon Bobek[2][iD], and Grzegorz J. Nalepa[2][iD]

[1] AGH University of Science and Technology, Krakow, Poland
m18.kuk@gmail.com

[2] Jagiellonian Human-Centered Artificial Intelligence Laboratory (JAHCAI) and Institute of Applied Computer Science, Jagiellonian University, Krakow, Poland

Abstract. Explainable Artificial Intelligence (XAI) aims at introducing transparency and intelligibility into the decision-making process of AI systems. In recent years, most efforts were made to build XAI algorithms that are able to explain black-box models. However, in many cases, including medical and industrial applications, the explanation of a decision may be worth equally or even more than the decision itself. This imposes a question about the quality of explanations. In this work, we aim at investigating how the explanations derived from black-box models combined with XAI algorithms differ from those obtained from inherently interpretable glass-box models. We also aim at answering the question whether there are justified cases to use less accurate glass-box models instead of complex black-box approaches. We perform our study on publicly available datasets.

Keywords: Explainable AI · Machine learning · Artificial intelligence · Data mining

1 Introduction

In recent years, the impact of machine learning on our daily life increased significantly, providing invaluable support to the decision making process in many domains. In insensitive areas, such as healthcare, industry, and law, where every decision may have serious consequences, the adoption of AI systems that cannot justify or explain their decisions is difficult and in many cases not desired. Such an observation stays in contradiction to the trend in the AI world, where the most progress is observed in the area of black-box models such as deep neural networks, random forests, etc. This duality led to the development of explainable AI methods, which allows introducing *transparency* and *intelligibility* to the decisions made by not interpretable black-box models. However, this transparency and intelligibility may serve different purposes, depending on the application area and the task that is to be solved with the AI method. In particular, we can define two main goals of XAI methods:

D. Groen et al. (Eds.): ICCS 2022, LNCS 13352, pp. 668–675, 2022.
https://doi.org/10.1007/978-3-031-08757-8_55

1) understand the mechanics of the ML model in order to debug the model, and possibly the dataset (i.e., what input drives the *model* to classify instance A as class C,

2) understand the phenomenon that is being modelled with AI methods and to build trust (i.e., what input makes the *instance A* to be classified as C).

While these goals might be indistinguishable at first glance, there is a fundamental difference in the assumptions that need to be fulfilled in both cases. In the first case, we assume that the model might be wrong, and we want to fix it, hence the information about the *model* is the most important. The model performance is an objective. In the second case, we assume that the model is correct and we want to use it to obtain information about the *class* or *instance* itself. The explanation itself is the main objective. In this paper, we focus on the second case. We provide a discussion on the performance of the glass-box models and black-box models explained with the use of XAI methods, in the situation when the objective is not to learn about the AI model, but about the phenomenon the model captures. We focused on rule-based explanations as one of the most understandable and widely applicable methods in industrial and medical cases. We performed a comparison of these two approaches on datasets from selected scikit-learn datasets and UCI Machine Learning Repository to see if a simple glass-box model can outperform complex XAI algorithms.

The rest of the paper is organised as follows: In Sect. 2 we describe a few papers which concern the explainable methods and we introduce our motivations. In Sect. 3 we present our approach to the performance comparison mentioned above. Next, Sect. 4 presents and discusses the results we obtained. Finally, in Sect. 5 we summary our work.

2 Related Works and Motivation

In this paper, we focused to evaluate what is the difference between the explanations obtained from black-box models combined with XAI algorithms and from those obtained based on interpretable glass-box models. To get such evaluation, firstly we verified the existing researches which concern glass-box and black-box models in the application of Explainable Artificial Intelligence.

In [10] the author makes a comparison of white and black box models. The author outlines that in some cases glass-box models could give as accurate results as the black box models. However, it strongly depends on the application domain and the data delivered. In the case of the black-box models, the author highlights that the experts do not need to understand the mathematical transformations behind them, but they proposed to deliver the output data in a similar form as input.

In [1] the authors pay attention to the fact that nowadays there is the need for XAI application due to commercial benefits, regulatory considerations, or in cases when the users have to effectively manage AI results. They outline that the black-box models do not disclose anything about internal design, structure, or implementation. On the other hand, the glass-box is completely exposed to the user.

In [6] the authors used glass-box and black-box models to predict the ambient black carbon concentration. They used several methods, whereas a neural network with LSTM layers gave the best results. However, they highlight that using black-box models like neural network or random forest complicates explanations.

In [9] the authors used the Anchor algorithm to obtain rules which could be explanations of each cluster of data. As a result, the proposed methodology is able to generate human-understandable rules which could be passed to the experts to support in the explainability process.

In [14] the authors used the black-box model to develop the structured attack (StrAttack), which is able to explore group sparsity in adversarial perturbation by sliding a mask through images. They demonstrate the developed method on datasets consisting of images. Furthermore, they outline that thanks to the sliding masks, they increase the interpretability of the model.

In [15] the authors also concentrate on image classification. They used a deep neural network to assign input to predefined classes. To interpret the models, they considered post-hoc interpretations. More specifically, they focus on the impact of the feature on the predicted result – they tried to uncover the casual relations between input and output.

In [11] the authors created an open-source Python package called InterpretML. They focused in most cases on the feature importance explanations, not on the human-readable rules.

I our work we focus on rule-based explanations, which according to our previous research [3] proves to be one of the most intelligible way of providing explanations to experts. Therefore, we mainly concentrate on the algorithms which can generate explanations which are represented as a logic implication (IF-THEN) by using a conjunction of relational statements. Such explanations can be executed with rule-based engines and verified according to selected metrics such as accuracy, precision, or recall. Having that, a research question arises: If the explanation is as much valuable as the model decisions, what kind of model should be applied to assure good interpretability along with high accuracy of explanations? Should it be 1) a glass-box model to directly generate explainable results, or 2) a complex black-box model and explain its results with Explainable Artificial Intelligence methods? To solve this research problem, we aim to apply XAI methods that are able to generate human-readable rules for complex black-box models and verify if these methods are needed to be applied in the case of simple tabular data or if we should make explanations directly with the use of the glass-box models.

3   Experimental Comparison

The scope of this work concentrates on the comparison of using simple glass-box models with complex black-box models explained with the XAI algorithm. In this work, we use a classification task as an exemplary problem to be solved. We considered the most popular classifiers available in the scikit-learn package [12].

For glass-box models, we selected: Decision Tree, Nearest Neighbors. For black-box models, we chose RBF SVM, Gauss Process, Random Forest, Neural Network, AdaBoost, Naive Bayes, QDA. We used a default models' settings as hyperparameter tuning was not the main goal of this paper. Our experiment considers two approaches for solving the classification task: 1) use directly explainable glass-box model, 2) use the black-box model, explain the predictions with XAI method, and solve the main problem based on the explanations obtained.

In the second approach, we took into consideration the XAI algorithms which generate explanations in the form of human-readable rules based on the trained classifier model. In this work, we considered three XAI methods: Anchor [13], Lux [4] and Lore [7]. Each of them generates instance-based explanations (rules), which subsequently were converted into XTT2 format [8]. To allow the results to be compared with glass-box models we used HEARTDROID inference engine [5] to predict the classification target (label) based on the obtained rules and data instances.

The schematic illustration of the considered approaches is presented in the Fig. 1.

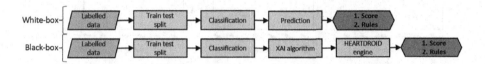

Fig. 1. Glass-box and black-box models approaches comparison.

To test the considered approaches and draw reliable conclusions, we chose data from different sources as an input to the experiments. In the work we used the following datasets: banknote and glass[1], cancer and iris[2], and titanic[3].

Each dataset has been divided into train and test datasets. For each considered classifier, we applied the same train instances to train the model and test instances to make predictions to maximize the reliability of the comparison results. For both considered approaches, we computed the accuracy, recall, and precision scores for each considered classifier to compare these two approaches.

4   Results and Discussion

In this section, we present the results of our experiments. Firstly, we compared the performance of all considered classifiers used directly to solve the classification problem. Then, we compared scores for the classification problem solved based on the rules generated for the black-box models with XAI methods (only the best results for each XAI method) vs glass-box model. Finally, we also considered the variance of scores that can be obtained for a selected XAI method depending on the black-box model explained. These results are presented in the following figures.

Figure 2 shows the results (scores) which were calculated for all classifiers used directly to predict the classification target (label). As can be seen, there are some datasets for which all classifiers perform with a high score (close to 1.0) such as iris and banknote that suggest that the problem to be solved in their case is relatively simple. For cancer and wine datasets, most classifiers also give high scores, but others (e.g. QDA, RBF, SVM) perform much worse. The most difficult problem to be solved is contained in the glass dataset for which the best score is lower than 0.7. Other difficult dataset is titanic for which scores are not greater than 0.8. Comparing different classifiers across

---

[1] See: https://archive.ics.uci.edu/ml.
[2] See: https://scikit-learn.org/stable/datasets.html.
[3] See: https://www.kaggle.com/datasets.

all considered datasets, it can be noticed that the glass-box Decision Tree model gives comparable results to black-box models. For the most difficult dataset (glass), it gives the highest precision, keeping recall and accuracy at a competitive level.

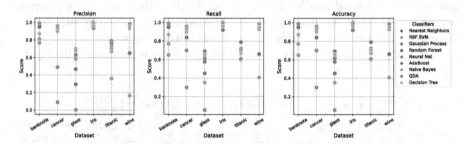

Fig. 2. Classifiers score comparison.

Figure 3 presents the comparison of the glass-box model results with the black-box models explanations obtained with XAI methods and executed with the use of HEARTDROID engine. In this figure, only the best scores for each dataset for each of the XAI methods are presented to compare the best possible results that can be obtained with a particular XAI method. In the case of using of XAI methods to generate rules, in Fig. 3 we can observe that the Anchor algorithm gives the best results in most datasets, but the Lux and Lore algorithms give noticeably worse results. In the case of the Lore algorithm, we noticed some of the bugs which resulted in scores equal to 0 which were marked on the charts. Only for the iris dataset (probably the easiest one), the Lux XAI method gives better results than the Anchor algorithm. Comparing the results from the exampled black-box models with Decision Tree, it can be noticed that for simple datasets like banknote, iris, or glass, the glass-box model gives better results. For slightly more difficult but still simple datasets (cancer and wine), slightly better results are obtained with the Anchor algorithm than the Decision Tree. However, the difference is not significant. In the case of more complex datasets like Titanic, glass-box model gives considerably worse results than the black-box model explained with the Anchor method.

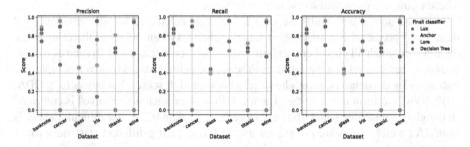

Fig. 3. XAI methods score comparison to glass-box model.

Figure 4 shows the variance of the performance obtained with the Anchor algorithm, as it gives the best results from the considered XAI methods, depending on the classification model used. We can observe the biggest score variance for cancer and wine datasets (relatively simple) for which the results strongly depend on the classifier model explained. The lowest variance is observed for glass and iris datasets, so the most and the least difficult datasets.

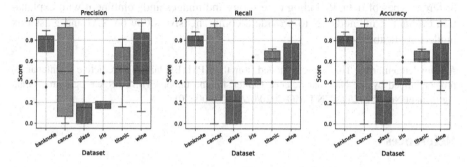

Fig. 4. XAI methods dependency on classifiers.

The obtained results allow us to compare how the explanations derived from black-box models combined with XAI algorithms differ from those obtained by interpretable glass-box models. Executed experiment proves that despite the fact that the black-box models are more complex and universal than the simple glass-box models, there is no need to apply them, especially for simple datasets. We found out that Decision Tree classifier gives competitive results and provides the model in an easily understandable format. However, in some examples, even in relatively simple datasets, it can be beneficial to apply more complex explanation methods (black-box model linked with XAI method) than simple glass-box models. Obtained results suggest also that when we need to consider more complex cases, better results can be obtained using the black-box models explained with XAI methods with human-readable rules. However, the final results strongly depend on the selected classifier. Hence, the properly chosen model which is treated as an input to the XAI method is important and has a significant impact on the final result.

5   Summary

In this paper, we made a glass-box and black-box classifiers comparison with the application in the explainable artificial intelligence area. The main goal of the work was to investigate if we should use the glass-box models to directly generate explanations or rather use a complex black-box model linked with XAI methods? We compared the classification scores for several classification methods and then we used the same trained models to obtain results with the use of XAI algorithm methods. We conducted our experiments based on the publicly available datasets. The results suggest that especially in the case of tabular data, it is worth investing resources into research on inherently explainable models, instead of relying on a combination of black-box and XAI algorithms. However, taking into account more complex analyses that concern e.g. embeddings or latent semantic analysis uses of glass-box models could be insufficient and then, black-box models with XAI methods could be applied. However, the choice should be made carefully, with additional evaluation of XAI results to select the most suitable approach. In future work, we plan to extend this analysis, taking into account different types of data, including time series and images and combine it with explanation evaluation methods [2] to provide a comprehensive study on XAI and glass-box models applicability.

Acknowledgements. This paper is funded from the XPM (Explainable Predictive Maintenance) project funded by the National Science Center, Poland under CHIST-ERA programme Grant Agreement No. 857925 (NCN UMO-2020/02/Y/ST6/00070)

References

1. Adadi, A., Berrada, M.: Peeking inside the black-box: a survey on explainable artificial intelligence (XAI). IEEE Access **6**, 52138–52160 (2018). https://doi.org/10.1109/ACCESS.2018.2870052
2. Bobek, S., Bałaga, P., Nalepa, G.J.: Towards model-agnostic ensemble explanations. In: Paszynski, M., Kranzlmüller, D., Krzhizhanovskaya, V.V., Dongarra, J.J., Sloot, P.M.A. (eds.) ICCS 2021. LNCS, vol. 12745, pp. 39–51. Springer, Cham (2021). https://doi.org/10.1007/978-3-030-77970-2_4
3. Bobek, S., Kuk, M., Brzegowski, J., Brzychczy, E., Nalepa, G.J.: KNAC: an approach for enhancing cluster analysis with background knowledge and explanations. CoRR abs/2112.08759 (2021), https://arxiv.org/abs/2112.08759
4. Bobek, S., Nalepa, G.J.: Introducing uncertainty into explainable AI methods. In: Paszynski, M., Kranzlmüller, D., Krzhizhanovskaya, V.V., Dongarra, J.J., Sloot, P.M.A. (eds.) ICCS 2021. LNCS, vol. 12747, pp. 444–457. Springer, Cham (2021). https://doi.org/10.1007/978-3-030-77980-1_34
5. Bobek, S., Nalepa, G.J., Ślażyński, M.: HeaRTDroid - rule engine for mobile and context-aware expert systems. Expert Syst. **36**(1), e12328 (2019)
6. Fung, P.L., et al.: Evaluation of white-box versus black-box machine learning models in estimating ambient black carbon concentration. J. Aerosol Sci. **152**, 105694 (2021)
7. Guidotti, R., Monreale, A., Ruggieri, S., Pedreschi, D., Turini, F., Giannotti, F.: Local rule-based explanations of black box decision systems. ArXiv abs/1805.10820 (2018)
8. Kaczor, K., Nalepa, G.J.: Critical evaluation of the XTT2 rule representation through comparison with CLIPS. In: KESE@ECAI (2012)
9. Kuk, M., Bobek, S., Nalepa, G.J.: Explainable clustering with multidimensional bounding boxes, pp. 1–10 (2021). https://doi.org/10.1109/DSAA53316.2021.9564220
10. Loyola-González, O.: Black-box vs. white-box: understanding their advantages and weaknesses from a practical point of view. IEEE Access **7**, 154096–154113 (2019). https://doi.org/10.1109/ACCESS.2019.2949286
11. Nori, H., Jenkins, S., Koch, P., Caruana, R.: Interpretml: a unified framework for machine learning interpretability (2019)
12. Pedregosa, F., Varoquaux, G., Gramfort, A., et al.: Scikit-learn: machine learning in Python. J. Mach. Learn. Res. **12**, 2825–2830 (2011)

13. Ribeiro, M.T., Singh, S., Guestrin, C.: Anchors: high-precision model-agnostic explanations. In: AAAI (2018)
14. Xu, K., et al.: Structured adversarial attack: towards general implementation and better interpretability (2019)
15. Zhang, X., Wang, N., Shen, H., Ji, S., Luo, X., Wang, T.: Interpretable deep learning under fire (2019)

Virtual Reality Prototype of a Linear Accelerator Simulator for Oncological Radiotherapy Training

Vei S. Chan[1]([✉]), Andrés Iglesias[2,3]([✉]), Habibollah Haron[1], Pedro J. Prada[4], Samuel Ruiz[4], Akemi Gálvez[2,3], Lihua You[5], Faezah M. Salleh[6], and Farhan Mohamed[1,7]

[1] School of Computing, Faculty of Engineering, University of Technology Malaysia, 81310 Johor Bahru, Malaysia
vschan2@live.utm.my, habib@utm.my
[2] Department of Applied Mathematics and Computational Sciences, University of Cantabria, 39005 Santander, Spain
{iglesias,galveza}@unican.es
[3] Faculty of Sciences, Toho University, 2-2-1 Miyama, Funabashi 274-8510, Japan
[4] Hospital Universitario Marqués de Valdecilla, 39008 Santander, Spain
jpedraja@hvvaldecilla.es, samuel.ruiz@scsalud.es
[5] National Center for Computer Animation, Faculty of Media and Communication, Bournemouth University, Poole BH12 5BB, UK
lyou@bournemouth.ac.uk
[6] Department of Biosciences, Faculty of Science, University of Technology Malaysia, 81310 Johor Bahru, Malaysia
faezah@utm.my
[7] Media and Game Innovation Centre of Excellence, 81310 Johor Bahru, Malaysia
farhan@utm.my

Abstract. Learning to operate medical equipment is one of the essential skills for providing efficient treatment to patients. One of the current problems faced by many medical institutions is the lack or shortage of specialized infrastructure for medical practitioners to conduct hands-on training. Medical equipment is mostly used for patients, limiting training time drastically. Virtual simulation can help alleviate this problem by providing the virtual embodiment of the medical facility in an affordable manner. This paper reports the current results of an ongoing project aimed at providing virtual reality-based technical training on various medical equipment to radiophysicist trainees. In particular, we introduce a virtual reality (VR) prototype of a linear accelerator simulator for oncological radiotherapy training. The paper discusses the main challenges and features of the VR prototype, including the system design and implementation. A key factor for trainees' access and usability is the user interface, particularly tailored in our prototype to provide a powerful and versatile yet friendly user interaction.

Supported by European Union Horizon 2020 project PDE-GIR (Ref. MSCA-RISE-778035), Malaysia's Fundamental Research Grant Scheme (No. 5F395) and project of Ref. TIN2017-89275-R, of the MCIN/AEI/10.13039/501100011033/FEDER, Spain.

Keywords: Virtual reality · Linear accelerator · Medical simulator training · Oncological radiotherapy · Head-mounted display · User interface

1   Introduction

With the increasing number of cancer cases worldwide, there is urging need for highly-skilled specialists from the various disciplines involved in the prevention, diagnosis, monitoring, and treatment of cancer. An illustrative example is given by oncological radiotherapy (ORT), a field where different types of oncologists work in close cooperation with radiophysicists, a type of medical physicists with the technical ability to operate medical radiation equipment efficiently [1,4].

Radiophysicists typically work with linear accelerators (LINAC), sophisticated devices used to speed up charged subatomic particles or ions through a series of oscillating electric potentials. Oncological radiotherapy linear accelerators (ORTLINAC) are used for procedures such as intensity-modulated radiation therapy (IMRT), a level-3 high-precision technique that combines the use of computer tomography (CT) imaging and multileaf collimators. In IMRT, CT is used to get a volumetric representation of the tumor, while the collimators use a set of individual "leaves" equipped with independent linear in/out movement (orthogonal to the radiotherapy beam) to fit the treatment volume to the boundary shape to the tumor and vary the radiation signal intensity accordingly. In this way, IMRT is used to deliver precise radiation doses at targeted areas within the tumor, thus reducing the radiation impact on healthy organs and tissues.

Unfortunately, IMRT requires a lot of expertise and considerable experience for optimal performance. For instance, an individual radiation treatment planning (RTP) must be set up for each patient before the therapy sessions. The plan needs to consider several factors and parameters, such as the region of interest (ROI) where the radiation beam will focus, the most suitable beam type, the energy to be applied, the appointment schedule of therapy sessions and many others. The RTP is intended at maximizing the treatment effectiveness while minimizing the physical strain upon the patients. Once all details of the RTP are agreed, radiation sessions are set on place. During the radiotherapy session, the practitioners arrange the patients on a motorised table with six degrees of freedom. Then, they use remote controls to move the table and match the tumor location and orientation to the radiation beam's focal point of ORTLINAC. The whole procedure is highly-demanding in terms of concentration and skills to get the precise position. As a result, intensive practising is required for the medical trainees to master the positioning of the patients to the precise radiation beam's focal point by using the remote controls. However, the ORTLINAC therapy schedule is often full due to its high demand [1]. The medical trainees have scarce time to access the facility for practising [5]. Moreover, it is not affordable to allocate ORTLINAC rooms for training purposes owing to their high costs [1,4]. In this context, virtual reality (VR) emerges as a suitable technology to simulate the real medical environment in the virtual world, allowing the trainees to conduct hands-on practice and let them accustomed to the environment.

Previous studies explored the usage of VR in training and education in various domains [14,19,20]. These works show that VR technology contributes to psychomotor or technical skills development, knowledge transfer, and social skills. In the radiotherapy field, there is also research work using VR technology for training the novice to operate the medical equipment [24]. VR is also used to inform the patients about the therapy session and reduce their anxiety [15,23]. Nevertheless, realism is still an issue because most studies visualise the virtual operation in the ORTLINAC room. In contrast, the practitioners in real-life situations are also involved in operating the equipment remotely in another control room. Therefore, both the ORTLINAC room and the control room must be fully integrated and coordinated in the VR simulator for realistic training. In addition, the user interaction in the VR simulator must be as similar as possible to that in the real-world setting. These are the goals of the present contribution.

This paper introduces a VR prototype of a unified system comprised of the ORTLINAC room and the control room along with their interactions. The paper describes the main tasks of the system design and implementation, including the research workflow to determine the user requirements and to address the user interaction issues. The structure of this paper is as follows: Sect. 2 reports previous work regarding VR for medical science. Section 3 describes the workflow for the design and implementation of the VR prototype introduced in this paper. Then, Sect. 4 shows the main results of the implementation. Lastly, Sect. 5 discusses the conclusions and future work in the field.

2   Previous Work

2.1   Virtual Simulation for Medical Science

The presence of virtual simulation is soundly significant in many areas of the medical science, such as radiation therapy [5], radiography [13], and surgery [24]. The 3D visualisation of a patient's body and internal organs is helpful for medical practitioners to make a treatment plan and discuss it with their peers. The large-size wall display can also help them present and collaborate effectively. In radiotherapy, most institutions use the virtual environment for radiotherapy training (VERT) system for such purposes [21]. Besides, the 3D view and virtual simulation can also benefit the teaching and learning process [9].

2.2   Oncology Radiotherapy Training Issues

To provide efficient treatment to the patients, quality and effective medical education are of utmost importance. Training is also essential in solving the shortage of qualified staff operating the medical equipment for radiotherapy [6,21]. Here we discuss some issues found in conventional ORTLINAC radiotherapy training.

Need to Learn Diverse Skills. Radiotherapy workflow involves many medical knowledge and skills [10]. According to [2,4], the workflow includes: (a) CT scanning to obtain the imaging data of patient's anatomy; (b) segmentation to extract the region of interest (organ, tumor); (c) treatment planning and evaluation; (d) quality assurance to avoid patient injury; (e) image-guided to place the patient on the ORTLINAC; and (f) perform the radiotherapy. Aside from the technical skills, the novice also needs to learn to communicate with patients and provide patient-care service [4]. These factors require the trainees to spend much time mastering these skills. However, the ORTLINAC equipment is often in high demand, giving the trainees sparse access for practising [1,5,13].

Patient Care Issues. The need to ensure patient safety and well-being during the ORT sessions is a big concern, as it may cause psychological pressure for the trainees [5,16]. Using phantoms might help minimize this issue [8]. However, the limited access to the ORTLINAC still affects the training progress.

Limitations of the 2D Medium. Conventional treatment planning and demonstration use 2D media to explain the concept, such as 2D imaging slice view and printed medium. Previous studies found it challenging to visualise and explain the spatial relationship between organs and anatomy and how the radiation beam affects these organs [9]. Therefore, 3D and immersive techniques can help effectively explain these spatial concepts to the trainees.

2.3 VR for Radiotherapy Education and Training

Several research works addressed the use of VR technology for RT training, including skin apposition application [3], medical imaging [13], breast cancer [15], prostate cancer [18], medical dosimetry [9,17], and brachytherapy [24].

The VERT System. Most of the VR training approaches make use of the VERT system, consisting of a wall-size display and an actual hand pendant controller to operate the virtual ORTLINAC. The studies show that VERT provides an optimal environment for hands-on practice [4,21]. The virtual simulation can offer a safe working environment where the trainees can practise by trial and error without the fear of injuring the actual patients [4,12]. The work in [5] reported that most of the trainees utilised VR simulation training significantly whenever it is available. This study is consistent with the high level of satisfaction and enjoyment among the trainees, as reported by [3,9,10,22]. These results showed that VR simulation could provide a conducive environment and a valuable opportunity for practice, which can help the trainees to master the skills effectively and in shorter time [9]. Besides, the large screen display of VERT helps the educators to demonstrate and explain therapy concepts in a classroom setting [17,21]. This display can help the trainees understand the spatial relationship between the radiation beam and the target organs [21] and allow more engagement and discussion of learned knowledge into the professional conversation [10]. These results provide evidence that VR-based training fosters the trainees' development and confidence in operating the radiotherapy equipment.

However, previous studies reported several limitations of the VERT system. First is the realism issue. Most users stated that the VERT lacks immersion and does not provide tactile feedback when the collision occurs [3,12]. A few

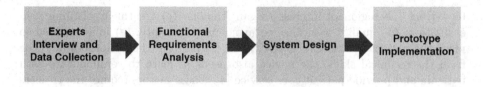

Fig. 1. Our design and implementation flowchart.

studies showed that the students obtained less performance and task accuracy in actual treatment planning after practice using virtual simulation units compared to conventional simulation practice by using the actual unit or the treatment planning software [12,17]. Some users also criticised the complex control of the VERT system as a limiting factor for their training and skill development [4].

Secondly, [10] revealed the lack of autonomous and self-directed learning in their blended learning framework. This issue is possibly caused by the large display unit where the student has less opportunity to conduct the practice by him/herself. Moreover, the COVID-19 pandemic also caused the cancellation of many on-site clinical practices [5], which further exacerbates the usual access limitation to the learning facility. Clearly, there is a need to explore an alternative immersive technology that can allow distance learning but without these issues.

Head-Mounted Displays. The advancement of VR technology allows the increased affordability of small-size equipment. The head-mounted display (HMD) is one such VR equipment that can solve the issues found in large 3D displays. Authors in [1] created an HMD VR application for ORTLINAC training, resulting in better learnability and effectiveness in training radiotherapy compared to VERT. The work in [24] utilised the room-scale VR headset HTC Vive for brachytherapy training to improve the trainees' technical skills. With the recent research trend in collaborative VR [11], HMDs can improve both autonomous and group learning environments.

Based on [10], medical practitioners and experts' involvement can help design the software and education curriculum to fulfil the real-life situation. Since the practitioners spend most of the time in the control room, there is a requirement to simulate the virtual embodiment in the control room where they have limited view and need to depend on the camera to operate the ORTLINAC. Accordingly, this paper emphasises the development of a virtual control room to train hand-eye coordination skills and spatial awareness in a limited viewing condition.

3   System Design and Implementation

This research work is part of an ongoing project aimed at providing VR-based technical training on various medical equipment to radiophysicist trainees. Although the full project is still a work in progress, we think that it has already reached significant results to justify publication. The work described in this paper concerns the development of a workable medical simulation training system to simulate the real-life working condition of ORTLINAC based on user requirements. In this context, this section focuses on the design and implementation workflow plan for creating a VR prototype of the ORTLINAC simulator for ORT training. Figure 1 shows the main steps of the design and implementation flowchart. They are described in detail in next subsections.

3.1   On-site Medical Facility Visits and Meetings with Experts

The first step of the process involves visits to medical facilities and meetings with medical practitioners to elicit the functional requirements of the system. Some authors visited the medical facilities in the oncology department of Hospital Universitario Marqués de Valdecilla, Santander, Spain, where they were presented the daily operation in both the ORTLINAC therapy room and the remote control room, including the features of ORTLINAC, how to control the ORTLINAC and some standard procedures in radiotherapy. The authors collected photos, videos and other materials to analyse and design the VR training system.

3.2   Functional Requirements

After the visits to the medical facilities and meetings with experts, the authors analysed the collected materials (video transcripts, photos, printed materials, and others) to extract the functional requirements for the VR ORTLINAC training system. Figure 2 shows the use case diagram of the VR medical simulation training system.

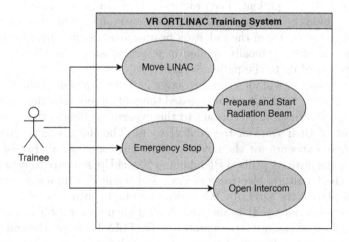

Fig. 2. The use case diagram of the VR ORTLINAC training system.

The trainee is the primary use case actor to use the VR ORTLINAC training system for practising. Table 1 shows the analysed functional requirements and their description.

Table 1. Functional requirements and their description.

ID	Requirement	Description
RQ1	Move LINAC	Trainee can move the position of table and gantry by pressing the button in the remote control panel
RQ2	Prepare and Start Radiation Beam	Trainee can choose, prepare, start and stop the radiation beam
RQ3	Emergency Stop	Trainee can perform an emergency stop to shut down the ORTLINAC immediately
RQ4	Open Intercom	Trainee can open the intercom and give instructions to the patient to adjust his/her position remotely during the radiotherapy session

3.3 System Design

As indicated above, there are two rooms for our VR-based training system: the ORTLINAC radiotherapy room and the control room. During the radiotherapy session, the practitioners in the radiotherapy room place and fasten the patient on the motorised table of ORTLINAC to stabilise and fix the patient' position. In some cases, an individually customized plastic mask is provided to the patient to wear during treatment. After this, any other person than the patient should leave the room to avoid the harmful effects of the radiation. The LINAC machine is operated from the control room, where the patient can be tracked through a window and/or one or several cameras. There is also an intercom for oral communication with the patient. The practitioners in the control room use different controls to guide the motion of the table and gantry and align the tumor's region of interest to the centre of the radiation beam's focal point. This external radiotherapy procedure is typically applied in several sessions distributed over days and weeks according to the patient's RTP.

In this paper, we will focus on the design and development of the virtual control room. Firstly, we designed and created the control room simulation according to the real-life situation. According to the experts' feedback, the practitioners spent most of their time in the control room. Therefore, this simulation can provide more exposure for the novices to the environment. To furnish the virtual scene, the authors utilised Blender and SketchUp software to create the 3D model of the furniture, electronic devices and medical equipment, such as the camera display of the ORTLINAC device room and control panels.

The user interface (UI) is also an essential element for interacting with the virtual world. For example, to operate the ORTLINAC using the control panel and view the camera display. We identified several design considerations, leading to different versions throughout the design process. The first design version relied on virtual buttons for user interaction, allowing the users to click on the buttons of the 3D model to perform different actions. However, this feature may cause navigation difficulties for the trainees because they may accidentally click on another nearby button. The alternative solution of increasing the size or changing the buttons' orientation may reduce the realism and familiarity to the actual control panel. Therefore, this work proposes sign-posting and annotation above the 3D models to attract the users to click on the button in the VR world. Once clicked, it will open a larger user interface panel that displays the camera view and control panel layout for easy viewing and selection, respectively. In this way, the 3D medical equipment can be displayed in VR at its original real-life size scale, thus improving the realism of the system.

3.4   Implementation

The VR ORTLINAC training system was developed in Unity3D with the Oculus Integration package. This package provides various templates and prefabs to develop a VR application in Unity3D, including an avatar framework and customised configurations. This work also used the Oculus Quest as the VR head-mounted display (HMD) with two Oculus Touch controllers for user interaction. The reason to use Oculus Quest is that it is a standalone system, requires fewer set-up procedures, and is very ubiquitous to carry around. In addition, Blender and SketchUp were used to create the 3D models of the medical equipment, electronic devices, furniture and room. Blender supports texture mapping on the 3D models to improve their visual realism. Table 2 shows the hardware and software specifications used in this work along with their versions.

Table 2. Hardware and software specifications.

Name	Category	Specification
Unity3D	Software -> Game Engine	Version: 2020.3.25f1
Oculus Integration	Software -> Unity Asset	Version: 37.0
Blender	Software -> 3D Modelling	Version: 3.0.0
SketchUp	Software -> 3D Modelling	Version: Pro 2022
Oculus Quest	Hardware -> HMD	Generation: 1
		Software version: 37.0

After creating the 3D models in Blender and SketchUp, they were imported in Unity3D to build the virtual control room scene according to the sketch and requirements. The Unity UI can implement the UI design and user interaction based on our design considerations. Furthermore, the VR system included the room-scale locomotion feature to let the users physically walk around the virtual room. Another VR feature is to show the 3D models of users' hands with a Touch controller and cast a laser pointer from the right-hand controller to allow the users to point and click on the virtual button. The inclusion of virtual hands can also improve the users' perception of presence in the VR world. Lastly, the VR training system was built as an Android application package (APK) file and deployed in the Oculus Quest. Additional visualization on smartphones and tablets has also been developed and is fully supported.

4 Results

This section presents the results of the VR ORTLINAC training prototype. There are three main components: the virtual remote control room, the ORTLINAC radiotherapy room, and the proposed UI and user interaction.

4.1 Remote Control Room

Figure 3 shows a scene comparison between the real (top) and virtual (bottom) environments. As the reader can see, the virtual simulation was created as similar as possible to the actual control room, including the furniture, electronic devices, and interior layout. Real-world textures were extracted from the photos and applied on the virtual surfaces to improve the overall realism of the scene.

Fig. 3. Comparison of actual (top) and virtual (bottom) control rooms.

Figure 4 shows the comparison of the 3D models of the camera display (left) and control panel (right). We also mapped the button icon and text annotation textures on the 3D models, based on their appearance in the actual equipment.

4.2 ORTLINAC Radiotherapy Room

For the ORTLINAC room, the current work focused on the 3D modelling of ORTLINAC equipment and the position of cameras. The authors edited and

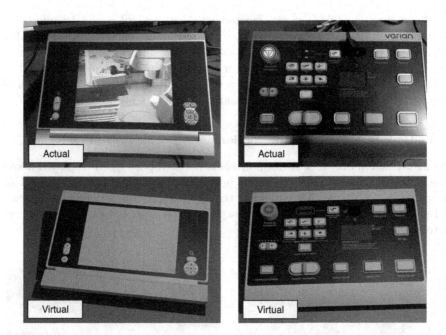

Fig. 4. The actual (top) and virtual (bottom) models of the camera display (left) and control panel (right).

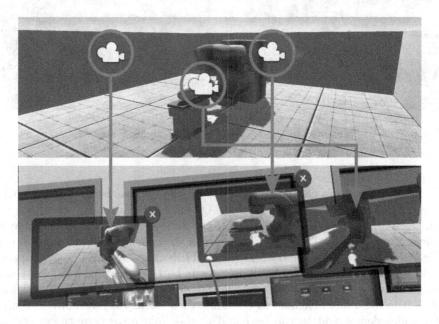

Fig. 5. The camera positioning in the ORTLINAC radiotherapy room (top) and their displays in the virtual control room (bottom).

modified the 3D LINAC model created by [7] in SketchUp and Blender, and shown in Fig. 5(top). Meanwhile, we set the positioning of cameras in the radiotherapy room in order to project the camera view into the camera displays in the virtual control room by using render texture mapping in Unity3D. Based on these features, this work simulates the remote control room successfully. Figure 5 shows the multiple camera positions and their displays in the control room.

4.3   User Interface and Interaction

As mentioned above, this work implements the sign-posting UI displayed above the 3D models to attract the users to point the controller's laser to the button, as shown in Fig. 6(left). After pressing the "A" button, this action opens a larger UI panel that displays the camera view or control panel layout for the users to interact, as shown in Fig. 6(right).

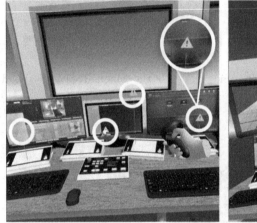

Fig. 6. The UIs presented in the VR ORTLINAC training system: sign-posting UIs (left); camera view and control panel layout UIs (right).

Besides, the users can point and click on the buttons in the control panel layout UI to control the movement of the ORTLINAC. Currently, this prototype only includes the functionalities to rotate the gantry and move the table linearly. The movement of ORTLINAC is reflected in the camera displays, as shown in Fig. 7. Hence, the users can observe the ORTLINAC position remotely. We also included the grab interaction by using the grip button in the left-hand controller when the virtual hand touches the UI. The users can grab any UI panel and place it in the desired location to customise their workspace, as shown in Fig. 8. The video demonstration can be found in this link: https://youtu.be/5YmY_0EsiLQ.

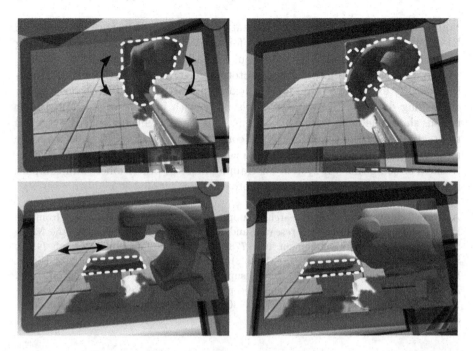

Fig. 7. The movement function for the rotation of the gantry and translation of the table before movement (left) and result after movement (right).

Fig. 8. The grab interaction to move the UI location: the grip button (left); the user activates his/her left hand on the UI panel (middle); using this feature to grab and drag the UI panel to other location (right).

5   Conclusions and Future Work

This paper presents a VR prototype of an ORTLINAC system for oncological radiotherapy training. Based on the experts' feedback, the ORTLINAC room and the control room are now integrated within a unified framework. Unity3D's render texture mapping functionalities are used to achieve the effect of the remote camera display for effective synchronization between both rooms. Also, several UI design considerations are proposed for the VR world, including sign-posting and displaying a larger UI button layout for easier user interaction.

We will continue improving the prototype in terms of functionalities, graphics, and user experience (UX) for the next step. Furthermore, 3D reconstruction of imaging data that shows a patient's body with internal organs and tumor regions should be included in this system. This feature can challenge the trainees to practice operating the ORTLINAC correctly and avoid collision between the patient and the gantry. In addition, we wish to evaluate the efficiency and effectiveness of using this VR system for training medical practitioners and compare it with the traditional pedagogical approach.

Acknowledgements. This research is supported by the PDE-GIR project from the European Union Horizon 2020 research and innovation programme under the Marie Skodowska-Curie grant agreement No. 778035. The Fundamental Research Grant Scheme (FRGS/1/2020/ICT06/UTM/02/2) under the Malaysias Ministry of Higher Education (MOHE) with vote No. 5F395 supported C.V. Siang for the doctoral study. A. Iglesias and A. Galvez also thank the projects TIN2017-89275-R and PID2021-127073OB-I00 funded by MCIN/AEI/10.13039/501100011033/FEDER "Una manera de hacer Europa".

References

1. Bannister, H., et al.: LINACVR: VR simulation for radiation therapy education. In: 25th ACM Symposium on Virtual Reality Software and Technology, pp. 1–10 (2019)
2. Boejen, A., Grau, C.: Virtual reality in radiation therapy training. Surg. Oncol. **20**(3), 185–188 (2011)
3. Bridge, P., Appleyard, R.M., Ward, J.W., Philips, R., Beavis, A.W.: The development and evaluation of a virtual radiotherapy treatment machine using an immersive visualisation environment. Comput. Educ. **49**(2), 481–494 (2007)
4. Bridge, P., Crowe, S.B., Gibson, G., Ellemor, N.J., Hargrave, C., Carmichael, M.: A virtual radiation therapy workflow training simulation. Radiography **22**(1), e59–e63 (2016)
5. Bridge, P., et al.: International audit of simulation use in pre-registration medical radiation science training. Radiography **27**(4), 1172–1178 (2021)
6. Burger, H., et al.: Bridging the radiotherapy education gap in Africa: lessons learnt from the cape town access to care training programme over the past 5 Years (2015–2019). J. Cancer Educ. 1–7 (2021). https://doi.org/10.1007/s13187-021-02010-5
7. Carina, C.: Linear Accelerator (2021). https://3dwarehouse.sketchup.com/model/u9d9c8922-fce6-4f94-8483-0c0b7bad6c39/Linear-Accelerator?hl=en
8. Chamunyonga, C., Burbery, J., Caldwell, P., Rutledge, P., Fielding, A., Crowe, S.: Utilising the virtual environment for radiotherapy training system to support undergraduate teaching of IMRT, VMAT, DCAT treatment planning, and QA concepts. J. Med. Imaging Radiat. Sci. **49**(1), 31–38 (2018)
9. Cheung, E.Y.W., Law, M.Y.Y., Cheung, F.: The Role of Virtual Environment for Radiotherapy Training (VERT) in Medical Dosimetry Education. J. Cancer Educ. **36**(2), 271–277 (2019). https://doi.org/10.1007/s13187-019-01622-2

10. Czaplinski, I., Fielding, A.L.: Developing a contextualised blended learning framework to enhance medical physics student learning and engagement. Physica Med. **72**, 22–29 (2020)
11. Ens, B., Bach, B., Cordeil, M., Engelke, U.: Grand challenges in immersive analytics. In: Conference on Human Factors in Computing Systems - Proceedings (2021)
12. Flinton, D.: Competency based assessment using a virtual environment for radiotherapy. Procedia Comput. Sci. **25**, 399–401 (2013)
13. Gunn, T., Rowntree, P., Starkey, D., Nissen, L.: The use of virtual reality computed tomography simulation within a medical imaging and a radiation therapy undergraduate programme. J. Med. Radiat. Sci. **68**(1), 28–36 (2021)
14. Isham, M.I.M., Haron, H.N.H., bin Mohamed, F., Siang, C.V.: VR welding kit: accuracy comparison between smartphone VR and standalone VR using RMSE. In: 2021 IEEE International Conference on Computing (ICOCO), pp. 341–346. IEEE (2021)
15. Jimenez, Y.A., Cumming, S., Wang, W., Stuart, K., Thwaites, D.I., Lewis, S.J.: Patient education using virtual reality increases knowledge and positive experience for breast cancer patients undergoing radiation therapy. Support. Care Cancer **26**(8), 2879–2888 (2018). https://doi.org/10.1007/s00520-018-4114-4
16. Kane, P.: Simulation-based education: a narrative review of the use of VERT in radiation therapy education (2018)
17. Leong, A., Herst, P., Kane, P.: VERT, a virtual clinical environment, enhances understanding of radiation therapy planning concepts. J. Med. Radiat. Sci. **65**(2), 97–105 (2018)
18. Marquess, M., et al.: A pilot study to determine if the use of a virtual reality education module reduces anxiety and increases comprehension in patients receiving radiation therapy. J. Radiat. Oncol. **6**(3), 317–322 (2017). https://doi.org/10.1007/s13566-017-0298-3
19. Na, K.S., Mohamed, F., Isham, M.I.M., Siang, C.V., Tasir, Z., Abas, M.A.: Virtual reality application integrated with learning analytics for enhancing english pronunciation: A conceptual framework. In: 2020 IEEE Conference on e-Learning, e-Management and e-Services (IC3e), pp. 82–87. IEEE (2020)
20. Novotny, J., et al.: Developing virtual reality visualizations for unsteady flow analysis of dinosaur track formation using scientific sketching. IEEE Trans. Vis. Comput. Graph. **25**(5), 2145–2154 (2019)
21. Phillips, R., et al.: Virtual reality training for radiotherapy becomes a reality. Stud. Health Technol. Inform. **132**, 366–371 (2008)
22. Ryan, E., Poole, C.: Impact of virtual learning environment on students' satisfaction, engagement, recall, and retention. J. Med. Imaging Radiat. Sci. **50**(3), 408–415 (2019)
23. Wang, L.J., Casto, B., Luh, J.Y., Wang, S.J.: Virtual reality-based education for patients undergoing radiation therapy. J. Cancer Educ. (3), 1–7 (2020). https://doi.org/10.1007/s13187-020-01870-7
24. Zhou, Z., Jiang, S., Yang, Z., Zhou, L.: Personalized planning and training system for brachytherapy based on virtual reality. Virt. Real. **23**(4), 347–361 (2018). https://doi.org/10.1007/s10055-018-0350-7

A Review of 3D Point Clouds Parameterization Methods

Zaiping Zhu[1]([✉]), Andres Iglesias[2,3], Lihua You[1], and Jian Jun Zhang[1]

[1] The National Center for Computer Animation, Bournemouth University, Poole, UK
s5319266@bournemouth.ac.uk
[2] Department of Applied Mathematics and Computational Sciences, University of Cantabria, 39005 Cantabria, Spain
[3] Department of Information Science, Faculty of Sciences, Toho University, 2-2-1 Miyama, Funabashi 274-8510, Japan

Abstract. 3D point clouds parameterization is a very important research topic in the fields of computer graphics and computer vision, which has many applications such as texturing, remeshing and morphing, etc. Different from mesh parameterization, point clouds parameterization is a more challenging task in general as there is normally no connectivity information between points. Due to this challenge, the papers on point clouds parameterization are not as many as those on mesh parameterization. To the best of our knowledge, there are no review papers about point clouds parameterization. In this paper, we present a survey of existing methods for parameterizing 3D point clouds. We start by introducing the applications and importance of point clouds parameterization before explaining some relevant concepts. According to the organization of the point clouds, we first divide point cloud parameterization methods into two groups: organized and unorganized ones. Since various methods for unorganized point cloud parameterization have been proposed, we further divide the group of unorganized point cloud parameterization methods into some subgroups based on the technique used for parameterization. The main ideas and properties of each method are discussed aiming to provide an overview of various methods and help with the selection of different methods for various applications.

Keywords: Parameterization · Organized point clouds · Unorganized point clouds · Mesh reconstruction

1 Introduction

3D point clouds parameterization, also called point clouds mapping, is the process of mapping a 3D point cloud onto a suitable (usually simpler) domain. It has many applications such as object classification, texture mapping and surface reconstruction [1–3]. In many situations, it is computationally expensive or difficult to work with 3D point clouds directly. Therefore, projecting them onto a lower-dimensional space without distorting their shape is necessary. Compared to mesh parameterization, 3D point clouds parameterization is more challenging in general because there is no connectivity information

D. Groen et al. (Eds.): ICCS 2022, LNCS 13352, pp. 690–703, 2022.
https://doi.org/10.1007/978-3-031-08757-8_57

between points, which hinders the direct extension of well-established mesh parameterization algorithms to point cloud parameterization. There are some survey papers on mesh parameterization [4, 5]. However, to the best of our knowledge, there are no survey papers about point clouds parameterization. In this paper, we will review the methods of parameterizing point clouds. Notice there are also some works on 2D point clouds parameterization. Since 2D point clouds parameterization is different from 3D point clouds parameterization in most cases, this paper will only focus on the methods of 3D point clouds parameterization.

Some methods have been proposed to parameterize point clouds. In this paper, we roughly divide them into two main groups according to whether point clouds are organized or not. For each of the two groups, we further divide it into some subgroups based on the property of the mapping process and review each of the methods.

2 Some Concepts

In this section, some concepts related to point clouds will be introduced to help readers understand the problem of point clouds parameterization. Since mesh parameterization has been well investigated in existing work and some ideas of mesh parameterization can be adopted by or adapted to point cloud parameterization, we will also introduce some concepts about mesh parameterization in this section.

1) **Organized and unorganized point clouds:** Generally, point clouds can be divided into organized and unorganized ones. Organized and unorganized point clouds are also called structured and unstructured point clouds, respectively. The division is determined by the way of storing point cloud data. For organized point clouds, the data are stored in a structured manner, while unorganized point cloud data are stored arbitrarily. Specifically, an organized point cloud is similar to a 2-D matrix and its data are divided into rows and columns according to the spatial relationships between the points. Accordingly, the spatial layout represented by the xyz-coordinates of the points in a point cloud decides the memory layout of the organized point cloud. Contrary to organized point clouds, unorganized point clouds are just a collection of 3-D coordinates, each of which denotes a single point.

2) **Global and local parameterization:** To parameterize point clouds, some methods map the whole point set of an underlying structure to a parameterization domain. In contrast, some other methods split the problem into several subproblems, each of which is called a local parameterization. The choice between global and local parameterization has impacts on mapping processes and results. Globally parameterizing the whole point set can guarantee the reconstructed mesh is a perfect manifold, meaning there are no seams, which may exist if the point cloud is partitioned and locally parameterized. However, processing the whole point cloud at the same time may be computationally expensive, especially for large structures.

3) **Topological shapes:** Topological shapes can be grouped based on the number of holes they own. Shapes with no holes such as spheres and bowls are treated as genus-0 shapes. Similarly, genus-1, genus-2 and genus-3 shapes have one, two and three holes in them, respectively, and so on.

4) **Bijective function:** also called bijection, invertible function, or one-to-one correspondence, pairs each element in one set exactly to one element in the other set, and
   vice versa.
5) **Isometric, conformal, and equiareal mappings:** Suppose f is a bijective function
   between a mesh S or a point cloud and a mapping domain S^*, then f is isometric
   (length preserving) if the length of any arcs on S is preserved on S^*; f is conformal
   (angle preserving) if the angle of intersection of every pair of intersecting arcs on S
   is preserved on S^*; f is equiareal (area preserving) if the area of an area element on S
   is preserved on S^*. Isometric mappings are equiareal and conformal. Any mappings
   that are equiareal and conformal are isometric mapping.

3   Parameterization Methods of Organized Point Clouds

To parameterize an organized point cloud, many methods iteratively obtain a topologi-
cally identical 2D triangulation from the underlying 3D triangulation of the point cloud,
and the 2D triangulation determines the parameter values of the vertices in the domain
plane. Depending on the ways of transforming from 3D to 2D, there are several methods, including Harmonic parameterization [6], Floater's barycentric mappings [7] and
the most Isometric parameterization [3]. For Harmonic parameterization in [6], the arc
length is regarded as the parameter value of a spline curve, which is used to minimize
the integral of the squared curvature with respect to the arc length for fairing the spline
curve. With regard to barycentric mappings in [7], a shape-preserving parameterization
method is applied for smooth surface fitting; the parameterization that is equivalent to a
planar triangulation can be obtained by solving a linear system based on the convex combination. In [3], Hormann and Greiner propose a method to parameterize triangulated
point clouds globally, the way of parameterizing inner point set is the same as that of
parameterizing boundary point set. However, they ignore the problem of parameterizing
triangulated point clouds with holes.

Energy function has also been defined to minimize the metric distortion in the transformation process from 3D to 2D. The methods described in [7, 8] follow the shared approach, which firstly parameterizes the boundary points, and then minimizes the following
edge-based energy function for the parameterization of inner points [3]:

$$E = \frac{1}{2} \sum c_{ij} \|P_i - P_j\|^2 \tag{1}$$

where c_{ij} is the edge coefficient that can be chosen in various ways, P_i and P_j are two
points at the same edge.

In order to reconstruct a tensor product B-spline surface from scattered 3D data with
specified topology, choosing a suitable way to parameterize the points is crucial in the
reconstruction process. The method adopted by Greiner and Hormann in [8] is called the
spring model. With this method, the edge of the 3D triangulation is replaced by a spring.
Then the boundary points are mapped first onto a plane and stay unchanged. Next, the
inner points are mapped onto this plane by minimizing the spring energy. The procedure
is repeated to improve the parameters until certain conditions are satisfied.

The above methods are mainly applicable to structured point clouds. They are not efficient when the number of points increases, and are likely to fail when holes and concave sections exist in the point clouds.

4 Parameterization Methods of Unorganized Point Clouds

In comparison with the parameterization of organized point clouds, many more methods have been proposed to parameterize unorganized point clouds. Table 1 lists these methods and gives the information about the category, parameter domain, local or global parameterization, topology, applications and publication year.

Table 1. Methods to parameterize unorganized point clouds.

Methods	Category	Parameter domain	Local/global parameterization	Topology	Applications	Year
"Simplicial" surface [10]	Base surfaces-based methods	Base surfaces	/	Arbitrary topology	Surface reconstruction	1992
Manually define [9]			Global	/	Least square fitting of B-spline curves and surfaces	1995
Minimizing quadratic function [11]			/	/	B-spline curves and surfaces approximation	2002
Recursive DBS [12]			Global/local	Disk	Efficient parameterization	2005
Recursive subdivision technique [13]			Global/local	Disk (With hole is ok)	Parameterizing point clouds	2007
Floater meshless parameterization [14–17]	Meshless parameterization	Plane	Global	Disk	Surface reconstruction	2000
Meshless parameterization for spherical topology [18]		Planes	Local	Genus-0	Surface reconstruction	2002
As-rigid-as-possible meshless parameterization [19]		Plane	Global	Disk	Denoising and parameterizing point clouds, mesh reconstruction	2010
Meshless quadrangulation by global parameterization [20]		Plane	Global	Arbitrary genus	Meshless quadrangulation	2011
Spherical embedding [23]	Spherical mapping	Sphere	Global	Genus-0	Mesh reconstruction	2004
3D point clouds parameterization algorithm [22]		Sphere	Global	Relatively simple models	Parameterizing point clouds	2008
Spherical conformal parameterization [21]		Sphere	Global	Genus-0	Mesh reconstruction	2016

(*continued*)

Table 1. (*continued*)

Methods	Category	Parameter domain	Local/global parameterization	Topology	Applications	Year
Discrete one-forms [24]	Adapt from mesh parameterization	Planes	Local	Genus-1	Mesh reconstruction	2006
Periodic global parameterization [25]		Plane	Global	Arbitrary genus	Direct quad-dominant meshing of point cloud	2011
PDE & SOM [26]	Neural networks-based methods	Adaptive base surface	Global	Complex sculptured surfaces	Surface reconstruction	2001
Adaptive sequential learning RBFnetworks [27]		/	Global	Freeform	Point-cloud surface parameterization	2013
Residual neural network [28]		/	Local	Fixed degree curve	Polynomial curve fitting	2021
A new parameterization method [29]	Other	/	/	/	NURBS surface interpolation	2000
Pointshop 3D [31]		/	/	/	Point-based surface editing	2002
Free-boundary conformal parameterization [30]		/	Global/local	/	Parameterizing point clouds for meshing	2022

According to the property of the mapping process, we divide the parameterization methods of unorganized point clouds into base surfaces-based methods, meshless parameterization, spherical mapping, methods adapted from mesh parameterization, neural networks-based methods, and other methods.

4.1 Base Surfaces-Based Methods

For parameterization of unorganized point clouds, base surfaces, which approximate the underlying structure of point clouds, have been widely applied to parameterize point clouds. Base surfaces can be a plane, a Coons patch, or a cylinder [2]. The parameter values of each point in a point cloud can be obtained by projecting the point cloud onto a base surface. The projection direction can either be perpendicular to the surface or based on a determined projection vector. According to [9], a base surface should own the following properties:

a) Unique local mapping: The uniqueness implies that any two different points on the underlying surface should be mapped onto two different locations on the mapping domain.

b) Smoothness and closeness of base surface: This indicates that a base surface should be as smooth and simple as possible, while still approximating the underlying surface as much as possible. The balance between these properties should be carefully considered.
c) Parameterization of base surface: This implies that how we parameterize a base surface has a direct effect on the parameterization of the fitting surface. We can choose a more suitable way to parameterize a base surface by referring to the underlying structure of the fitting surface.

To get access to such base surfaces, some approaches have been proposed. For example, Hoppe et al. [10] propose a method to produce so-called "simplicial" surfaces. They first define a function to estimate the signed geometric distance to the underlying surface of the point clouds, then a contouring algorithm is applied to approximate the underlying surface by a "simplicial" surface. Their method is capable of reconstructing a surface with or without boundary from an unorganized point set. However, there is no formal guarantee that the reconstructed result is correct and the space required to store the reconstruction is relatively large. In [9], users can also manually define some section curves and four boundary curves to get a base surface of a point cloud, as some characteristic curves approximating the underlying structure of the point cloud are sufficient in defining a base surface. But it is also necessary to take advantage of the interior characteristic curves when the geometry is complex, even though just four corner points can be used to create a base surface in some cases. A base surface can also be obtained by iteratively minimizing a quadratic objective function [11]. With this method, a linear system of equations is solved in each step. To parameterize unstructured point clouds, Dynamic Base Surfaces (DBS) are also proposed by Azariadis [2]. As its name implies, a BDS is gradually improved regarding its approximation to the underlying structure of a point cloud, and the parameter value of each point in the point cloud is obtained by projecting it orthogonally to the DBS. Different from existing methods, no restrictions are required for the density and the homogeneity of point clouds. The limitation of this method is that it is only applicable to the point clouds where a closed boundary consisting of four curves exists. Azariadis and Sapidis [12] present a method to parameterize a point cloud globally and/or locally using recursive dynamic base surfaces. Their method can handle arbitrary point clouds of disk topology. Figure 1 shows the local parameterization of one subset of several point clouds using this method. The same authors [13] extend the DBS concept and use a recursive subdivision method to improve the accuracy of point clouds parameterization, especially for some small regions of the point clouds, where the approximation error by the DBS is not acceptable. They divide such regions into smaller parts and the points on these parts are approximated by c^0 composite surface based on recursive DBS subdivision to increase the approximation error, then to make the point clouds parameterization more accurately.

4.2 Meshless Parameterization

Meshless parameterization, first proposed by Floater and Reimers in [14], is also a widely used method to parameterize and mesh point clouds. As shown in Fig. 2, the main idea of meshless parameterization is to map the points in a point cloud onto a plane, where the

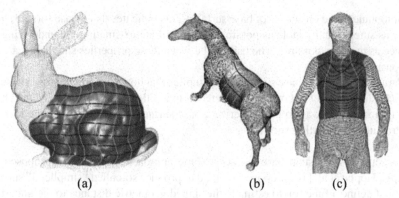

(a)   (b)   (c)

Fig. 1. Local parameterization of: (a) "bunny" point cloud, (b) "horse" point cloud, and (c) "human" point cloud [12].

mapping points are triangulated using an appropriate triangulation method, and then the original point cloud is meshed with the same triangulation edge structure as the mapping points. In order to make sure the reconstructed mesh has high quality, the mapping points should preserve the local structure of the original point clouds as much as possible. Therefore, the shape distortion ought to be minimized in the parameterization process. This is formulated as the problem of solving a sparse linear system [14, 15]. Since the mapping does not depend on the topological structure of point clouds, this method is called meshless parameterization. After the projection, the corresponding triangulation of the point clouds before mapping can be obtained by triangulating the projecting points in the planar parameter domain. This method has some limitations. First of all, solving a large linear system using their method is not efficient. Secondly, the reconstructed 3D triangles may distort and intersect each other due to the artificial convex boundary, which is also a problem when there are concave holes and the convex combination is not well defined along the concave parts of the hole boundary. To improve the efficiency of solving the linear system more efficiently, Volodine et al. [16] show that it can be done by an appropriate reordering of the matrix, which enables the linear system to be solved efficiently by deploying a direct sparse solver. To overcome the second problem, the same authors [17] extend the method to avoid distortion in the vicinity of concave boundaries by inserting virtual points to the concave neighbourhood, which can make sure the convex combination mapping is always defined. The methods described in [17] are only applicable to disk shape point clouds. To make the method presented in [17] more general, Hormann and Reimers [18] present an algorithm that can handle genus-0 topology as well by dividing the problem of triangulating point clouds into subproblems, each of which can be solved using the method in [17]. To improve the reconstructed result, Zhang et al. [19] apply an "as-rigid-as-possible" meshless parameterization method to parameterize a disk topology point cloud onto a plane while denoising the point cloud. Since their method can preserve local distances in the point cloud, a more regular 3D mesh can be obtained. Li et al. [20] present a meshless global parameterization method to parameterize point clouds and use the obtained parameterization to mesh the point clouds automatically.

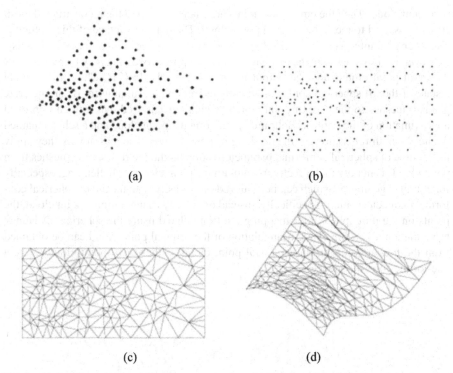

Fig. 2. (a) Point set. (b) meshless parameterization. (c) Delaunay triangulation of the mapping points. (d) surface triangulation [15].

4.3 Spherical Mapping

When the underlying structure of the point clouds is closed, which means there are no boundaries of the structure, "spherical mapping" is normally applied to parameterize the point clouds. The reason why "spherical mapping" is applied under such conditions can be partly explained by the uniform theorem [21], which states that every genus-0 closed surface is conformally equivalent to S^2. Thus, mapping from a genus-0 surface to the unit sphere is natural. The same idea is also applied to genus-0 point clouds. One such example is shown in Fig. 3. The problem of forming a spherical mapping given a point cloud model P can be formulated as [22]:

$$s = o + r_s \frac{p - o}{||p - o||} \tag{2}$$

where s are the spherical mapping points, p is the original point set, o is the centre of the original point set and r_s represents the largest distance between the original point set and the centre with the radius of the sphere.

Spherical parameterization is mostly used to mesh point clouds. For example, Zwicker and Gotsman [23] present a method to reconstruct a manifold genus-0 mesh from a 3D point cloud by using spherical embedding of a k-nearest neighbourhood graph of a point cloud. Then the embedded points are triangulated and the reconstructed mesh structure is used to mesh the original point cloud. The main advantage of this method is that it can guarantee a closed manifold genus-0 mesh, even the input point cloud is noisy. However, its drawbacks are that pre-processing and post-processing may be required for the input point clouds and the output mesh, respectively. In [21], Choi et al. extend a state-of-the-art spherical conformal parameterization algorithm used to parameterize genus-0 meshes to the case of point clouds, which are achieved by using an improved approximation of the Laplace-Beltrami operator on the point cloud and a scheme named the north-south reiteration for the meshing of point clouds. The reason why they apply the method of spherical conformal parameterization method to reconstruct meshes from point clouds is mainly that directly triangulating a point cloud is challenging, especially for complex geometry, which can be achieved more easily with the aid of spherical conformal parameterization. Specifically, instead of directly triangulating a point cloud, the points on the unit sphere after mapping are triangulated using the spherical Delaunay triangulation algorithm. Then triangulation of the original point cloud can be obtained from the triangulation on the spherical point cloud as these two point clouds have a one-to-one correspondence.

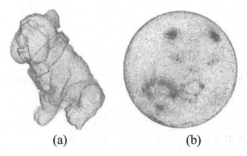

(a)   (b)

Fig. 3. (a) A bulldog point cloud. (b) the spherical conformal parameterization of the bulldog point cloud [21].

4.4 Methods Adapted from Mesh Parameterization

There are also some methods that are adapted from parameterizing meshes to parameterizing point clouds. For example, Tewari et al. modify the harmonic one-form method used in parameterizing manifold meshes to parameterize genus-1 point clouds that are sampled from such meshes [24]. They locally parameterize the subsets of a point cloud and the way they parameterize the point cloud can guarantee the consistency between the pieces. Even though the reconstructed results using their method are not much better than other reconstruction techniques, their method presents some new tools to the surface reconstruction problem and is very simple to implement. Li et al. [25] present a new method to reconstruct quad-dominant mesh from unorganized point clouds using the adapted periodic global parameterization method, which is modified from the periodic global parameterization method that is used to parameterize a triangle mesh. The local

Delaunay triangulation is used to design the parameterization of the point cloud. Their method can be used to deal with noisy point clouds without global connectivity. But it suffers from close-by structures because topological errors may be raised from the local Delaunay triangulation method by connecting two nearby surfaces.

4.5 Neural Networks-Based Methods

With the rapid development of neural network techniques, they have been applied to three main tasks of point cloud processing, i.e., 3D shape classification, 3D object detection and tracking, and 3D point cloud segmentation [26]. Besides their applications in the three main tasks, some researchers have investigated neural network-based point cloud parameterization. For example, Barhak and Fischer [27] adopt a self-organizing map (SOM) for the parameterization of small sets of clean points with low-frequency spatial variations, which can be used to reconstruct smooth surfaces. There are mainly two steps in the parameterization process: In the first step, Partial Differential Equation (PDE) and SOM are applied where the former technique can yield a parametric grid without self-intersection and the latter one makes sure all the sampled points have an impact on the grid, which guarantees the uniformity and smoothness of the reconstructed surface. In the second step, an adaptively modified 3D base surface is created for point clouds parameterization. Meng et al. [28] proposed a method to parameterize larger, noisy and unoriented point clouds by using adaptive sequential learning RBFnetworks. The network adopts a dynamic structure by adaptive learning and the neurons are adjustable regarding their locations, widths and weights, thus making it more powerful compared to other methods that apply RBFs at determined locations and scales. What is more, multi-level parameterization and multiple level-of-details (LODs) can be achieved in two ways. When multiple LODs meshes are required, parameterizing the point clouds with the best resolution and the points and surfaces can be computed at degrading sampling level to get the required LODs. In the second case where only one downgraded LOD is required, downgraded parameterization can be applied to obtain the result. Scholz and Juttler [29] apply residual deep neural networks to parameterize point clouds for polynomial curve fitting. Since the network approximates the function that assigns a suitable parameter value to a sequence of data points, optimal curve reconstruction from point clouds can be obtained. However, their method is only applicable to a small number of sample points and the proposed neural networks do not consider discrete surface point data. Figure 4 shows the layout of their proposed residual neural network.

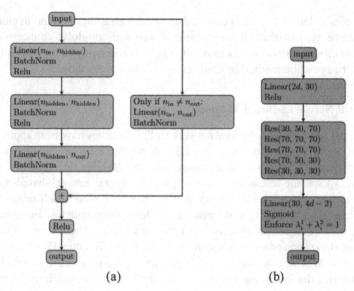

Fig. 4. (a) The layout of a building block. (b) the layout of the whole residual neural network [29].

4.6 Other Methods

Some other methods cannot be easily grouped. Therefore, we refer to them as other methods in this subsection and review them below.

As Ma and Kruth discuss in [9], three methods are usually adopted to parameterize digitized points for performing least squares fitting of B-spline curves and surfaces. These three methods are uniform parameters, cumulative chord length parameters and centripetal parameterization parameters. Since all these methods assume that the points are scattered in a special pattern, like chain points for curves and grid points for surfaces, these methods are very likely to fail when the points are irregularly spaced. To address this issue, Ma and Kruth [9] propose a simple technique, which parameterizes the irregularly spaced points by projecting them onto a base surface and obtaining their parameters from the parameters of the projected points. Jung and Kim [30] propose a new method to parameterize data points for NURBS surface interpolation, which is more powerful than the existing point clouds parameterization methods. With this method, the parameter value at the maximum of each rational B-spline basis function is treated as the parameter value of the corresponding data point. The empirical results show that their method outperforms the other methods as aforementioned in [10] regarding interpolation surfaces. In addition, many works consider mapping them onto a simple domain with a fixed boundary shape such as a sphere, a circle or a rectangle. However, some undesirable distortion may occur during the parameterization process due to the fixed boundary shape. To overcome such a problem, Choi et al. [31] develop a free-boundary conformal parameterization technique to parameterize disk-shape point clouds, which leads to high quality of the reconstructed mesh. By free boundary, it means that the positions of only two boundary points are fixed, and the left boundary points are parameterized to a suitable location automatically based on the structure of the original point clouds. To make the parameterization of point clouds more flexible, Zwicker et al. [32] present a system in which interactively parameterizing point clouds can be done. During the mapping process, an objective function is applied to minimize distortions automatically. Furthermore, the user can adjust the mapping intuitively at the same time.

5   Conclusion

In this paper, we have reviewed various methods used to parameterize 3D point clouds. These methods are grouped into organized point parameterization and unorganized point cloud parameterization ones and unorganized point cloud parameterization methods are further divided into some subgroups according to the property of the point clouds and the mapping technique. We discussed each of these methods.

It should be pointed out that there is no "best" parameterization method applicable to all point clouds, as one method may succeed in parameterizing some point clouds but fail in parameterizing other point clouds. Therefore, for a given point cloud, it is necessary to choose a suitable method to parameterize the 3D point cloud according to the desirable properties of low distortion and high computing efficiency in parameterizing the point cloud.

Acknowledgements. This research is supported by the PDE-GIR project which has received funding from the European Union Horizon 2020 research and innovation programme under the Marie Skodowska-Curie grant agreement No 778035. Zaiping Zhu is also sponsored by China Scholarship Council.

References

1. Meng, T.W., Choi, G.P.-T., Lui, L.M.: Tempo: feature-endowed teichmuller extremal mappings of point clouds. SIAM J. Imag. Sci. **9**, 1922–1962 (2016)
2. Azariadis, P.N.: Parameterization of clouds of unorganized points using dynamic base surfaces. Comput. Aided Des. **36**, 607–623 (2004)
3. Hormann, K., Greiner, G.: MIPS: an efficient global parametrization method. Curve and Surface Design: Saint-Malo, pp. 153–162 (2000)
4. Floater, M.S., Hormann, K.: Surface parameterization: a tutorial and survey. Adv. Multiresolution Geometric Modelling **2005**, 157–186 (2005)
5. Sheffer, A., Hormann, K., Levy, B., Desbrun, M., Zhou, K., Praun, E., Hoppe, H.: Mesh parameterization: theory and practice. ACM SIGGRAPPH, course notes **2007**, 10
6. Eck, M., Hadenfeld, J.: Local energy fairing of B-spline curves. In: Hagen, H., Farin, G., Noltemeier, H., (eds.) Proceedings of the Geometric Modelling, pp. 129–147. Springer, Vienna, (1995)
7. Floater, M.S.: Parametrization and smooth approximation of surface triangulations. Comput. Aided Geometric Design **14**(3), 231–250 (1997)
8. Greiner, G., Hormann, K.: Interpolating and approximating scattered 3D-data with hierarchical tensor product B-splines. In: Proceedings of the in Surface Fitting and Multiresolution Methods, pp. 163–172. Vanderbilt University Press (1997)

9. Ma, W., Kruth, J.-P.: Parameterization of randomly measured points for least squares fitting of B-spline curves and surfaces. Comput. Aided Des. **27**, 663–675 (1995)

10. Surface Reconstruction from Unorganized Points | Proceedings of the 19th Annual Conference on Computer Graphics and Interactive Techniques Available online: https://dl.acm.org/doi/abs/https://doi.org/10.1145/133994.134011. Accessed 19 Feb 2022

11. Pottmann, H., Leopoldseder, S., Hofer, M.: Approximation with active B-spline curves and surfaces. In: Proceedings of the 10th Pacific Conference on Computer Graphics and Applications, 2002. Proceedings, pp. 8–25. IEEE (2002)

12. Azariadis, P., Sapidis, N.: Efficient parameterization of 3D point-sets using recursive dynamic base surfaces. In: Panhellenic Conference on Informatics, pp. 296–306. Springer, Heidelberg (2005)

13. Azariadis, P., Sapidis, N.: Product design using point-cloud surfaces: a recursive subdivision technique for point parameterization. Comput. Ind. **58**, 832–843 (2007). https://doi.org/10.1016/j.compind.2007.03.001

14. Floater, M.S., Reimers, M.: Meshless parameterization and surface reconstruction. Comput. Aided Geometric Design **18**, 77–92

15. Floater, M.S.: Meshless parameterization and B-spline surface approximation. In: Proceedings of the Mathematics of Surfaces IX, pp. 1–18. Springer (2000)

16. Volodine, T., Roose, D., Vanderstraeten, D., Volodine, T., Roose, D., Vanderstraeten, D.: 65F05, 65M50. Efficient Triangulation of Point Clouds Using Floater Parameterization 2004

17. Volodine, T., Vanderstraeten, D., Roose, D.: Experiments on the Parameterization of Point Clouds with Holes. TW Reports, volume TW432 **2005**, 12

18. Hormann, K., Reimers, M.: Triangulating point clouds with spherical topology. Curve and Surface Design: Saint-Malo (2002), pp. 215–224 (2002)

19. Zhang, L., Liu, L., Gotsman, C., Huang, H.: Mesh reconstruction by meshless denoising and parameterization. Comput. Graph. **34**, 198–208 (2010)

20. Li, E.: Meshless quadrangulation by global parameterization. Comput. Graph. **35**(5), 992–1000 (2011)

21. Choi, G.P.-T., Ho, K.T., Lui, L.M.: Spherical conformal parameterization of genus-0 point clouds for meshing. arXiv:1508.07569 [cs, math] **2016**. https://doi.org/10.1137/15M1037561

22. 3D Point Clouds Parameterization Alogrithm | IEEE Conference Publication | IEEE Xplore Available online: https://ieeexplore.ieee.org/abstract/document/4697396. Accessed 19 Feb 2022

23. Zwicker, M., Gotsman, C.: Meshing point clouds using spherical parameterization. In: PBG, pp. 173–180 (2004)

24. Tewari, G., Gotsman, C., Gortler, S.J.: Meshing genus-1 point clouds using discrete one-forms. Comput. Graph. **30**, 917–926 (2006). https://doi.org/10.1016/j.cag.2006.08.019

25. Li, E., Che, W., Zhang, X., Zhang, Y.-K., Xu, B.: Direct quad-dominant meshing of point cloud via global parameterization. Comput. Graph. **35**, 452–460 (2011). https://doi.org/10.1016/j.cag.2011.03.021

26. Guo, Y., Wang, H., Hu, Q., Liu, H., Liu, L., Bennamoun, M.: Deep learning for 3D point clouds: a survey. IEEE Trans. Pattern Anal. Mach. Intell. **43**, 4338–4364 (2021)

27. Barhak, J., Fischer, A.: Parameterization and reconstruction from 3D scattered points based on neural network and PDE techniques. IEEE Trans. Visual Comput. Graphics **7**, 1–16 (2001)

28. Meng, Q., Li, B., Holstein, H., Liu, Y.: Parameterization of point-cloud freeform surfaces using adaptive sequential learning RBFnetworks. Pattern Recogn. **46**, 2361–2375 (2013). https://doi.org/10.1016/j.patcog.2013.01.017

29. Scholz, F., Jüttler, B.: Parameterization for polynomial curve approximation via residual deep neural networks. Comput. Aided Geometric Design **85**, 101977 (2021). https://doi.org/10.1016/j.cagd.2021.101977

30. Jung, H.B., Kim, K.: A new parameterisation method for NURBS surface interpolation. Int. J. Adv. Manuf. Technol. **16**, 784–790 (2000). https://doi.org/10.1007/s001700070012
31. Choi, G.P.T., Liu, Y., Lui, L.M.: Free-Boundary Conformal Parameterization of Point Clouds. arXiv:2010.15399 [cs, math] 2021
32. Zwicker, M., Pauly, M., Knoll, O., Gross, M.: Pointshop 3D: an interactive system for point-based surface editing. ACM Trans. Graph. (TOG) **21**, 322–329 (2002)

Machine Learning and Data Assimilation for Dynamical Systems

Statistical Prediction of Extreme Events from Small Datasets

Alberto Racca[1(✉)] and Luca Magri[1,2,3]

[1] Department of Engineering, University of Cambridge, Cambridge, UK
ar994@cam.ac.uk
[2] Aeronautics Department, Imperial College London, London, UK
l.magri@imperial.ac.uk
[3] The Alan Turing Institute, London, UK

Abstract. We propose Echo State Networks (ESNs) to predict the statistics of extreme events in a turbulent flow. We train the ESNs on small datasets that lack information about the extreme events. We asses whether the networks are able to extrapolate from the small imperfect datasets and predict the heavy-tail statistics that describe the events. We find that the networks correctly predict the events and improve the statistics of the system with respect to the training data in almost all cases analysed. This opens up new possibilities for the statistical prediction of extreme events in turbulence.

Keywords: Extreme events · Reservoir computing · Heavy tail distribution

1 Introduction

Extreme events arise in multiple natural systems, such as oceanic rogue waves, weather events and earthquakes [1]. A way to tackle extreme events is by computing their statistics to predict the probability of their occurrence. Because extreme events are typically rare, information about the heavy tail of the distribution that describes the events is seldom available. This hinders the performance of data-driven methods, which struggle to predict the events when extrapolating from imperfect datasets [8]. In this work, we assess the capability of a form of reservoir computing, the Echo State Network [5], to predict the statistics of extreme events in a turbulent flow [6]. In particular, we analyse the ability of the networks to improve the prediction of the statistics of the system with respect to the available training data. The paper is organised as follow. Section 2 introduces the turbulent flow model. Section 3 describes the Echo State Network. Section 4 analyses the statistical prediction of extreme events. We summarize the work and present future developments in Sect. 5.

A. Racca is supported by the EPSRC-DTP and the Cambridge Commonwealth, European & International Trust under a Cambridge European Scholarship. L. Magri is supported by the ERC Starting Grant PhyCo 949388.

D. Groen et al. (Eds.): ICCS 2022, LNCS 13352, pp. 707–713, 2022.
https://doi.org/10.1007/978-3-031-08757-8_58

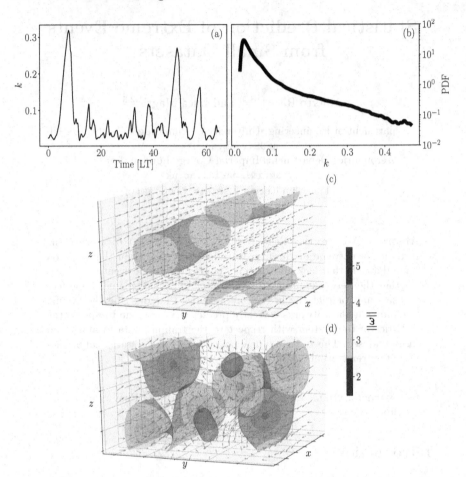

Fig. 1. One time series of the kinetic energy, (a), and Probability Density Function of the kinetic energy computed from the entire dataset, (b). The time in panel (a) is normalized by the Lyapunov time. Vorticity isosurfaces, $\boldsymbol{\omega} = \nabla \times \mathbf{v}$, and velocity flowfield before, (c), and after, (d), an extreme event. The laminar structure, (c), breaks down into vortices, (d).

2   A Low-Dimensional Model for Turbulent Shear Flow

We study a nine-equation model of a shear flow between infinite plates subjected to sinusoidal body forcing [6]. The incompressible Navier-Stokes equations are

$$\frac{d\mathbf{v}}{dt} = -(\mathbf{v} \cdot \nabla)\mathbf{v} - \nabla p + \frac{1}{\text{Re}}\Delta\mathbf{v} + \mathbf{F}(y), \tag{1}$$

where $\mathbf{v} = (u, v, w)$ is the velocity, p is the pressure, Re is the Reynolds number, $\mathbf{F}(y) = \sqrt{2}\pi^2/(4\text{Re})\sin(\pi y/2)\mathbf{e}_x$ is the body forcing along x, y is the direction of the shear between the plates and z is the spanwise direction. We solve the flow in the domain $L_x \times L_y \times L_z$, where the boundary conditions are free slip at $y \pm L_y/2$, and periodic at $x = [0; L_x]$ and $z = [0; L_z]$. Here, we set $L_x = 4\pi, L_y = 2, L_z = 2\pi$ and Re $= 400$ [9]. We project (1) on compositions of Fourier modes, $\hat{\mathbf{v}}_i(\mathbf{x})$, so that the velocity is $\mathbf{v}(\mathbf{x},t) = \sum_{i=1}^{9} a_i(t)\hat{\mathbf{v}}_i(\mathbf{x})$. The projection generates nine nonlinear ordinary differential equations for the amplitudes, $a_i(t)$, which are the state of the system [6]. The system displays a chaotic transient that converges to the laminar solution $a_1 = 1, a_2 = \cdots = a_9 = 0$. In the turbulent transient, the kinetic energy,

$$k = 0.5 \sum_{i=1}^{9} a_i^2, \qquad (2)$$

shows intermittent large bursts, i.e. extreme events, panel (a) in Fig. 1, which generate the heavy tail of the distribution [8], panel (b). In the figure, time is expressed in Lyapunov Times (LT), where a LT is the inverse of the Lyapunov exponent, $\Lambda \simeq 0.0163$. The Lyapunov exponent is the average exponential rate at which arbitrarily close trajectories diverge, which is computed with the QR algorithm [2,3]. Each extreme event is an attempt of the system to reach the laminar solution. During an extreme event, the flow slowly laminarizes, panel (c), but the laminar structure violently breaks down into vortices, panel (d). To study only the transient, we (i) generate 2000 time series series of length of 4000 time units through a 4th order Runge-Kutta scheme with $dt = 0.25$, (ii) discard all the time series that laminarized, i.e. the ones with $k \geq 0.48$, and (iii) use the remaining time series as data. The different time series are obtained by randomly perturbing a fixed initial condition [9].

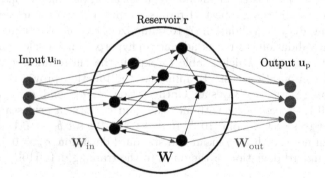

Fig. 2. Schematic representation of the echo state network.

3   Echo State Networks

As shown in Fig. 2, in an Echo State Network [5], at the i-th time step the high-dimensional reservoir state, $\mathbf{r}(t_i) \in \mathbb{R}^{N_r}$, is a function of its previous value and the current input, $\mathbf{u}_{\text{in}}(t_i) \in \mathbb{R}^{N_u}$. The output, $\mathbf{u}_{\text{p}}(t_{i+1})$, which is the predicted state at the next time step, is a linear combination of $\mathbf{r}(t_i)$:

$$\mathbf{r}(t_i) = \tanh\left(\mathbf{W}_{\text{in}}[\tilde{\mathbf{u}}_{\text{in}}(t_i); b_{\text{in}}] + \mathbf{W}\mathbf{r}(t_{i-1})\right); \quad \mathbf{u}_{\text{p}}(t_{i+1}) = \mathbf{W}_{\text{out}}[\mathbf{r}(t_i); 1] \quad (3)$$

where ($\tilde{}$) indicates normalization by the range component-wise, $\mathbf{W} \in \mathbb{R}^{N_r \times N_r}$ is the state matrix, $\mathbf{W}_{\text{in}} \in \mathbb{R}^{N_r \times (N_u+1)}$ is the input matrix, $\mathbf{W}_{\text{out}} \in \mathbb{R}^{N_u \times (N_r+1)}$ is the output matrix, b_{in} is the input bias and $[\,;\,]$ indicates vertical concatenation. $\mathbf{W}_{\text{in}}$ and $\mathbf{W}$ are sparse, randomly generated and fixed. These are constructed in order for the network to satisfy the echo state property [5]. The input matrix, $\mathbf{W}_{\text{in}}$, has only one element different from zero per row, which is sampled from a uniform distribution in $[-\sigma_{\text{in}}, \sigma_{\text{in}}]$, where σ_{in} is the input scaling. The state matrix, $\mathbf{W}$, is an Erdős-Renyi matrix with average connectivity $\langle d \rangle$. This means that each neuron (each row of $\mathbf{W}$) has on average only $\langle d \rangle$ connections (non-zero elements). The value of the non-zero elements is obtained by sampling from an uniform distribution in $[-1, 1]$; the entire matrix is then scaled by a multiplication factor to set its spectral radius, ρ. The only trainable weights are those in the output matrix, $\mathbf{W}_{\text{out}}$. Thanks to the architecture of the ESN, training the network by minimizing the Mean Square Error (MSE) on $N_t + 1$ points consists of solving the linear system

$$(\mathbf{R}\mathbf{R}^T + \beta\mathbf{I})\mathbf{W}_{\text{out}}^T = \mathbf{R}\mathbf{U}_{\text{d}}^T, \quad (4)$$

where $\mathbf{R} \in \mathbb{R}^{(N_r+1) \times N_t}$ and $\mathbf{U}_{\text{d}} \in \mathbb{R}^{N_u \times N_t}$ are the horizontal concatenation of the reservoir states with bias, $[\mathbf{r}; 1]$, and of the output data, respectively; $\mathbf{I}$ is the identity matrix and β is the Tikhonov regularization parameter [5].

The input scaling, σ_{in}, spectral radius, ρ, and Tikhonov parameter, β, are selected using Recycle Validation [7] to minimize the MSE of the kinetic energy. The Recycle Validation is a recent advance in hyperparameter selection in Recurrent Neural Networks, which is able to exploit the entire dataset while keeping a small computation cost. To minimize the function provided by the validation strategy, we use Bayesian Optimization for σ_{in} and ρ in the interval $[0.1, 10] \times [0.1, 1]$ seen in logarithmic scale and perform a grid search in each $[\sigma_{\text{in}}, \rho]$ point to select β from $[10^{-6}, 10^{-9}, 10^{-12}]$. We set $b_{\text{in}} = 0.1$, $d = 20$ and add gaussian noise with zero mean and standard deviation, $\sigma_n = 0.01\sigma_u$, where σ_u is the standard deviation of the data, to the training data [10].

4   Statistical Prediction of Extreme Events

We study the capability of the networks to predict the statistics of the system through long-term predictions. Long-term predictions are closed-loop predictions, i.e. predictions where we feed the output of the ESN as an input for the next time step, which lasts several tens of Lyapunov Times. These predictions diverge from the true trajectory due to the chaotic nature of the signal, but remain in the region of phase space of the chaotic transient. In doing so, they replicate the statistics of the true signal. The long-term predictions are generated in the following way: (i) from 500 different starting points in the training set, we generate 500 different time series by letting the ESN evolve each time for 4000 time units ($\simeq$ 65LTs); (ii) we discard the laminarized time series and (iii) use the remaining ones to compute the statistics as done for the data, see Sect. 2. To quantitatively assess the prediction of the statistics, we use the Kantorovich metric [4], $\mathcal{K}$, also known as Earth mover's distance, and the Mean Logarithmic Error (MLE) with respect to the true Probability Density Function of the kinetic energy, $\mathrm{PDF}_{\mathrm{True}}(k)$,

$$\mathcal{K} = \int_{-\infty}^{\infty} |\mathrm{CDF}_{\mathrm{True}}(k) - \mathrm{CDF}_j(k)| dk, \tag{5}$$

$$\mathrm{MLE} = \sum_{i=1}^{n_b} n_b^{-1} |\log_{10}(\mathrm{PDF}_{\mathrm{True}}(k)_i - \log_{10}(\mathrm{PDF}_j(k)_i)|, \tag{6}$$

where CDF is the Cumulative Distribution Function, j indicates the PDF we are comparing with the true data and n_b is the number of bins used in the PDF. When a bin has a value equal to zero and the logarithm is undefined, we saturate the logarithmic error in the bin to be equal to 1. On the one hand, we use the Kantorovich metric to assess the overall prediction of the PDF of the kinetic energy. On the other hand, we use the MLE to assess the prediction of the extreme events, as the logarithm highlights the errors in the small values of the tail.

In Fig. 3, we compare the statistics of the training data and an ensemble of 10 networks of 2000 neurons. We do so because the objective of predicting the statistics is to improve our knowledge, by employing the networks, with respect to the already available knowledge, the training data. Panel (a) shows the PDF of the kinetic energy in the training set for different sizes of the training set, from 1 time series to the entire data (1440). The prediction of the PDF improves with the size of the datasets, and values of the tail up to laminarization are observed only after 100 time series. The unresolved tail due to lack of data is a signature problem of data-driven analysis of extreme events [8]. Panels (b)–(c) show the Kantorovich metric and the MLE of the training sets and networks as a function of the training set size. The networks improve the prediction of the PDF with respect to the available data in all figures of merits analyzed, except for one outlier. The MLE of the training set improves more than the Kantorovich metric as the dataset becomes larger. This happens because a small amount of data is needed to accurately describe the peak of the PDF, which affects more the Kantorovich metric, while many time series are needed to describe the tail, which affects more the Mean Logarithmic Error. The results indicate that the networks are able to extrapolate from an imperfect dataset and improve the prediction of the overall dynamics of the system.

Fig. 4 shows the statistics of the square of the normal vorticity to the midplane, $\omega_y = \frac{\partial u}{\partial z} - \frac{\partial w}{\partial x}$. We plot the square of the vorticity, ω_y^2, because the symmetry of the problem causes the time-average of the vorticity to be equal to zero. Panel (a) shows the flowfield of the time-average, $\overline{(\ )}$, for the entire data, while panels (b) and (c) show the error with respect to (a) for an Echo State Network and the ten time series training set, respectively. All networks in the ensemble decrease the average error, up to values 7 times smaller than the training data (results not shown). This means that the Echo State Networks are able to extrapolate the statistics of the flowfield in addition to the statistics of the kinetic energy.

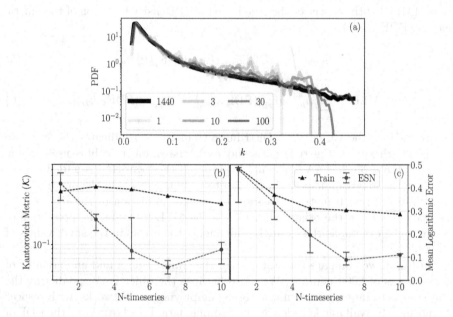

Fig. 3. PDF of the kinetic energy for different number of time series up to the entire data of 1440 time series, (a). For example, 3 means that the PDF is computed from 3 out of 1440 time series. 25th, 50th and 75th percentiles of Kantorovich Metric, (b), and MLE, (c), for the training set (Train), and the networks (ESN) as a function of the number of time series in the training set (N-timeseries). For example, N-timeseries means that the training set consists of N out 1440 time series.

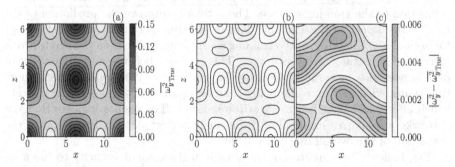

Fig. 4. Time average of the square of the midplane vorticity, $\omega_y^2(x,0,z)$ for the entire data, (a), error for a 2000 neurons network, (b), and training set, (c).

5   Conclusions and Future Directions

We propose Echo State Networks to predict the statistics of a reduced-order model of turbulent shear flow that exhibits extreme events. We train fully data-driven ESNs on multiple small datasets and compare the statistics predicted by the networks with the statistics available during training. We find that the networks improve the prediction of the statistics of the kinetic energy and of the vorticity flowfield, sometimes by up to one order of magnitude. This means that the networks are able to extrapolate the statistics of the system when trained on small imperfect datasets. Future work will consist of extending the present results to higher-dimensional turbulent systems through the combination of Echo State Networks and autoencoders.

The code is available on the github repository MagriLab/ESN-MFE.

References

1. Farazmand, M., Sapsis, T.P.: Extreme events: mechanisms and prediction. Appl. Mech. Rev. **71**(5) (08 2019). https://doi.org/10.1115/1.4042065, 050801
2. Ginelli, F., Poggi, P., Turchi, A., Chaté, H., Livi, R., Politi, A.: Characterizing dynamics with covariant lyapunov vectors. Phys. Rev. Lett. **99**(13), 130601 (2007)
3. Huhn, F., Magri, L.: Stability, sensitivity and optimisation of chaotic acoustic oscillations. J. Fluid Mech. **882**, A24 (2020). https://doi.org/10.1017/jfm.2019.828
4. Kantorovich, L.V.: On the translocation of masses. In: Dokl. Akad. Nauk. USSR (NS), vol. 37, pp. 199–201 (1942)
5. Lukoševičius, M.: A practical guide to applying echo state networks. In: Montavon, G., Orr, G.B., Müller, K.-R. (eds.) Neural Networks: Tricks of the Trade. LNCS, vol. 7700, pp. 659–686. Springer, Heidelberg (2012). https://doi.org/10.1007/978-3-642-35289-8_36
6. Moehlis, J., Faisst, H., Eckhardt, B.: A low-dimensional model for turbulent shear flows. New J. Phys. **6**(1), 56 (2004)
7. Racca, A., Magri, L.: Robust optimization and validation of echo state networks for learning chaotic dynamics. Neural Netw. **142**, 252–268 (2021)
8. Sapsis, T.P.: Statistics of extreme events in fluid flows and waves. Ann. Rev. Fluid Mech. **53**, 85–111 (2021)
9. Srinivasan, P.A., Guastoni, L., Azizpour, H., Schlatter, P., Vinuesa, R.: Predictions of turbulent shear flows using deep neural networks. Phys. Rev. Fluids **4**, 054603 (2019)
10. Vlachas, P.R., Pathak, J., Hunt, B.R., Sapsis, T.P., Girvan, M., Ott, E., Koumout-sakos, P.: Backpropagation algorithms and reservoir computing in recurrent neural networks for the forecasting of complex spatiotemporal dynamics. Neural Netw. **126**, 191–217 (2020)

Outlier Detection for Categorial Data Using Clustering Algorithms

Agnieszka Nowak-Brzezińska[(✉)] and Weronika Łazarz

Institute of Computer Science, Faculty of Science and Technology, University of Silesia,
Bankowa 12, Katowice 40-007, Poland
agnieszka.nowak-brzezinska@us.edu.pl

Abstract. Detecting outliers is a widely studied problem in many disciplines, including statistics, data mining and machine learning. All anomaly detection activities are aimed at identifying cases of unusual behavior when compared to the remaining set. There are many methods to deal with this issue, which are applicable depending on the size of the dataset, the way it is stored and the type of attributes and their values. Most of them focus on traditional datasets with a large number of quantitative attributes. While there are many solutions available for quantitative data, it remains problematic to find efficient methods for qualitative data. The main idea behind this article was to compare categorical data clustering algorithms: *K-modes* and *ROCK*. In the course of the research, the authors analyzed the clusters detected by the indicated algorithms, using several datasets different in terms of the number of objects and variables, and conducted experiments on the parameters of the algorithms. The presented study has made it possible to check whether the algorithms detect the same outliers in the data and how much they depend on individual parameters such as the number of variables, tuples and categories of a qualitative variable.

Keywords: Qualitative data · Outlier detection · Data clustering · *K-modes* · *ROCK*

1 Introduction

The article deals with the clustering of qualitative data to detect outliers in these data. Thus, in the paper, we encounter two research problems: clustering qualitative data and detecting outliers in such data. We look at outliers as atypical (rare) data. If we use clustering algorithms for this purpose, outliers are data that are much more difficult to include in any group than the typical (normal) data. Clustering qualitative data is a more extensive research problem than clustering quantitative data. We count the distance between the numeric values on each attribute that describes the objects. Quantitative data can be normalized which allows us to interpret the differences between the compared objects properly. Assessing the similarity between two objects described by qualitative attributes is a challenging task. Let us take *eye color* as an example of a qualitative attribute. Now, let us take into account three persons: A with *blue* eyes, B with *brown* and C with *gray* eyes. There are various methods to measure their similarity. We may say that *blue* is more similar to *gray* than *brown*. In fact, we know that *gray* is much

D. Groen et al. (Eds.): ICCS 2022, LNCS 13352, pp. 714–727, 2022.
https://doi.org/10.1007/978-3-031-08757-8_59

more similar to *blue* than *brown*. But we may also want to compare them as a plain text and then *blue* and *brown* share the same initial letter which makes them more similar than pairs *blue-gray* or *brown-gray*. It all depends on the method we use to compare the objects. It is also worth remembering that the comparing of the objects in the set will significantly impact the structure of the groups that we create.

By default, clustering algorithms, known in the literature for years, are based on the concept of data distances in a metric space, e.g., in Euclidean space. The smaller the distance between the objects, the greater the probability that they will form one group. If the distance between a given object from all created groups is too great, then we should consider the object as an *outlier* in the data. This idea seems logical. In the context of qualitative data: when a given object shows no similarity to the created groups, then it can be considered an outlier in the data.

In the study, we have made use of real datasets from various fields. This type of data very often contains some unusual pieces of data. They are not the result of a measurement error, but they actually differ from most of the data in the set. It is not always the case that one or more objects stand out significantly from the rest, and we can easily see it. Sometimes, it is also the case that certain subsets of objects differ to the same extent from the majority of data. The problem becomes even more complicated when we take into account the fact that these objects in the sets may be more or less differentiated by the specificity of the domain they come from, but also by the method of describing these data (the number of attributes, the number of possible values of these attributes, the number of objects). When objects are described on a categorical scale, the effectiveness of their correct clustering and outlier detection is necessary for a deeper study. In this paper, we analyze clustering algorithms from two types of clustering: hierarchical ($ROCK$) and non-hierarchical ($K - modes$). In case of quantitative data, the clustering process works as follows. Hierarchical algorithms in each iteration look for a pair or groups of objects with a smallest distance and combine them into a group. The process is repeated until an expected number of clusters is reached or until all groups have merged into one group. On the other hand, non-hierarchical algorithms (like the most popular clustering algorithm $K - means$), search for the best partition for a predetermined number of groups so that the distances inside the clusters are small and the clusters are as large as possible. In qualitative data, we should modify the algorithms to be suitable for operating on data for which we cannot explicitly measure distances. In case of non-hierarchical algorithms, we cannot use the $K - means$ algorithm because it forms its representative by determining the value of the so-called center of gravity of the group. For quantitative data, it is simply an arithmetic mean of the attribute values describing the features that make up the group. For qualitative data, we cannot derive a mean value. However, we can find a most common value. And this is the concept behind the $K - modes$ algorithm we chose for our research. In case of hierarchical algorithms, where two objects with the shortest distance are combined into a group iteratively, for datasets with qualitative data we cannot rely on the notion of distance. Instead, we use measures to determine the similarity of objects and, at each step of the algorithm, we connect the objects or groups of objects with the greatest similarity. This is the main idea of the $ROCK$ algorithm - a hierarchical clustering algorithm for qualitative data. We group the data to explore it better. Exploration has to do with the fact that apart from its obvious task, which is discovering patterns or rules in data, we can also discover unusual data, outliers in the data.

Therefore, in this study, we decided to investigate the effectiveness of the two selected clustering algorithms: $K - modes$ and $ROCK$, in outlier detecting. We want to compare how consistent the algorithms are in this respect. If they are consistent, then they should designate the same objects for potential outliers. In the research, we will change the clustering parameters to find the optimal results. We will repeat the experiments for 5%, 10%, and 15% outliers in the dataset. We expect that the more outliers we identify, the greater the coverage of the analyzed methods may be. We present the results in the section on experiments and research results.

2   State of Art

The methods of outlier detecting in datasets can be divided into formal and informal. Most formal tests require test statistics to test hypotheses and usually rely on some well-behaved distribution to check whether the extreme target value is out of range. However, real-world data distributions may be unknown or may not follow specific distributions. That is why it is worth considering other solutions, for example, clustering algorithms. In addition to the distribution-based methods, cluster-based approaches are also welcome. These approaches can effectively identify outliers as points that do not belong to the created clusters or the clusters distinguished by a small number of elements [6,9]. So far, numerous works have been published focusing on detecting outliers and good data clusters in a quantitative dataset. The most well-known algorithm is the LOF (Local Outlier Factor) algorithm proposed by Breunig in [2], in which local outliers are detected. Based on the ratio of the local density of a given object and the local density of its nearest neighbors, the LOF factor is calculated. Then, the objects with the highest LOF values are considered as outliers. Another method that isolates outliers and normal objects is the $IsolationForest$ method based on the construction of a forest of binary isolation trees. Then outliers are observations with shortest average path lengths from the root to the leaf [8]. The indicated algorithms are widely used in IT systems, both to clean datasets from noise so that they do not interfere with the system operation, and to detect unusual observations in the data for a further analysis. The presence of outliers in qualitative data can significantly disrupt the effectiveness of machine learning algorithms that try to find patterns in the data, such as rules, decision rules or association rules. Dividing the objects into groups in which the objects are as similar to each other as possible and thus detecting objects that do not match the groups is a very efficient solution to explore the outliers. We decided to choose two clustering algorithms, $K - modes$ and $ROCK$ - as they are the representatives of both hierarchical and non-hierarchical clustering algorithms. We found them very simple to interpret and implement on real data. So far, no papers describing the application of the indicated algorithms on a large scale or comparing the results with the distinction as to the type of data processed and the time of execution have been published. This has become the direct motivation of the authors of this paper to analyze those two selected clustering algorithms $K - modes$ and $ROCK$ in the context of their efficiency in detecting outliers in the qualitative data.

3 Data Clustering

The problem of clustering is one of the most researched issues in social sciences, psychology, medicine, machine learning, and data science. In addition to the standard benefits of data clustering, it has found a wide application in dataset processing with categorical domains, both in the course of preparation for mining and in the modeling process itself. Here, data clustering was used to find outliers in qualitative datasets. The two algorithms described in this section differ in terms of data clustering and outliers detection. The $K - modes$ algorithm, most frequently used in research and real IT systems, creates groups of clusters from objects closest to selected centroids and defines outliers as objects farthest from the cluster center. The $ROCK$ algorithm calculates the similarity measures between objects and groups of objects, creating data clusters containing objects that should not belong to any other cluster.

When dealing with quantitative data, we can easily use descriptive statistics, using quantities such as mean, median, standard deviation, and variance. When we handle qualitative data, it is not possible. We only know the most common value - a dominant. In such a case, clustering algorithms will cluster objects with the same value of a given attribute into groups. Of course, large clusters will be created by objects with a value equal to the dominant for a given attribute. For the clusters to be of good quality, we must effectively detect unusual data not to disturb the coherence and separation of the created data structures. We do not make assumptions that our sets contain outliers. We want our model to deal with any given dataset. If there are no outliers in the set, the cluster quality indicators will be very close to the values expected for the sets without outliers.

3.1 K-modes Clustering

The $K - modes$ clustering algorithm was proposed as an alternative to the popular $K - means$ algorithm, the most used centroid-based non-hierarchical algorithm [5]. The modifications made to the $K - means$ algorithm include using a simple measure of matching dissimilarity for qualitative features, replacing the group averages with vectors composed of the most common values at individual coordinates of the objects (modes), and using a frequency-based method to modes update. Let $X = \{x_1, \ldots, x_n\}$ be a set of n-objects x, such that $x = (x_1, \ldots, x_m)$. The dissimilarity measure of x_1, x_2 objects is defined as $d(x_1, x_2) = \sum_{i=1}^{m} \sigma(x_{1i}, x_{2i})$, where $\sigma(x_{1i}, x_{2i}) = 0$ if $x_{1i} = x_{2i}$ and 1 otherwise. Having $A = \{A_1, \ldots, A_m\}$ - set of the attributes of the objects in X it is possible to define $S \subseteq X$ - a cluster of data. The mode of $S = \{x_1, \ldots, x_p\}, 1 \leq p \leq n$ is the vector $q = (q_1, \ldots, q_m)$ which minimizes the function $D(S, q) = \sum_{i=1}^{n} d(x_i, q)$ called the cost function. A cluster center is called a mode and is defined by considering those values of the attributes that appear most frequently in the data points which belong to that cluster. The $K - modes$ (Algorithm 1) algorithm begins with a random selection of k objects (centroids) which are the central objects of k clusters. Then, the dissimilarity measure is calculated and the closest centroid is determined for each object. When all objects are assigned to individual clusters, the centroids are updated by creating new modes from objects present in the cluster. The calculations are repeated until the differences in the generated clusters in the following steps cease to exist.

Algorithm 1. $K - modes$ algorithm

input: X-dataset, k-expected number of clusters
output: a set of k clusters

1. Randomly select k items (modes) from the dataset.
2. For each pair (mode, object), calculate the dissimilarity measure.
3. For each object that is not a mode, find the mode closest to the object.
4. Join objects with the corresponding modes to create clusters.
5. For all clusters, recalculate the modal vectors containing in successive coordinates the most common values on attributes of cluster objects.
6. Perform steps 3-5 until the generated clusters do not repeat themselves.

The $K - modes$ algorithm is the easiest to implement and the most popular among the categorical data clustering algorithms because it is linearly scalable concerning the size of the dataset. The disadvantage of the algorithm is that it selects random initial modes, leading to unique structures around objects that are undesirable in the set. A method to prevent such situations is to draw the initial set of modes multiple times and assign each object to the cluster with the greatest number of times. The output clusters generated by the $K - modes$ algorithm have a similar cardinality, which does not have to reflect the actual data clusters on the sets having atypical distributions of variables. As with most categorical clusters, clusters containing a tiny number of elements or a single element can be considered outliers. The specifics of $K-modes$ clustering show that we will create single-element clusters only if the initially drawn object is an outlier. If we want to obtain a reliable mapping in small individual clusters, we can run the algorithm multiple times, each time randomizing a different set of initial $K - modes$ and finish the work when the variability is low in the final set of clusters. Finding the similarity between a data object and a cluster requires n operations, which for all k clusters is nk. Assigning objects to the appropriate k clusters and updating mods also require nk operations. Assuming the algorithm is run I times for different starting objects, the algorithm will have a linear complexity of $O(nkI)$.

3.2 ROCK Clustering

The $ROCK$ algorithm (RObust Clustering using linKs) [4], is a hierarchical clustering algorithm for categorical data. The algorithm introduces notions of neighbors and links. A point's neighbours are those points that are considerably similar to it. A similarity function between points defines the closeness between pairs of points. A user defines the threshold for which the pairs of points with a similarity function value greater than or equal to this value are considered to be neighbors. The number of links between pairs of points is defined to be the number of common neighbors for the points. The larger the number of links between a pair of points, the greater the likelihood is that they belong in the same cluster. Starting with each point in its own cluster, the algorithm repeatedly merges the two closest clusters till a desired number of clusters remain or when a situation arises in which no two clusters can be merged.

Algorithm 2. $ROCK$ algorithm

input: sample set of objects. Number of k clusters to be found. The similarity threshold: $\theta \geq 0.4$
output: A group of objects - a cluster
Do for All Data {

1. Initially, place each object into a separate cluster.
2. Construction of a Similarity Matrix with similarity for each pair of objects (A,B) using measure $Similarity(A, B) = \frac{|A \cap B|}{|A \cup B|}$
3. Computation of an Adjacency Matrix (A) using a similarity threshold $\theta \geq 0.4$ if $similarity(A, B) \geq \theta$ then 1; else 0
4. Compute a Link Matrix by multiplying an Adjacency Matrix by itself to find the number of links.
5. Calculation of a Goodness Measure for each pair of objects by using the g function
6. Merge the two objects with the highest similarity (goodness measure).
7. When no more entry exists in the goodness measure table then stop the algorithm which by now should have returned k number of clusters and outliers (if any), otherwise go to step 4.

}

The following features of this algorithm are necessary to define:

- Links - the number of common neighbors between two objects.
- Neighbors - if a similarity between two points exceeds certain similarity threshold, they are neighbors: if $similarity(A, B) \geq \theta$ then two points A, B are neighbors, for θ being a user-specified threshold.
- Criterion Function - the objective is to achieve a good cluster quality by maximizing the sum of links of intra cluster point pairs and minimizing the sum of links of inter cluster point pairs.
- Goodness Measure to maximize the criterion function and identify the best pair of clusters to be merged at each step of the $ROCK$ clustering algorithm.

$ROCK$ is a unique algorithm because it assumes that an attribute value, in addition to its frequency, must be examined based on the number of other attribute values with which it occurs. Due to its high computational complexity, $ROCK$ is good at detecting outliers in small datasets, and its computational time increases as the records in the set increase. This is because each record must be treated as a unique data cluster. If the user does not have a comprehensive knowledge about the dataset, the appropriate selection of the θ value and the minimum number of clusters generated on the output is a challenging task. The $ROCK$ algorithm is very resistant to outliers and can successfully identify outliers that are relatively isolated from the rest of the points. The ones with very few or no neighbors in one- or several-member clusters will be considered outliers. The overall computational complexity will depend on the number of neighbors of each facility. In most cases, the order of complexity will be $O(n^2 \log n)$. If a maximum and an average number of neighbors are close to n, then the algorithm's complexity increases to $O(n^3)$.

4   Conducted Research

The algorithms described in Sect. 3 were implemented in the Python language (version 3.8.8). We used the JupyterHub (version 6.3.0) environment available at *https://jupyter.org/hub* for the implementation and visualization of the data. Jupyter-Hub runs in the cloud or on hardware locally and supports a preconfigured data science environment for each user. We used Anaconda package containing most of the libraries, enabling machine learning models and visualization of results. The existing models of the Scikit-Learn library were used to implement the $K-modes$ algorithm. The $ROCK$ algorithm due to a lack of previous implementation was implemented by the authors. We used the Matplotlib library and the Pandas Dataframe structure for data visualization. Most of the computation is based on the Pandas data structures that hold the results.

The computer program described by the authors has been divided into sections containing:

- Importing Python libraries SciPy (1.6.2), Scikit-learn (0.24.1), NumPy, Pandas (1.2.4), Matplotlib (3.3.4) and libraries to perform operations related to time.
- Implementing algorithms: $ROCK$ with the parameters: k denoting the expected number of clusters and *theta* being a parameter of a function that returns an estimated number of neighbors and $K-modes$ with k parameter denoting the expected number of clusters and *threshold* parameter denoting the percentage of expected outliers.
- Data preprocessing: dealing with missing values (function that completes missing fields with the most common value in a column and removes columns that contain more than 60 empty values), coding the variables (encoding text values into numerical values), decoding encoded text variables.
- Uploading all datasets (reading, calculating the descriptive statistics, encoding text variables for the selected dataset to visualize the result).
- Execution of $ROCK$ and $K-modes$ algorithms on datasets. Presentation of the algorithms' computation time in relation to the type of the algorithm.
- Presentation of the algorithms' computation time in relation to the number of variables, the number of records, and data diversity.
- Listing the numbers of individual clusters obtained by the $ROCK$, $K-modes$ algorithms.
- Showing the selected dataset with assigned cluster numbers for the $ROCK$ and $K-modes$ algorithms and flags that indicate whether a record has been classified as an outlier. If the flag is -1, the object is an outlier. If it is 1, the object is considered normal.
- Presentation of the matrix of similarities and differences in classifying values as outliers for the $ROCK$ and $K-modes$ algorithms when compared in pairs.
- Identification of common outliers generated by the $ROCK$ and $K-modes$ algorithms.

The source of the software was placed in the GitHub repository: https://github.com/wlazarz/outliers2. It contains the implementation of the $K-modes$ and $ROCK$ algorithms and six datasets on which the experiments were conducted. The sequence of steps performed to compare the clustering and outliers detection algorithms is presented in

Fig. 1. The equipment specification on which we conduct our research is as follows: MacBook Pro Retina (15-inch, Mid 2015), macOS Catalina (10.15.7), 2,2 GHz processor four-core Intel Core I7, RAM 16 GB 1600 MHz MHz DDR3, GPU Intel Iris Pro 1536 MB. GPU acceleration and XAMPP were not used.

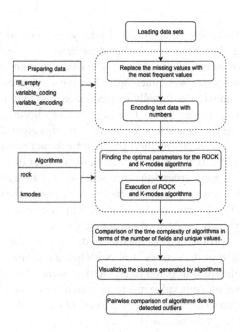

Fig. 1. Scheme of the program comparing algorithms clustering data and detecting outliers.

4.1 Data Description

We used six qualitative datasets to compare the algorithms that detect outliers in the data, each with a different structure of the variables matched to the clustering-based algorithms which support the detection of outliers in the qualitative datasets. The sets have different sizes and consist of a different number of categorical variables. The characteristics of the selected datasets are presented in Table 1. All analyzed datasets are real datasets, four of which relate to the domain of medicine ($Primary\ Tumor$ [10], $Lymphography$ [11], $SPECT\ Heart$ [12], $Covid-19$ [13]). In addition to the medical databases, two others were also analyzed: BM_attack [16] and $wiki$ [15]. The set $wiki$ contains the highest number of objects (913) and attribute values (285 unique values).

The first step in the project was to load datasets and prepare them properly before clustering commences. In all datasets, we filled empty fields with the most common value on a given variable. Categorical variables were encoded into numeric variables on *Primary Tumor Dataset* and *Lymphography Dataset*. Despite reducing the dataset to a numerical form, algorithms working on qualitative sets treat numbers as categories of

722   A. Nowak-Brzezińska and W. Łazarz

variables. The process of numerical encoding of the test values was intended to reduce
a long execution time of the algorithms resulting from the need to compare each sign
of the test value.

4.2 Methodology

We conducted the experiments empirically. Initially, we tried to automate the experiments by launching the execution of the algorithms: $K - modes$ and $ROCK$,
and changing the parameter values of these algorithms iteratively. However, several
lenghtly multi-hour processes were interrupted by an excessive memory consumption.
As a result, the experiments were finally carried out empirically for the gradually and
consciously changed parameter values (e.g., number of clusters). The *elbow method*
was used while looking for parameters for the $K - modes$ algorithm [14]. If the number of clusters selected with this method generated substantial outliers (many objects
were on the border of 5%, 10%, 15% of outliers), the number of clusters was increased
or decreased, still oscillating around the threshold point. The authors checked a cluster relevance using the *Silhouette method*, but the structure of created clusters was not
always satisfactory [17]. In case of the $ROCK$ algorithm we took into account the
number of clusters (already established during the execution of the $K - modes$ algorithm) and and initial epsilon value (a maximum distance at which elements can be in
one cluster) = 0.6. Most of the sets we dealt with had a reasonable number of outliers
within the epsilon value of 0.6. If too many outliers were obtained, the epsilon value
was increased. If increasing this value results in even more outliers, the number of clusters was decreased. Conversely, for too few outliers obtained, the epsilon was reduced,
or the number of clusters was increased.

5   Experiments

This section covers the results of the comparison of the two algorithms described in the
previous section: $ROCK$ and $K - modes$. We compared the algorithms in terms of
their time complexity. At the very beginning, it is worth emphasizing that in this paper,
we present the results obtained as a result of optimization of clustering parameters.
Thus, by diligently changing the clustering parameters of both algorithms, we checked
which combination of the values of these parameters gives optimal results. These optimal results (as one of many obtained) are presented below.

5.1   Time Complexities of Clustering Algorithms

Based on the sets described in Sect. 4.1, we performed an analysis of time complexity of
the algorithms described in this work. The execution time of the algorithms is given in
seconds. The study was conducted in the $Jupyter Hub$ environment installed locally on
$MacBook Pro$ hardware with $Intel Core i7$ quad-core processor and 16 GB RAM. The
datasets are characterized by a different number of objects and variables and represent
different types of data. The results are included in Table 1.

Table 1. Time complexity for $ROCK$, $K - modes$ and $K - means$ clustering algorithms

Dataset	Rows	Columns	Values	Time Complexity [s]		
				ROCK	K-modes	K-means
BM_attack	322	6	20	5,81	1,4	0,11
SPECT	267	23	46	3,67	2,91	0,47
primary-tumor	339	18	58	6,91	3,18	0,77
lymphography	148	19	62	0,72	1,52	0,22
covid	204	16	91	1,64	1,84	0,29
wiki	913	53	285	**141,96**	**26,57**	**3,06**

The $K - modes$ algorithm has an average linear or near-square complexity when diagnosed with many clusters. Regardless of the number of records, variables, and values, the execution time for the $K - modes$ algorithm is the lowest for each dataset. We can observe that the complexity of the $ROCK$ algorithm increases rapidly with the increase in the number of data.

5.2 Outlier Detection for Clustering Results

Algorithms working on qualitative datasets require the indication of individual parameters for the dataset: the number of generated clusters in case of the $K - modes$ algorithm and a minimum number of generated clusters and in case of the $ROCK$ algorithm the estimated number of neighbors between objects in the clusters. Implementing the $ROCK$ algorithm became a tough challenge due to a very high computational complexity and unusual parameters. We selected the $ROCK$ algorithm parameters on a trial and error basis. While the $ROCK$ algorithm analyzes the similarities not only between objects but also between clusters that should be merged into a single cluster, the $K - modes$ algorithm arranges objects from a dataset between clusters so that each cluster contains a similar amount of data and focuses only on the similarities between individual objects in the data. As mentioned earlier, the definition of an outlier generated by the $ROCK$ algorithm, taken from [4] indicates one-element classes. The records marked as anomalies by the $K - modes$ algorithm are the records from the farthest neighborhood of the centroid in which cluster the object is located. All datasets used in this research were taken from the *UCI Machine Learning Repository* database and represent real data collected during research on real data objects with different distributions, possibly containing a small number of deviations, which results in significantly different sizes of clusters generated by the $ROCK$ algorithm. The results of the outlier detection analysis for the lymphography set are presented in Fig. 2.

Data clustering algorithms do not have a natural definition of outliers and do not return points considered as variances in the data. The problem of marking objects that differ the most from the others due to the calculations characteristic of the algorithm was solved by generating an additional column for the dataset containing the values -1 or 1, where the value -1 means that the object was considered an outlier and 1 means that the object is normal. In most cases, the analyzed algorithms returned completely

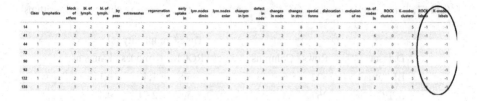

Fig. 2. The results of the outlier detection analysis for the lymphography set

different results. Large differences in outliers selection are the results of the different nature of those algorithms. The $ROCK$ algorithm is the most diligent in detecting outliers. It focuses on inter-object and inter-cluster connections, tying them together until well-defined clusters are obtained with the number of common neighbors below a certain threshold. Thus, single-member clusters contain far-away objects from every other cluster and every data object. In case of the $K - modes$ algorithm, due to randomness during the selection of an initial set of cluster centroids, outliers are considered as the objects whose distance from the centroids in the clusters they belong to, is the greatest. Due to a very different approach to determining good clusters and detecting outliers by these two algorithms, the anomaly classification result will also be different for each of the algorithms. We can design the anomaly search process in a qualitative set in two steps. Initially, all algorithms for the low anomaly threshold can search for common anomalies. If the process does not return results, you can increase the threshold and see if there are common outliers in the set this time.

5.3 Detection of Common Outliers

We should notice the relationship between the number of outliers and the degree of coverage of clustering algorithms in the context of outliers detection. Table 2 presents some interesting results. For each of the analyzed knowledge bases and the three analyzed levels of the number of outliers (5%, 10%, and 15%, respectively), the table presents the number of clusters for each of the algorithms: $ROCK$ and $K - modes$, number of outliers detected by each of these algorithms separately, and then the number of common outliers detected by these algorithms and a percentage that these common outliers represent concerning the entire analyzed set. One of the more essential conclusions is that, the more outliers we look for (5%, 10%, or 15%), by running each of the two analyzed algorithms separately, the more common outliers are found by these algorithms. For example, we found 3, 6, and 8 common outliers in the lymphography dataset, respectively, for the 5%, 10%, and 15% outliers we searched. There are also interesting results in the BM_attack dataset. In regard to the number of outliers we searched for, the number of actually found outliers and common outliers did not change (2 common outliers no matter how many outliers we were looking for). It is worth looking at the structure of this data set. It contains the fewest attributes and possible values of these attributes when compared to the rest of the sets, which brings about difficulties with regards to distinguishing objects from each other and detecting a greater or lesser number of outliers. In general, when analyzing all sets, one can notice a specific influence the number of attributes and their values have on the efficiency of outlier detection. The more attribute values there are, the greater the coverage of commonly detected outliers. This is easily explained. With a greater number of features describing the objects, we achieve a greater differentiation, so it is easier for us to correctly (not accidentally) determine the outliers.

Table 2. The results of % of common outliers obtained for 5%, 10%, and 15% of outliers in each of the datasets

Dataset	%	Clusters ROCK	K-modes	Outliers ROCK	K-modes	common outliers	% of common outliers
lymphography	5%	2	5	5	5	3	0,020000
	10%	3	5	16	11	6	0,040500
	15%	3	5	16	21	8	0,054100
covid	5%	3	6	7	6	1	0,004900
	10%	5	6	34	25	13	0,063700
	15%	5	6	34	25	13	0,063700
SPECT	5%	6	4	8	10	4	0,014980
	10%	1	4	23	21	10	0,037450
	15%	6	4	50	47	31	0,116100
BM_attack	5%	20	3	3	44	2	0,006200
	10%	20	3	3	44	2	0,006200
	15%	20	3	3	44	2	0,006200
primary-tumor	5%	3	5	12	24	8	0,023599
	10%	6	5	40	24	19	0,056000
	15%	6	5	40	53	28	0,082596
wiki	5%	2	9	28	24	13	0,014240
	10%	1	9	78	96	44	0,048193
	15%	5	9	143	143	74	0,081100

5.4 Evaluation of the Proposed Methods

As part of this work, a vast number of experiments were performed. We changed the values of individual parameters to observe changes in the cluster structure, the number of generated outliers, and most importantly, in assessing whether the analyzed clustering algorithms return similar results in terms of outliers. In the study, we considered real datasets which frequently contain unusual data. They are not the result of a measurement error, but they differ from most data in the set. It is not always the case that one or more objects stand out significantly from the rest, and we can easily see it. Sometimes, it is also the case that specific subsets of objects differ to the same extent from most of the data. The problem becomes even more complicated when we take into account the fact that these objects in the sets may be more or less differentiated by the specificity of the domain they come from, but also by the method of describing these data (the number of attributes, the number of possible values of these attributes, the number of objects). When objects are described on a categorical (qualitative) scale, the effectiveness of their proper clustering and outlier detection is necessary for a deeper study. Hence, in this paper, we analyze selected clustering algorithms which exemplify two types of clustering: hierarchical ($ROCK$) and non-hierarchical ($K - modes$). Analysis of the results allows us to conclude that if we care about the speed of calculations or have a large dataset, a good choice will be to use the $K - modes$ algorithm. The algorithm is recommended to be used in datasets that we know are divided into a small number of large clusters. Then the initially drawn centroids will have less influence on clustering quality. In most cases, the most reasonable approach is to use the $ROCK$ method because it performs an exhaustive analysis of the dataset in search of outliers - it approaches object variables individually. It looks for relationships between objects and variables (attributes and their values). The main disadvantage of this algorithm is a very high computational complexity, which in extreme cases may be close to the cube of the number of objects in the set. For this reason, the algorithm is a good choice if we have small datasets, up to 1000 records. Another difficulty is the selection of the distance between the clusters and the minimum number of clusters. The algorithm execution time and clustering quality are improved by knowing an estimated number of clusters in the set and how far the elements should be apart from each other to not be included in a common cluster. Let us suppose that we do not have an exhaustive knowledge about the dataset. In that case, it is worth running the algorithm many times and analyzing the generated clusters to assess the quality of the parameters.

6   Conclusions

This paper focuses on searching for outliers in qualitative data sets depending on the type and the number of variables. Section 3 describes relatively novel approaches to qualitative clustering data. The results presented in this paper are based on six datasets characterized by a different structure. While there is a multitude of solutions related to quantitative data, clustering data containing only qualitative variables remains a challenge for data scientists. The authors attempted to compare the effectiveness of cluster and outlier detection in qualitative datasets, between which there is no explicit comparison so far. Algorithms based on quantitative data generally tend to have better mathematical properties. This does not apply to qualitative sets, so it is difficult to determine which algorithm works better on the data, and it is difficult to detect natural groups. We define the performance of algorithms in terms of their scalability and cluster generation time. We can draw a primary conclusion from the research that the data structure significantly impacts the algorithm's time complexity. The $K - modes$ algorithm defines clusters and outliers as objects far away from modes if we have visible modes in a data set. Otherwise, the optimal number of clusters can be very large or very small, and objects that should be in separate clusters will be in one due to a small distance from central modes. Then, it is better to use the $ROCK$ algorithm, which is less efficient and has a much greater computation complexity but is not sensitive to unusual data distribution. We should adequately select the algorithm for a dataset. Each algorithm classifies outliers differently and the results will differ. Algorithms based on categorical data clustering are relatively new methods of detecting outliers in data, having no implementation in commonly used programming languages. The discussed *ROCK* and *K-modes* algorithms introduce different methods to solve this problem and give different solutions in terms of their performance concerning the time needed to execute the algorithms when the number of records and dimensions change. The quality of the created clusters is measured by the user's knowledge and the examination of the results. The user sets basic parameters of clustering, which require an extensive knowledge of the data [1].

References

1. Carletti, M., Terzi, M., Susto, G.A.: Interpretable anomaly detection with DIFFI: depth-based feature importance for the isolation forest, 1–12. IEEE, US (2000). arXiv preprint arXiv:2007.11117
2. Breunig, M.M., Kriegel, H.P., Ng, R.T., Sander, J.: LOF: identifying density-based local outliers. In: Proceedings of the 2000 ACM SIGMOD International Conference on Management of Data, pp. 93–104 (2000)
3. Gibson, D. and Kleinberg, J. and Raghavan, P.: Clustering categorical data: an approach based on dynamical systems. In: Proceedings of the 24th International Conference on Very Large Data Bases, the VLDB Journal, pp. 222–236 (2000)
4. Guha, S., Rastogi, R., Shim, K.: ROCK: a robust custering algorithm for categorical attributes. Inf. Syst. **25**(5), 345–366 (2000)
5. Huang, Z.: A fast clustering algorithm to cluster very large categorical data sets in data mining. Data Min. Knowl. Discov. **2**(3), 283–304 (1998)
6. Jiang, M.F., Tseng, S.S., Su, C.M.: Two-phase clustering process for outliers detection. Pattern Recogn. Lett. **22**(6–7), 691–700 (2001)
7. Kaufman, L., Rousseeuw, P.J.: Finding Groups in Data: an Introduction to Cluster Analysis. John Wiley & Sons (2005)
8. Liu, F.T., Ting, K.M., Zhou, Z.-H.: Isolation forest. In: IEEE International Conference on Data Mining, pp. 413–422 (2009)
9. Loureiro, A., Torgo L., Soares, C.: Outlier detection using clustering methods: a data cleaning application. In: Proceedings of KDNet Symposium on Knowledge-Based Systems for the Public Sector (2004)
10. Primary Tumor, UCI Machine Learning Repository. https://archive.ics.uci.edu/ml/datasets/Primary+Tumor. Accessed 23 May 2020
11. Lymphography, UCI Machine Learning Repository. https://archive.ics.uci.edu/ml/datasets/Lymphography. Accessed 23 May 2020
12. SPECT, UCI Machine Learning Repository. https://archive.ics.uci.edu/ml/datasets/spect+heart. Accessed 9 June 2021
13. COVID, UCI Machine Learning Repository. https://www.kaggle.com/anushiagrawal/effects-on-personality-due-to-covid19. Accessed 9 June 2021
14. Elbow method. https://en.wikipedia.org/wiki/Elbow_method_(clustering). Accessed 9 June 2021
15. Wiki, UCI Machine Learning Repository. https://archive.ics.uci.edu/ml/datasets/wiki4HE. Accessed 9 June 2021
16. BM_attack, UCI Machine Learning Repository. https://archive.ics.uci.edu/ml/datasets/Dishonest+Internet+users+Dataset. Accessed 9 June 2021
17. Sillhouette method. https://en.wikipedia.org/wiki/Silhouette_(clustering). Accessed 9 June 2021

Augmenting Graph Inductive Learning Model with Topographical Features

Kalyani Selvarajah$^{(\boxtimes)}$ and Jae Muzzin

School of Computer Science, University of Windsor, Windsor, ON, Canada
{kalyanis,muzzin4}@uwindsor.ca

Abstract. Knowledge Graph (KG) completion aims to find the missing entities or relationships in a knowledge graph. Although many approaches have been proposed to construct complete KGs, graph embedding methods have recently gained massive attention. These methods performed well in transductive settings, where the entire collection of entities must be known during training. However, it is still unclear how effectively the embedding methods capture the relational semantics when new entities are added to KGs over time. This paper proposes a method, AGIL, for learning relational semantics in knowledge graphs to address this issue. Given a pair of nodes in a knowledge graph, our proposed method extracts a subgraph that contains common neighbors of the two nodes. The subgraph nodes are then labeled based on their distance from the two input nodes. Some heuristic features are computed and given along with the adjacency matrix of the subgraph as input to a graph neural network. The GNN predicts the likelihood of a relationship between the two nodes. We conducted experiments on five real datasets to demonstrate the effectiveness of the proposed framework. The AGIL in relation prediction outperforms the baselines both in the inductive and transductive setting.

Keywords: Knowledge graphs · Graph neural networks · Subgraph

1 Introduction

In recent years, significant progress has been made in the construction and deployment of knowledge graphs (KGs) [29]. KGs represent structured relational information in the form of subject-predicate-object (SPO) triples, e.g.,$\langle Justin\ Trudeau, fatherOf, Xavier James\rangle$. Freebase [4], YAGO [23], DBPedia [1], ConceptNet [22], and Never-ending language learning (NELL) [6] are a few prominent examples of large KGs. Recently, KGs have gained widespread attention because of their benefits in a variety of applications, including question answering [14], dialogue generation [13], information retrieval [30], entity linking [10] and recommendation systems [37].

Despite their usefulness and popularity, KGs are often noisy and incomplete because it is challenging to incorporate all information in the real world, and these data are typically dynamic and evolving, making it difficult to generate accurate and complete KGs [28]. Therefore, automating the construction

D. Groen et al. (Eds.): ICCS 2022, LNCS 13352, pp. 728–741, 2022.
https://doi.org/10.1007/978-3-031-08757-8_60

of a complete KG is a tedious process. Various techniques have been proposed for knowledge graph completion, such as the traditional Statistical Relational Learning (SRL) methods and Knowledge graph embedding methods. Building a complete KG is possible by predicting objects (known as link prediction) and relations.

A relation or logical induction prediction problem discovers probabilistic logical rules from a given KG. Induction can be learned in several ways such as from examples [16] and from interpretations [7]. For example, let's say, "the 23rd prime minister of Canada, Justin Trudeau lives in Ottawa, and is married to Sophie Trudeau." The first-order logic of the above sentence would be $LivesIn(Justin$ $Trudeau, Ottawa) \wedge MarriedTo(JustinTrudeau, SophieTrudeau)$. Therefore, a logical rule can be derived based on the concept that a married couple lives together (generally); $LivesIn(X, Y) \wedge MarriedTo(X, Z) \rightarrow LivesIn(Z, Y)$. This rule can be used to find the relation or possible hypothesis $LivesIn(SophieTrudeau, Ottawa)$. Here, the known logical rules have been generalized to derive a new rule or relationship which is true most of the time. Additionally, this rule predicts the relation for the entities which did not exist when KGs were trained. In reality, KGs evolve with time and new entities will join. Most of the existing embedding-based methods are highly successful in predicting the relations if the entities were seen when KGs were trained, which is transductive reasoning. Generalizing relational semantics is a challenging task and important to see the relationships in unseen entities, which is inductive reasoning. However, these embedding methods have some limitations in explicitly capturing the relational semantics when new entities are added to KGs over time.

This paper proposes an Augmenting Graph Inductive Learning (AGIL) framework to learn relational semantics in a given KG, and predict relations $(s, ?, o)$. Since much of the existing machine learning methods suffer from scalability issues, recently, PLACN [17] and GraIL [25] applied subgraph-based methods in link prediction and relation prediction, respectively, to overcome this problem. GraIL used a Graph Neural Network (GNN) based relations prediction method to learn relational semantics even if the entities were unseen during training. However, GraIL operated strictly on subgraphs and utilized no additional information. PLACN, on the other hand, successfully used local features as additional information for link prediction. So, our proposed model exploits both PLACN and GraIL to derive AGIL, which includes three primary steps. First, the subgraph is extracted with common neighbors of the target link between nodes i and j. The common neighbors in the enclosed subgraph are collected till k number of hops. Then the subgraph is labeled using the Double-Radius Node Labeling [35] method. In the final steps, the heuristic features of nodes for the entire subgraph are extracted and fed into GNN along with the adjacency matrix of the subgraph, which aggregates the feature vectors into a scoring function for the prediction.

Our Contribution: The followings are the summary of our contributions:

1. We propose an Augmenting Graph Inductive Learning (AGIL) framework based on common neighbors-based subgraphs for relations prediction in both transductive and inductive settings.
2. We extract heuristic features of nodes from the entire subgraph and model a new prediction framework based on Graph Neural Networks (GNN);

The rest of the paper is organized as follows. Section 2 discusses related existing work. Our framework is presented in Sect. 3. Following that, Sect. 4 presents the experimental setup and the corresponding results. Finally, Sect. 5 concludes the research idea of this paper with directions for future work.

2   Related Work

Multiple methods have been proposed to construct a complete knowledge graph. Graph Embedding is one of the most broadly used solutions for Knowledge-Graph Completion challenges. Translation-based approach [5,5,24] Bilinear-based approach [27,33] and Neural-Network-based approach [2,8] are well-known graph embedding approaches.

Traditional approaches on the KG embedding methods are in a transductive manner. They require all entities during training. However, many real-world KGs are ever-evolving by adding new entities and relationships. Several inductive KG embedding approaches are proposed to address the issue of emergent entities. Graph2Gauss [3] is an approach to generalize to unseen nodes efficiently on large-scale attributed graphs using node features. Then Hamilton et.al. [11] proposed a generic inductive framework, GraphSAGE, that efficiently generates node embeddings for previously unseen data in a graph by leveraging node feature information. Node features are, however, not available in many KGs. In addition to these inductive embedding methods, DRUM [18], NeuralLP [34], and RuleN [15] are few models which learn logical rule and predict relations in KGs.

Recently, GraIL [25] was proposed to generalize inductive relation based on subgraph reasoning. Since GraIL shows comparatively better performance than the state-of-the-art methods, we consider extending it. Additionally, SEAL [35], PLACN [17] and DLP-LES [21] are few recent approaches that successfully extracted subgraphs from a given networks and applied heuristic features to train the model. Motivated by their high performance, we incorporate these heuristic features with our model.

3   Problem Definition and Proposed Approach:

Given a KG, $G = \langle V, E, R \rangle$ is a directed graph, where V is the set of vertices, E is the set of edges and R represents the set of relations. The edges in E connect two vertices to form triplets (h, r, t), where h is a head entity in V, t is a tail entity in E and r is a relation in R, i.e., $E = \{(h, r, t) | h \in V, r \in R, t \in V\}$.

In a given KG, there is a high chance of missing relations $(h, ?, t)$, head entity $(?, r, t)$ and tail entity $(h, r, ?)$. Knowledge graph completion in a given KG, G is defined as the task of predicting missing triplets, $E' = \{(h, r, t) | h \in V, r \in R, t \in V, (h, r, t) \notin E\}$ in both transductive and inductive settings. In the transductive setting, the entities in a test triple are considered to be in the set of training entities. Predicting missing triplets in the transductive setting is defined as $E'' = \{(h, r, t) | h \in V, r \in R, t \in V, (h, r, t) \notin E\}$. In the inductive setting, the entities in a test triple are never seen in the set of training entities. Predicting missing triplets in the inductive setting is defined as $E''' = \{(h, r, t) | h \in V' \text{ or } t \in V', r \in R, (h, r, t) \notin E\}$, where $V' \cap V = \emptyset$ and $V' \neq \emptyset$.

Our primary objective is to predict the relation between two nodes. We employ Graph Neural Network (GNN) [19] to learn the knowledge graph's structural semantics. The proposed model has the following steps;

1. Subgraph extraction.
2. Node labeling.
3. Feature matrix construction.
4. Scoring the subgraph using GNN.

3.1 Subgraph Extraction

For each triple in the knowledge graph, the subgraph is extracted with the goal of isolating the connecting nodes between the two target nodes u and v. We wish to isolate only the nodes which are found along every possible path between the head and tail of the knowledge triple, referred to as the target nodes of the subgraph. A few approaches in the existing literature [17,35] have been proposed for subgraph extraction from a given graph. AGIL extracts subgraphs using common neighbors of any targeted nodes u and v because sufficient information of entire nodes of subgraphs can be taken for the training process [17]. Moreover, having additional information about the shared neighbours of both nodes u and v allows to determine the future existence of a relationship between them. We set a number k for the number of hops to collect the nodes in the subgraph, which can be defined as given below.

Definition 1. *Subgraph:* *For a given knowledge graph* $G = \langle V, E, R \rangle$*, let* $\Gamma_k(x)$ *be the neighbors of* x *within* k *hops. The subgraph of a target link between nodes* u *and* v *is given by the function* $S : V^2 \rightarrow 2^V$*, the function that returns the set of common neighbor nodes connecting* u *and* v*,*

$$S = \bigcup_{i=1}^{k} (\Gamma_i(u) \cap \Gamma_i(v)); \; for \; some \; m > 1 \tag{1}$$

where $\{u, v\} \in V'$*,* $V' \subseteq V$ *and* V' *is a set of common neighbors for the targeted nodes, and* $|V'| = \emptyset$*.*

3.2   Subgraph Node Labeling

Generally, GNN takes both feature matrix X and adjacency matrix A as input, (A, X). To construct a feature matrix X of a subgraph, the position of nodes are really important to maintain the consistency of the structural information. GNN learns the existence of target links for prediction. So, we exploit the Double-Radius Node Labeling method, which was proposed by SEAL [35] to label the subgraphs.

Each label is a 2-tuple. The first element is the distance from the first target node, the second element is the distance from the second. The target node labels are always $(0, 1)$ and $(1, 0)$. Figure 1 is an example of a subgraph for target nodes $\langle University, ?, ComputerScience \rangle$, and the labeled subgraph.

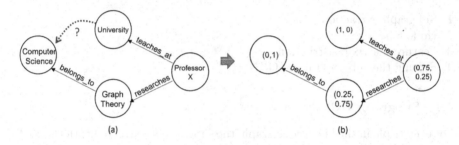

(a)   (b)

Fig. 1. (a) Subgraph of target nodes. (b)Labeled Subgraph.

3.3   Feature Matrix Construction

The GraIL [25] graph neural network architecture considers only the structural node feature X for predicting triplets of a given KG. However, we believe that in addition to the structural node feature, incorporating explicit features of the subgraph to the feature matrix X send additional information to the graph neural network training model. Since a knowledge graph does not always have any explicit feature information about a node, we decided to use the topological feature of the subgraph to see the importance of topological features in relation prediction. We belive that topological heuristics are useful in knowledge graph completion because entities are less likely to form relationships with entities that are farther away. Similarly with social networks, people tend to create new relationships with people that are closer to them. The motivation to apply topological heuristics to knowledge graphs was inspired by research in social networks. Our model uses the proximity measures taken from the topology as a heuristic in link prediction.

The specific heuristics used in this research were chosen to give a precise measurement of the notion of proximity of entities within the knowledge graph. In research done by Liben-Nowell and Kleinberg [12] with neighbour-based proximity measures, it was found that predictions outperformed chance by factors of

40 to 50, which led the authors to concluded that topology does indeed contain latent information which can be used to predict missing or future relationships.

We chose to use multiple proximity measurements as each has their own characteristics. AGIL uses the following five simple heuristics as used in PLACN model. Here $\Gamma(v)$ and $\Gamma(u)$ specify the set of neighbors within k hops for nodes v and u respectively.

Common Neighbors (CN) counts how many neighbours any two vertices v and u have in common.

$$CN_{u,v} = |\Gamma_k(v) \cap \Gamma_k(u)| \tag{2}$$

Jaccard Coefficient (JC) produces the normalized form of CN.

$$JC_{u,v} = \frac{|\Gamma_k(v) \cap \Gamma_k(u)|}{|\Gamma_k(v) \cup \Gamma_k(u)|} \tag{3}$$

Adamic-Adar (AA) is a modified version of JC, which gives a higher priority to the common neighbors with lower degree.

$$AA_{u,v} = \sum_{w \in |\Gamma_k(v) \cup \Gamma_k(u)|} \frac{1}{log|\Gamma_k(w)|} \tag{4}$$

Preferential Attachment (PA) The idea behind PA is that a node with a higher degree has a better probability of forming new connections.

$$PA_{u,v} = |\Gamma_k(v).\Gamma_k(u)| \tag{5}$$

Resource Allocation (RA) is much more similar to AA, but gives higher priority to low-degree common neighbors.

$$RA_{u,v} = \sum_{w \in |\Gamma_k(i) \cup \Gamma_k(j)|} \frac{1}{|\Gamma_k(w)|} \tag{6}$$

Let $f : V^2 \to R^5$ be the function which returns the set of above five heuristic features for the pair of nodes u, v. So, $f(u, v)$ returns a vector of five components, each holding the CN, JC, AA, PA, and RA Value. Let S be the set of nodes in the enclosing subgraph of target nodes u and v, and $\{u, v\} \in S$ and $S \subseteq V$. Then for every node $i \in S$, we can evaluate the heuristic features of x and each of x's neighbors $y \in \Gamma(x)$ using $f(x, y)$.

Here, we discuss how we calculate the feature vector of node x. Let P_x be the matrix whose columns are label of nodes in S and rows are five feature vectors, i.e., $P_x = [R_0, R_1, \ldots, R_n]$, where $R_y = f(x, y) \ \forall y \in S$ and $n = |S|$. The matrix P_x contains all five heuristic features of every possible pair of nodes in the subgraph. The number of rows and columns of P_x are 5 and $|S|$ respectively.

Example 1. *Consider a subgraph* $S = \{u, v, w, x, y, z\}$, *and* u *and* v *are the target nodes of the subgraph* S, *then*

$$P_x = \begin{bmatrix} CN_u & CN_v & CN_w & 0 & CN_y & CN_z \\ AA_u & AA_v & AA_w & 0 & AA_y & AA_z \\ JC_u & JC_v & JC_w & 0 & JC_y & JC_z \\ RA_u & RA_v & RA_w & 0 & RA_y & RA_z \\ PA_u & PA_v & PA_w & 0 & PA_y & PA_z \end{bmatrix}_{5 \times n}$$

where the column for x *is zero, because it is pointless to compare* x's *topology to itself.*

Since the size of P_x depends on the size of the subgraph, $|S|$, which is variable, this is not suitable for scaling in training on when the node degree of the graph is very high. A very large feature vector can cause critical performance issues in the model. This is where the Fixed Sized Subgraph and Variable Sized Subgraph models diverge. Each take a different approach in deriving a feature vector F_x from the topology matrix P_x.

PLACN uses a constant value k, which is the absolute maximum size a subgraph may reach. k is derived in a way to be large enough for most node pairs. The value of k is a function of the number of edges and nodes in the complete graph.

$$k \approx \left\lceil \frac{2|E|}{|V|} \left(1 + \frac{2|E|}{|V|(|V| - 1)} \right) \right\rceil \tag{7}$$

The theoretical analysis of GraIL determined that any logical rule R derived from the topology of a knowledge graph uniquely corresponds to a set of nodes connected through a sequence of relations, and that GraIL can learn this rule if the nodes and relations are present in the graph neural network.

To examine the differences between fixed and variable size sub graph, we constructed our feature matrix and sent it to GNN.

Topology Information in Variable Sized Subgraphs: In order to have a feature vector of constant size, we take a statistical analysis of each heuristic feature, across all of the nodes in the subgraph. For each heuristic feature $R \in R^5$, we can take the mean, median, standard deviation, minimum, maximum and variance across all of the nodes in the subgraph. In other words, we can apply the statistical functions to the rows of the topology matrix P_x.

Let $F_x = Stat(P_x)$, where $Stat(P_x)$ replaces each row of P_x with $\langle Mean(r),$ $Median(r), Variance(r), Min(r), Max(r), Std(r)\rangle$, So the resulting feature matrix has 30 elements,

$$F_x = \begin{bmatrix} Mean(CN) & Med(CN) & Var(CN) & Min(CN) & Max(CN) & STD(CN) \\ Mean(AA) & Med(AA) & Var(AA) & Min(AA) & Max(AA) & STD(AA) \\ Mean(JC) & Med(JC) & Var(JC) & Min(JC) & Max(JC) & STD(JC) \\ Mean(RA) & Med(RA) & Var(RA) & Min(RA) & Max(RA) & STD(RA) \\ Mean(PA) & Med(PA) & Var(PA) & Min(PA) & Max(PA) & STD(PA) \end{bmatrix}_{5 \times 6}$$

Therefore, F_x sees the rows of P_x replaced by the statistical results which are rows of fixed size 5, since we consider five heuristic values and columns of fixed size 6 since there are 6 statistical functions. F_x will always have a total of 30 elements, suitable to be encoded into a node x's feature vector for training in the Graph Neural Network. We simply list all 30 elements as components of the final feature vector.

Topology Information in Fixed Sized Subgraphs: As PLACN used in it's architecture, the topological feature matrix P_x of subgraphs is fixed for a given KG. The columns correspond to the fixed subgraph size and the rows correspond to each heuristic function. Therefore the size of P_x is always $5 \times |S|$. Thus, for fixed sized subgraphs, we can directly encode P_x,

$$F_x = P_x \tag{8}$$

In practice, this has led to very large vectors, when the fixed size of the subgraphs is large.

3.4   Scoring Subgraph Using GNN

This section explains the importance of GNN in our framework.

GNN Message Passing: In a GNN, a hidden embedding h_u^k for each node $u \in V$ is updated on each message-passing iteration based on information gathered from u's graph neighbor $\Gamma(u)$. In other terms, the representation of the node u is iteratively updated by aggregating its neighbors' representations [32]. So basically, GNN works based on two functions: Aggregation function passes information from $\Gamma(u)$ to u, and update function update features of u based on the information to form an embedded representation.

In AGIL model, each enclosed subgraph has a network of k-hop neighborhood nodes. So, after aggregating for k iteration, the kth layer of GNN is represented as,

$$m_u^k = \text{AGGREGATE}^k(\{h_v^{k-1} : v \in \Gamma(u)\}) \tag{9}$$

$$h_u^k = \text{UPDATE}^k(h_u^{k-1}, m_u^k) \tag{10}$$

where h_u^k is the feature vector of node u at kth iteration, Initially, $h_u^0 = X_u$, and m_u^k is the message aggregated from $\Gamma(u)$ at kth iteration.

In Eq. 9, there are various approaches proposed for message AGGREGATE function. Motivated by these architectures, GraIL adopts the method proposed by [20]. The following function defines the message aggregated function in a relational multi-graph:

$$h_u^k = \sigma\left(\sum_{r \in R}\sum_{v \in \Gamma_i^r} \alpha_{rr_u vu} W_r^{k-1} h_v^{k-1} + W_0^{k-1} h_u^{k-1}\right) \tag{11}$$

where k is the current layer of the neural network, u is the node being aggregated, R is the set of relationship types, $\alpha_{rr_u vu}$ is the attention value for layer k, r is

Table 1. The statistical information of datasets for inductive setting, where R, V and E are relations, vertices and edges respectively.

		WN18RR			FB15k-237			NELL-995		
		# R	# V	# E	# R	# V	# E	# R	# V	# E
v1	Train	9	2746	6678	183	2000	5226	14	10915	5540
	ind-test	9	992	1991	146	1500	2404	14	225	1034
v2	Train	10	6954	18968	203	3000	12085	88	2564	10109
	ind-test	10	2923	4863	176	2000	5092	79	4937	5521
v3	Train	11	12078	32150	218	4000	22394	142	4647	20117
	ind-test	11	5084	7470	187	3000	9137	122	4921	9668
v4	Train	9	3861	9842	222	5000	33916	77	2092	9289
	ind-test	9	7208	15157	204	3500	14554	61	3294	8520

any relationship , r_t is a target relationship between nodes v and u, W_r^k is the transformation matrix for r and layer k, and h_v^k is the feature vector of the node v.

The GNN uses an aggregation function to distribute features of nodes into their neighbors, for each layer of the neural network. The aggregation function used by GraIL uses the node's labels as the feature vector. We append the elements of the feature matrix F_x to the h vector used in formulation 11. The feature vector h in the GraIL model uses only the node labels $(L1, L2)$, for example $(1, 0)$, or (25.75). However, the node structured information (i.e., node labels) are limited information for training GNN. Therefore, we extend the feature vector with the 30 elements from F_x. So, in AGIL model, the feature vector h for node x would incorporate the statistical analysis of heuristic features as below;
$\langle L1, L2, Mean(CN), Med(CN), Var(CN), Min(CN), Max(CN), STD(CN),$
$Mean(AA), Med(AA), Var(AA), Min(AA), Max(AA), STD(AA), Mean(JC),$
$Med(JC), Var(JC), Min(JC), Max(JC), STD(JC), Mean(RA), Med(RA),$
$Var(RA), Min(RA), Max(RA), STD(RA), Mean(PA), Med(PA), Var(PA),$
$Min(PA), Max(PA), STD(PA)\rangle.$

At each layer, the graph neural network continuously combines feature vectors of nodes with the aggregates of their 1-hop neighborhoods.

4   Experiments

We perform experiments to demonstrate the efficiency and effectiveness of our framework, AGIL. Experiments are carried out on benchmark datasets, WN18RR [9], FB15k-237 [26], and NELL-995 [31] which were originally developed for transductive settings. To conduct inductive relation prediction, we use 4 versions of inductive datasets and 2 versions of transductive datasets, which are prepared by the GraIL authors and identical to the data used in their experiments. They constructed fully-inductive benchmark datasets by sampling disjoint subgraphs from the KGs. These datasets consist of two set of graphs: Train-graph and Ind-test-graph. Table 1 represents the statistical information on how benchmark datasets are split for inductive setting.

All the experiments are performed on a Intel(R) Core(TM) i7-3770 CPU computer @3.40 GHZ speed and 24 GB of RAM.

Table 2. Inductive Setting Experimental Result (AUC-PR)

	WN18RR				FB15k-237				NELL-995			
	v1	v2	v3	v4	v1	v2	v3	v4	v1	v2	v3	v4
Neural-LP	86.02	83.78	62.90	82.06	69.64	76.55	73.95	75.74	64.66	83.61	87.58	85.69
DRUM	86.02	84.05	63.20	82.06	69.71	76.44	74.03	76.20	59.86	83.99	87.71	85.94
RuleN	90.26	89.01	76.46	85.75	75.24	88.70	91.24	91.79	84.99	88.40	87.20	80.52
GraIL	94.32	94.18	85.80	92.72	84.69	90.57	91.68	**94.46**	86.05	92.62	93.34	87.50
AGIL (F-Subgraph)	**96.38**	**95.77**	**89.28**	**95.66**	73.5	84.56	76.4	NA	90.56	93.7	94.18	NA
AGIL (V-Subgraph)	94.76	94.92	86.46	93.65	**87.42**	**91.20**	**93.44**	93.52	**91.21**	**96.84**	**97.04**	**95.42**

4.1 Inductive Relation Prediction

We test our model, AGIL on inductive datasets to determine if it can generalise relations when the entities aren't visible during GNN training. AGIL is trained on Train-graph and tested on Ind-test-graph.

To evaluate the performance, we compare AGIL against the following state-of-the-art methods.

1. NeuralLP [34]: an end-to-end differentiable model for inductive relation prediction.
2. DRUM [18]: a scalable and differentiable approach for mining first-order logical rules from KG.
3. RuleN [15]: statistical rule mining method, and the current state-of-the-art in inductive relation prediction on KGs.
4. GraIL [25]: inductive relation prediction by subgraph reasoning, and highly similar to AGIL.

We use the original source code by the authors for the implementation of above methods, NeuralLP[1], DRUM[2], RuleN[3] and GraIL[4]. For AGIL framework, the implementation is built upon the Python code base provided by [25] in their GraIL implementation. It uses the Deep Graph Learning library to implement a graph neural network.

Results and Discussion: The performance of the experimental setup for AGIL is represented in Table 2 against baseline methods. The Precision Recall

---

[1] https://github.com/fanyangxyz/Neural-LP.
[2] https://github.com/alisadeghian/DRUM.
[3] https://web.informatik.uni-mannheim.de/RuleN/.
[4] https://github.com/kkteru/grail.

Area Under Curve (AUC-PR) is used to evaluate the model's accuracy. The AGIL model is tested based on fixed sized subgraph (F-Subgraph) as proposed in PLACN, and variable sized subgraph (V-Subgraph). We observed that the model with fixed sized subgraph fails to perform better in some dataset such as v4-FB15k-237 and v4-NELL-995. The poor performance might be due to the absence of critical nodes and relations in the subgraph with fixed size neighbours. But, AGIL model with variable sized subgraph outperforms most of the standard baseline methods. In the NELL-995 dataset, the improvement is most significant compared to the other datasets. In WN18RR dataset, AGIL performs significantly better when we use fixed sized subgraph extraction. This indicates that for any knowledge graph of realistic size, fixed sized subgraphs are not always suitable for Inductive Graph Neural Network models.

If k value is sufficiently large enough, it may include all connecting paths. The recommended calculation to derive k by PLACN was shown to be too low for certain data sets, such as FB15k-237. If a knowledge graph has a high number of cycles, there may be many alternative paths between target nodes. Due to the truncation of the subgraph size to k, only a subset of possible paths will be analyzed by the neural network. Therefore, only a subset of the possible inductive rules will be learned by the GNN. When those inductive rules are applied to link prediction, they fail to produce accurate results. Both AGIL and GraIL provide a limiting factor to prevent excessively large subgraphs. It limits the number of hops from each node to a maximum, in all experiments, this maximum was 3 hops.

Moreover, GraIL outperforms on v4 of FB15k-237. However, the performance of AGIL is still close to GraIL on this dataset.

4.2   Transductive Relation Prediction

Most existing embedding based KG completion methods consider transductive setting for the prediction. Basically, all the existing KGs including WN18RR, FB15k-237, and NELL-995 are originally developed for the transductive setting. We test AGIL on transductive setting to determine it can predict the links accurately. We then compare AGIL against GraIL and RuleN.

Table 3. Transductive Setting Experimental Result (AUC-PR)

	WN18RR		FB15k-237		NELL-995	
	v1	v2	v1	v2	v1	v2
RuleN	81.79	83.97	87.07	92.49	80.16	87.87
GraIL	89.00	90.66	88.97	93.78	83.95	92.73
AGIL	**92.77**	**92.80**	**90.03**	**95.56**	**92.44**	**93.84**

Results and Discussion: The experimental results on transductive setting is represented in Table 3, which compares AGIL with GraIL and state-of-the-art method RuleN. In all the cases, AGIL outperforms the other two methods.

For the time being, we could not compare AGIL with other embedded-based methods. We will compare this in the future.

During the experiments it was shown that use of feature vectors would cause the model to overfit the training data, and loose some generality when applied to the test triples. To resolve this, we utilized the NodeNorm function [36] to normalize the feature vector. This gives the effect of making each feature vector have the save variance. Zhou et.al [36] have observed that GNNs perform poorly when the variance of features of nodes is very high. The normalization replaces each component in the feature vector with the difference from the mean divided by the variance.

The code is available in the GitHub link:
https://anonymous.4open.science/r/agil2021/README.md.

5   Conclusions

This paper examines an augmenting graph inductive learning framework based on GNN, named AGIL. Since many real-world KGs evolve with time, training very large networks with GNN is a challenging task. Therefore, we used a common neighbor-based subgraph to solve the scalability issue. Although AGIL is highly similar to the recently proposed model GraIL, AGIL incorporates topological heuristic features as additional information when GNN trains. Experimentally, we can see that the additional feature information gives better accuracy in both transductive and inductive settings. We also proved experimentally that fixed-sized subgraphs are not always suitable for Inductive Graph Neural Network models. Overall, our model, AGIL, outperforms most of the baseline methods. In the future, we are planning to examine the importance of individual topological features for the relation prediction.

References

1. Auer, S., Bizer, C., Kobilarov, G., Lehmann, J., Cyganiak, R., Ives, Z.: DBpedia: a nucleus for a web of open data. In: Aberer, K., et al. (eds.) ASWC/ISWC -2007. LNCS, vol. 4825, pp. 722–735. Springer, Heidelberg (2007). https://doi.org/10.1007/978-3-540-76298-0_52

2. Balažević, I., Allen, C., Hospedales, T.M.: Hypernetwork knowledge graph embeddings. In: Tetko, I.V., Kůrková, V., Karpov, P., Theis, F. (eds.) ICANN 2019. LNCS, vol. 11731, pp. 553–565. Springer, Cham (2019). https://doi.org/10.1007/978-3-030-30493-5_52

3. Bojchevski, A., Günnemann, S.: Deep gaussian embedding of graphs: unsupervised inductive learning via ranking. arXiv preprint arXiv:1707.03815 (2017)

4. Bollacker, K., Evans, C., Paritosh, P., Sturge, T., Taylor, J.: Freebase: a collaboratively created graph database for structuring human knowledge. In: Proceedings of the 2008 ACM SIGMOD International Conference on Management of Data, pp. 1247–1250 (2008)

5. Bordes, A., Usunier, N., Garcia-Duran, A., Weston, J., Yakhnenko, O.: Translating embeddings for modeling multi-relational data. In: Advances in Neural Information Processing Systems 26 (2013)

6. Carlson, A., Betteridge, J., Kisiel, B., Settles, B., Hruschka, E.R., Mitchell, T.M.: Toward an architecture for never-ending language learning. In: Twenty-Fourth AAAI Conference on Artificial Intelligence (2010)

7. De Raedt, L., Džeroski, S.: First-order JK-clausal theories are PAC-learnable. Artif. Intell. **70**(1–2), 375–392 (1994)

8. Demir, C., Ngomo, A.-C.N.: Convolutional complex knowledge graph embeddings. In: Verborgh, R., Hose, K., Paulheim, H., Champin, P.-A., Maleshkova, M., Corcho, O., Ristoski, P., Alam, M. (eds.) ESWC 2021. LNCS, vol. 12731, pp. 409–424. Springer, Cham (2021). https://doi.org/10.1007/978-3-030-77385-4_24

9. Dettmers, T., Minervini, P., Stenetorp, P., Riedel, S.: Convolutional 2D knowledge graph embeddings. In: Thirty-second AAAI Conference on Artificial Intelligence (2018)

10. Hachey, B., Radford, W., Nothman, J., Honnibal, M., Curran, J.R.: Evaluating entity linking with wikipedia. Artif. Intell. **194**, 130–150 (2013)

11. Hamilton, W.L., Ying, R., Leskovec, J.: Inductive representation learning on large graphs. In: Proceedings of the 31st International Conference on Neural Information Processing Systems, pp. 1025–1035 (2017)

12. Liben-Nowell, D., Kleinberg, J.: The link-prediction problem for social networks. J. Am. Soc. Inf. Sci. Technol. **58**(7), 1019–1031 (2007)

13. Liu, S., Chen, H., Ren, Z., Feng, Y., Liu, Q., Yin, D.: Knowledge diffusion for neural dialogue generation. In: Proceedings of the 56th Annual Meeting of the Association for Computational Linguistics (Volume 1: Long Papers), pp. 1489–1498 (2018)

14. Lukovnikov, D., Fischer, A., Lehmann, J., Auer, S.: Neural network-based question answering over knowledge graphs on word and character level. In: Proceedings of the 26th International Conference on World Wide Web, pp. 1211–1220 (2017)

15. Meilicke, C., Fink, M., Wang, Y., Ruffinelli, D., Gemulla, R., Stuckenschmidt, H.: Fine-grained evaluation of rule- and embedding-based systems for knowledge graph completion. In: Vrandečić, D., et al. (eds.) ISWC 2018. LNCS, vol. 11136, pp. 3–20. Springer, Cham (2018). https://doi.org/10.1007/978-3-030-00671-6_1

16. Muggleton, S.: Inductive logic programming. New Gene. Comput. **8**(4), 295–318 (1991)

17. Ragunathan, K., Selvarajah, K., Kobti, Z.: Link prediction by analyzing common neighbors based subgraphs using convolutional neural network. In: ECAI 2020, pp. 1906–1913. IOS Press (2020)

18. Sadeghian, A., Armandpour, M., Ding, P., Wang, D.Z.: Drum: end-to-end differentiable rule mining on knowledge graphs. arXiv preprint arXiv:1911.00055 (2019)

19. Scarselli, F., Gori, M., Tsoi, A.C., Hagenbuchner, M., Monfardini, G.: The graph neural network model. IEEE Trans. Neural Netw. **20**(1), 61–80 (2008)

20. Schlichtkrull, M., Kipf, T.N., Bloem, P., van den Berg, R., Titov, I., Welling, M.: Modeling relational data with graph convolutional networks. In: Gangemi, A., et al. (eds.) ESWC 2018. LNCS, vol. 10843, pp. 593–607. Springer, Cham (2018). https://doi.org/10.1007/978-3-319-93417-4_38

21. Selvarajah, K., Ragunathan, K., Kobti, Z., Kargar, M.: Dynamic network link prediction by learning effective subgraphs using CNN-LSTM. In: 2020 International Joint Conference on Neural Networks (IJCNN), pp. 1–8. IEEE (2020)

22. Speer, R., Chin, J., Havasi, C.: Conceptnet 5.5: An open multilingual graph of general knowledge. In: Thirty-first AAAI Conference on Artificial Intelligence (2017)

23. Suchanek, F.M., Kasneci, G., Weikum, G.: Yago: a core of semantic knowledge. In: Proceedings of the 16th International Conference on World Wide Web, pp. 697–706 (2007)

24. Sun, Z., Deng, Z.H., Nie, J.Y., Tang, J.: Rotate: knowledge graph embedding by relational rotation in complex space. arXiv preprint arXiv:1902.10197 (2019)
25. Teru, K., Denis, E., Hamilton, W.: Inductive relation prediction by subgraph reasoning. In: International Conference on Machine Learning, pp. 9448–9457. PMLR (2020)
26. Toutanova, K., Chen, D., Pantel, P., Poon, H., Choudhury, P., Gamon, M.: Representing text for joint embedding of text and knowledge bases. In: Proceedings of the 2015 Conference on Empirical Methods in Natural Language Processing, pp. 1499–1509 (2015)
27. Trouillon, T., Welbl, J., Riedel, S., Gaussier, É., Bouchard, G.: Complex embeddings for simple link prediction. In: International Conference on Machine Learning, pp. 2071–2080. PMLR (2016)
28. Wang, M., Qiu, L., Wang, X.: A survey on knowledge graph embeddings for link prediction. Symmetry 13(3), 485 (2021)
29. Wang, Q., Mao, Z., Wang, B., Guo, L.: Knowledge graph embedding: a survey of approaches and applications. IEEE Trans. Knowl. Data Eng. 29(12), 2724–2743 (2017)
30. Xiong, C., Callan, J.: Esdrank: connecting query and documents through external semi-structured data. In: Proceedings of the 24th ACM International on Conference on Information and Knowledge Management, pp. 951–960 (2015)
31. Xiong, W., Hoang, T., Wang, W.Y.: Deeppath: a reinforcement learning method for knowledge graph reasoning. arXiv preprint arXiv:1707.06690 (2017)
32. Xu, K., Hu, W., Leskovec, J., Jegelka, S.: How powerful are graph neural networks? arXiv preprint arXiv:1810.00826 (2018)
33. Yang, B., Yih, W.t., He, X., Gao, J., Deng, L.: Embedding entities and relations for learning and inference in knowledge bases. arXiv preprint arXiv:1412.6575 (2014)
34. Yang, F., Yang, Z., Cohen, W.W.: Differentiable learning of logical rules for knowledge base reasoning. arXiv preprint arXiv:1702.08367 (2017)
35. Zhang, M., Chen, Y.: Link prediction based on graph neural networks. Adv. Neural. Inf. Process. Syst. 31, 5165–5175 (2018)
36. Zhou, K., Dong, Y., Wang, K., Lee, W.S., Hooi, B., Xu, H., Feng, J.: Understanding and resolving performance degradation in graph convolutional networks. arXiv preprint arXiv:2006.07107 (2020)
37. Zhu, F., Wang, Y., Chen, C., Liu, G., Orgun, M., Wu, J.: A deep framework for cross-domain and cross-system recommendations. arXiv preprint arXiv:2009.06215 (2020)

Generative Networks Applied to Model Fluid Flows

Mustapha Jolaade[✉], Vinicius L. S. Silva, Claire E. Heaney, and Christopher C. Pain

Applied Modelling and Computation Group, Imperial College London, London, UK
moj20@ic.ac.uk

Abstract. The production of numerous high fidelity simulations has been a key aspect of research for many-query problems in fluid dynamics. The computational resources and time required to generate these simulations can be so large and impractical. With several successes of generative models, we explore the performance and powerful generative capabilities of both generative adversarial network (GAN) and adversarial autoencoder (AAE) to predict the evolution in time of a highly nonlinear fluid flow. These generative models are incorporated within a reduced-order model framework. The test case comprises two-dimensional Gaussian vortices governed by the time-dependent Navier-Stokes equation. We show that both the GAN and AAE are able to predict the evolution of the positions of the vortices forward in time, generating new samples that have never before been seen by the neural networks.

Keywords: Generative adversarial networks · Adversarial autoencoder · Two-dimensional turbulence · Spatial-temporal predictions · Deep learning

1 Introduction

The study of fluid dynamics has involved massive amounts of data generated either from controlled experiments, field measurements or large-scale numerical simulations. The high volume of data, amongst other reasons, means these methods can be relatively slow and require a great deal of computational power to be able to model the underlying physics. While advancements in high performance computing research has boosted speed and accuracy of numerical simulation, obstacles still remain [2]. Thus, the development of computational frameworks that are accurate, robust, cheap and fast enough to model fluid dynamics remains a key aspect of computational science and engineering research.

In this paper, generative models, a branch of machine learning, is applied to a two-dimensional turbulent fluid problem for the purposes of rapidly predicting forward in time while avoiding the high computational cost of traditional numerical methods.

Generative models have garnered a huge amount of interest in recent years [11]. The main idea behind generative models is to build a statistical model

© The Author(s), under exclusive license to Springer Nature Switzerland AG 2022
D. Groen et al. (Eds.): ICCS 2022, LNCS 13352, pp. 742–755, 2022.
https://doi.org/10.1007/978-3-031-08757-8_61

around a given dataset that is capable of generating new sample instances that appear to be taken from the original dataset. These new samples can further be used for tackling problems related to the case under study. When the building process is based on deep networks (artificial neural networks such as convolutional neural networks (CNN) [12]) that use multiple layers to capture how patterns/features of the dataset are organised or clustered, the resulting model is termed a deep generative model. Once a deep generative model has learned the structure of the training dataset, by being fed a random vector as input, its networks can generate desired samples from complex probability distributions in high-dimensional spaces [8]. In building deep generative networks two main methods have been widely used. The first is a variational autoencoder that uses stochastic variational inference to minimize the lower bound of the data likelihood [11]. The second is a generative adversarial network (GAN) whereby two players (neural network) play a zero-sum game. The game seeks to minimize the distribution divergence between the model output and the real/training dataset by using real samples as a proxy for optimization. A novel third method born out of the amalgamation of these two methods is the use of adversarial autoencoder (AAE) [14]. In this project, attention is given to both GAN and AAE as data-driven methods for prediction and modelling of spatial-temporal turbulent fluid flow.

Although reduced order models have been used for time-dependent turbulent fluid modelling in areas such as subsurface flow [3] and for the solution of the Navier-stokes equation [19]. In this project, for the first time, we use generative models in a reduced-order model framework to carry out efficient predictions in time of a two-dimensional turbulent fluid flow problem.

The rest of this paper is structured as follows: the next section provides a description of the methodology adopted from [16] for spatial-temporal prediction with GANs. Here, we also include the methodology for prediction using the AAE. Section 3, introduces the test fluid system and a relevant discussion about the transformation carried out to make the data suitable for use. The obtained results from predicting single and multiple time levels are also presented in Sect. 3. Finally, conclusion and remarks about possible future work are provided in Sect. 4.

2   Methodology

The use of GANs for time series prediction and data assimilation of real world dynamical systems has been proven to be successful for the spatial-temporal spread of COVID-19 using SEIRS type models [15,16]. Particularly, the method in [16] has been shown to be independent of the underlying system, thus this project will apply the same method for the two-dimensional turbulent fluid model.

In this project, we start by building a reduced model of the turbulent fluid flow, going from a high-fidelity spatial domain to a lower dimensional representation. Then, a generative model is built and trained to learn a mapping between a input latent vector and the lower dimensional representation. Finally, we apply the processes of simulating forward in time using the capabilities of the generative models. The aim is then for the generative networks to serve as surrogate models that can reproduce the high-fidelity numerical model.

2.1 POD-based Non-Intrusive Reduced Order Modelling

The connection between physics-based machine learning and dimensionality reduction has been substantially studied and well-documented [18]. Results of these studies have shown that many methods used to obtain a low-dimensional subspace of a system are related to machine learning methods. In modern computational research, Reduced Order Modelling (ROM) is a well-known technique for dimensionality reduction [3]. By constructing reduced-order models that encapsulate the original features of the fluid systems while maintaining its underlying physics, it is possible to seek solutions to a model in an efficient and much less expensive way [20]. The Non-Intrusive Reduced Order Modelling (NIROM) is a type of ROM so named due to its non-dependent on the system under study. This model reduction approach can use proper orthogonal decomposition (POD) [17] to derive a physics-inspired low-dimensional parameterization that represents the high dimension of the high-fidelity spatial domain of the fluid model (i.e. state of snapshots). POD is closely related to the principal component analysis (PCA) method in statistics and was first used for turbulent flows by [13].

In this project, the dimensionality reduction aspect of our methodology is set within a NIROM framework that involved computing the POD basis vectors (via PCA) using the snapshots of the input data [17].

Consider a three-dimensional field ω, which is dependent on some input parameter and varies in space and time. We can define its function as $\omega : \mathcal{X} \times \mathcal{T} \times \zeta \to \mathbb{R}$ where $\mathcal{X}$ is the spatial domain, $\mathcal{T}$ is the time domain, and an input domain ζ of initial parameters/condition. The aim of data-driven/non-intrusive dimensionality reduction is to find an approximate model for ω from the data

$$\mathcal{D} \subset \{\omega(\boldsymbol{x}, t, \boldsymbol{z}) \mid \boldsymbol{x} \in \mathcal{X}, t \in \mathcal{T}, \boldsymbol{z} \in \zeta\} \tag{1}$$

which, in this case, are snapshots in time of the field. The desired approximate model of the field can be expressed as a linear expansion in the POD basis. This POD basis would be computed from many snapshots data developed as solutions of a high-fidelity model that describes the field. To compute the POD basis, we consider a snapshot data to be $\omega(t; \boldsymbol{z}) \in \mathbb{R}^{n_x}$ where n_x is the dimension of the spatial domain (from finite discretization). Thus, the set $\{\omega(t_i; z_j) \mid i = 1, \cdots, n_t; \ j = 1, \cdots, n_z\}$ of snapshots at n_t different time levels/steps of $t_1, t_2, \cdots, t_{n_t} \in \mathcal{T}$ and n_z different initial input conditions of $z_1, z_2, \cdots, z_{n_z} \in \zeta$ comprises of $n_s = n_t n_z$ snapshots. The snapshot matrix can be defined as $\boldsymbol{S} \in \mathbb{R}^{n_x \times n_s}$ with each row corresponding to a spatial location and each column representing a snapshot in the set. At this stage, PCA can then be introduced for dimensionality reduction.PCA seeks a transformation $\boldsymbol{T}$ that maps each vector $\{\boldsymbol{\omega}(t_i; \boldsymbol{z}_j)$ in $\boldsymbol{S}$ (i.e. each snapshot) from the original dimensional space of n_x to a new space that only keeps the first r principal components using the first r eigenvectors of the transformed matrix [10].

The idea is to maximize the variance of the original data while minimising the total least squared errors in the representation of the snapshots. The size of the POD basis/principal component r is chosen by specifying a tolerance in this error calculation. This user-specified tolerance, k also indicates how much information/energy of the data is captured by the resulting snapshot representation. We chose r such that:

$$\frac{\sum\limits_{k=1}^{r} \sigma_k^2}{\sum\limits_{k=1}^{n_s} \sigma_k^2} > k, k = 0.999 \tag{2}$$

This means given a snapshot field we can compute its original state, using the POD coefficients, with 99.9% reconstruction accuracy. Hence, once the dimension reduction is completed, the POD expansion coefficients $\boldsymbol{\theta}_{k=1,\dots,r}(t; \boldsymbol{z})$ denote the model approximation and parameterization of a snapshot field $\boldsymbol{\omega}(t; \boldsymbol{z})$ at time t and input conditions $\boldsymbol{z}$. The coefficients are then employed in the training of generative models for the time series prediction. Results of the POD-based compression are shown and discussed in Sect. 3.2.

2.2   Generative Models

Generative modelling is the process of training a machine learning model with specific data to produce 'fake' data from a distribution that mimics the probability distribution of the original training set. Here, we produce two generative models to perform time series prediction of a turbulent fluid flow. The two models utilized are: a generative adversarial network (GAN) [7] and an adversarial autoencoder (AAE) [14]. The choice of these models was based on their proven successes in the use of nonlinear fluid modelling. In the result section, a comparison between outputs of the two models is presented.

Generative Adversarial Network: A GAN is an artificial learning technology that is composed of two neural networks as shown in Fig. 1. GANs have been adopted widely in several research areas, showing huge successes in practical applications including simulating fluid models [5]. The training process is essentially a game between two models competing as adversaries. While the generator module (G) generates fake samples from an input random distribution, a discriminator module (D) tries to distinguish between real samples drawn from the original distribution and the sample output from the generator. D does this by estimating a score which serves as the probability that a particular sample came from the original distribution i.e. $D(G(\boldsymbol{\theta}_r)) = 1$. The training process of a GAN is a minimization-maximization problem that is based on a cross-entropy loss function

$$\boldsymbol{J}(D, G) : \min_{G} \max_{D} E_{\theta_r \sim p_{data}(\theta_r)}[log D(\theta_r)] + E_{\mathbf{z} \sim p_z(\mathbf{z})}[log(1 - D(G(\theta)))] \tag{3}$$

where $p_{data}(\theta_r)$ is the probability data distribution of the target output of real samples θ_r and $p_z(z)$ is the prior distribution for the random latent vector $\mathbf{z}$. The training process involves:

- Updating D with gradients that maximize the discriminator function by differentiating with respect to parameters of the discriminator.
- Updating G with gradients that minimize the generator function by differentiating with respect to parameters of the generator.

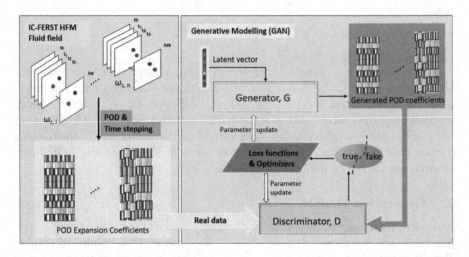

Fig. 1. Generative modelling using GAN. In this workflow, real samples obtained from POD-based NIROM are utilized as training data for the discriminator module of a GAN. Fake data produced by the generator, G from an input latent vector is simultaneously used in the training process, with loss back propagated through both neural network modules.

A common problem in the use of GANs for sample generation is mode collapse. Typically, a GAN is trained to produce a wide variety of outputs that mimic the training data distribution. For example, if a GAN is trained with pictures of different dog breeds, we want a different dog for every random input to the dog generator. However, it is possible that the generator only produces a small set of realistic outputs and learns to generate only that seemingly credible output (or small set of outputs) to the discriminator.

The Wasserstein GAN (WGAN) [1] is a type of GAN that avoids this problem of mode collapse by circumventing the issue of vanishing gradients. This implies that the discriminator is trained to optimality, learning to reject any output/set of outputs the generator tries to stabilize on. The WGAN method introduces a new loss function that alternatively minimizes an approximation of the Earth Mover distance between the distributions completely avoiding mode collapse.

In developing a GAN for the generative modelling of this project, a typical Deep Convolutional GAN (DCGAN) was developed and trained to produce the target output. Following evidence of mode collapse however, an Improved WGAN [9] was also developed by altering the loss functions of the original DCGAN. The WGAN also included a gradient penalty term that led to more diverse output from the generator.

The WGAN loss function uses a Earth mover distance criteria to enforce match of a prior data distribution. The loss function for this type of GAN is given as follows

$$L = \mathbb{E}_{\hat{\mathbf{x}} \sim \mathbb{P}_g} \left[D\left(\tilde{\mathbf{x}}\right)\right] - \mathbb{E}_{\mathbf{x} \sim \mathbb{P}_r} \left[D\left(\mathbf{x}\right)\right] + \lambda \mathbb{E}_{\hat{\mathbf{x}} \sim \mathbb{P}_{\hat{\mathbf{x}}}} \left[\left(\left\|\nabla_{\tilde{\mathbf{x}}} D\left(\tilde{\mathbf{x}}\right)\right\|_2 - 1\right)^2\right] \quad (4)$$

where the second term is a gradient penalty that replaces weight clipping to achieve Lipschitz continuity (gradient with norm at most 1 everywhere). The discriminator in this GAN works as a critic.

The generative model (GAN and/or WGAN) developed and trained using the presented workflow can be used for time-series/forward prediction without any changes to its structure. This is also the case when the model is utilized for the assimilation of given observation/sensor data [16].

Adversarial Autoencoder: A second type of generative model built and implemented in this project is an AAE (Fig. 2). Similar to a GAN, the AAE was proposed as a generative model that seeks to match an aggregated posterior distribution of its hidden latent vector with a prior distribution. To be able to function as a deep generative model, the AAE is trained to perform variational inference that enables its decoder to learn a statistical model that maps between the imposed prior and the data distribution. The AAE has a wide range of applications including semi-supervised classification, unsupervised clustering and data visualization [14]. In the field of computational fluid dynamics, Cheng et al. [4] studied the capability of an advanced deep-AAE for parameterizing nonlinear fluid flow and utilized it in the prediction of a water collapse test case. Here, we develop an AAE and test it for prediction of nonlinear turbulent flow.

2.3 Space-time Predictions Using Generative Models

The goal of this project is to show that generative models such as GANs and AAEs can be utilized for the time-series prediction of nonlinear turbulent fluid models. This section discusses the time-series prediction and an algorithm for its implementation. The methodology proposed by [16] is further tested on a two-dimensional turbulent fluid model to obtain a surrogate model that is accurate and computationally cheap. The next subsections discuss the this method and its components.

Prediction Using GANs: The ability of a GAN to produce realistic samples that seem to belong to a prior distribution is leveraged in this project. To

748   M. Jolaade et al.

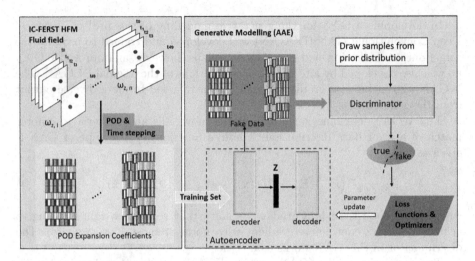

Fig. 2. Generative modelling using AAE. The training process of this workflow attempts to match output of the autoencoder with a prior distribution. While the encoder generates fake samples that matches this distribution, the discriminator attempts to critic against the generated samples.

predict forward in time, an algorithm, Predictive GAN (PredGAN) algorithm [16] is implemented in this project on two-dimensional turbulent flow data. The PredGAN algorithm begins with training a GAN to generate a data sequence of $p+1$ time levels from an input latent vector. To achieve this, the GAN is trained with $p+1$ consecutive time levels of compressed variables/POD coefficients concatenated to form a trajectory. Once the training is completed, the generator of the GAN is capable of producing fake snapshots at multiple time levels, $n-p$ to n where $n \geq p$. In order to complete prediction with the trained GAN, the first p time levels of a known trajectory/given solution is matched with corresponding time levels of the output of the GAN through loss optimization. Once convergence has been reached, the additional time step $p+1$, in the output of the generator serves as the forward prediction of the trajectory. This process can be repeated by using the predicted $p+1$ solution as a known solution while similar optimization is carried out to predict time level $p+2$. Ultimately, all time steps can be predicted by replicating the process and obtaining a new time step for each iteration of the PredGAN algorithm.

Prediction Using Adversarial Autoencoder: To predict with an autoencoder, the following steps were followed:

1. Since the autoencoder does not require a latent variable as input, we use the first $p-1$ time levels of the known solution as input.
2. To predict forward, the $p-1$ time level is used as an initial guess for the desired p time level. and passed into the autoencoder to give a prediction.

Following successful training, the autoencoder then attempts to match the true snapshot at time p from the input initial guess.
3. The output prediction from a single iteration through the autoencoder is further re-used as input guess for the time level p and the time series is passed through the autoencoder till convergence is reached.
4. The final output is the snapshot prediction at time p.
5. For multiple time level predictions, the process is repeated from steps 1–4 using the last p time levels as initial guesses for subsequent time levels.

3   Implementation and Results

3.1   Case Study: Parameter-varying Flow in a Periodic Box

In order to train a GAN capable of time-series prediction of the two-dimensional turbulent fluid problem, a dataset comprising two velocities component (x, y) and the pressure for each discretized node of a two-dimensional incompressible Navier-Stokes simulations has been obtained. Figure 3 shows the spatial properties of each snapshot. This dataset represents a parameter-varying flow in a fixed-wall box. Given that the convolutional layers of a neural network are designed to detect object/features anywhere in an image, it can be used in this project since the large-parameter variations implicit in the dataset generation is of a similar nature as object randomly located in an image [6].

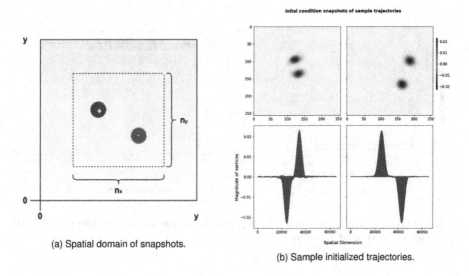

(a) Spatial domain of snapshots.

(b) Sample initialized trajectories.

Fig. 3. Two-dimensional Gaussian vortices in a square domain. The positive and negative vortices are of equal strength and each snapshot $S \in \mathbb{R}^{y \times y}$ where $y = 256$, are randomly initialized within a predefined subdomain $n_x \times n_y$. The images on the bottom right show the magnitude of the vortices projected over a 1-D domain.

The simulations were run on Imperial College-Finite Element Reservoir Simulator (IC-FERST) using the following criteria: turbulent flow with Re=5000, constant viscosity and no slip walls boundary conditions. The first step in the project is to transform the velocity dataset into vorticity data so it represents initial Gaussian (randomly initialized) vortices that decay due to viscosity changes. Following this transformation, the vorticity data are then compressed by carrying out a POD.

The training set for this project included snapshots from 300 separate trajectories. Snapshots from trajectories were obtained such that each trajectory included 50 snapshots - a total of 15K snapshots. Prior to actual training, the data is prepared for time series prediction by concatenating successive snapshots, 5 s apart, into a time series of 7 instances (i.e. each time series represents vortices' evolution over a period of 30 s). This sums up to 6K distinct time series - one trajectory can be split into a maximum of 20 time series of 7 snapshots. In predicting with generative models post-training, we were able to forecast 1–3 additional instances (evolution over a period of ≤15 s) for never before seen time series (30 trajectories). See 3.3 for more details.

3.2   POD Compression and Order Reduction

Each snapshot is no longer a 256 by 256 array but now represented using 292 features (POD coefficients). The cumulative information/energy retained measured using explained variance is over 99.99% as shown in Fig. 4. A visual comparison of the compression is shown in Fig. 5.

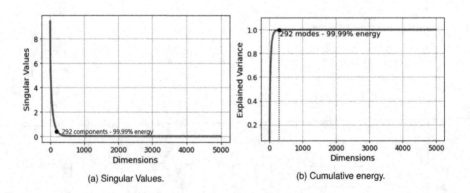

(a) Singular Values.   (b) Cumulative energy.

Fig. 4. POD singular values and relative cumulative energy for the two-dimensional Gaussian vorticity filed snapshot set.

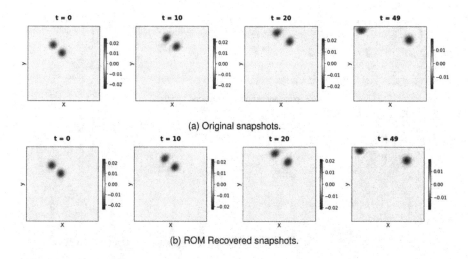

(a) Original snapshots.

(b) ROM Recovered snapshots.

Fig. 5. Original and recovered snapshots following POD-based NIROM. The reduction was specified to retain 99.99% information from the original snapshots. The reduction decreased the dimension from 256-by-256 to 292 POD coefficients.

3.3 Prediction Using GAN and AAE

Single Time Level Prediction: In this section, we apply the PredGAN algorithm to a sample trajectory to predict a single time level forward. Following the training of a WGAN-GP and an AAE with 7 time levels from sample trajectories, we proceed to predict a single step forward in the test set using the PredGAN and PredAAE algorithms. In this application, the first 6 time levels (t=0 to t=25) are considered known while the 7th time level (t=30) is the predicted time step. Results of the single time level prediction can be found in Fig. 6. A sample trajectory (Fig. 6a) serves as the input to the generative models (WGAN-GP and AAE). Snapshots of each generative model shows output following convergence of the loss between generated sample and known solution (Fig. 6bc). The second row visualizes the mismatch between the magnitude and location of the true data (in blue) and generated prediction (in orange). The vertical axis represents the magnitude of the vortices while the horizontal axis is a one-dimensional projection of the two-dimensional domain. Given these results, AAE is shown to have a better performance both for predicting forward and matching known solutions with samples generated from a random input latent vector.

Multiple Time Levels Prediction: The results from predicting multiple time levels, shown in Fig. 7, follows a similar pattern as that of the single time level prediction. Following results for the single time level prediction (t=30), we proceed to predict multiple time levels from t=35 to t=45. It is worth mentioning that this data was not present in the training set. The first row of snapshots

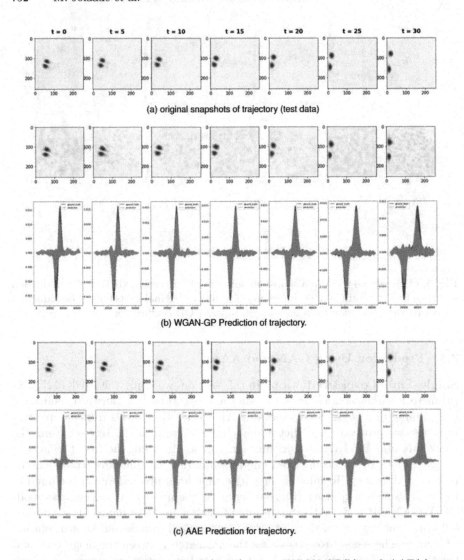

(a) original snapshots of trajectory (test data)

(b) WGAN-GP Prediction of trajectory.

(c) AAE Prediction for trajectory.

Fig. 6. Prediction of one time level (t=30) using WGAN-GP(b) and AAE(c) on a sample trajectory from train dataset.

(Fig. 7a.) shows the true snapshot of the trajectory at times t=35 to t=45. This is the known/given solution form the high fidelity simulation. Figure 7b. shows predicted output for these time levels using the same WGAN-GP. Here, we see that while the the WGAN-GP is able to predict the spatio-temporal distribution, the prediction ability reduces with forward time. The AAE (Fig. 7c), however, shows no such sign.

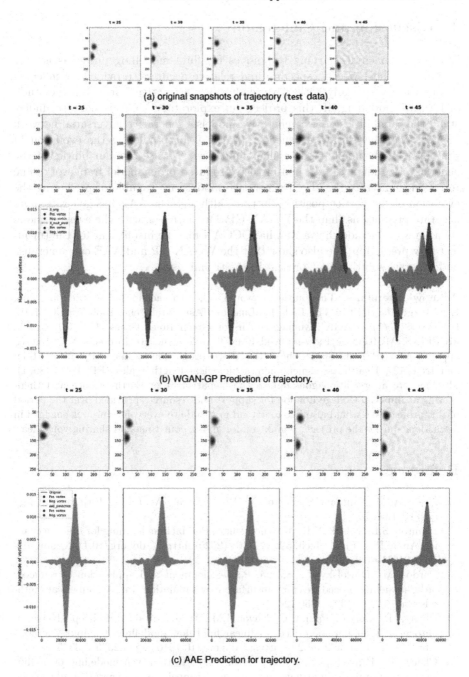

(a) original snapshots of trajectory (test data)

(b) WGAN-GP Prediction of trajectory.

(c) AAE Prediction for trajectory.

Fig. 7. Multiple time level prediction (t=35 to t=45) using WGAN-GP(b) and AAE(c) on a sample trajectory.

## 4	Discussion and Conclusion

The use of machine learning techniques for fluid modelling problems is very promising. The low cost, speedup and relative accuracy provided by machine learning tools, specially generative models, are attractive features in the study of forward modelling. In this project, an exploratory study is done to understand the capabilities of two generative models - generative adversarial network (GAN) and adversarial autoencoder (AAE) - for predicting the evolution in time of a highly nonlinear turbulent fluid flow. We use the capabilities of the generative models within a non-intrusive reduced order model framework. The results demonstrate that both generative models are capable of predicting the evolution of the vortice positions in time, although the AAE has generate more accurate predictions than the WGAN-GP. Furthermore, with the event of mode collapse, we conclude that a 'vanilla' DCGAN may be insufficient for the turbulent flow prediction. We also show that the WGAN-GP and AAE can generalise and generate solutions not present in the training set.

Acknowledgements. The authors would like to acknowledge the following EPSRC grants: RELIANT, Risk EvaLuatIon fAst iNtelligent Tool for COVID19 (EP/V036777/1); MAGIC, Managing Air for Green Inner Cities (EP/N010221/1); MUFFINS, MUltiphase Flow-induced Fluid-flexible structure InteractioN in Subsea applications (EP/P033180/1); the PREMIERE programme grant (EP/T00 0414/1); and INHALE, Health assessment across biological length scales (EP/T00 3189/1). Most sincere appreciation goes out to the Department of Earth Science and Engineering at Imperial College London for support and resources provided over the period of this project. We would also like to extend gratitude to everyone who was engaged in discussions during the project. Thank you for giving your time and sharing your ideas.

References

1. Arjovsky, M., Chintala, S., Bottou, L.: Wasserstein GAN (2017), http://arxiv.org/abs/1701.07875
2. Brunton, S.L., Noack, B.R., Koumoutsakos, P.: Machine learning for fluid mechanics. Annu. Rev. Fluid Mech. **52**, 477–508 (2020). https://doi.org/10.1146/annurev-fluid-010719-060214
3. Cardos, M., Durlofsky, L., Sarma, P.: Development and application of reduced-order modeling procedures for subsurface flow simulation. Int. J. Numerical Methods Eng. **77**(9), 1322–1350 (2009)
4. Cheng, M., Fang, F., Pain, C.C., Navon, I.M.: An advanced hybrid deep adversarial autoencoder for parameterized nonlinear fluid flow modelling. Comput. Methods Appl. Mech. Eng. 372 (2020). https://doi.org/10.1016/j.cma.2020.113375
5. Cheng, M., Fang, F., Pain, C.C., Navon, I.M.: Data-driven modelling of nonlinear spatio-temporal fluid flows using a deep convolutional generative adversarial network. Comput. Methods Appl. Mech. Eng. 365 (2020)
6. Gonzalez, F.J., Balajewicz, M.: Deep convolutional recurrent autoencoders for learning low-dimensional feature dynamics of fluid systems (2018), http://arxiv.org/abs/1808.01346

7. Goodfellow, I.: NIPS 2016 Tutorial: Generative Adversarial Networks (2016), http://arxiv.org/abs/1701.00160
8. Goodfellow, I., Bengio, Y., Courville, A.: Deep Learning. MIT Press (2016), http://www.deeplearningbook.org
9. Gulrajani, I., Ahmed, F., Arjovsky, M., Dumoulin, V., Courville, A.: Improved training of wasserstein GANs. Adv. Neural Inf. Process. Syst. 2017-Decem, 5768–5778 (2017)
10. Hotelling, H.: Analysis of a complex of statistical variables into principal components. J. Educ. Psychol. **24**(6), 417 (1933)
11. Kingma, D.P., Welling, M.: Auto-encoding variational bayes (2014)
12. Lecun, Y., Bengio, Y., Hinton, G.: Deep learning. Nature **521**(7553), 436–444 (2015). https://doi.org/10.1038/nature14539
13. Lumley, J.: Atmospheric turbulence and radio wave propagation. In: Atmospheric Turbulence and Radio Wave Propagation, pp. 166–178 (1967)
14. Makhzani, A., Shlens, J., Jaitly, N., Goodfellow, I., Frey, B.: Adversarial Autoencoders (2015), http://arxiv.org/abs/1511.05644
15. Quilodrán-Casas, C., Silva, V.L., Arcucci, R., Heaney, C.E., Guo, Y., Pain, C.C.: Digital twins based on bidirectional LSTM and GAN for modelling the covid-19 pandemic. Neurocomputing **470**, 11–28 (2022)
16. Silva, V.L.S., Heaney, C.E., Li, Y., Pain, C.C.: Data Assimilation Predictive GAN (DA-PredGAN): applied to determine the spread of COVID-19 (2021), http://arxiv.org/abs/2105.07729
17. Sirovich, L.: Turbulence and the dynamics of coherent structures. III. Dynamics and scaling. Q. Appl. Math. **45**(3), 583–590 (1987). https://doi.org/10.1090/qam/910464
18. Swischuk, R., Mainini, L., Peherstorfer, B., Willcox, K.: Projection-based model reduction: formulations for physics-based machine learning. Comput. Fluids **179**, 704–717 (2019)
19. Xiao, D., Fang, F., Buchan, A.G., Pain, C.C., Navon, I.M., Muggeridge, A.: Nonintrusive reduced order modelling of the Navier-Stokes equations. Comput. Methods Appl. Mech. Eng. **293**, 522–541 (2015). https://doi.org/10.1016/j.cma.2015.05.015
20. Xiao, D., Yang, P., Fang, F., Xiang, J., Pain, C.C., Navon, I.M., Chen, M.: A nonintrusive reduced-order model for compressible fluid and fractured solid coupling and its application to blasting. J. Comput. Phys. **330**, 221–244 (2017). https://doi.org/10.1016/j.jcp.2016.10.068

Towards Social Machine Learning
for Natural Disasters

Jake Lever[1,2,3(✉)] and Rossella Arcucci[1,2,3]

[1] Department of Earth Science and Engineering, Imperial College London,
Exhibition Rd, London SW7 2BX, UK
{j.lever20,r.arcucci}@imperial.ac.uk
[2] Data Science Institute, Imperial College London William Penney Laboratory,
South Kensington, London SW7 2AZ, UK
[3] Leverhulme Centre for Wildfires, Environment and Society, Imperial College
Department of Physics, South Kensington, London SW7 2BW, UK

Abstract. We propose an approach for integrating social media data
with physical data from satellites for the prediction of natural disasters.
We show that this integration can improve accuracy in disaster management models, and propose a modular system for disaster instance
and severity prediction using social media as a data source. The system is designed to be extensible to cover many disaster domains, social
media platform streams, and machine learning methods. We additionally present a test case in the domain context of wildfires, using Twitter
as a social data source and physical satellite data from the Global Fire
Atlas. We show as a proof of concept for the system how this model can
accurate predict wildfire attributes based on social media analysis, and
also model social media sentiment dynamics over the course of the wildfire event. We outline how this system can be extended to cover wider
disaster domains using different types of social media data as an input
source, maximising the generalisability of the system.

Keywords: Natural disasters · Machine learning · Sentiment analysis

1 Introduction

As the climate begins to change, the severity and frequency of natural disasters is
increasing yearly. Between 1998 and 2017, climate-related/geophysical disasters
killed 1.3 million people, and left a further 4.4 billion injured, homeless, displaced
or in need of emergency assistance [7]. In the last 30 years, the number of climate
related disasters has tripled, and this increase is almost certain to continue into
the future as climate records continue to be broken all over the world.

This increase in natural disaster activity due to climate change is additionally
being compounded by human activity. Modern agricultural practices increase the
risk of natural disaster through deforestation, and air pollution and the emissions
of water soluble particles into the atmosphere also increase the risk of extreme
weather. These disasters are also often highly linked, in ways which are still being

D. Groen et al. (Eds.): ICCS 2022, LNCS 13352, pp. 756–769, 2022.
https://doi.org/10.1007/978-3-031-08757-8_62

studied; such as the recent arctic heatwave in 2020, which destabilized the polar vortex and allowed a cold front of air to move down over North America, causing sub-zero temperatures in Texas, freezing the power grid and leaving 210 people dead. Increasing world population and housing, water, food and health crises in many highly populated areas are forcing more people into living at the wildland urban interface, which is at a raised risk of disasters due to these reasons. As a result, more individuals & infrastructure will be increasingly exposed to more frequent and intense natural disasters, raising the overall potential cost to society of these destructive events. Since 1980, the US has sustained 310 weather events and climate disasters where the overall damages & costs reached or exceeded $1 billion USD (adjusted for CPI to 2021). 20 of these events occurred in 2021 alone, leading to the deaths of 688 people in this year. The total cost of these 310 events alone exceeds $2.155 trillion USD [17].

Advancements in computational models in the past 20 years have allowed humans to predict localised environmental conditions with increasing accuracy. This is due to increased computational resources, advancements in understanding of these systems, and the inclusion of better data sources. However, numerical computational models during natural disasters often suffer in terms of accuracy due to their extreme and unpredictable nature, meaning often not enough useful training data is available. One massive source of data which is increasingly being used in models is social media.

Social media plays an ever increasing role in society. In 2020, 23% of US adults reported getting their news from social media often [19], up from 18% in 2016 [19]. Additionally, 82% of US adults are using social media as of 2021 [6], representing a massive audience and huge amounts of shared information. This paper proposes a more simple, socially focused model which aims to avoid the common pitfalls of modern commonly used models by combining social media data as an input source. In the advent of social media in the last decade, more studies developing social and physical models are including these types of data sources as an input [1,9,14,22]. These studies introduce the concept of the 'human sensor', where social media users are considered to be noisy sensors, posting a subjective account of their localised conditions. By analysing the response from these sensors, we can infer a model of the disaster as it unfolds in real time. We propose a system for collecting and analysing social media data to improve current natural disaster models, by monitoring online discussions and sentiments with the aim of ultimately identifying areas of actionable interest to disaster management teams. We show that online social discussions and sentiment are often linked to disaster activity, and that models can be trained to predict these shifts in sentiment over the course of the disaster. We also propose methods for information extraction & analysis of textual tweet data published during natural disasters, and discuss how this could be incorporated into a real-time model for public alerts.

The paper is structured as follows; Sect. 2 outlines the need for more socially conscious natural disaster models, and discusses the benefits of this. Following this, Sect. 3 defines the domain specific, modular architecture of the system we are proposing, describing the function of each of these modules. In Sect. 4 we present a test case of the system in the context of north American wildfires from 2016. Finally, Sect. 5 summarises our contribution and further work.

2   Background

Extensive work has been put into climate modelling with increasing success, and managing and mitigating climactic effects has been achieved in the past with good results [13], such as flood mitigation with dams or drought management using reservoirs [16]. However, it has been shown that during natural disasters, the coordination of teams from a crisis management perspective becomes challenging [15]. Often emergency operations are given from a centre and coordinate multiple organisations including local government, police, fire, hospital, utility, and Red Cross representatives. These teams are often 'flown in' ad hoc, and are expected to collaborate to deliver optimal crisis management at often very short timescales. This can very easily lead to poor communication and coordination of disaster management teams at a time when quick, coordinated and effective action is often key to saving lives.

Inherent properties of social media have been shown to lend themselves towards crisis management during natural disasters from both sides of the public / disaster management coin [23]. On the one hand, social media allows people in different locations to post a subjective description of their surroundings & immediate dangers as a disaster unfolds, which again employs the concept of the human sensor. By in-taking and analysing this data during disasters, management teams could build up a geographic picture of these noisy accounts, and use predictions to quickly identify areas of interest/danger from public accounts posted in real-time.

Conversely, social media can also be effectively used by these management teams to relay operational messages, warnings, and updates back to the public once they have been authorized. An example of this has already been implemented by Google using satellite data for a number of crisis alerts including flood forecasting, wildfire boundary lines, earthquakes, and more [10]. Operational updates posted on social media by local authorities could also be used in models to coordinate information/response strategies between the different organisations, such as the ones mentioned previously, involved in crisis management. This could ease the communication strains between organisations and help coordinate a quicker, more direct response.

Social media data is increasingly being analysed using Natural Language Processing (NLP) methods. NLP is a set of broad analytical techniques for computationally interpreting human language [12]. The field has undergone rapid recent development, with increasingly more information such as topics, intent, themes, entities, and sentiments being able to be inferred by models from text. Due to the vast amounts of text information generated by social media posts, it has been shown that analysis of these posts show insights [1,2,5,9,22] into processes and events. This paper aims to yield similar insights in the context of natural disasters by performing information extraction & sentiment analysis (SA). APIs exist for accessing data from all main social media platforms, including Twitter [20], Facebook [8], Instagram [8], and Reddit [18].

By implementing a machine learning (ML) based information system such as the one described in this paper, we can improve the streamlining of information at emergency operations centres, hereby allowing disaster management teams to make real-time, bottom-up decision as an event unfolds. Previous work has shown that there is a link between social media expression/sentiment, and natural disaster activity in the context of hurricanes [22].

This paper proposes a system, shown in Fig. 1, for collecting and analysing social media data in real time, and uses models trained on historical data to predict instances and characteristics of natural disasters. We show that gathering information on social media given the current state of the art of language models is feasible and efficient, and discuss the domain specific training using a test case in the context of wildfires for forecasting. We outline a system which adds a social aspect to disaster models, and contributes to the advancement of their capabilities by providing additional information to the overall system.

3   System Overview

The main concept behind the system is the live extraction of important information and sentiments from social media channels regarding natural disasters as they unfold. This system could be implemented initially in a predictive manner, predicting whether there is a natural disaster unfolding, and subsequently monitor and track the disaster by analysing social media discussions and extracting information from this live text data.

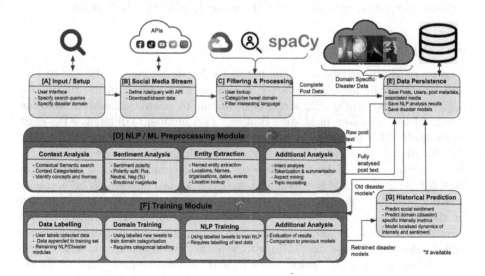

Fig. 1. Modular system diagram outlining flow of processes.

The system implements a live data stream based on search queries and using a prediction model which is specific to the disaster domain for the model. Here we mean the domain to be the category of natural disaster in the scope of this study, for example; Wildfires (this domain will be the demonstration case presented in Sect. 4), Earthquakes, Flood forecasting, or Hurricanes. With more work, the system could also further still be extended to include modelling and prediction on a wider range of types of events such as protests and political unrest, as well as political events and elections, health and food crises, and crime. This would require further development, primarily a more complex language model.

As mentioned, the prediction model would need to be domain specific to the type of disaster it is making predictions on. A model is made to be domain specific by training on a historical social media dataset which is again domain/disaster specific. For example, a model specific to the domain of wildfires would need to be trained on wildfire specific social media data. The entire system is designed to be modular and so can be adapted to be domain specific for each type of natural disaster, as well as different platforms of social media. This section will outline the systems component modules, and how they differ in terms of domains. The operational modular system is shown in Fig. 1. Each modules function is now briefly explained.

3.1 [A] System Input/Setup

The system is designed to take a series of input queries, or rules on which to set up a live real time stream for social media posts which are published satisfying one of the given rules. This is shown in Module A of Fig. 1. The rules must be formatted conforming to the social media platforms query syntax, e.g. Twitter's syntax rules [20]. The set of queries must be domain specific to the disaster, and include keywords specifically used in language specific to this domain.

3.2 [B] Social Media Data Streaming

This system implements real time social media streaming for the use of this live data in predictions. The function of this module is the implementation of a social media platform specific streaming function to download live data from the given website, as shown in Module B of Fig. 1. A number of different APIs exist to facilitate this for Twitter [20], Facebook [8], Instagram [8], and Reddit [18]. The function sends a set of queries, outlined with the domain specific language in the syntax outlined by the individual APIs documentation, as outlined in Sect. 3.1. This starts the live data stream. Here, we can also search on accounts which are publishing posts, and part of the training phase for this system will be the creation of a list of reputable accounts to monitor. These accounts will have been shown to be useful in providing information about disaster instance and development. These may include local authority service accounts, local and environmental journalists, rescue workers, and local government and authorities.

3.3   [C] Filtering and Text Pre-processing

Following collection of the social media data, filtering of results removes false positives (posts which do not mention the natural disaster domain when the model detects that they have) from the collected data. Filtering by misuse/misleading phrases, e.g. 'Corruption is spreading like wildfire' makes the dataset cleaner, resulting in more accurate models. This is represented by Module C of Fig. 1. Text is also pre-processed in this module, removing punctuation, hyperlinks, correcting spelling mistakes and formalising words etc., reducing noise in the text.

3.4   [D] Information Extraction and Sentimental Analysis

The function of this module is for the generation of social sentiment data from posts on natural disasters, and the extraction of published information which may be of help to the prediction module of the system. This is shown in Module D of Fig. 1. Sentiment Analysis is the process of computationally extracting a numerical value corresponding to the overall emotional leaning of the text. This is achieved using analysis of words used, word patterns and part-of-speech (POS). This is useful because it allows us to convert the qualitative text data into quantitative metrics which can be used for further analysis. The system can then be trained to predict the aggregated sentimental values for each natural disaster in Sect. 3.7. SA was performed using Google's NLP API for SA [11], and yields two metrics:

- Firstly, the **Sentiment Score** ranges between -1.0 (negative) and 1.0 (positive) and corresponds to the overall emotional leaning of the text.
- Secondly, tweet **Magnitude** indicates the overall strength of emotion (both positive and negative) within the given text, between 0.0 and +inf. [11]

This defines two numerical sentimental variables; S and M, which are the social sentiment variables for each natural disaster, observing the constraints $S \in \mathbb{R} : 0 \leq S \leq 1$ and $M \in \mathbb{R} : 0 \leq M$ respectively.

A domain specific language model could be used to infer greater insight into the text data collected by the collection module outlined in Sect. 3.2. Models like this can be trained using domain specific language datasets, which can be the textual social media data previously collected as part of the system setup in Sect. 3.6. The NLP model will be similar to the one implemented for hurricanes in [22], and will additionally aim to extract domain specific disaster characteristics from the noisy text data. These may include but are not limited to; disaster duration, total area damage, and number of people displaced, in need of emergency assistance, injured or killed. This is achieved through the sentimental & linguistic analysis of the words used in the posts, for example; contextual, entity, and intent analysis, results of which are also saved to the database. Tokenisation of the text is also stored.

3.5 [E] Data Persistence

Data persistence and model iteration & integration are a key aspect of the evolutionary aspect of this system. Due to the systems online data collection module outlined in Sect. 3.2, data is constantly collected and analysed surrounding different types of natural disaster domains. The function of this module is to save analysed social media data, including metadata on posts, users etc. and from Sect. 3.2, NLP and sentimental analysis results from Sect. 3.4. Additionally stored will be the models trained in Sect. 3.6 and the downloaded natural disaster data, as shown in Module E of Fig. 1.

3.6 [F] System Training (Pre-Operational)

Before the system can be used in real time for operational use, a model needs to be trained in order to make the model domain specific for the type of natural disaster it will be used for. Each type of disaster discussed in Sect. 3 will have different types of expression on social media and the dynamics and relationship between social media activity and disaster activity will vary from disaster to disaster. Thus, it is necessary to implement individual models for each type of disaster.

To train a domain specific model, we need a similarly domain specific historical social media dataset related to this type of disaster. The system is set up using historical natural disaster data for the domain specific disaster data in order to implement a base model.

This module also supports the labelling of training data gathered from the data collection and analysis modules of the system. The reason for this is for the iterative improvement of the ML models in Modules D & G, described in Sects. 3.4 & 3.7 respectively.

An important aspect of this training phase is that it can be achieved offline. That is, models can be iteratively trained while the system is online, and swapped out as they are improved. Due to the system constantly collecting and saving tweet data surrounding these types of natural disasters, the database used to train these models will only get larger and more diverse, both geographically and between disaster domains, as the system is used more. This new data can be labelled and then used to iteratively train improved systems in a supervised manner, or remain unlabelled for models to be trained using unsupervised methods. To summarise, the system is designed to have an offline training programme which iteratively retrains and improves the model as it runs, leading to evolutionary improvement and in turn greater accuracy, and the rapid development & deployment of improved predictive models.

3.7 [G] ML Prediction Model

This part of the system utilises the model trained historical data outlined in Sect. 3.6. We take the model trained on historical, domain specific disaster data, and use this to make predictions about the occurrences of new disaster instances. The aims of this module are as follows;

- To detect instances and categorise types of new disasters, and predict the times and locations of these new events.
- To predict the severity of disasters over the course of the crisis period. By severity we mean the physical domain specific disaster variables from the historical data supplied in the training dataset outlined in Sect. 3.6, e.g. size of area affected, number of injuries etc. We hypothesise that there exists some function of f which allows us to predict disaster intensity from social media post data.

The ML method which we utilise for this system is the Gradient Boosted Random Forest, implemented in python with the *XGBoost* package [21]. Random Forests are an ensemble method extension of Classification and Regression Trees (CART) [4]. In the training phase, these methods start with a simple model, often a single tree (or weak learner), and then additive training occurs where trees are generated and added to build up a forest of trees which is used for the final prediction. For new data, predictions are then made by averaging the majority vote of all trees in the forest. The GBRFs are built and evaluated using the MAE, and the Gini coefficient [4]. The Gini coefficient is a scoring metric which is a measure of the degree to which a particular element is wrongly classified when randomly chosen and it is expressed by the formula:

$$\sum_{i=1}^{n} p_i(1 - p_i) \tag{1}$$

where p_i denotes that the physical event i happens (or that the sentiment $i \in S$ is evaluated) and n is the number of possible events (or possible sentiments). Gradient boosting Random Forest algorithms are particular Random Forests which begin with a base (weak) learner (tree), and consecutively add more weak learners to the ensemble with the goal of minimising a loss function [4]. Given a training sample $\{x_i, y_i\}_1^n$, the goal is to find function of $F'(x)$ such that the expected value of the loss function $\psi(y, F(x))$ is minimised [4]. The gradient boosting problem is then expressed as:

$$F'(x) = \arg \min_{F(x)} \mathbb{E}_{x,y} \psi(y, F(x)) \tag{2}$$

where $\mathbb{E}$ denotes the expected value. The boosting algorithm then approximates $F'(x)$ by additive expansion, which can be summarised as:

$$F(x) = \sum_{m=0}^{s} \beta_m h(x, a_m) \tag{3}$$

where the functions $h(x, a)$ are known as the 'weak learners'.

Over a series of s steps, the weak learners are sequentially added, and the expansion coefficients $a = \{a_1, a_2, ...\}$ and $\{\beta\}_0^s$ are jointly fit to the current models pseudo-residuals. For $m = 1, 2, \ldots, s$, this gives

$$(\beta_m, a_m) = \arg \min_{\beta, a} \sum_{i=1}^{n} \psi(y_i, F_{m-1}(x_i) + \beta h(x_i; a)) \tag{4}$$

and

$$F_m(\mathbf{x}) = F_{m-1}(\mathbf{x}) + \beta_m h(\mathbf{x}; a_m) \tag{5}$$

which shows the step-wise optimisation of F . Equation (4) is then optimised to solve the loss function ψ by fitting least squares on $h(\mathbf{x}, a)$. This replaces the optimisation problem presented in Eq. (2) with one based on reducing least squares in Eq. (4).

The models were trained on a wide grid search covering the hyper-parameters denoted by:

- *ETA* (Learning Rate): Step size shrinkage used in update.
- *maxDepth*: Maximum depth of a tree.
- *minChildWeight*: Minimum sum of instance weight needed in a child.
- *subsample*: Subsample ratio of the training instances.
- *colSampleByTree*: Subsample ratio of columns when constructing each tree.
- *nEstimators*: Number of trees in the ensemble.

The output of this module is the overall system output: predictions of domain, instances and severity of the natural disaster. This represents a socialised model of the disaster, from detection to evolution and finally resolution. Once a disaster has been detected and registered in the application database, it follows the process flow shown in Fig. 1, being continuously updated as more information on this suspected disaster is streamed through by the data collection stream. A disaster is monitored until information is passed through that this crisis is resolved. We now demonstrate in Sect. 4 a prototype of the model described in Sect. 3, applied with a wildfire domain.

4   Experimental Test Case: Wildfires - Satellites and Twitter

We now present a prototype version of the domain specific model applied to satellite wildfire data using twitter as the social media data source. We choose wildfire events occurring in Australia and North America (United States and Canada) in 2016 for the creation of our training dataset. We chose to use this type of disaster domain in this geographic area due to the abundance of both wildfire activity in this area and active Twitter users. We now outline the implementation of the system for this particular context.

Satellite & Twitter Data Collection: For the combination of social media data with wildfire data in our model component of the system, we chose to use twitter for the historical social media source, and wildfire data from the Global Fire Atlas [3] for the 2016 wildfire data. The physical wildfire characteristics taken from this data are: Latitude ($\circ$), Longitude ($\circ$), Size (km2), Perimeter (Per) (km), Duration (d) (days), Speed(km/day), Expansion (Exp) (km2/day), and Start (S_{DOY}) and End (E_{DOY}) . This defines in the physical vector $\mathbf{x}$ of our model:

$$\mathbf{x} = [Lat, Lon, Size, Per, d, Speed, Exp, S_{DOY}, E_{DOY}, Pop_{Density}] \tag{6}$$

Twitter's V2 API [20] with an academic research product track was used for the collection of the twitter data associated with historical wildfires, using a query implementation in line with Twitter's query syntax [20]. Queries were designed to search for tweets mentioning locations of the burn as well as a set of wildfire domain specific keywords to search on; Fires OR Wildfires OR Bushfires OR "Landscape Burn" OR "Wildland Burn" , as well as generated hashtags included in the query. Tweets which contained certain misuse phrases such as "like wildfire" were removed to reduce noise. Tweets were saved to a database with meta and user data and media, along with a Fire ID. The result is a dataset of US wildfires in 2016 and tweets associated with these events.

Twitter Data Analysis: We are now able to analyse this text data for wildfires and generate social sentiment variables from this data using SA. Recalling Sect. 3.4 where the two numerical sentimental variables S and M are defined, Tweets are grouped by day for each fire and social sentiment values from the S and M are averaged and summed to generate the following social sentiment variables for each wildfire:

- S_{mean} : Average Daily Sentiment Score
- M_{mean} : Average Daily Magnitude Score
- S_{ovr}: Overall Sentiment Score
- M_{ovr}: Overall Magnitude Score
- Tot_{tweets}: Total number of Tweets for Wildfire

These variables constitute the sentimental vector

$$\mathbf{y} = [S_{mean}, M_{mean}, S_{ovr}, M_{ovr}, Tot_{tweets}]. \tag{7}$$

The generation of these social sentiment variables for each wildfire represents the completion of our two datasets in both the Australian (AUS) and North American (US) domains.

The data can now be viewed from a temporal perspective, by plotting heatmaps for the online social sentiment across the year of 2016. Fig. 2 shows how online sentimental activity matches the fire seasons in the two geographic domains, which is a positive indication of the quality of the data.

Wildfire & Social Modelling Results: Two types of ML models were implemented using the *XGBoost* package [21] in python as outlined in Sect. 3.7. The first type of model takes as an input the physical vector $\mathbf{x}$, and is trained using this data on the target vector $\mathbf{y}$. The 5 variables which were described in this section were used to train 5 models, one predicting each social sentiment variable. These models are called the sentimental prediction models. The second type of model predicted the 10 physical wildfire characteristics vector $\mathbf{x}$, using the social sentiment vector $\mathbf{y}$, predicting in the opposite direction to the sentimental prediction model. This type of model was called the physics prediction model, and there were 10 of these models trained, one for each variable, meaning total of 15 models were trained on the combined AUS + US dataset. The results are shown in Tables 1

(a) Heatmap of US online social sentiment for 2016

(b) Heatmap of AUS online social sentiment for 2016

Fig. 2. US (a) and AUS (b) online social Sentiment heatmaps for 2016

& 2 below. As the last column in Tables 1 & 2 shows, the execution time (Exe_T) for running the model on a 2020 Macbook Pro 2.3GHz 8 core i7 Intel processor is very low, whch is an important condition for real time operational predictions.

Table 1. Results from predicting social sentiment variables from physical wildfire characteristics.

Variable	MAE	RMSE	Gini	$Gini_N$	Score	Exe_T (msecs)
S_{mean}	6.34	17.83	0.367	0.846	38.7%	2.94
M_{mean}	10.88	40.85	0.316	0.822	38.3%	5.75
M_{ovr}	107.27	407.36	0.338	0.872	24.4%	3.37
S_{ovr}	54.75	364.76	0.381	0.865	15.3%	2.28
tot_{tweets}	458.927	2777.28	0.323	0.863	10.6%	2.94

Table 1 demonstrates that the models for predicting average Sentiment and Magnitude (S_{mean} and M_{mean}) show good results when attempting to predict these variables, with low MAEs of 6–10. Overall Sentiment and Magnitude (S_{ovr} and M_{ovr}) models also preformed well. The most notable result shown in Table 2 is the model predicting fire Duration (d). This model showed very positive results, with an MAE of 0.84. This shows that the resulting model was able to predict wildfire duration to within one day from social sentiment values/social media data alone. Additionally, *Speed* and *Exp* both performed well.

Table 2. Results from predicting physical variables from social sentiment data.

Variable	MAE	RMSE	Gini	$Gini_N$	Score	Exe_T (msecs)
Lat	3.72	5.2	0.344	0.868	0.75	4.90
Lon	6.185	7.40	1.73	0.929	0.79	6.11
$Size$	9.66	33.89	0.354	0.835	0.14	1.95
Per	7.22	15.89	0.227	0.764	0.24	3.69
d	0.84	2.01	0.207	0.961	0.87	4.22
$Speed$	0.533	0.86	0.143	0.659	0.25	2.32
Exp	0.88	3.24	0.226	0.652	0.16	2.09
$Pop_{Density}$	69.05	575.18	0.397	0.424	0.02	2.22
S_{DOY}	40.73	60.90	0.100	0.736	0.56	2.83
E_{DOY}	39.62	61.15	0.095	0.734	0.56	5.48

5   Conclusion and Future Work

This paper outlines a system to facilitate the integration of social media data into physical models, discusses the benefit of this, and demonstrates a prototype test case with current wildfire models. As shown, social media data is being adopted increasingly in scientific studies and specifically disaster management, yielding benefits which could be transferred to natural disaster relief efforts. This work attempts to bridge this gap by developing a modular social media alarm system for natural disasters which predicts instances and localised severity of the event based on analysis of social media posts.

Having successfully trained and implemented a retrospective historical social media wildfire model using Twitter as a data source, the next step for testing this type of model would be the integration/coupling of the system with a real time wildfire model. The modular design of the system allows for rapid updating of the system as future work allows. This will primarily be focused on the development of ML methods for information extraction of the post text data and modelling disaster activity. The benefit of this will be improved accuracy of these models which will allow localised modelling of disaster conditions.

Social media data represents a near limitless supply of real time data on almost any large event. If disaster models do not consider this data, then this represents a loss of information, as there is data available via these networks which is not accessible elsewhere. Systems which analyse this data in real time are able to recoup this loss, which ultimately leads to the development of improved models.

References

1. Alkouz, B., Aghbari, Z.A., Abawajy, J.H.: Tweetluenza: predicting flu trends from twitter data. Big Data Mining Anal. **2**(4), 273–287 (2019). https://doi.org/10.26599/BDMA.2019.9020012
2. Alrashdi, R., O'Keefe, S.: Automatic Labeling of Tweets for Crisis Response Using Distant Supervision, p. 418–425. Association for Computing Machinery, New York, NY, USA (2020), https://doi.org/10.1145/3366424.3383757
3. Andela, N., et al.: The global fire atlas of individual fire size, duration, speed and direction. Earth Syst. Sci. Data **11**(2), 529–552 (2019)
4. Bishop, C.M.: Pattern recognition and machine learning (2006)
5. Cui, R., Gallino, S., Moreno, A., Zhang, D.J.: The operational value of social media information. Prod. Oper. Manage. **27**(10), 1749–1769 (2018)
6. Department, S.R.: Social media usage in U.S. November 2021, https://www.statista.com/statistics/273476/percentage-of-us-population-with-a-social-network-profile/
7. for Research on the Epidemiology of Disasters & The UN Office for Disaster Risk Reduction, T.C.: Economic Losses, Poverty & Disasters 1998–2017. Tech. rep., The Centre for Research on the Epidemiology of Disasters & The UN Office for Disaster Risk Reduction (2018)
8. Facebook: Facebook graph API, https://developers.facebook.com/docs/graph-api/
9. Gallagher, R.J., Reagan, A.J., Danforth, C.M., Dodds, P.S.: Divergent discourse between protests and counter-protests: #blacklivesmatter and #alllivesmatter. PLOS ONE **13**, 1–23 (2018). https://doi.org/10.1371/journal.pone.0195644
10. Google: Forecasting & alerts: Google's crisis alerts provide access to trusted safety information across search, maps, and android (2017). https://crisisresponse.google/forecasting-and-alerts/. Accessed 17 Jan 2021
11. Google: Analyzing sentiment (2021), https://cloud.google.com/natural-language/docs/analyzing-sentiment
12. Hirschberg, J., Manning, C.D.: Advances in natural language processing. Science **349**(6245), 261–266 (2015). https://www.science.org/doi/abs/10.1126/science.aaa8685, https://doi.org/10.1126/science.aaa8685
13. Dullaart, J.C.M., Muis, S., Bloemendaal, N., Aerts, J.C.J.H.: Advancing global storm surge modelling using the new ERA5 climate reanalysis. Climate Dyn. 1007–1021 (2019). https://doi.org/10.1007/s00382-019-05044-0
14. Kryvasheyeu, Y., et al.: Rapid assessment of disaster damage using social media activity. Sci. Adv. **2**(3) (2016). https://advances.sciencemag.org/content/2/3/e1500779, https://doi.org/10.1126/sciadv.1500779
15. Laura G. Militello, Emily S. Patterson, L.B.R.W.: Information flow during crisis management: challenges to coordination in the emergency operations center. Cogn. Technol. Work **9**(1), 25–31 (2007). https://doi.org/10.1007/s10111-006-0059-3
16. Lempérière, F.: Dams and floods. Engineering **3**(1), 144–149 (2017)
17. National Centers for Environmental Information: U.s. billion-dollar weather and climate disasters, 1980 - present (NCEI accession 0209268) (2021). https://www.ncdc.noaa.gov/billions/overview. Accessed 17 Jan 2021
18. Reddit: Reddit API documentation, https://www.reddit.com/dev/api/
19. Shearer, E.: 86% of Americans get news online from smartphone, computer or tablet, January 2021, https://www.pewresearch.org/fact-tank/2021/01/12/more-than-eight-in-ten-americans-get-news-from-digital-devices/

20. Twitter: Twitter API v2, https://developer.twitter.com/en/docs/twitter-api
21. XGBoost: Dart booster (2020), https://xgboost.readthedocs.io/en/latest/tutorials/dart.html
22. Yao, F., Wang, Y.: Domain-specific sentiment analysis for tweets during hurricanes (DSSA-H): a domain-adversarial neural-network-based approach. Comput. Environ. Urban Syst. **83**, 101522 (2020)
23. Young, C., Kuligowski, E., Pradhan, A.: A review of social media use during disaster response and recovery phases. https://doi.org/10.6028/NIST.TN.2086. Accessed 31 Jan 2020

Correction to: GPU Accelerated Modelling and Forecasting for Large Time Series

Christos K. Filelis - Papadopoulos ⓘ, John P. Morrison ⓘ,
and Philip O'Reilly ⓘ

Correction to:
Chapter "GPU Accelerated Modelling and Forecasting
for Large Time Series" in: D. Groen et al. (Eds.):
Computational Science – ICCS 2022 **LNCS 13352,**
https://doi.org/10.1007/978-3-031-08757-8_33

The chapter "GPU Accelerated Modelling and Forecasting for Large Time Series" was previously published non-open access. It has now been changed to open access under a CC BY 4.0 license and the copyright holder updated to 'The Author(s)'. The book has also been updated with this change. In addition to this, the acknowledgement section of this chapter has been updated.

The updated original version of this chapter can be found at
https://doi.org/10.1007/978-3-031-08757-8_33

D. Groen et al. (Eds.): ICCS 2022, LNCS 13352, p. C1, 2023.
https://doi.org/10.1007/978-3-031-08757-8_63

Author Index

Printed in the United States
by Baker & Taylor Publisher Services